OCEANOGRAPHY AND MARINE BIOLOGY

AN ANNUAL REVIEW

Volume 23

OCEANOGRAPHY AND MARINE BIOLOGY

AN ANNUAL REVIEW

Volume 23

HAROLD BARNES, *Founder Editor*

MARGARET BARNES, *Editor*
The Dunstaffnage Marine Research Laboratory
Oban, Argyll, Scotland

ABERDEEN UNIVERSITY PRESS

FIRST PUBLISHED IN 1985

British Library Cataloguing in Publication Data

Oceanography and marine biology: an annual review.—Vol. 23
1. Oceanography – Periodicals
2. Marine biology – Periodicals
551.46′005 GC1

ISBN 0 08 030397-8
ISSN 0078-3218

PRINTED IN GREAT BRITAIN
in 10 *point Times Roman type*
BY ABERDEEN UNIVERSITY PRESS
ABERDEEN

PREFACE

Once again scientists from many parts of the world have responded to invitations to contribute to this series of Annual Reviews. It is always a pleasure to acknowledge the willing co-operation of contributors and to thank them for acceding to editorial requests. As usual, I am fortunate in having the advice of many friends and colleagues including, in particular, Drs A. D. Ansell, R. N. Gibson, and T. H. Pearson. Good relations with the publishers also make the task of producing this publication a pleasure.

CONTENTS

Oceanogr. Mar. Biol. Ann. Rev., 1985, **23,** 11–103
Margaret Barnes, Ed.
Aberdeen University Press

THE AGE OF TIDES[1]

T. S. MURTY
Institute of Ocean Sciences, Department of Fisheries and Oceans, Sidney, B.C., Canada V8L 4B2

and

M. I. EL-SABH
Département d'Océanographie, Université du Québec à Rimouski, 300 avenue des Ursulines, Rimouski, Québec, Canada G5L 3A1

INTRODUCTION

Spring tides usually occur a day or two after full moon and new moon, and this delay is referred to as the age of the tide (Garrett & Munk, 1971). In other words, the age of tide is a term used to denote the interval between the time of new moon or full moon (when the equilibrium semi-diurnal tide is maximum) and the time of the local spring tide (Webb, 1973). Whewell (1883) was the first to have used the terminology "age of tide"; on the other hand, Defant (1961) referred to it as "spring retardation". Wood (1978) used the terms "age of the phase inequality" and "age of the diurnal inequality" to denote, respectively, the ages of the semi-diurnal and diurnal tides.

A localized increase in the age of the tide is a good indication of resonances at that location (Webb, 1973). It is with this view that we study the geographical distribution of the age of the tide. The present review attempts to bring together all the published, and most of the unpublished work on the age of tides. The review will begin by considering the previous work on the age of tides followed by a brief description of the geography of the global oceans and the smaller water bodies selected for a detailed study. A brief astronomical background and the tide-generating forces will then be considered. Tidal analysis and tidal constituents, followed by the tides in the global oceans are then examined. Next, age of semi-diurnal tide in the global oceans, the negative ages, the age of parallax, and the age of diurnal tides will be considered. The age of parallax involves the semi-diurnal tidal constituents M_2 and N_2. Unlike the semi-diurnal tidal constituent S_2 (which along with M_2 is involved in the calculation of the age of the semi-diurnal tide), N_2 is not contaminated by radiational effects (Garrett & Munk, 1971). Some of these details will be included in this review.

[1] Contribution du Laboratoire Océanologique de Rimouski.

Next we go into detailed studies for some selected water bodies. We start with a description of the tides in the St Lawrence Gulf and Estuary (in eastern Canada), followed by a discussion on the age of tides in this system. Descriptions of tides and age of tides in the Arabian Gulf (also called Persian Gulf), the Gulf of Oman, the Red Sea, the Gulf of Aden, waters around Australia and New Zealand (which represent, respectively, an island of continental size and a reasonably large island system), are discussed. Following Webb (1973), we shall examine the distribution of the age of tide near the heads of estuaries. The final section provides a summary and some suggestions for future work.

REVIEW OF LITERATURE

The two classical papers on the topic of the age of tide are those of Garrett & Munk (1971) and Webb (1973). We shall start, however, with a brief description of some earlier works.

Newton (1687) attributed the age of tide to the inertia of the water. Laplace (1799) pointed out that the age of tide is expressible in terms of the difference in phase lag of the lunar and solar tides behind their forcing functions. Young (1823) was among the first to point out the analogy between the age of the tide and the increase with frequency of the phase lag of a forced linearly damped harmonic oscillator. As noted earlier Whewell (1883) was the first to have used the term "age of tide". According to Harris (1894–1907) the frequency of the semi-diurnal tide is close to the frequencies of one or more normal modes of the global oceans. Lamb (1932) provided a mathematical derivation of the idea of Laplace mentioned above. Also with reference to the suggestion of Young (1823), Lamb expressed doubt whether the phase lag or its increase with frequency could be sufficiently large to agree with the observations, noting that it is only near resonance that a large increase of phase lag with frequency occurs.

Proudman (1941) showed analytically that energy dissipation in coastal seas is necessary to produce predominantly positive age in a hemispherical frictionless ocean. Defant (1961) used the terminology "spring retardation" instead of "age of tide", in his study in the German Bight of the North Sea. The two semi-diurnal tidal constituents involved in estimating the age of the tide are the M_2 and S_2. Munk & Cartwright (1966) and Cartwright (1968) showed that the S_2 tide is contaminated by radiational effects; the details will be given later. Easton (1970) studied the age of tides in and around Australia. Farquharson (1970) explained the age of tide as due to the variations which occur in the M_2S_2 amphidromic system, with the semi-monthly change in the moon's phase.

Garrett & Munk (1971) raised the question: why is the phase lag (for the semi-diurnal tide, a phase lag of 1·016° corresponds to an age of one hour) of the solar semi-diurnal tide S_2 typically greater than that of the lunar semi-diurnal tide M_2, and also, how can a difference of several tens of degrees occur for a difference in frequency of less than 4%? They suggested that friction is in some manner responsible for the age of the tide, particularly if there is some resonant frequency of the oceans close to the semi-diurnal tidal frequencies. They explained qualitatively this phenomenon by the phase change with frequency, by expressing the response of the ocean to tidal forcing as an expansion in terms of the normal modes of the ocean. They also showed that such an expansion allows for the age of tide to be large, positive, and vary from place to place.

If the semi-diurnal tide were to be dominated by one mode, then the age of tide would be positive and would have the same value everywhere. The variation of the age from one place to another is due to the spatial dependence of the normal modes. By studying in detail the spatial variation in the age of the tide, one can understand better the normal modes of the oceans. According to Garrett & Munk (1971), unusual combination of the normal modes could give rise to large positive values, and also negative values for the age of the tide. They also pointed out one of the weaknesses in their mathematical formulation of the problem, namely that the forcing functions for the M_2 and S_2 tides differ only in frequency. Earlier we referred to the work of Munk & Cartwright (1966) and Cartwright (1968) in which they showed that the S_2 tide is contaminated by radiational effects. According to Garrett & Munk (1971), this difficulty with radiational effects could be avoided by studying the "age of parallax inequality" (which is computed as the difference in phase between M_2 and N_2 divided by the difference in their speeds) instead of the age of the semi-diurnal tide. Note that neither M_2 nor N_2 is contaminated by radiational effects. The N_2 tide is due to the fluctuating distance of the moon from the earth (its frequency is one cycle per month less than that of M_2, and three cycles per month less than that of S_2). There is no published literature on the age of the diurnal tide (between constituents K_1 and O_1).

Webb (1973) studied the geographical distribution of the age of the tide to learn about the geographical distribution of the localized resonances in the world oceans. He pointed out that the regions of the global oceans where the age of tide is large happen to be the same where dissipation is concentrated (Miller, 1966) and in some situations localized resonances are expected (such as Hudson Bay, Coral Sea, and Gulf of Patagonia, *etc.*). Making use of this concept of localized resonance, Webb questioned whether an overall "Q" factor for the global oceans, as used by Garrett & Munk (1971) is justified. Instead, Webb (1973) suggested that one should consider the dynamics of tidal motions in terms of decay times (approximately equal to the age of the tide) and the time required for the energy entering the deep ocean to be transported to the dissipation areas of resonance. Cartwright (1978) recognizes that this is a controversial subject to which it is difficult to apply precise arguments.

Some of the important results from the study of Webb (1973) are: (1) over most of the ocean, the age of the tide lies between 0 and 60 h, with a typical variation of about 10 h between adjacent stations; (2) because of the random nature of much of this variation, it is difficult to produce contour maps of the age of the tide; (3) most of the regions with an age of the tide of greater than 60 h are associated with strong topographic features (most common feature being a large continental shelf); (4) there are some regions with a large positive age of tide which are not associated with obvious topographic features; (5) there are a few regions where a negative age of tide is found; and (6) occasionally the extreme values of the age of tide are associated with regions where the amplitude of the S_2 constituent is less than 5 cm. In these cases the extreme values of the age of the tide may occur due to the noise present in the tidal records. In addition, the S_2 may be affected by radiational effects.

Following Webb (1973) we write the tidal response function as

$$R(\omega) = A(\omega)\cdot e^{\theta(\omega)} \tag{1}$$

where A and θ are real. Then $\partial\theta/\partial\omega$ at a particular frequency is equal to the length of time that a signal at that frequency is delayed by the system (this delay is the age of the tide). If one represents the physical system as a set of resonant states, then a long time delay anywhere in the system will correspond to the dominant effect of one or more resonant states with a long decay time. Hence the localized regions with a large value of age of tide correspond to localized resonances of the ocean. Often the topographic features are large enough to cause resonances. Away from the resonances, the age of the semi-diurnal tide is usually less than 50 h. Large values of age of tide are found near estuary heads, probably due to shallow-water effects. There is a misprint in the equation for the age of the semi-diurnal tide, in Webb's (1973) paper; this does not, however, affect any of his results. The equation for the age α of the semi-diurnal tide should be

$$\alpha = \frac{(\text{phase of } S_2 - \text{phase of } M_2)}{1{\cdot}016} \text{ h} \tag{2}$$

The phases being expressed in degrees with reference to the Greenwich meridian. In Webb's paper, on page 847, the terms in the numerator are reversed.

Heath (1981a,b) studied the age of the tide around New Zealand and found that large variations occur. He attributed the large negative ages to the differing spatial distributions of the M_2 and S_2 tides. Bye & Heath (1975) suggested that it is misleading in the case of New Zealand to associate changes in the age of tide with local resonance effects (Webb, 1976).

Pugh (1981) offered an explanation for anomalous values of the age of tide. He suggested that in seas where the spring amphidrome is not degenerate, the smallest semi-diurnal tides of the spring–neap cycle near the amphidromic position will occur, when the rest of the water-body is experiencing spring tides. This will lead to an anomalous tidal age to the local tides. Pingree (1983) studied the dissipation mechanism by examining the spring–neap tide envelope.

GEOGRAPHY OF THE AREA STUDIED

We shall present results for the age of the tide in the global oceans (Fig. 1) as well as in some selected smaller water-bodies. The water-bodies that will receive detailed scrutiny are waters around Australia (representing continental scale), waters around New Zealand (representing a reasonably large island system). Three other water-bodies also will be studied in detail. These are: (a) the Gulf of St Lawrence and the St Lawrence Estuary (Fig. 2) in eastern Canada (representing a large marginal sea and estuary system), (b) the Arabian Gulf and the Gulf of Oman (Fig. 3) (representing a combination of two gulfs with some estuarine effects), and (c) Red Sea and the Gulf of Aden (Fig. 4) (representing a combination of a long, narrow and deep gulf and a wide gulf with no estuarine effects).

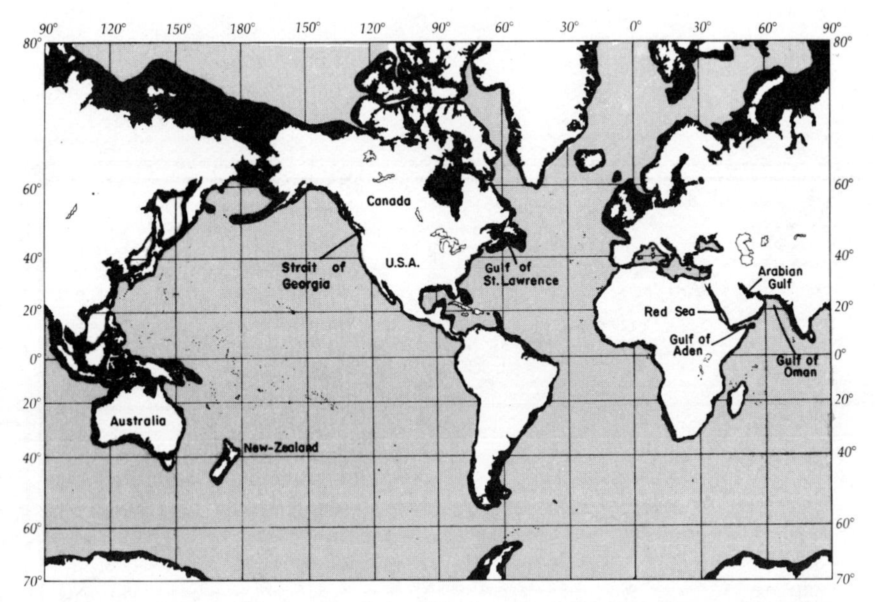

Fig. 1.—Geographical map of the global oceans, showing the continental shelves and selected water-bodies mentioned in the text.

The Red Sea is connected to the Gulf of Aden through the narrow Strait of Bab-el-Mandab (Fig. 4); the Arabian Gulf is connected to the Gulf of Oman through the narrow Strait of Hormuz (Fig. 3). No rivers flow into the Red Sea, hence estuarine effects are minimal to non-existent in the Red Sea. The Euphrates and Tigris Rivers flow into the Arabian Gulf at Shatt-al-Arab and thus some estuarine effects exist in the Arabian Gulf–Gulf of Oman system. On the other hand, many rivers flow into the Gulf of St Lawrence and St Lawrence Estuary (Fig. 2) and in this system strong estuarine effects exist.

Important topographical variations also occur in the St Lawrence system. The Laurentian Channel is very deep; and in contrast the Magdalen Shallows are very shallow. In the St Lawrence Estuary, the depths are greater downstream of the Saguenay barrier, and upstream of this barrier it is shallow. Important topographic variations also occur in the Arabian Gulf–Gulf of Oman system. The Arabian Gulf is shallow in the southwestern part. The Red Sea–Gulf of Aden system is somewhat similar to the Arabian Gulf–Gulf of Oman system in the sense that the narrow Strait of Hormuz connects the Arabian Gulf with the Gulf of Oman, just as the narrow Strait of Bab-el-Mandab connects the Red Sea with the Gulf of Aden.

ASTRONOMICAL FACTORS

The tides in the oceans on the planet Earth are caused exclusively by the gravitational attraction of the moon and sun, the influence of other celestial bodies being completely negligible. Hence for computing the tides we have to

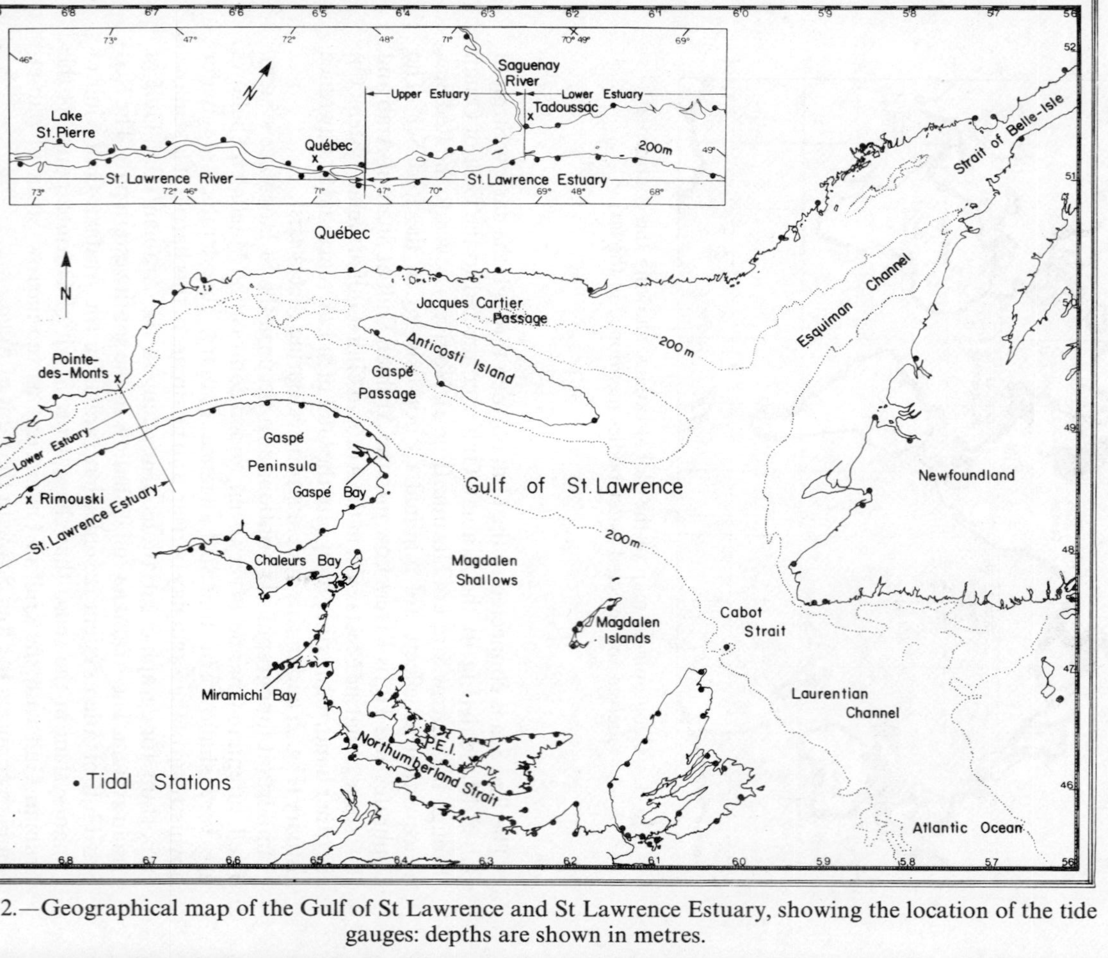

Fig. 2.—Geographical map of the Gulf of St Lawrence and St Lawrence Estuary, showing the location of the tide gauges: depths are shown in metres.

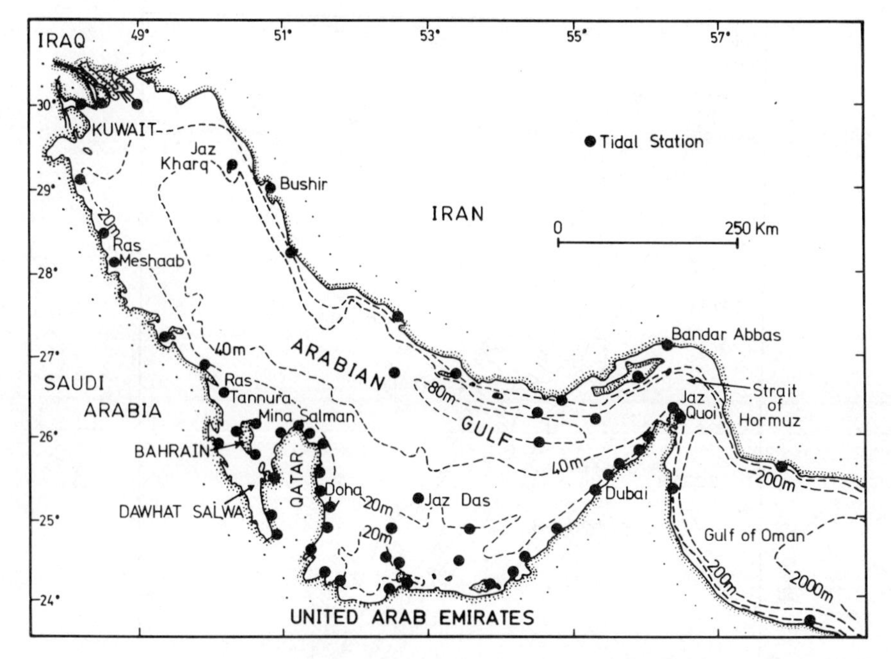

Fig. 3.—Geographical map of the Arabian Gulf and the Gulf of Oman, showing the location of tide gauges: depths are shown in metres.

know the changing interrelationships of the earth, moon, and sun, in terms of their motions and distances. In the subsequent discussion in this section our major source of information is the excellent monograph by Wood (1978).

Although one can specify the positions of astronomical bodies through at least five different coordinate systems, for tides, only three of these five systems are relevant. Table I summarizes the features of these three different systems, while Table II summarizes the mathematical relations among these coordinate systems.

According to Wood (1978) the earth is subject to the following seven astronomical motions: (1) the diurnal rotation of the earth, (2) the annual revolution of the earth around the sun, (3) the precision of the equinoxes, (4) the nutational motion of the polar axis, (5) the space motion of the earth with the sun towards the apex of the sun's way, (6) the rotation of the galaxy around its own centre, and (7) the irregular shifting of the earth's crust, as the geographic pole of the earth, is displaced at a non-uniform rate, with respect to its pole of figure; this motion causing "variation of latitude".

For tidal studies, we need to consider only two of the above seven motions, namely the diurnal rotation of the earth and the annual revolution of the earth around the sun. One has to consider also the monthly revolution of the moon around the earth. In addition, we have to include the effect of the moon's position (in declination) upon its rate of change (in right ascension). Although these four types of motions are real, the actual measurement of each of these motions is made in terms of an apparent change in position, or

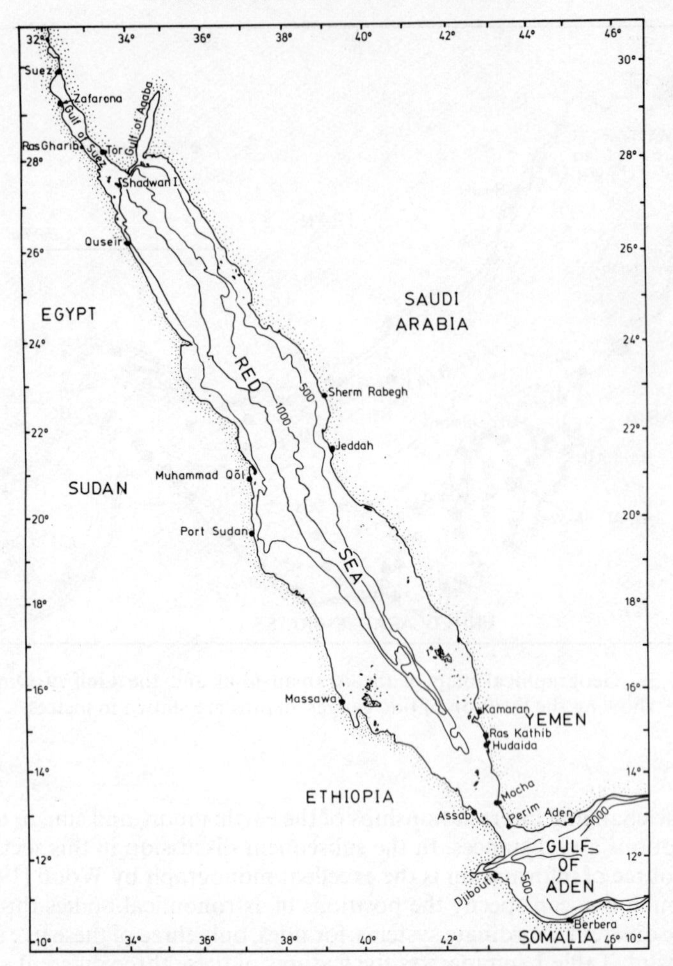

Fig. 4.—Geographical map of the Red Sea and the Gulf of Aden, showing the location of tide gauges: depths are shown in metres.

displacement of the sun and moon on the celestial sphere. Before we discuss the implications of these four types of motions, we shall introduce certain terms following Wood (1978).

Twice a month at new moon and full moon (respectively referred to as conjunction and opposition), the earth, moon and sun arrange themselves into direct alignment in celestial longitude, and because of the combination of their gravitational forces, enhanced tide-raising forces occur (tides generated at such times are known as spring tides). Because of the elliptic shape of the lunar orbit, once during a revolution the moon reaches its closest monthly approach to the earth, and this position is referred to as "perigee". Usually the passage of the moon through the perigee and the alignment of the sun, moon and earth at new or full moon (either position is referred to as syzygy) do not occur at the same time. This occurs infrequently (due to commensurable relationships

TABLE 1

Coordinate systems to specify positions of astronomical bodies relevant for tides (*based on Wood, 1978*)

Coordinate system	Features of the system
Equatorial system	The basic reference circle is the earth's equator projected upon the celestial sphere. The north and south celestial poles are located by an axis projected through the north and south geographic poles of the earth and extended beyond the two points of intersection with the celestial sphere. The celestial equator is an extension of the earth's equator outwards to intersect with the celestial sphere. The "astronomical declination" (δ) of a body is measured in degrees, minutes and seconds of arc perpendicularly north or south from the celestial equator through 90° to the north or south celestial pole. The coordinate corresponding to geographical longitude is the "right ascension" (α) which is only measured from west to east in hours, minutes and seconds of time through 24 h, rather than in units of arc. The astronomical position for 0 (or 24) h is the point of intersection of the celestial equator with the ecliptic on the celestial sphere (the ecliptic is the apparent path of the sun around the celestial sphere as the earth revolves around the sun, making the sun to appear to revolve around the earth in the same direction).
Ecliptic system	The fundamental plane is the plane of the earth's annual revolution around the sun. The coordinate of celestial longitude (λ) is measured eastward through 360° through vernal equinox (the celestial equator and the ecliptic intersect at two points, referred to as the vernal equinox and the autumnal equinox), but in the plane of, or parallel to the ecliptic rather than the celestial equator, and in degrees, minutes and seconds of arc, rather than time. The celestial longitude is measured in a counterclockwise sense of rotation, as viewed from the north pole of the ecliptic. The other coordinate is the celestial latitude (β) which is measured in degrees, minutes and seconds of arc directly north or south of (*i.e.* along ecliptic meridians perpendicular to) the plane of the ecliptic. At the north and south poles it reaches 90° (for the moon β ranges only from 4°56′ to 5°20′). The advantage of the ecliptic system is that it eliminates the inclination of the earth's axis of rotation with reference to the ecliptic, and is especially useful when the sun's apparent annual motion and the motions of the moon with respect to this motion are considered.
Horizon system	This is a localized system and involves the position of a celestial body as seen from a specified point on the earth's surface. The azimuth and altitude of the celestial body are topocentric and not geocentric as in the other two systems. The astronomical horizon is a plane perpendicular to the local direction of gravity. The zenith is the point where the local direction of gravity intersects the celestial sphere. The coordinate of azimuth (Z) is measured in the plane of the horizon, in degrees, minutes and seconds of arc through 360° in a counterclockwise (westerly) direction as viewed from the local zenith of the place. Vertical angular distances are measured in one direction only (in degrees, minutes and seconds of arc) above the astronomical horizon through 90° towards the astronomical zenith, and this coordinate is known as astronomical altitude (H). Small circles of equal altitude parallel to the horizon are called almucantars; great circles perpendicular to the horizon and passing through the north and south points on the horizon, together with the zenith, nadir, and north and south celestial poles are known as the celestial meridians. This system is particularly useful in computing the tides due to the moon, noting that the moon is relatively close to the earth compared with other celestial bodies. Since the earth is an irregular spheroid, the moon's parallax and distance from the earth's surface vary significantly at any angular altitude above the horizon. Thus, the horizon system is useful in tidal calculations in determining the local zenith distance of the moon, which enters into the computation of parallax as a measure of distance of the moon from the earth's surface, as well as in the lunar augmentation.

TABLE II

General equations for transformation of coordinates (based on Wood, 1978): α = apparent right ascension; δ = declination; λ = celestial longitude; β = celestial latitude; ε = obliquity of the ecliptic (angle of inclination between the celestial equator and the ecliptic); Z = zenith distance; A = azimuth; H = altitude; ϕ = latitude of the place of observation

Equatorial system to ecliptic system or reverse	Equatorial system to the horizon system or reverse
$\sin\beta = \sin\delta\cdot\cos\varepsilon - \cos\delta\cdot\sin\alpha\sin\delta$	$\sin z\cdot\sin A = -\cos\delta\cdot\sin H$
$\cos\beta\cdot\sin\lambda = \sin\delta\cdot\sin\varepsilon + \cos\delta\sin\alpha\cdot\cos\delta$	$\cos z = \delta\sin\phi + \cos\delta\cos H\cdot\cos\phi$
$\cos\beta\cdot\cos\lambda = \cos\delta\cdot\cos\alpha$	$\sin z\cdot\cos A = \sin\delta\cos\phi - \cos\delta\cdot\cos H\cdot\sin\phi$
and	and
$\sin\delta = \sin\beta\cdot\cos\varepsilon + \cos\beta\sin\lambda\cdot\sin\varepsilon$	$\cos\delta\cdot\sin H = -\sin z\cdot\sin A$
$\cos\delta\cdot\sin\alpha = \cos\beta\cdot\sin\lambda\cos\varepsilon - \sin\beta\sin\varepsilon$	$\sin\delta = \cos z\cdot\sin\phi + \sin z\cdot\cos A\cdot\cos\phi$
$\cos\delta\cdot\cos\alpha = \cos\beta\cdot\cos\lambda$	$\cos\delta\cos H = \cos z\cdot\cos\phi - \sin z\cdot\cos A\cdot\sin\phi$

between synodic and anomalistic months); when the time separation between these two events is not greater than 36 h, the astronomical situation is referred to as perigee-syzygy, and the tides (with increased range) are known as "perigean spring tides".

According to Wood (1978), whenever the perigee and syzygy occur with a separation of few hours, augmented dynamic influences act to increase the eccentricity of the lunar orbit, the lunar parallax and thus the orbital velocity of the moon. These solar induced perturbations also reduce the moon's perigee distance; at certain times, lunar passage through perigee involves a particularly close approach of the moon to the earth. To distinguish these cases of unusually close perigee, Wood (1978) coined a new term "proxigee" and the associated tides of proportionately increased amplitude are designated as "proxigean spring tides".

Twice each month (at new moon and full moon) the sun and the moon in their real as well as apparent revolution (with reference to the earth) arrange themselves into direct alignment with the earth in celestial longitude (see Fig. 5A,B). According to Wood (1978) the moon may either lie along a straight line connecting the earth and the sun (as happens at new moon or conjunction) or on the far side of the earth from the sun (as occurs at full moon or opposition). Note that, in either case, if the moon simultaneously crosses the plane in which the earth revolves around the sun (or comes within a limiting angular distance thereof) a solar or lunar eclipse occurs.

The alignment of the sun and moon with the earth in celestial longitude occurs twice in a duration of 29·53 days; these are the syzygy periods. The moon revolves round the earth in an elliptic orbit, with the earth occupying one of the two foci (C in Fig. 5A). At least once a month (the moon takes 27·55 days to complete one revolution round the earth, and this makes it possible to

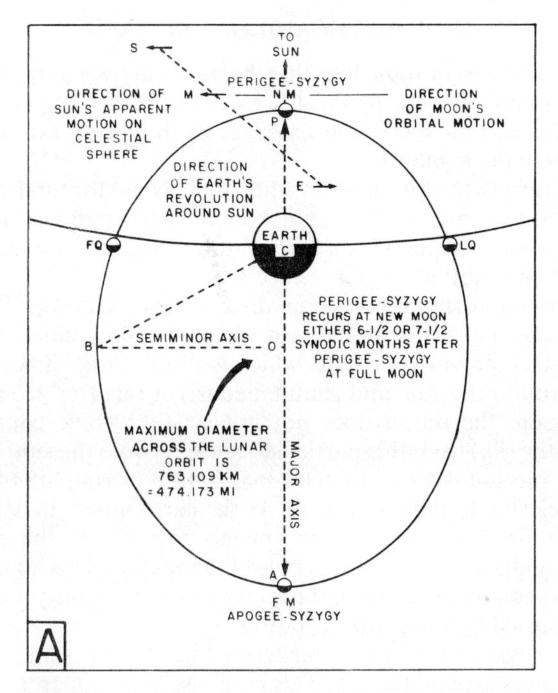

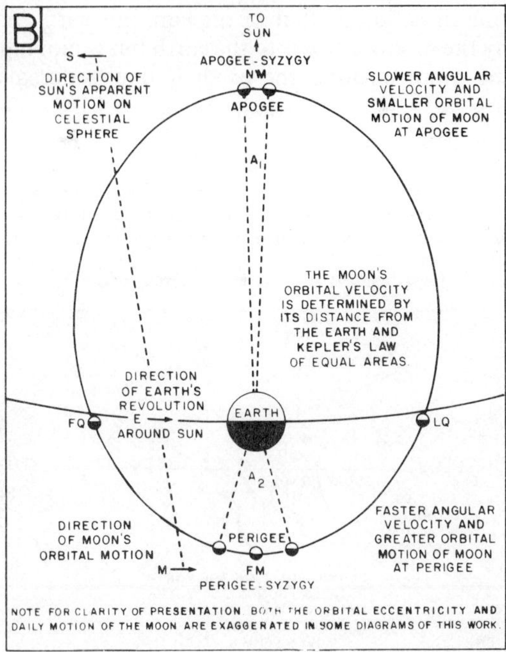

Fig. 5—A, syzygy alignment of moon and sun at new moon, with the moon between the earth and the sun: a near coincidence of perigee and syzygy can also occur at full moon; from Wood (1978). B, revolution of the moon around the earth in an elliptic orbit brings the moon to perigee each anomalistic month, averaging 27·555 days: the moon then reaches maximum orbital velocity; from Wood (1978).

have two occurrences in some months) the moon arrives in the position of its closest approach to the earth, which we defined earlier as perigee. The time duration between two successive passages of the moon through perigee is known as anomalistic month.

Now we return to the four motions that are relevant for tidal computations. These are: (1) the diurnal rotation of the earth, (2) the earth's annual revolution around the sun, (3) the moon's revolution around the earth, and (4) declinational and right ascensional effects.

The following information is taken directly from Wood (1978). The earth revolves around the sun in a direction which is counterclockwise as viewed from the north pole of the ecliptic, which is in the same direction as earth's revolution around the sun, and simultaneously it revolves around the earth. For this reason, the moon does not appear to revolve counterclockwise (eastward in the sky) from this particular cause, as does the sun. The apparent annual solar motion is equal to 360°/365·25 days or roughly one degree per day, in the celestial longitude (Fig. 6). As the earth moves in a counterclockwise sense around the sun, the sun appears to move in the same sense of revolution as judged from the north pole of the ecliptic. This apparent easterly motion has to be subtracted from the apparent daily westerly motion of rising and setting caused by the earth's rotation.

The moon revolves around the earth (see Fig. 6) in the same sense in which the earth revolves around the sun. This gives rise to an apparent motion of the moon in the same direction as that of the sun, but with greater magnitude because, not only the moon is closer to the earth but is moving faster. The true revolution of the moon around the earth is one "sidereal month" and is

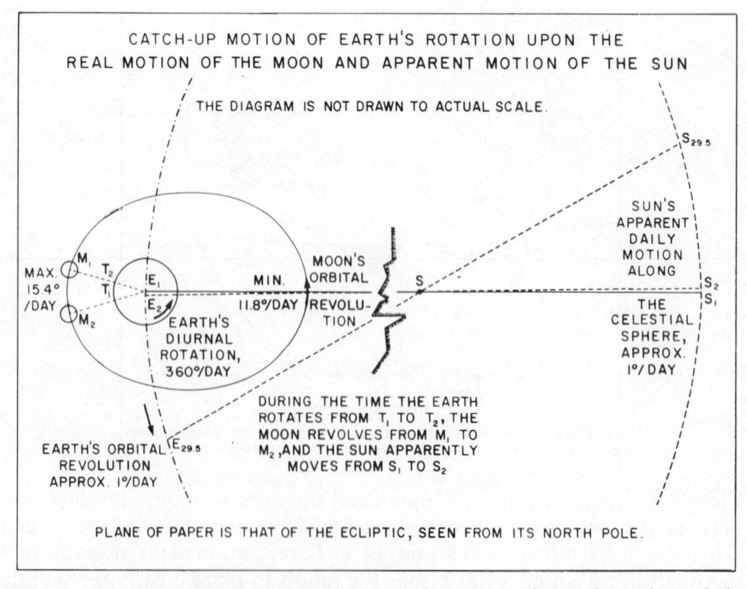

Fig. 6.—Catch-up motion of earth's rotation upon the real motion of the moon and apparent motion of the sun (from Wood, 1978).

27·321661 days. The "synodic month" is the period between alignments of the sun and moon in identical longitudes (or right ascensions) and the next occurrence of this same syzygy position. Thus it is the interval between either two full moons or between two new moons and is 29·530589 days.

As the moon revolves around the earth in its monthly orbit, at one position in the orbit referred to as "perigee", the moon will reach its closest monthly approach to the earth, and roughly half a month later, the moon will reach its greatest monthly distance from the earth, known as "apogee". Similarly, in the earth's annual motion around the sun, the earth will pass through a position of closest approach (between 2 and 4 January) to the sun, known as "perihelion", and about six months later (between 3 and 6 July) the earth will pass through a position of greatest distance, referred to as "aphelion".

Because of the increased gravitational forces involved at the time of closest approach, the speed of the secondary object (for the revolution of the moon round the earth, the secondary object is the moon, for the revolution of the earth around the sun, obviously the secondary object is the earth) in its orbit will be greater. This happens at perigee (for the moon) and at perihelion (for the earth). Exactly opposite effects occur at apogee and aphelion. Instead of specifying the angular velocity for the earth around the sun, we, however, treat it as angular velocity of the sun round the earth. The maximum and minimum daily angular velocities of the moon are 15·4° and 11·8°, respectively, whereas for the sun these are 1·016° and 0·983°, respectively.

TIDE-GENERATING FORCES

The tides at a given location in the ocean are due to the super-position of the long gravity waves of tidal frequences generated in the ocean by the tide-raising forces of the sun and the moon. In this section our main source of information is the excellent manual by Forrester (1983). It will be shown that the tide-generating forces are portions of the gravitational attractions of the sun and the moon that are not balanced by the centripetal acceleration of the earth in its orbital motion. Note that only at the centre of the earth, are the gravitational attractions exactly balanced by the centripetal accelerations, which in fact is a required constraint for orbital motions.

The centrifugal force due to the rotation of the earth is included in earth's gravity; since earth's gravity is contant in time, it cannot make any contribution to the tide-generating forces. Following Forrester (1983) we shall first discuss the tide-raising force of the sun because the orbital parameters for the sun–earth system are easier to understand, and apply similar principles to the earth–moon system.

To derive an expression for the sun's tide-generating force, we start with Newton's second law of motion which states that the acceleration of a body equals the force acting on it per unit mass, or

$$\text{acceleration} = \frac{\text{force}}{\text{mass}} \tag{3}$$

Next we invoke the law of universal gravitation which states that the gravitational attraction F_g on a unit mass due to a body of mass M is

$$F_g = \frac{GM}{r^2} \tag{4}$$

where r is the distance between the body and the unit mass and G is the universal gravitational constant. For a body moving with a velocity v along an orbit with a radius of curvature r, the centripetal acceleration (*i.e.* acceleration towards the centre of curvature) is given by

$$A_c = \frac{v^2}{r^2} \tag{5}$$

The ratio of the gravitational attractions of the sun and the moon on the earth is

$$\frac{F_g(\text{sun})}{F_g(\text{moon})} = \frac{27 \times 10^6}{390^2} = 178 \tag{6}$$

In arriving at the ratio 178 we made use of Equation 4 and also the fact that the sun is 390 times further than the moon from the earth, and also the mass of the sun is 27 million times the mass of the moon. The interesting fact is that, although the sun's gravitational attraction on the earth is 178 times greater than the gravitational attraction of the moon on the earth, yet it is the moon that produces larger tides on the earth than the sun. This apparent paradox can easily be resolved by noting that it is only that part of the gravitational force that is not balanced by the centripetal acceleration in the orbital motion that generates tides. It will be shown that this unbalanced portion is inversely proportional to the cube of the distance from the earth, but is directly proportional to the mass of the attracting body. Hence the tide-raising force of the sun on the earth is approximately 178/390 = 0·46 times that of the tide-raising force of the moon on the earth.

With reference to Figure 7, we note that, since the acceleration due to the earth's diurnal rotation along its own axis is already included in earth's gravity, for the following discussion we assume that the earth has a fixed orientation in space during its revolution around the sun; hence the centripetal acceleration is the same everywhere on the earth. The centripetal acceleration at the centre O of the earth is exactly balanced by the gravitational attraction of the sun at that point (this is a necessary constraint required by the orbital motion). Thus, the centripetal acceleration (which is uniform on the earth is given by GS/r^2 where S is the mass of the sun, and r is the distance of the sun to the earth's centre (although we say here that the earth is revolving round the sun, in reality the sun and the earth are orbiting around a common centre of mass, which is less than 500 km from the centre of the sun).

Consider the balance of forces at the four points A, A′, B, B′, in Figure 7. At point A, the gravitational attraction of the sun is greater than at the earth's centre O; the difference between the gravitational force and the centripetal force (*i.e.* the unbalanced component) tries to accelerate a mass at A, away from O and towards the sun. At A′, the gravitational force is less than at O, and the unbalanced component accelerates a mass at A′ away from O and away from the sun. At points B and B′, the gravitational force is not much different from that at O; its direction, however, is slightly different. Thus the unbalanced

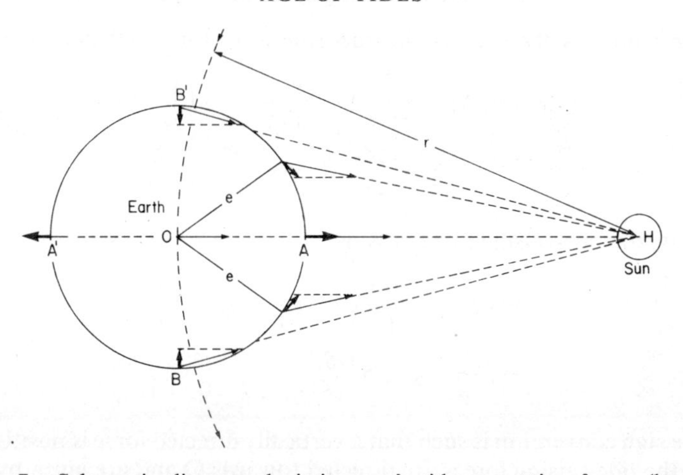

Fig. 7.—Schematic diagram depicting sun's tide-generating forces (heavy arrows) which are the differences between sun's gravitational force on the earth and the earth's centripetal acceleration in solar orbit: note that this diagram which shows a portion of the earth's orbit around the sun is not drawn to scale; in particular the cross-section through the earth is greatly exaggerated with reference to the size and distance of the sun; from Forrester (1983).

components at B and B′ act towards O. Basically, the sun's tide-generating forces are these unbalanced components of the gravitational attraction. It can be seen that at the four points A, A′, B, B′, the tide-raising forces are vertical, and at other points the forces are inclined to the vertical, the angle of inclination depending upon the location of the point.

At four intermediate locations the tide-raising forces are completely horizontal; these horizontal tide-raising forces are referred to as the tractive forces. These forces tend to accelerate the water away from B and B′ towards A and A′ in such a way that the water surface everywhere is perpendicular to the vectorial sum of the tide-generating force and earth's gravity. Such an ideal surface, referred to as the mean sea level (determined by gravity alone) is known as the equilibrium tide.

Now we shall derive expressions for the tide-raising forces of the sun. Let H be the centre of the sun (Fig. 7) and e the radius of the earth. The sun's gravitational attraction is:

$$\begin{aligned} \text{at O:} &\quad GS/r^2 \\ \text{at A:} &\quad GS/(r-e)^2 \\ \text{at A}': &\quad GS/(r+e)^2 \\ \text{at B and B}': &\quad GS/(r+e)^2 \end{aligned} \tag{7}$$

Note that all these attractions are directed towards the centre H of the sun. For the expressions $1/(r-e)^2$ and $1/(r+e)^2$ we use binomial expansions and ignore the higher powers of e/r. Thus, the tide-generating force at A due to the sun is

$$\begin{aligned} F_t(\mathrm{A}) &= \frac{GS}{r^2}\left(1+2\frac{e}{r}+\cdots-1\right) \\ &\simeq 2\frac{GSe}{r^3}. \end{aligned} \tag{8}$$

Similarly, the tide-raising force at A′ is

$$\begin{aligned} -F_t(\mathrm{A}') &= \frac{GS}{r^2}\left(1-2\frac{e}{r}+\cdots-1\right) \\ &\simeq -2\frac{GSe}{r^3} \end{aligned} \tag{9}$$

here the sign convention is such that a vertically directed force is positive. At B and B′ the tide-raising forces are directed towards O and are given by

$$\begin{aligned} -F_t(\mathrm{B}) &= -F_t(\mathrm{B}') \\ &= \left|\frac{GS}{r^2}\left(1+\frac{e^2}{r^2}\right)^{-1}\right|\sin\beta \simeq \frac{GSe}{r^3} \end{aligned} \tag{10}$$

where β is the angle OHB or OHB′. In deriving Equation 10, e^2/r^2 was ignored and $\sin\beta$ was approximately equated to e/r. Thus, the tide-raising forces due to the sun are directly proportional to the mass of the sun and inversely proportional to the cube of the distance between the earth and the sun. Next we shall consider the tide-raising forces due to the moon.

The earth and the moon orbit about a common centre of mass which is inside the earth (about 1700 km below the surface). One can use the same arguments (as for the sun's tide-raising forces) to derive the tide-raising forces due to the moon, with the understanding that now H is the centre of the moon and r is the distance of the moon to the centre of the earth. The expansional forces at the points for which the moon is in the zenith and the nadir are

$$F_t(\mathrm{A}) = F_t(\mathrm{A}') \simeq 2\frac{GMe}{r^3}. \tag{11}$$

The compressional forces on the great circle around the earth's surface midway between these two points are

$$F_t(\mathrm{B}) = F_t(\mathrm{B}') \simeq -\frac{GMe}{r^3}. \tag{12}$$

Now following Forrester (1983) we shall show that the tide-raising forces are very small in magnitude. The ratio of the tide-generating force of the moon (the tide-generating force of the sun is about half that of the moon, as was shown earlier) to earth's gravity is about 10^{-7}. Although the tidal forces are very small, they act on every water particle in the ocean and accelerate those particles towards the sublunar (or subsolar) point on the other side of the earth, and towards its antipode on the far side. Thus, although the perturbations set up in this manner in the deep ocean are small, they get amplified in shallow coastal areas.

Now we shall consider the tidal potential and the equilibrium tide. The tide-raising forces due to the moon and the sun can be written in the form of the negative gradient of the so-called tidal potential. The tidal potential can be combined with the potential for gravity (referred to as the geopotential) and the equi-potential surfaces of the combined field can be interpreted as level surfaces. In other words, one of these equi-potential surfaces is the surface of the equilibrium tide, *i.e.* the surface which the water would assume, if it can respond instantaneously to the tidal forces. The geopotential, however, is time-invariant. We are interested only in the variations in time of the tidal potential.

The tidal potentials, p_t, at an arbitrary point P on the earth's surface are given by

$$-p_t(\text{moon}) = \frac{GMa^2}{2r_m^3}(3\cos^2\alpha_m - 1) \tag{13}$$

and

$$-p_t(\text{sun}) = \frac{GSa^2}{2r_s^3}(3\cos^2\alpha_s - 1) \tag{14}$$

where M and S are, respectively, the masses of the moon and the sun, a is the distance from earth's centre to P ($a = e$, the radius of the earth if P is on the earth's surface), r_m and r_s are distances of the moon and the sun from the earth, and α_m and α_s are their zenith angles (*i.e.* colatitudes). The vertical component of the tidal force can be obtained by differentiating Equation 14 with respect to a. With $\cos\alpha = 1$, this reproduces the expression 8 for the tidal force at A (Fig. 7), and with $\cos\alpha = 0$, it reproduces the expression 10 for the tidal force at B.

The height of the equilibrium tide with reference to the mean sea level can be calculated by recognizing that the equilibrium tide surface is an equi-potential surface for the combined gravity and tidal fields. Thus an increase in the tidal potential results in a decrease in the geopotential (and *vice versa*) which leads to a fall in the water surface level. Since an equilibrium tide height of h means an increase of g.h. in the geopotential, the tidal potential p_t has to be (to maintain an equi-potential surface) such that

$$\Delta h = \frac{-p_t}{g}. \tag{15}$$

Substituting Equations 13 and 14 into 15 and noting that the universal constant of gravitation G is related to the acceleration due to gravity g through

$$G = \frac{g \cdot e^2}{E} \tag{16}$$

where e is the radius of the earth and E is its mass, we get for the heights of the lunar and solar equilibrium tides

$$\Delta h(\text{moon}) = \frac{Me^4}{2Er_m^3}(3\cos^2\alpha_m - 1) \tag{17}$$

$$\Delta h(\text{sun}) = \frac{Se^4}{2Er_s^3}(3\cos^2\alpha_s - 1). \tag{18}$$

In deriving this, we took $a = e$ because we are concerned with equilibrium tide for points on the earth's surface.

Forrester (1983) gives the following values: radius of the earth $e = 6400$ km, the ratio of the mass M of the moon to the mass E of the earth is 0·012, the ratio of the radius of the earth e to the distance r_m between the earth and the moon is 0·017, and the corresponding ratio for the sun is $4{\cdot}3 \times 10^{-5}$. The ratio of the mass of the sun S to the mass of the earth E is $3{\cdot}3 \times 10^{-5}$. Since the extreme values for the equilibrium tide occur for $\alpha = 0°$ and $90°$ we get, for the heights of the extreme equilibrium tide

$$\Delta h(\text{moon}) = 0{\cdot}38 \text{ m, and } -0{\cdot}19 \text{ m} \tag{19}$$

and

$$\Delta h(\text{sun}) = 0{\cdot}17 \text{ m, and } -0{\cdot}08 \text{ m.} \tag{20}$$

As for the ratio of the tide-raising forces, the ratio of the extreme solar equilibrium tide to the extreme lunar equilibrium tide is again 0·46.

There is no location on the earth where the observed tide is identical to the equilibrium tide at that location. With reference to Figure 8 the equilibrium spring tide occurs on the day when the sun's and the moon's high waters (HW) occur at the same meridian (at new and full moon). These high waters occur around local noon and midnight and are naturally greater than the average value because of the combination of the equilibrium tides due to the sun and

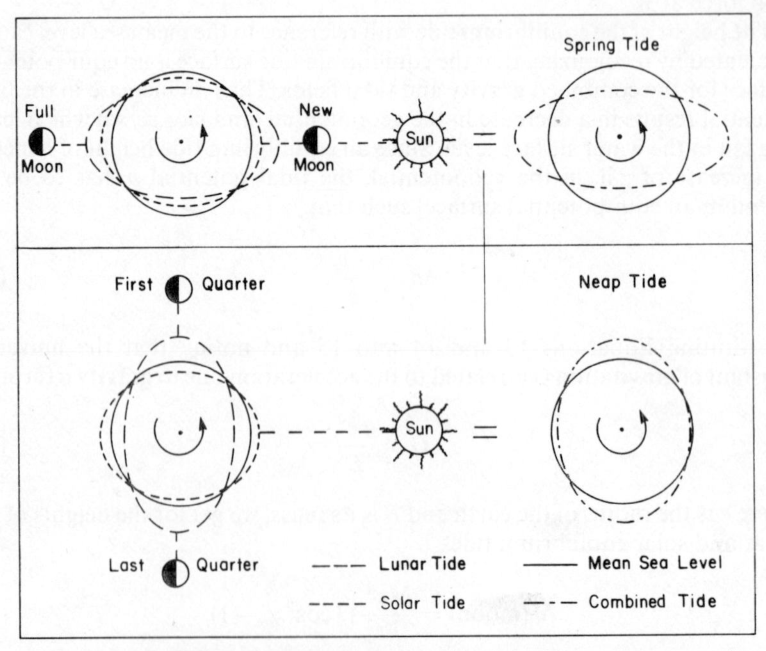

Fig. 8.— Combination of lunar and solar equilibrium tides to generate spring tides at new moon and full moon and neap tides at first and last quarters (from Forrester, 1983).

the moon. Similarly, the two low waters (LW) also reinforce each other and the combined value is smaller than the average value. Hence there is a greater than average range of equilibrium semi-diurnal tide associated with spring tides. During the moon's first and last quarter, the equilibrium neap tide occurs because the HW's due to the sun and the moon almost coincide in time with the other LW's (Fig. 8). Hence the tidal range is smaller than average at neap tides.

TIDAL CONSTITUENTS

In this section, following Forrester (1983) we shall discuss how the important tidal constituents arise. Figure 9 shows the equilibrium tide due to the sun superimposed on the equi-geopotential surfaces of the mean sea level, for three different positions of the sun. One can see from the top part of Figure 9, which corresponds to the sun on the Equator (zero declination), an observer on the Equator rotates with the earth once each solar day with respect to the sun's equilibrium tide, passing through LW at the points B and B′, where B′ is on the opposite side of the earth from B; the observer passes through HW at points A and A′. In other words, an observer at any latitude would have equilibrium LW as the meridian on which he is located passes through the points B and B′, and he has HW as his meridian passes through A and A′. Note that the amplitude of the HW would be smaller with the increase of latitude in both the Northern and Southern Hemispheres. The passage through HW and LW gives rise to the semi-diurnal solar constituent S_1 with a frequency of two cycles per day (frequency of 30° per hour). The same argument can be used for

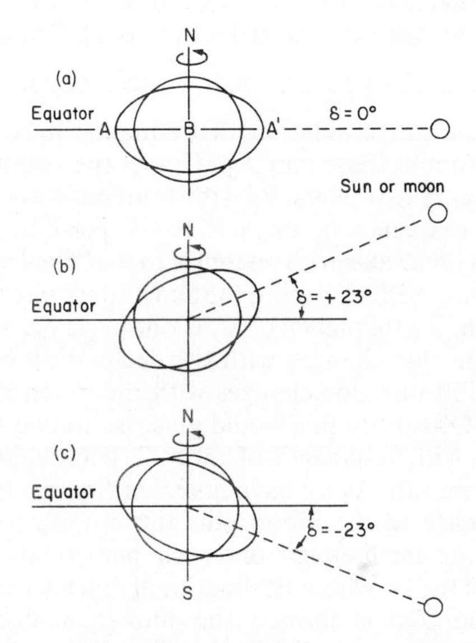

Fig. 9.—Equilibrium tidal surfaces (AA′) for sun or moon (a) on equator, (b) north of equator, and (c) south of equator (from Forrester, 1983).

the moon, and thus we obtain the semi-diurnal lunar constituent M_2, with a frequency of two cycles per lunar day. The lunar day is approximately 50 minutes longer than the solar day due to the fact that the moon advances about 12·5° in its orbit every day with reference to the position of the sun. Thus, the frequency of the M_2 constituent works out to be about 28·98° per hour.

Next we shall consider the origin of the diurnal constituents, K_1 and O_1. With reference to the middle and bottom parts of Figure 9, when the sun is north and south of the Equator, one centre of HW for the solar equilibrium tide is north and the other centre is south of the Equator. Then an observer at the Equator would still notice two equal HW's and two equal LW per day, with the difference that the HW's would not be as high as in case (a). For case (b) shown in Figure 9, an observer at a northern latitude would have HHW at noon and LHW at midnight; an observer at a southern latitude experiences HHW at midnight and LHW at noon. For case (c) of Figure 9 the situation is the opposite of case (b). The height difference between HHW and LHW is known as the diurnal inequality; it increases with the declination of the sun, as well as with the latitude of the observer, for the case of the equilibrium tide.

It can be shown that, the zone of low water around the earth for the equilibrium tide cannot extend any farther north or south than $(90° - \delta)$ where δ is the sun's declination. This means, at latitudes greater than $(90° - \delta)$ one can only observe a distorted diurnal tide, with only one HW and extended period of low water. One can regard a semi-diurnal tide with a diurnal inequality as the combination of a semi-diurnal and a diurnal tide. Note that the diurnal tide must augment the noon HW for the case where the sun and the observer are on the same side of the Equator; the diurnal tide must reduce to zero when the sun is on the Equator, and must reinforce the HW at midnight when the sun is on the opposite side of the Equator. To be able to do this, we need more than one diurnal constituent.

Let n_1 be the angular speed of 360° per solar day (i.e. 15° per hour) and n_0 is the angular speed 360° per year (*i.e.* 0·04° per hour). One can write

$$\cos(n_1 + n_0)t + \cos(n_1 - n_0)t = 2(\cos n_0 t)(\cos n_1 t). \tag{21}$$

This results in a diurnal oscillation of frequency n_1 whose amplitude is modulated at the annual frequency n_0. This is the origin of the two solar declinational diurnal constituents P_1 with frequency $n_1 - n_0$ (*i.e.* 14·96° per hour) and K_1 with frequency $n_1 + n_0$ (*i.e.* 15·04° per hour). Since the rate of rotation of the earth on its axis with reference to the "fixed stars" is equal to the sum of its rotation rate with respect to the sun and its rate of revolution in the orbit around the sun, the frequency of K_1 is one cycle per sidereal day. Just as the solar equilibrium tide changes with the declination of the sun over any year, the lunar equilibrium tide changes with the moon's declination over a period of a month. Obviously this would give rise to two lunar declinational diurnal constituents with frequencies of one cycle per lunar day plus and minus one cycle per lunar month. As we have noted earlier, the frequency of earth's rotation with reference to the moon plus the moon's frequency of orbital revolution around the earth equals one cycle per sideral day. Thus K_1 also happens to be one of the two lunar declinational diurnal constituents. Because of this double rôle, K_1 is termed the luni-solar declinational diurnal constituent. The other lunar declinational diurnal constituent is O_1 with an angular speed of 13·94° per hour.

So far we have discussed the origin of the two semi-diurnal constituents S_2 and M_2 and the two diurnal constituents K_1 and O_1. In the present study we need to consider one more semi-diurnal constituent, namely N_2. We shall now consider its origin, again following Forrester (1983). We have seen that the orbit of the moon around the earth and that of the earth about the sun are ellipses with the earth as one of the foci for the first case and the sun as one for the second. Because of the elliptical nature of the orbit, the distances between the earth and the moon and between the earth and the sun change during the period of the orbit (the orbital periods being one year for the earth around the sun and one month for the moon around the earth). We also saw that the orbital points at which the earth is closest and farthest from the sun are respectively referred to as perihelion and aphelion. Similarly for the moon in the orbit around the earth, the corresponding terms are perigee and apogee. Since the tidal potential varies inversely as the cube of the distance r_m or r_s according to Equations 13 and 15, the shapes of the solar equilibrium tides change at perihelion and aphelion, and the shape of the lunar equilibrium tide changes at perigee and apogee. Because the tidal potential due to the moon is greater, these changes are much more pronounced for the lunar equilibrium tide. Thus, the amplitude of the solar equilibrium tide is modulated with a period of one year, and the amplitude of the lunar equilibrium tide is modulated with a period of one month.

According to Forrester (1983) the combined effect of the amplitude and phase modulations can be simulated by adding to each constituent two satellite constituents with frequencies equal to that of the main constituents plus and minus the orbital frequency, but with the amplitude of one satellite greater than the other. One can regard the tidal constituents as rotating vectors, because they have a fixed amplitude and a uniformly increasing phase angle. In Figure 10 OR is the main constituent with angular speed n, RS is the

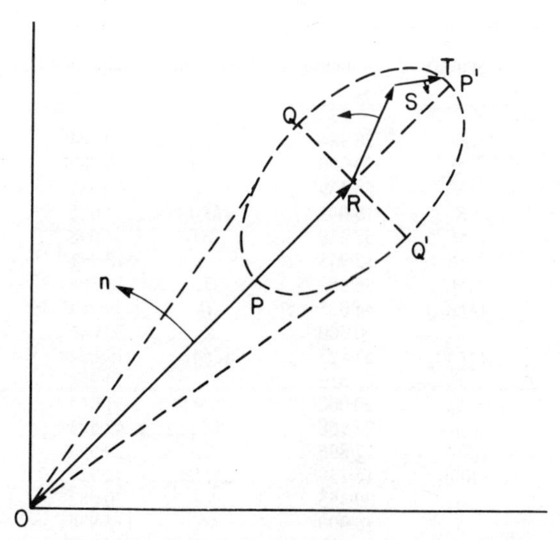

Fig. 10.—Amplitude and phase modulation produced by vector sum of main constituent plus two counter-rotating satellite constituents (from Forrester, 1983).

first satellite constituent with angular speed $n_1 = n+n'$, and ST is the second satellite constituent with angular speed $n_2 = n-n'$. The sum of these three constituents is OT; relative to the rotating vector OR, the point T traces an ellipse. The centre of this ellipse is R, its semi-major axis is equal to the sum of the satellite amplitudes, and its semi-minor axis is equal to the difference in their amplitudes.

It can be seen from Figure 10 that as the point T moves around the ellipse once for each cycle of the main constituent, the amplitude of the vector sum OT oscillates between OP and OP′; the phase oscillates about that of OR through the angle QOQ′. Note that if the satellite amplitudes are equal, the ellipse degenerates into a straight line, and there is no phase modulation (only amplitude modulation). This is the origin of the larger and smaller lunar elliptic semi-diurnal constituents N_2 (with an angular speed of 28·44° per hour) and L_2 (with an angular speed of 29·53° per hour). The corresponding larger and smaller solar elliptic semi-diurnal constituents are T_2 (angular speed of 29·96° per hour) and R_2 (angular speed of 30·04° per hour). The amplitude of R_2 is so small that for all practical purposes it can be ignored.

Several other tidal constituents arise due to various causes. Since these other tidal constituents are not relevant for the present study we will not go into any detailed discussions about these, but refer the reader to Forrester (1983) and Godin (1972). Table III lists the tidal constituents (along with their frequencies) commonly used by the National Ocean Survey of the U.S.A. (Harris, 1981).

TABLE III

Tidal constituents commonly used by the National Ocean Survey, U.S.: the frequency is in degrees per hour; from Harris (1981)

Symbol	Frequency	Symbol	Frequency
M_2	28·984	Mm	0·544
S_2	30·000	Ssa	0·082
N_2	28·439	Sa	0·041
K_1	15·041	Msf	1·015
M_4	57·968	Mf	1·098
O_1	13·943	ρ_1	13·471
M_6	86·952	Q_1	13·398
(MK_3)	44·025	T_2	29·958
S_4	60·000	R_2	30·041
$(MN)_4$	57·423	$(2Q)_1$	12·854
ν_2	28·512	P_1	14·958
S_6	90·000	$(2SM)_2$	31·015
μ_2	27·968	M_3	43·476
$(2N)_2$	27·895	L_2	29·528
$(00)_1$	16·139	$(2MK)_3$	42·927
λ_2	29·455	K_2	30·082
S_1	15·000	M_8	115·936
M_1	14·496	$(MS)_4$	58·984
J_1	15·585		

A tide wave gets distorted as it travels in shallow water, and such a distortion could be represented by superposition of the harmonic tidal frequencies. This is the origin of the shallow water tidal constituents, also known as over-tides. They simply are used as a mathematical convenience for the simulation of the distortion of the tide wave, and they do not occur, at least directly due to the tide-generating forces.

TIDES IN THE GLOBAL OCEANS

Following Murty (1984) we shall review the tidal regimes in the global oceans. The tidal regimes in the world oceans are usually described through cotidal charts, which show lines of simultaneous occurrence of high water in various regions. Whewell (1833) appears to have produced the first cotidal chart for the semi-diurnal tides in the global oceans. These charts are not very accurate because they are produced by simply interpolating from the coastal data. The next charts to appear were those of Harris (1894–1907). This work also suffered many drawbacks, notable ones being neglect of the earth's rotation and not giving absolute values of amplitudes of the tides. Sterneck (1920) produced charts for the semi-diurnal tides, mainly based on interpolation of observed data. Dietrich (1944a,b) gave charts for the constituents M_2, S_2, K_1, and O_1, Villain (1952) gave a chart for the M_2 tide in the global oceans. Defant (1961) appears to be among the first to have produced a cotidal chart based on a numerical model. Since then, dozens of numerically produced cotidal charts for various water-bodies on the globe have appeared in the literature.

Proudman (1944) studied the distribution of the M_2 tide in a section of the Atlantic Ocean between 35° S and 45° N including the effects of the earth's rotation but ignoring coastal energy dissipation. He prescribed arbitrary values for the current and water level along the 35° S latitude as boundary conditions. The motion he considered consisted of an independent tide due to the tidal potential and four free oscillations in the form of northward and southward propagating Kelvin and Poincaré waves. Linear combinations of these have been made to agree with observed M_2 at different coastal locations.

Accad & Pekeris (1978) studied M_2 and S_2 tidal regimes in the global oceans by solving the Laplace tidal equations, based on a knowledge of the tidal potential alone. There are about 1300 tidal stations in the open ocean at which the tide is measured by recording the pressure with an instrument located at the bottom. The topography of the global ocean model is made up of 2° arcs of latitude and longitude on a Mercator projection. It was assumed that tidal dissipation occurred only at the coast where a portion of the incident tidal energy is assumed to be absorbed. To understand the influence of sharp corners introduced in the numerical model by approximating the continuous coastline with arcs of latitude and longitude, a model with a smooth coastline was also run. The influence of the sharp corners appears to be insignificant in the results. The M_2 and S_2 tidal regimes in the global oceans as computed by Accad & Pekeris (1978) are shown in Figures 11 and 12, respectively. Observed and calculated tides at selected island stations in the oceans are compared in Table IV.

Harris (1911) appears to be among the first to give a cotidal chart for the Arctic Ocean. He showed that the tide entering from the Atlantic Ocean takes

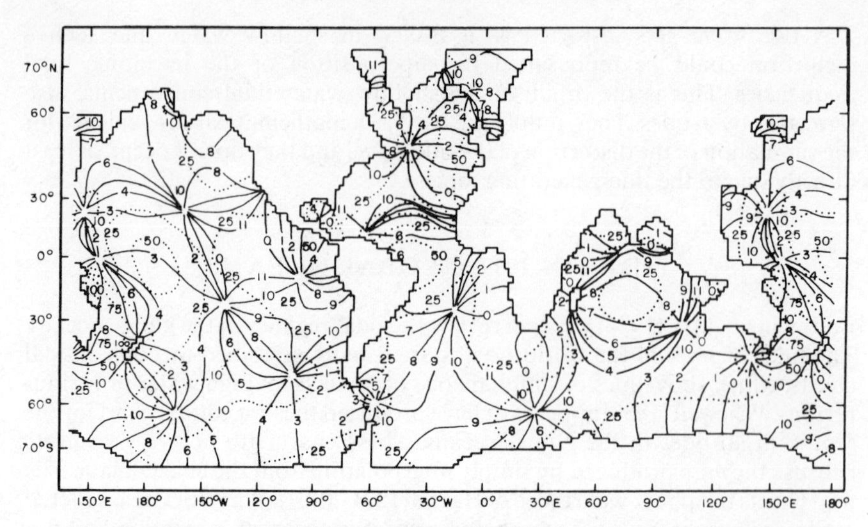

Fig. 11.—Distribution of the M_2 tide in the global oceans: solid lines show phases in Greenwich hours and dotted lines show range in centimetres (from Accad & Pekeris, 1978).

about 20 h to cross the Arctic Ocean. Following the observations taken during the Maud expedition, Defant (1924) and Fjeldstad (1929a,b) showed that the tide wave takes only 12 h (instead of 20 h) to cross the Arctic Ocean. Goldsbrough (1913) developed an analytical model for the tides in the Arctic Ocean. His study, which was limited to the region between 60° and 75°30′ N, showed that the semi-diurnal tides in the Arctic Ocean are very small.

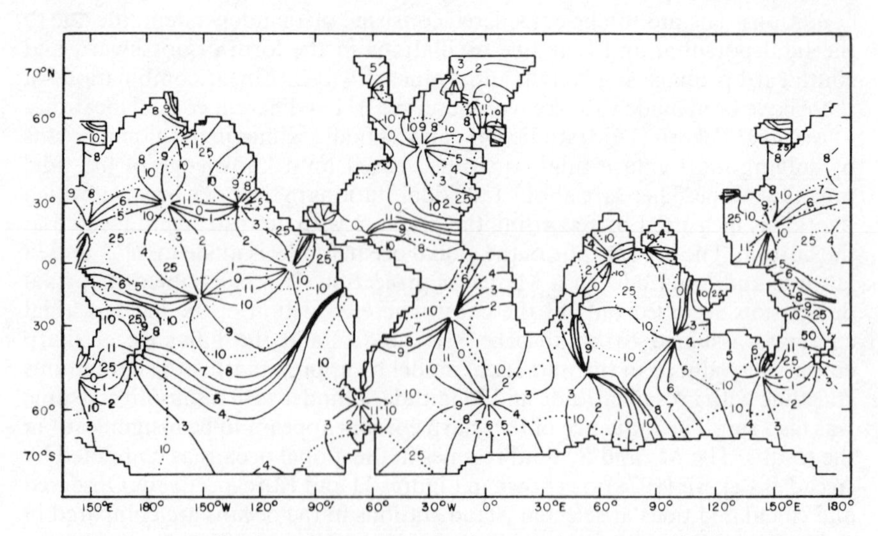

Fig. 12.—Distribution of the S_2 tide in the global oceans: solid lines show phases in Greenwich hours and dotted lines show tidal range in centimetres; from Accad & Pekeris (1978).

TABLE IV

Comparison of observed and computed M_2 tidal amplitudes and phases at island stations in the oceans (from Accad & Pekeris, 1978)

Station	Region	Amplitude (cm)		Phase lag (h)	
		Observed	Computed	Observed	Computed
North Atlantic Ocean					
Flores Island	Azores	39	34	2·0	1·8
Santa Maria Island	Azores	51	48	2·0	2·1
Funchal	Madeira	72	71	1·5	1·8
Tenerife	Canary Islands	69	70	1·0	1·3
Santo Antao	Cape Verde Island	30	21	8·7	9·2
Eleuretha Island	Bahama Island	32	20	0·7	11·7
St George's Island	Bermuda	37	24	0·0	11·5
South Atlantic Ocean					
Ascension Island		33	27	5·9	6·0
Fernando de Noronha		79	70	6·9	7·0
St Helena Island		32	27	2·7	2·4
Isla Trinidade		33	33	7·0	6·8
Tristan da Cunha		23	25	0·4	0·3
Stanley Harbour	Falkland Island	45	31	9·1	9·1
Elsehul	South Georgia	27	22	9·0	9·1
Scotia Bay	South Orkney Island	46	30	8·7	8·8
Indian Ocean					
Port Victoria	Seychelles Island	40	35	0·4	0·4
Port Louis	Mauritius	13	43	8·9	8·9
Addu Atoll		29	36	8·4	8·5
Port Refuge	Cocos Island	27	31	10·4	9·5
St Paul Island		38	36	7·7	8·0
Port-aux-Français	Kerguelen	51	16	6·5	6·8
North Pacific Ocean					
Pagan Island	Marianas Island	17	2	9·8	4·1
Kusail Island	Caroline Island	42	49	4·3	3·3
Port Rhin	Marshall Island	57	55	4·4	3·6
Midway Island		11	6	3·0	4·2
Johnston Island		27	23	3·5	3·0
Honolulu	Hawaiian Islands	16	16	2·1	0·1
Kahului	Hawaiian Islands	18	18	0·4	0·1
Hilo	Hawaiian Islands	21	20	1·0	0·3
South Pacific Ocean					
Lord Howe Island	Tasman Sea	59	33	10·1	9·0
Paagoumene Bay	New Caledonia	45	47	9·0	8·3
Vila Harbour	New Hebrides	34	34	6·9	7·3
Kingston	Norfolk Island	57	79	8·8	8·3
Nukualofa	Tonga Island	52	45	6·4	5·9
Funafuti	Ellice Island	57	50	5·1	4·5
Aitutaki	Cook Island	17	32	7·0	5·6
Hanga Piko	Easter Island	21	12	0·5	10·0
Caleta Aeolian	Galapagos Island	72	49	8·2	7·3
Cambier Island	Polynesian Archipelago	27	12	11·6	9·0
Ahe		12	4	2·9	4·1
Nukuhiva		47	21	1·2	0·6

Goldsborough concluded that the independent tide in the Arctic Ocean is insignificant and the observed tide is a co-oscillation with the Atlantic Ocean tide.

Defant (1924) used a one-dimensional numerical model for the combined Atlantic and Arctic oceans and showed that for the M_2 tide in the Arctic Ocean there is an amphidromic point north of Canada. Nekrasov (1962) used Defant's method for the Greenland Sea and the Norwegian Sea. Dvorkin, Kagan & Kleshyava (1972), used a two-dimensional numerical model with a grid spacing 1°. Actually, Zahel's (1973) model is for the global oceans and also includes the Arctic Ocean. Kowalik & Untersteiner (1978) developed a two-dimensional numerical model with a grid size of 75 km for the Arctic Ocean. One novel feature of this model is the manner in which these authors avoided the difficulty associated with the integration in a spherical polar coordinate system near the pole (the North Pole in this case). Kowalik & Bich Hung (1977) used a stereographic polar coordinate system modified by means of a scale factor m, and this was adapted by Kowalik & Untersteiner (1978).

Take the origin at the North Pole and let x and y, be coordinates along 0 and 90° E longitudes. Let M and N be the transport components along the x- and y-axes and η be the free surface elevation. Then the equations of motion and continuity are

$$\frac{\partial M}{\partial t} - m^2 A \nabla^2 M + \frac{k}{H^2} M|M| - fN + mgH \frac{\partial \eta}{\partial x} = F_x \tag{22}$$

$$\frac{\partial N}{\partial t} - m^2 A \nabla^2 N + \frac{k}{H^2} N|M| + fM + mgH \frac{\partial \eta}{\partial y} = F_y \tag{23}$$

$$\frac{\partial \eta}{\partial t} + m \frac{\partial M}{\partial x} + m \frac{\partial N}{\partial y} = 0. \tag{24}$$

Here, A is the horizontal eddy viscosity (taken as 10^9 cm^2·s^{-1}), k is a bottom friction coefficient (3×10^{-3}), f is the Coriolis parameter, $m = (M,N)$, $H(x,y)$ is the water depth, and F_x and F_y are the components of the tidal potential (taken to be zero). The gravity g is determined from

$$g = 978{\cdot}05\,(1 + 0{\cdot}0053 \sin^2 \phi)\ \text{cm·s}^{-2} \tag{25}$$

where ϕ is the latitude (of a given grid point). The scale factor m relates the surface elements δS on a stereographic map to those on a sphere as follows:

$$m = \frac{\delta S_{\text{max}}}{\delta S_{\text{sphere}}} = \frac{1 + \sin \phi_0}{1 + \sin \phi} \tag{26}$$

where ϕ_0 is the latitude through which a parallel plane of the stereographic projection passes. Near the pole, m is unity.

At the closed boundary

$$M = N = 0 \tag{27}$$

and at the open boundary

$$\eta = \eta(t) \tag{28}$$

is prescribed. Conservation of mass is assured by requiring that the integral taken over one tidal period T_p along the open boundary be zero:

$$\int_0^{T_p}\int_0^L (M\cos\alpha + N\sin\alpha)\,dt\,dS = 0 \tag{29}$$

where α is the angle between the perpendicular direction to the open boundary and the x-axis and L is the length of the open boundary.

A staggered grid in space (Hansen, 1962) and a leapfrog scheme in time are used. The stability criterions is

$$\Delta t \leq \frac{\Delta}{m\sqrt{2gH}} \tag{30}$$

where Δ is the grid size (taken as 75 km). In practice the following stability criterion is, however, more appropriate (Phillips, 1960):

$$\Delta t \leq \frac{R + 2m^2 \dfrac{A}{\Delta^2}}{2f^2} \tag{31}$$

where R is related to r, H, M, and N (see Kowalik & Untersteiner, 1978).

Topography plays a major rôle in the Arctic Ocean tidal model because the depth varies from 5·1 to 0 km. The north Siberian Shelf is the widest continental shelf on the globe and hence, lateral friction is also important. One might expect that at a critical latitude (where the tidal and inertial periods agree with each other), resonance would occur. Flattery (1967) showed, however, that this does not happen. Kagan (1968) showed that the almost permanent ice cover in the Arctic Ocean has negligible effect on the tide. For a general discussion of the influence of an ice layer on long waves, see Murty & Polavarapu (1979) and Murty & Holloway (in press).

The cotidal and corange lines for the M_2 tide in the Arctic Ocean are shown in Figures 13 and 14, respectively. These cotidal and corange maps agree with those of the U.S. Navy Hydrographic Office (Anonymous, 1958) and those of Zahel (1977). There is an amphidromic point at latitude 81°30′ N and longitude 133° W. This is in the deep water of the Canadian Basin off Prince Patrick Island. The M_2 tide from the Atlantic Ocean mainly enters through the Greenland Sea. At Spitsbergen its amplitude is about 40 cm and during its propagation northwards decreases to about 2 cm near the East Siberian and Chukchi Seas. In the region of the East Siberian Shelf, Chukchi Sea, and Beaufort Sea, the cotidal lines are somewhat parallel to the depth contours. At the entrance to the East Siberian Sea the amplitude is 10–15 cm falling to 2–3 cm near the coast.

Another branch of the Atlantic Ocean tide enters the Arctic Ocean between Spitsbergen and Norway. This causes higher amplitudes in the southern part of the Barents Sea and in the White Sea and begins to dissipate towards Novaya Zemlya. The model of Kowalik & Untersteiner (1978) did not reproduce the tide in the area of the White Sea, western part of the Barents Sea, and around Novaya Zemlya.

Platzman (1975, 1979) calculated the normal modes of the Atlantic and Indian Oceans, including their topography and the effect of the earth's rotation in a finite-element framework and paying particular attention to the

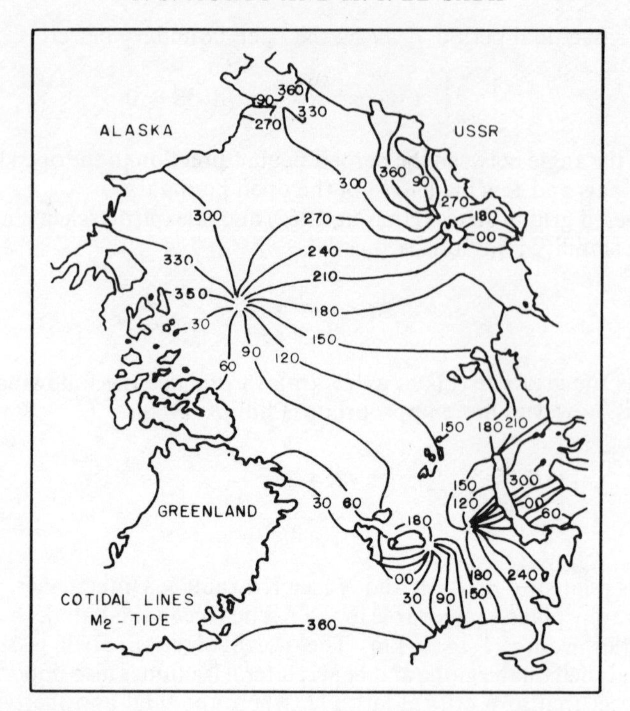

Fig. 13.—Cotidal lines for M_2 in the Arctic Ocean: phase angles (with reference to Greenwich) are in degrees; from Kowalik & Untersteiner (1978).

effects of multiple connectivity due to the presence of islands. Tidal regimes can then be developed using the results of the normal modes. Several authors (*e.g.* Fairbairn, 1954) used the so-called Proudman tidal theorem to calculate the tides in the oceans or in small water-bodies. This theorem permits one to determine the tidal regime across the open boundary of a water-body, knowing the tides around its shore. Foreman, Delves, Barrodale & Henry (1980) showed, however, that the mathematical problem is ill-posed, and attempts to use the theorem numerically have failed.

Thacker (1980) simulated tidal motion on a sphere using a geodesic finite-difference method. This method permits variable resolution of the grid, and the use of a three-dimensional artesian co-ordinate system, rather than a two-dimensional curvilinear spherical co-ordinate system, makes computations on irregular surfaces somewhat easier. Clarke & Battisti (1980) considered the influence of the continental shelf on coastal tides. Their coastal boundary layer theory suggests that semi-diurnal tides will be amplified on wide continental shelves in middle and low latitudes, but diurnal tides will not be amplified. Observations bear out these theoretical results.

Zahel (1973) used a numerical model to derive the cotidal chart for the diurnal tidal constituent K_1 in the global oceans. Figure 15 shows this chart. By far the most comprehensive study of global ocean tides is by Schwiderski (1978a,b, 1980a,b,c, 1981a,b,c,d,e, 1982a,b, 1983) who included several novel

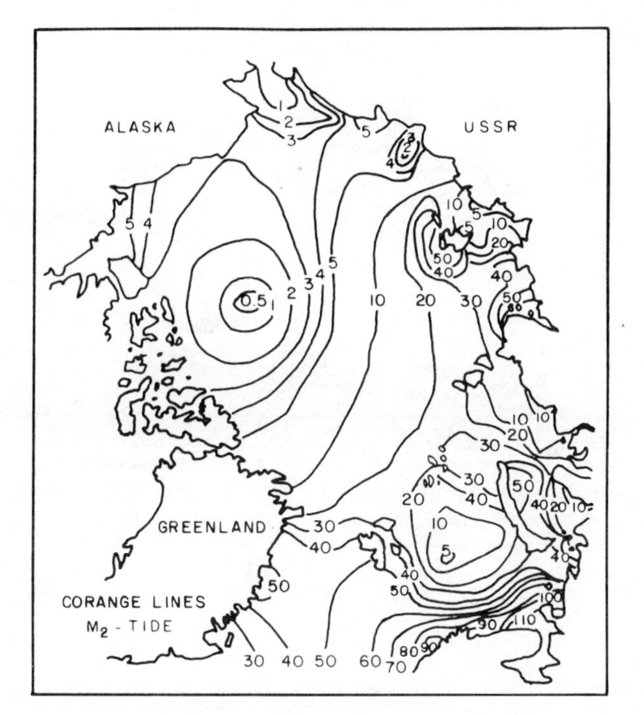

Fig. 14.—Corange lines (centimetres) of M_2 in the Arctic Ocean (from Kowalik & Untersteiner, 1978).

features in his numerical models. First of all, he derived the ocean tidal equations in continuous as well as discrete forms. In these equations the Boussinesq linear eddy dissipation law was used and the eddy viscosity was made to depend on the lateral mesh area (which is derived from the grid size and the local water depth). The bottom friction coefficient was also made to depend on the mesh area. In addition to the tidal potential, secondary factors such as the influence of the oceanic and terrestrial tides on the tidal potential were included. Figures 16 and 17, respectively, show the co-amplitude and co-phase lines of the semi-diurnal constituent N_2 for the Arctic Ocean based on Schwiderski (1981b).

AGE OF THE SEMI-DIURNAL TIDES IN THE GLOBAL OCEANS

Following Garrett & Munk (1971) and Webb (1973) we shall discuss the geographical distribution of the age of the semi-diurnal tide in the global oceans. The formula for determining the age of the semi-diurnal tide is given in Equation 2. Garrett & Munk (1971) computed the age of the tide at 647 of the world's ports. In this computation they made use of the tidal constants published by the International Hydrographic Bureau (1966); the ports on the

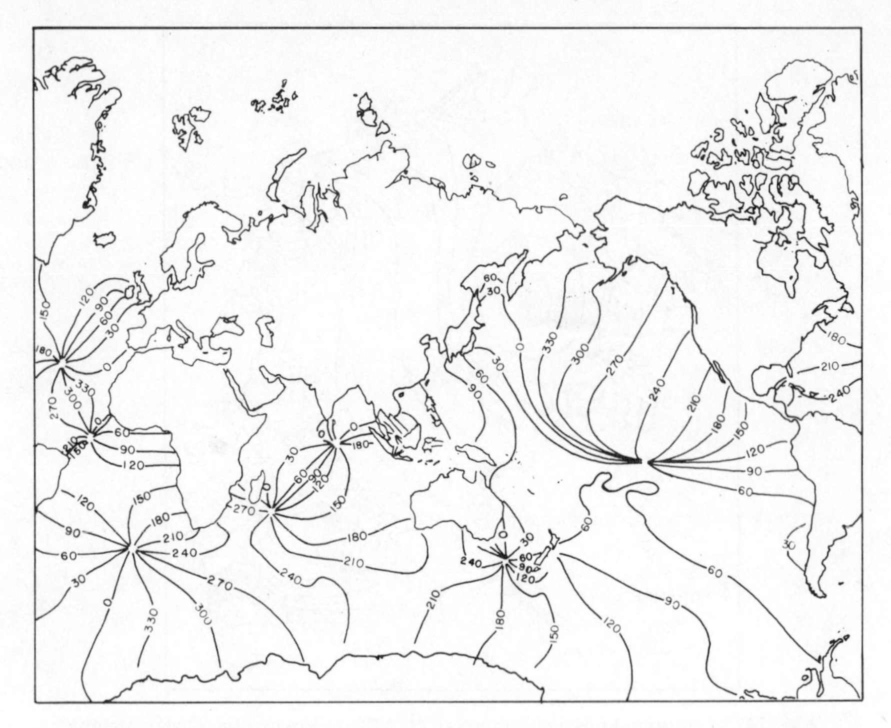

Fig. 15.—Corange and cotidal lines for the K_1 tide in a 4° world-ocean-model (from Zahel, 1973).

Baltic Sea, the Mediterranean Sea, the Black Sea, and the Great Lakes are excluded in these computations. Table V shows the distribution of the age of the tide based on these calculations.

Garrett & Munk (1971) suggest that friction is the cause of the age of the tide, particularly when there is some resonant frequency of the oceans close to the semi-diurnal tidal frequencies. They wrote the response of the ocean to the tide-generating forces as an expansion in terms of the damped normal modes of the ocean, and showed that the age of the tide not only varies from place to place but also can assume large positive values, and sometimes even negative values.

Let ζ be the water surface elevation at location x at time t, ω_n the frequency of the nth normal mode of the oceans, and $S_n(x)$ the corresponding eigenfunction (normalized). Then the response of the oceans to tidal forcing at frequency ω is given by

$$\zeta(x,t) = \mathrm{Re}\sum_n \frac{B_n(\omega)}{\omega_n - \omega - \frac{1}{2}iQ_n^{-1}\omega_n} S_n(x)\mathrm{e}^{-i\omega t} \tag{32}$$

Here the dissipation is represented by the imaginary term. The parameter Q_n is defined such that the nth normal mode dissipates a fraction $2\pi/Q_n$ of its energy in each cycle. At this stage, these authors make the following three

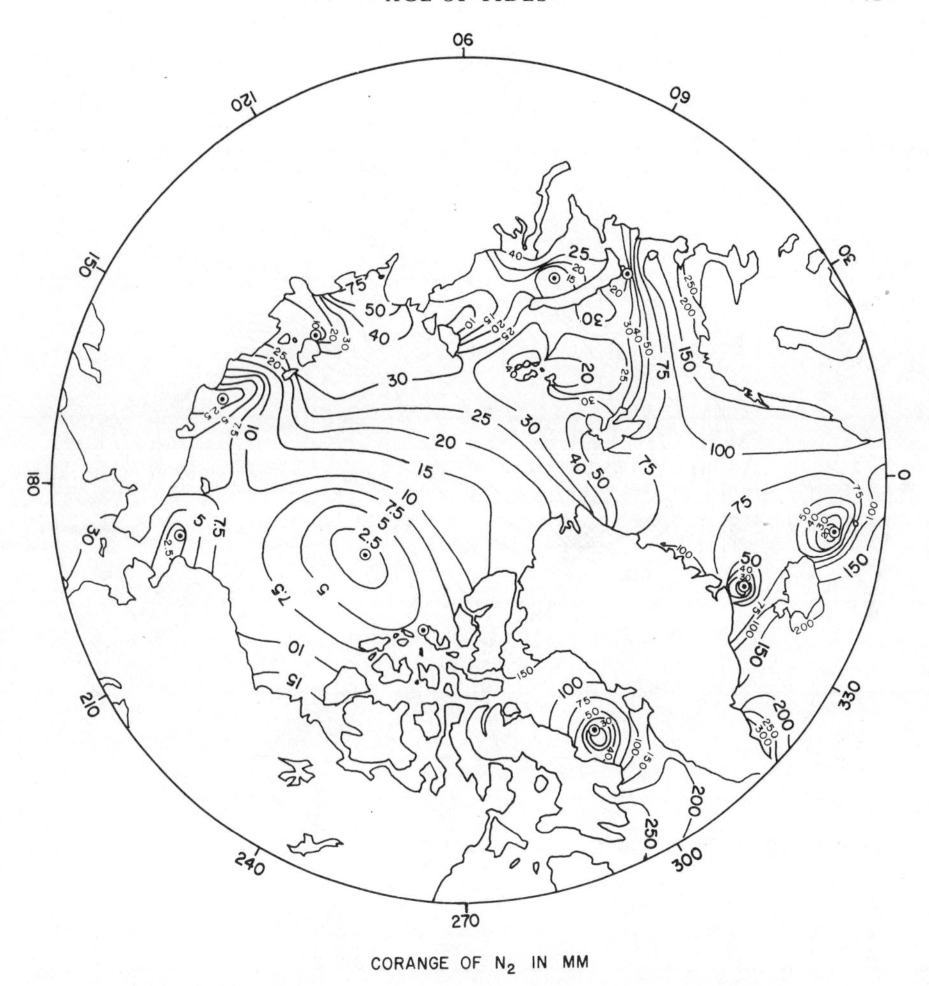

Fig. 16.—Corange map of N_2 ocean tide in the Arctic Ocean: amplitudes are in mm; from Schwiderski (1981b).

assumptions: (a) the set of functions $S_n(x)$ is complete, (b) the problem is linear, as implied in Equation 32, and (c) for ω close to ω_n, the phase of $B_n(\omega)$ changes much less rapidly than the phase of the denominator in Equation 32. Let us examine the justification for these assumptions.

Assumption (c) requires that the forcing function has the same spatial dependence for the two neighbouring frequencies one wishes to examine. This assumption holds well for the solar and lunar tides. The first assumption is also quite reasonable. Garrett & Munk (1971) are honest in admitting that assumption (b) is probably not justifiable. The importance of the non-linearities in the dissipation of tidal energy in shallow water cannot be ignored. Garrett & Munk, however, provided some justification for the linearity assumption.

Fig. 17.—Cotidal map of N_2 ocean tide in the Arctic Ocean: phases angles (with reference to Greenwich) are in degrees; from Schwiderski (1981b).

Equation 32 can be written for any location x as

$$\zeta = \mathrm{Re} \sum_n a_n(\omega) e^{i\theta n(\omega)} e^{-i\omega t} = \mathrm{Re}\, A(\omega) e^{i\theta(\omega)} e^{-i\omega t}. \tag{33}$$

Now we shall consider the variation of the phase lag θ with frequency. Following Garrett & Munk (1971), we assume that a single mode is dominant, identified by subscript 0. Also assume that $B_0(\omega)$ is more or less constant. Let

$$B_0(\omega) \cdot S_0(x) = C_0. \tag{34}$$

Then one can write

$$A e^{i\theta} = a_0 e^{i\theta_0} = \frac{C_0}{\omega_0 - \omega - \frac{1}{2} i Q_0^{-1} \omega_0}. \tag{35}$$

TABLE V

The distribution of the age of the semi-diurnal tide (from Garrett & Munk, 1971)

Age (days)	Atlantic	Pacific and Indian	Total
−7	1	2	3
−6	0	0	0
−5	0	4	4
−4	0	5	5
−3	2	3	5
−2	4	4	8
−1	3	6	9
0	34	53	87
1	111	186	297
2	63	103	166
3	19	24	43
4	3	6	9
5	0	4	4
6	2	3	5
7	1	1	2
Total	243	404	647

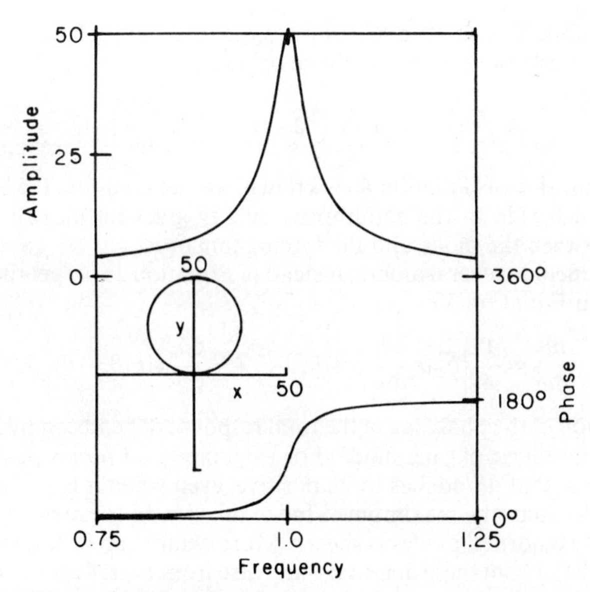

Fig. 18.—Response of one normal mode, $\omega = 1$, $Q = 25$ (from Garrett & Munk, 1971).

Figure 18 shows the variation of the amplitude and phase with the frequency, for the case $\omega_0 = 1$ and $Q_0 = 25$. The insert plot between the two curves is the imaginary part Y *versus* the real part X of $A\mathrm{e}^{i\theta}$. Note that θ changes from 0° to 180° and passes through 90° at $\omega = \omega_0$. The following relation gives the gradient, whose maximum value of $2Q_0/\omega_0$ occurs at $\omega = \omega_0$,

$$\frac{\mathrm{d}\theta}{\mathrm{d}\omega} = \frac{\frac{1}{2}Q_0^{-1}\omega_0}{(\omega_0-\omega)^2+(\frac{1}{2}Q_0^{-1}\omega_0)^2} \tag{36}$$

hence

$$Q_0 \geq \tfrac{1}{2}\omega_0\frac{\mathrm{d}\theta}{\mathrm{d}\omega}. \tag{37}$$

The inference is that if the semi-diurnal tide were dominated by one mode, the age of the tide would be positive, and would be uniform (*i.e.* same value everywhere). Also this value will provide a lower bound for the Q of that mode (Q being a measure of the rate of loss of energy).

Let $\mathrm{d}S$ be an element of surface area, then energy, E of this mode can be approximated as

$$E = \int \tfrac{1}{2}\rho g|\zeta|^2\,\mathrm{d}S = \frac{\frac{1}{2}\rho g|B_0|^2 \int |S_0|^2\,\mathrm{d}S}{(\omega^\circ-\omega)^2+(\frac{1}{2}Q_0^{-1}\omega_0)^2}. \tag{38}$$

Let D be the rate of energy dissipation, then

$$E = \frac{Q_0 D}{\omega_0}. \tag{39}$$

From Equations 38 and 39, one can write the following relation which holds whether the mode is close to resonance or not.

$$\omega^{-2}\rho g|B_0|^2 \int |S_0|^2\,\mathrm{d}S = \omega^{-2}D\left(\frac{\mathrm{d}\theta}{\mathrm{d}\omega}\right)^{-1}. \tag{40}$$

Since the right side of Equation 40 is known, we can compute the left side; the ratio of the left side to the equilibrium energy gives an idea of the spatial coupling between the mode and the forcing function.

When we include several modes, instead of Equation 36 we get the following relation from Equation 33

$$\frac{\mathrm{d}\theta}{\mathrm{d}\omega} = \frac{1}{A}\left\{\sum a_n\frac{\mathrm{d}\theta_n}{\mathrm{d}\omega}\cos(\theta_n-\theta)+\sum\frac{\mathrm{d}a_n}{\mathrm{d}\omega}\sin(\theta_n-\theta)\right\}. \tag{41}$$

Then variation of the phase lag of the total response depends on the changes in amplitude and phase of each mode. For each mode $\mathrm{d}\theta_n/\mathrm{d}\omega$ is positive, but it does not mean that $\mathrm{d}\theta/\mathrm{d}\omega$ has to be positive; even when it is positive, it need not be smaller than the maximum value of $\mathrm{d}\theta_n/\mathrm{d}\omega$. In Figures 19 and 20 the response of two normal modes is shown. More details can be found in Garrett & Munk (1971). From these diagrams (and also from Figs. 4 and 5 of Garrett & Munk, not shown here) one can deduce that the variation of the phase with frequency is extremely variable. Generally, $\mathrm{d}\theta_n/\mathrm{d}\omega$ is positive and is usually

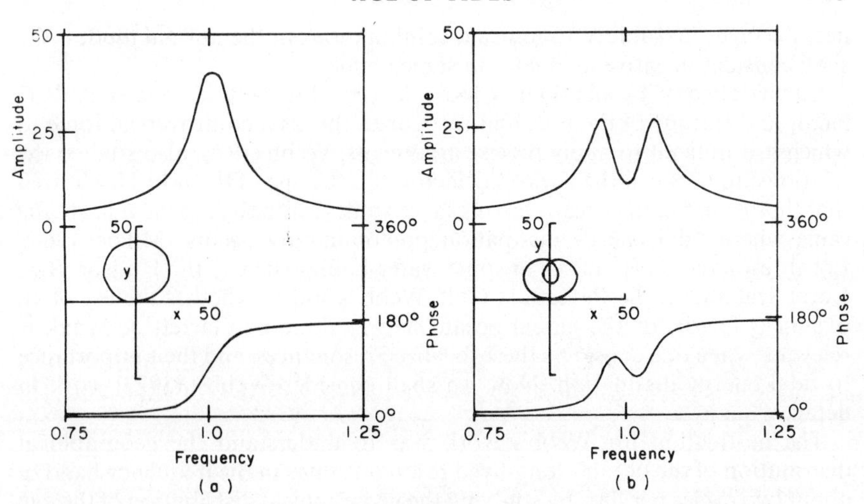

Fig. 19.—Response of two normal modes, $C_1 = C_2 = 1$, $Q_1 = Q_2 = 25$: (a) $\omega_1 = 0{\cdot}99$, $\omega_2 = 1{\cdot}01$; (b) $\omega_1 = 0{\cdot}96$, $\omega_2 = 1{\cdot}04$; from Garrett & Munk (1971).

less than the maximum $d\theta_n/d\omega$; larger gradients are found, however, in some situations and also a rapid phase change of greater than 360° could occur. The behaviour is expected to be generally similar when we include more modes.

Thus, the origin of the variation of the age of the tide from one place to another is due to the spatial dependence of the normal modes $S_n(x)$ in Equation 32. This is why Garrett & Munk (1971) suggest that one can better understand the normal modes of the ocean by studying the spatial distribution of the age of the tides. These authors attributed the large positive ages, and the

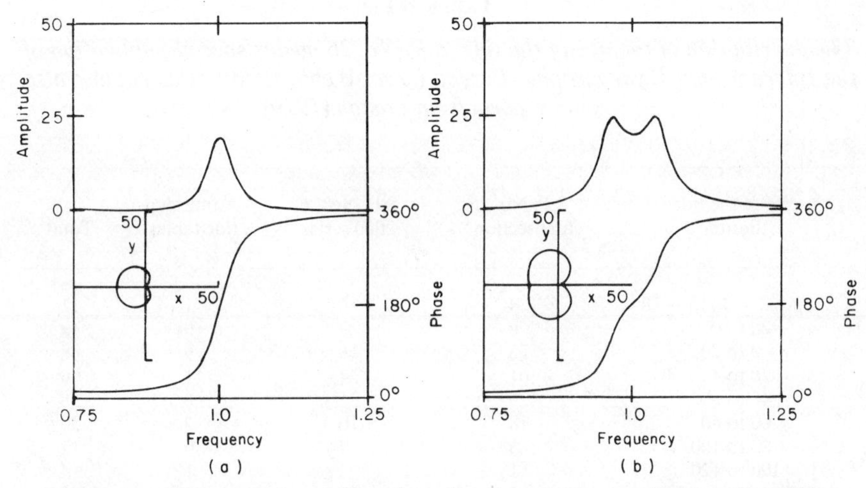

Fig. 20.—Response of two normal modes, $C_1 = 1$, $C_2 = -1$, $Q_1 = Q_2 = 25$: the values of ω_1 and ω_2 are similar to Figure 19; from Garrett & Munk (1971).

negative ages in Table V to unusual combinations of the normal modes. We shall consider negative ages later in some detail.

Cartwright (1978) mentions a loose relationship between the ocean's Q factor and the age of the tide, and recognizes this as a controversial topic to which it is difficult to apply precise arguments. Webb (1973) also studied the relationship between the ocean's Q factor and the age of the tide. He showed that those parts of the oceans having large values for the age of the tide are the same where tidal energy dissipation predominantly occurs (Miller, 1966). Local resonances also occur in some water-bodies such as the Hudson Bay, Coral Sea, and in the Patagonia Gulf. Webb wonders whether the use of an overall Q factor for the global oceans as was done by Garrett & Munk is relevant, when one considers these localized resonances, and their importance to tidal energy dissipation. Now we shall consider Webb's (1973) work in detail.

The motivation for Webb's work was to understand the geographical distribution of the ocean's long-lived resonant states in the frequency band of about two cycles per day, by studying the geographical distribution of the age of the semi-diurnal tide. Webb found that these resonances are localized geographically and are often associated with the broad continental shelves which were found to be important for tidal dissipation (Miller, 1966). Inclusion of these localized resonances gives a decay time of about 30 to 35 h for the tidal energy in the ocean.

Webb (1973) calculated the age of the semi-diurnal tide at ports (distributed as uniformly as possible) around the Atlantic, Indian, and Pacific Oceans. These locations were selected to be approximately 800 km apart, except along the coast of Antarctica where the data was sparse. Table VI shows the distribution of the age of the tide.

Figure 21 shows the distribution of the age of the semi-diurnal tide around

TABLE VI

*The distribution of the age of the tide: S.P. No. 26 means special publication of the International Hydrographic Bureau (from Webb, 1973); *this column also includes ports from Easton (1970)*

Age of the tide (hours)	S.P. No. 26 first section*	S.P. No. 26 2000 series	Admiralty tide tables	Total
Approx. −90 to −20	4	6	1	11
−20 to 0	6	7	0	13
0 to 20	23	16	9	48
20 to 40	61	24	11	96
40 to 60	25	25	6	56
60 to 80	18	10	2	30
80 to 100	5	7	0	12
100 to 120	1	2	1	4
120 to approx. 264	7	13	0	20

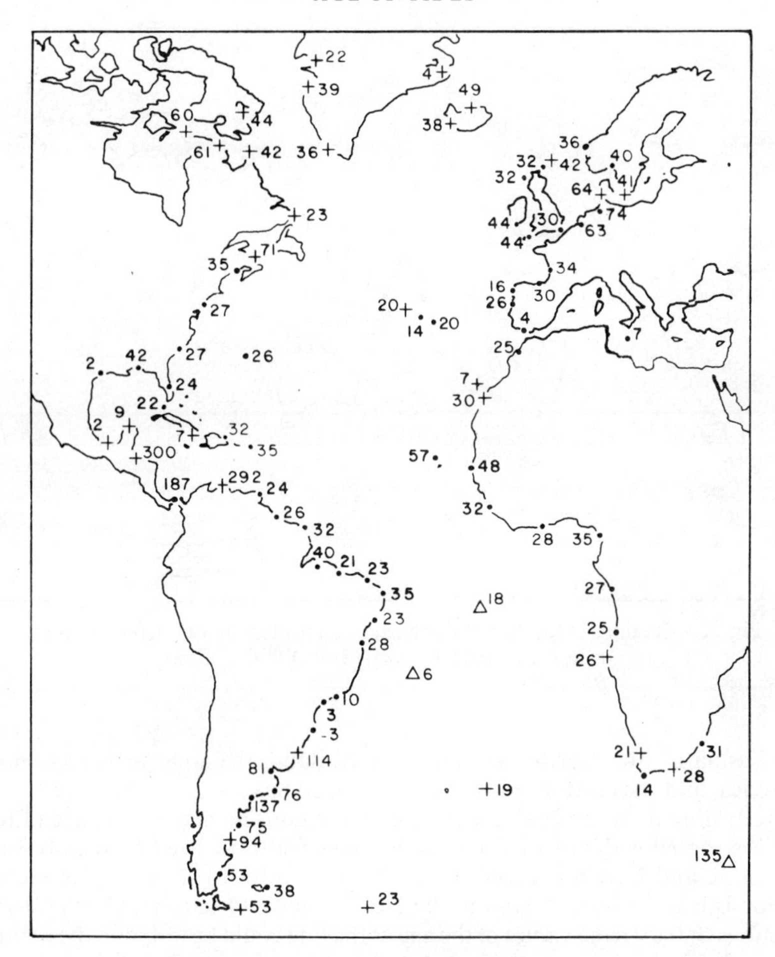

Fig. 21.—The age of the semi-diurnal tide in hours for Atlantic ports: ●, S.P. no. 26 (first section); +, S.P. no. 26 (2000 series); △, Admiralty Tide Tables; from Webb (1973).

the Atlantic Ocean. It can be seen that, over most of the coastlines the age of tide has values in the range of 0 to 60 h with a typical variation of about 10 h between adjacent ports. It is not convenient to draw contours representing the age of the tide because of the random nature of the variation. Figure 22 shows the regions with age of tide greater than 60 h or having negative ages. Most of the regions with an age of tide greater than 60 h are associated with significant topographical features, such as large continental shelves. Large values of the age of tide are found on the shelf off Argentina, in the North Sea, Hudson Strait, Bering Sea, Sea of Okhotsk, Arafura Sea, and the Ross Sea. Note that Miller (1966) found such topographical features important for tidal dissipation (as can be seen from an examination of Fig. 3 in Webb, 1973, not shown here). Other topographic features associated with large values of the age of tide

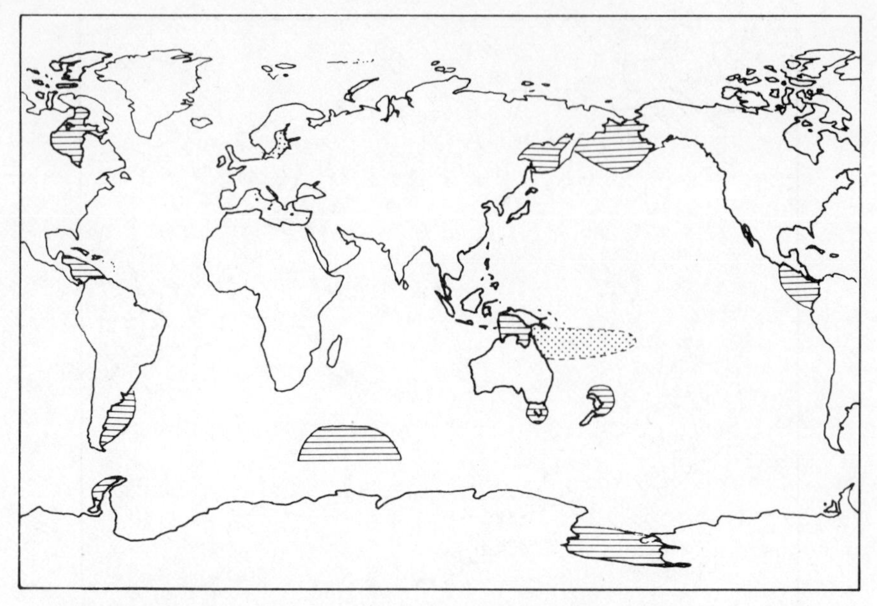

Fig. 22.—Regions of the ocean with an age of tide of greater than 60 h (striped) or less than 0 h (stippled) (from Webb, 1973).

are Tasmania, the North Island of New Zealand, the areas around Central America, and the eastern side of the Caribbean Sea.

According to Webb (1973) there are two regions having a large age of tide not associated with any obvious topographic features. The first region is the Antarctic and includes Kerguelen and Prince Edward Islands. The second region is near Midway Island in the North Pacific Ocean. Webb (1973) also cautions that extreme values of the age of the tide could be found in the regions where the S_2 tide has an amplitude less than 5 cm. As mentioned earlier the S_2 tide could be contaminated by radiational effects; we shall consider this later. Some of the areas are the Aleutian Islands, Midway Island, Kerguelen Island, Easter Island, the Ross Sea, and the East coast of Central America. Negative ages are found in the Baltic Sea, and in a rather extensive region extending from Northern Queensland to the Samoan Islands.

Webb (1976) suggests that a localized increase in the age of the tide is a good indication of resonance. One cannot account for the observed increases in the age of the tide in terms of the delay due to the propagation of the tide as a long gravity wave. Webb (1976) suggests that the increase in the age is due to resonances on the continental shelves, since the selves are usually of widths comparable to a quarter wavelength of the semi-diurnal tide. He developed a simple analytical model to show that resonance on the continental shelf leads to an increase in the age of the tide and also in enhanced absorption of tidal energy.

Heath (1981b) considered the age of tide and resonances in the northern parts of the Atlantic and Pacific Oceans. He starts with the hypothesis that for understanding the oceanic response to tidal forcing, one has to calculate the periods and rate of energy dissipation of the normal modes of the ocean. Earlier we have seen that if Q is a measure of the rate of loss of energy, the system loses $2\pi/Q$ of its energy during every cycle. Furthermore, as was shown earlier, Garrett & Munk (1971), based on an average worldwide value for the age of the semi-diurnal tide, suggested that a lower bound for Q (for normal modes at frequencies in the semi-diurnal tidal band) is about 25. We have seen that Webb (1973) argued that regions with large values of the age of the tide represent localized resonances associated with topographical features, and these localized resonances are not representative of the open ocean conditions. He estimated that the overall decay time of tidal energy is 30 to 35 h, which is equivalent to a Q of about 18. Using a numerical model, Hendershott (1972) obtained a Q of about 34.

Heath (1981b) starts with the hypothesis that in an ocean where only a single normal mode dominates a tidal band, one would not expect much spatial variation either in the ratio of the amplitudes of the tidal constituents or in their phase difference. An examination of this for the northern part of the Atlantic Ocean, making use of the pelagic tidal data (Cartwright, Zetler & Hamon, 1979) appears to bear out Heath's hypothesis (see Table VII and Fig. 23).

Next Heath (1981b) makes use of the calculations of Platzman (1975) who computed the normal modes of the Atlantic and Indian Oceans. Platzman showed that the period of the dominant normal mode near the semi-diurnal tidal band is 12·8 h. The next dominant mode has a period of 14·4 h, quite removed from the semi-diurnal tidal band. Hence, the idea that in the semi-diurnal tidal band in the North Atlantic, a single mode is dominant, appears to be reasonable. Heath (1981b) then refers to the work of Garrett & Greenberg (1977) in which they analysed the response of the semi-diurnal tidal constituents at two locations in the western part of the North Atlantic Ocean, by assuming that the response is dominated by one mode. Table VIII summarizes these results. Making use of the observed pelagic tides, Heath (1981b) made more estimates for the resonant period and Q in the semi-diurnal tidal band.

The situation in the Pacific Ocean is quite different, according to Heath (1981b). The amphidromic points of the main tidal constituents are well separated (see Figs 24 and 25) and the ratio of the amplitudes and phases of the tidal constituents vary significantly. In deducing these results, Heath made use of the cotidal charts of Luther & Wünsch (1975) and Accad & Pekeris (1978). Because of this variation, one has to include at least two normal modes. Heath estimated the resonant periods and Q of the two assumed normal modes in the semi-diurnal tidal band for the western part of the Pacific Ocean from the observed spatial separation of the area of maximum amplitude for the three semi-diurnal tidal constituents M_2, S_2 and N_2, the related relative amplitude of the maxima, and the spatial variation in the age of the tide. For the Northern Atlantic, using a single normal mode, Heath obtained a resonant period lying between 12·6 and 12·8 h and a Q having a value between 10 and 12. For the Pacific Ocean, using two normal modes, he obtained resonant periods of 10·2 and 12·8 h and a Q of about 5. Heath recognizes that his estimate of 5 for Q is quite a bit lower than the estimates by others (Garrett & Munk, 1971; Hendershott, 1972; Webb, 1973). Heath justifies his value by suggesting that, if we agree with his calculation of the frequency separation of the modes, the two

TABLE VII

Resonant period and Q for fits of the semi-diurnal tidal constituents in the deeper part of the North Atlantic Ocean, to a simple oscillator response: here Q is the same for M_2, S_2 and N_2 tidal constituents; from Heath (1981b)

	Amp. ratio S_2/M_2	Phase diff. $M_2–S_2$	Calculated normal period (h)	Mode Q	Amp. ratio N_2/M_2	Phase diff. $M_2–N_2$	Calculated normal period (h)	Mode Q	Amp. ratio S_2/N_2	Phase diff. $S_2–N_2$	Calculated normal period (h)	Mode Q
Eastern North Atlantic Ocean												
No. of sites 15												
Vector average	0·33	−32	12·55	11·4	0·21	21	12·7	10·3	1·56	53	12·63	11·4
Scalar S.D.	0·03	7			0·012	3			0·22	9		
Extreme phase	0·31	−19	12·85	11·4	0·22	15	12·9	8·6	1·60	66	12·55	13·0
difference	0·37	−40	12·4	11·5	0·21	26	12·65	12·6	1·55	35	12·90	8·8
Extreme amplitude	0·39	−39	12·35	11·4	0·23	20	12·85	11	1·98	62	12·45	11·4
ratio	0·28	−29	12·65	15	0·20	23	12·6	10·5	1·19	51	12·75	15
Western North Atlantic												
No. of sites 13												
Vector average	0·19	−29	12·6	31	0·23	22	12·8	12	0·84	51	12·76	23·1
Scalar S.D.	0·18		4		0·01	3			0·10	4		
Extreme phase	0·16	−23	12·6	52	0·23	17	12·9	10·7	0·77	58	12·75	26
difference	0·21	−33	12·6	26	0·22	29	12·65	15	0·90	43	12·8	21
Extreme amplitude	0·16	−23	12·6	52	0·25	18	12·95	14	1·04	46	12·8	17
ratio	0·22	−23	12·7	28	0·21	23	12·7	10·3	0·70	48	12·75	29

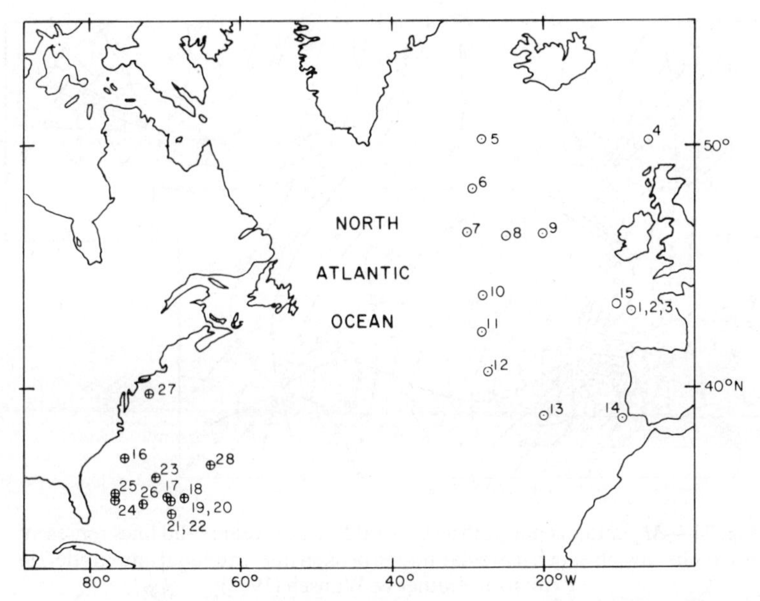

Fig. 23.—Location of the pelagic tide-gauge observations in the deeper part of the North Atlantic Ocean: site numbers 1 to 15 are in the eastern North Atlantic and 16 to 28 in the western North Atlantic.

modes have to be evenly distributed on either side of the semi-diurnal tidal band. Otherwise the *Q* has to be small to allow contributions from the modes to the semi-diurnal tides.

NEGATIVE AGES

One aspect of the age of tides problem that has not been satisfactorily explained is the existence of negative ages and the cause or causes for it. Garrett & Munk (1971) examined the age of tides in terms of the normal modes

TABLE VIII

Resonant period and Q in the North Atlantic Ocean (from Garrett & Greenberg, 1977)

Location of observation point	Resonant period (h)	*Q*
Bermuda	12·82	16·4
Halifax	12·79	17·1

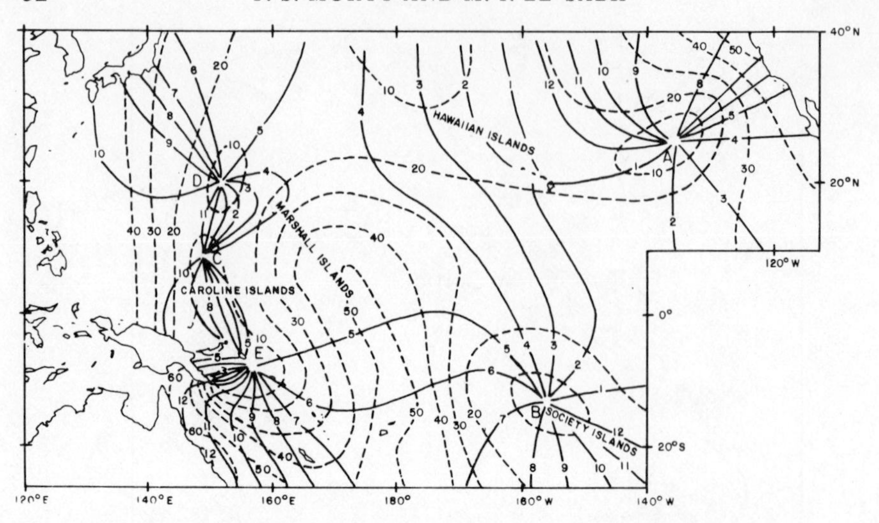

Fig. 24.—M_2 cotidal chart for the Central Pacific Ocean: solid lines represent equal Greenwich epoch g in solar hours; broken lines are equal amplitude h in cm; from Luther & Wünsch (1975).

of the oceans and attributed the negative ages to the effect of unusual combination of the modes. It can be shown that negative values for the age of the tide may be erroneous at locations where the amplitude of the S_2 tidal constituent is smaller than about 5 cm, because the S_2 tide may be contaminated by radiational effects.

Heath (1981a) attributed large negative ages around New Zealand to the

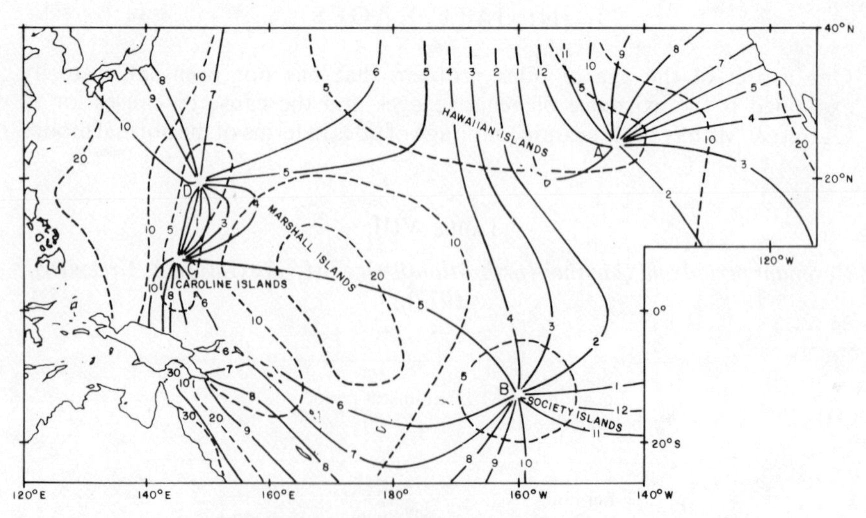

Fig. 25.—S_2 cotidal chart for the Central Pacific Ocean: same as for Figure 24; from Luther & Wünsch (1975).

differing spatial distributions of the M_2 and S_2 tides. Farquharson (1970) accounted for negative ages due to the fact that the amphidromic points for the two constituents (involved in the age calculation) do not coincide. The co-phase lines of the amphidrome of one constituent interacting with the co-phase lines of the amphidrome of the other constituent, can produce negative ages. Pugh (1981) offered an explanation for anomalous values of the age of the tide, including negative values. He suggests that in seas where the spring amphidrome is not degenerate, the smallest semi-diurnal tides of the spring–neap cycle near the amphidromic position will occur, when the rest of the water-body is experiencing spring tides. This will lead to an anomalous tidal age to the local tides.

Wood (1978) identified two astronomical factors that influence the local heights and times of arrival of the tides (besides the usual hydrodynamic and topographic causes). These are the lunar retardation and the interaction between the tidal force envelopes of the moon and sun. Following Wood (1978) we shall briefly discuss these factors and examine their relevance to negative ages. Earlier we noted that the earth rotates on its axis from one meridian transit of the mean sun to the next in 24 h, which is referred to as the mean solar day. We also saw that the time interval between two successive upper transits of the moon across the local meridian of the place under discussion is known as the lunar or tidal day. We shall see now, how due to astronomical causes, the lunar or tidal day exceeds the mean solar day.

The angular velocity of the moon around the earth is roughly 12·2° per day, whereas the angular velocity of earth's rotation in the same direction is 360° per day. Thus, during a given day, any point on the rotating earth has to execute a rotation of $360° + 12{\cdot}2° = 372{\cdot}2°$ to keep up with the moon. Since a rotation through 360° corresponds to 24 h (or one hour equals 15°) the extra amount of rotation of 12·2° every day requires an extra time of 48·8 min. In estimating this value the following assumptions, however, are implied: (a) the moon revolves around the earth in a circular orbit, and (b) the moon's speed of revolution does not change. It can be shown that, removal of these two assumptions gives on the average an extra time of 50·415 min each day for a sublunar point on the rotating earth to regain this position along the major axis of the moon's tidal force envelope (note that the tide-generating force is a maximum here). Thus, due to the influence of the lunar retardation, the recurrence of a tide of the same phase and similar height would occur at intervals of 24 h and 50 min, and this time interval has been established as the tidal day.

The second astronomical factor that can influence the time of arrival of tides of a given phase at any location is due to the interaction between the tidal force envelopes of the moon and the sun. This interaction causes a priming of the tides (*i.e.* high waters occurring before the moon reaches the local meridian of the tide gauge location) due to a displacement of the force components. This acceleration in the arrival times of the tides happens between new moon and first-quarter phase and between full moon and third-quarter phase. On the other hand, between first-quarter phase and full moon, and between third-quarter phase and new moon, an opposite displacement of force components takes place which causes a delay in the arrival of tides, this phenomenon being known as the lagging of tides. In this case high water occurs several hours after the moon has reached the local meridian.

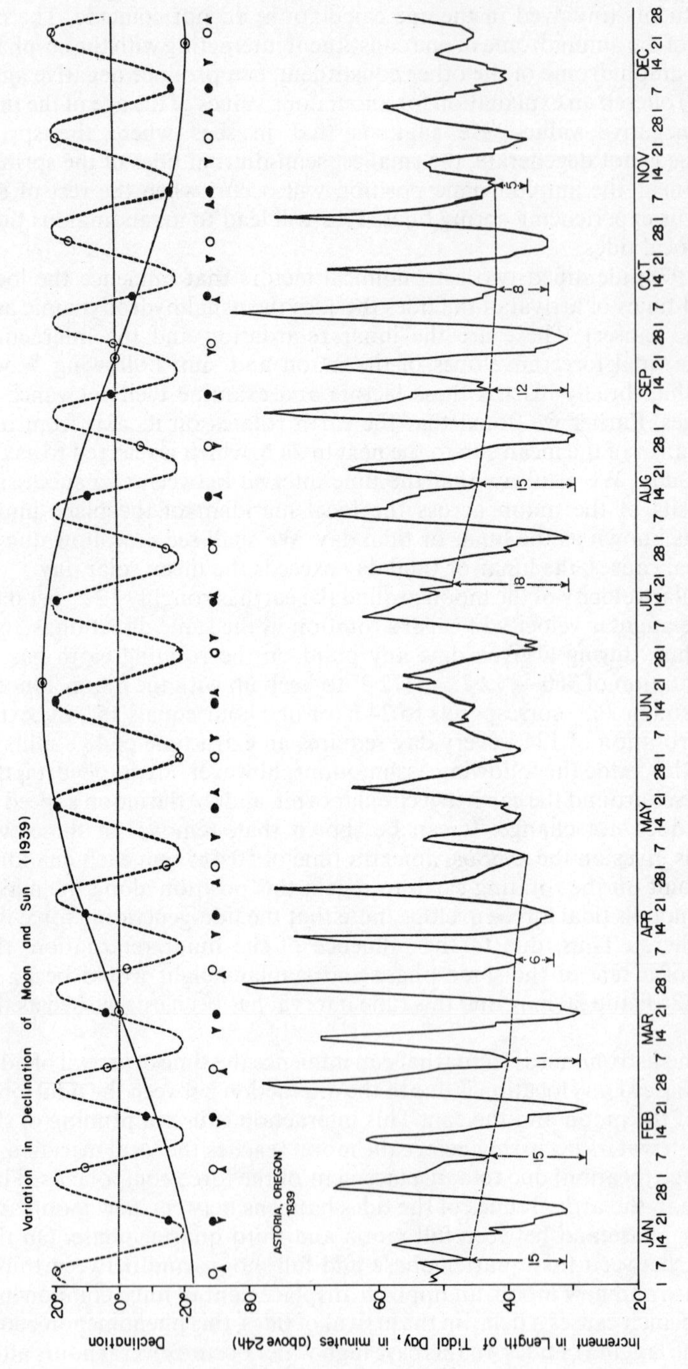

Fig. 26.—Variation in the length of the tidal day during the year 1939 at Astoria (Oregon, U.S.A.) shown as function of the increment to be added to the time of the previous day's higher high water to establish the time of occurrence of H.H.W. on the current day (modified from Wood, 1978).

Following Wood (1978) we shall show an example of tidal priming and lagging at Astoria, Oregon, U.S.A., for the year 1939 (Fig. 26). The maxima and minima in the tide curves in this diagram show the variations in the length of the true tidal day and these variations can be explained in terms of the changing phase relationships of the moon with reference to the sun. It can be shown that (see Wood, 1978, for details) from the last-quarter phase to new moon, the tidal day gets continually shorter and reaches its minimum value at new moon. A similar situation occurs between first-quarter phase and full moon. This phenomenon of lunar priming accounts for the successive minima in the length of the tidal day depicted in Figure 26. The increase in the length of the tidal day between new moon and first-quarter and between full moon and last-quarter phases can be accounted for in terms of the greater relative speed of the moon in its orbit as compared with the apparent daily motion of the sun.

Negative ages occur not only for the semi-diurnal tides, but also for diurnal tides. Negative values occasionally also occur for the age of parallax. Wood (1978) mentions that negative values for the parallax age occur in the Gulf of Mexico. In later sections we shall give detailed distributions of negative ages in some selected water-bodies. We conclude this section by suggesting that several explanations can be offered for the existence of negative ages. None of these explanations are totally satisfactory. Probably a combination of all these factors accounts for negative ages. In any case, negative ages of tide need further study and clarification.

AGE OF PARALLAX AND AGE OF DIURNAL TIDE

Earlier we mentioned that the S_2 tide is contaminated by radiational effects, and Garrett & Munk (1971) suggested that this contamination could be avoided by computing the so-called "age of parallax inequality" making use of the M_2 and N_2 tidal constituents. Wood (1978) indeed has done this, and we also computed the age of parallax (for convenience we shall drop the word inequality) for several locations, the details of which will be given in later sections. Garrett & Munk (1971) also suggested that one can compute the age of the diurnal tide making use of the constituents K_1 and O_1. We computed the age of the diurnal tide at several locations in some selected water-bodies.

Now we shall consider some details of the contamination of the S_2 tide by radiational effects. To do this, first we define tidal admittances, following Cartwright (1978). The tide-generating potential is usually expressed for convenience as a series expansion of low order spherical harmonics. Then to understand the response of the ocean to the tidal potential, all one has to do is study the response to these spatial harmonics individually. Cartwright (1978) argues that the oceanic tidal response is strongly linear. In justification of this he mentions that the spectra of tidal records at locations near the deep ocean show strong lines at the frequencies of the major harmonic constituents present in the tidal potential, and very little energy at multiples of these frequencies. Cartwright (1978), however, recognizes that non-linear effects are important in shallow seas and coastal waters. Since these have little effect on the global ocean tides, one can ignore non-linear effects, and express the response as follows:

$$\zeta(t) = \mathrm{Re}\int_0^{\infty} [A(t-\tau)+iB(t-\tau)]R(\tau)\,\mathrm{d}\tau. \qquad (42)$$

Here ζ is the deviation of the sea surface from its mean level at a given location, A and B are the time-dependent parts of the tidal potential for a particular spherical harmonic and R is a response function. Note that here, for convenience the subscripts m, n on the spherical harmonic functions are omitted. The following expression is equivalent to Equation 42

$$H(f) = Z(f)\cdot G(f) \qquad (43)$$

where $H(f)$ and $G(f)$ are the complex spectra respectively of ζ and A in the frequency domain f. The function $Z(f)$ is referred to as the complex admittance of the ocean to the equilibrium tide; it is the Fourier transform of $R(t)$.

Starting from a set of tidal data, when admittances to the gravitational tidal potential are determined, one usually discovers a small but certain anomaly at the frequencies of the solar tides near one cycle per day, two cycles per day, and one cycle per year. These anomalies were ascribed to meteorological effects (Cartwright, 1978). Schureman (1941) pointed out that even though the tidal constituents S_a, S_{2a} and S_1 (which represent, respectively, the annual, semi-annual and solar diurnal periods) occur in the gravitational tide-producing forces, tidal variations at these periods are related to meteorological causes (*e.g.* changes in air temperature and pressure, and periodic land and sea breezes). According to Zetler (1971) since the physical mechanisms creating S_a, S_{2a} and S_1 remain fixed in phase, these are traditionally included in tides.

Based on a standard tidal analysis, if the phases of the semi-diurnal constituents are plotted as a function of frequency, one can see an irregularity for S_2 (30° per solar hour). Similar results can be seen in plots of semi-diurnal amplitudes divided by the appropriate theoretical coefficients (Zetler, 1971). Zetler (1971) mentions that although the tidal mathematicians have been aware of this for many years, it had not been generally realized that the analysed S_2 constituent consisted of a trigonometric combination of gravitational and radiational tides. What this means to practical tide prediction is that small errors are introduced into tidal predictions by inferring additional constituents based on the assumption that the analysed S_2 is solely due to gravitational causes.

Chapman & Lindzen (1970) showed that the atmospheric tides are almost entirely at the solar frequencies owing to the thermal effect of radiation. The fact that the gravitational tidal effects have negligible effect on the atmosphere can be seen from the observation that the lunar atmospheric tides are negligibly small compared with the solar atmospheric tides. Hollingsworth (1971) showed that a portion of this small lunar atmospheric tide is generated by oceanic coupling. Munk & Cartwright (1966) hypothesized that the solar radiation is the cause of the anomalies in the solar ocean tides. Cartwright (1978) suggests the following physical mechanisms: (a) direct radiation pressure, which is quite small, (b) direct surface heating and cooling, (c) onshore winds caused by heating and cooling of coastal land, and (d) coupling with the atmospheric tide through surface pressure.

Munk & Cartwright (1966) defined a radiational potential whose gradient is equal to the amount of radiation received per unit area at a given point on the earth's surface, ignoring transient atmospheric losses. Using this radiation potential, one can deal with the various physical mechanisms suggested above. The radiation potential $U(\theta,\lambda,t)$ is defined as

$$U(\theta,\lambda,t) = \begin{cases} S\left(\dfrac{\xi}{\bar{\xi}}\right)\cos\alpha & \text{for} \quad 0 \leq \alpha \leq \dfrac{\pi}{2}, \quad \text{day} \\ 0 & \text{for} \quad \dfrac{\pi}{2} < \alpha \leq \pi, \quad \text{night} \end{cases} \tag{44}$$

Here S is the solar constant (taken as unity for convenience), ξ is the sun's parallax, $\bar{\xi}$ is the average value of ξ, α is the zenith angle of the sun at the point P in Figure 27, θ is the latitude, λ is the longitude of the location under discussion and t is time.

Similar to the gravitational potential, one can also expand the radiational potential U in terms of spherical harmonics. Because of the asymmetrical nature of Equation 44, harmonics P_1^0, P^1 appear. Due to the night-time cut off of the solar radiation, harmonics of degree 2, 4, 6, . . . have a slow convergence. According to Cartwright (1978) the harmonic P_1^0 has a strong annual term, otherwise weakly represented in the gravitational tide. This provides what is usually called the "seasonal variation in mean sea level". Pattullo *et al.* (1955) studied the global distribution of this, and its physical cause is a combination of the last three of the four physical mechanisms listed above.

The term P_1^1 which is proportional to $e^{i\lambda}\cos\theta$ contains a term of solar daily period, and this is also weakly represented in the gravitational tide. On the other hand, the terms P_2^0, P_2^1, P_2^2 are similar to the gravitational terms, but one can distinguish these in the ocean tide by the absence of the contributions from the moon at their respective frequencies (particularly in the tidal constituents K_1 and K_2). Cartwright & Tayler (1971) tabulated the leading terms in the complete harmonic development of the radiation potential.

The following information is taken directly from Cartwright (1978). Analysis of tidal records using both gravitational and radiational tidal potentials by Cartwright (1968), Zetler (1971), and Cartwright & Edden (1977)

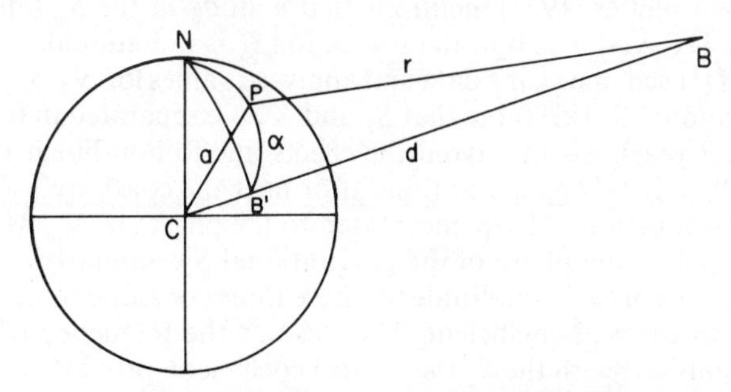

Fig. 27.—Centre of earth is C, North Pole is N, sun is B, distance d from C at Zenith angle α relative to point P on the earth's surface (from Cartwright, 1978).

shows that the major component of the radiational anomaly is the forcing from the atmospheric tides. Note that the diurnal anomaly S_1 is much weaker compared with the semi-diurnal anomaly S_2. This may appear as a paradox at the outset, for a response to the diurnal radiation pattern, but such a result is well established for the surface atmospheric pressure. According to Chapman & Lindzen (1970) the tide in the upper atmosphere is mainly diurnal with higher harmonics, but the wave form in the vertical direction is such that the diurnal component is suppressed at the ground level and the first harmonic S_2 predominates. In the oceans the typical amplitude of S_1 is about one centimetre, whereas the amplitude of the radiational S_2 (with values of about 5 to 10 cm) is about 17% of the gravitational S_2.

Another point to be made in support of the statement that the radiational S_2 tide is related to the atmospheric tide is the observation that the phase of the radiational S_2 tends to lead the phase of the gravitational S_2 by about 240°. This agrees with the well-established fact for the semi-diurnal atmospheric tide, that its minimum value (which corresponds to a positive static rise in the sea level) occurs at four o'clock local time, at every point on the earth. Cartwright (1978) cautions that since the amplitude of the atmospheric tide is only about one millibar, which corresponds statically to only about one centimetre rise in sea level, the dynamics of the transfer mechanism is not properly understood.

Next we shall discuss how to determine the radiational S_2 tide in practice. Munk & Cartwright (1966) developed the so-called response method of tidal analysis making use of the smoothness of the admittances of the species frequency bands, to separate the contributions from the gravitational, radiational potentials and also from non-linear effects. This method, however, requires a large amount of computation. Hence Zetler (1971) devised a somewhat simpler method, which allows us to determine the radiational S_2 from traditionally obtained tidal harmonic constituents.

Following Zetler (1971), the solar diurnal S_1 tide is flanked on either side (*i.e.* in frequency) by K_1 and P_1 which are just one cycle per year different from S_1. Note that P_1 and K_1 are two of the three largest diurnal components (the third one being O_1). One year's observed tidal data are needed to obtain the harmonic constants for S_1. Since the amplitude of S_1 is small, not much attention is paid as to its source (gravitational or radiational). Using data at Honolulu, Munk, Zetler & Groves (1965) showed that the line stands above the continuum. Zetler (1971) mentions that a study of the S_1 tide along the coasts of the U.S.A. shows that the source for S_1 is radiational.

Zetler (1971) used one year's data and analysed values for N_2, M_2, S_2 and K_2 to infer radiational S_2 tide (note that S_2 and K_2 are separated in frequency by two cycles per year). Also to avoid the effects due to non-linear interactions (Zetler, 1969), only tide gauge stations at or near the coast are used.

A parabolic function of frequency, fitted to the phases of N_2, M_2 and K_2 is used to interpolate the phase of the gravitational S_2. Similarly a parabola is fitted to the ratios of each amplitude for these three constituents divided by the appropriate theoretical coefficient. The ratio at the frequency of S_2 on this parabola, combined with the S_2 theoretical coefficient, provides an amplitude for the gravitational S_2. The phase and amplitude of the gravitational S_2 thus determined are subtracted vectorially from the observed S_2 to obtain a phase and amplitude for the radiational S_2. Zetler (1971) pointed out some weaknesses in this approach and how they may be corrected. Table IX lists the amplitudes of the gravitational and radiational S_2 at some selected stations on the western and eastern coasts of the U.S.A.

We have already mentioned that the trouble with the radiation contamination of the S_2 tide can be avoided by computing the age of parallax, using

$$\text{Age of parallax} = \frac{(\text{phase of } M_2 - \text{phase of } N_2)}{0{\cdot}544}. \quad (45)$$

The age is in hours provided that the phases are expressed in degrees. Wood

TABLE IX

Amplitudes (cm) of gravitational and radiational S_2 at some stations on the coast of U.S.A. (based on Zetler, 1981)

Tide-Gauge Station	Amplitude (cm) of S_2 Gravitational	Radiational
Western Coast		
Neah Bay	21·9	4·3
Aberdeen	29·6	6·7
Toke Point	29·0	7·0
Astoria	26·8	6·4
Newport, South Beach	25·3	4·3
Marshfield, Coos Bay	19·2	4·9
Crescent City	17·4	4·0
Humboldt Bay	16·2	2·1
San Francisco	15·5	4·0
Santa Barbara	19·2	2·1
Rincon Island	19·8	2·1
Santa Monica	20·7	1·8
Los Angeles, Outer Harbor	23·2	3·0
La Jolla, Pier Gage	22·0	0·3
La Jolla, Bottom	19·8	1·4
San Diego	22·6	1·8
Eastern Coast		
East Port	58·8	18·6
Boston	29·6	7·6
New York, the Battery	15·2	2·4
Sandy Hook	15·2	1·8
Atlantic City	12·5	0·6
Cape May	15·2	2·1
Breakwater Harbor	14·0	2·7
Old Point Comfort	9·8	2·7
South Port	13·7	3·7
Myrtle Beach	14·3	1·5
Charleston	12·5	0·6
Daytona Beach	9·8	0·6
Patrick Air Force Base	9·4	1·2
Miami	7·6	1·2
Key West	6·1	0·9

(1978) points out that in parts of the Gulf of Mexico the age of parallax is negative. He also compared the age of parallax and age of semi-diurnal tide (which he refers to as the age of phase inequality) at Sandy Hook, New Jersey, U.S.A., which have values, respectively, of 33 and 22 h. Also following Wood (1978) we define the age of the diurnal tide (which he refers to as the age of the diurnal inequality) as:

$$\text{Age of diurnal of} = \frac{(\text{phase of } K_1 - \text{phase of } O_1)}{1{\cdot}11}. \qquad (46)$$

Again the age is in hours, provided the phases are in degrees. In later sections we shall present detailed distributions of the age of parallax and age of diurnal tide in some selected water-bodies.

TIDES IN THE ST LAWRENCE GULF AND ESTUARY

The Gulf of St Lawrence (see Fig. 2, p. 16) is a triangular-shaped semi-enclosed sea on the eastern coast of Canada. It has a surface area of about 226 000 km^2 with a mean depth of 152 m. There are two connections with the ocean. The larger, Cabot Strait on the southeast, is 104 km wide with a maximum depth of 480 m. The smaller, Strait of Belle Isle at the northeastern corner, is 16 km wide with a sill depth of only 60 m. The Gulf of St Lawrence receives the discharge from a major river the St Lawrence, which enters its northwestern corner and has a very important influence on its dynamics. In addition, the Gulf is fed, especially along its northern coast, by a large number of intermediate and small rivers. The mean annual freshwater discharge into the Gulf was estimated by El-Sabh (1977) to be $19{\cdot}1 \times 10^3$ m^3/s, two-thirds of which comes from the St Lawrence River.

The St Lawrence Gulf and Estuary are not large enough to generate independent tides (*i.e.* due to tidal potential). Hence the tides in this system are co-oscillating tides with the Atlantic Ocean. In the Gulf of St Lawrence and its approaches there are nearly two hundred sites at which tides have been systematically observed and analysed (see Fig. 2, p. 16). The co-oscillating tide in the Gulf is mainly through the Cabot Strait while the co-oscillation through the Belle Isle Strait has insignificant influence on the tides in the Gulf. In this section we shall summarize the major characteristics of the semi-diurnal and diurnal tides in the St Lawrence system. For more details the reader is referred to Farquharson (1970), Godin (1979, 1980), and Levesque, Murty & El-Sabh (1979).

Figure 28A shows the co-amplitude and co-phase lines for the semi-diurnal tidal constituent M_2. As can be seen, there exists a principal amphidromic point within the Gulf just westward of the Magdalen Islands. There is a second amphidromic point at the northern end of Northumberland Strait and a degenerate amphidromic system in the Belle Isle Strait. Similar patterns are also found for the two other semi-diurnal constituents N_2 and S_2 (Godin, 1980).

The movement of the semi-diurnal oscillation in the Gulf of St Lawrence, because of the presence of the Magdalen amphidrome, consists of counter-clockwise motion of the crest of high water which enters Cabot Strait, then

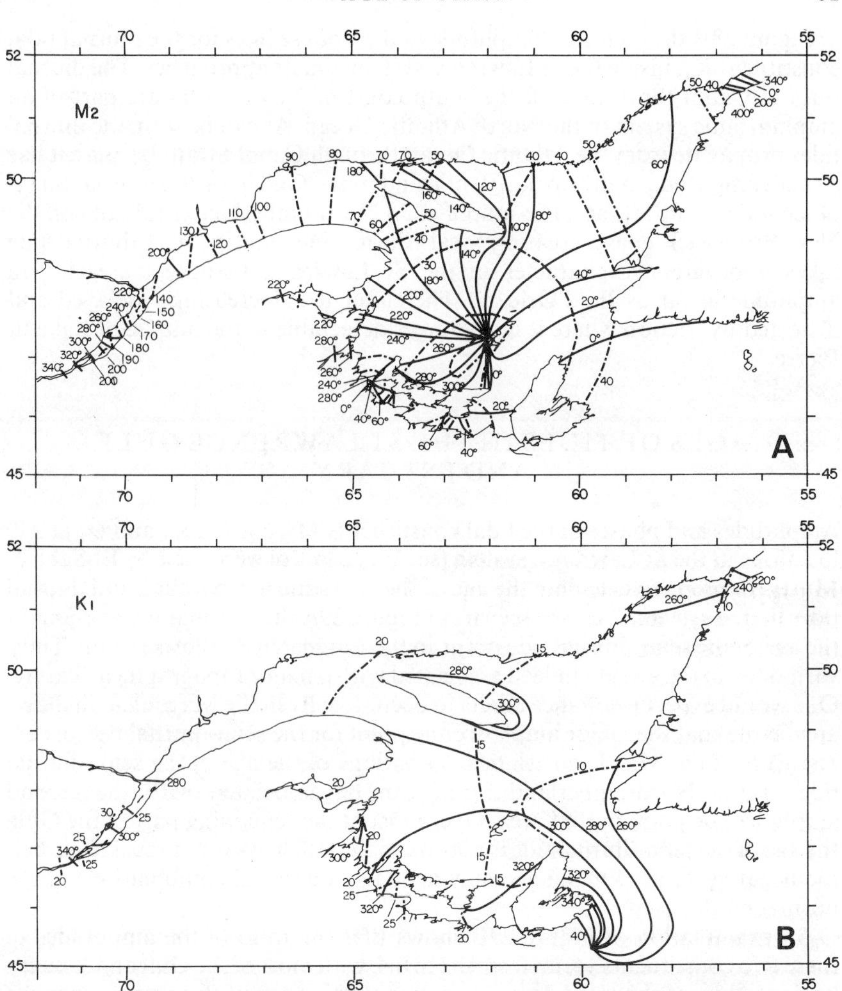

Fig. 28.—Cotidal and corange lines for the M_2 tide (A) and K_1 tide (B) in the St Lawrence Gulf and Estuary: phase angles (with reference to Greenwich) are in degrees and amplitudes are in centimetres; from Godin (1979).

progresses along the west coast of Newfoundland, the Quebec north shore, Gaspé Peninsula and the northeastern coast of Prince Edward Island. Inside the Northumberland Strait, the motion consists of a wave entering from the south progressing up the strait with increasing amplitude until it meets the region of the amphidromic point, where there is a rapid decrease in amplitude and marked delay of the tide. Past the Gaspé Peninsula, the semi-diurnal tide takes a progressive character and travels towards Québec with continuous amplification. The maximum amplitude is found between Isle-aux-Coudres and the eastern side of Ile d'Orléans. It takes eleven hours to progress from Québec to the western extremity of Lake St Pierre beyond which the tide is no longer detectable.

Figure 28B shows the co-amplitude and co-phase lines for the diurnal tidal constituent K_1, inside the St Lawrence system and its approaches. The diurnal tides at Cabot Strait and off the south coast of Nova Scotia are part of an amphidromic system in the North Atlantic Ocean. As can be seen, the diurnal tides propagate from the Atlantic Ocean through Cabot Strait. Because it has a small amplitude in the open Atlantic and in the Gulf area, the diurnal tide is of secondary importance. It is amplified in Northumberland Strait and off the New Brunswick coast, creating local diurnal inequalities. The diurnal tide takes a progressive character in the St Lawrence Estuary increasing in amplitude as far as Ile d'Orléans. Thereafter it is increasingly delayed and distorted by friction where it is no longer detectable in the middle of Lake St Pierre.

AGES OF TIDES IN THE ST LAWRENCE GULF AND ESTUARY

Amplitudes and phases of the tidal constituents M_2, S_2, N_2, K_1 and O_1 at 229 locations in the St Lawrence system (see Fig. 2, p. 16) were used by El-Sabh & Murty (in prep.) to calculate the age of the semi-diurnal, parallax and diurnal tides in that system. It can be seen from Figure 29A that the major variations in the age of the semi-diurnal tide occur in the Magdalen Shallows region. Thus, variations in the age are indeed associated with a major topographical feature. One would expect resonance effects to occur locally in the Magdalen Shallows area. Note that the major amphidromic point for the semi-diurnal tide occurs just west of the Magdalen Islands. Variations of the age of the semi-diurnal tide in the Northumberland Strait can be associated with the second amphidromic point in that area. Over most of the remaining part of the Gulf, the age of the semi-diurnal tide is approximately 30 h. As was discussed earlier, the negative values for the age may arise from unusual combinations of the normal modes.

An examination of Figure 29B shows that the ratio of the amplitudes of these two constituents varies from 0·2 to 0·4, with most of the Gulf and Estuary having values of about 0·3. Large variations (ratios up to 3·0) occur again in the Magdalen Shallows area where the semi-diurnal tide has an amphidromic point.

Figure 29C and D show, respectively, the contours of the age of parallax and the N_2/M_2 amplitude ratio. It can be seen that the pattern for the age of parallax is quite similar to that of the age of the semi-diurnal tide. Unlike the ratio S_2/M_2, the ratio N_2/M_2 does not, however, vary much and for the St Lawrence Gulf and Estuary it has values ranging from 0·2 to 0·3. Thus, the suggestion by Garrett & Munk (1971) that the age of parallax might be more suitable than the age of the semi-diurnal tide appears to be correct.

One very interesting result for the age of the diurnal tide in the St Lawrence system (Fig. 29E) is that it is negative everywhere. Variation from minus eighty (−80) hours to minus twenty-three (−23) hours occurs between the eastern end of Anticosti Island and the Cabot Strait. Relatively large negative ages occur outside the St Lawrence system on the Atlantic coast of the Cape Breton Island. This picture is consistent with the cotidal charts for the diurnal constituents K_1 and O_1 which show no amphidromic points in the Gulf of St

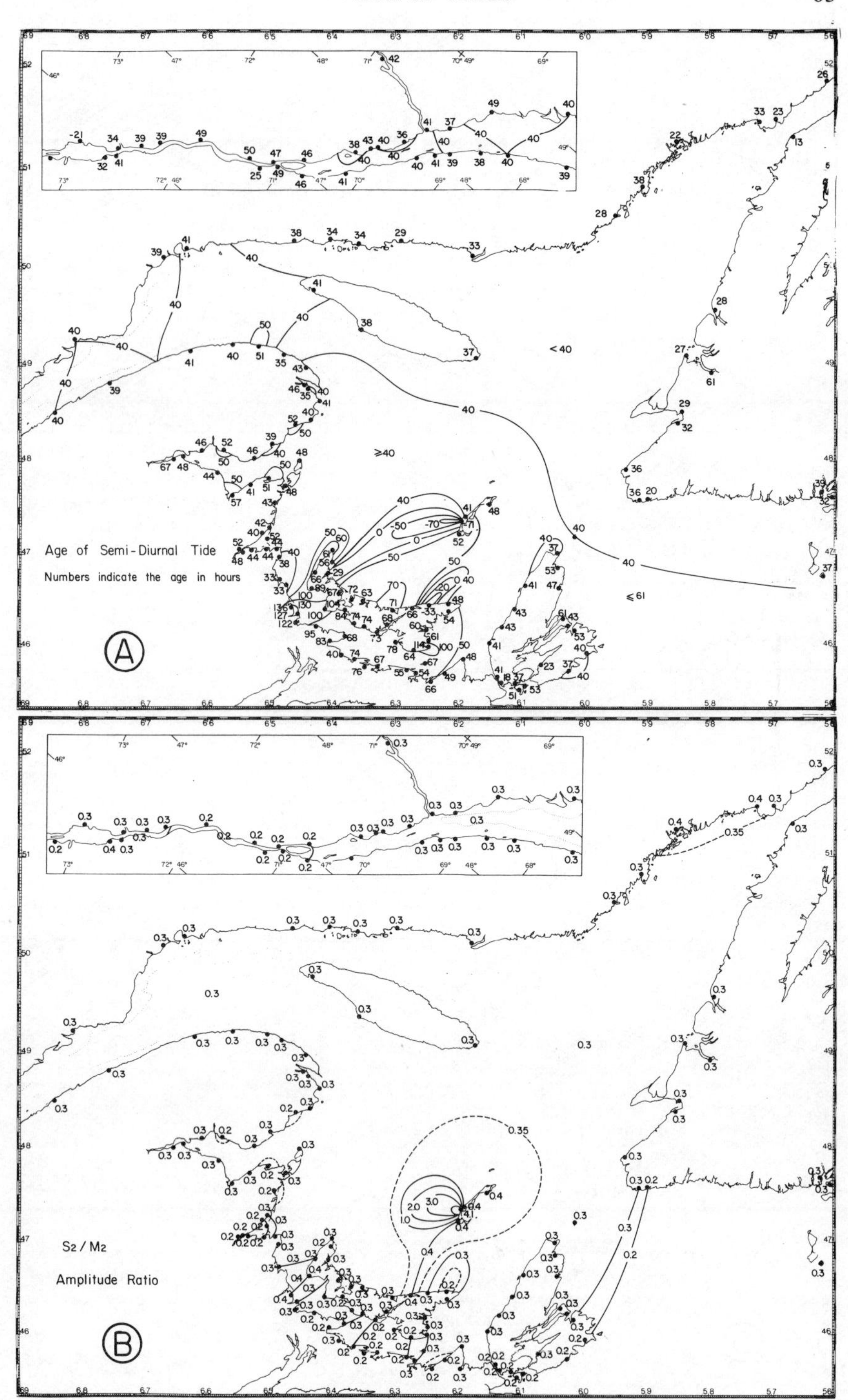
Age of Semi-Diurnal Tide
Numbers indicate the age in hours
A
S2 / M2
Amplitude Ratio
B

Fig. 29A, B.

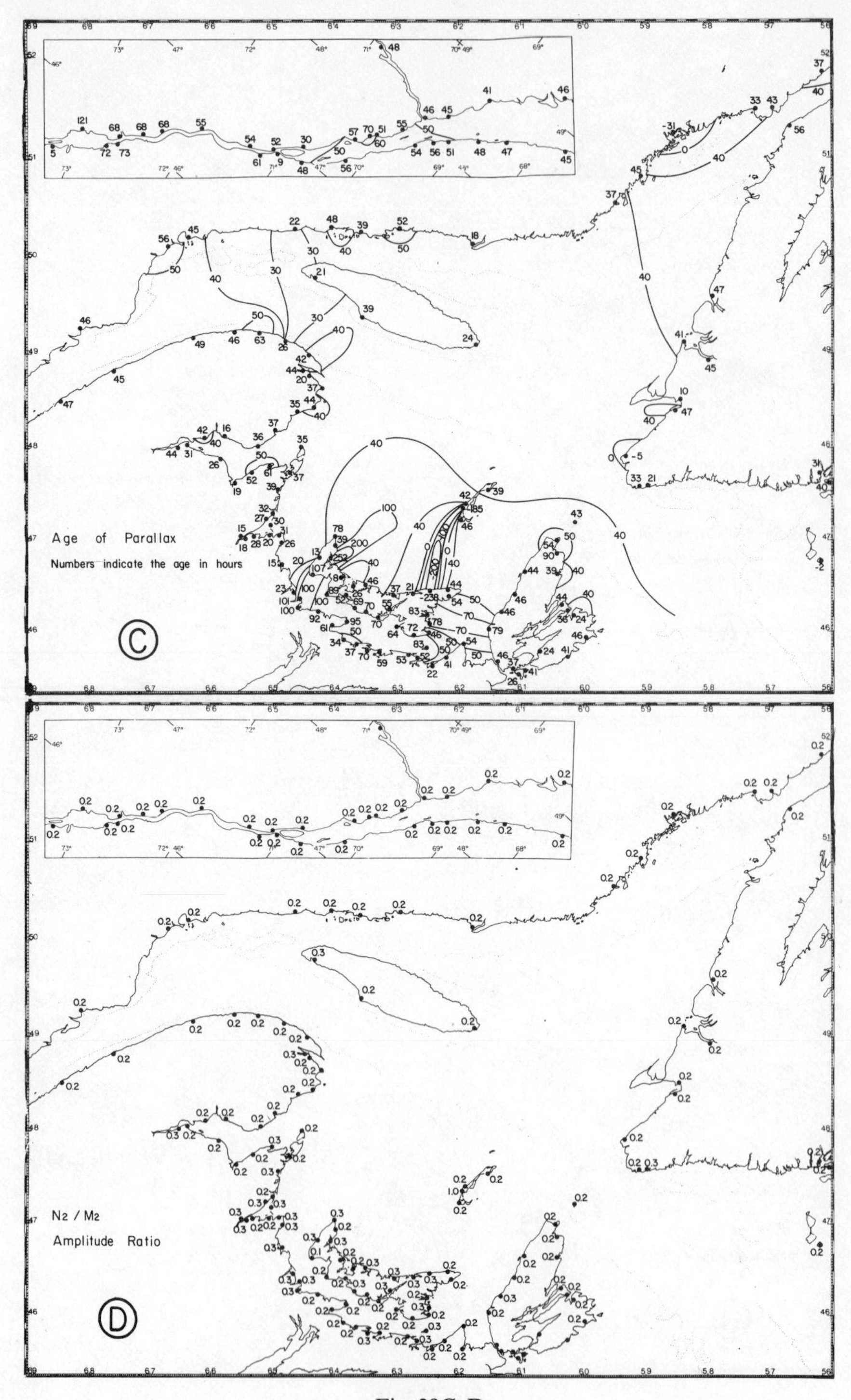

Age of Parallax
Numbers indicate the age in hours
C
N2 / M2
Amplitude Ratio
D

Fig. 29C, D.

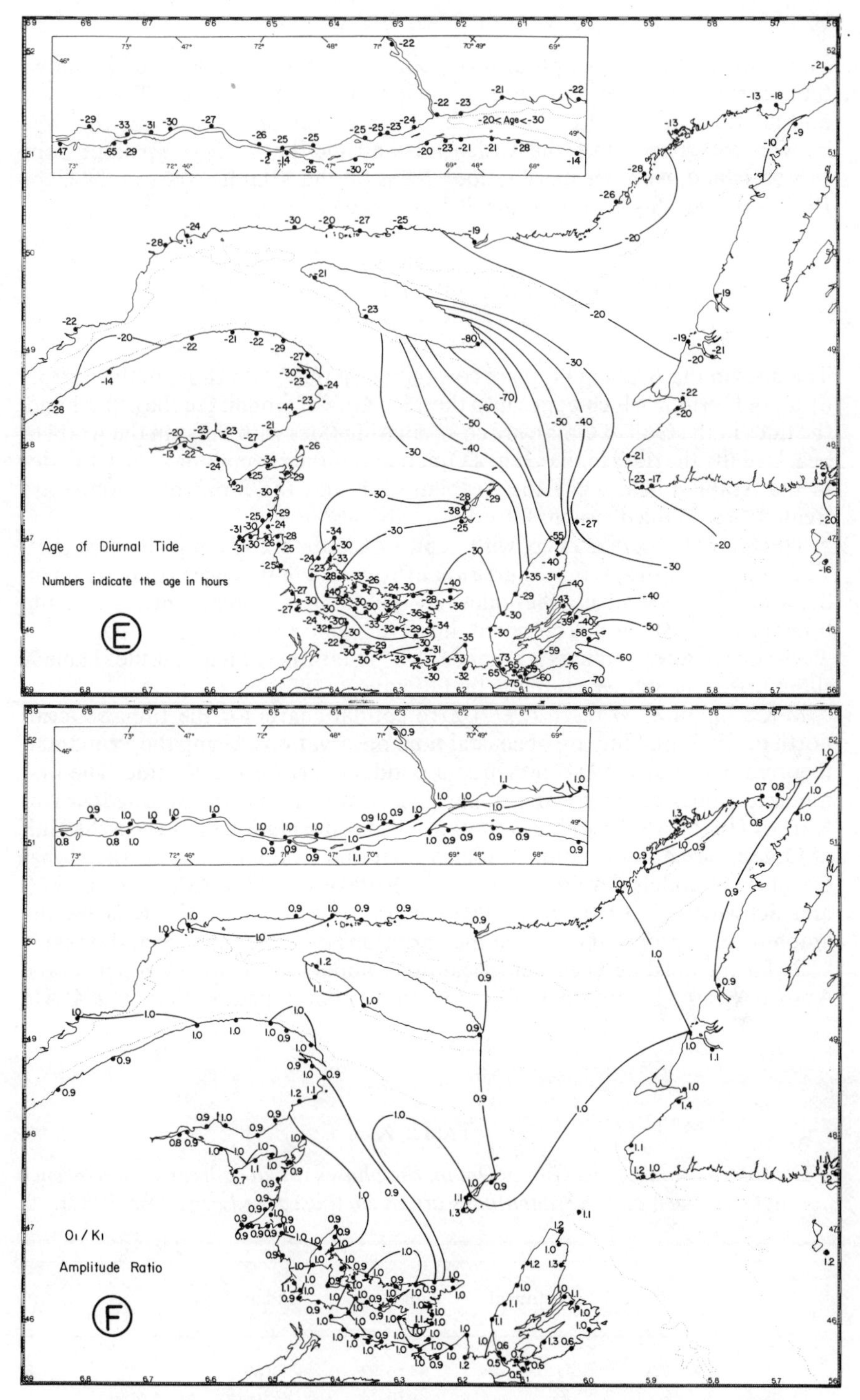

Fig. 29.—Distribution of the age of (A) semi-diurnal tide, (C) age of parallax and (E) age of diurnal tide in the St Lawrence Gulf and Estuary: (B), (D) and (F) show, respectively, the distribution of S_2/M_2, N_2/M_2, and O_1/K_1 amplitude ratios; from El-Sabh & Murty (in prep.).

Lawrence but have an amphidromic point outside the Gulf. Figure 29F shows that the ratios of O_1/K_1 are fairly uniform throughout the system. Thus, for the diurnal tide, one could look for important resonance effects in the area between the eastern tip of the Anticosti Island and the northern tip of the Cape Breton Island. Another place to look for is on the Atlantic coast of the Cape Breton Island (this is outside the St Lawrence system).

TIDES IN THE ARABIAN GULF AND GULF OF OMAN

The tides in the Arabian Gulf are co-oscillating tides with those in the narrow Strait of Hormuz which opens into the deep Gulf of Oman. On the other hand, the tides in the Gulf of Oman are co-oscillating tides with those in the Arabian Sea. Usually the tides in the Gulf of Oman are studied in conjunction with tides in the Arabian Sea, while the Arabian Gulf and the Strait of Hormuz are treated as a coupled system for tidal computations.

The Gulf of Oman is deep, with depths exceeding 2 km in its eastern part. The continental shelf is quite narrow. LeProvost (1984) suggests that, based on Schwiderski (1980a,b,c), the tides in the Gulf of Oman must be quite homogeneous. It should be noted, however, that Schwiderski's model is for global ocean tides and uses a grid of $1° \times 1°$ in latitude and longitude. Table X summarizes the inferences made by LeProvost (1984).

McCammon & Wünsch (1977) gave cotidal charts for the Indian Ocean north of 15° S, making use of coastal tidal observations. Using the Proudman theorem, Fairbairn (1954) prepared a cotidal chart for the K_2 tide. The first numerical tidal model for the Indian Ocean was developed by Bogdanov & Karkov (1975). All these charts contain a little information on tides in the Gulf of Oman. Some information can also be found in the results for the global ocean tidal models developed by Accad & Pekeris (1978), Zahel (1970, 1973) and Schwiderski (1980a,b). In these works, the main emphasis is on the amphidromic points and not on the variations of the amplitude and phase.

By far the most detailed numerical tidal model for the northern part of the Arabian Sea including the Gulf of Oman was developed by Elahi (1984). He

TABLE X

Tidal information for the Gulf of Oman; the phases are in degrees with reference to Greenwich and the amplitudes are in cm (based on LeProvost, 1984)

Tidal constituent	Amplitude	Phase
M_2	63 to 72	160
S_2	22 to 24	194 to 199
K_1	32 to 37	339 to 348

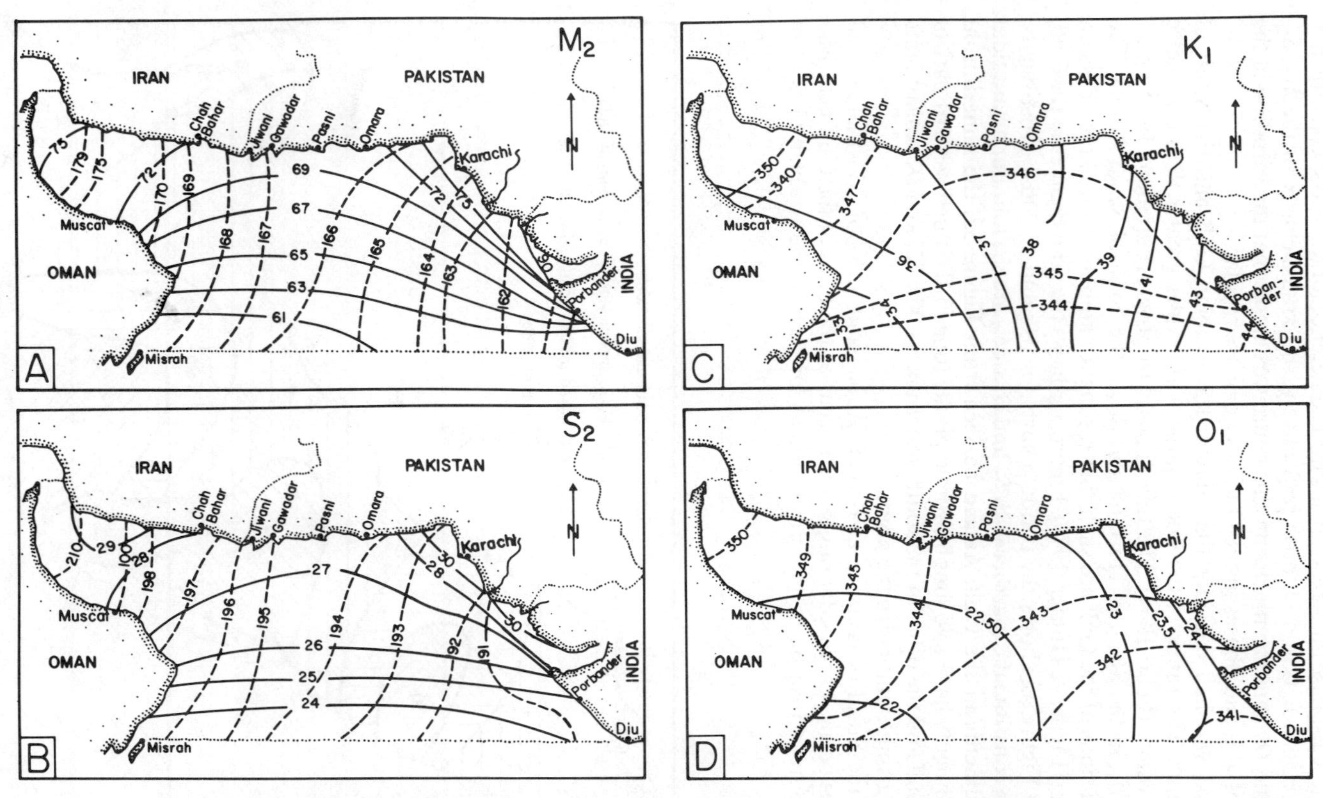

Fig. 30.—Distribution of cotidal (solid lines) and corange (broken lines) for (A) M_2 tide, (B) S_2 tide, (C) K_1 tide and (D) O_1 tide in the Northern Arabian Sea (from Elahi, 1984).

gave cotidal charts in which the co-amplitude lines are drawn with a successive difference of 2 cm for the M_2 tide, 1 cm for the S_2 and K_1 tides and 5 cm for the O_1 tide. Co-phase lines are drawn with 1° difference for these four constituents. Figure 30A, B, C and D show, respectively, the cotidal charts for the M_2, S_2, K_1, and O_1 constituents. From this figure we can see that the variation of the amplitudes and phases in the Gulf of Oman are regular.

Next we shall consider the tides in the Arabian Gulf and the Strait of Hormuz. The range of the tide is large with values greater than one metre everywhere (Lehr, 1984). The tide appears to progress up along the Iranian coast from the Strait of Hormuz and down the coast of Saudi Arabia. The dimensions of the Gulf are such that resonance amplification of the tides can occur (Hughes & Hunter, 1979). Figure 31 shows the tidal regimes in the Gulf, while Figure 32 shows the cotidal charts for M_2 and K_1 tidal constituents. These constituents, together with S_2 and O_1 tides, are the four important tidal constituents in the Gulf. As can be seen from Figure 32, the semi-diurnal constituents have two amphidromic points (one in the northwestern part of the Gulf and the other in the southwestern part). The K_1 and O_1 constituents have a single amphidromic point.

In the Strait of Hormuz at Bandar Abbas and at Khor Kawi large tidal ranges occur. Large tidal ranges also occur at the head of the Gulf near Fao and Khor Musa Bab. Numerical models for tides in the Arabian Gulf were

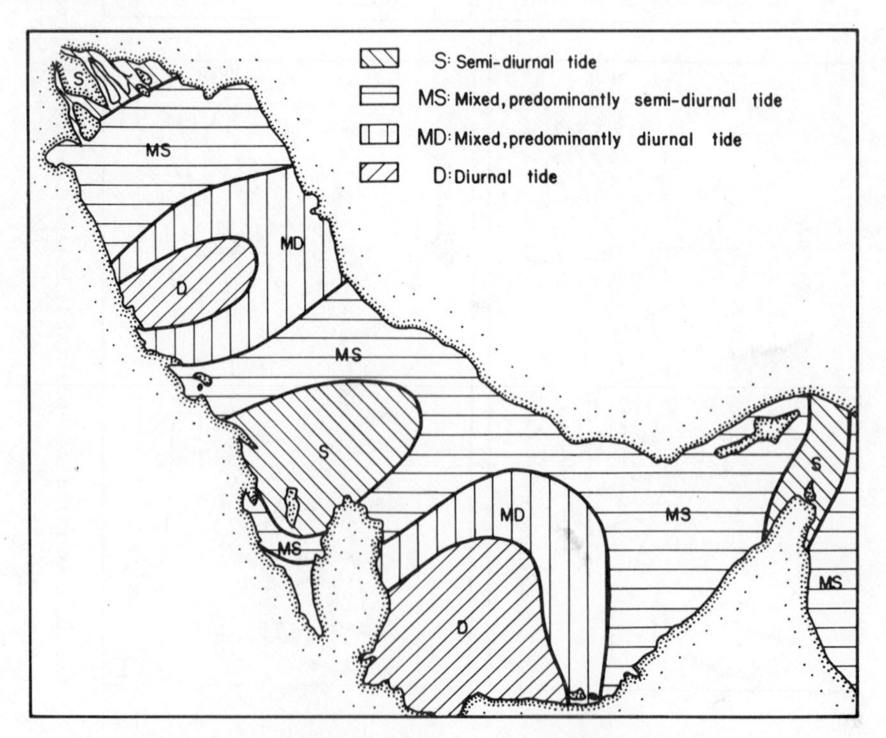

Fig. 31.—Classification of tides in the Arabian Gulf (from El-Sabh, 1982).

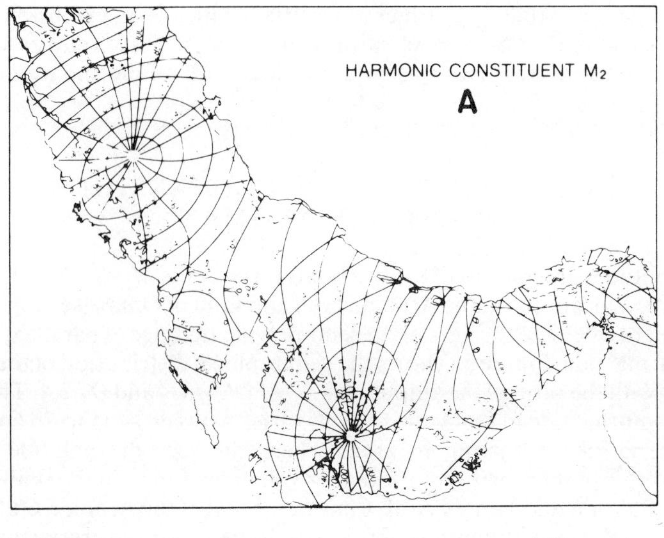

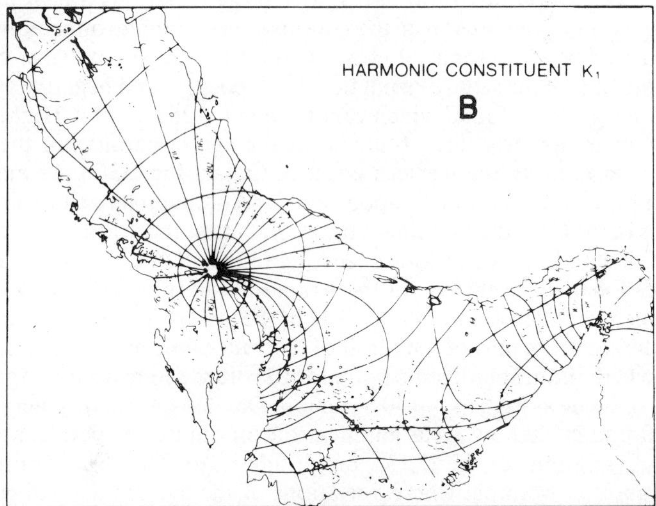

Fig. 32.—Cotidal and corange lines for (A) M_2 tide and (B) K_1, S_2 tide in the Arabian Gulf (from British Admiralty Chart, 1967).

developed by Von Trepka (1968), Evans-Roberts (1979), Lardner, Belen & Cekirge (1982), LeProvost (1984), and Murty & El-Sabh (1984).

The area around Bahrain is of particular interest because, important semi-diurnal tidal variations occur here and the topography is quite complex. According to LeProvost (1984) the Danish Hydraulic Institute developed four numerical tidal models with successively finer grids for the following areas: a model of the Gulf, with 8 km grid, a secondary model (Dawhat Salwa) with a grid of 3 km, a regional model, between Bahrain and Saudi Arabia coast with a grid mesh of 740 m, and a local model going down to 247 m. Another example of a local model is that of Lardner *et al.* (1982) which is an extension of their multi-block model. This model, with a grid size of 5 km, covers the area between the Saudi Arabian coast south of Ras Tanura and the west coast of Qatar.

AGE OF TIDES IN THE ARABIAN GULF AND GULF OF OMAN

The amplitudes and phases of the five tidal constituents M_2, S_2, N_2, O_1 and K_1 at eighty-six locations on the Arabian Gulf and Gulf of Oman (see Fig. 3, p. 17) were used to calculate the ages of the semi-diurnal tide, age of parallax, and age of the diurnal tide. Figure 33 shows the geographical distribution of these ages, together with the ratio of the amplitudes S_2/M_2, N_2/M_2 and O_1/K_1. The age of the semi-diurnal tide in the Arabian Gulf varies between −35 to 76 h with the major variations occurring in areas where the semi-diurnal tide has an amphidromic point. Furthermore, higher values are found in the shallow area between Bahrain and United Arab Emirates. Most of the Arabian Gulf has an average S_2/M_2 amplitude ratio of 0·3, with large variations (ratios up to 0·7) occurring again in the area near the southwestern amphidromic point.

Comparison between Figure 33A and C shows a similar pattern of the age of parallax and age of the semi-diurnal tide. Both positive and negative values for the age of parallax are found which vary between −26 and 77 h. Higher values of up to 195 h, however, are obtained for the age of parallax in the area of Dawhat Salwa, along the western coast of Qatar (Fig. 33D). As mentioned earlier, this area has a complex topography and is characterized by important variations of the semi-diurnal tide. The N_2/M_2 amplitude ratio has an average value of 0·2 with large variations occurring in the southern shallow area.

All ages for the diurnal tide in the Arabian Gulf (Fig. 33E) show negative values which vary between −4 and −92 h; the lower value is found at Jazirat Quoi in the southern part of the Strait of Hormuz while higher values occur in the area where the amphidromic point for the diurnal tide occurs. Again, ages of the diurnal tide as high as −140 h occur in Dawhat Salwa. In general, higher ages are found in shallow areas and near the amphidromic points, associated with large variations of the O_1/K_1 amplitude ratio. Thus, the Arabian Gulf presents another example where variations of the age are associated with a major topographic feature.

One interesting result obtained is that all ages in the Gulf of Oman show positive values: age of semi-diurnal tide at Muscat equals 29 h, age of parallax is 13 h, and age of diurnal tide is 6 h. Note that there are no amphidromic points either for the semi-diurnal or for the diurnal tides in the Gulf of Oman.

TIDES IN THE RED SEA AND GULF OF ADEN

The geography of the Red Sea and the Gulf of Aden has been shown earlier in Figure 4 (see p. 18). This diagram also shows the locations of the tide-gauge stations. First, we shall give a brief description of the water-bodies under discussion.

The Red Sea lies between the latitudes 12° N and 30° N and between the

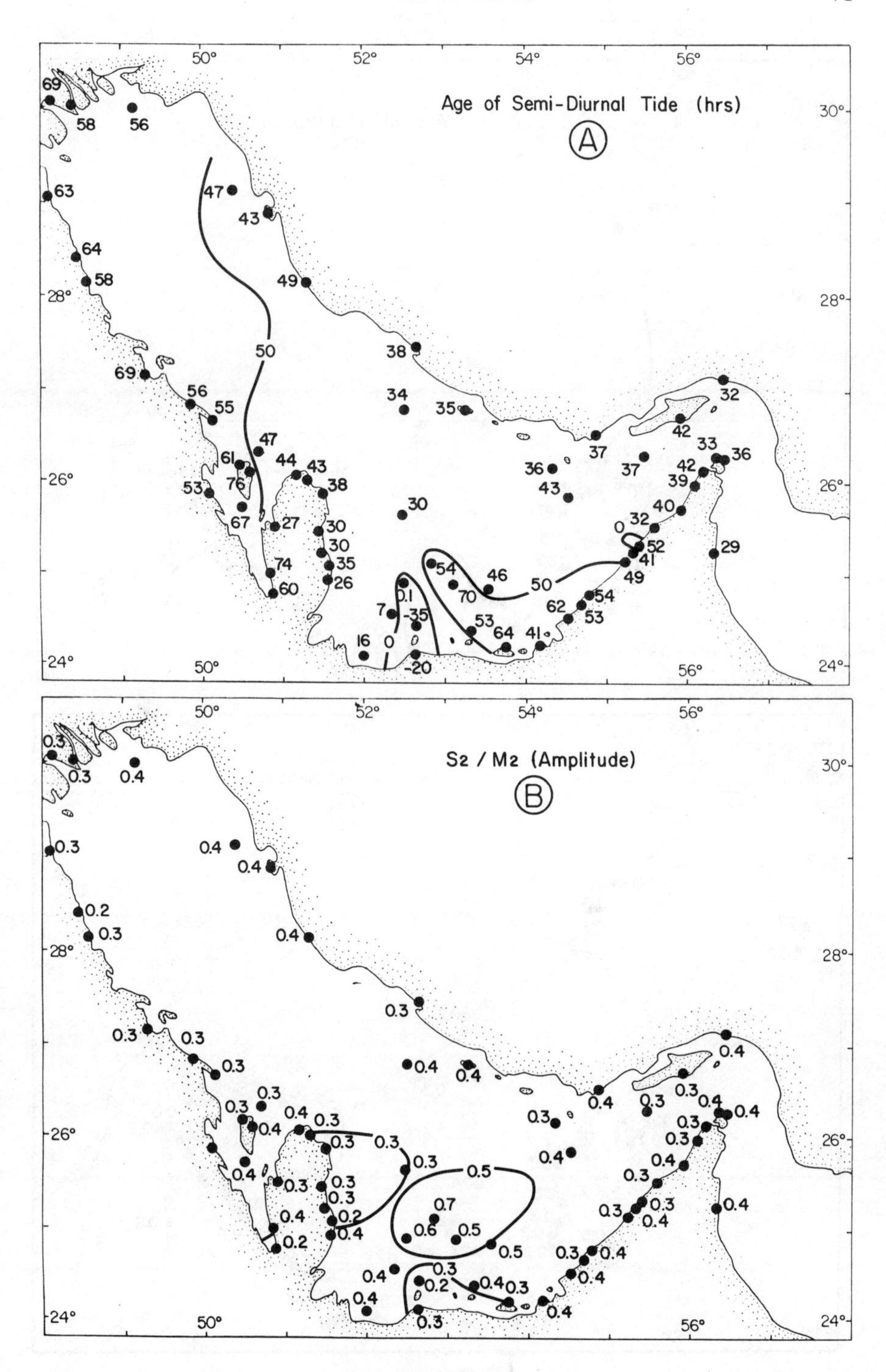

Fig. 33A, B.

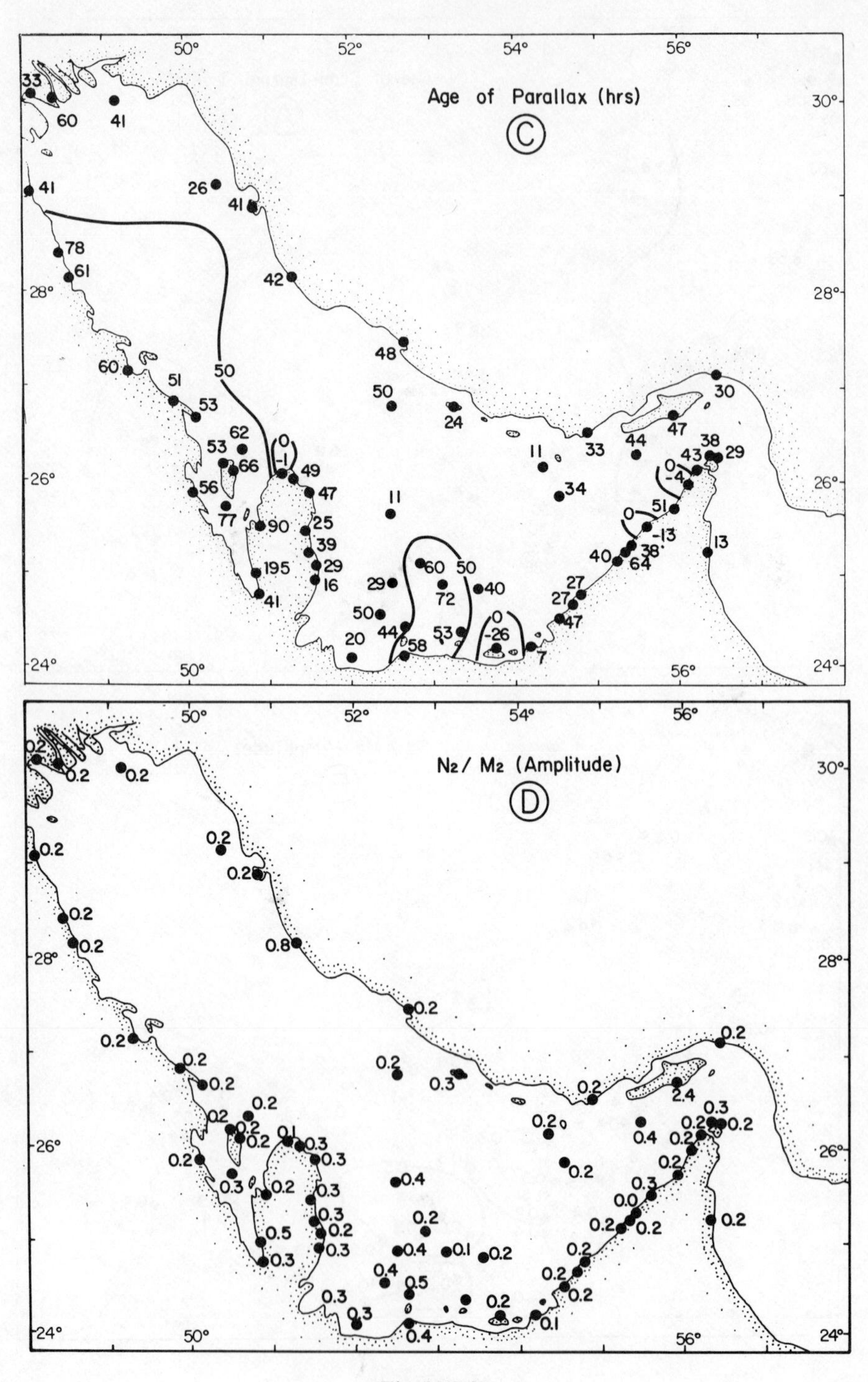

Fig. 33C, D.

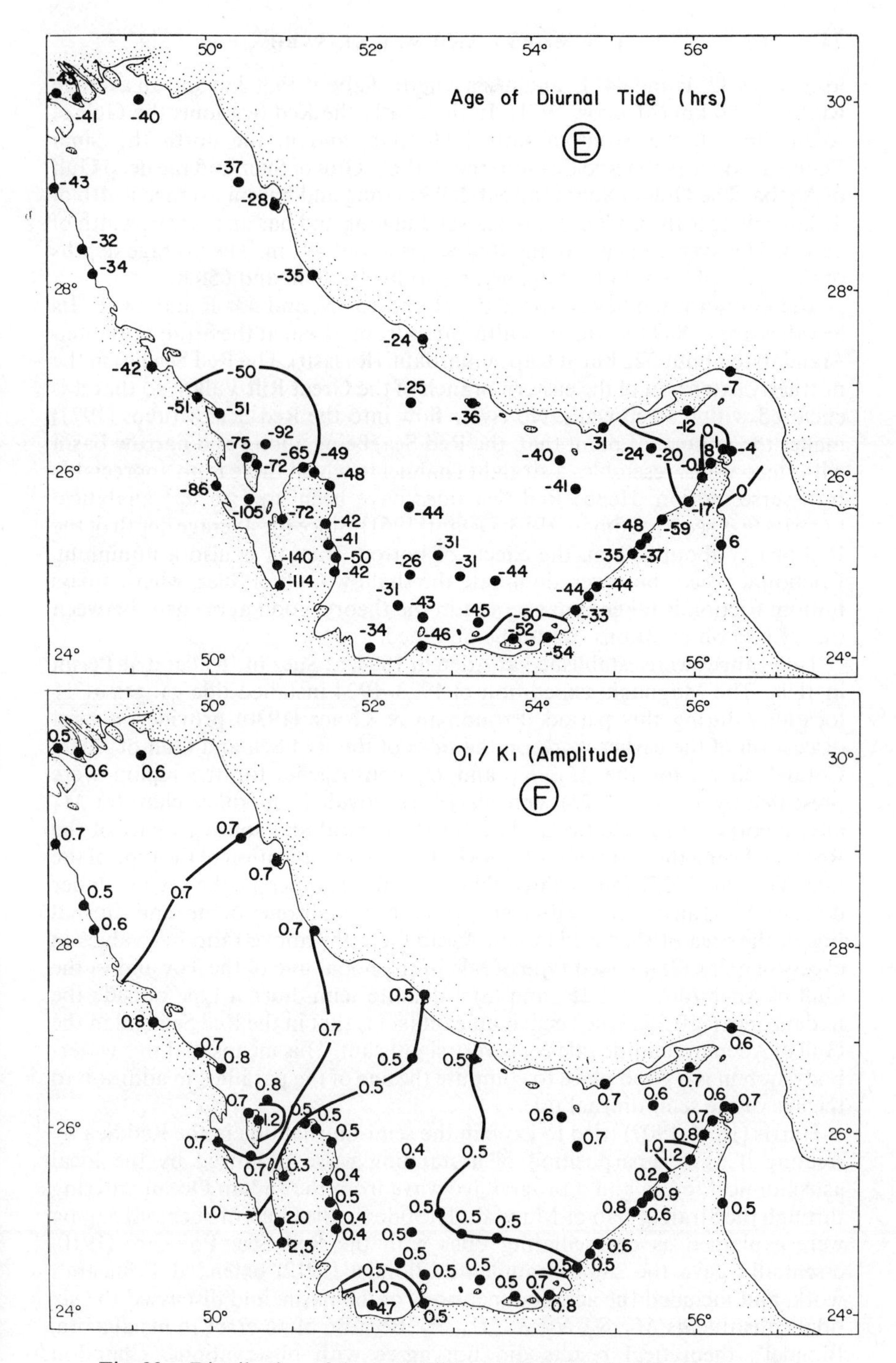

Fig. 33.—Distribution of the age of (A) semi-diurnal tide, (C) age of parallax and (E) age of diurnal tide in the Arabian Gulf: (B), (D), and (F) show, respectively, the distribution of S_2/M_2, N_2/M_2, and O_1/K_1 amplitude ratios.

longitudes 32° E and 44° E, and has a length of about 1930 km and an average width of 280 km (Morcos, 1971). In the south, the Red Sea joins the Gulf of Aden through the Strait of Bab-el-Mandab, and in the north the Sinai Peninsula divides the Red Sea into the shallow Gulf of Suez and the deep Gulf of Aqaba. The Gulf of Suez is about 250 km long and has an average width of 32 km, whereas the Gulf of Aqaba is 150 km long and has an average width of 16 km. The average depth of the Red Sea is about 491 m. The average depths of the Gulfs of Suez and Aqaba are, respectively, 36 m and 650 m.

The Gulf of Aden lies between 10° N and 15° N, and 43° E and 52° E. Its length is about 889 km and its width varies from 21 km at the Strait of Bab-el-Mandab to about 322 km at Cape Guardafui (Rasasir). The Red Sea lies in the northerly extension of the eastern branch of the Great Rift Valley, so that it is enclosed within steep walls. No rivers flow into the Red Sea. Morcos (1971) makes the important point that, the Red Sea, being a long and narrow basin with steep walls, resembles a straight channel in which there is no appreciable transverse motion. Hence Red Sea tides have been used to test analytical theories (*e.g.* see Proudman, 1953; Defant, 1961). Since the average depth of the Red Sea is about 491 m, the effects of bottom friction is also a minimum. Frictional effects, however, dominate the shallow Gulf of Suez, where unless bottom friction is taken into account in the theory, good agreement between theory and observations cannot be obtained.

Tide gauges were established at Aden in 1894, at Suez in 1897 and at Perim in 1898. The Magnaghi expedition of 1923–1924 installed tide gauges at 11 locations during this period. Proudman & Grace (1930) provided a brief discussion of the earlier work on the tides of the Red Sea and Gulf of Aden. Cotidal charts for the M_2, K_1 and O_1 constituents for the region were presented by Vercelli (1925). Vercelli (1931) provided a modified chart for M_2 after incorporating additional data for the central and southern part of the Red Sea. Using the ratio $(K_1+O_1)/(M_2+S_2)$ as an indication of the type of the tide, Vercelli (1925) found three different tidal regimes in the region under discussion: (1) an extremely diurnal type in the nodal zones of the semi-diurnal tide in the area of Port Sudan and Assab (here the above ratio has values in excess of 1·25), (2) a mixed type of tide in the nodal line of the Tor and in the Gulf of Aden (0·25 to 1·25), and (3) a definite semi-diurnal type outside the nodal zones ($<0{\cdot}25$). One very significant fact is that in the Red Sea and in the Gulf of Aden the elliptic tide N_2 is very significant. This means that this water-body system is a good place to compute the age of the parallax, in addition to the age of the semi-diurnal tide.

Harris (1894–1907) tried to explain the semi-diurnal tide in the Red Sea by treating it as a superposition of a standing wave produced by the local astronomical forces and a progressive wave from the Indian Ocean entering through the Strait of Bab-el-Mandab. The tides in the Gulfs of Suez and Aqaba were explained as co-oscillating tides with the Red Sea. Poincaré (1910) essentially gave the same arguments. Blondel (1912) extended Poincaré's work, and included the actual dimensions of the basin, and discussed the six tidal constituents M_2, S_2, N_2, K_1, O_1, P_1. Because of an error in his algebra, Blondel's theoretical results did not agree with observations. Chandon (1930a,b) corrected this error and she obtained better agreement with observations for M_2 except at Perim.

Defant (1919) gave a theoretical discussion of the spring tides of the Red Sea by including the co-oscillating and independent tides. Making use of the results of the Magnaghi expedition, Defant (1926) computed the tides in the Red Sea and concluded that they are essentially co-oscillating with those of the Gulf of Aden. Sterneck (1927) suggested that the ratio of the amplitude of the independent tide to that of the co-oscillating tide is 1:3 for M_2 and 1:4 for S_2 for most of the Red Sea. Figure 34 shows the M_2 tide in the Red Sea at various locations. The co-oscillating and the independent tides are shown separately and the observed values are shown by dots.

The tides in the Red Sea are mainly semi-diurnal (Fig. 35). There is a difference of six hours between the times of high water in the north and south, so that it is high water at the southern end when it is low water at the northern end and *vice versa*. An average spring tide range of about 0·5 m at both the northern and southern ends decreases towards the central part of the Red Sea. Near Port Sudan and Jeddah, the semi-diurnal tide is negligible. Here there is an anti-clockwise amphidromic system (Fig. 35). The tidal range is also insignificant just north of Bab-el-Mandab (between Assab and Mocha). From here southward, the time of high water changes by several hours, and the spring tidal range increases to about 1·0 m near Perim.

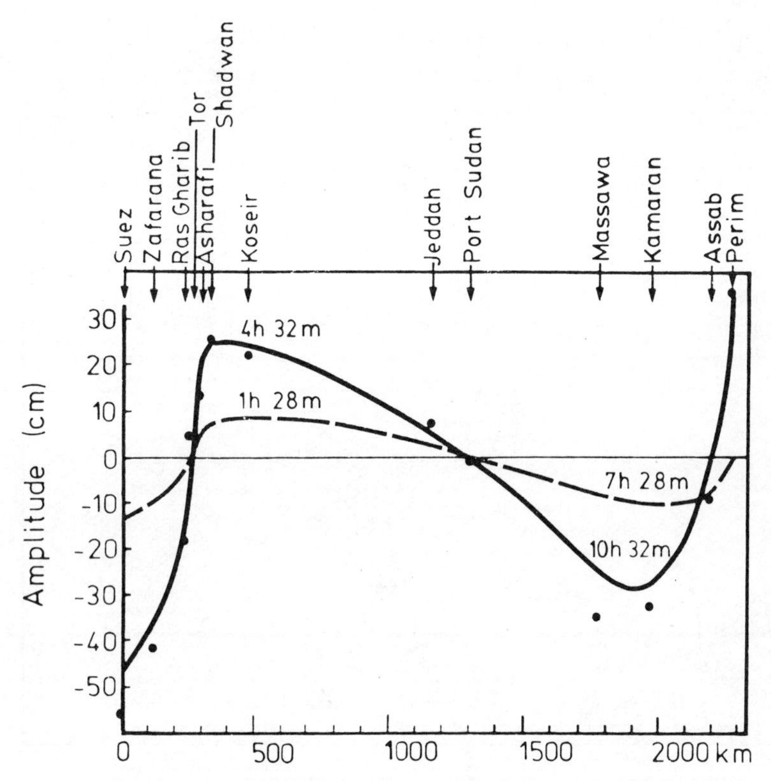

Fig. 34.—A plot of the semi-diurnal tide of the Red Sea and the Gulf of Suez: solid line shows the co-oscillating tide; broken line shows the independent tide; dots denote the observations; from Grace (1930).

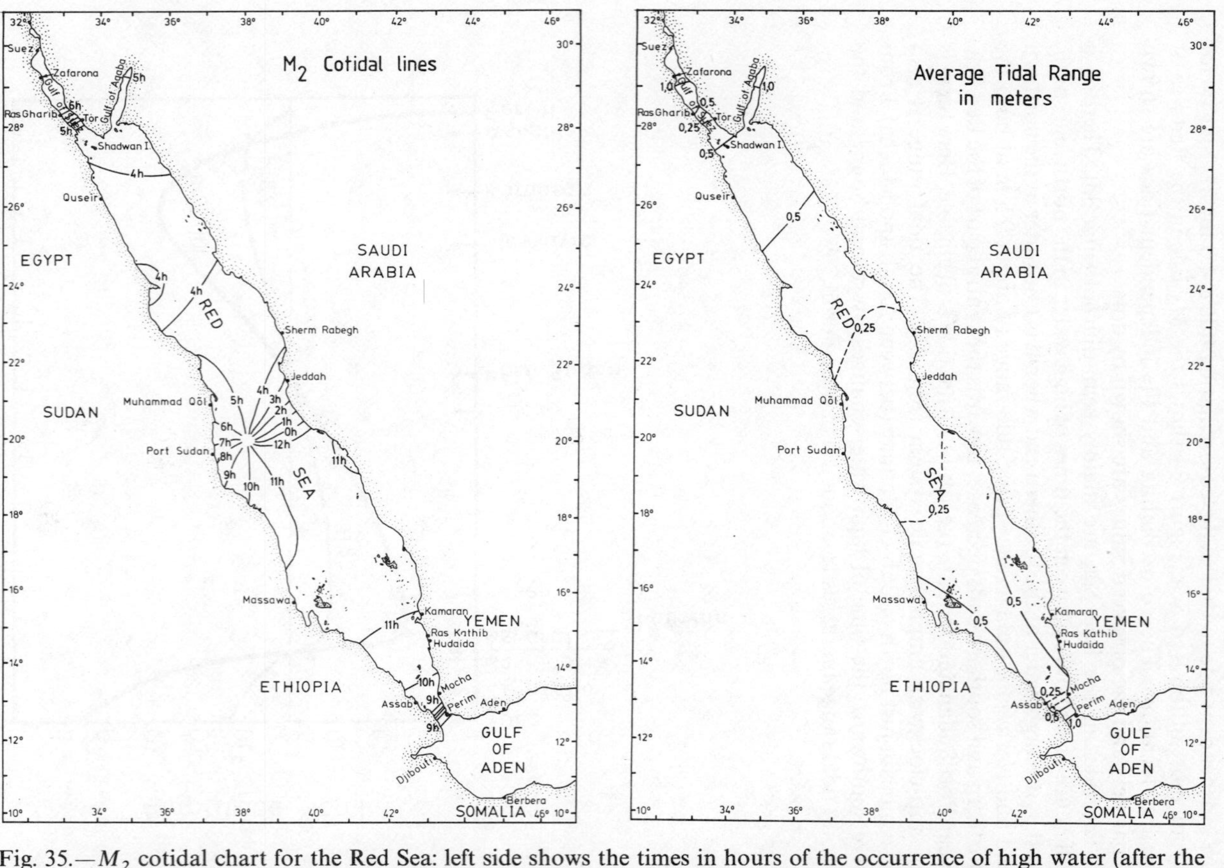

Fig. 35.—M_2 cotidal chart for the Red Sea: left side shows the times in hours of the occurrence of high water (after the transit of the moon at Greenwich); right side shows the average tidal range in metres; from Anonymous (1963).

The diurnal tides are discernible where the semi-diurnal tides are weak. Except at Perim and in the Strait of Bab-el-Mandab, the diurnal tide is small compared with the semi-diurnal tide. The diurnal tide is significant in the area between Perim and the archipelagos of Dahalak and Farasan. Note that at Aden, there is strong diurnal inequality, and occasionally there is only one tidal cycle per day.

In the Gulf of Suez, the tidal range decreases from the entrance to the Bank of Tor, and then increases to about 1·5 m at Suez. There is a difference of six hours between the high waters at the northern and southern ends of the Gulf. Hence the tide has the shape of a standing wave, with high water in the south when it is low water in the north and *vice versa*. At Suez there is a difference in the two semi-diurnal tidal cycles per day.

In the Gulf of Aqaba, high water occurs almost simultaneously over the whole Gulf. High water occurs at the northern end of the Gulf about an hour after high water occurs in the northern part of the Red Sea. The spring tidal range increases from about 0·5 m at the entrance to about 1·7 m at the head of the Gulf of Aqaba.

AGE OF TIDES IN THE RED SEA AND GULF OF ADEN

The amplitudes and phases of the tidal constituents M_2, S_2, N_2, O_1 and K_1 at twenty-two locations on the Red Sea, Gulf of Aden, and Gulf of Suez were taken from the Admiralty Tide Tables for 1982 and this data is supplemented with additional data from the International Hydrographic Bureau (1966). The phases of the constituents at all these locations are expressed in terms of the Greenwich mean phase. Based on these five constituents El-Sabh & Murty (in prep.) calculated the age of the semi-diurnal and diurnal tides in addition to the age of parallax. Table XI summarizes these calculations while Figure 36 shows their geographical distributions. In the parentheses the ratio of the amplitudes of S_2/M_2, N_2/M_2 and O_1/K_1 are also shown in Figure 36.

With reference to Figure 36A it can be seen that the ratio of the amplitudes of S_2 to M_2 varied from 0·1 to 0·4 at all the stations, except for Port Sudan where it is 1·1. This anomaly could be easily accounted for because of the presence of an amphidromic point for M_2 near Port Sudan. The lowest values of the ratio (0·1) occur at Jeddah, Muhammad Qol, and Port Sudan and are probably associated with resonance phenomenon in the deepest part of the Red Sea where the depths are about 3000 m.

The only negative values for the age of the semi-diurnal tide occur at Assab and Mocha which are in the Strait of Bab-el-Mandab. These negative values either could be erroneous, because the amplitudes of the S_2 tide at the two stations are, respectively, 3 and 4 cm. As pointed out earlier, for amplitudes less than about 5 cm, the S_2 tide could be contaminated by radiational effects, and hence the age of the semi-diurnal tide computed may not be taken seriously. On the other hand, if indeed these negative values are real, probably they could be accounted for by an unusual combination of the normal modes (Garrett & Munk, 1971). It is interesting to note that the Strait of Bab-el-Mandab is shallower and much narrower compared with the Red Sea and the Gulf of Aden, and most likely has its own distinct normal modes.

TABLE XI

Amplitude (cm), phases (GMT Zone) for the major tidal constituents and age of the semi-diurnal and diurnal tides together with age of parallax in the Red Sea and Gulf of Aden (modified from El-Sabh & Murty, in prep.)

Station	Amplitude (cm)					Phase (GMT)								Age of tide (h)		Age of parallax
	S_2	M_2	N_2	O_1	K_1	S_2	M_2	N_2	O_1	K_1	S_2/M_2	O_1/K_1	N_2/M_2	Semi-diurnal	Diurnal	(h)
Aden	21	48	13	20	40	158	134	132	352	350	0·4	0·5	0·3	18	2	9
Perim	17	37	10	18	35	159	136	139	352	350	0·5	0·5	0·3	17	2	−0·06
Mocha	04	11	07	08	07	194	237	218	352	350	0·4	1·1	0·6	−48	2	41
Hudaida	06	30	10	01	01	351	305	268	047	337	0·2	1·0	0·3	39	64	74
Ras Khathib	07	26	05	01	04	339	294	284	082	069	0·3	0·3	0·2	38	12	25
Kamaran	09	33	09	01	02	334	300	283	140	034	0·3	0·5	0·3	27	96	37
Jeddah	01	07	02	02	04	326	293	279	056	089	0·1	0·5	0·3	27	−30	30
Sherm Rabegh	02	11	03	01	01	165	124	062	123	178	0·2	1·0	0·3	34	−50	120
Tor	02	08	03	02	04	232	203	176	159	164	0·2	0·4	0·3	25	−5	53
Suez	14	56	18	01	05	306	276	248	170	153	0·3	0·3	0·3	26	15	55
Zafarana	13	42	14	01	22	301	278	250	199	180	0·3	0·0	0·3	19	18	55
Ras Ghan'b	07	18	06	02	02	304	272	244	157	150	0·4	0·9	0·3	28	6	55
Quseir	06	22	07	02	02	139	110	082	192	157	0·3	1·0	0·3	25	31	55
Muhammad Qol	01	06	02	02	03	185	130	082	175	166	0·1	0·6	0·3	50	8	92
Port Sudan	01	01	01	02	02	258	192	174	178	170	1·1	0·9	0·3	61	7	37
Massawa	12	33	09	02	02	335	324	295	182	164	0·4	0·8	0·3	5	17	59
Assab	03	08	02	06	18	173	268	265	222	335	0·4	0·3	0·3	−99	−102	11
Djbouti	21	46	13	19	39	163	139	137	357	354	0·4	0·5	0·3	17	3	09
Berbera	20	48	13	19	46	160	135	131	356	349	0·4	0·4	0·3	18	7	14
Shadwan Island	04	25	08	01	02	146	115	087	178	168	0·2	0·3	0·3	27	9	55
Harmil Island	04	13	04	01	02	337	315	293	199	166	0·3	0·6	0·3	21	30	47
Seba Island	20	38	09	21	41	162	140	139	007	347	0·5	0·5	0·2	22	−310	07

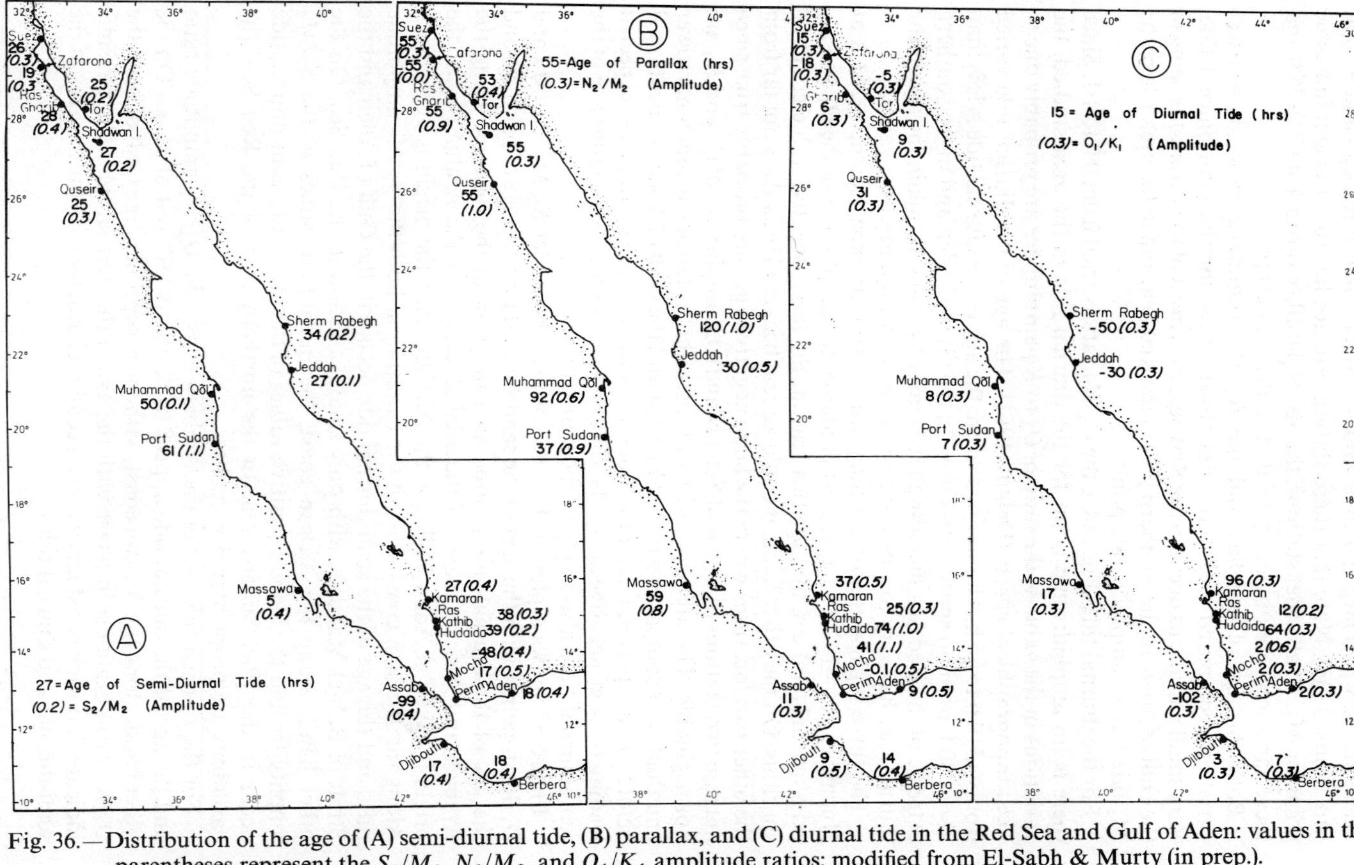

Fig. 36.—Distribution of the age of (A) semi-diurnal tide, (B) parallax, and (C) diurnal tide in the Red Sea and Gulf of Aden: values in the parentheses represent the S_2/M_2, N_2/M_2, and O_1/K_1 amplitude ratios; modified from El-Sabh & Murty (in prep.).

Figure 36B shows the results for the age of parallax. There is only one negative value, −0·06 h, at Perim. This value can practically be taken as zero. There are only two large positive values, 120 h at Sharm Rabegh and 92 h at Muhammad Qol. Note that these stations are not far from the amphidromic point for M_2. At all other stations, the age of parallax varies from 7 to 75 h and these variations are obviously related to the topography.

Both the age of parallax and the N_2/M_2 amplitude ratio show fewer variations compared with those described above for the semi-diurnal tide. Very small variations occur in the Red Sea, with most of the variations being in the Gulf of Aden, in the southern part of the Red Sea, and in the area where the M_2 tide has an amphidromic point.

For the diurnal tide (Fig. 36C), the O_1/K_1 ratio varied from 0·0 to 1·1. Since there is no amphidromic point for the diurnal tide in the area studied, the variations in the value of the ratio of O_1 to K_1 amplitudes are probably due to local topographical effects (Defant, 1961). The age of the diurnal tide varied from −310 h to 96 h. The stations where the age of the diurnal tide is less than about 30 h need no special attention. In the Gulf of Aden and in the southern extreme of the Red Sea including the Strait of Bab-el-Mandab, the age of the diurnal tide has very low positive values or very high negative values.

The three highest negative values are −310 h at Seba Island, −263 h at Hudaida and −102 h at Assab. At Hudaida, the amplitudes of the O_1 and K_1 tides are only 1·0 cm, and for this reason, the large negative value could be spurious. On the other hand, it could be real because Hudaida is not far from the other two locations where the large negative ages are probably true, since at these two stations (Assab and Seba Island) the amplitudes of O_1 and K_1 are not negligible. The area comprised of these three stations is much narrower and shallower compared with the Red Sea and the Gulf of Aden, as mentioned before. Thus, peculiar combinations of normal modes could be expected leading to large negative ages. The same arguments for the negative age of the semi-diurnal tides also apply to the diurnal tides.

Figure 37 shows a plot of the ratio of the amplitudes S_2/M_2, N_2/M_2 and O_1/K_1, together with the ages of the semi-diurnal tide, the age of parallax, and the age of the diurnal tide plotted as a function of the distance along the Arabian and African coasts of the area studied. For the Arabian coast the origin is taken at Aden while for the African coast the origin is at Berbera. Along the Arabian coast most of the variation in both the S_2/M_2 amplitude ratio and the age of the semi-diurnal tide occurs in the Gulf of Aden and the Strait of Bab-el-Mandab, with only small variation in the Red Sea. On the other hand, along the African coast, the maximum values of the S_2/M_2 amplitudes and the highest positive values of the age of the semi-diurnal tide occur in the Port Sudan area. In the northern part of the Red Sea, the variations are proportionally smaller.

For the diurnal tide, along the Arabian coast, the O_1/K_1 amplitude ratio and the age of the diurnal tide appear to be about 180° out of phase. On the other hand, along the African coast, with the exception of the Gulf of Suez, the O_1/K_1 ratio is almost in phase with the age of the diurnal tide. These results indicate that each coast has its own peculiar characteristics for the ages of the semi-diurnal and diurnal tides.

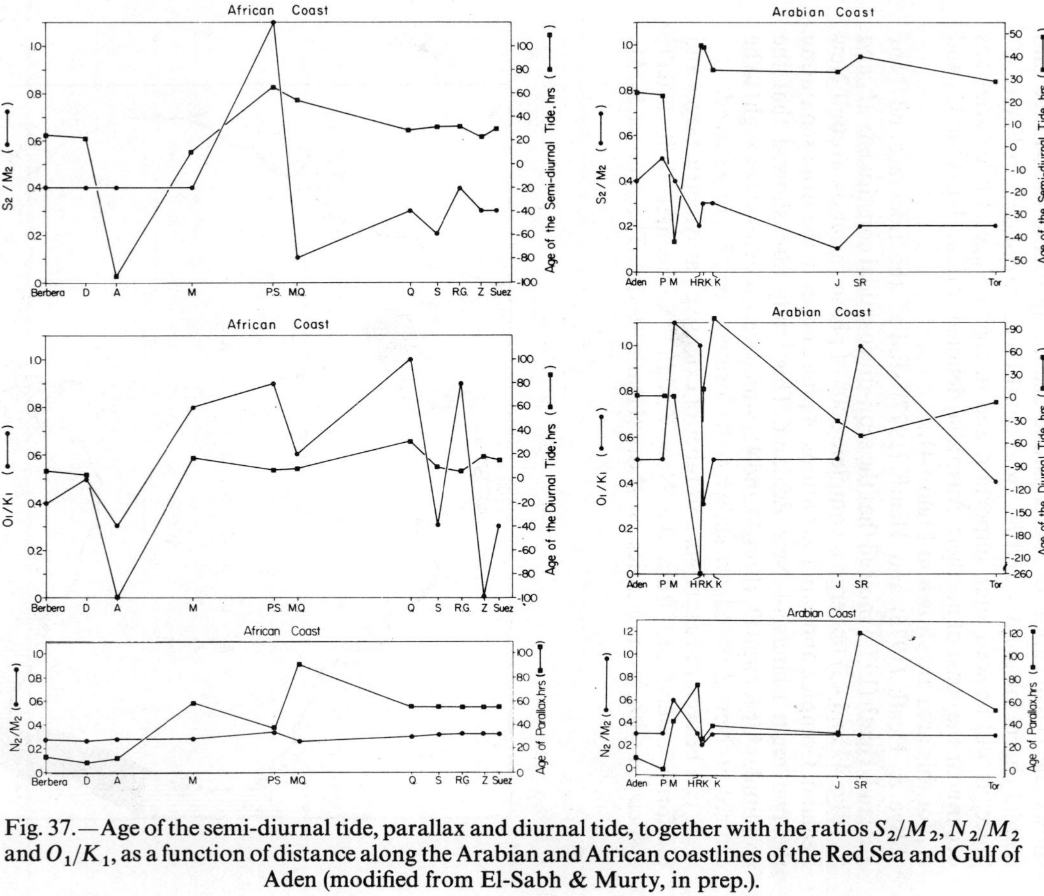

Fig. 37.—Age of the semi-diurnal tide, parallax and diurnal tide, together with the ratios S_2/M_2, N_2/M_2 and O_1/K_1, as a function of distance along the Arabian and African coastlines of the Red Sea and Gulf of Aden (modified from El-Sabh & Murty, in prep.).

TIDES IN AND AROUND AUSTRALIA AND NEW ZEALAND

Easton (1970) provided a good review of the tides around the Australian coastlines. Figure 38 categorizes the coastline into four types: semi-diurnal, mixed mainly semi-diurnal, mixed mainly diurnal, and diurnal. Figure 39 shows the classification of the coastline according to the tidal range and cotidal charts for the M_2 and K_1 tidal constituents are shown in Figure 40. Since, as will be seen later, important and interesting age of tide variations occur near the coast of northern Australia, detailed cotidal charts for M_2 and K_1 for this area are shown in Figure 41.

Bye & Heath (1975) and Heath (1977) studied the tides around New Zealand. Heath (1977) showed that the semi-diurnal tidal constituents M_2 and S_2 (Figs 42 and 43) undergo a complete 360° of phase variation around New Zealand. Complex and rapid variations of phase occur in the strait separating the two main islands of New Zealand. This study also showed that the amplitudes for K_1 and O_1 (Figs 44 and 45) which were previously thought to be centred on New Zealand, are shown to be located east of New Zealand.

As can be seen in Figure 46, a number of extensive bathymetric ridges and platforms extend out from the New Zealand continental shelf, usually associated with promontories on shore (Heath, 1981a). The ratio of S_2/M_2

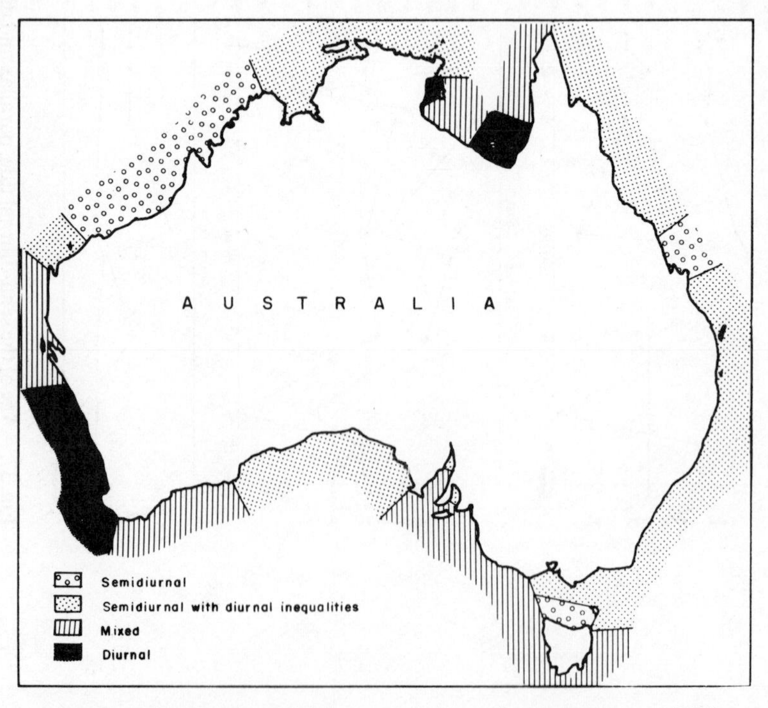

Fig. 38.—Classification of tides around Australia (from Easton, 1970).

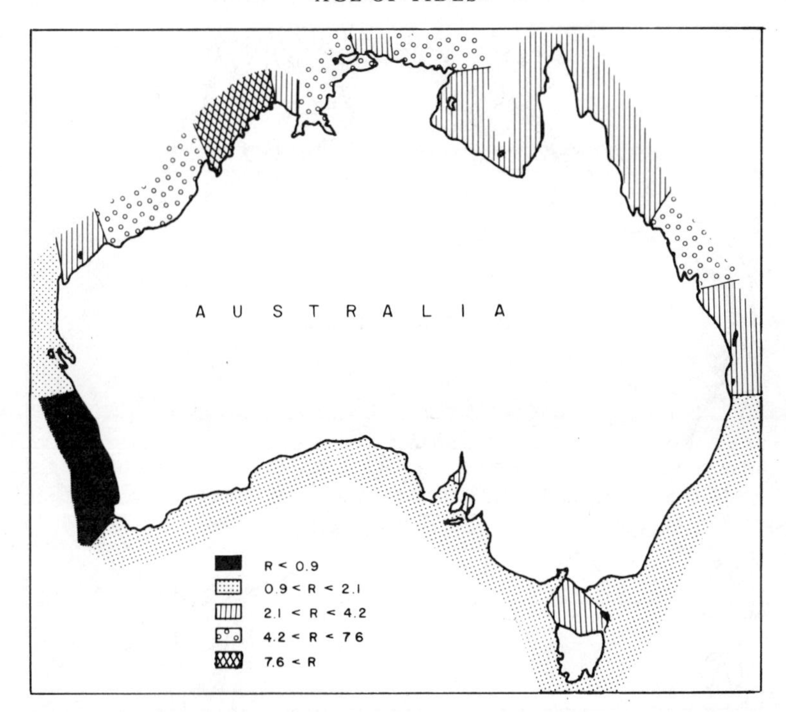

Fig. 39.—Tidal ranges in metres around Australia (from Easton, 1970).

amplitude (Fig. 47) decreases from the western coast to the eastern coast. Note that on the eastern coast the amplitudes of S_2 are only 0·03 to 0·05 m whereas on the western coast the amplitudes are 0·2 to 0·4 m. Although the amplitudes of the M_2 tide also decrease from the western coast to the eastern coast, the decrease is not as drastic as for the S_2 constituent. The fact of the rapid spatial change of the M_2 phase relative to S_2 on the western coast of New Zealand, and the small decrease in the amplitude of the S_2 tide relative to M_2 across the Campbell Plateau south from New Zealand, suggests that the relative amplitudes of the standing and progressive waves differ between M_2 and S_2 (Heath, 1981a).

For the diurnal tides Heath (1981a) found a somewhat different picture from that described by Defant (1961). Defant described an amphidrome for K_1 centred on New Zealand with a complete rotation of 360° (clockwise) around both islands. The cotidal chart given by Heath (1977) does not, however, agree with this. Heath's results suggests a K_1 amphidrome located in the Hikurangi Trench east of New Zealand. Large phase variations for K_1 occur in the Cook Strait, particularly in its narrows. The constituent O_1 has also an amphidrome located east of New Zealand. The amphidrome for O_1 appears to be located further north than that for K_1 (east of Hawk Bay as opposed to Cook Strait). The constituent O_1 has low amplitudes around New Zealand and has no drastic phase variation in the narrows of Cook Strait (Heath, 1977).

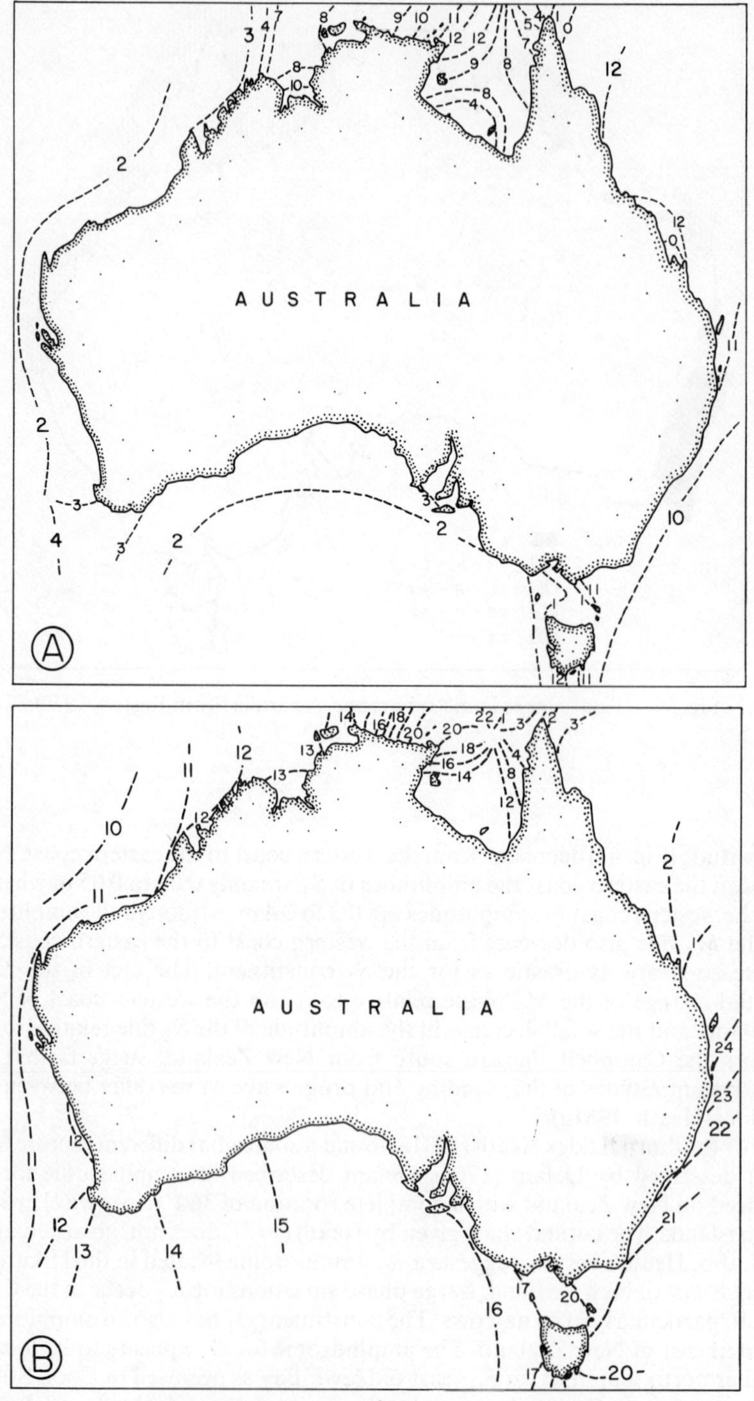

Fig. 40.—Cotidal charts around Australia for (A) M_2 and (B) K_1 tide (from Easton, 1970).

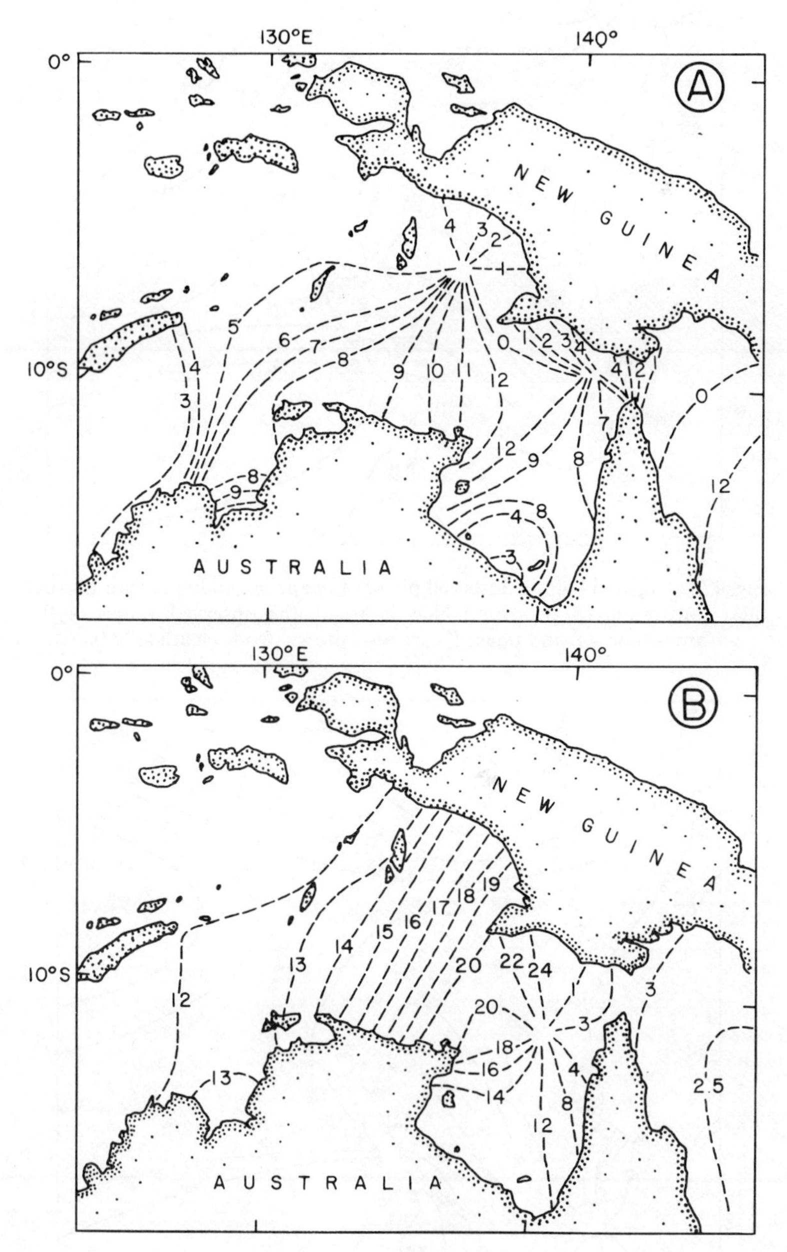

Fig. 41.—Cotidal charts at northern Australia for (A) M_2 and (B) K_1 tide (from Easton, 1970).

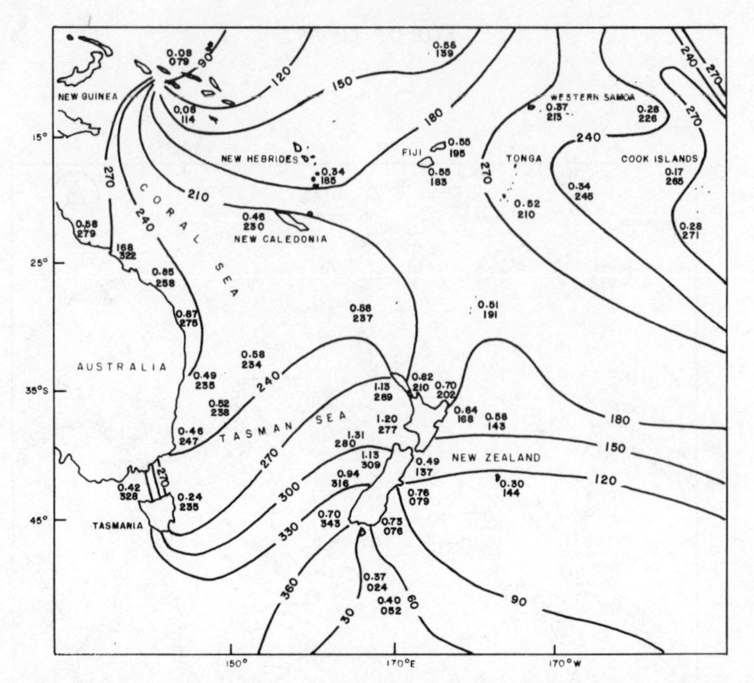

Fig. 42.—Contours of the observed phase of the principal lunar semi-diurnal tidal constituent (M_2) around New Zealand: the observed values of the amplitude (m) and phase (°) are also shown; from Heath (1977).

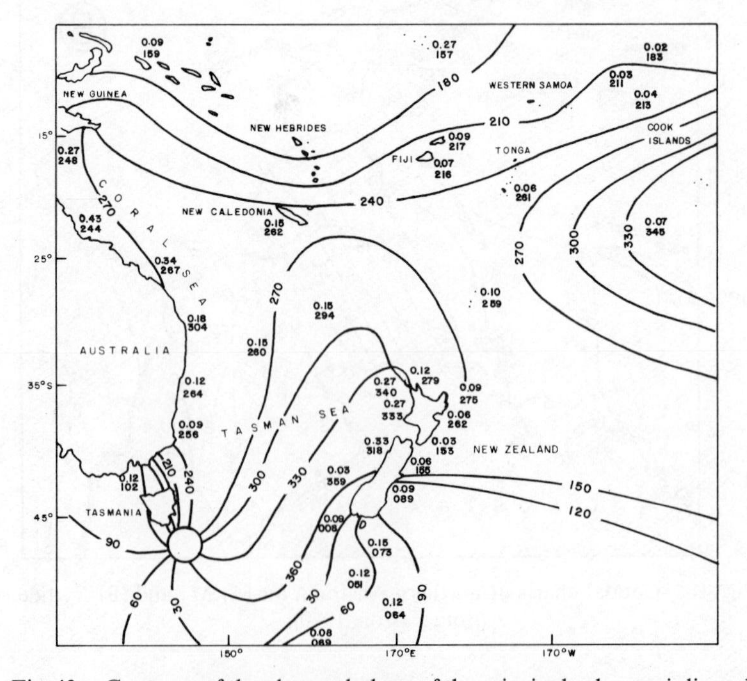

Fig. 43.—Contours of the observed phase of the principal solar semi-diurnal tidal constituent (S_2) around New Zealand: the observed values of the amplitude (m) and phase (°) are also shown; from Heath (1977).

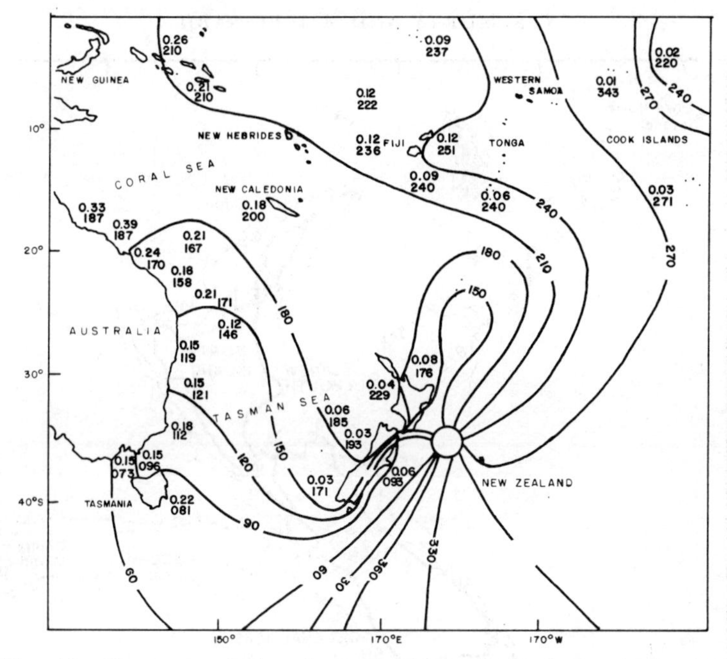

Fig. 44.—Contours of the observed phase of the lunar declinational constituent (K_1) around New Zealand: the observed values of the amplitude (m) and phase (°) are also shown; from Heath (1977).

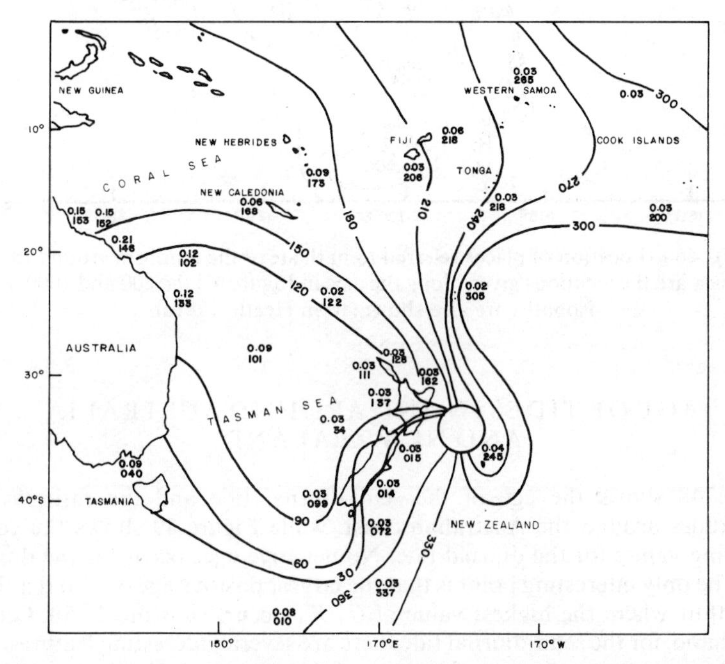

Fig. 45.—Contours of the observed phase of the lunar declination constituents (O_1) around New Zealand: the observed values of the amplitude (m) and phase (°) are also shown; from Heath (1977).

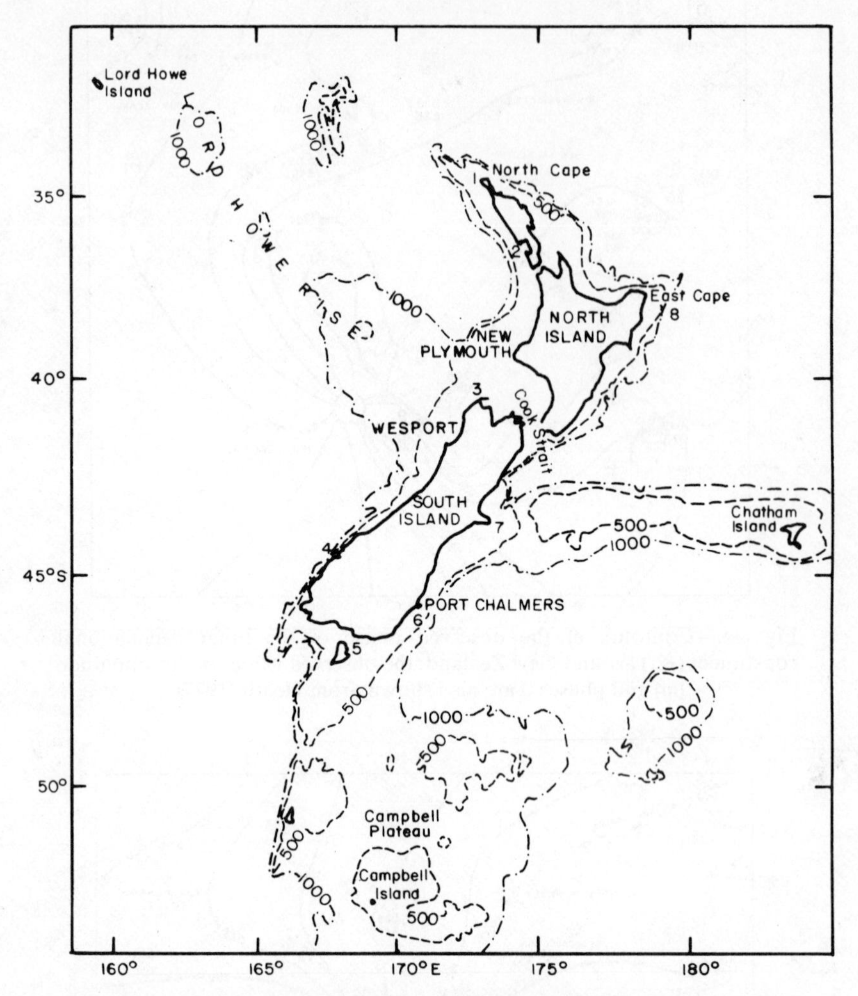

Fig. 46.—Location of places referred to in the text: the numbers around the coast are the locations given along the axis in Figure 47; the 500 and 1000 m isobaths are also shown; from Heath (1981a).

AGE OF TIDES IN AND AROUND AUSTRALIA AND NEW ZEALAND

Figure 48 shows the age of the semi-diurnal tide and the ratio S_2/M_2 amplitudes around the Australian coast, while Figure 49 shows the corresponding values for the diurnal tide. No negative ages occur for the diurnal tide. The only interesting point is that the largest positive age of 71 h is at Port McArthur where the highest value of O_1/K_1 occurs (Easton, 1970). On the other hand, for the semi-diurnal tide there are several interesting features. The three largest positive values of the age of the semi-diurnal tide occur at the three smallest ratios of S_2/M_2 (excepting one very small value). There are three

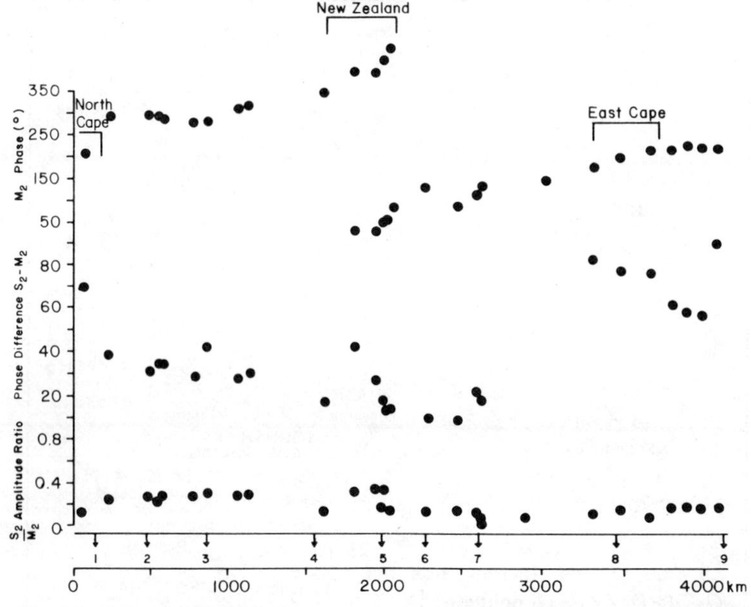

Fig. 47.—Phase of the principal lunar semi-diurnal tide (M_2), the phase difference between the principal solar semi-diurnal tide (S_2) and the M_2 tide, and the S_2/M_2 amplitude ratio around New Zealand coast: the numbers along the axis refer to the locations shown in Figure 46; from Heath (1981a).

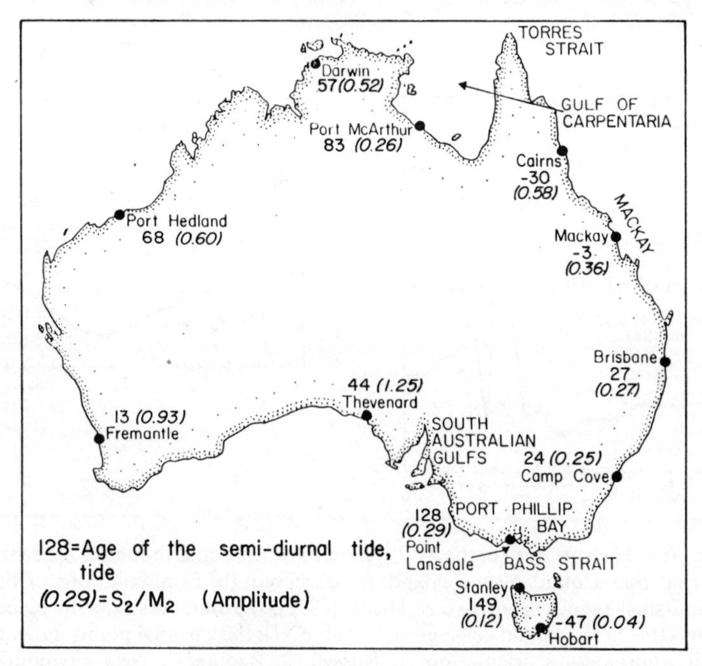

Fig. 48.—Distribution of the age of semi-diurnal tide around Australia: values in the parentheses represent the S_2/M_2 amplitude ratio; based on data from Easton (1970).

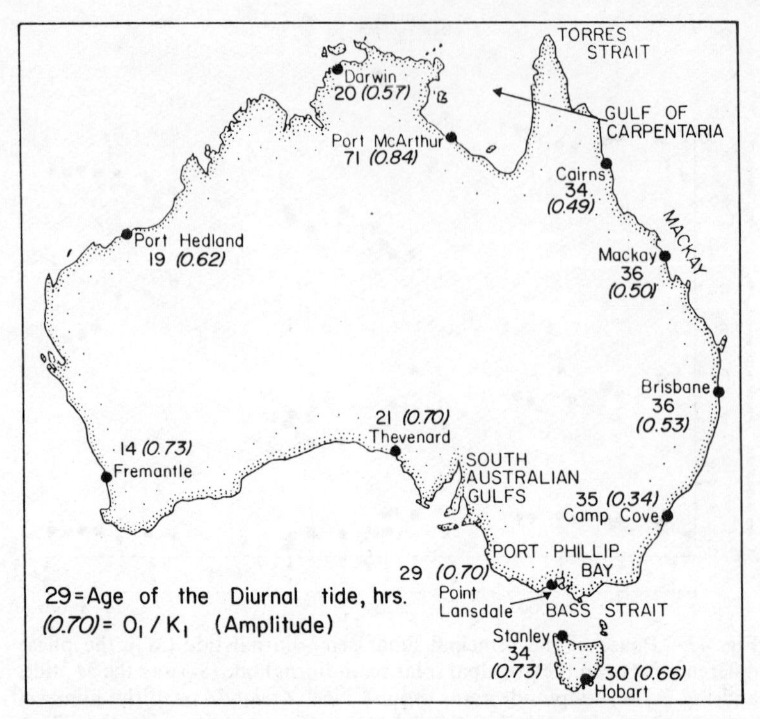

Fig. 49.—Distribution of the age of diurnal tide around Australia: values in the parentheses represent the O_1/K_1 amplitude ratio; based on data from Easton (1970).

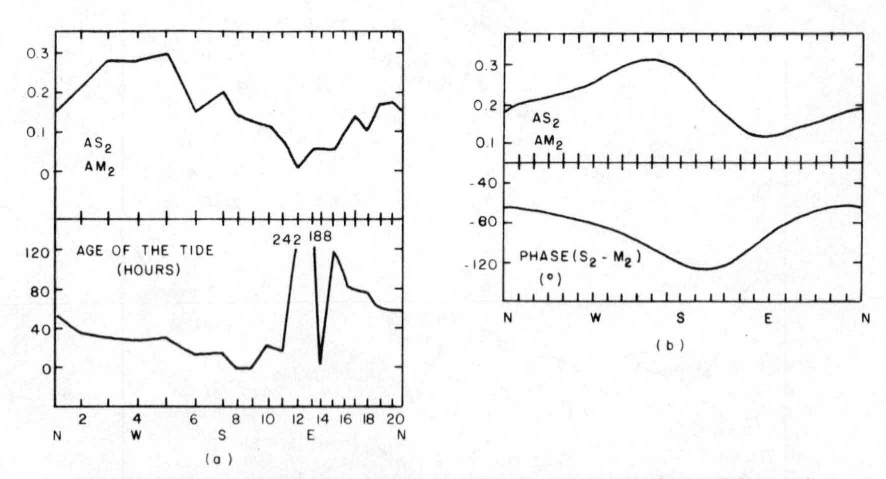

Fig. 50.—The variation of the M_2/S_2 amplitude ratio and the age of the semi-diurnal tide around New Zealand: (a) observed; (b) from solutions of the island-shelf model; from Bye & Heath (1975); the numbers refer to places round the New Zealand coast given in Table XII; the cardinal points indicate their approximate orientation; 1, Russell; 2, Raglan; 3, New Plymouth; 4, Westport; 5, Greymouth; 6, Deep Cove; 7, Bluff; 8, Port Chalmers; 9, Timaru; 10, Akaroa; 11, Lyttelton; 12, Kaikoura; 13, Wellington; 14, Castlepoint; 15, Napier; 16, Gisborne; 17, Hicks Bay; 18, Tauranga; 19, Auckland; 20, Whangarei.

TABLE XII

Phase ($g°$, referred to Greenwich Mean Time) and amplitude (Am) of the principal (M_2) and solar (S_2) semi-diurnal tidal constituents taken from the Admiralty Tide Tables 1964 (Hydrographic Department, 1963): the quantity ($GS_2 - GM_2$)/1·016 is an estimate of the semi-diurnal age of the tide (from Bye & Heath, 1975); note that the published phases g relative to GMT have been converted to local standard time, i.e. $G = g - 12\,\omega$, where ω is the corresponding angular frequency

Place	Location Lat. °S	Location Long. °E	gM_2	AM_2	gS_2	AS_2	$\frac{AS_2}{AM_2}$	$\frac{GS_2 - GM_2}{1{\cdot}016}$
A. Western coast								
Norfolk	29°04′	167°56′	237	0·6	294	0·2	0·26	44
Lord Howe	31°32′	159°04′	234	0·6	260	0·2	0·26	14
New Caledonia	22°02′	166°08′	235	0·5	273	0·1	0·27	25
New Plymouth	39°04′	174°03′	277	1·2	318	0·3	0·275	28
Kawhia	38°04′	174°49′	285	1·1	332	0·3	0·27	34
Raglan	37°48′	174°53′	289	1·1	336	0·2	0·21	34
Manukau	37°03′	174°31′	289	1·1	333	0·3	0·26	31
Onechunga	36°56′	174°47′	304	1·3	358	0·3	0·26	41
Hokianga	35°24′	173°31′	289	1·1	340	0·3	0·24	38
Westport	41°44′	171°36′	309	1·1	349	0·3	0·27	27
Greymouth	42°26′	171°13′	316	0·9	359	0·3	0·29	30
Cook Strait Tasman Bay								
Makara	41°13′	174°37′	255	0·3	346	0·2	0·7	78
Porirua	41°06′	174°52′	289	0·4	355	0·2	0·54	53
Nelson	41°16′	173°16′	280	1·3	335	0·4	0·30	42
Greville	40°52′	173°48′	272	1·2	324	0·5	0·39	39
B. Northeastern coast								
Russell	35°16′	174°07′	210	0·8	279	0·1	0·15	56
Whangarei	35°49′	174°30′	213	0·9	283	0·2	0·17	57
Auckland	36°51′	174°46′	204	1·1	278	0·2	0·16	61
Great Barrier	36°09′	175°19′	199	0·9	284	0·1	0·14	72
Tauranga	37°39′	176°11′	202	0·7	290	0·1	0·09	75
Hicks Bay	37°34′	178°19′	186	0·7	275	0·1	0·13	76
C. Eastern coast (North of Chatham Rise)								
Gisborne	38°41′	178°02′	168	0·6	262	0·1	0·1	81
Napier	39°29′	176°55′	166	0·6	297	0·0	0·05	117
Castlepoint	40°55′	176°13′	143	0·6	153	0·0	0·05	−2
Wellington	41°17′	174°47′	137	0·5	340	0·0	0·06	118
Cape Campbell	41°44′	174°15′	151	0·6	288	0·0	0·05	123
Kaikoura	42°24′	173°42′	140	1·0	038	0·0	0·00	242
D. Eastern coast (South of Chatham Rise)								
Lyttelton	43°36′	172°44′	126	0·9	155	0·1	0·07	17
Akaroa	43°48′	172°55′	106	0·9	141	0·1	0·11	22
Timaru	44°24′	171°15′	079	0·8	089	0·1	0·13	−2
Port Chalmers	45°49′	170°38′	109	0·7	120	0·1	0·13	−1
Nugget Point	46°27′	169°49′	079	0·7	094	0·1	0·13	6
Deep Cove	45°27′	167°10′	343	0·7	008	0·1	0·13	13
Bluff	46°36′	168°20′	047	0·9	073	0·2	0·17	14
Patterson Island	46°54′	168°07′	044	0·8	058	0·2	0·19	2
Macquarie Island	54°31′	158°58′	028	0·3	069	0·1	0·3	28
Campbell Island	52°33′	169°13′	052	0·4	064	0·1	0·31	0
Auckland Island	50°52′	166°05′	024	0·4	051	0·1	0·33	15

negative values for the age of the semi-diurnal tide. Of these, one is small (−3 h). The other two are −47 h at Hobart and −30 h at Cairns. At Hobart the ratio of S_2/M_2 is insignificantly small. The large values of the age of the semi-diurnal and diurnal tides at Port McArthur appear to be related to the complex cotidal chart for M_2 and K_1 shown earlier.

Bye & Heath (1975) found large variations in the age of tides from place to place around the coastlines of New Zealand (Fig. 50). The ratio of the amplitude of the semi-diurnal constituents S_2 and M_2 (see Table XII) shows four distinct geographical groups: a group on the western coast with a ratio of about 0·25; a group on the northeastern coast with a ratio of about 0·15; a group on the eastern coast north of the Chatham Rise with a ratio of about 0·05; and finally a group on the eastern coast south of the Chatham Rise with a ratio of about 0·13. The Cook Strait region and the islands south of New Zealand do not fit into this category.

It can be seen from Table XII that the age of the semi-diurnal tide around New Zealand varies from 5 to 75 h, except near Cook Strait where S_2 tide has very small amplitudes. Bye & Heath (1975) explained that the main cause for the observed variation in the age of the semi-diurnal tide is due to the fact that the primary M_2 and S_2 progressive waves approach New Zealand from different directions, with the circular asymmetry for each component being introduced by the presence of more than one mode.

A superposition of the M_2 tide from the northeast and the S_2 tide from the northwest would give a variation in the age of the tide as observed in Table XII. Heath (1981a) cautions that it may be misleading, at least in the case of New Zealand, to relate variations in the age of tide with local resonance effects. Webb (1973) found it significant with regard to the age of the semi-diurnal tide in the region between the Coral Sea and Samoa. Bye & Heath (1975) gave some importance to the large positive ages off the northeast coast of New Zealand. Heath (1981a) suggests that the unusually large positive ages or negative ages (large or small) occur due to the different spatial distributions of the M_2 and S_2 tidal regimes. Large values of the age of the semi-diurnal tide occur on the eastern coast of New Zealand near Cook Strait. According to Heath (1981a) the reason for this is the direct entry of S_2 tidal energy into the Cook Strait from the western coast. Note that the S_2 tide in the western part of the Cook Strait is almost completely out of phase with the small amplitude northward directed S_2 tide on the eastern coast. Destructive interference makes the S_2 tidal amplitudes small in the eastern part of the Cook Strait; also large errors could result in estimating the phases.

AGE OF TIDES AT ESTUARY HEADS

Examination of Figure 51, which shows the geographical distribution of the age of the semi-diurnal tide off Argentina, indicates that not only the age of tide varies over even quite small distances but also there is a tendency for largest values to be found near the head of estuaries. Similar behaviour can be found in the North Sea as can be seen in Figure 52. Webb (1973) suggests that shallow-water effects may be causing this increased value of age of tide near estuary heads.

From Figure 52 it can be seen that the age of the semi-diurnal tide (Defant,

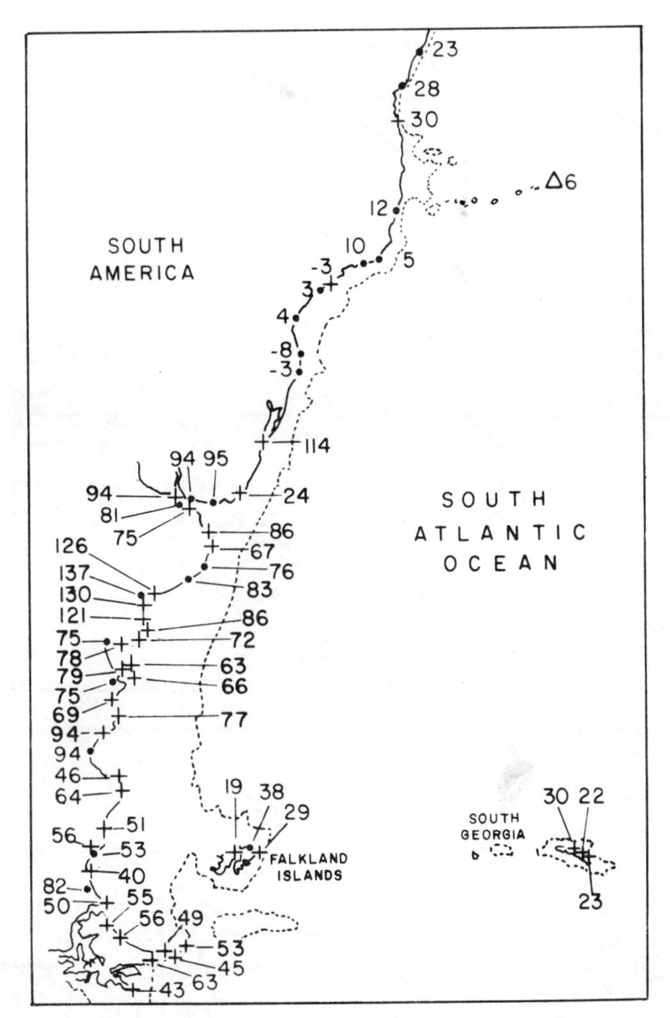

Fig. 51.—The age of the semi-diurnal tide (hours) for ports in the south-western Atlantic Ocean (from Webb, 1973).

1961, calls this the spring retardation) increases in the corner of the Deutsche Bucht up to almost three days. In the Kattegat there is a great change from a lag of about 30 h to a priming of about 40 h. Defant (1961) recognizes the important rôle played by topography in determining these variations. He also remarks that the ratio of the ranges of the neap tide to the spring tide which should have a theoretical value of 36% increases from north to southeast in the North Sea and achieves values as high as 89% near Hamburg.

Heath (1981a) mentions that the phase of the semi-diurnal tides does not show a gradual variation around the New Zealand coast except on the open coast. He makes the important point that at the major promontories

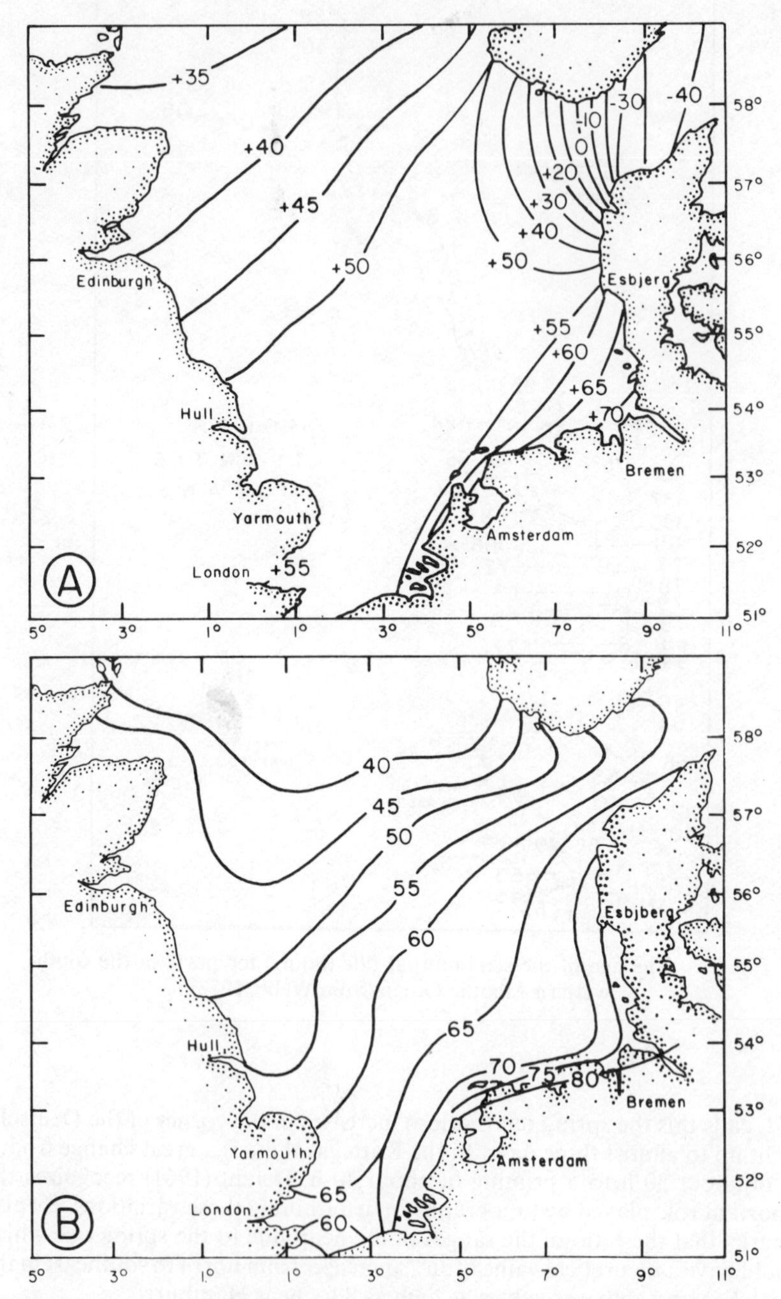

Fig. 52.—Age of semi-diurnal tide (hours) in the North Sea (A), and contours of the ratio of spring tide to neap tide in per cent (B) (from Defant, 1961).

TABLE XIII

Variation of the age of tide (hours) from the mouths to the heads of estuaries

Water-body	Estuary	Age of semi-diurnal tide		Age of parallax		Age of diurnal tide	
		At mouth	At head	At mouth	At head	At mouth	At head
St Lawrence Estuary	St Lawrence	39	46	45	48	−22	−25
Gulf of St Lawrence	Gaspé Bay	41	46	37	44	−24	−30
	Chaleurs Bay	48	67	35	44	−26	−13
	Miramichi Bay	44	48	31	18	−28	−31
Arabian Gulf	Shatt-al-Arab	69	78	33	53	−43	−30

(especially at the southern end of the South Island and near North Cape and East Cape) the phase varies rather abruptly.

We have also examined the detailed distributions of the age of semi-diurnal tide, age of diurnal tide and age of parallax near estuary heads. In the Red Sea and Gulf of Aden, although there are interesting geographical variations, they cannot be related to any estuarine effects because no rivers flow into these two water-bodies. Our study for the St Lawrence Gulf and Estuary and for the Arabian Gulf is summarized in Table XIII. Apart from some irregular variations, for twelve of the fifteen cases listed in Table XIII, there is indeed a general tendency for the age to increase from the mouth towards the head. The fact that the O_1/K_1 ratio at the head of Shatt-al-Arab Estuary equals 0·1 and appears to be the lowest value in the Arabian Gulf water-body, may explain the results obtained for the age of diurnal tide in that area. We have no explanation, however, for the two other exceptions (age of parallax for Miramichi Bay and age of diurnal tide for Chaleurs Bay). The N_2/M_2 and O_1/K_1 ratios at these two locations appear to vary regularly in much the same way as in the other estuaries within the St Lawrence system. In general, our study appears to support the conjecture of Webb (1973), but some more estuarine cases should be examined.

DISCUSSION

We have examined the geographical distribution of the ages of the semi-diurnal tide, parallax and diurnal tide for (a) the Gulf of St Lawrence and St Lawrence Estuary, (b) Red Sea and Gulf of Aden, (c) Arabian Gulf and Gulf of Oman, and (d) Strait of Georgia on the west coast of Canada (Fig. 53). In addition to these computations, the ages of the semi-diurnal and diurnal tides were examined for the waters around Australia and New Zealand. For the global oceans, only the age of the semi-diurnal tide was examined.

There is a tendency for the age of the diurnal tide to be either positive or negative in a given water-body. Only in the Red Sea and Gulf of Aden and around New Zealand, does the age of the diurnal tide have both positive as well as negative values. It has only positive values in the Strait of Georgia and in the Gulf of Oman and on the Australian coast. Negative values are only obtained in the Gulf of St Lawrence, St Lawrence Estuary and in the Arabian Gulf. One tentative conclusion one can draw is that these water-bodies, where negative values totally occur, have large areas of shallow water. On the other hand, in the deeper coastal water-bodies such as the Strait of Georgia and the Gulf of Oman, the age of the diurnal tide is everywhere positive.

Another similarity between the Strait of Georgia and the Gulf of Oman is that not only the age of the diurnal tide is positive but also the age of the semi-diurnal tide and the age of parallax are positive. Furthermore, the range of variation for each age is quite small. For example, in the Strait of Georgia, the age of the semi-diurnal tide has the range of 19 to 29 h, the age of parallax has the range of 48 to 72 h and the age of the diurnal tide varies between 19 and 25 h.

Next consider the age of parallax. It is completely positive in the Red Sea, Gulf of Aden, Gulf of Oman and Strait of Georgia, which are all deep water-bodies. It has both positive and negative values in the Gulf of St Lawrence, St

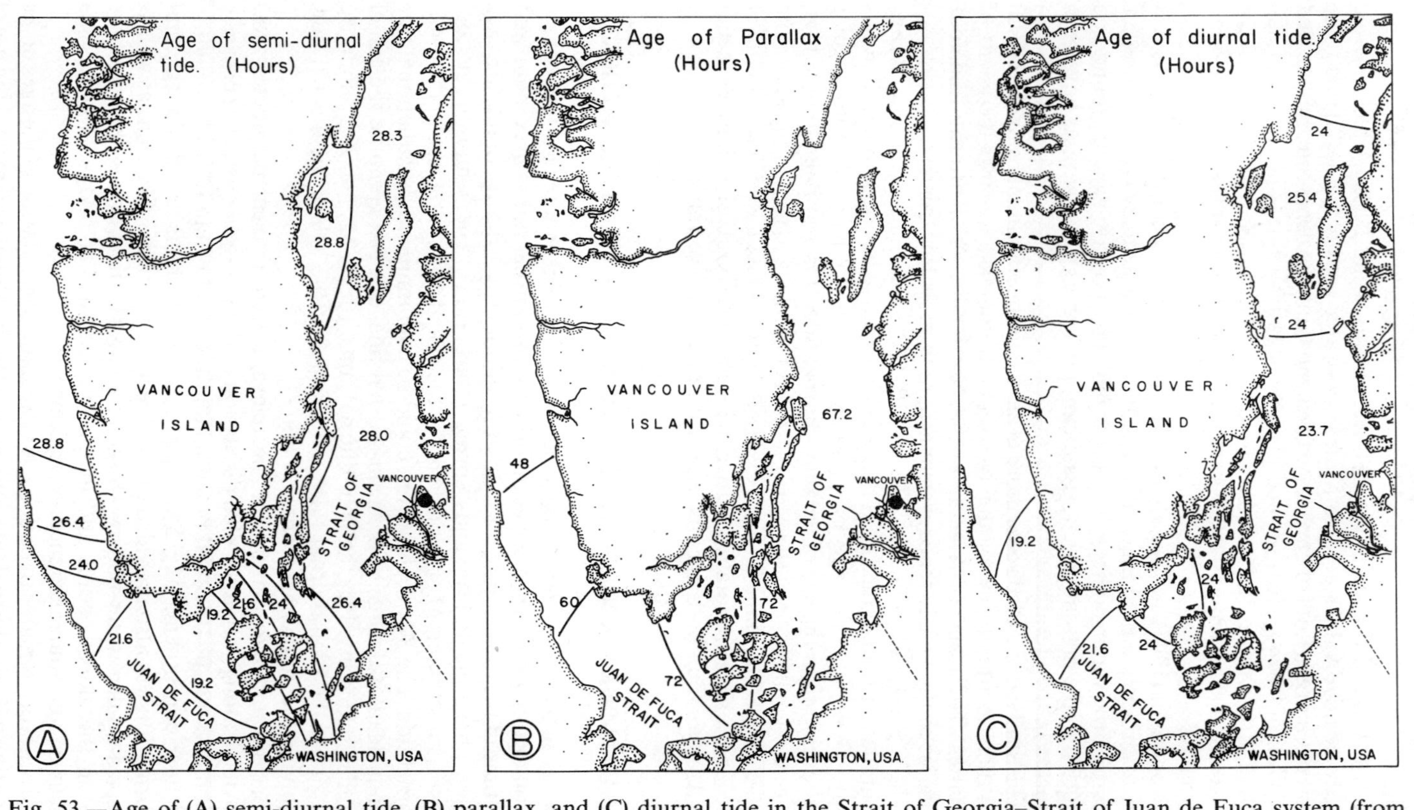

Fig. 53.—Age of (A) semi-diurnal tide, (B) parallax, and (C) diurnal tide in the Strait of Georgia–Strait of Juan de Fuca system (from Anonymous, 1982).

Lawrence Estuary and the Arabian Gulf, which have extensive shallow-water regions.

For the age of the semi-diurnal tide, the only water-bodies which showed totally positive values are again the deep-water coastal bodies, namely the Strait of Georgia, Gulf of Oman, Red Sea and Gulf of Aden. Note that the Red Sea–Gulf of Aden system showed just one negative value. The Arabian Gulf, Gulf of St Lawrence, the waters around Australia and New Zealand, and the global oceans showed positive as well as negative values for the age of the semi-diurnal tide.

Proudman (1941) showed that energy dissipation in coastal seas is necessary to produce a predominantly positive age in a hemispherical frictionless ocean. In the water-bodies where predominantly positive ages are found and reported in the present review, significant tidal energy dissipations have been reported in the literature.

This study also added some more evidence to the statement by Webb (1973) that there is a tendency for the age of tide to increase towards the heads of estuaries. The coast of Argentina, the North Sea, the Kattegat, and four estuaries in the St Lawrence system generally show this trend.

The suggestion by Garrett & Munk (1971) that one might compute the age of parallax instead of the age of the semi-diurnal tide has been followed in this study. Generally, the age of parallax appears to have a more regular distribution than the age of the semi-diurnal tide. It is gratifying to note that, the general behaviour of the age of parallax conforms to that of the age of the semi-diurnal tide; in the Strait of Georgia, Gulf of Oman, Red Sea and Gulf of Aden, both ages are totally positive, whereas in the Gulf of St Lawrence and the Arabian Gulf both ages have positive and negative values.

SUMMARY AND SUGGESTIONS FOR FUTURE WORK

The age of the tide is a term used to denote the interval between the time of new moon or full moon when the equilibrium semi-diurnal tide is maximum, and the time of the local spring tide. Other terminology for the age of tide are "age of the phase inequality" and "spring retardation". The semi-diurnal tidal constituents S_2 and M_2 are relevant for the calculation of the age of the semi-diurnal tide. The S_2 tide, however, could be contaminated by radiational tide and, instead of the age of the semi-diurnal tide, one can compute the age of parallax which involves the semi-diurnal constituents M_2 and N_2. Similar to the age of the semi-diurnal tide, one can define the age of the diurnal tide, involving the tidal constituents K_1 and O_1.

Background information on the relevant astronomical parameters was provided, along with a detailed discussion of the tide-generating forces of the moon and the sun, the tidal potential and the equilibrium tide. The origin of the tidal constituents M_2, S_2, N_2, K_1, and O_1 was discussed. Background information was also provided on tidal regimes in the global oceans as well as selected smaller water-bodies. The geographical distribution of the age of the tide in the global oceans and in the selected smaller bodies of water was discussed and some inferences were drawn.

Much work remains to be done. A systematic computation of the ages of the semi-diurnal tide, diurnal tide, and parallax in the global oceans, making use of all available relevant tide-gauge data is useful. This should be supplemented with data on the M_2, S_2, N_2, K_1, and O_1 tidal regimes computed from sophisticated numerical tidal models such as those of Schwiderski referred to earlier. One should follow up the suggestion of Webb (1973) and pay particular attention to the geographical distribution of the age of tide near estuary heads. Two problems require more clarification: the existence of the negative ages and its causes has no completely satisfactory explanation, and the original suggestion of Garrett & Munk (1971) about using the geographical distribution of the age of the tide as an indicator of resonances in the ocean needs further examination, in view of the criticism of Webb (1973) and others.

The authors are at present involved in deducing the amplitudes and phases of the radiational S_2 at several Canadian tide stations. As was mentioned earlier, the analysis of Garrett & Munk (1971) applies to linear systems. One might be able to look at the dynamics of non-linear systems through numerical models. It should also be kept in mind that local resonances cannot exist on scales which are less than a quarter of a tidal wavelength.

ACKNOWLEDGEMENTS

The authors wish to express their appreciation to Dr Margaret Barnes for her encouragement during the preparation of this review. We also thank Drs Andrew Bennett and Roger Flather for useful discussions. The Marine Environmental Data Services, Ottawa, have kindly provided us with tidal harmonics in the selected water-bodies dealt with in the present study. Sincere thanks are also due to Guy Gendron for computer assistance, Laval Hotton and Coralie Wallace for preparing the figures and Chantal Parisé for typing the manuscript. This study was supported by grants from the Natural Sciences and Engineering Research Council of Canada and the Université du Québec à Rimouski to M. I. El-Sabh.

REFERENCES

(For the sake of completeness some references, marked with an asterisk, have been included although they are not referred to in the text.)

Accad, Y. & Pekeris, C. L., 1978. *Phil. Trans. R. Soc. Ser. A*, **290**, 235–266.

Admiralty Tide Tables, 1982. Vols I, II, III. The Hydrographer of the Navy, U.K.

*Airy, G. B., 1845. *Encyclopedia Metropolitana*, **3.**

Anonymous, 1958. *Oceanographic Atlas of the Polar Seas.* U.S. Navy Hydrographic Office, Washington, D.C., U.S.A., 149 pp.

Anonymous, 1963. *Handbuch für das Rote Meer und den Golf von Aden.* Deutsches Hydrographisches Inst., Hamburg, 569 pp.

Anonymous, 1982. *Tidal and Residual Currents in the Strait of Georgia–Strait of Juan de Fuca system.* Beak Consultants Ltd, Vancouver, 71 pp.

Blondel, A. 1912. *Annls Fac. Sci. Univ. Toulouse*, **3,** 151–207.

Bogdanov, K. T. & Karkov, B. V., 1975. *Oceanology*, **15,** 156–160.

*Bogdanov, K. T. & Magarik, V. A., 1967. *Dokl. Akad. Nauk SSSR*, **172,** 1315–1317.

British Admiralty Chart, 1967. *Persian Gulf Tidal Constituents*, chart no. 5081. The Hydrographer of the Navy, London.

*Brown, P. J., 1973. *J. mar. Res.* **31**, 1–10.

Bye, J. A. T. & Heath, R. A., 1975. *J. mar. Res.*, **33,** 423–442.
Cartwright, D. E., 1968. *Phil. Trans. R. Soc. Ser. A*, **263,** 1–55.
*Cartwright, D. E., 1971. *Phil. Trans. R. Soc. Ser. A*, **270,** 603–649.
*Cartwright, D. E., 1974. *Accademia Nazionale dei Lincei, Roma*, **206,** 91–95.
*Cartwright, D. E., 1977. *Phil. Trans. R. Soc. Ser. A*, **284,** 537–546.
Cartwright, D. E., 1978. *Int. hydr. Rev.*, **55,** 35–84.
Cartwright, D. E. & Edden, A. C., 1977. *Ann. Geophys.*, **33,** 179–182.
Cartwright, D. E. & Tayler, R. J., 1971. *Geophys. J. R. astr. Soc.*, **23,** 45–74.
Cartwright, D. E., Zetler, B. D. & Hamon, B. V., 1979. *Pelagic Tidal Constants.* Compiled by I.A.P.S.O. Advisory Committee on Tides and Mean Sea Level, I.U.G.G., No. 30, 65 pp.
Chandon, E., 1930a. Thèse Fac. Sci. de Paris, Gauthier-Villars, Paris, 101 pp.
Chandon, E., 1930b. *Int. hydrogr. Rev.*, **7,** 176–177.
Chapman, S. & Lindzen, R., 1970. *Atmospheric Tides.* Reidel, Dordrecht, 200 pp.
Clarke, A. K. & Battisti, D. S., 1980. *Ocean Model.*, **51,** 7 only.
*Darwin, G. H., 1898. *The Tides and Kindred Phenomena of the Solar System.* Republished in 1962, Greenman and Cooper Publ. Comp., San Francisco, 378 pp.
Defant, A., 1919. *Denkschr. Akad. Wiss., Wien*, **96,** 110–137.
Defant, A., 1924. *Ann. Hydrogr. mar. Meteorol.*, **52,** 153–166, 177–184.
Defant, A., 1926. *Annln Hydrogr. Berl.*, **54,** 185–194.
Defant, A., 1961. *Physical Oceanography, Vol. 2.* Pergamon Press, Oxford, 598 pp.
Dietrich, G., 1944a. *Veröffentlichungen des Inst. Meereskunde an der Univ. Berlin*, A41, 7–68.
Dietrich, G., 1944b. *Z. ges. Erdk.*, **3,** 69 only.
*Doodson, A. T. & Warburg, H. D., 1941. *Admiralty Manual of Tides.* H.M. Stationery Office, London, Reprinted 1952, 270 pp.
Dvorkin, E. N., Kagan, B. A. & Kleshyava, G. P., 1972. *Izv. Akad. Nauk S.S.S.R. Ser. Fiz. Atmos. Okean.*, **8,** 298–306 (in Russian).
Easton, A. K., 1970. *The Tides of the Continent of Australia.* Horace Lamb Centre, Flinders University, South Austr. Res. Paper 37, 326 pp.
Elahi, K. Z., 1984. In, *Oceanographic Modelling of the Kuwait Action Plan Region*, edited by M. I. El-Sabh, UNESCO Mar. Sci. Reports No. 28, p. 68 (Abstract only).
El-Sabh, M. I., 1977. *J. Fish. Res. Bd Canada*, **34,** 516–548.
El-Sabh, M. I., 1982. *Numerical modelling of tides in the Arabian Gulf.* Fac. Mar. Sci., King Abdulaziz Univ., Jeddah, Unpubl. Rep., 113 pp.
Evans-Roberts, D. J., 1979. *Consulting Engineer*, June 1979.
Fairbairn, L. A., 1954. *Phil. Trans. R. Soc. Ser. A*, **247,** 191–212.
Farquharson, W. I., 1970. *Tides, Tidal Streams and Currents in the Gulf of St Lawrence.* Atlantic Oceanogr. Lab., Bedford Inst. Oceanogr., Dartmouth, Canada, 78 pp.
Fjeldstad, J. E., 1929a. *Gerlands Beitr. Geophys.*, **9,** 237–247.
Fjeldstad, J. E., 1929b. *Sci. Res. Norweg.*, **4,** 3–80.
Flattery, T. W., 1967. Ph.D. thesis, University of Chicago, U.S.A., 168 pp.
*Foreman, M. G. G., 1977. *Manual for Tidal Heights Analysis and Prediction*, Pacific Mar. Sci. Rep. 77-10, Inst. Ocean Sci., Sidney, B.C., Canada, 97 pp.
Foreman, M. G. G., Delves, L. M., Barrodale, I. & Henry, R. F., 1980. *Geophys. J. R. astr. Soc.*, **63,** 467–478.
Forrester, W. D., 1983. *Canadian Tidal Manual*, Dept Fish. & Oceans, Canada, Ottawa, 138 pp.
*Garrett, C. J. R., 1972. *Nature, Lond.*, **238,** 441–443.
Garrett, C. J. R. & Munk, W. H., 1971. *Deep-Sea Res.*, **18,** 493–503.
Garrett, C. J. R. & Greenberg, D. A. 1977. *J. phys. Oceanogr.*, **7,** 171–181.
Godin, G., 1972. *The Analysis of Tides.* University of Toronto Press, Toronto, 264 pp.
Godin, G., 1979. *Nat. Can. (Que.)*, **106,** 105–121.
Godin, G., 1980. *Cotidal Charts for Canada.* Mar. Sci. & Inform. Dir., Dept Fish. & Oceans, Canada, MS Rep. Ser., No. 55, 93 pp.

Goldsbrough, G. R., 1913. *Proc. Lond. math. Soc.*, **14,** 31–66.
*Gotlib, V. Y. & Kagan, V. Y., 1983. *Oceanology*, **23,** 401–404.
Grace, S. F., 1930. *Mon. Not. R. astr. Soc. geophys.* Suppl., No. 2, 273–296.
Hansen, W., 1962. In, *Proc. Symp. Mathematical-Hydrodynamical Methods of Phys. Oceanogr.*, Sept. 1961, Inst. für Meereskunde, Univ. Hamburg, 25–34.
Harris, D. L., 1981. *Tides and Tidal Datums in the United States.* U.S. Army Corps Engineers, Coastal Engineer. Res. Centre, Fort Belvoir, U.S.A., Special Rep. No. 7, 382 pp.
Harris, R. A., 1894–1907. *Manual of Tides, Parts I–V*, U.S. Coast and Geodetic Survey Rep., Washington, D.C., 363–364, 1904.
Harris, R. A., 1911. *Arctic Tides.* U.S. Coast and Geodetic Survey, Washington, D.C., 103 pp.
Heath, R. A., 1977. *N.Z. J. mar. freshw. Res.*, **12,** 87–97.
Heath, R. S., 1981a. *Deep-Sea Res.*, **28A,** 847–858.
Heath, R. S., 1981b. *Deep-Sea Res.*, **28A,** 451–493.
Hendershott, M. C., 1972. *Geophys. J. R. astr. Soc.*, **29,** 389–402.
*Hendershott, M. C., 1972. *J. R. astr. Soc.*, **29,** 238–403.
*Hendershott, M. C., 1973. *Trans. Am. geophys. Un.*, **54,** 76–86.
*Hendershott, M. C. & Munk, W. H., 1970. *Ann. Rev. Fluid Mech.*, **2,** 205–224.
*Hendershott, M. C. & Speranza, A., 1971. *Deep-Sea Res.*, **18,** 959–980.
Hollingsworth, A., 1971. *J. atmosph. Sci.*, **28,** 1021–1044.
Hugues, P. & Hunter, J. R., 1979. *Physical Oceanography and Numerical Modelling of the Kuwait Action Plan Region*, UNESCO, Div. Mar. Sci., Rep. MARINF/278.
Hydrographic Department, 1963. *The Admiralty Tide Tables, Vol. III, Pacific and Adjacent Seas*, 1964. London.
International Hydrographic Bureau, 1966. *Harmonic Constants*, Special Publ. No. 26, Int. Hydrogr. Bur., Monaco.
*Jeffreys, H., 1968. *Geophys. J. R. astr. Soc.*, **16,** 253–258.
Kagan, B. A., 1968. *Gidrometeorolog. Izd.*, **218,** 5 m, Abb. Leningrad, U.S.S.R.
Kowalik, A. & Untersteiner, N., 1978. *Dt. hydrogr. Z.*, **31,** 216–229.
Kowalik, Z. & Bich Hung, N., 1977. *Oceanology*, **7,** 5–20.
Lamb, H., 1932. *Hydrodynamics.* Cambridge University Press, Cambridge, 6th edition, 738 pp.
Laplace, P. S. Marquis de, 1799. *Traité de Mécanique Céleste*, Imprimerie de Crapelet, Paris.
Lardner, R. W., Belen, M. S. & Cekirge, H. M., 1982. *Comput. Math. with Appls*, **8,** 425–444.
*LeBlond, P. H. & Mysak, L. A., 1978. *Waves in the Ocean.* Elsevier Scientific Publ. Comp., New York, 602 pp.
Lehr, W. J., 1984. In, *Oceanographic Modelling of the Kuwait Action Plan Region*, edited by M. I. El-Sabh, UNESCO Mar. Sci. Rep. No. 28, pp. 4–11.
LeProvost, C., 1984. In, *Oceanographic Modelling of the Kuwait Action Plan Region*, edited by M. I. El-Sabh, UNESCO Mar. Sci. Rep. No. 28, pp. 25–36.
Levesque, L., Murty, T. S. & El-Sabh, M. I., 1979. *Int. hydrog. Rev.*, **56,** 117–132.
*Lisitzin, E., 1974. *Sea-level Changes*, Elsevier Scientific Publ. Comp., New York, 286 pp.
*Longuet-Higgins, M. S., 1968. *Phil. Trans. R. Soc. Ser. A*, **262,** 511–607.
*Longuet-Higgins, M. S., 1971. *J. geophys. Res.*, **76,** 3517–3522.
*Longuet-Higgins, M. S. & Pond, G. S., 1970. *Phil. Trans. R. Soc. Ser. A*, **266,** 193–223.
Luther, D. S. & Wünsch, C., 1975. *J. phys. Oceanogr.*, **5,** 222–230.
McCammon, C. & Wünsch, C., 1977. *J. geophys. Res.*, **82,** 5993–5998.
*Miller, G. R., 1964. Ph.D. thesis, University of Calif., San Diego, 120 pp.
Miller, G. R., 1966. *J. geophys. Res.*, **71,** 2485–2489.
*Mofjeld, H. O. & Rattray, Jr, M., 1971. *J. mar. Res.*, **29,** 281–305.
Morcos, S. A., 1971. *Oceanogr. Mar. Biol. Ann. Rev.*, **8,** 73–202.

Munk, W. H. & Cartwright, D. E., 1966. *Phil. Trans. R. Soc. Ser. A*, **259,** 533–581.
*Munk, W. H. & MacDonald, G. J. F., 1960. *The Rotation of the Earth.* Cambridge University Press, Cambridge, 323 pp.
Munk, W. H., Zetler, B. & Groves, G. W., 1965. *Geophys. J.*, **10,** 211–219.
*Murty, T. S., 1977. *Bull. Fish. Res. Bd Canada*, No. 198, 337 pp.
Murty, T. S., 1984. *Storm Surges-Meteorological Ocean Tides.* Canadian Fish. & Aquatic Sci., Bull. No. 212, 897 pp.
Murty, T. S. & El-Sabh, M. I., 1984. In, *Oceanographic Modelling of the Kuwait Action Plan Region*, edited by M. I. El-Sabh, UNESCO Mar. Sci. Rep. No. 28, pp. 12–24.
Murty, T. S. & Holloway, G., in press, *J. Waterway Port cstl Engng.*
Murty, T. S. & Polavarapu, R. J. 1979. *Mar. Geodesy*, **2,** 99–125.
Nekrasov, A. V., 1962. *Tr. Leningr. Gidrometeorol. Inst.*, **16,** 49–57 (in Russian).
*Neumann, G. & Pierson, W. J., 1966. *Principles of Physical Oceanography*, Prentice-Hall Comp., Englewood Cliffs, N.J., 545 pp.
Newton, I., 1687. *Principia mathematica* (Trans. Andrew Motte) D. Adee, New York (1948).
Pattullo, J., Munk, W. H., Revelle, R. & Strong, E. 1955. *J. mar. Res.*, **14,** 88–155.
*Pekeris, C. L. & Accad, Y., 1969. *Phil. Trans. R. Soc.*, **265,** 413–436.
Phillips, N. A., 1960. In, *Advances in Computers, Vol. 1*, edited by F. K. Alt, Academic Press, New York, pp. 43–90.
Pingree, R. D., 1983. *Deep-Sea Res.*, **30,** 929–944.
Platzman, G. W., 1975. *J. phys. Oceanogr.*, **5,** 117–128.
Platzman, G. W., 1979. *J. phys. Oceanogr.*, **9,** 1276–1283.
Poincaré, H., 1910. *Leçons de Mécanique Céleste. T.3. Théories des marées*, Gauthier-Villars, Paris, pp. 386–387.
*Proudman, J., 1928. *Mon. Not. R. astr. Soc., Geophys. Suppl.*, No. 2, 32–43.
Proudman, J., 1941. *Mon. Not. R. astr. Soc., Geophys. Suppl.*, No. 59, 23–26.
Proudman, J., 1944. *Mon. Not. R. astr. Soc., Geophys. Suppl.*, No. 104, 244–256.
Proudman, J., 1953. *Dynamical Oceanography*, Methuen & Co. Ltd, London, 409 pp.
Proudman, J. & Grace, S. F., 1930. *UGGI, Section d'Océanographie, Comm. des Marées*, Bull. No. 15, 16 pp.
Pugh, D. T., 1981. *Jl R. astr. Soc.*, **67,** 515–527.
Schureman, P., 1941. *Manual of Harmonic Analysis and Prediction of Tides.* Special Publ. 98, U.S. Coast and Geodetic Survey, 317 pp.
Schwiderski, E. W., 1978a. *Global Ocean Tides. Part I. A Detailed Hydrodynamical Interpolation Model*, Rep. N.S.W.C./D.L. TR-3866, Naval Surface Weapons Center, Dahlgren, Virginia, U.S.A., 88 pp.
Schwiderski, E. W., 1978b. *Hydrodynamically Defined Ocean Bathymetry*, Rep. N.S.W.C./D.L. TR-3888, Naval Surface Weapons Center, Dahlgren, Virginia, U.S.A., 57 pp.
Schwiderski, E. W., 1980a. *Mar. Geodesy*, **3,** 161–217.
Schwiderski, E. W., 1980b. *Mar. Geodesy*, **3,** 219–255.
Schwiderski, E. W., 1980c. *Rev. Geophys. Space Phys.*, **18,** 243–268.
Schwiderski, E. W., 1981a. *The Semi-diurnal Principal Solar Tide* (S_2). Atlas of tidal charts and maps, Rep. N.S.W.C./D.L. TR 81-122, Naval Surface Weapons Center, Dahlgren, Virginia, U.S.A., 12 pp.
Schwiderski, E. W., 1981b. *Global Ocean Tides. Part VI. The Semi-diurnal Lunar* (N_2). Atlas of tidal charts and maps, Rep. N.S.W.C. TR 81-218, Naval Surface Weapons Center, Dahlgren, Virginia, U.S.A., 11 pp.
Schwiderski, E. W., 1981c. *Global Ocean Tides. Part VII. The Diurnal Principal Solar Tide* (P_1). Atlas of tidal charts and maps, Rep. N.S.W.C. TR 81-220, Naval Surface Weapons Center, Dahlgren, Virginia, U.S.A., 11 pp.
Schwiderski, E. W., 1981d. *Global Ocean Tides, Part VIII. The Semi-diurnal Luni-Solar Declinational Tide* (K_2). Atlas of tidal charts and maps, Rep. N.S.W.C. TR 81-222, Naval Surface Weapons Center, Dahlgren, Virginia, U.S.A., 11 pp.

Schwiderski, E. W., 1981e. *Global Ocean Tides, Part IX. The Diurnal Elliptical Lunar Tide* (Q_1). Atlas of tidal charts and maps, Rep. N.S.W.C. TR 81-224, Naval Surface Weapons Center, Dahlgren, Virginia, U.S.A., 11 pp.
Schwiderski, E. W., 1982a. *Global Ocean Tides, Part X. The Fortnightly Lunar Tide* (*Mf*). Atlas of tidal charts and maps, Rep. N.S.W.C. TR 82-100, Naval Surface Weapons Center, Dahlgren, Virginia, U.S.A., 10 pp.
Schwiderski, E. W., 1982b. *Exact Expansions of Arctic Ocean Tides*, Rep. N.S.W.C. TR 81-494, Naval Surface Weapons Center, Dahlgren, Virginia, U.S.A., 23 pp.
Schwiderski, E. W., 1983. *Mar. Geodesy*, **6,** 219–265.
Sterneck, R., 1920. *Stiz. Ber. Akad. Wiss. Wien*, **129,** 131–150.
Sterneck, R., 1927. *Annln Hydrogr. Berl.*, **55,** 129–134.
*Taylor, G. I., 1921. *Proc. Lond. math. Soc.*, **20,** 148–181.
Thacker, W. C., 1980. *J. Comput. Phys.*, **37,** 355–370.
Vercelli, F., 1925. *Annali idrogr.*, **11,** I–VII + 188 pp. (and 1927, 13–208).
Vercelli, F., 1931. *Annali idrogr.*, **12,** 74 pp.
Villain, C., 1952. *Ann. Hydrogr.*, **3,** 269.
Von Trepka, L., 1968. *Inst. für Meereskunde der Univ. Hamburg*, **10,** 59–63.
Webb, D. J., 1973. *Deep-Sea Res.*, **20,** 847–852.
*Webb, D. J., 1974. *Rev. Geophys. Space Phys.*, **12,** 103–116.
Webb, D. J., 1976. *Deep-Sea Res.*, **23,** 1–15.
*Wheeler, W. W., 1960. *A Practical Manual of Tides and Waves*, Longmans & Green, London, 201 pp.
Whewell, W., 1833. *Phil. Trans. R. Soc. Ser. A*, 147–236.
Wood, F. J., 1978. *The Strategic Role of Perigean Spring Tides in Nautical History and North American Coastal Flooding, 1635–1976.* N.O.A.A., U.S., Dept. Commerce, Washington, D.C., 529 pp.
Young, T., 1823. *Tides*, In, *Encycl. Britannica, Vol. 21*, 8th edition, Little and Brown, Boston (1853).
Zahel, W., 1970. *Mitteilungen des Inst. für Meereskunde der Univ. Hamburg*, No. 17, 50 pp.
Zahel, W., 1973. *Pure appl. Geophys.*, **109,** 1819–1825.
Zahel, W., 1977. In, *Tidal Friction and Earth's Rotation*, edited by P. Brosche & J. Sündermann, Springer-Verlag, New York, pp. 98–124.
Zetler, B. D., 1971. *J. phys. Oceanogr.*, **1,** 34–38.
*Zetler, B. D. & Munk, W. H., 1975. *J. mar. Res.* **33,** 1–13.
*Zetler, B. D., Munk, W. H., Mofjeld, H., Brown, W. & Dormer, F., 1975. *J. phys. Oceanogr.*, **5,** 430–441.

Oceanogr. Mar. Biol. Ann. Rev., 1985, **23**, 105–182
Margaret Barnes, Ed.
Aberdeen University Press

THE BENGUELA ECOSYSTEM PART I. EVOLUTION OF THE BENGUELA, PHYSICAL FEATURES AND PROCESSES

L. V. SHANNON
Sea Fisheries Research Institute, Private Bag X2, Rogge Bay 8012, Cape Town, South Africa

INTRODUCTION

The Benguela is one of the four major eastern boundary current regions of the World ocean and the oceanography of the western coast of Africa south of about 15° S, like that off California, Peru and North West Africa is dominated by a coastal upwelling system. Research in the area during the nineteenth and first part of the twentieth century was directed mainly at taxonomy, at improving navigational safety, and the development of fisheries. In spite of the observations of Ross (1847) early workers such as Muhry (1862) and Petermann (1865) viewed the Benguela Current as a northward extension of the West Wind Drift and it was only during the 1920s that Meyer (1923) showed conclusively that they are separated by a well-defined convergence zone, the Subtropical Convergence. After World War II a detailed study of the oceanography of southern African west coast waters commenced. The literature prior to 1970 was largely descriptive in nature, and it is only during the last fifteen years that mesoscale processes (*i.e.* processes occurring over spatial scales of tens of kilometres to a few hundred kilometres with time scales of hours to a few days) have been given serious attention.

This review will take a broad view of the status of the Benguela ecosystem and the main thrust will be aimed at reviewing the present state of knowledge of the main processes governing the system. It has been divided into parts, *viz.* Part I which deals with the evolution and physics of the Benguela system, and which is the subject of this paper, Part II (Chapman & Shannon, 1985, also appearing in this volume) which deals with chemical processes, and Part III will address the various biological processes. Part III is in preparation and will appear in Volume 24 of *Oceanography and Marine Biology: An Annual Review* (1986) with L. V. Shannon, J. G. Field and W. R. Siegfried as authors. Part IV is also in preparation. Part I has been subdivided into a number of sections covering the origin of the Benguela system, large scale features, meteorology, seasonal and inter-annual variability and mesoscale physical processes. Two important papers on the physical oceanography of the region appeared during 1983 (Nelson & Hutchings, 1983; Parrish, Bakun, Husby &

Nelson, 1983) and the present article is intended to complement and build on the work of these authors. Processes in the nearshore region, *e.g.* in the intertidal zone and the surf zone, will not be considered in this review except where they are important for the broader understanding of the system as a whole.

In a review of a complex ecosystem such as that of the Benguela it is difficult, if not impossible to cite all relevant literature without making the paper cumbersome and unreadable. Accordingly the list of references, while being reasonably comprehensive, is by no means exhaustive. It should, however, be adequate to lead into the remainder of the available literature on the Benguela ecosystem and into the more specialized disciplinary literature. Readers are also referred to the comprehensive bibliography on the physical oceanography of the South East Atlantic compiled by Lutjeharms, van Ballegooyen & Valentine (1981).

EVOLUTION OF THE BENGUELA CURRENT AND UPWELLING SYSTEM

The South Atlantic Ocean began to form about 130 Myr (million year) ago (Kennett, 1982; Dingle, Siesser & Newton, 1983) following the rifting of Gondwanaland and the separation of the South American and African plates. Van Zinderen Bakker (1975) has suggested that it may have taken until mid Tertiary times before the ocean had sufficient zonal width to permit the development of its present circular wind and oceanic circulations. Kennett (1982) has indicated that the anticyclonic gyre was well established in the South Atlantic by the Eocene, which, if correct, implies that an equatorward flow (at the surface) along the west coast of southern Africa has existed at least for the past 50 Myr. This early current, the precursor of the South East Atlantic Drift, had however little in common with the eastern boundary current regime off southwestern Africa as we know it today—fossil evidence from the Eocene marine transgression implies that warm water conditions prevailed in the South East Atlantic Ocean during the early to middle Tertiary (Ward, Seely & Lancaster, 1983). A number of important geological and climatic events were necessary before the present thermohaline circulation in the World ocean could be established (see Fig. 1). During the Paleocene (65 to 55 Myr ago) Australia and Antarctica were joined. They began to drift apart 55 Myr ago (Kennett, 1978) but circum-Antarctic flow was blocked by the South Tasman Rise and Tasmania. During the Eocene the Southern Ocean was relatively warm and Antarctica largely non-glaciated, and according to Kennett (1978) a major climatic boundary was crossed at the Eocene-Oligocene boundary (38 Myr ago) when Antarctic sea ice began to form and a drop of 5 °C in the temperature of bottom waters resulted. Kennett (1978) has suggested that this threshold was crossed because of the gradual isolation of Australia and Antarctica and perhaps the opening of the Drake Passage. Furthermore, by the early Oligocene the opening of the Atlantic had progressed to a point where the South Atlantic was open to Antarctic bottom waters (Deacon, 1983). This implies that the thermohaline circulation as we know it today could not have been initiated until after the onset of the Oligocene period, *i.e.* more recently than 38 Myr ago. Other major events which must have significantly

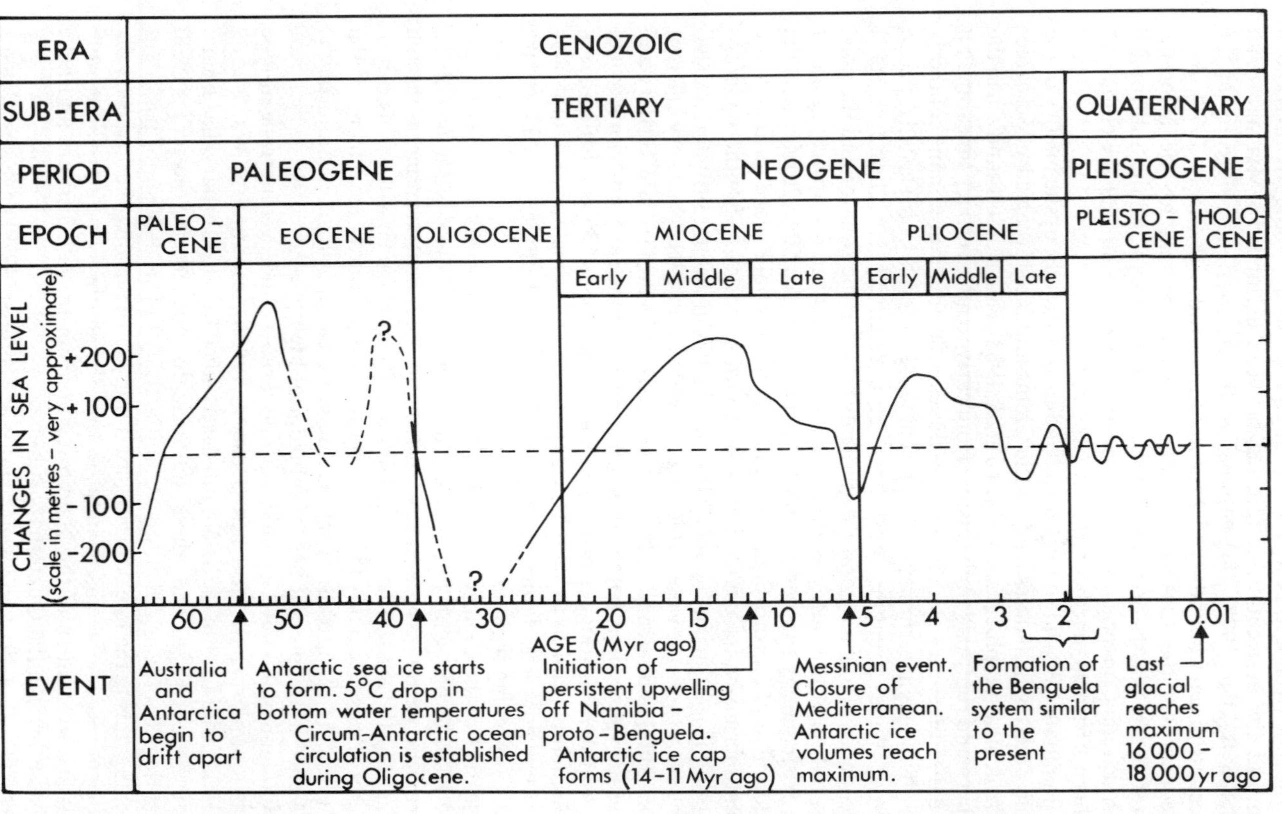

Fig. 1.—Cenozoic time scales, important events and sea level fluctuations around southern Africa from work of Hendey (1983), Deacon (1983), Kennett (1978) and others.

influenced circulation in the Atlantic Ocean were the closure of the mid-American seaway and the formation of the Antarctic ice-cap during the mid to late Miocene (Kennett, 1978) and the complete closure of the Mediterranean basin and the desiccation of this sea between 6·2 and 5·3 Myr ago (Van Zinderen Bakker, 1978)—ice volumes in the Antarctic were greater at the Miocene-Pliocene boundary (5 Myr ago) than today (Kennett, 1978). As a consequence of the substantial cooling during the mid to late Miocene, separate water masses originated in the Southern Ocean and the Subtropical and Antarctic Convergences were formed, with the result that the climate of southern Africa changed completely (Coetzee, 1978).

A dominance of arid to semi-arid conditions throughout the history of the Namib desert, which dates back to the Cretaceous, are revealed by Mesozoic-Cenozoic stratigraphic records (Ward *et al.*, 1983), with a desert sand sea in the southern and central Namib existing from early to mid-Tertiary times, and an internal sedimentary structure of the dunes reflecting a dominant southerly palaeo-wind regime, similar to the present wind patterns, which lasted for 20 to 30 Myr. Wind strengths must have been relatively low, however, as a consequence of low temperature gradients between the equator and the pole. Thus, although this suggests that a driving mechanism for upwelling might have existed along the western coast of southern Africa 50 to 20 Myr ago, the development of a proto-Benguela system could not take place until after the present thermo-haline circulation in the South Atlantic and Southern Oceans was established during the late Miocene. While upwelling probably existed during early times the water which was upwelled would have had very different temperature, salinity, and nutrient characteristics to those of late Tertiary and Quaternary upwelling waters. It is considered that the early upwelling would have appeared as tongues rather than as a continuous Ekman drift process. Consequently prior to the late Miocene, the Benguela region would not have supported flora and fauna characteristic of present-day eastern boundary current upwelling regimes.

A feature of the Cenozoic era has been the equatorwards movement of the westerly wind belt with important consequences for the contraction of the subtropical ocean gyre systems and meridional heat transfer, although throughout the era southern Africa did not at any stage lie within this belt of westerly winds (Deacon, 1983). It was the equatorwards movement of the westerlies and the growth of the Antarctic ice-cap that focused atmospheric subsidence in the present belt of subtropical high pressure cells—which was the fundamental cause of mid-latitude aridity and Mediterranean-type climates bordering the arid belt (Deacon, 1983). According to Ward *et al.* (1983) the aridification of the Namib desert since the late Tertiary, has been a progressive process, and these authors do not accept Siesser's (1978) contention that the aridification was initiated by the onset of major persistent upwelling during the late Miocene, for a number of reasons. Nevertheless the geological evidence seems to indicate that the southerly palaeo-winds were intensified over much of the Namib region from the late Tertiary and have persisted through the Quaternary sub-era.

Siesser (1978, 1980) and Diester-Haass & Schrader (1979) have postulated a late Miocene origin for strong persistent Benguela upwelling off northern Namibia, based on Deep-Sea Drilling Project cores near to the abutment of the Walvis Ridge with the African continent (site 362/362A—approximate latitude 20° S). More recent studies by Meyers *et al.* (1983) on cores taken by the GLOMAR CHALLENGER in a similar area at DSDP 530 and 532, support the earlier findings of Siesser (1980) and others that date the onset of upwelling as late Miocene. Diester-Haass & Schrader (1979) considered that the greater abundance of siliceous microfossils in the Pleistocene might have been due to an intensification of cold water current patterns and to higher wind velocities (the latter being deduced from the high abundance of displaced continental opal remains). The view of Seisser (1978) was that upwelling off northern Namibia was weak and spasmodic from the mid or late Oligocene to mid Miocene period, and intensified during the late Miocene (12 Myr ago). Siesser (1980) presented a plausible argument for the late Miocene origin of northern Benguela upwelling, and suggested that the enormous blooms of a single nanoplanktonic organism, *Braarudosphaera*, reflected in the high accumulation rates during Oligocene times were more probably a response to a regional South Atlantic event rather than to local Benguela upwelling. The sediment deposit rates and organic carbon contents from the core analysis are shown in Figure 2. Although these data related only to the northern Namibian Benguela system, Siesser (1980) argued intuitively that it was probable that upwelling would begin all along the coast with only a slight lag from place to place, once the forcing mechanisms were established. The palaeo-climatic and palaeo-ecological records of the southern part of the Benguela system (around the South West Cape) strongly suggest, however, that persistent upwelling was not established in the region until the late Pliocene, about which time, from the work of Siesser (1980) and Meyers *et al.* (1983), productivity in the northern Benguela reached a maximum.

In a recent article on fossil sea-birds from early Pliocene (5 Myr ago) deposits in the South Western Cape, Olson (1983) found that the region had a more sub-Antarctic marine environment then than at present, and that the marine avifauna of the southern Benguela area has changed drastically since the early Pliocene. Studies by Tankard & Rogers (1978), Coetzee (1978),

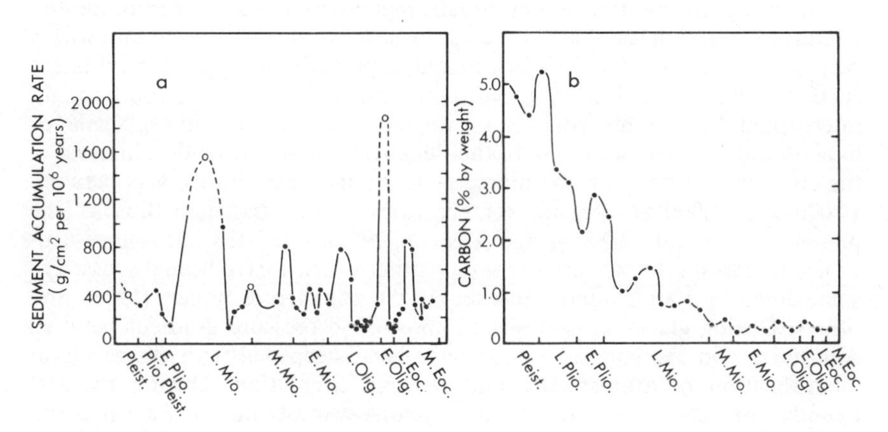

Fig. 2.—a, total corrected sediment accumulation rates for DSDP site 362/362A (after Siesser, 1980). b, organic carbon content in cores from DSDP site 362/362A (after Siesser, 1980).

Hendey (1983) and Coetzee, Scholtz & Deacon (1983) on pollens and vertebrate fauna from the late Tertiary and Quaternary fossil deposits in the South Western Cape indicate that the climate of the region during the early Pliocene was in an intermediate stage, between that of the Miocene and the present. Hendey (1983) considered that a summer wet–winter dry rainfall pattern was characteristic of the early Pliocene in the South Western Cape. The present arid summer–wet winter Mediterranean type climate in the southern Benguela region only appears to have been fully established towards the end of the Pliocene, *i.e.* about 2 Myr ago (Tankard & Rogers, 1978; Hendey, 1983). At this time there was a sharp increase in the organic carbon content of marine sediment cores from northern Namibia (Siesser, 1980, see also Fig. 2). Also by this time the Agulhas Current was similar to its present form, its quasi-modern flow patterns having been established about 5 Myr ago in the early Pliocene (Martin, 1981). Thus, although there is evidence of equatorward flow in the South East Atlantic and sporadic coastal upwelling off Namibia since the Oligocene epoch, and although a proto-Benguela system appears to have been initiated off Namibia during the late Miocene (Seisser, 1980), the Benguela system as we know it today is clearly of more recent origin, dating from the late Pliocene.

Marine transgressions and regressions during the history of the Benguela system (*e.g.* Siesser & Dingle, 1981) would have had a significant impact on the upwelling regimes and current patterns. Sea level fluctuations since the Paleocene epoch from Hendey (1983) have been included in Figure 1. High sea levels during the mid Miocene and early to mid Pliocene times would have resulted in the present continental shelf being 100 to 200 m deeper, with the drowning of lower lying coastal areas and the formation of coastal islands (*e.g.* the Cape Peninsula) and an irregular coast having numerous embayments. Conversely, at the Miocene–Pliocene boundary and also during the mid to late Pliocene the lower sea levels would have exposed large areas of shelf, *e.g.* St Helena Bay. These bottom topographic and orographic changes would obviously have had a major effect on the positions of upwelling centres, the oceanic front and the general dynamics of the Benguela system.

A feature of the modern global climatic regimes initiated in the Pliocene and continuing through the Pleistocene has been the characteristic rhythm with a periodicity of about 100 000 years, linked to perturbations of the orbit of the earth relative to the sun, of cooler climates, *i.e.* glacials or hypothermals, interrupted by shorter periods of warmer climates, *i.e.* interglacials or hyperthermals (Deacon, 1983). In the Benguela regime the coldest interval of the late Pleistocene, with the most severe climatic conditions, was between 16 000 and 18 000 yr ago, and temperatures approximated to those of the present from about 12 000 yr ago (Deacon, 1983). The late Cenozoic palaeoenvironments on the western coast of southern Africa have been discussed in some detail by Van Zinderen Bakker (1975), Tankard & Rogers (1978), and others. During glacial episodes the atmospheric pressure gradient between Antarctica and the equator would have been steepened with the resultant intensification of atmospheric and oceanic circulation. During the last hypothermal, the South Atlantic anticyclone was situated 5° further north (Van Zinderen Bakker, 1975) and cold polar air could penetrate the interior of southern Africa up to about 24° S (Coetzee *et al.*, 1983). The consequence of this would have been an intensification of upwelling and for the Benguela system to shift further north. Van Andel & Calvert (1971) suggested in fact that the Benguela Current was intensified during glacials in order to account for the increased erosion they observed on the Namibian shelf. In support of the concept of a northward shift, Bornhold (1973) suggested that there was a northward extension of the Benguela regime along the coast of Angola and an onset of upwelling off the Congo and Gabon, as deduced from the analysis of Angola Basin sediments. To what extent the Walvis Ridge would have acted as a barrier to this hypothesized northward extension of the system is a matter for speculation. The inferred atmospheric and oceanic systems around southern Africa during glacials and interglacials are illustrated in Figure 3, which is

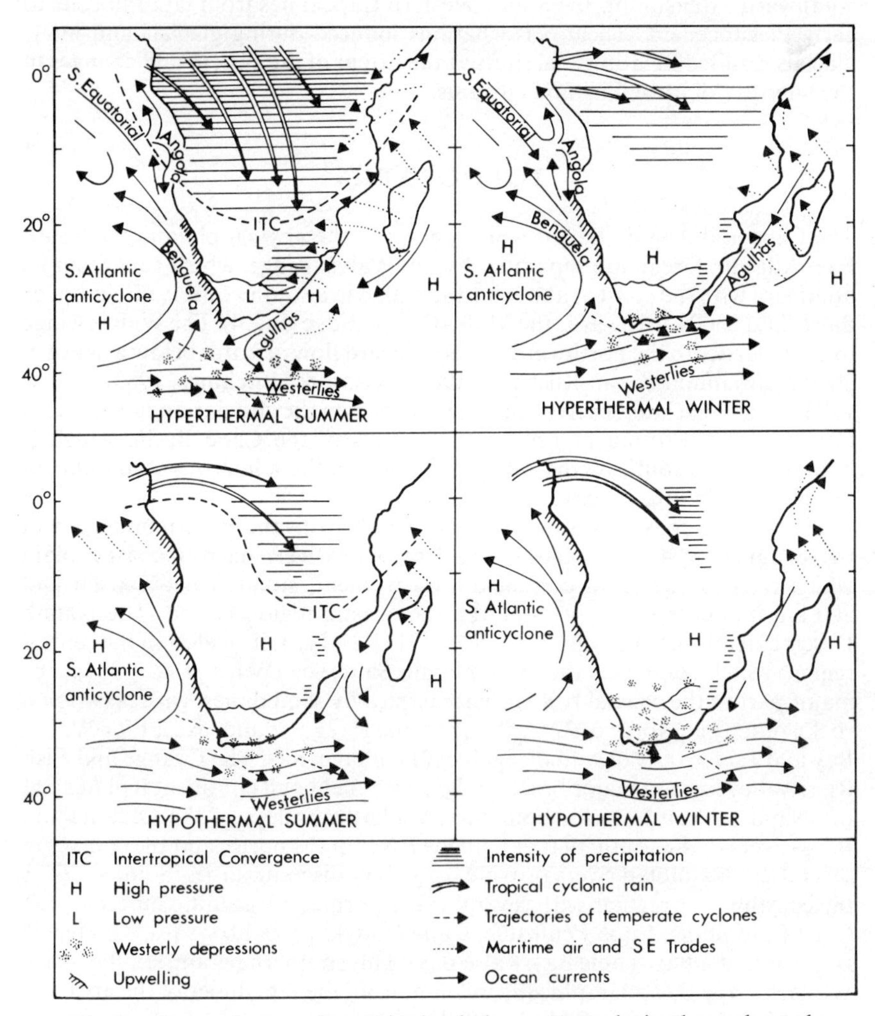

Fig. 3.—Atmospheric and oceanic circulation patterns during hyperthermal (interglacial) and hypothermal (glacial) times (after Van Zinderen Bakker, 1976; Tankard & Rogers, 1978).

from Van Zinderen Bakker (1976). It should be noted, however, that in his most recent interpretation of African palaeo-environments during the last glacial (Van Zinderen Bakker, 1982) this author ascribed the changes as being a consequence of worldwide cooling and a strengthening of the circulation systems rather than to a latitudinal shift in climatic belts. Deep-sea cores from the northern Benguela show a distinct layering of organic carbon concentrations corresponding to cycles of 30 000 to 50 000 years (Meyers *et al.*, 1983) which reflect changes in upwelling intensity in the area and/or of sea level.

To summarize, therefore, while there is evidence of sporadic upwelling occurring during the Oligocene, the proto-Benguela system off Namibia was not initiated until the late Miocene. Upwelling was intensified progressively during the Pliocene, and the full development of the Benguela system and its southward extension to the South Western Cape dates from late Pliocene to early Pleistocene. Cyclical perturbations induced during glacials and interglacials resulted in latitudinal shifts in the extent of the system and changes in the intensity of upwelling and currents.

TOPOGRAPHY

The Cape and Angola Basins which comprise the abyssal plain in the South East Atlantic Ocean are separated by the Walvis Ridge, which runs from its abutment with the coast at about latitude 20° S in a southwesterly direction for more than 2500 km towards the Mid-Atlantic Ridge (Fig. 4). The Walvis Ridge forms a barrier to the northward and southward flow of water below a depth of 3000 m (Shannon & van Rijswijck, 1969; Nelson & Hutchings, 1983) and, as will be discussed later, exerts a major influence on the circulation in the South East Atlantic. Prominent geological features of the Cape Basin, which is bounded in the south by the Agulhas Ridge, are the numerous seamounts of volcanic origin (*e.g.* Discovery, Vema).

The western coast of southern Africa, which forms the eastern boundary of the Benguela system, is characterized by a relatively narrow coastal plain which rises to the main continental escarpment, situated between 50 and 200 km inland (see Fig. 4). Much of the coastal region is arid. The Namib Desert extends between about 14° S and 31° S and is at its widest in the central region, which comprises the main Namib Sand Sea (Ward *et al.*, 1983). The major part of the coastal belt is characterized by sand dunes, with occasional rocky outcrops. North of 32° S the coastline is regular and, except for Walvis Bay and Lüderitz, is devoid of significant embayments. The Orange and Fish River valley forms a major break in the escarpment and continental plateau at the Namibia–South Africa boundary, while the valleys of the Olifants and Berg Rivers in the south, that of Kunene River in the north, and the courses of several dry Namibian rivers provide secondary discontinuities. South of 32° S the coastline is irregular with several capes (formed by granitic outcrops, *e.g.* Cape Columbine, Cape Peninsula, Cape Hangklip) and bays (*e.g.* St Helena Bay, Saldanha Bay, Table Bay, False Bay). This southern region, *viz.* the South Western Cape is topographically different from the remainder of the area and has a Mediterranean type climate and vegetation.

The bathymetry of the western continental margin of southern Africa is variable, with narrow parts of the continental shelf situated off southern

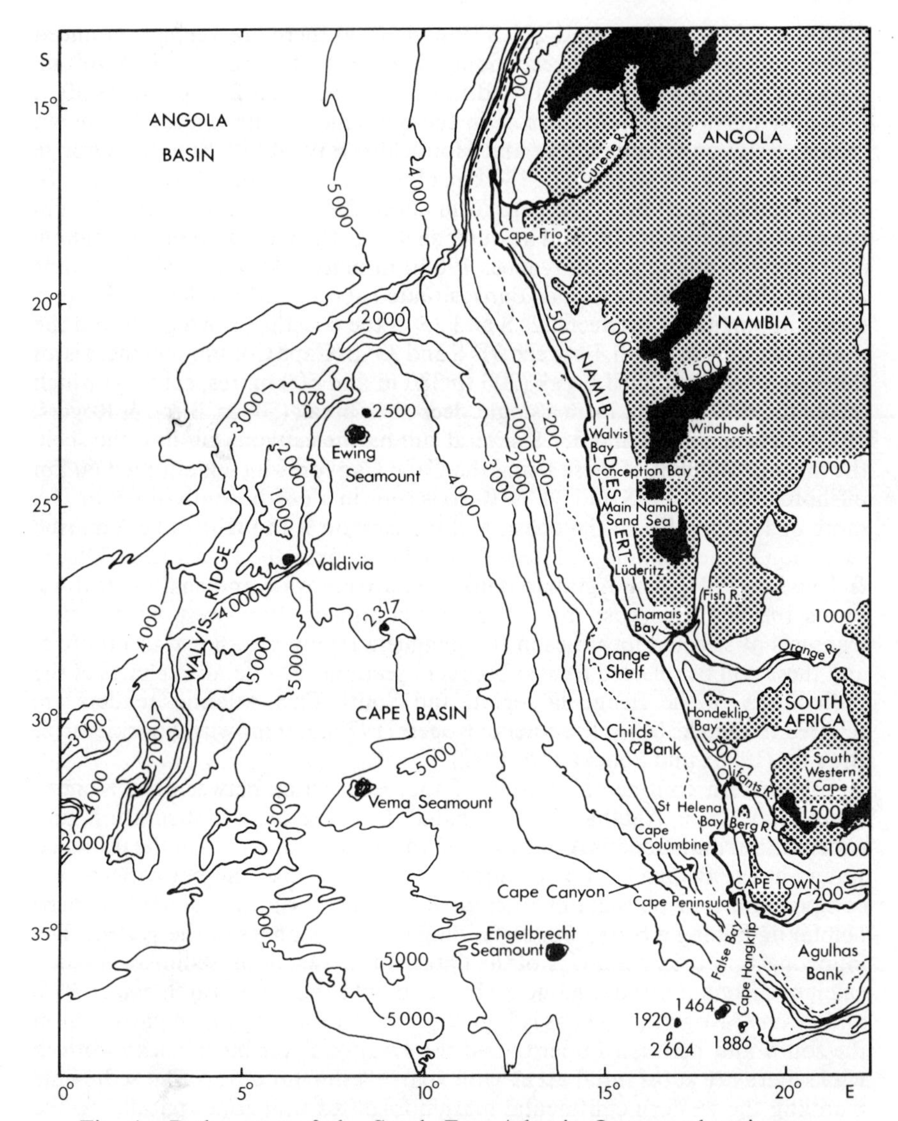

Fig. 4.—Bathymetry of the South East Atlantic Ocean and main orographical features of southwestern Africa.

Angola (20 km), south of Lüderitz (75 km) and off the Cape Peninsula (40 km) and the widest zones off the Orange River (180 km) and in the extreme south (Agulhas Bank). In the Kunene margin the shelf is narrow (about 45 km), about 200 m deep and the continental slope is relatively steep (Bremner, 1981). Between Cape Frio (18° S) and Chamais Bay (28° S), the Walvis Shelf, which is typically 140 km wide, is relatively deep on average, with the shelf break being at about 350 m on average (Birch, Rogers, Bremner & Moir, 1976). Double shelf breaks are, however, common off the west coast (Siesser, Scrutton &

Simpson, 1974), and around 23° S (Walvis Bay) there are very pronounced inner and outer breaks corresponding to depths of about 140 and 400 m, respectively (see Fig. 5). The broad inner shelf between 22° S and 23°30′ S probably plays an important rôle in the dynamics of the central Namibian region. Between Chamais Bay and Hondeklip Bay (30°30′ S) is the Orange Shelf which is at its widest, *viz.* 180 km, off the mouth of the Orange River. In this region the shelf break deepens from about 200 m in the north to 500 m in the south (Birch *et al.*, 1976). The outer shelf, *viz.* the Orange Bank, is shallow (160 m) while the mid shelf reaches 190 m in places. At about 31° S there is another shallow feature, Childs Bank, situated about 150 km offshore. Further south, in particular between 32° S and 35° S the coastline is irregular and the shelf is variable in width. Between 31° S and 33° S (Cape Columbine) there is an inner and an outer shelf break (200 to 380 m and 500 m, respectively) which merge south of 33° S to form a single, deep shelf break (500 m, Birch & Rogers, 1973). Between 31° S and 35° S several submarine canyons cut into the shelf, the most prominent of these being the Cape Canyon which is situated 60 km offshore between 33° S and 34° S. Its axis runs in a north–south direction, *i.e.* more or less parallel to the coast, and the canyon is thought to be a marine extension of the Olifants and Berg Rivers dating from the Palaeogene (Dingle & Hendey, 1984). The Agulhas Bank, a relatively wide and shallow feature, forms the southernmost margin of the continent. East–west bathymetric transects at selected localities in the Benguela region are shown in Figure 5, and these will be referred to in subsequent sections. For detailed charts of the bathymetry of the Benguela region and South East Atlantic, readers are referred to Dingle, Moir, Bremner & Rogers (1977), Rabinowitz, Shackleton & Brenner (1980), and Shackleton (1982).

The western continental margin of southern Africa between the Kunene River and Cape Agulhas is dominated by biogenic sedimentation and concomitant authigenesis resulting from the high productivity of the upwelled waters in the Benguela system (Birch *et al.*, 1976). Like the bathymetry, the composition and physical characteristics of the surficial sediments can be helpful in gaining a broad understanding of the dynamics of the system. The following very brief summary of the nature of the sea floor, sediment texture, calcium carbonate, and organic carbon content is based on Birch *et al.* (1976) and Birch & Rogers (1973), as is Figure 6. Much of the region, in particular in the south and between Lüderitz and the Orange River has a rocky bottom while there are substantial areas with sparse sediment cover. The sediments mantling the western continental margin form textural zones parallel to the

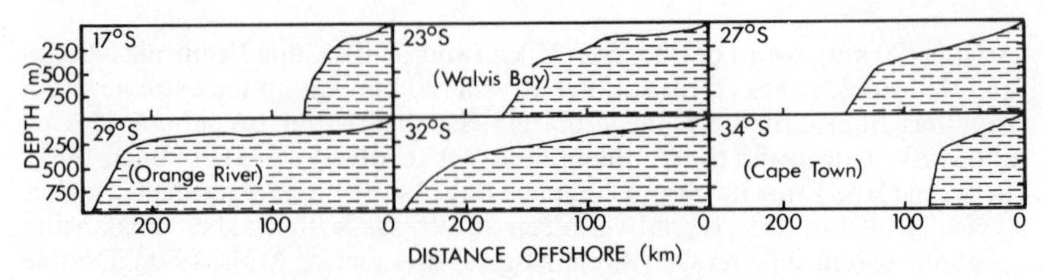

Fig. 5.—Shelf profiles at selected latitudes.

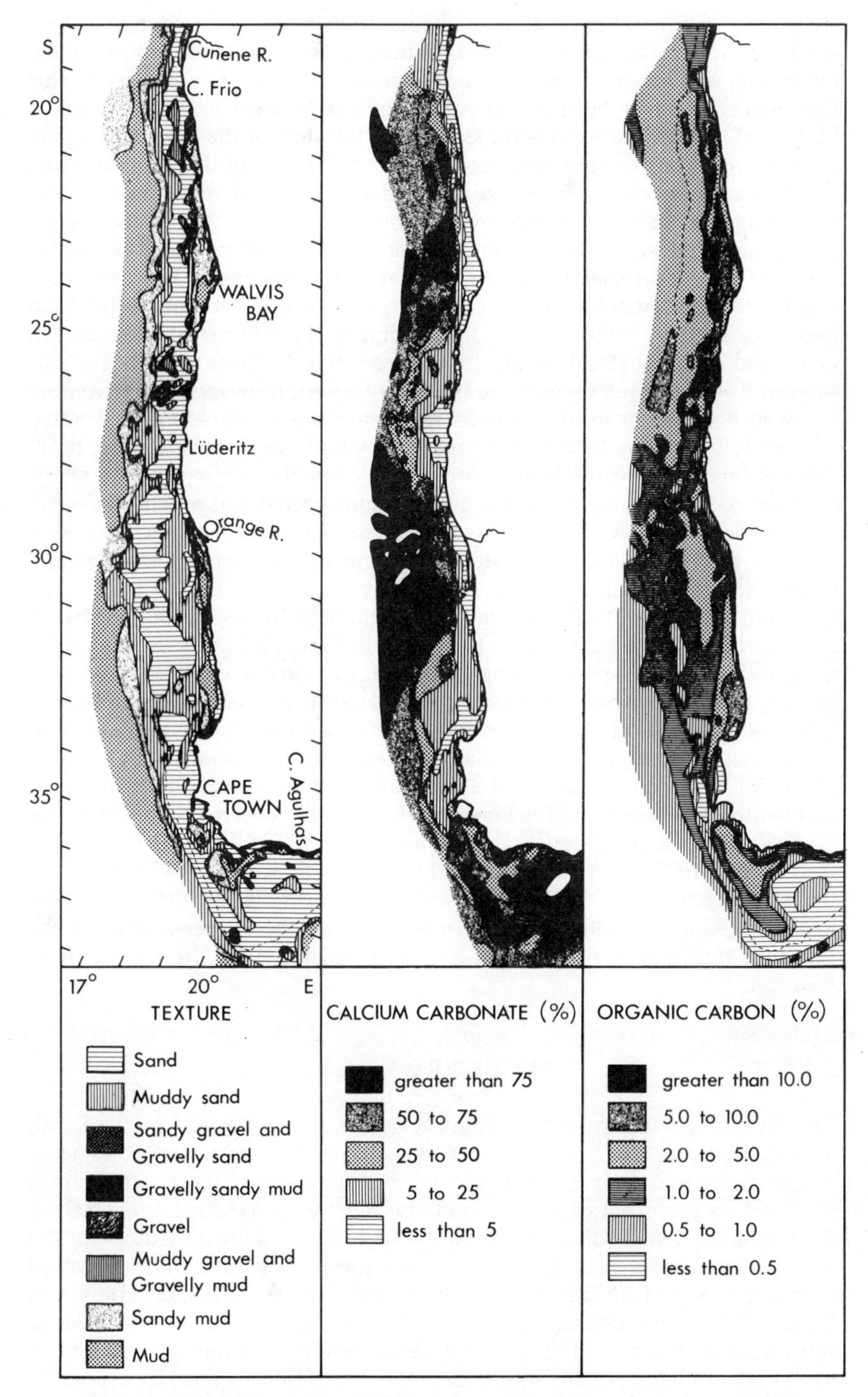

Fig. 6.—Sediment texture, calcium carbonate and organic carbon content along the west coast (from Birch *et al.*, 1976).

coast, and in general the sediments become finer seawards, changing from sands on the inner and middle shelves to muddy sands and sandy muds on the outer shelf and muds on the outer continental slope (Birch *et al.*, 1976). This pattern has, however, been considerably modified by localized river input and biological deposition. Significant features of the shelf in the Benguela region are the two extensive mud belts, each almost 500 km long. The southern belt which extends between the Orange and Olifants Rivers, up to 40 km wide and averaging 15 m thick, is situated over the outer edge of the middle shelf and is mainly of terrigenous (river) origin (Birch *et al.*, 1976). The northern belt which lies over the middle shelf between Cape Frio and Conception Bay comprises organic rich diatomaceous oozes ($>5\%$ organic carbon content). Other than two large deposits of dominantly fine sediment, the outer shelf is covered in sand-sized material (Birch *et al.*, 1976). According to these authors the calcareous flora and fauna which are the most important biogenic contributors to the shelf sediments in the Benguela region are mainly planktonic or benthic foraminiferans, while locally, siliceous phytoplankton dominate some nearshore sediments and polychaete worms have converted some of the terrestrially derived muds on the western coast into sand-sized faecal pellets. The outer and middle parts of the shelf are covered by carbonate-rich sediment ($>50\%$ $CaCO_3$). Birch *et al.* (1976) have shown that the highest organic carbon values on the shelf are related to the diatomaceous muds between Lüderitz and Cape Cross (15% C_{org}) and to the organic rich faecal pellet muds on the shelf break north of Lüderitz ($>5\%$ C_{org}), on the inner shelf and off Lamberts Bay ($>5\%$ C_{org}), and south of Cape Hangklip ($>2\%$ C_{org}). Sediment, depleted in organic carbon, in some regions is associated with deposits of glauconite and apatite mixed as pellets (Birch, 1979). The manner in which the organic-rich sediment curves eastward around the Agulhas Bank (see Fig. 6) is regarded by Birch *et al.* (1976) as an indication that this southernmost region lies within the sedimentological regime of the Benguela. Readers are referred to the following for further information: van Andel & Calvert (1971); Calvert & Price (1971a); Birch (1975, 1977, 1978); Scrutton & Dingle (1975); Rogers (1977); Bremner (1978, 1980a,b, 1981, 1983); Summerhayes, Bornhold & Embley (1979); Robson (1983). A synthesis of information about the sedimentology of the Benguela region is in preparation by Drs J. Rogers, J. M. Bremner and R. Johnson.

METEOROLOGY

The prevailing winds over the Benguela region are determined by the South Atlantic high pressure system (anticyclone), the pressure field over the adjacent subcontinent and by eastward moving cyclones across the southern part produced by perturbations on the subtropical jet stream (Schell, 1968; Nelson & Hutchings, 1983). The South Atlantic high is maintained throughout the year but undergoes seasonal shifts in position (it is approximately centred around 30° S : 5° E in summer and 26° S : 10° E in winter, Schell, 1968; van Loon, 1972a) and intensity (3 to 4 mb). The pressure over the African subcontinent changes radically from a well-developed low during summer to a weak high in winter as the continental heat low and Intertropical Convergence Zone moves northwards, and consequently the pressure gradient along the western coast is seasonally variable. The curved anticyclonic flow associated with the South Atlantic high is guided by the coastline owing to the desert-like nature of the coastal plain acting as a thermal barrier to cross flow (Nelson & Hutchings, 1983) and by the orography of the continental escarpment. As a result the winds along the western coast of southern Africa are predominantly southerly and upwelling favourable. The Ekman transport computed from an extensive set of data by Parrish *et al.* (1983) for 1° latitude and longitude rectangles during January to February (summer) and July to August (winter) is illustrated in Figure 7a, and the effect of the seasonal migration of the pressure systems is clearly evident. There are four features apparent in Figure 7a which are worth highlighting. First, the Lüderitz area is the principal potential upwelling centre in the system, while at Cape Frio a secondary region of winds highly favourable for upwelling exists; secondly, there is a pronounced offshore divergence zone north of 28° S, which supports the earlier findings of Stander (1964), of a progressive anticyclonic rotation in the wind field with increasing distance offshore; thirdly, the wind stress maximum is situated in a band away from the coast, evident particularly during winter; the fourth feature is that the difference between the summer and winter nearshore Ekman transport north of Lüderitz is not large (*cf.* Stander's 1964 wind data), but is substantial in the south. What is less evident in Figure 7a but is apparent in the work of Berrit (1976), from the plots of wind speed cubed in the paper of Parrish *et al.* (1984), of wind stress and wind stress curl in Duing, Ostapoff & Merle (1980), and of nearshore wind stress in Boyd, Husby & Norton (in prep.)—see Figure 7b—is that the 15° S parallel is the approximate northern boundary of the highly upwelling favourable wind field. North of this latitude winds tend to be lighter and more onshore. It should be noted that Figure 7a was intended by Parrish *et al.* (1983) to facilitate the inter-regional comparison of four upwelling systems, and probably Figure 7b is more suited to a discussion of the longshore variation of wind stress near the coast in the Benguela region as the 1° rectangles are taken adjacent to the coast. Figure 7a tends to over-emphasize the winter wind stress near 27° S (Lüderitz) and under-emphasize the summer upwelling favourable winds in the southern part of the system. In Figure 7b the wind stress maxima near 33° to 34° S and 31° S during spring and summer are shown quite clearly.

The essential differences in the seasonal wind regime between the northern and southern parts of the Benguela region were illustrated by Hart & Currie (1960) from a consideration of four coastal sites. Their diagrammatic representation is shown here in Figure 8. In winter, with the northward shift of the pressure systems, the effect is much more pronounced in the south where the frequency of winds with westerly components—*i.e.* non-upwelling favourable—is significant. In this southern region of the Benguela, wind-induced upwelling is highly seasonal and reaches a maximum during spring and summer (Shannon, 1966; Andrews & Hutchings, 1980) and the upwelling season extends from September to March (Andrews & Hutchings, 1980). North of about 31° S the macroscale wind field exhibits relatively less seasonal variation. Upwelling is perennial here, but with a spring–summer maximum and autumn minimal as far north as 25° S and a late winter–spring maximum north of this latitude (Stander 1964; Schell, 1968). While the wind off northern and central Namibia shows relatively little seasonal variation, there are nevertheless slight maxima in the upwelling favourable wind during April to

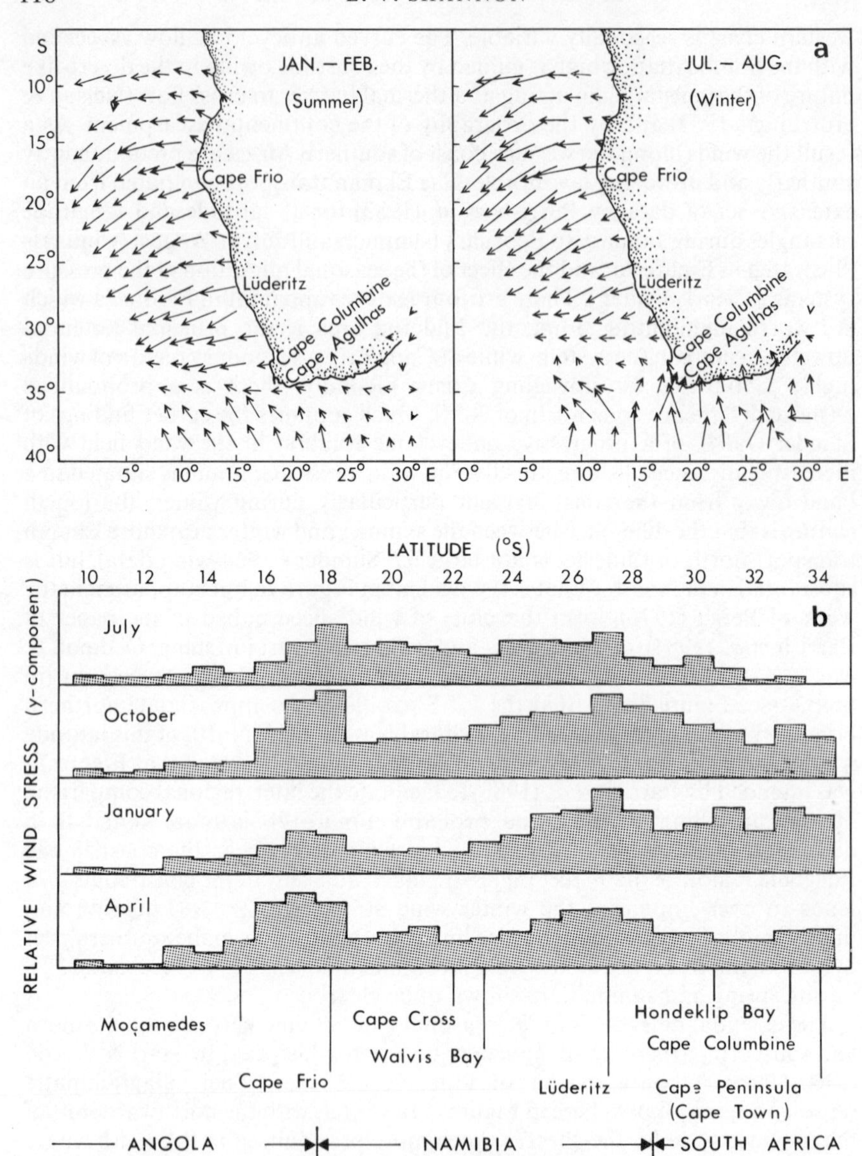

Fig. 7.—a, Ekman transport during summer and winter (after Parrish *et al.*, 1984). b, Y-component of wind stress per 1° rectangle adjacent to the coast (after Boyd *et al.*, in prep.).

May and October (Stander, 1958, 1963; Berrit, 1976; Boyd *et al.*, in prep.). Thus the Benguela region can be divided into two distinct regimes.

Land-sea breezes are common along the coast north of Cape Columbine (Jackson, 1947), and the diurnal modulation of the coastal wind has been described very adequately by Hart & Currie (1960) and Stander (1958, 1963,

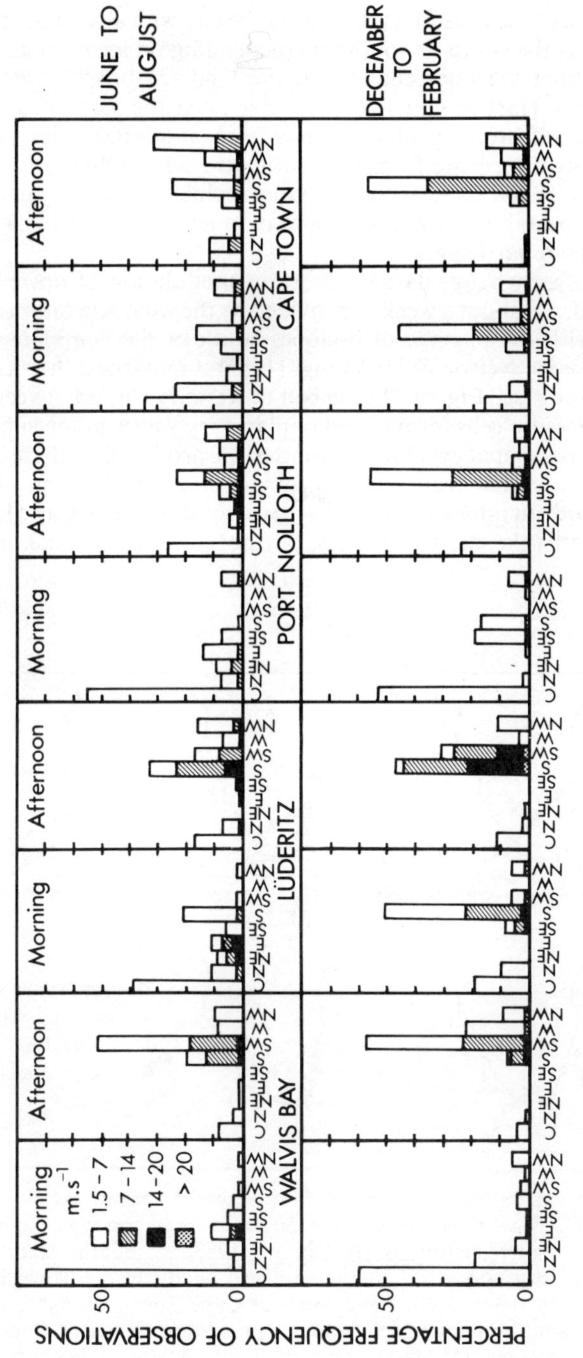

Fig. 8.—Percentage frequency and speeds of coastal winds in winter and summer (after Hart & Currie, 1960).

1964). The contrast between the morning and afternoon winds, in particular those at Walvis Bay, is illustrated in Figure 8. In this figure the marked intensification of the coastal winds during the day with a veering or backing of winds towards the south or southwest (depending on location and season) is evident. Distinct seasonal changes in the land-sea breezes were noted by Stander (1964). Hart & Currie (1960) have cited the view of S. P. Jackson "... that the sea-breeze probably has a fetch of 80–100 miles over the sea, that is from its divergence from the southeast trade". Obviously this diurnal pulsing of winds and the seasonal variation exhibited therein must be important for the coastal upwelling dynamics of much of the Benguela region, the northern part in particular.

In the southern Benguela an important modulation of upwelling with a longer period, *viz.* about a week, is provided by the wind relaxation or reversals associated with the passage of cyclones south of the continent during the upwelling season. Nelson & Hutchings (1983) summarized the situation very neatly as follows (see Fig. 9): "In the belt of westerly winds between 35° S and 45° S, low pressure cells form ahead of planetary waves in the subtropical jet stream. The associated cyclonic rotation of air produced as these cells advect eastwards, causes the wind field as far north as the Olifants River to be modulated with an intensity which increases southwards to Cape Point. In the summer months the effect is usually, but not necessarily, weak, manifesting

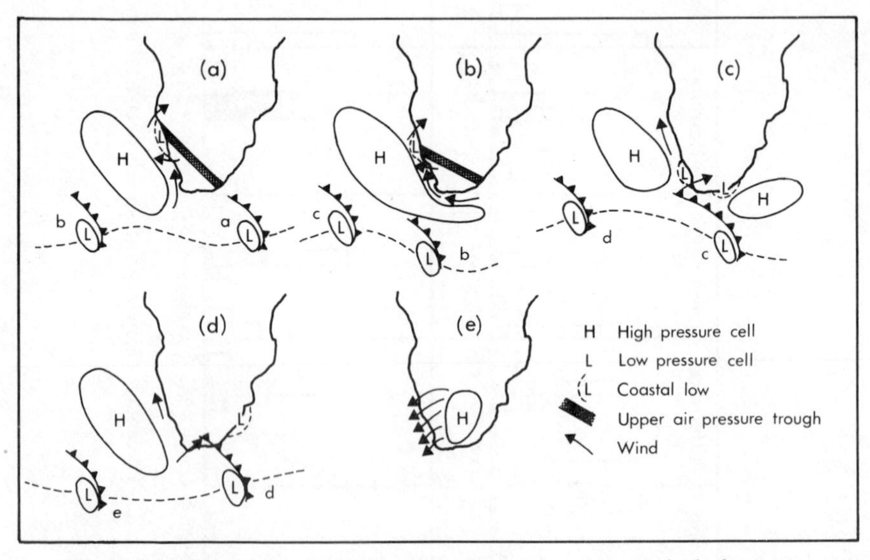

Fig. 9.—Cyclic weather pattern over the Benguela system typical of summer conditions (after Nelson & Hutchings, 1983). a, South Atlantic high established, coastal low at Lüderitz, southerly winds at Cape Town; b, South Atlantic high ridging, gale force winds at Cape Town, coastal low moves south; c, South Atlantic high weakens, northwest winds at Cape Town, following passage of coastal low; d, South Atlantic high strengthens, southerly winds along west coast; e, berg wind conditions.

Fig. 10.—Aerosol plumes of sand and dust due to a katabatic wind event: NIMBUS 7 CZCS, 670 nm band, 9 May 1979; (after Shannon & Anderson, 1982).

itself as a periodic weakening of the South Atlantic High and slackening or abatement of south easterly winds along the coast. In the winter months, the effect may be, but is not necessarily, strong, bringing in its extreme form gale force north westerly to south westerly winds of some hours duration, in cycles of three to six days.... Associated with the approach of cyclonic systems, is the appearance of cells of low pressure which form near Luderitz (Taljaard, Schmidt and Van Loon, 1961) and travel round the subcontinent as trapped waves at a speed of about 750 km day^{-1}. . . . The migration of these cells southward along the coast seems to occur as a precursor to the approach of the cyclonic systems to the south (Nguyen Ngoc Anh and Gill, 1981) and is best observed in the summer months under conditions of weak modulation of the South Atlantic High pressure cell. The cyclonic rotation of air about these cells suppresses upwelling locally as the wave travels along the coast and the relaxation in the wind at the centre causes suitable conditions for the generation of inertial motions and possibly shelf waves."

Another large-scale feature of the meteorology of virtually the entire Benguela region is the occurrence of "berg" winds during autumn and winter (Jackson, 1947; de Wet, 1979). These occasional katabatic wind events are associated with the formation of a large high pressure system (a precursor to the coastal lows) over or just south of the southern or southeastern part of the subcontinent of several days duration. The anticyclonic circulation around the high results in a strong (up to 15 $m{\cdot}s^{-1}$ or more) easterly to northeasterly flow off the plateau of dry adiabatically heated air. Satellite imagery suggests that these winds are locally intensified by topographic features such as river valleys, and they may transport substantial quantities of sand and dust out to sea (Shannon & Anderson, 1982). This aeolian transport to a distance of 150 km offshore between 18° S and 30° S during a single berg wind event is shown in Figure 10 which emphasizes the directional coherency of the wind over a distance of nearly 1500 km. It has been suggested by Hart & Currie (1960) that these berg winds have little effect over the sea owing to their divergence above the marine boundary layer. On the macroscale, however, they appear to suppress upwelling in the southern region (Nelson & Hutchings, 1983). A. J. Boyd (pers. comm.) has indicated, however, that these winds can produce localized upwelling, but that this effect seems to be limited to within 10 km of the coast.

Very little has been published on the interannual variations in the Benguela wind regime. An upwelling index based on the integrated longshore component of the wind at a single site (Cape Point) since 1961 has been cited in Hutchings, Nelson, Horstman & Tarr (1983), Nelson & Hutchings (1983), Crawford, Shelton & Hutchings (1984) and Hutchings, Holden & Mitchell-Innes (1984), and shows minima during 1966 and 1983 and a maximum in 1975, but no clear trends. The Cape Point site is, however, not characteristic of the whole Benguela region. Early work by Rawson (1908) has indicated a 19-year periodicity in the latitudinal position of the subtropical high pressure belt, which he tentatively suggested may be linked to the lunar period of 18·6 years, while Dyer & Tyson (1975) and Tyson (1981) have indicated a distinct periodicity in summer rainfall over the interior of South Africa. It seems probable that there may well be an interannual wind cycle over the Benguela regime, but long reliable sets of data are not readily available to verify this. It is also probable that the northern and southern Benguela regions will be affected differently, as the recent studies on the southern Benguela warm event of 1982–1983 suggest (*e.g.* Shannon, 1983; Boyd & Agenbag, 1984b; Walker, Taunton-Clark & Pugh, 1984).

As indicated earlier, much of the western coast is arid and has a low rainfall. Rainfall over the sea in the Benguela region ranges from less than 10 cm·yr^{-1} in the north to 50 cm·yr^{-1} in the south, while the respective annual evaporation ranges from less than 75 to 125 cm (Albrecht, 1960 as cited in Van Loon, 1972b). On average, the only significant inputs of fresh water into the Benguela system are *via* the Orange and Kunene rivers (summer) in the north and Olifants and Berg rivers (winter) in the south. The last three have mean annual run-offs of 7300×10^6, 708×10^6, and 528×10^6 m^3, respectively (Department of Environment Affairs, South Africa). During periods of heavy rainfall in the catchment areas the discharge is, however, manifest as a thin lens of low salinity water with mesoscale dimensions (Shannon, 1966). The impact of the episodic floods which occur in the Namib with a time scale of decades (*e.g.* during 1933–1934; Ward, Seely & Lancaster, 1983) on the oceanography of the Benguela has not been well documented.

Fog is common over much of the region north of 32° S, in particular around Walvis Bay. Diagrams in van Loon (1972b) indicate the existence of a deep centre of dew point depression at 850 mb along the western coast with a summer maximum (16 °C) centred around 27° S and in winter at 22° S (18 °C). Total cloud cover generally increases offshore and from south to north, with inshore values ranging from about 40 to 70% (Van Loon, 1972b; Parrish *et al.*, 1983). The summer minimum zone indicated by Parrish *et al.* (1983) lies between 25° S and 34° S while in winter it shifts north to between 22° S and 32° S in sympathy with seasonal movement of wind belts. The relatively cloud free nature of the southern Benguela region makes it amenable to investigation using satellites (Shannon & Anderson, 1982).

WATER MASSES

The large scale hydrology of the South East Atlantic has been described by a number of authors. Clowes (1950) synthesized much of the early German (Meteor) and British (Discovery) work, while authors such as Fuglister (1960), Hart & Currie (1960), Darbyshire (1963), Stander (1964), Shannon (1966), Shannon & van Rijswijck (1969), Visser (1969a,b), Moroshkin, Bubnov & Bulatov (1970), Welsh & Visser (1970), and Henry (1975) have reported the results of several cruises undertaken in the region during the 1950s and 1960s.

There are several water masses present off the western coast of southern Africa, including *inter alia* tropical and subtropical surface waters, South Atlantic central, Antarctic Intermediate, deep and bottom water. The principal water masses are annotated on the suite of *T-S* plots for the Benguela area between latitudes 15° S and 35° S shown in Figure 11 and the general similarity between the *T-S* curves is evident. According to Clowes (1950), Stander (1964) and Shannon (1966) the water which upwells to the surface and to subsurface depths along the coast between Cape Frio (about 18° S) and Cape Point (about 34° S) is central water. This water mass which corresponds to the linear portion of the *T-S* curve connecting the approximate points 6 °C, 34·5‰ and 16 °C, 35·5‰ is found throughout the South East Atlantic, either

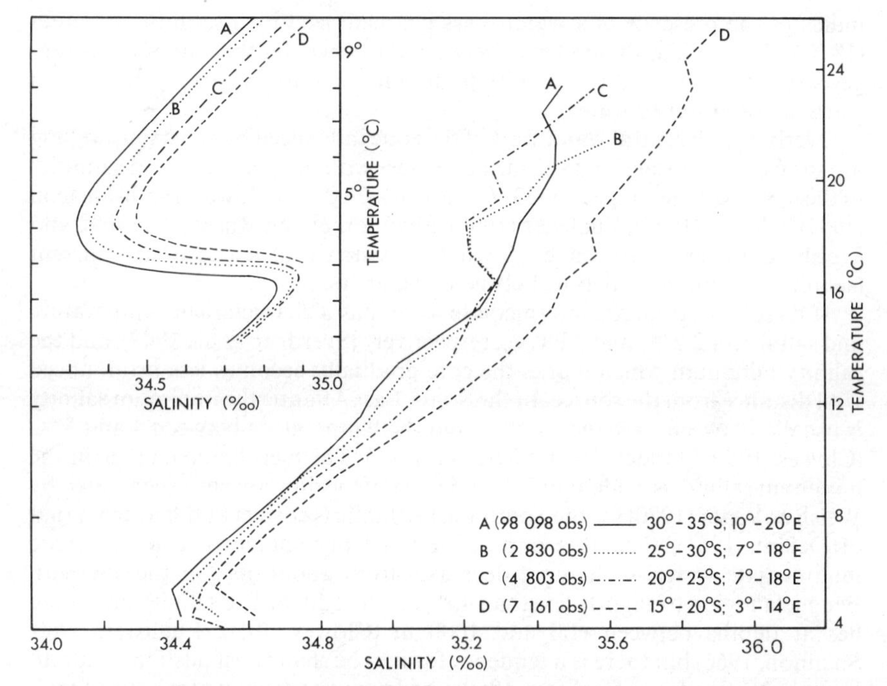

Fig. 11.—Mean salinity per 1 °C temperature interval between 15° S and 35° S, with the lower portions of characteristic *T-S* curves, drawn from scatter plots, inset.

as a layer separating the surface and Antarctic intermediate layers in the oceanic region, or as the sole or main water mass present over the Benguela continental shelf. Central water is formed in the Subtropical Convergence region by the sinking and northward spreading of mixed subtropical and sub-antarctic water masses (Sverdrup, Johnson & Fleming, 1942; Orren, 1963, 1966). Central waters in both the South East Atlantic and South West Indian Oceans have very similar *T-S* characteristics (Orren, 1963; Shannon, 1966) and it is difficult to quantify the contribution of the latter to the Benguela system. Clowes (1950) and Shannon (1966) have suggested that some Indian Ocean central water enters the Benguela region at subsurface depths, although its contribution is probably small. Dietrich (1935) and Darbyshire (1963), however, have shown that the dynamic topography is favourable for the entrainment of this water around the Agulhas Bank. Clowes (1950), Shannon (1966), and Visser (1969a), by considering the properties of central water at the 26·70 to 26·75 sigma-*t* and 140 cl·t^{-1} isanosteric surfaces (approximately 10° to 11 °C; 34·9 to 35·0‰) deduced that in the oceanic region the flow of central water was predominantly northerly to northwesterly as far as latitude 20° S. North of 20° S the central water *T-S* curve is laterally displaced by about 0·1‰ (see Fig. 11) suggesting that in this region the central water is not of Benguela origin but has been advected southwards from lower latitudes. The lower upper portion of the *T-S* curves for the region between 20° and 30° S indicates the presence of a water mass characterized by a salinity minimum (18 °C, 35·2°/₀₀). The dissimilarity between the curves for the four 5° areas suggests that the origin of the salinity minimum may not be simply due to sun warming of upwelled water.

Overlying the central water west of the zone influenced by coastal upwelling are subtropical surface and subsurface waters with temperatures and salinities in the approximate ranges 15–23 °C and 35·4–36·0°/₀₀ (Clowes, 1950). Deacon (1937), Clowes (1950), Fuglister (1960), Shannon & van Rijswijck (1969), and Welsh & Visser (1970) have shown the existence of a subsurface current (salinity maximum) at depths between 50 and 200 m.

At its source Antarctic intermediate water has a characteristic temperature and salinity of 2·2 °C and 33·8°/₀₀, respectively (Sverdrup *et al.*, 1942), and the salinity minimum which marks the core gradually becomes less pronounced with distance from the source. In the South East Atlantic the minimum salinity is usually between 34·3 and 34·5°/₀₀ and the temperature between 4 and 5 °C (Clowes, 1950; Stander, 1964; Shannon, 1966). The meridional change in the minimum salinity is evident in Figure 11, and if the dilution curve generated by Welsh & Visser (1970) for the South East Atlantic is correct (it differs from that of Defant, 1961), then this implies that the percentage of true Antarctic intermediate water in the core decreases from about 50% in the southern Benguela region to slightly less than 40% north of 20° S. The salinity minimum lies at depths between 600 and 1000 m (Clowes, 1950; Fuglister, 1960; Shannon, 1966) but there is a tendency for it to be shallowest offshore at about latitude 24° S (refer to Fuglister, 1960) and immediately west of the shelf break in the southern Benguela region (*e.g.* Fig. 12). Isentropic analysis by Clowes (1950) and Shannon (1966) using sigma-*t* surface 27·25 suggested that there is a

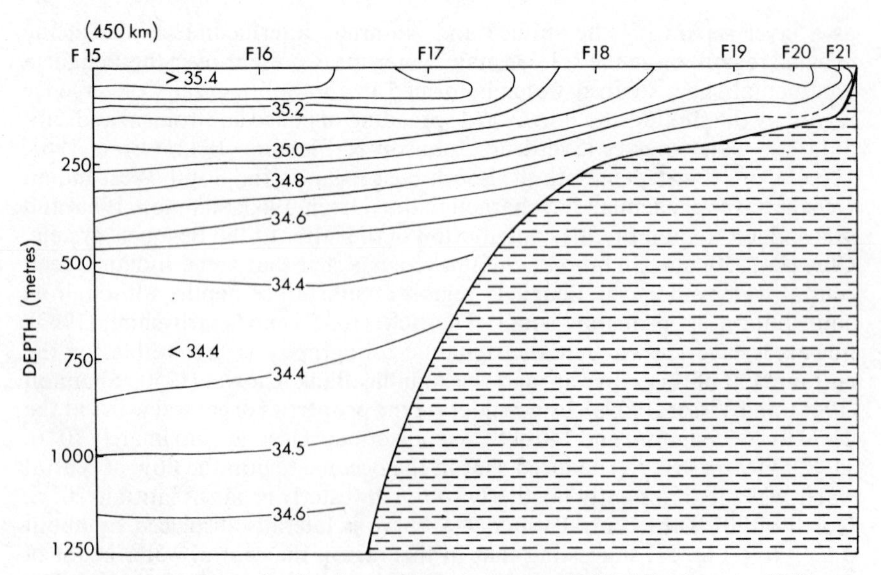

Fig. 12.—Salinity distribution off Roodewal Bay (30°21′ S), January 1959 (after Shannon, 1966).

northwesterly flow of Antarctic intermediate water south of latitude 25° S, a finding which was supported by Visser (1969a). Recent direct current measurements by G. Nelson (pers. comm.), however, show substantial poleward movement of this water mass in the Benguela region. Some Antarctic intermediate water may be advected from the Indian Ocean into the Atlantic around the Agulhas Bank (Clowes, 1950; Shannon, 1966). North of 25° S Visser's (1969a) data indicated a southward movement at the 80 cl·t^{-1} isanosteric surface (equivalent to sigma-*t* 27·28) within 500 km of the coast.

North Atlantic deep water is detected as a high salinity layer lying beneath the Antarctic intermediate water, and according to Clowes (1950) can be traced as far south as 56° S. Shannon & van Rijswijck (1969) noted the presence of the deep water core at depths ranging from about 2000 to 3000 m in the South East Atlantic, it being deepest at about latitude 32° S. Its temperature and salinity decreased from about 3·1 °C, 34·93‰ at 16° S and 2·9 °C and 34·89‰ at 24° S to 2·4 °C and 34·87‰ at 32° S, suggesting slow southward movement. At 24° S in the Angola Basin the deep water is characterized by a slight salinity minimum, with the relatively uniform warm and saline North Atlantic bottom water lying below it. In the Cape Basin, however, the deep layer overlies the colder and less saline Antarctic bottom water which has a typical temperature and salinity of $<1{\cdot}5$ °C and $<34{\cdot}77$‰, respectively, and is present at depths deeper than 4000 m (Shannon & van Rijswijck, 1969). These authors considered that the Walvis Ridge effectively blocked the southward penetration of North Atlantic bottom water into the Cape Basin and northward flow of Antarctic bottom water into the Angola Basin, although they did detect some leakage at the break in the Walvis Ridge at 31° S; 2° E. The temperature profile along latitude 24° S (Fig. 13, from Fuglister, 1960) illustrates this blocking effect. Subsequent to Shannon & van Rijswijck's (1969) study the question of the penetration of Antarctic bottom water from the Cape into the Angola Basins was considered in more detail by Connary & Ewing (1974).

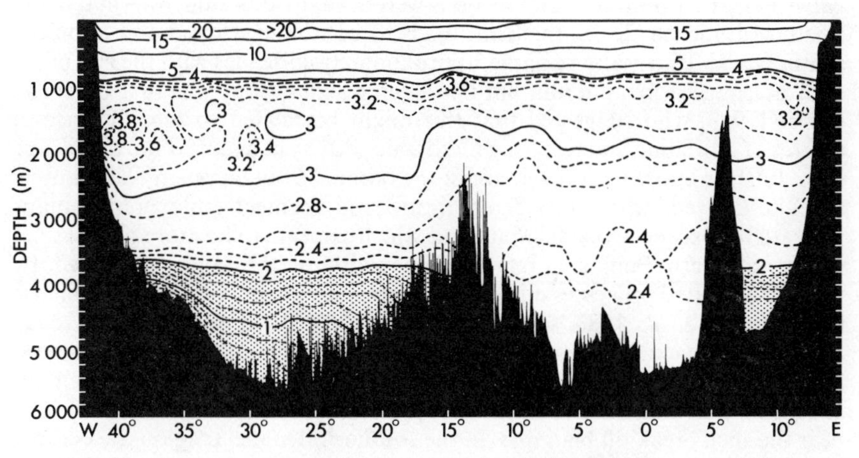

Fig. 13.—Temperature profile across the South Atlantic Ocean at 24° S, October 1958 (modified from Fuglister, 1960).

BOUNDARIES OF THE BENGUELA SYSTEM AND LARGE SCALE FRONTAL FEATURES

The "Benguela Current" was defined by Hart & Currie (1960) as the name applying "... to the region of cool upwelled coastal water along the South-west coast of Africa", *i.e.* water characterized by a pronounced negative surface temperature anomaly found mainly between 15° S and 34° S within 185 km of the coast, which forms the eastern periphery of the anticyclonic gyre in the South Atlantic. This definition was followed by *inter alia* Shannon (1966) while others have preferred to define the Benguela Current in terms of generally northward setting currents. Bang (1971) questioned both definitions; if considered on the basis of upwelled water the character of the "current" becomes absurdly discontinuous in both time and space; if defined in terms of surface flow, then the Benguela cannot be differentiated from the southeast Trade Wind Drift. Bang (1971) proposed that the term "Benguela Current" should be considered as "... that area east of the offshore divergence within which, as has been well established, oceanic processes are dominated by short-term atmospheric interactions. The divergence is thus, at least partly, a hydrodynamic discontinuity reflecting the geomorphological discontinuity of the continental slope and separating the weather-dominated Benguela system from what might be termed the climatic flywheel of the southeast Atlantic deep sea circulations." Several subsequent workers, *e.g.* Lutjeharms (1977), Nelson & Hutchings (1983) and Parrish *et al.* (1983) have tended to refer to the area as the Benguela current system, Benguela upwelling area, Benguela system or Benguela region. Throughout this review it is referred to as Benguela system or region. As processes taking place both seawards and shorewards of the shelf break and oceanic front are important for the understanding of the ecosystem as a whole, the western boundary of the system will, for the purpose of this review, be considered as being fairly open-ended.

While the question of definition of the seaward boundary of the Benguela system is somewhat academic, there generally exists over much of the area between Cape Point and Cape Frio a well-developed oceanic thermal front. South of Lüderitz (27° S) the front tends to be well developed and although spatially and temporally variable, approximately coincides with the run of the shelf break. The meandering nature of the oceanic front was first noted by Currie (1953) who suggested that this might be related to the existence of centres of upwelling and resulting mesoscale eddy systems as far north as Cape Frio. Little is known, however, about the oceanic frontal system off Namibia. Satellite-derived, sea-surface temperature and pigment (chlorophyll) maps indicate the existence of a frontal band which appears to be more diffuse than in the southern Benguela region. The width of the zone influenced by upwelling related processes off Namibia varies seasonally (see Fig. 14, which is from Parrish *et al.*, 1983; see also Stander, 1964; Boyd & Agenbag, 1984a). While the surface features, however, change in time and space, the oceanographic station spacing and sampling frequency has generally been inadequate to establish the existence and persistence of a baroclinic frontal zone near the shelf break off Namibia. In the southern Benguela region the oceanic frontal system is better documented, particularly between Cape Point and Cape Columbine thanks to the pioneering work of the late Dr Nils Bang

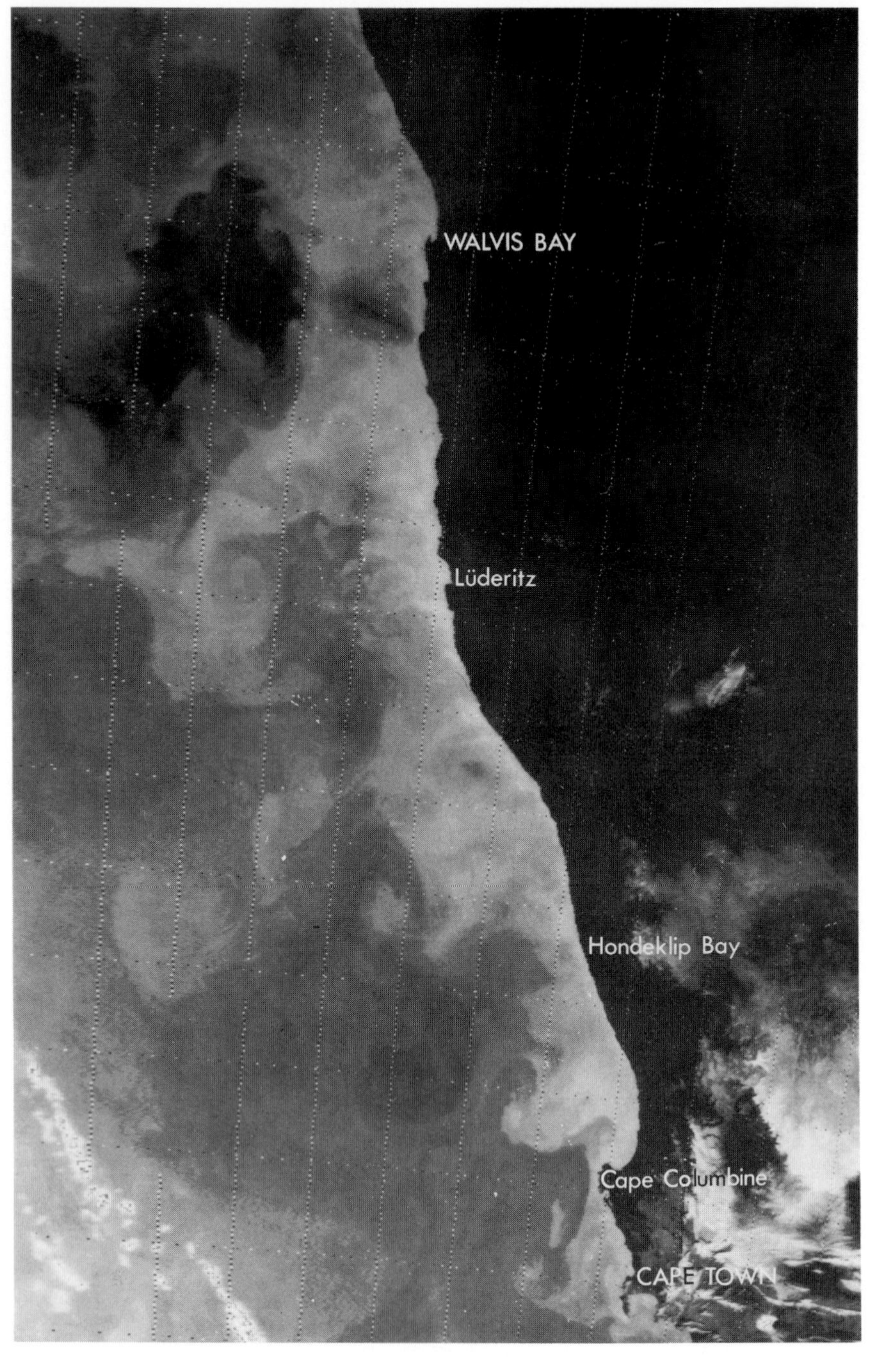

Fig. 15.—NOAA 1 enhanced infrared image of the Benguela, 15 June 1979, showing frontal features (from Van Foreest *et al.*, in press).

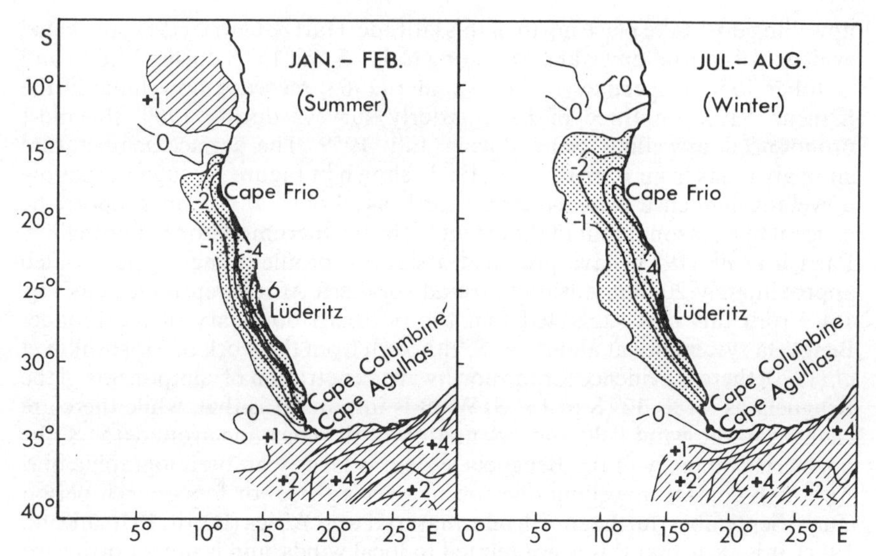

Fig. 14.—Sea surface temperature anomaly (from Parrish *et al.*, 1983).

(Bang, 1971, 1973, 1974; Bang & Andrews, 1974). Bang's mesoscale observations will be discussed later; at this stage it will suffice to record that Bang drew a clear distinction between the inshore frontal system and the strongly baroclinic offshore frontal zone (divergence) close to the shelf break. Another phenomenon which was observed by Bang (1971) in the vicinity of the shelf edge divergence between the Orange River and Hondeklip Bay was the existence of slicks, which he suggested might be related to internal wave activity. These features have been observed in some satellite images further south (Apel, Bryne, Proni & Charnell 1975; Nelson & Shannon, 1983), while they are commonly sighted off Namibia during research cruises (A. J. Boyd, pers. comm.).

Pronounced waves or meanders in the main oceanic thermal or pigment front have been illustrated by various authors, *inter alia* Currie (1953), Bang (1971), Lutjeharms (1981a), Shannon, Walters, Mostert & Anderson (1983), and Shannon, Hutchings, Bailey & Shelton (1984). In a recent paper Van Foreest, Shillington & Legeckis (in press) have suggested that the large-scale stationary features which have been observed at the thermal front from satellite imagery of the Benguela (see Fig. 15) may be related to the existence of a barotropic shelf wave with a 2·2-day period. These authors have pointed out, however, that while the wavelength of such a wave fits the observed structural features well, it does not explain their baroclinic and stationary nature.

The northern and southern boundaries of the Benguela system are reasonably well defined, although in the literature there is some disagreement as to what constitutes the boundaries. While Copenhagen (1953), Shannon, Nelson & Jury (1981), and Nelson & Hutchings (1983) have cited Cape Frio (18° S) as the effective northernmost boundary of the upwelling system, Hart & Currie (1960), Stander (1964), and Parrish *et al.* (1983) have shown that upwelling does take place north of this latitude. Hart & Currie (1960) provided evidence of coastal upwelling extending to 17° S and 15° S during March and October 1950, respectively, while Stander (1964) recorded upwelling off the Kunene River on three of his quarterly surveys during 1959, the most pronounced upwelling being during July 1959. The surface temperature anomaly maps from Parrish *et al.* (1983), shown in Figure 14, suggest that the upwelling influence extends as far north as about 15° S, and support the concept of a seasonal shift in the extent of the northernmost zone of upwelling. Parrish *et al.* (1983) have provided a sigma-*t* profile along a line situated approximately 200 km offshore around southern Africa (reproduced as Fig. 16). From this it is suggested that the northern boundary of the broader Benguela system lies at about 16° S, although from the work of Moroshkin *et al.* (1970) there is evidence for the northward penetration of components of the Benguela as far as 12° S to 13° S. What is important is that, while there are changes in the wind field and orientation of the coastline around 15° S, the northern boundary of the Benguela system is largely an oceanographic one. Although coastal upwelling does occur over a three- or four-month period (June–September) further north off equatorial west Africa (Berrit, 1976; Picaut, 1981), it is clear that this is not related to local winds, and is not a northward extension of the Benguela system. While the wind field north of 15° S is not highly favourable for upwelling, some uplift of warm saline waters close inshore from above the intense pycnocline is nevertheless possible.

Hart & Currie (1960) regarded the southernmost extent of the Benguela upwelling area as about 34° S and Andrews & Hutchings (1980) endorsed the view that the Cape Peninsula is the southernmost significant upwelling site. During the summer the upwelling zone can, however, extend as far south and east as Cape Agulhas (35° S: 20° E) which is regarded by Harris (1978) and Shannon *et al.* (1983) as a more appropriate boundary of the western coast system than Cape Point. At both of these capes the orientation of the coastline changes by about 45°. In addition, there is a marked change in the wind field

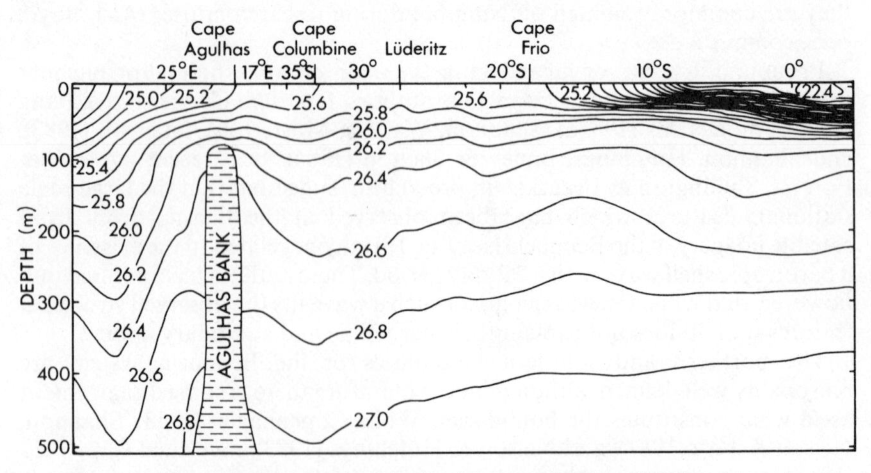

Fig. 16.—Sigma-*t* compared with depth approximately 200 km from the coast around southern Africa (after Parrish *et al.*, 1983).

south of about 35° S (see Fig. 7). Shannon *et al.* (1981) broadly interpreted the southern boundary of the Benguela system as the Agulhas retroflection area, which implied the inclusion of the Agulhas Bank region within the system (see also Fig. 16). That water of Agulhas origin contributes to the offshore part of the Benguela is well established (Clowes, 1950; Darbyshire, 1963; Shannon, 1966; Bang, 1973). As Bang (1973) stated, ". . . vestiges of Agulhas Bank or Agulhas Current water are almost always found off the Cape upwell cell". Reference to Figure 6 shows that organically rich sediments occur over the western Agulhas Bank, which is ecologically an integral part of the productive western coast regime. Thus, the southern boundary of the Benguela is produced by a combination of meteorological, oceanographical, and topographical factors.

The mesoscale processes associated with the northern and southern boundaries of the Benguela system and the oceanic front are discussed in more detail in subsequent sections.

MACROSCALE CIRCULATION

The following is a brief discussion of the large scale circulation in the upper layers of the South East Atlantic between latitudes 13° S and 38° S and the 0° and 20° E meridians. Readers are referred to the section (pp. 122–125) on water masses for information on the movement of the deeper water masses, while the literature pertaining to the highly variable and complex dynamics of the continental shelf region will be reviewed later.

SURFACE CURRENTS

Prior to the work undertaken during the Meteor and Discovery expeditions earlier this century, knowledge of the surface currents in the region was based on ships' drift measurements. Defant (1936) as cited in Hart & Currie (1960) analysed an extensive set of data of Dutch current observations and averaged them seasonally into one degree rectangles. His charts show the existence of a well-defined current in a band 200–300 km wide running in a north northwesterly direction, adjacent to the coast at 34° S and moving progressively offshore northwards. North of 20° S the streamlines bend westwards. Between 35° S and 20° S Defant (1936) indicated a one-sided divergence line, west of which the flow was predominantly westwards. It is significant that two satellite-tracked spar buoys (FGGE type) with drogues set just below the surface, released west of Cape Town during March 1977 (Harris & Shannon, 1979) and February 1979 (Nelson & Hutchings, 1983) followed similar general paths to that indicated by Defant's "Benguela current" streamlines. Nelson & Hutchings (1983) suggested that the current is topographically steered, accelerating in areas of steep topography and meandering over the planes. Harris & Shannon (1979) noted a predominantly westerly movement between 25° S and 18° S in accordance with the dynamic topography of the area (Stander, 1964; Shannon & van Rijswijck, 1969; Moroshkin, Bubnov & Bulatov, 1970). It is perhaps also significant that the drifter released in March 1977 crossed into the Angola Basin over the gap in the Walvis Ridge at about 22° S. In this respect Shannon & van Rijswijck noted that the Ridge appeared to influence local currents significantly. According to Harris & Shannon (1979) their drifter indicated a mean velocity of 17 $cm \cdot s^{-1}$, calculated over the shortest distance between the first and final fixes, which is similar to the mean velocity of drift cards passing through the region (Stander, Shannon & Campbell, 1969; Shannon, Stander & Campbell, 1973). Harris and Shannon (1979) recorded a 15° deviation of drifter trajectory to the left of the cumulative wind stress vectors, a finding supported by the inferred drift card trajectories in the region (see Shannon *et al.*, 1973). The implications of this are that, away from the coast and in the absence of strong gradient currents, the surface currents will be closely related to the prevailing wind. The anticyclonic curvature of both wind and surface current fields in the area tends to support this (compare Figs 7 and 17).

From drift card returns, Shannon, Stander & Campbell (1973) estimated a rotational period of the South Atlantic gyre of about 38 months. Although relatively few drift cards released in the Atlantic between 30° S and 50° S have been recovered along the west coast of southern Africa, those which have, suggest a mean speed in the West Wind Drift of about 15 $cm \cdot s^{-1}$. On the basis of drift card data Stander *et al.* (1969) speculated on the probable transport of *Jasus tristani* larvae from Tristan da Cunha to the Vema Seamount, and from their work the meandering nature of the currents between 30° S and 40° S and 10° W and 10° E can be inferred, and which has subsequently been confirmed by the trajectories of three satellite-tracked drifters through the area (Lutjeharms & Heydorn, 1981; Lutjeharms & Valentine, 1981). Their oscillating paths which abruptly change direction from eastwards to northwards on approaching 10° E were in agreement with the dynamic topography of Dietrich (1935).

The work of Stander (1964), Moroshkin *et al.* (1970) and Dias (1983a) has greatly contributed to the understanding of the complex flow patterns in the region north of 23° S. Moroshkin *et al.* (1970) showed a substantial westward flow between 23° S and 15° S at the surface with a large cyclonic gyre (r = 300 km) separating this "west (main) branch of the Benguela current" and the eastward flowing South Equatorial Countercurrent further north. These authors proposed the existence of three branches of the Benguela north of 20° S with a "Benguela divergence" separating the main (west) branch from the more easterly branches. They also showed the merging of the swift southerly flowing Angola Current inshore and the Benguela. It should be noted that the study of Moroshkin *et al.* (1970) related to autumn, a season which, as will be seen in the subsection on seasonal and interannual variability, is not typical of 'average' conditions in the Benguela region.

Different authors have measured or inferred surface or near-surface currents using various techniques, including drift cards, ships drift, satellite-tracked drifters, dynamic topography and isentropic analysis, and an attempt to synthesize the data has been made in Figure 17. It is based on the following: Rennell (1832); Dietrich (1935); Defant (1936); Clowes (1950); Hart & Currie (1960); Darbyshire (1963); Stander (1964); Shannon (1966); Duncan & Nell (1969); Shannon & van Rijswijck (1969); Stander *et al.* (1969); Visser (1969a); Moroshkin *et al.* (1970); Shannon *et al.* (1973); Harris & Shannon (1979); Lutjeharms & Heydorn (1981); Lutjeharms & Valentine (1981); Boyd & Agenbag (1984a); Nelson & Hutchings (1983); Dias (1983a); and Parrish *et al.* (1983).

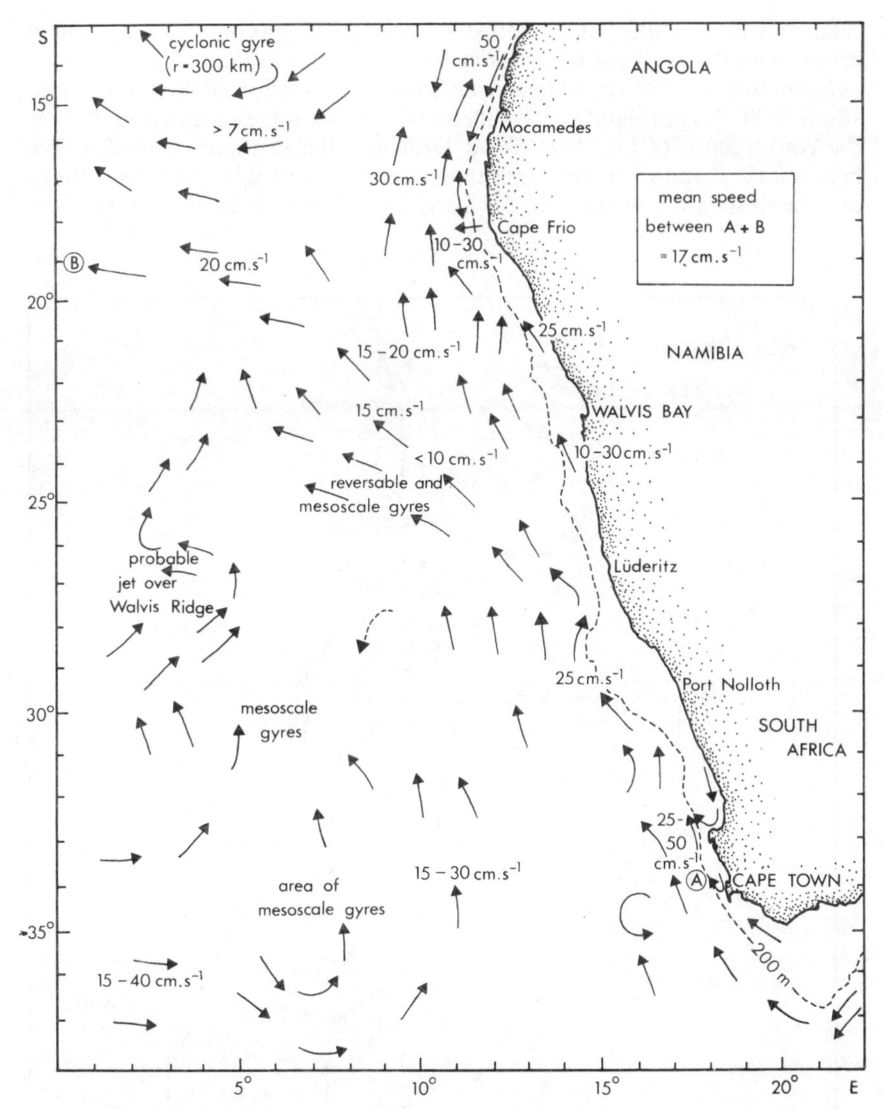

Fig. 17.—Surface currents—composite of work of several authors (refer to text).

CURRENTS BETWEEN 200 AND 300 M

The circulation of South Atlantic central water between 200 and 300 m is illustrated schematically in Figure 18. This figure is a composite based on the dynamic topography of Dietrich (1935), Stander (1964), Shannon & van Rijswijck (1969), and Moroshkin *et al.* (1970) and the isentropic analysis of Shannon (1966) and Visser (1969a). A note of caution must, however, be sounded. The papers of Shannon & van Rijswijck (1969) and Visser (1969a) relate to winter, while that of Moroshkin *et al.* (1970) was based on an autumn cruise, with the resultant bias of some of the features in Figure 18.

Excluding the 250 km wide coastal band, the direction of flow in the area south of 20° S is not significantly different from that at the surface (Fig. 17), *e.g.* the convergence of the West Wind Drift and the main equatorward flow between 10° E and 15° E; the topographic control exerted by the Walvis Ridge and the westward flow between 20° S and 25° S. North of 20° S the circulation

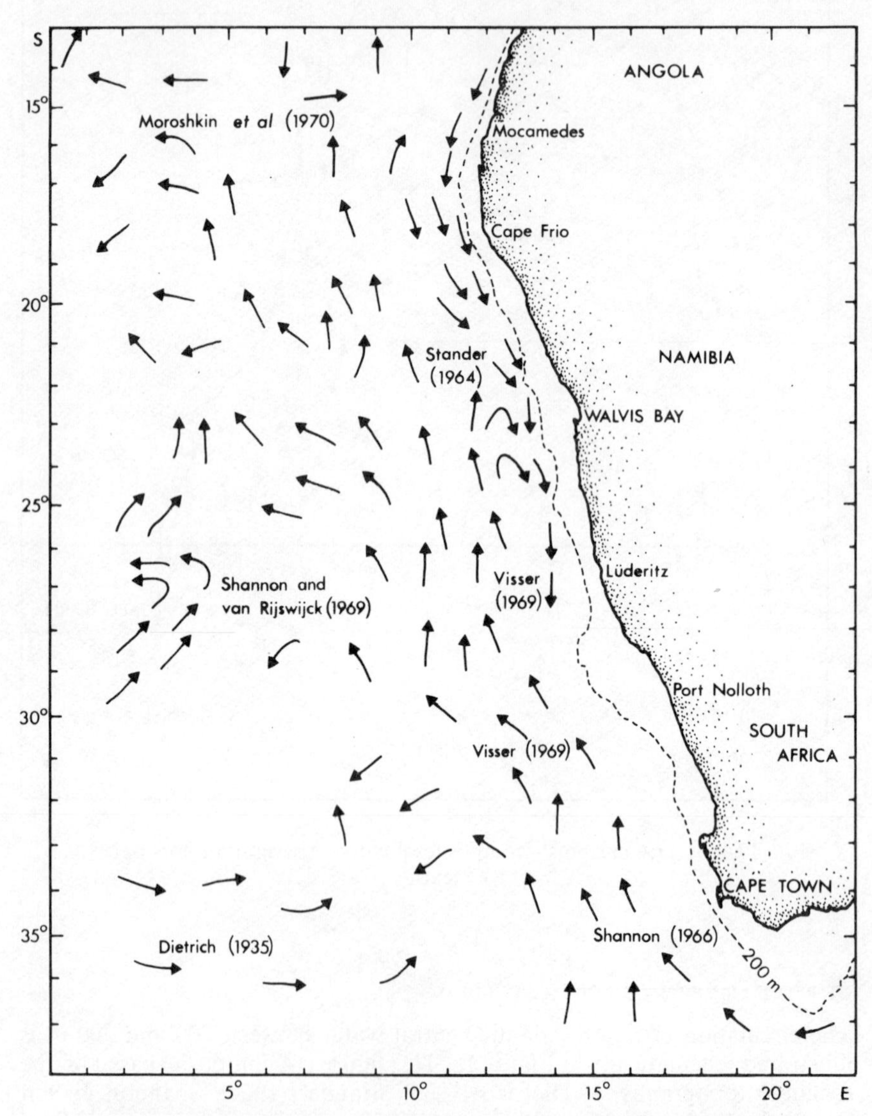

Fig. 18.—Probable movement of central water between 200 m and 300 m—composite based on work of Dietrich (1935), Stander (1964), Shannon (1966), Visser (1969a), Shannon & van Rijswijck (1969), and Moroshkin *et al.* (1970).

is evidently more complex. Moroshkin *et al.*'s (1970) data suggest a predominantly westward meandering movement west of 10° E and south of the anticyclonic gyre which is centred around 13° S: 4° E. Various authors, including *inter alia* Hart & Currie (1960), Stander (1964), Visser (1969a), De Decker (1970), Moroshkin *et al.* (1970), Bailey (1979), and Nelson & Hutchings (1983) have shown the existence of a poleward undercurrent flowing parallel to the coast west of the shelf break and penetrating as far south as Lüderitz (Visser, 1969a) and the Cape of Good Hope (De Decker 1970, Andrews & Hutchings, 1980). This undercurrent is at times characterized by its low dissolved oxygen content—typically $<2\ ml \cdot l^{-1}$, and $<1\ ml \cdot l^{-1}$ at the core—and was postulated by Hart & Currie (1960) as a deep compensation current, which according to Stander (1964) results in the poleward extension of the equatorial oxygen-poor zone. Nelson & Hutchings (1983) have suggested that a weak cyclonic gyral motion which exists in the region with the Walvis Ridge as its northern boundary would tend to trap the oxygen-deficient water. The available literature on the formation and advection of the oxygen-depleted layers is reviewed in Chapman & Shannon (1985).

VOLUME FLUXES

TABLE I

Volume fluxes in the Benguela system

Author	Latitude (°S)	Equatorward flux (Sv)	Lateral input flux (Sv)	Lateral output flux (Sv)	Comments
Defant & Wüst (1938)	25	14·9			Above 1000 db
Defant & Wüst (1938)	28·5	15·7			" "
Sverdrup *et al.* (1942)	30	16			Upper water (above intermediate)
Bang & Andrews (1974)	34	7			Shelf-edge jet only
Bang (1976)	32–34·5	10*	0·5–0·7		*Shelf-edge jet only, at 32·5° S
Carmack & Aagaard (1977)	20–25		1·7+	1·7	+1 Sv Ekman transport plus 0·7 Sv baroclinic transport above 200 m
Dias (1983b)	12	−1·2 to −3·7			Angola system above 400 m relative to 800 db, between coast and 9° E

The equatorward volume transport in the Benguela system appears to be of the order of 15 Sv (1 Sv = 10^6 $m^3 \cdot s^{-1}$), a figure which is comparable with the other major eastern boundary currents (Wooster & Reid, 1963). Bang & Andrews (1974) and Bang (1976) estimated fluxes of 7 Sv and 10 Sv, respectively, for the equatorward shelf-edge jet in the southern Benguela which represents a substantial proportion of the gross transport in the South Atlantic. Nelson & Hutchings (1983) cited recent unpublished work which suggests that in the vicinity of the shelf-edge jet the flux may be locally more intense.

The (few) published estimates of the meridional and lateral volume fluxes in the Benguela system are summarized in Table I.

SEASONAL CHANGES AND INTER-ANNUAL VARIABILITY

The seasonal distribution of temperature and salinity along the western coast has been described by several authors *inter alia*, Clowes (1954), Buys (1957, 1959), Stander (1958, 1963, 1964), Shannon (1966), Schell (1970), Wooster (1973), Christensen (1980), O'Toole (1980), Boyd & Agenbag (1984a), Parrish *et al.* (1983), Stetsjuk (1983), and Strogalev (1983). Readers are also referred to the section on macroscale meteorology.

Christensen (1980) showed the mean monthly surface temperatures around southern Africa south of 25° S from data provided by commercial shipping during the period 1968–1978, while more recently Boyd & Agenbag (1984a) have discussed the seasonal surface structure of the Benguela between 17° S and 34° S and between 37 and 300 km offshore using a similar set of data but from a slightly longer period (1968–1980). Boyd & Agenbag have compared their seasonal (three months per season, *viz.* December–February, *etc.*) maps with those generated from a more extensive data set, but with a coarser grid spacing, by Parrish *et al.* (1984) for the two two-month periods, January to February and July to August. The seasonal averages from Boyd & Agenbag (1984a) are shown in Figure 19. These authors drew attention to the general similarity between the winter and spring distribution with water cooler than 16° C along the entire coast between 18° S and 34° S extending up to 300 km offshore. During summer and autumn (very similar—see Fig. 19) the area of cool water contracts meridionally and zonally. Longshore temperature gradients are weak (1 °C or less per 1° of latitude) compared with the offshore gradients, the latter being strongest in summer and autumn in the area south of Walvis Bay (Boyd & Agenbag, 1984a). North of Walvis Bay the offshore gradients weaken slightly and the isotherms bend more towards the coast. The large-scale seasonal surface temperatures broadly reflect changes in insolation, upwelling, vertical mixing and horizontal advection. They do not, however, show the intense mesoscale coastal upwelling events.

The changes in temperature and salinity in the upper 50 m based on monthly sampling by research vessels in areas off Namibia (21° S–24° S) and off the Cape (32° S–33° S) have been described by Buys (1957, 1959), Stander (1958, 1963), Stander & De Decker (1969) and Boyd & Agenbag (1984a,b). The

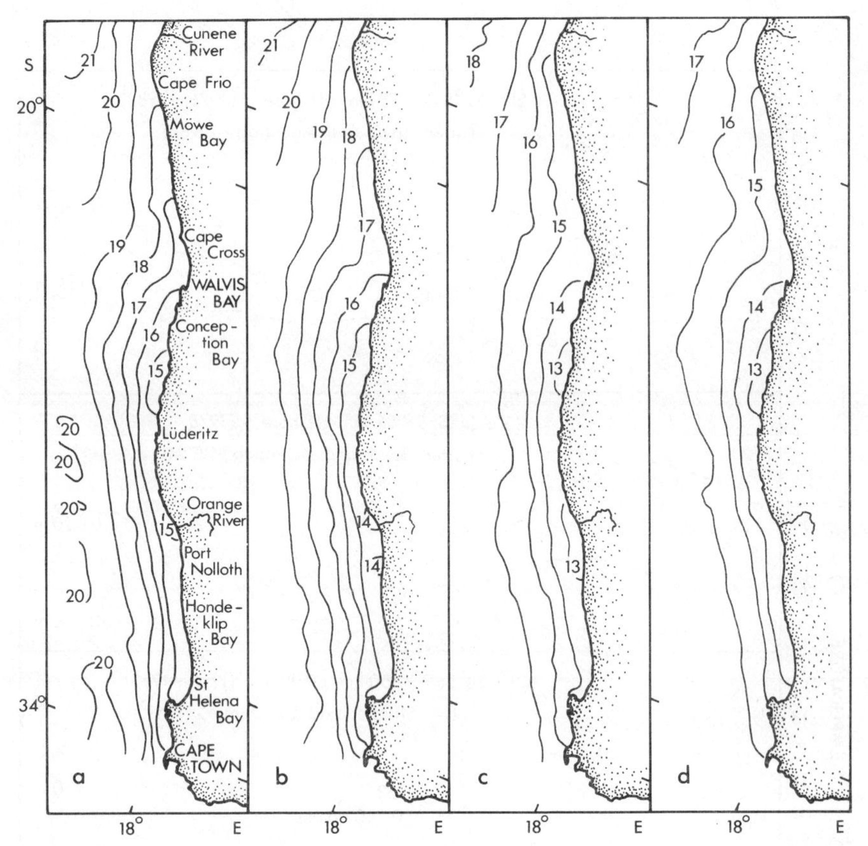

Fig. 19.—Average seasonal sea surface temperature (°C) (after Boyd & Agenbag, 1984a): a, summer; b, autumn; c, winter; d, spring.

monthly variation in temperature in the two regions is shown in Figure 20, and indicates lowest values during August and close agreement between the 20 m and 0–50 m averaged temperatures. The seasonal signal appears to be more pronounced off Namibia than off the Cape where, excluding May, there is little change in the temperature at 20 m and 0–50 m during the year. Both Namibia and the Cape curves show peak temperatures during late summer to autumn with a two-month lag between the two regions (March and May, respectively).

Off Namibia the salinity tends to follow the temperature curve but with a lag of about one to two months (see Stander & De Decker, 1969), whereas off the Cape the situation is more complex. (A note of caution: the area between 32° S and 33° S encompasses a strong baroclinic frontal zone and is highly variable. The variability off the Cape is evident from the time series of monthly measurements made by Andrews & Hutchings, 1980, near Cape Town—see Fig. 21.)

Two studies on the occurrence of thermoclines in the Benguela system have been made, one by Duncan (1964) and the other by du Plessis (1967) for the

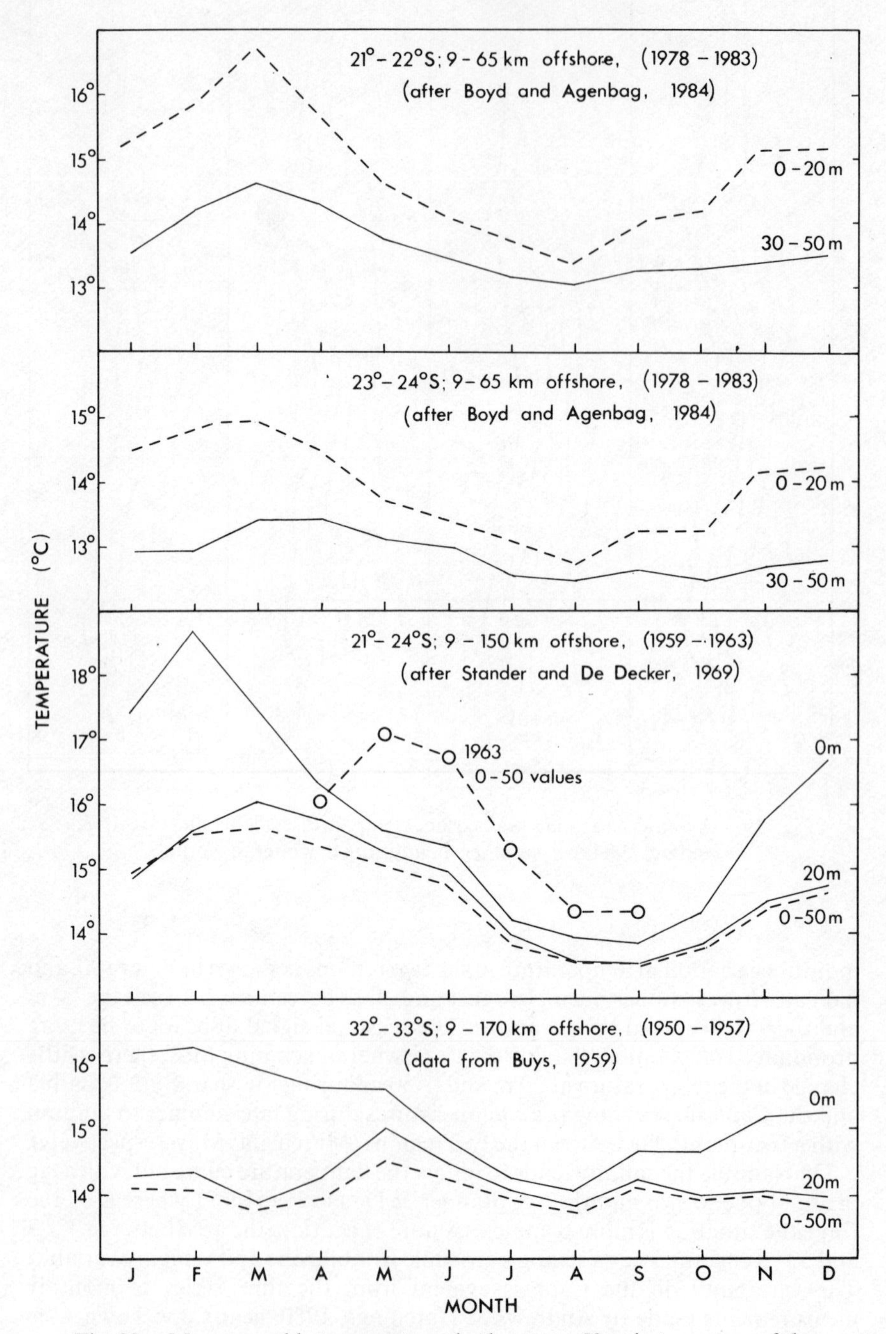

Fig. 20.—Mean monthly temperatures in the upper 50 m in two areas of the Benguela system.

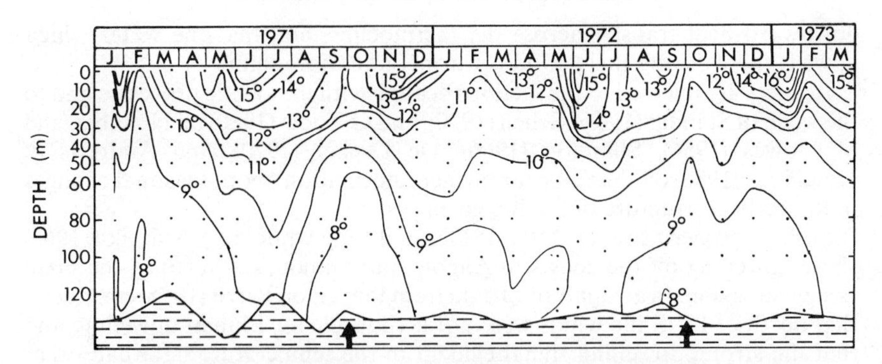

Fig. 21.—Monthly temperature distribution 25 km west of Cape Town showing intrusions of cold water near the bottom during September–October (after Andrews & Hutchings, 1980).

areas between 32° S and 33° S and 21° S and 24° S, respectively. The results are summarized in Table II. Off Namibia thermoclines tended to be shallower and except during summer were less frequent than in the south, and were associated on the average with slightly warmer less stratified water. Boyd & Agenbag (1984a) have suggested that the warming of the layer below 30 m off Namibia during late summer to early autumn is not primarily caused by

TABLE II

*Occurrence of thermoclines at two sites in the Benguela system: *estimated from diagrams*

Area, author, etc.	Season	Frequency occurrence (%)	Depth of top of thermocline (m)	Temperature at top of thermocline (°C)	Temperature drop across thermocline (°C)
21° S–24° S, 9–150 km offshore	Summer	86	10	17·5	3·1
1959–1965,	Autumn	52	22	16·8	2·2
from du Plessis	Winter	9	29	15·4	1·5
(1967)	Spring	24	8	15·5	2·0
32° S–33° S 9–170 km	Summer	87	17	15·6*	4·2*
offshore 1955–1961,	Autumn	87	22	15·0*	4·0*
from Duncan	Winter	58	39	13·8*	3·0*
(1964)	Spring	60	33	14·2*	3·2*

downward heat transfer across the thermocline, and that the water which upwells there during this period is warmer and more saline.

Readers are referred to the sections dealing with mesoscale processes and to the work of Stander (1964), Schell (1970), and O'Toole (1980) off Namibia, and of Clowes (1954), Shannon (1966), De Decker (1970), and Andrews & Hutchings (1980) off the Cape for further information about seasonal changes in the vertical structure of the Benguela system.

Surface isotherms in the Atlantic Ocean during winter (*e.g.* Mazeika, 1968) show upwelling off the coasts of Angola and Gabon, *i.e.* north of the main Benguela region. In an analysis of data from the region Berrit (1976) concluded that north of 15° S the winds were not favourable for Ekman upwelling and that the strong upwelling signal evident in the temperature data had other causes. Upwelling occurs in the eastern equatorial region of the Atlantic and along the Gulf of Guinea coast during the austral winter, and Moore *et al.* (1978) attributed part of this upwelling to an internal Kelvin wave generated by increased easterly winds off northern Brazil. This wave travels along the equatorial wave guide and reflects at the African coast polewards and westwards in the form of coastal Kelvin and Rossby waves. Subsequent studies by Picaut (1981) have indicated that the wave propagates polewards between 1° S and 13° S with a phase velocity of about $0{\cdot}7\ m{\cdot}s^{-1}$ and then propagates further south, but with some distortion. If Picaut's results are extrapolated then it would imply that the wave could reach Namibia during August and Cape Town (34° S) about a month or two (allowing for changes in stratification) later. Although there is no definite proof that this does happen, the data below 30 m in Boyd & Agenbag (1984a) and the appearance on the shelf near Cape Town of cold (<8 °C) water during September to October (refer to Fig. 21—from Andrews & Hutchings, 1980) seem to support the wave concept. A comparable rate of progression of a "warm pulse" could possibly be inferred from a comparison of the diagrams presented by Berrit (1976) and Figure 20. It should also be noted that a warm pulse is evident off Abijan during April which is approximately the same time as the Benguela subsurface temperature maximum. Furthermore, Hirst & Hastenrath (1983) postulated a link between the (austral) summer relaxation of the westward wind stress in the equatorial western Atlantic, the subsequent warm pulse off Angola and the Angolan rainy season (March to April). It is tempting to speculate whether the upwelling system throughout the Benguela is not perhaps 'primed' by this coastal trapped Kelvin wave which then facilitates Ekman upwelling. The lack of a strong seasonal wind signal off Namibia and the oceanographic data from both the Cape and Namibia seem to support the idea of priming. If the Benguela upwelling is primed and then terminated by Kelvin waves, then this implies that a major causative mechanism of the inter-annual variability of the system would have its origin in the equatorial region of the Atlantic. Variations due to changes in the local wind field would be superimposed on this.

Very little has been published on the inter-annual variability in the Benguela system, and it is evident from what literature is available that long term records, where they exist, have yet to be adequately analysed. That perturbations occur on the time scale of years and decades seems probable from the biological record (Shannon, Crawford & Duffy, 1984), but the inherent short term and spatial variability of the system has presented problems for monitoring. Buys (1957, 1959), Stander (1958, 1963), Stander & De Decker (1969), Strogalev (1983), and Boyd & Agenbag (1984b) have all analysed relatively short oceanographic sets of data in the northern and southern Benguela regions and have suggested certain trends. In the Cape, the work of Buys (1959) indicated that for the period 1950 to 1957 there was a cool interval during 1954 to 1955 and 1955 to 1956, and he suggested the possibility of a seven-year cycle, a view supported by Stander (1963) in his analysis of the 1954 to 1961 Namibian records, where a cool period was noted during 1957 to 1958. Strogalev's (1983) data indicated that 1972 to 1974 and 1976 were warmer than normal years off central Namibia. The recent study of the 1982 to 1983 southern Benguela warm event (reported in *S. Afr. J. Sci.*, Vol. 80, No. 2, February 1984) has highlighted the inter-annual variability in the system since 1950. Walker, Taunton-Clark & Pugh (1984), see Table III, reported on various sea surface temperature data for the southern Benguela for the post 1956 period and concluded that, while warm events in this region corresponding to Pacific events peaked during the summer, they had a distinctly different character to the El Niño, being related to variations in the wind field rather than to the advection of warm water into the system from the north. These authors showed that the early 1960s were 1 to 2 °C warmer than average years with 1963 (see last paragraph) being the warmest. 1967, 1971, and 1979 were cool years in the southern region and Walker *et al.* (1984) indicated a possible relationship between the position of the Subtropical Convergence and local sea temperatures (see also Gillooly & Walker, 1984). While the summer of 1982 to 1983 was abnormal in the southern Benguela with respect to temperature (Walker *et al.*, 1984; Duffy, Berruti, Randall & Cooper, 1984; Gillooly & Walker, 1984), wind (Nelson & Walker, 1984; Schulze, 1984; Hutchings, Holden & Mitchell-Innes, 1984), sea level (Brundrit, de Cuevas & Shipley, 1984), and biological characteristics (Shannon, Crawford & Duffy, 1984; Shannon, Brundrit *et al.*, 1984; Branch, 1984; Duffy *et al.*, 1984; Shannon & Chapman, 1983b; Hutchings *et al.*, 1984), off Namibia the work of Boyd & Agenbag (1984b) has shown no significant anomaly, although the preceding autumn and winter (1982) were characterized by cooler than usual conditions.

Probably the most satisfactory approach to the question of inter-annual variability is the examination of long-term tidal records. Recent studies on sea level, adjusted for atmospheric pressure and tides, by Brundrit, Shipley, de Cuevas & Brundrit (1983) and Brundrit *et al.* (1984) have shown coherency in the monthly records from nine sites along the coast (22°57′ S to 34°35′ S), with the inter-annual contribution, which has a very large spatial structure, revealing the same trend at each site. Their results suggested a decline in sea level since 1979 with 1982 showing a lower level in the inter-annual cycle. Anomalously high values were observed by Brundrit *et al.* (1984) at the southern sites during the 1982 to 1983 spring and summer. On the longer term, Brundrit *et al.* (1983) showed peaks during 1963 and 1968 to 1969 and troughs during 1965 to 1966 and 1971 (Table III).

Since the early 1950s, although there have been several warm and cool periods in the Benguela, only two events approximating to major El Niño type situations have occurred *viz.* in 1963 and 1984. During the 1963 event temperatures 2–4 °C and salinities 0·1–0·2‰ above normal were recorded in the upper 50 m off Namibia (Stander & De Decker, 1969). These authors noted that the southward intrusion of warm saline Angolan water was not accom-

TABLE III

*Comparison of upwelling indices for the southern Benguela region: *, 1960s were generally warm; ‡, units of relative wind displacement*

Year	Upwelling index/reference: Sea temperature — Shannon (1976) St Helena Bay area	Sea temperature — Walker *et al.* (1984) Table Bay & west coast*	Wind — Nelson & Hutchings (1983)	Wind — Nelson & Walker (1984)	Sea level — Brundrit *et al.* (1983, 1984)
1953–1954	Warm				
1955–1956	Cool				
1957	Warm	Warm			
1958	Cool				
1959	}				Low (2nd half of 1959)
1960	}	Warm			
1963	Warm }	} Very warm			High (mid-1963)
1964	Warm }	}			
1966			Low (14)‡		Low (also 1965)
1967		Cool			
1968					High
1969			Low (18)‡		High
1970		Cool			
1971		Cool			Low (early 1971)
1972		Warm			
1973		Slightly warm			
1974		Warm }	1974–1976		
1975		}	High (22)‡		
1976–1977		Summer warm }			
1978		Cool			
1979		Cool			
1982–1983		Warm		Abnormally low	Low in 1982, summer of 1982 to 1983 anomalous

panied by a decrease in upwelling favourable winds. A recent event of similar magnitude occurred during the late summer of 1984 (Boyd & Thomas, 1984). The effect of the 1963 event was felt subsequently in the Cape (Brundrit *et al.* 1983, and Table III). From data presented by Walter (1937), it is evident that events of similar or greater magnitude occurred during 1934, when monthly mean sea temperatures were 2–3 °C above the long-term average from March through July at Swakopmund. The anomaly was accompanied by a reversal or slackening of the usual northerly flow of the Benguela, and the flood waters from the Orange River were reported as moving southwards instead of northwards. Although not yet confirmed by the physical data, biological records (*e.g.* Shannon, Crawford & Duffy, 1984), suggest that Benguela El Niño may also have occurred in 1950 to 1951 and around the turn of the century. These perturbations do, however, seem to be less frequent in the Benguela system, than in the eastern Pacific.

THE NAMAQUA–LÜDERITZ UPWELLING AREA

The principal upwelling centre of the Benguela, as borne out by the work of several authors, *inter alia* Copenhagen (1953), Stander (1964), Boyd & Agenbag (1984a), Parrish *et al.* (1983), and Stetsjuk (1983) is in the vicinity of Lüderitz (27° S)—approximately equidistant from the northern and southern boundaries of the system. Defant (1936) noted that the zone of greatest negative surface temperature anomaly was situated between 23° S and 31° S an observation which, although based on relatively few data, has been substantiated by subsequent analyses by Wooster (1973), Parrish *et al.* (1984) and Stetsjuk (1983). Copenhagen (1953) identified on the basis of temperature and bathymetry three main centres of upwelling in the Benguela, *viz.* Lüderitz, Saldanha Bay, and Cape Point with secondary centres at Hondeklip Bay and Walvis Bay. Boyd & Cruickshank (1983) indicated maximum negative surface temperature anomalies 37 km offshore at 25° S and 29° S, which corresponded to the positions of two cool tongues noted by Hart & Currie (1960) on both their cruises, by Stander (1964), and by Bang (1971). These tongues which are evidently related to both the bathymetry and the wind field have as their bases Lüderitz and Hondeklip Bay and, together comprise a major environmental barrier in the Benguela, effectively dividing the system into two. Relatively few oceanographic stations have been occupied in the central Benguela, probably because most of the research has been focused on the important pelagic fishing areas which lie to the north and to the south of the 'cold' region. In the following paragraphs published work relating to mesoscale upwelling processes in the Namaqua (28° S–31° S) and Lüderitz (24° S–28° S) zones is discussed.

NAMAQUA ZONE

A cool wedge-shaped zone extending northwards and broadening from Hondeklip Bay to the Orange bight is evident from satellite thermal infrared imagery, while Stander (1964) and Shannon (1966) have shown the presence of cold water in the bight during most seasons. Nelson & Hutchings (1983) and Taunton-Clark (in press) considered that this tongue was largely determined

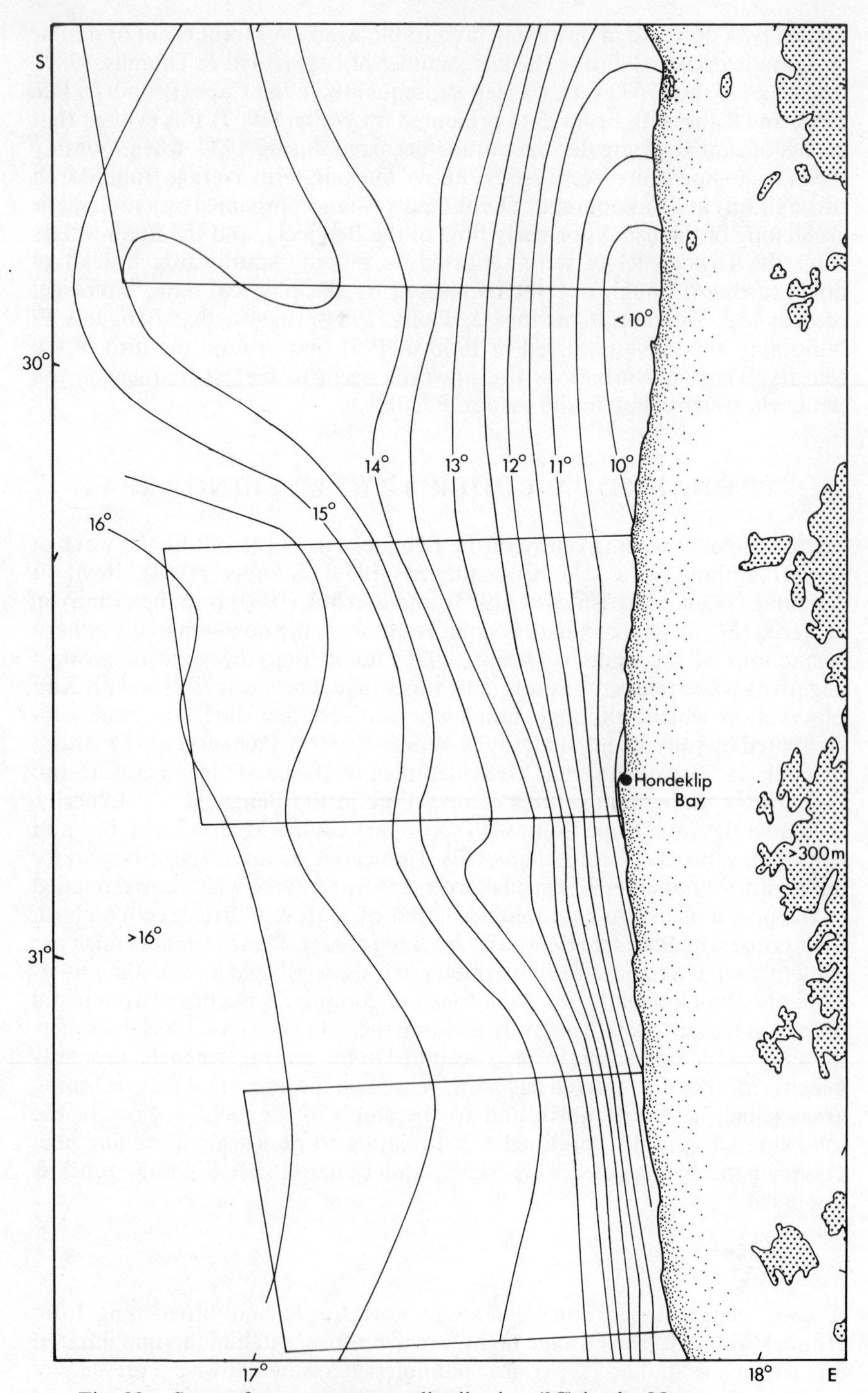

Fig. 22.—Sea surface temperature distribution (°C) in the Namaqua zone, 10th November 1980 (from Taunton-Clark, in press).

by the bathymetry. Southwest of Hondeklip Bay the shelf is narrow and deep, and in this region cold water would be readily available close inshore. Further north the shelf broadens and the inner shelf is marked by a progressive shallowing towards the coast (Taunton-Clark, in press). Moreover the mean orientation of the coast changes slightly at Hondeklip Bay. The progressive northward shoaling of the deeper (8–12 °C) isotherms over the mid-shelf between 31° S and 29° S is evident from Bang's (1971) sequence of sections. Winds in the area are predominantly longshore with a diurnal sea-breeze modulation and a strong perturbation as coastal lows pass through and the South Atlantic high weakens (Nelson & Hutchings, 1983). Nelson, Kamstra & Walker (1983) have examined the mean annual winds and surface temperature on a half degree rectangle grid and have shown the existence of an area of maximum negative wind stress curl adjacent to the coast at Hondeklip Bay, with the coolest water to the north in the Orange bight. Taunton-Clark (in press) has recently reported on the results of a series of meteorological and surface temperature measurements made on aerial surveys of the Namaqua zone during the last quarter of 1980. Figure 22 illustrates what this author described as a typically developed upwelling tongue. Taunton-Clark (in press) recorded a wind speed maximum offshore, with a local maximum northwest of Hondeklip Bay coinciding with the plume base and the local thermal low pressure area. Lowest temperatures were recorded north of Hondeklip Bay and the author felt that the upwelling response in the area was markedly slower than off the Cape Peninsula. Drift data in Clowes (1954) is consistent with the configuration of the Namaqua upwelling tongue.

Although Stander (1964) and Shannon (1966) found no evidence to support the classical cellular structure (Fig. 23) proposed by Hart & Currie (1960),

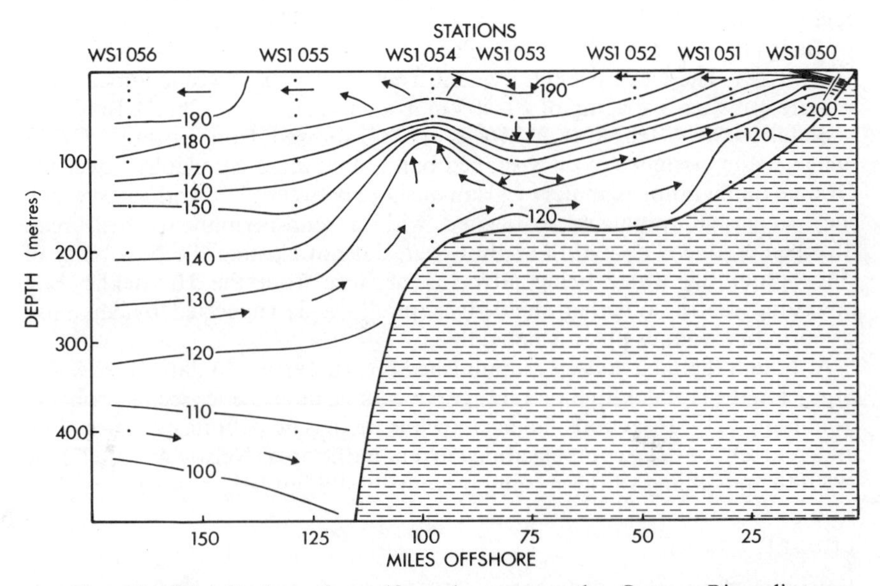

Fig. 23.—Distribution of specific volume anomaly, Orange River line, September 1950, showing cellular structure (from Hart & Currie, 1960).

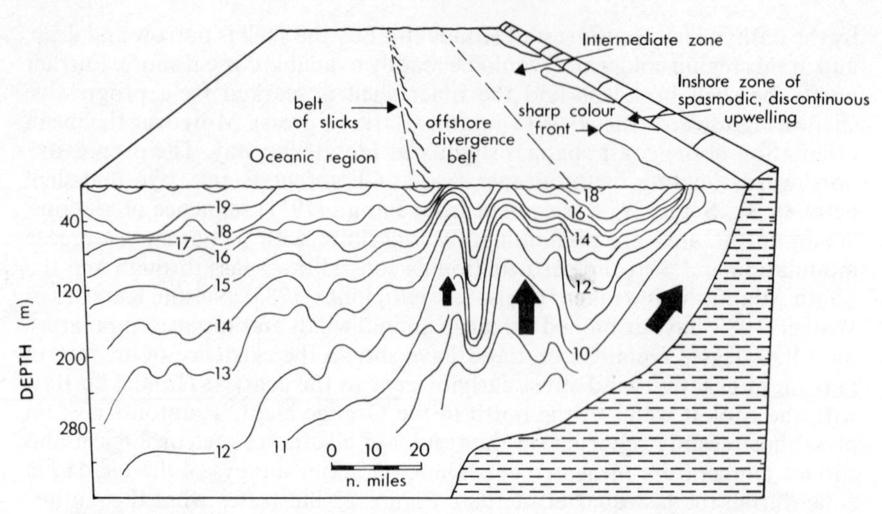

Fig. 24.—Main structural features of the Benguela region between 29° S and 32° S identified by Bang (1971).

probably on account of their wide station separations, Bang's (1971) vertical temperature sections of the area between 29° S and 31° S during February 1966 show marked wave-like features over the outer shelf and shelf break, which he suggested were indicative of surface divergence. This author's annotated schematic representation of the main structural features of the Benguela between 29° S and 32° S is shown in Figure 24. The presence of surface slicks, noted by Bang (1971) over the shelf break, and which were probably related to internal waves, were subsequently observed in LANDSAT MSS imagery of the region between 30° S and 32° S by Apel *et al.* (1975). The slick spacing recorded by Bang (1971) *viz.* 1–2 km, was similar to that of Apel *et al.* (1975) offshore, *viz.* 1·6–1·8 km. The wave packets evident from the LANDSAT scene appeared to radiate out with a spacing of 20–40 km from a source near Child's Bank.

Shannon, Schlittenhardt & Mostert (1984) showed, from NIMBUS-7 CZCS imagery, the existence of an S-shaped band of elevated chlorophyll over the Orange Banks (approximately 150 km offshore between 28° S and 30° S) which they suggested might be associated with a semi-permanent shelf-break divergence zone. These authors, however, did not exclude the possibility of chlorophyll-rich water being advected offshore from the Hondeklip Bay upwelling centre. Whether shelf-edge upwelling, as suggested by Shannon (1966) actually takes place is debatable.

Although there have been no published measurements to date of the shelf-edge jet in the Hondeklip Bay–Orange River area, its existence seems probable from Bang's (1971) sections. Likewise, while there is no definite evidence of an undercurrent or deep compensation current in the area (Nelson & Hutchings, 1983) the possibility of these being present should not be excluded.

LÜDERITZ ZONE

Following the work of Copenhagen (1953), Currie (1953), and Hart & Currie (1960) which identified the region around Lüderitz as an important upwelling site the subsequent studies by Stander (1964), Bailey (1979) and Stetsjuk (1983) demarcated the spatial and seasonal extent of upwelling in the area. According to Stander (1964), "Low surface temperatures are most conspicuous between 26° and 28° S. It would undoubtedly appear that the area from the Orange River mouth to Lüderitz is a region where upwelling occurs repeatedly, probably more frequently and intensely than elsewhere along this coast." Stander's (1964) work which was based on nine quarterly surveys (1959 to 1961) showed persistent upwelling throughout the year with a slight maximum in spring and a minimum during autumn, which is in agreement with the trend evident in Bailey's (1979) analysis of the mean monthly upwelling favourable winds at Lüderitz from 1971 through 1977. Bailey described the wind field in some detail and showed that the southerly coastal winds were fairly consistent throughout the year with a tendency towards a maximum during the last quarter and a minimum (half to two-thirds of average speed) between May and July. (The wind speed maximum is situated 50 km or more offshore.) This is in general agreement with Stetsjuk's (1983) "cold advection" values which indicate minimum upwelling in the region 24° S to 28° S between July and August and maximum upwelling between October and December. Bailey (1979) also found good agreement between the monthly average sea-surface temperatures within 80 km of the coast between 24° S and 29° S and the winds at Lüderitz from 1969 through 1977, and he considered that the upwelling event scale for the Lüderitz zone was relatively long—more analogous to the Peru and North West African situations rather than to those off Oregon and in the southern Benguela. The sea temperature and wind speed records from Walter (1937) for Lüderitz during November 1927, which showed a good inverse correlation between the two variables, suggest, however, a fairly rapid response (Fig. 25). Somewhat curiously Bailey (1979) observed that during certain years in January to March warmer water appeared following stronger southerly winds (recorded at Lüderitz) in December to January. This response may be due to the large eddy, centred around 26° S to 27° S, which is characteristic of the Lüderitz zone upwelling, noted by Hart & Currie (1960), Stander (1964), Bang (1971), and Bailey (1979) and which is associated with a southward flow north of 25° S and an eastward intrusion of oceanic water between 27° S and 28° S, possibly as some form of compensation.

The temperature and salinity distribution at 0, 200, and 400 m off Namibia during January 1960 is shown in Figure 26 (from Stander 1964), and the spatial scale and impact of the Lüderitz upwelling site on the Benguela system is immediately evident as is the convergence zone between the tongue and the northern Namibian regime at 22° S to 24° S. Nelson & Hutchings (1983) have suggested that the coherency in the tongues of water between 200 and 400 m moving slightly offshore during the summer, points to the possibility of wind-induced upwelling being enhanced in the area by the bottom topography. Stander commented that cold water (7–10·5 °C) was consistently present in the region adjacent to the shelf at these depths. As in the Namaqua zone, the availability of this cold water near to the coast between 27° S and 28° S (the shelf is deep here with the shelf break at about 500 m—see Fig. 5, p. 114) coupled with the orientation of the coast and the shelf break appear to be important considerations for the upwelling dynamics.

Although unfortunately Stander (1964) did not show vertical profiles along his Lüderitz line, seasonal mean vertical sections from the same set of data are

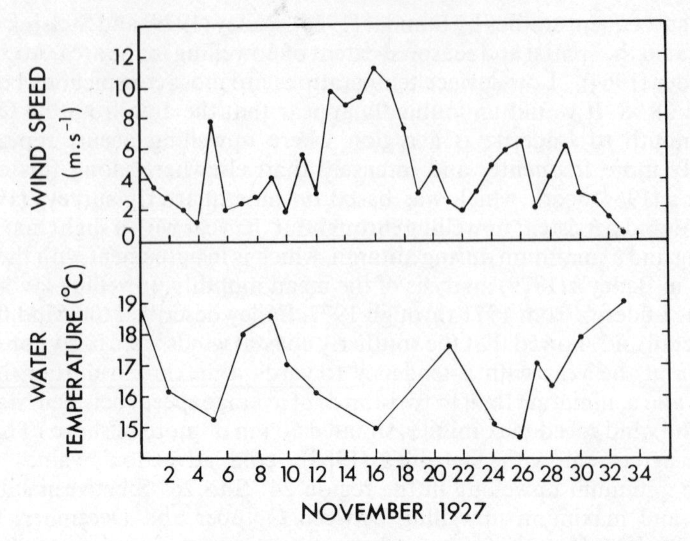

Fig. 25.—Variations in wind speed and water temperature at Lüderitz during November 1927 (after Walter, 1937).

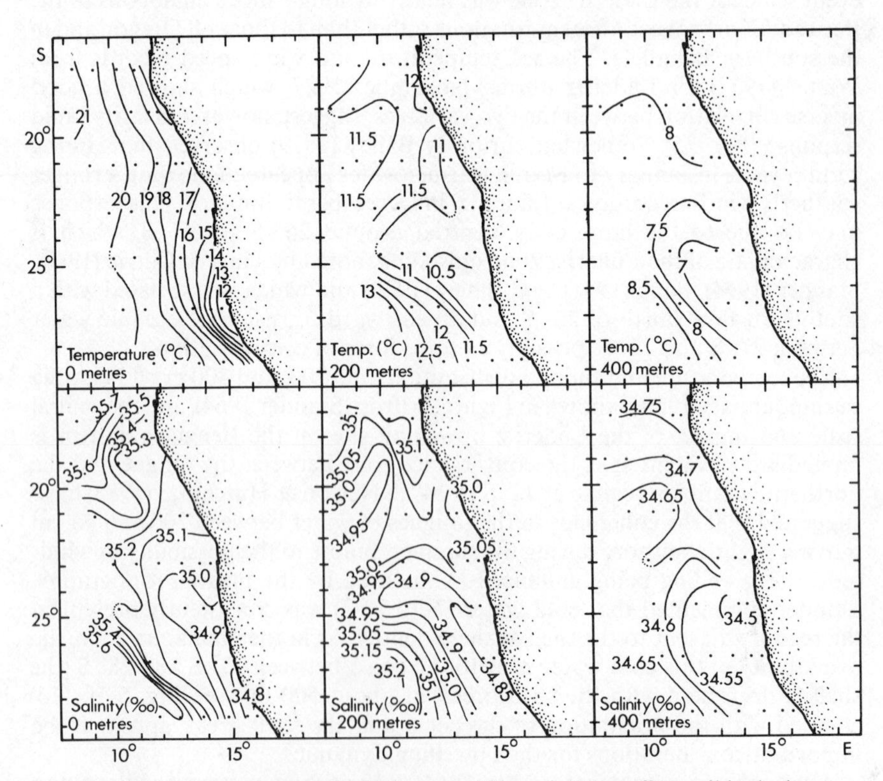

Fig. 26.—Temperature and salinity at 0, 200, and 400 m off Namibia during January 1960 (after Stander, 1964).

available in Schell (1970) and these confirm the pronounced uplift of the isolines over the shelf. Hart & Currie (1960) and Stander (1964) found that the upwelled water off southern Namibia originated from a depth of 200–300 m: Calvert & Price (1971b) have quoted a depth of 220 m for their Sylvia Hill line (25° S—their southernmost line and one on which maximum active upwelling was noted) during October 1968, while Bailey (1979) recorded depths of 300 and 330 m for lines off Chamais Bay (27°56′ S) and Marshall Rocks (26°21′ S), respectively, during an upwelling event in November 1976. It should be noted also that Latun (1962), using a two-layer steady state model, obtained good agreement with Hart & Currie's (1960) observations of upwelling on their Orange River line. He found strongest upwelling ($3{\cdot}5$ to $4{\cdot}7 \times 10^{-4}$ cm·s^{-1}) close to the shelf break and that water could not rise to the surface layer from levels below 325 m. Bailey (1979) has compared this maximum upwelling

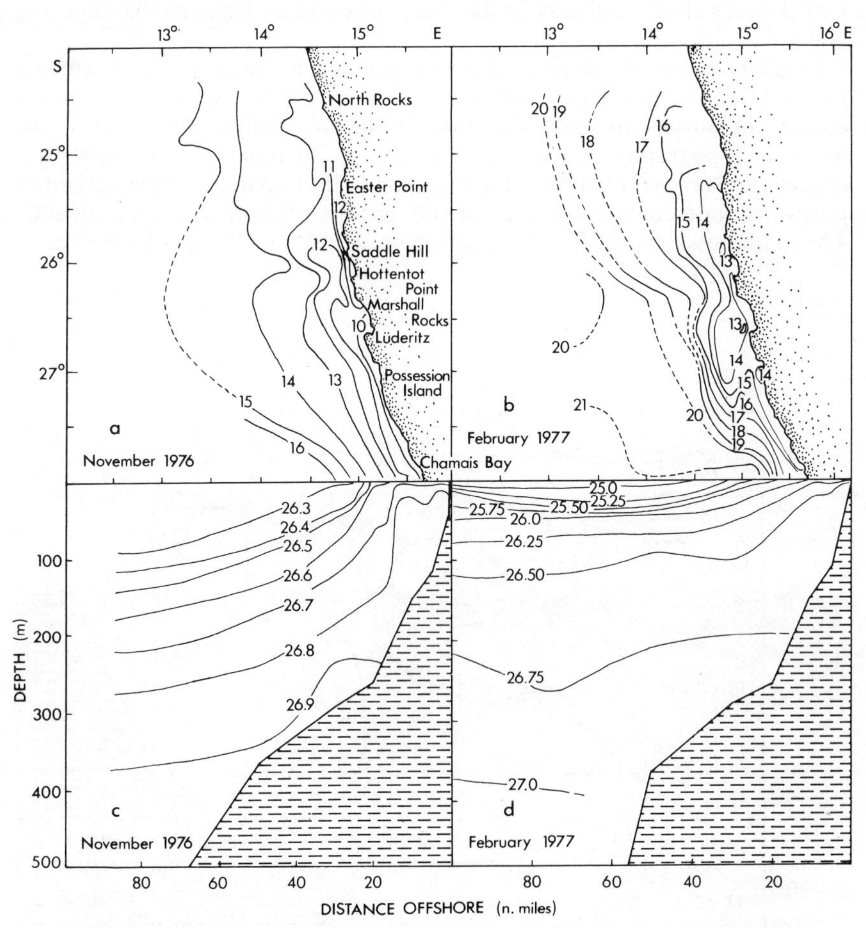

Fig. 27.—Surface temperature (°C) distribution in Lüderitz zone and sigma-*t* profile along the Marshall Rocks line (26°21′ S) during upwelling period (a,c) and quiescent period (b,d) (from Bailey, 1979).

condition with the quiescent case (February 1977) and his surface temperature charts of the area and vertical sigma-*t* profiles along the Marshall Rocks line for these two extreme cases are shown dramatically in Figure 27.

The only study of the response of the Lüderitz system to a mesoscale wind event is that by Bang (1971) who illustrated the effect of 3–6 days of gale force southeasterly wind. (His dramatic diagram is shown in Figure 28.) Bang noted that the mixing effects of the gale were smaller than anticipated, the thermocline depth having increased by 5–15 m. The 16 °C isotherm was displaced up to 80 km seawards over a few days, and temperature changes of 0·5–1 °C well below the thermocline at depths between 100 and 250 m were evident. Bang postulated the existence of a "pivot" or point of little change in the system at about 27°40′ S: 14°55′ E from his and Stander's (1964) data. Bang felt that, while upwelling occurs to the north and to the south, the vicinity of the pivot was one of comparative stability. The position of the pivot coincides with the central Benguela "environmental basin" proposed by Boyd & Cruickshank (1983) and may be the effective boundary between the Cape and Namibian subsystems.

Evidence for the existence of a deep compensation—poleward undercurrent in the Lüderitz zone was examined by Bailey (1979). Although no direct current measurements have been made in the area, Bailey deduced from the dynamic topography of the region (*e.g.* Fig. 29) together with isentropic analysis and an examination of the oxygen distribution that, during autumn, a poleward undercurrent was present 100–150 km offshore at a depth of 200–400 m, and over the shelf in the south between 50 and 100 m. Visser (1969a)

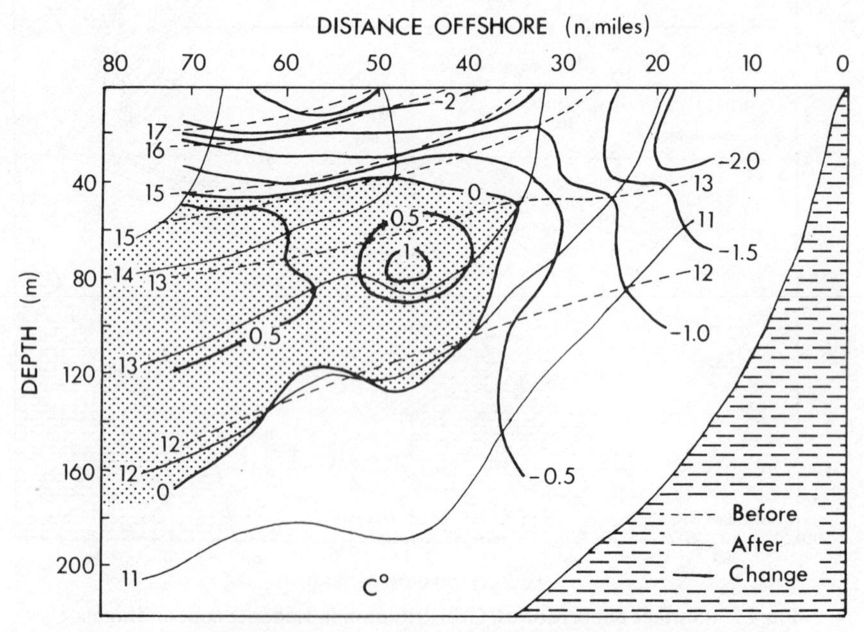

Fig. 28.—Temperature structure (°C) at 26° S (from Bang, 1971) before and after a southeasterly gale during February 1966.

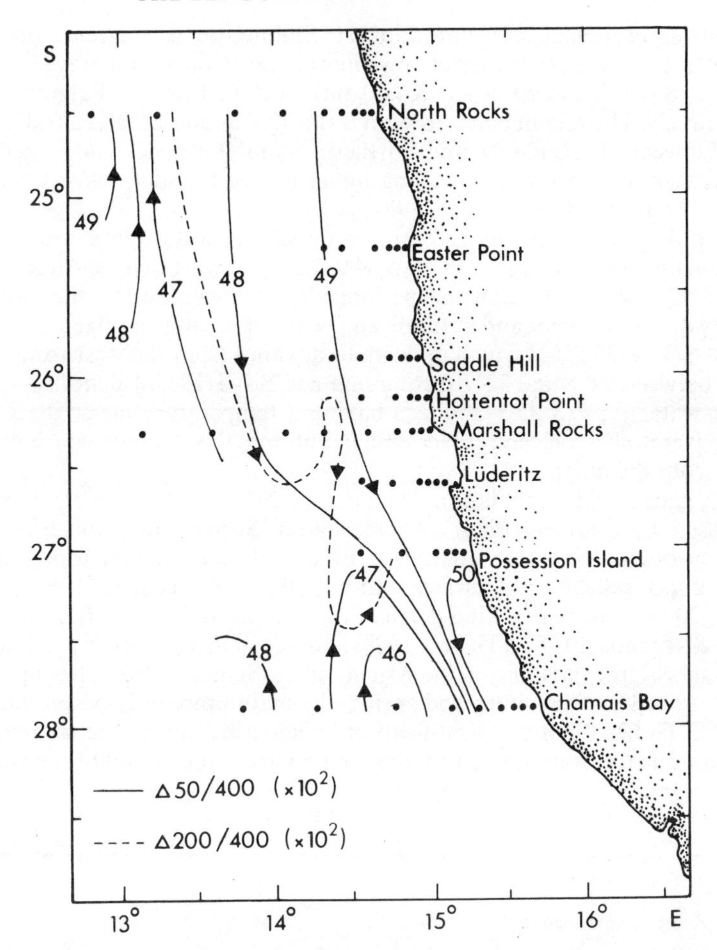

Fig. 29.—Water movement inferred from dynamic topography, May 1976 (from Bailey, 1979).

also showed an undercurrent 100–150 km west of Lüderitz at a depth of about 100 m during July 1959. The fact that it has not been detected during other seasons should not exclude it as a permanent feature, but may rather be symptomatic of the station spacings and measurement techniques.

MESOSCALE PROCESSES IN THE NORTHERN BENGUELA

The literature on the northern Benguela region is mainly of a descriptive nature (*e.g.* Currie, 1953; Hart & Currie, 1960; Stander, 1964; Calvert & Price, 1971b) and relatively little is known about the mesoscale dynamics off central and northern Namibia. Reference to the wind field and bathymetry and the work of Hart & Currie (1960), Stander (1964), O'Toole (1980), Boyd (1983a),

Parrish *et al.* (1983) and Van Foreest, Shillington & Legeckis (in press) suggests that there are centres of upwelling around Conception Bay (24° S) and near 18° S–19° S and 20° S–21° S, *i.e.* south of Cape Frio and Palgrave Point, respectively. The region between Walvis Bay (23° S) and 21° S is a transitional area between the Lüderitz and northern Namibian zones and is generally characterized by a lower upwelling intensity (see Stander, 1964; Calvert & Price, 1971b; Boyd & Agenbag, 1984a).

O'Toole (1980) characterized three main surface water types in the region, *viz.* cool upwelled water (12–18 °C, *S* 34·9–35·2$^\circ/_{oo}$), warm, saline Angola water (17–22 °C, *S* 35·5–35·9$^\circ/_{oo}$) which periodically advances towards the southeast, mainly during summer and autumn, and water of oceanic or mixed origin (16–20 °C, *S* 35·2–35·5$^\circ/_{oo}$) which appears to advance from the west towards the coast between 19° S and 22° S during summer. Boyd (1983a) identified a fourth saline water type (*S* 35·3–35·5$^\circ/_{oo}$) having a temperature lower than 15 °C which is upwelled off central Namibia in autumn, the origin of which appears to be from the north.

The geostrophic circulation during nine cruises off Namibia has been discussed by Stander (1964) in some detail. Surface and subsurface flow patterns often differ substantially, *e.g.* Figure 30. The dynamic topography is not a good indication of surface currents (Boyd & Agenbag, 1984a) as the upper 20 m is primarily wind driven (Moroshkin, Bubnov & Bulatov, 1970; Boyd & Agenbag, 1984a; Hagen, 1984) and it is in this surface layer that most of the short-term variability in the system occurs (Stander, 1963). The upper 50 m is well mixed during winter and spring, the main upwelling season (Stander, 1964; O'Toole, 1980) but stratification is increased during the summer and autumn due to insolation, advection, and a partial relaxation of the wind.

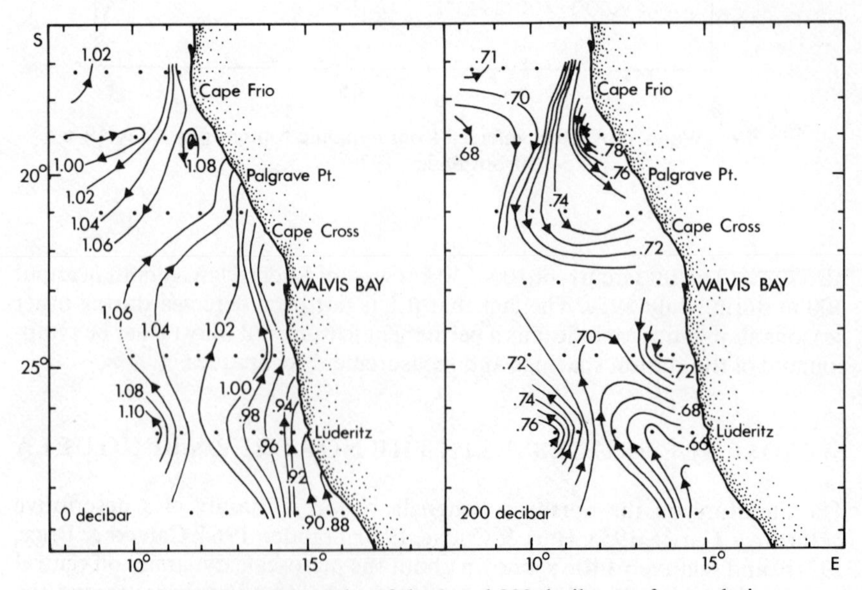

Fig. 30.—Dynamic topography of the 0 and 200 decibar surfaces relative to 1000 decibars, October 1959 (from Stander, 1964).

Except near the northern boundary of the Benguela or during wind reversals, the surface currents are predominantly longshore towards the northwest. Drogue studies by Boyd & Agenbag (1984a) at 10 m depth 46 km offshore (approximately over the 200 m isobath) at 21° S and 24° S showed relatively consistent motion to the northwest (10–30 $cm{\cdot}s^{-1}$) in accordance with the prevailing winds. These authors suggested that this flow probably extended as far as about 18° S where the drogues indicated a pronounced offshore movement near to the Angola front. The work of Stander (1964), Yelizarov (1967), and Filippov & Kolesnikov (1971) revealed a fairly complex system of currents between 50 and 300 m. Yelizarov (1967) showed pronounced poleward flow west of the shelf break between 20° S and 28° S and he postulated the existence of a mesoscale anticyclonic gyre centred around 20°30′ S: 12°30′ E. Several of Stander's (1964) diagrams indicate two large eddies, one centred around Lüderitz and the other around 19° S to 20° S, with eastward flow between 21° S and 23° S (see also Fig. 30). The latter feature is difficult to explain—the area between Walvis Bay and Cape Cross is not a main upwelling site—unless a narrow jet is present over the shelf break. Nelson & Hutchings (1983) postulated the existence of a large cyclonic gyre between 18° S and 25° S, with poleward flow near the shelf break while Stander (1964) indicated a subsurface cyclonic gyre of smaller dimensions. Whether or not this gyre extends to the surface is, however, a matter of speculation, as is its permanence. That there is a perennial poleward jet or subsurface compensation current extending down to 300–400 m in the northern Benguela region, often but not always associated with low oxygen water, seems to be well established (Hart & Currie, 1960; Stander, 1964; Yelizarov, 1967; De Decker, 1970; Moroshkin *et al.*, 1970). The sediment texture and composition maps (Bremner, 1981; also Fig. 6, p. 115) show the existence of a narrow band of relatively coarse material deficient in organic carbon on the outer part of the inner shelf (*i.e.* west of the diatomaceous mud belt) extending between Cape Frio and Walvis Bay, and which may be associated with a subsurface jet. In this respect Hagen (1979) and Hagen *et al.* (1981) have shown evidence of an equatorward jet in the upper 100 m over or near the upper shelf break at 20° S. Its speed and persistence, however, have yet to be confirmed by direct measurement.

Data presented by Hart & Currie (1960), Stander (1964), Schell (1970), and Calvert & Price (1971b) indicated that the maximum depth affected by upwelling off central Namibia was generally little more than 200 m. Stander (1964) showed that the depth tended to increase through the upwelling season, occasionally reaching about 300 m near the end of the season. At 17° S (Kunene River) it seems that the upwelled water originates from a depth of 100 m or shallower, probably on account of the increased stratification in the vicinity of the northern boundary of the Benguela (see Fig. 16, p. 128). The only published estimate of the rate of upwelling off central Namibia is Stander's value of 1·9 $m{\cdot}day^{-1}$ ($2{\cdot}2 \times 10^{-3}$ $cm{\cdot}s^{-1}$) for periods of vigorous upwelling.

THE CENTRAL NAMIBIAN REGION

The central Namibian region (21° S to 24° S) can broadly be divided into two areas. Between Conception Bay (24° S) and Walvis Bay (23° S) the coastline runs approximately from north to south, while north of about 22°30′ S the orientation of the coast is about 330°. A double shelf break is characteristic of much of the region (see Fig. 5, p. 114) and is most marked west of Swakopmund where the upper break occurs at 140 m some 100 km offshore and the lower break at 400 m about 160 km offshore. The broad shallow Swakop shelf probably plays an important rôle in the dynamics of the region. Except near Conception Bay and at 21° S where the shelf is narrow, water from depths greater than 150–200 m is not readily available close inshore. The extremities of the central region seem to be upwelling centres (see figures in Calvert & Price, 1971b; O'Toole, 1980; Boyd, 1983a) although A. J. Boyd (pers. comm.) feels that they only appear as such in view of the reduced upwelling intensity in the area between 21°45′ S and about 23° S.

Diagrams in Stander (1962, 1964), O'Toole (1980), Parrish *et al.* (1983), Boyd (1983a), and Boyd & Agenbag (1984a) strongly suggest the existence of a semi-permanent convergence zone off central Namibia between the Lüderitz and northern Namibian systems. It is most pronounced during the summer between latitudes 22° S and 23° S (*e.g.* Fig. 25), and is at times more evident at subsurface depths (*e.g.* Fig. 30). Also the mean annual "cold advective values" of Stetsjuk (1983) showed a well-defined maximum (warm) at 22° S. A. J. Boyd (pers. comm.) prefers to consider the central region as a transitional zone with local bathymetry influencing conditions rather than an area whose dynamics are determined by convergence. That the configurations of the large upwelling tongue which emanates from Lüderitz to an extent determines the structure of the region nevertheless seems probable from Stander (1964), Bang (1971), and Parrish *et al.* (1983). Bang (1971) noted southward motion of the front just south of 24° S and inshore south of Walvis Bay following a gale, indicating cyclonic circulation and possibly nearshore compensation flow around the main Lüderitz upwelling tongue. Bang's data also seem to suggest a secondary centre of upwelling off Conception Bay.

The thermohaline characteristics of the central Namibian region are not well understood. The salinity fluctuations lag the temperature by about one to two months (Stander & De Decker, 1969) and this complicates the interpretation of upwelling (Boyd, 1983a). During autumn warmer, higher salinity water (15 °C; *S* 35·2 to 35·4‰) is upwelled in the region (Boyd, 1983a; Boyd & Agenbag, 1984a), and Boyd (1983a) suggested that this upwelled water had a more northerly or offshore origin than at other times of the year. Alternatively, instead of upwelling, the surface water which is removed through Ekman transport may be replaced to an extent by horizontal advection of warm saline water from the north during periods when stratification is at a maximum, but this would tend to contradict the observations of equatorward currents at 10 m by Boyd & Agenbag (1984a). (These authors suggested a poleward compensation just below the thermocline when stratification was maximum.) Nevertheless surface compensation may occur during periods of weak coastal winds, and whatever the origin of the water upwelled, the topography of the region between 22° S and 23° S would not tend to favour upwelling of water from deeper than 150 to 200 m.

Temperatures reach a maximum during March (see Fig. 20, p. 136; Strogalev, 1983), and during late summer water of oceanic or Angolan origin tends to move in from the north or west and then to retreat as the upwelling is intensified later through autumn when stratification decreases. This process may be what happened in the sequence shown in Figure 31. What should not

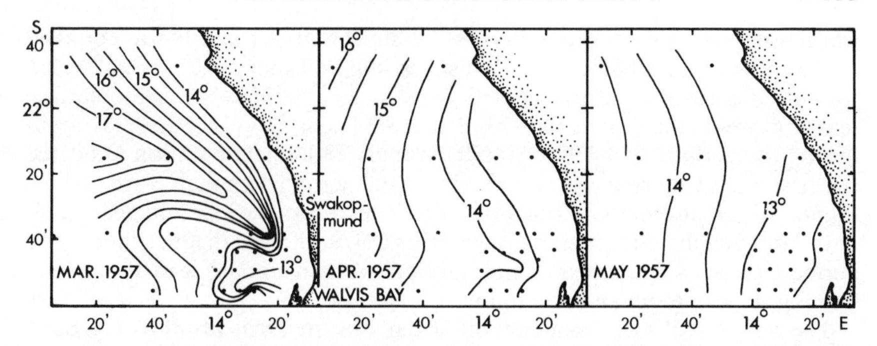

Fig. 31.—Temperature distribution (°C) at 20 m near Walvis Bay (after Stander, 1962).

be lost sight of is that the Tropic of Capricorn passes through Walvis Bay and during summer insolation is substantial (Copenhagen, 1953; du Plessis, 1967; Boyd & Agenbag, 1984a).

Although the station spacing in most published studies is inadequate to deduce the nature of the frontal structure over the shelf breaks, the work of Hart & Currie (1960), Stander (1962, 1964), Visser (1969a), Schell (1970), and Hagen *et al.* (1981) suggest the existence of a baroclinic jet. A subsurface salinity maximum at 50–100 m west of the upper shelf break is well defined during summer (Stander, 1962), but whether this is indicative of an undercurrent or merely an advanced stage of upwelling is not clear. Another feature which may be characteristic of summer is the presence of high salinity water (>35·0‰) at the bottom on the mid-upper shelf off Walvis Bay (Stander, 1962) and also noted by Hart & Currie (1960) during September to October 1950.

In an analysis of winds at Pelican Point and temperatures at selected sites near Walvis Bay, Stander (1963) found a good inverse relationship between the southerly wind component and 0 and 20 m temperatures (Fig. 32) and fair agreement between downwelling favourable winds and the occurrence of

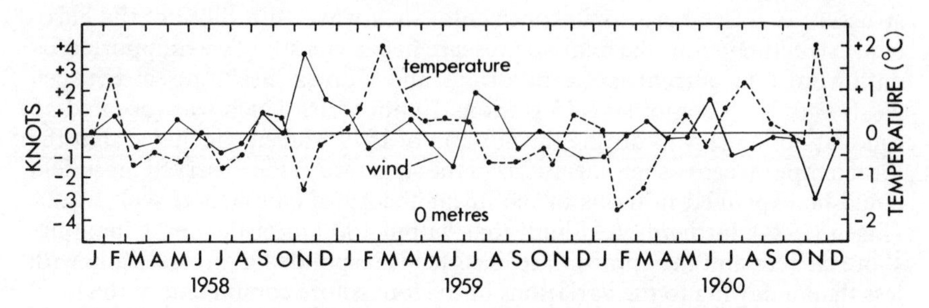

Fig. 32.—The relationship between sea water temperature anomaly (broken lines) at selected stations and the southerly wind component (solid lines) at Pelican Point, Walvis Bay (after Stander, 1963).

warm water. His view was that most of the short term variability was confined to the upper 20 m. What is somewhat surprising, considering the spatial scales, was the degree of coherency noted by Stander (1963) between the surface temperatures in the Walvis Bay (23° S) and St Helena Bay (32° S to 33° S) areas during the period 1954–1957. Stander's single 23-h anchor station indicated the existence of internal waves with a half-tidal period. Although Boyd (1983a) found little change in the vertical structure over a two-week period near 22° S, data from Schultz, Schemainda & Nehring (1979) showed distinct short-term periodicity, possibly related to internal tides with an amplitude of about 10 m (see Fig. 17 in Chapman & Shannon, 1985).

The only direct measurements of currents in the areas are those of Boyd (1983b), Boyd & Agenbag (1984a), Boyd, Potgieter & Buys (1983), and Hagen (1984). The drogue tracking study of Boyd (1983b) over the shelf around 22° S suggested that diurnal land–sea breezes control the currents in the upper 5 m, with a response time of a few hours, but that deeper currents were not directly influenced by the wind. The average surface current (10–15 $cm{\cdot}s^{-1}$) was 1·7% of the wind speed. Boyd noted marked shear between the surface currents and those at 20 and 30 m. At the latter depth Boyd observed a meandering poleward undercurrent with an average velocity of 3·5 $cm{\cdot}s^{-1}$. Boyd *et al.* (1983) identified three main time scales from a current meter mooring near Walvis Bay between October 1981 and March 1982, *viz.* 2–4 days; tidal or diurnal; those with periods of less than half a day. These authors obtained a fair correlation between the mean daily southerly wind component and the onshore subsurface currents, the latter evidently compensating for the cyclonic circulation and net surface flow out of the Bay noted by Pieterse & van der Post (1967).

Hagen (1979) suggested that the rhythms with periods of several days in the current field over the shelf were due mainly to the dynamics of these waves, while Hagen (1981) concluded that the energy-rich barotropic shelf waves occasionally impressed their space–time structures upon that of the local baroclinic mass field. His model, which ignored stratification, indicated that, for the shelf configuration at 20°30′ S there was a northward propagation of barotropic mesoscale eddies of 9 $km{\cdot}day^{-1}$, while the second and third modes of the solution yielded wavelengths and periods of 600 km and 350 km and 5 days and 7·2 days, respectively, *i.e.* similar longshore spacings to the features observed by Van Foreest *et al.* (in press). In a follow-up investigation Hagen (1984) compared the space–time patterns arising from the cross-shore modal structure of a free barotropic continental shelf wave with those of the baroclinic structures from the relative pressure fields. His study was supported by data from four current meter moorings and a cross-shelf transect between 20° S and 21° S comprising 15 stations 10 km apart which was repeated 15 times at 36-h intervals during the autumn of 1979. Hagen concluded that the local temporal cross-shelf variations in the structure of the observed mass field could be explained in terms of the linear theory of continental shelf waves. Hagen (1984), furthermore, found that within 100 km of the coast the long-shore current and the appropriate mass fields responded hydrostatically with less than a day lag to the variations in the long-shore component of the local wind. These events had a mean period of 5·6 days. Seawards of the 100-km wide coastal zone Hagen indicated the possible existence of an internal Rossby wave with a period of 13 days.

INTERACTION BETWEEN THE BENGUELA AND ANGOLA SYSTEMS

The region between 15° S and 20° S is extremely complex and not well understood (see Currie, 1953; Stander, 1964; Filippov & Kolesnikov, 1971; Moroshkin *et al.*, 1970; O'Toole, 1980; Parrish *et al.*, 1983; Nelson & Hutchings, 1983; Boyd & Thomas, 1984). Usually evident from the temperature and salinity distribution is a well-defined convergence which demarcates the approximate boundary between the surface waters of the Angola and Benguela systems. This front which is positioned approximately over the Walvis Ridge abutment extends from close to the coast near 17° S offshore in a southwesterly direction and appears to migrate seasonally over about 3° of latitude (18° S in March, 15° S during winter). The surface temperature and salinity changes associated with it are about 4 °C and 0·4°/₀₀, respectively—the distinct differences in the thermohaline characteristics of Angola water and that of Benguela origin are illustrated by the upper portions of the mean *T-S* curves (see Fig. 12, p. 124).

The interpretation of the mesoscale oceanographic processes in the region are complicated by the fact that the majority of cruises tended to terminate near the Angola–Namibia geographic boundary, and by the generally inadequate spatial and temporal scales of measurement. The timing of the cruises, many of which have taken place during autumn when the southward flowing Angola Current (Fig. 33) is near a maximum and upwelling off

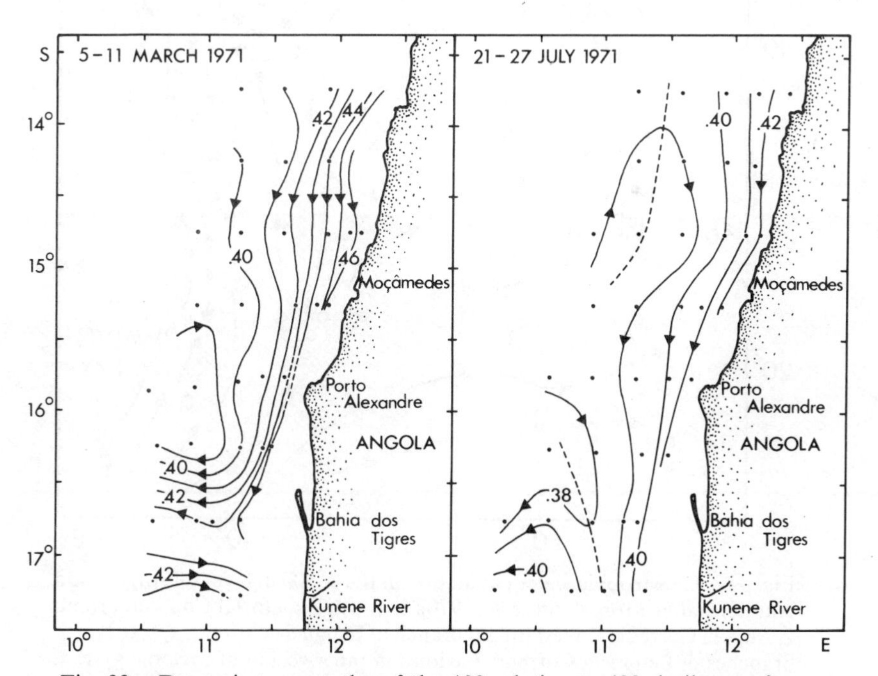

Fig. 33.—Dynamic topography of the 100 relative to 400 decibar surface during March and July 1971 off Angola (after Dias, 1983b).

Namibia near a minimum may also have led to some bias in the published interpretations.

Except for the work of Stander (1964) little appears to be known about the seasonal changes at 17° S. His figures indicate, that at this latitude, stratification reaches a maximum during summer west of the 100-km wide coastal zone, and a minimum during winter. Stander (1964) showed clear evidence of upwelling off the Kunene River at times. His diagrams suggest that during summer, when it occurs, it does so very close inshore and the upwelled water originates from depths shallower than 50 m. During winter the depth affected by upwelling may extend to 100 m. From the wind stress (Wooster, 1973; Berrit, 1976; Duing, Ostapoff & Merle, 1980; Parrish *et al.*, 1983) and topography it might be expected that the area north of Cape Frio would be a key area for upwelling throughout the year (with slight maxima during May and October), but this does not seem to be the case. Either the stratification is such, particularly during summer, that relatively warm saline water is recycled at very shallow depths or otherwise the surface water transported offshore (*cf.* drogue studies

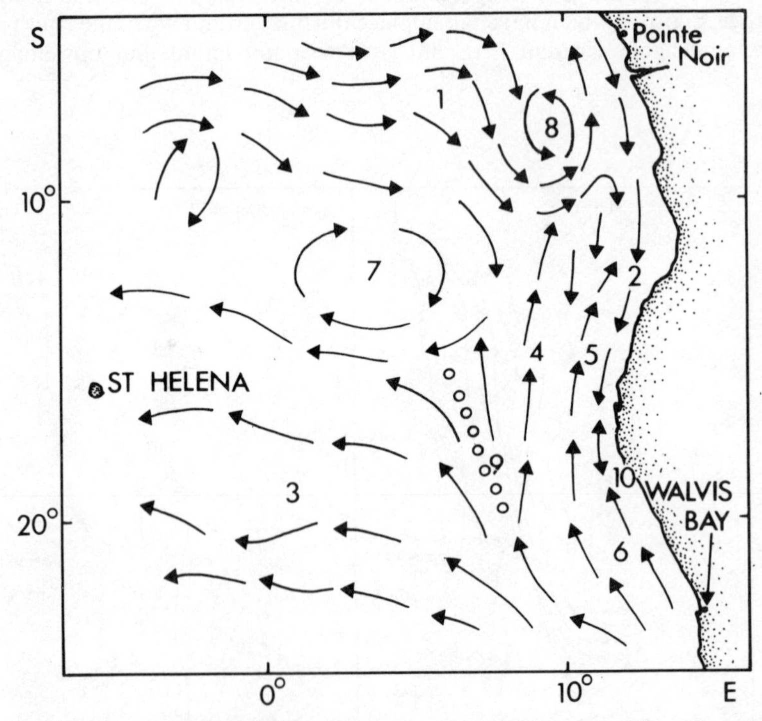

Fig. 34.—Geostrophic water circulation in the 0–100 m layer off Angola and Namibia (after Moroshkin *et al.*, 1970): 1, South Equatorial Countercurrent; 2, Angola Current; 3, West (main) branch of Benguela Current; 4, 5, 6, North branches of Benguela Current; 7, eddies in inner region of cyclonic gyre; 8, anticyclonic curl; 9, Benguela Divergence; 10, merging zone of Angola Current and north littoral branch of Benguela Current.

of Boyd & Agenbag, 1984a) is replaced by the southward horizontal advection of Angola water, as intimated by Hart & Currie (1960). The latter would explain the strong coastal downwelling and southward flow observed by Stander (1964, his plate 69) during October 1959 when upwelling was taking place further south (see also Fig. 30).

Stander (1964) noted that the surface flow becomes more complex and variable with decreasing latitude off Namibia, in particular between 23° S and 17° S, and his dynamic topography also shows similar complexity and variability at subsurface depths (100 m). The most comprehensive single investigation of the circulation off Angola and northern Namibia was undertaken between April and June 1968 (Moroshkin *et al.*, 1970). These authors identified northward filaments of the Benguela west of the coastal Angola Current, extending to 12° S to 13° S, and they postulated the existence of a divergence zone between these northerly and the main (westerly) branch of the Benguela north of 20° S. The schematic representation of flow in the upper 100 m proposed by Moroshkin *et al.* (1970) is shown in Figure 34 which, although for autumn, is in substantial agreement with the studies of Dias (1983a,b) off southern Angola and Stander (1964) and Robson (1983) off Namibia. The surface southerly flow inshore is most marked north of 17° S, while at this latitude Stander (1964) indicated a semi-permanent northerly flow between 100 and 200 km offshore and more variable flow near the "divergence". Filippov & Kolesnikov (1971) have, however, questioned whether the equatorward intrusions off Angola are of Benguela origin and whether the poleward flow off Namibia is an extension of the Angola Current. Nevertheless the available evidence does seem to suggest that subsurface meridional interaction between the Angola and Benguela systems can occur over a distance of about 1000 km.

MESOSCALE PROCESSES IN THE SOUTHERN BENGUELA

The general features of the oceanography around the South West Cape have been described by Isaac (1937), Clowes (1950), Darbyshire (1963, 1966), Shannon (1966), Bang (1973), Andrews & Hutchings (1980), and others, while Nelson & Hutchings (1983) have reviewed recent work on mesoscale upwelling processes in the region.

The dynamics of the southern Benguela upwelling system are governed largely by three factors, *viz.* mesoscale atmospheric perturbations, the topography, and the influence of the Agulhas Current system. The periodic relaxation in upwelling caused by the free zonal passage of easterly moving cyclones to the south of the continent, described by Nelson & Hutchings (1983), was summarized earlier in this review and readers are referred to Figure 9 (see p. 120). Typically wind reversals modulate the system with a period of about six days. These modulations are superimposed on the seasonal cycle, which is much more pronounced in the Cape Peninsula and Agulhas Bank areas than in the north. The seasonal wind stress curl diagrams in Kamstra (in press) illustrate this quite dramatically. In winter a broad convergence zone exists south and east of the Cape Peninsula with smaller zones in Table and St Helena Bays, whereas in summer the wind field around the South West Cape is predominantly divergent and strongly so immediately west of the Cape Peninsula. (These features are also evident from the study of Shannon, Chapman, Eagle & McClurg, 1983.) The topography of the southern Benguela region is complex. The coastline is irregular with several bays and capes and changes its general orientation markedly at Cape Point, just south of 34° S, while in places mountain ranges are present close to the coast (see Fig. 4, p. 113). These topographic features shear the wind stress field, giving rise to alongshore variability in the coastal upwelling (Jury, 1984, in press). Intrusions of water of Agulhas Current origin, which appear off the western coast either as shallow tongues or as advected eddies (sheared off from the current in the Agulhas retroflection area) provide a modulation with time scales ranging from a few days to several months.

Although Copenhagen (1953) identified both the Cape Peninsula and Cape Columbine areas as important upwelling centres, the tongue-like nature of the upwelling was first noted by Andrews & Cram (1969) in their classic aerial and shipboard study. Subsequent investigations have shown localized upwelling off other capes (*e.g.* off Cape Hangklip—Cram, 1970; Jury, 1980, 1984, in press). The mesoscale structure of the upwelling tongues is particularly noticeable from satellite thermal infrared and ocean colour imagery (*e.g.* Harris, 1978; Lutjeharms, 1981a; Shannon & Anderson, 1982; Shannon, Mostert, Walters & Anderson, 1983; Shannon & Lutjeharms, 1983). Figure 35 (from Harris, 1978) shows two broad but discrete tongues emanating from the Cape Peninsula and Cape Columbine systems during an advanced stage of upwelling, and well-developed upwelling can be seen extending as far east as Cape Agulhas.

Prior to the work of Bang & Andrews (1974) knowledge of the circulation around the South West Cape was based on data from drift card returns (Clowes, 1954; Duncan, 1965; Duncan & Nell, 1969), Pisa tubes (Duncan, 1966), ships' drift (*e.g.* Rennell, 1832; Defant, 1936; Tripp, 1967) and movement inferred from isentropic and *T-S* analysis (Clowes, 1950; Shannon, 1966; Darbyshire, 1966; De Decker, 1970) and the dynamic topography (Dietrich, 1935; Darbyshire, 1963). From the above a relatively simple picture emerged with cyclonic curving flow around the Cape, predominantly wind driven at the surface during much of the year, current reversals during periods of northwesterly winds, intermittent countercurrents close inshore, and an ill-defined poleward undercurrent. Much of this work has been synthesized by Harris (1978). The complexity of the current field was only fully appreciated after the advent of relatively recent current metering and buoy tracking programmes.

The area has a high degree of variability. Lutjeharms (1977) addressed the question of time variations in the spatial scales and intensities of ocean circulation around the South West Cape, and noted changes in the amplitude of the 26·4 sigma-*t* surface of up to 50 m. His energy distribution over the spatial spectrum showed two configurations; the most common one having an abrupt change in slope at 150 km (*i.e.* a predilection for disturbances with dimensions of less than 150 km) which compared well with the characteristic dimensions (120 km) of frontal eddies between 30° S and 34° S (Lutjeharms, 1981a). The second configuration, which showed a monotonic increase for all distances measured, was tentatively attributed by this author to sporadic extensive upwelling.

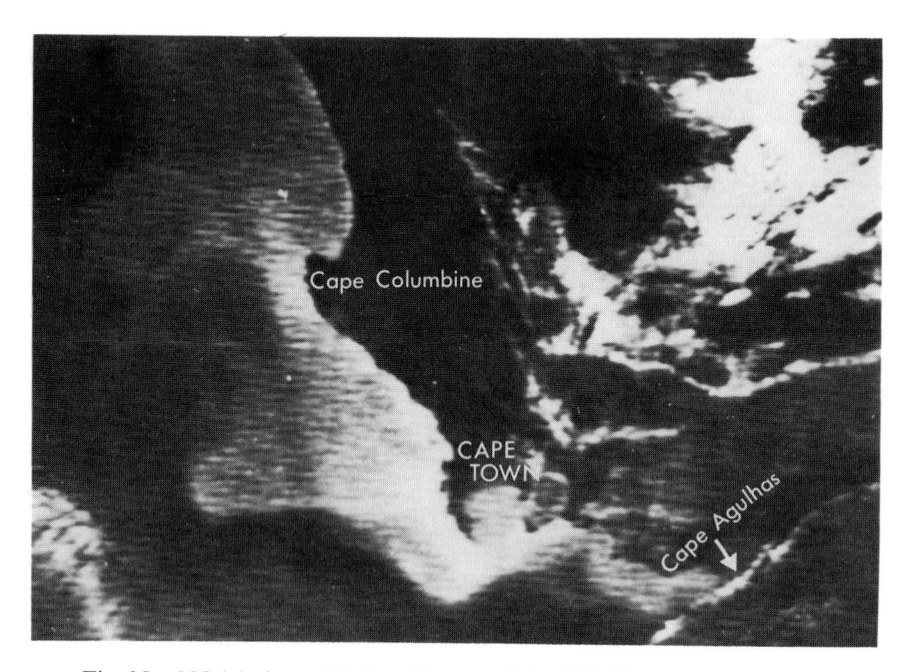

Fig. 35.—NOAA thermal infrared image of the S.W. Cape on 15th March, 1977 showing well-developed upwelling off Cape Columbine and the Cape Peninsula and extending as far east as Cape Agulhas (after Harris, 1978).

In the following paragraphs an attempt will be made to synthesize the present knowledge of mesoscale upwelling processes in the Cape Peninsula and Cape Columbine areas as well as the influence of the Agulhas system on the Benguela regime. Fine-scale investigations undertaken in semi-enclosed bays or close inshore and which relate primarily to tidal or wave driven regimes (*e.g.* Shannon & Stander, 1977; Gunn, 1977; Bain, 1983) have not been included.

INFLUENCE OF THE AGULHAS CURRENT SYSTEM

The southernmost part of the Benguela system, like the northern extremity, is influenced by a warm water regime. In many ways both the northern and southern boundaries can be considered as mirror images, pulsating seasonally, but out of phase by a few months (Shannon, 1977). The warmer surface waters present offshore in the south (>20 °C in summer, >16 °C in winter at 20° E) have properties of both South Atlantic and South Indian Subtropical surface water as well as Agulhas Current water. These waters are advected northwards around the Cape during the upwelling season (Darbyshire, 1963, 1966; Shannon, 1966; Duncan & Nell, 1969) and may result in the intensification of the horizontal frontal thermal gradients in the Cape Peninsula and Cape Columbine upwelling areas (Bang & Andrews, 1974; Lutjeharms & Valentine, 1981).

There has been some difference in opinion as to the extent of the penetration of the Agulhas Current water into the Benguela region (*e.g.* Shannon, 1966; Darbyshire, 1966; Orren & Shannon, 1967; Jones, 1971), although this can in part be attributed to the definitions of the water types. The surface water of the Agulhas Current is warm (generally >21 °C, but somewhat cooler in winter and typically >23 °C during summer—Shannon, 1970; Pearce, 1977) with a salinity of 35·1–35·4‰ (Pearce, 1977) which is lower than South Indian Subtropical surface water. It has a subsurface salinity maximum of about 35·5‰ (Clowes, 1950; Darbyshire, 1964; Pearce, 1977) which corresponds to a temperature of around 18 °C—*i.e.* similar *T-S* to that of much of the subtropical surface water in the South East Atlantic between 30° S and 35° S. Central water in the Agulhas likewise has similar *T-S* characteristics to that in the South Atlantic (Clowes, 1950; Orren, 1963; Shannon, 1966). Thus, except at the surface, it is extremely difficult to assess the degree of interaction between the Agulhas Current and Benguela on the basis of their thermohaline characteristics. Relatively pure Agulhas Current water does, however, penetrate into the South East Atlantic at times (usually during summer months when the wind regime facilitates this advection) in the form of shallowing warm filaments, generally less than 50 m deep or as eddies. Generally the tongues of warm Agulhas surface water are situated 180 km or further away from the coast (Shannon, 1966) and seldom penetrate much north of 33° S. One such tongue moving around the Agulhas Bank is evident in Figure 36a (from Bang, 1973, 1976).

In the area south of the Agulhas Bank the Agulhas Current executes an abrupt turn (Dietrich, 1935; Duncan, 1968; Bang, 1970; Gründlingh & Lutjeharms, 1979) and it is in this retroflection zone, which is one of high shear, that various independent circulation features may be spawned (Lutjeharms, 1981b). Lutjeharms & Valentine (1981) and Lutjeharms (1981c) have shown that at least two mechanisms exist for the advection of pockets of warm Agulhas Current water into the South Atlantic, *viz.* first, the growth of instabilities and the shedding of eddies on the northern border of the Agulhas Current and their subsequent advection across the Agulhas Bank and around the Cape and secondly, the formation of rings at the retroflection point southwest of the continent (similar to Gulf Stream rings). A drifting buoy placed at the centre of one such ring during May 1979 followed a meandering northward path west of, but in sympathy with, the Benguela thermal front, with the ring slowly losing its character due to mixing with colder surrounding water (Lutjeharms & Valentine, 1981). A conceptual image of the formation of eddies and rings is shown in Figure 36(b). Nelson & Hutchings (1983) suggested that the advection of vortices into the Benguela system could significantly alter the current flow patterns on a time scale of a few days.

Several authors, *inter alia* Dietrich (1935), Clowes (1950), Shannon (1966) and Darbyshire (1963, 1966) have shown that the movement of subsurface and central water components from the Agulhas Current around the Agulhas Bank and into the Benguela region is probable, and it has been suggested (Shannon, 1966) that modified Agulhas central water may upwell along the western coast at least as far as 30° S, and probably further north.

While there is some uncertainty as to the contribution of the Agulhas Current *per se* to the Benguela system, there is general agreement that, that of Agulhas Bank water is substantial, and that the region is ecologically of utmost

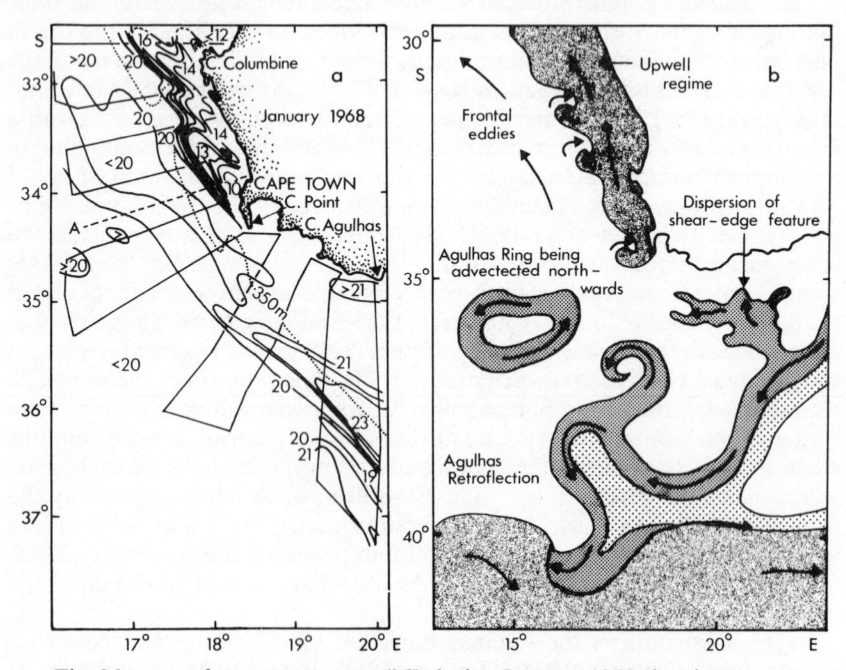

Fig. 36.—a, surface temperatures (°C) during January 1968 showing strong frontal feature pivoted on Cape Point and filament of Agulhas Current in the south (after Bang, 1973); b, conceptual image of Lutjeharms (1981b,c) showing formation of shear-edge eddies and rings.

importance to the Benguela. The formative mechanism of Agulhas Bank water is not properly understood but it appears that it comprises modified Agulhas Current and Atlantic water types. The Agulhas Bank is characterized by very strong winds during winter (Parrish *et al.*, 1983) which together with reduced stability facilitate mixing to a depth of at least 75 m (Pugh, 1982) and occasionally throughout the water column. During summer and autumn a strong thermocline is established at about 50 m which requires wind speeds greater than 20 $m \cdot s^{-1}$ for its erosion (Pugh, 1982). Net currents over the Bank are weak but indicate a strong diurnal activity (Welsh, 1964; Schumann & Perrins, 1983), and Shannon & Chapman (1983a) have suggested that a residence time of water on the Bank may be as long as several weeks. On the western side, cold water moves up onto the shelf during spring (Tromp, Lazarus & Horstman, 1975; Nelson & Hutchings, 1983), and remains on the shelf until autumn (the pronounced uplift in spring and early autumn evident in Tromp *et al.*, 1975, may be related to the bimodality in the southeasterly wind maxima in the area) and upwelling of central water can occur close inshore west of Cape Agulhas (Tromp *et al.*, 1975; Harris, 1978) and occasionally as far east as Algoa Bay (Schumann, Perrins & Hunter, 1982). Recent studies by Shannon & Chapman (1983a) on surface currents and winds and Schumann & Beekman (1984) on the thermal structure suggest that there is a divergence or transitional zone over the Bank around 21° E, which coincides with a break in the bathymetry, changes in sediment structure and composition (see Fig. 6, p. 115) and composition of fauna. It is also possible that internal waves of tidal character which are indicated by the strong surface signals in satellite images over the western Agulhas Bank (Nelson & Shannon, 1983) may play an important rôle in the mixing of subsurface and deeper waters in this region. Ships' drift (Harris, 1978), isopycnal and isentropic analysis (Shannon, 1966; Tromp *et al.*, 1975), drift cards (Duncan & Nell, 1969; Shelton & Kriel, 1980; Shannon *et al.*, 1983), *Physalia* (Shannon & Chapman, 1983b), and fish eggs and larvae (Shelton & Hutchings, 1982; Shelton, 1984) indicate a net westward, then northward movement around the Cape of surface Agulhas Bank water in the area west of 20° E during the summer (October to March) and an intermittent eastward movement in the inshore region further east. During winter and during periods of prolonged westerly winds the drift card returns suggest a net inshore eastward flow of surface water in the area between 18° E and 27° E (Algoa Bay), which implies that South Atlantic Subtropical surface water is seasonally important along much of the southern coast of the continent.

From the above it appears that, as in the northern Benguela–Angola system zone, the interactions between the Benguela and Agulhas systems can extend over a substantial distance, *i.e.* around 1000 km. Whether Agulhas water can maintain its identity as far north as 21° S on the western coast as suggested by Darbyshire (1966) is, however, questionable. Perhaps studies using stable isotope ratios or radionuclides (*e.g.* Shannon, Cherry & Orren, 1970) would be useful here.

THE CAPE PENINSULA UPWELLING AREA

The Cape Peninsula must surely rate as one of the most spectacular upwelling sites in the World. It is near the southern extremity of a subcontinent whose eastern boundary current system is bounded on its poleward end by a warm water regime. It is an area where bottom topography and orographically induced wind curl are at least as important as Ekman divergence in governing the upwelling process. It is a region where the wind field is modulated on a time scale of days as well as seasonally and where upwelling occurs on spatial scales from less than one kilometre to tens of kilometres, responding in a matter of hours to changes in the wind. It is also an area of immense natural beauty. The upwelling season off the Cape Peninsula extends from September to March (Andrews & Hutchings, 1980), with maxima in upwelling favourable winds in November and March at Cape Point. The temperature and salinity time series of Andrews & Hutchings (1980) showed that the properties of both the offshore water and subsurface (shelf) water followed a simple seasonal cycle. Distinct bimodality in the annual cycle was, however, exhibited by the nearshore surface (upper 30 m) waters, which may be related to the corresponding bimodality in the longshore winds.

The shelf break which is convoluted by the Cape Point Valley (Nelson, in press) is deep and the shelf itself is narrow with the result that cold water is available at subsurface depths close inshore, in particular near Cape Point, which Bang (1971, 1973) regarded as the southern pivot point of the Benguela system. The Table Mountain range which stands isolated to the flow of southerly wind creates local zones of strong wind curl and divergence (Nelson, in press). According to Nelson (in press) there are four definable classes of atmospheric events, *viz.* deep southeaster, shallow southeaster, northwester, and coastal low. Following on from the work of Andrews & Hutchings (1980), the wind field around the Cape Peninsula was studied during CUEX (Cape Upwelling Experiment) by means of aerial surveys augmented by coastal weather stations and data from shipping (Jury, 1980, 1981, 1984, in press; Nelson, Kamstra & Walker, 1983; Kamstra, in press). The characteristics of this wind field have been described in some detail by Jury (1980) and these have been highlighted in Shannon, Nelson & Jury (1981) and Nelson & Hutchings (1983).

The Cape Peninsula upwelling tongue or "plume" (see Nelson, 1981) which is characteristic of the alongshore variability in upwelling in the region was first documented by Andrews & Cram (1969), and was subsequently shown by Andrews & Hutchings (1980) to be a semi-permanent feature of the area during the summer months. Its surface features are masked during northwesterly winds but it only disappears entirely during the winter months (Andrews & Hutchings, 1980). These authors demonstrated that the pronounced short-term variability of the tongue and in the subsurface structure is imposed on the seasonal cycle, which they suggested was also wind-generated. Changes in sea surface temperature distribution in response to mesoscale wind events were examined by Jury (1980, 1984). His work has resulted in a substantially improved understanding of the dynamics of the southern Benguela, and the essential features of his study of the Cape Peninsula are briefly as follows.

The onset of southerly winds with wind speed maxima north of the Cape Peninsula, following the passage of a cold front, initiates upwelling off the northwestern Peninsula and entrains oceanic water and compresses the oceanic thermal front against the coast in the south. The travelling anti-cyclones passing south of the region create deep (2000 m) air flow from 160° when they are supported by an upper air ridge. This "deep southeaster" becomes topographically accelerated over the Cape Peninsula in a classic cape effect resulting in the growth of upwelling and the formation of pronounced tongues below the wind jets in Table Bay and immediately south of the Table Mountain chain (*i.e.* at Bakoven) as well as at other localities further south along the 60 km Peninsula mountain range. With the South Atlantic high ridging around south of the country the orographic influences are magnified due to the capping effect of the depressed inversion layer which results in the formation of a calm zone (wind shadow) north of the Peninsula and cyclonic wind vorticity. During this "shallow southeaster" phase the upwelling tongue and thermal oceanic front progress westwards and southwards. Finally, with the establishment of a coastal low and approach of the next cold front, wind reversals occur and northwesterly winds cause a relaxation in the upwelling and a compaction of the oceanic thermal front over the shelf region.

Jury (1984) has demonstrated that vertical wind shear controls the interaction of topography and winds in the area and his results indicated that vertical shears of $-2 \times 10^{-2} \cdot s^{-1}$ produced horizontal wind vorticities of $-6 \times 10^{-4} \cdot s^{-1}$ and alongshore sea surface temperature gradients of 1 °C per 10 km during the summer upwelling season. Two of Jury's (in press) case studies which illustrated the growth and decay of upwelling tongues off the Cape Peninsula are shown diagrammatically in Figure 37. Using data collected during CUEX, Taunton-Clark (in press) attempted to quantify the growth and decay of the Cape Peninsula upwelling tongue by comparing the areas encompassed by selected isotherms (measured by airborne radiation thermometry) with wind records from coastal weather stations. His results, although showing a fair degree of scatter, indicated a linear relationship between the plume area and the wind displacements, with a response time of between 12 and 24 h. The latter should be compared with the "almost immediate" response noted by Andrews & Hutchings (1980), and the rapid changes in superficial frontal gradients noted by Bang (1973) over 12 and 16 h. On a larger scale, Lutjeharms (1981a) obtained a fair correlation ($r = 0{\cdot}73$) between wind stress and the area of upwelled water as inferred from infrared satellite images.

The existence of a well-developed front positioned approximately over the shelf break west of the Cape Peninsula and which forms the seaward boundary of the coastal upwelling region has been noted by various authors *inter alia* Duncan (1966), Andrews & Cram (1969), Bang (1973), Bang & Andrews (1974), Andrews & Hutchings (1980), and Shannon *et al.* (1981). According to Brundrit (1981) it could be established in perhaps five days of vigorous upwelling, but once established, its main features could persist over an entire upwelling season, particularly at subsurface depths. Indeed Duncan (1966), Shannon *et al.* (1981) and Hutchings, Holden & Mitchell-Innes (1984) have shown that the subsurface front is maintained during winter and during periods of sustained downwelling. Off the Cape Peninsula water of 9–10 °C is almost always present shorewards of the front at relatively shallow depths (50 to 100 m or less) even during winter, and this dynamic priming of the system implies that upwelling can be rapidly switched on and off by the wind. Andrews & Hutchings (1980) nevertheless show that cooler, lower salinity water is available at these depths in summer than in winter. Associated with the front and the shelf break off the Cape Peninsula is a well-developed equatorward jet (Duncan, 1966; Bang & Andrews, 1974) which appears to be

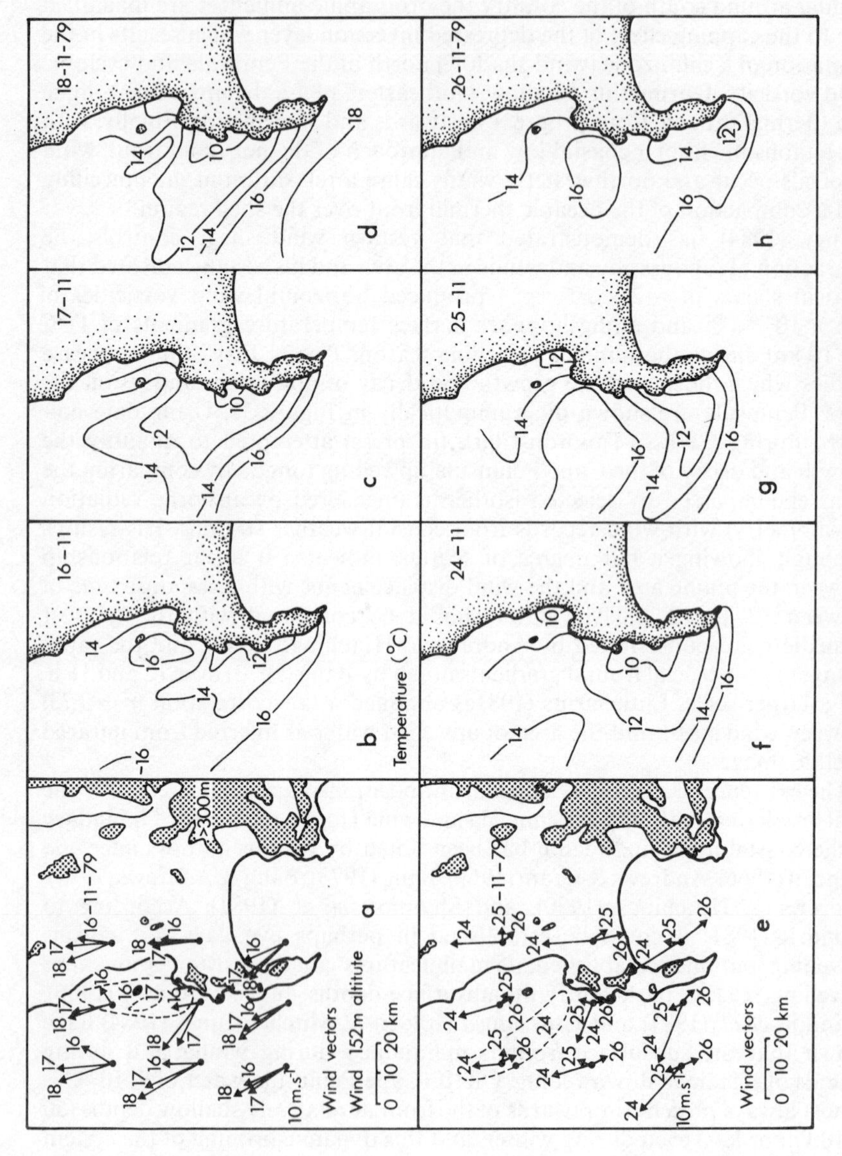

Fig. 37.—Growth and decay of upwelling off the Cape Peninsula (after Jury, in press).

more intense below the surface layer. Bang & Andrews (1974) have described how the front gradually moves seawards during periods of prolonged favourable wind stress as a result of potential energy being accumulated faster than the frontal jet can utilize it, and once it reaches the edge of the shelf it can expand downwards—*i.e.* an equilibrium position exists in the vicinity of the shelf break. These authors have suggested that the influx of warm water from the Agulhas Bank outside the Cape upwelling cell, by intensifying frontal gradients, favours the maintenance of both front and jet. The intense front and jet measured by Bang & Andrews (1974) during January 1973 is shown in Figure 38. Bang (1973), who first described the front in detail, drew a

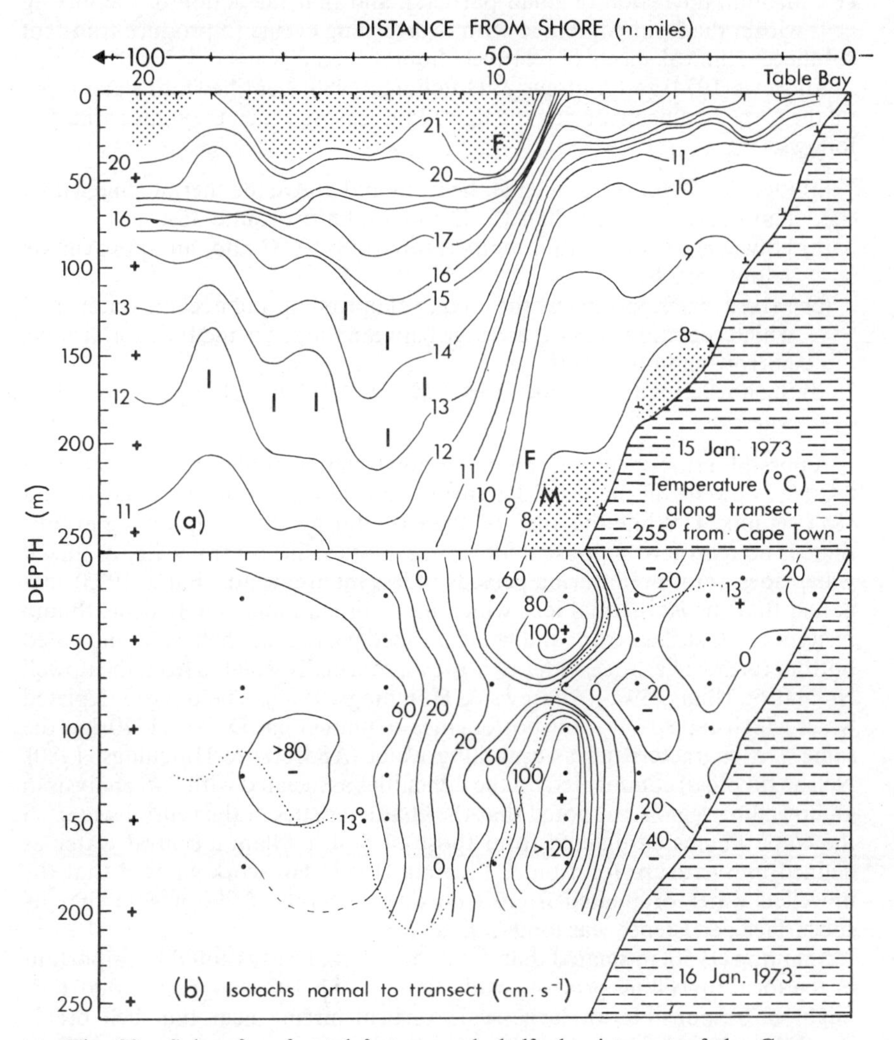

Fig. 38.—Subsurface frontal feature and shelf-edge jet west of the Cape Peninsula at about 34° S during January 1973 (after Bang & Andrews, 1974).

distinction between the inshore frontal zone, formed by the outcropping of the seasonal thermocline 18–28 km offshore and which was dominated by the activity of mixing cells (shallow pockets imbedded in the front), and the main frontal system further offshore—his "offshore upwell region"—which had strong subsurface thermal gradients. Brundrit (1981) showed that the frontal jet is close to instability, and it has been suggested that small perturbations in the front can rapidly grow into fully fledged mesoscale eddies, as evident in satellite imagery of the region (Lutjeharms, 1981a; Shannon & Lutjeharms, 1983). These authors suggested furthermore that the cascading of potential energy at an intensive front caused by baroclinic instabilities could provide a potent mechanism for cross-frontal mixing. In this respect, Bang (1973) postulated that the observed mixing cells could provide a mechanism for cross-frontal advection of small particles, and that the action of the mixing cells within the front combined with overturning events to produce transient subthermoclinical sheets of relatively homogeneous water.

Andrews (1974) and Andrews & Hutchings (1980) identified four water types off the Cape Peninsula during the upwelling season, which they categorized as follows.

(a) Oceanic water lying outside the front and above the thermocline with a temperature greater than 18 °C and a salinity around 35·4‰.
(b) Upwelling water with a temperature of 8–10 °C and, an approximate salinity of 34·7‰.
(c) Mixed water, formed by the mixing of upwelling and oceanic waters and which generally exists as a wedge between the coast and the frontal zone, narrowing in the south.
(d) Shelf water with a temperature less than 8 °C underlying the upwelling water.

A possible criticism of this categorization is that it implies that only water of 8–10 °C upwells off the Cape Peninsula and that the water between 10 and 18 °C is mixed. While this may be the case during strong upwelling events, there is no reason to suppose that warmer more saline central water or mixed water does not upwell during periods of less intense winds. Bang (1973) suggested that the isothermal lens which he recorded immediately beneath and inshore of the surface front and which formed part of the shelf water indicated bottom turbulence. Shelf water is, however, normally isolated from the upwell circulation (Bang, 1973; Andrews & Hutchings, 1980). The oxygen-depleted water which appears near the bottom in late summer (De Decker, 1970) has the same *T-S* characteristics as upwelling water (Andrews & Hutchings, 1980). Nelson (in press) commented on the difficulties associated with *T-S* analysis in an upwelling region but noted that the characteristics of the central water off the Cape Peninsula differed from those of South Atlantic central water as defined in Sverdrup, Johnson & Fleming (1942). His work showed that the upwelled water originated from a maximum depth of 200 m and that its subsurface persistence was longer than two days.

Shannon (1966) intimated that breaking internal waves could be important in the local upwelling process, while Bang (1971, 1973) suggested that they might be responsible for large scale vertical mixing near the shelf break. Recently, internal tides with large amplitudes have been measured near Cape Point, and the question of these waves breaking in areas of steep bathymetry such as exist off the Cape Peninsula has been addressed by Nelson (in press). His work also showed evidence of turbulent mixing between 250 and 400 m west of the front.

Assuming a steady wind of 5 $m \cdot s^{-1}$ Shannon (1966) showed from theoretical considerations that water could upwell off the Cape at a rate of 11 $m \cdot day^{-1}$. From the displacement of isohalines Andrews, Cram & Visser (1970) estimated a rate of around 21 $m \cdot day^{-1}$ with winds of about 10 $m \cdot s^{-1}$, while Andrews (1974) and Andrews & Hutchings (1980) noted a linear relationship ($r = 0{\cdot}95$) between the rate of upwelling and the longshore wind speed, with a mean uplift during active upwelling conditions of 21 $m \cdot day^{-1}$ and a maximum of 32 $m \cdot day^{-1}$. It should be noted, however, that these rates refer to part of the system for part of the time and as such are not true averages. From nutrient data Andrews & Hutchings (1980) obtained a mean upwelling volume flux for the Cape Peninsula cell of 9×10^9 $m^3 \cdot day^{-1}$ which equates to an upwelling rate of 4·5 $m \cdot day^{-1}$ over the whole study area (2000 km^2). By comparison, Bang's (1976) estimate of a flux of $0{\cdot}5 \times 10^6$ $m^3 \cdot s^{-1}$ for the Cape Columbine and Cape Peninsula zones (15 000 km^2) would give a flux of 6×10^9 $m^3 \cdot day^{-1}$ for the Cape Peninsula. Bang (1976) also suggested that localized springs or fountains of upwelling of crystal clear ice-blue water of 8–9 °C with displacements of 30 $m \cdot h^{-1}$ might exist in the system. Andrews & Hutchings (1980) felt that coastal downwelling was a minor feature off the Cape Peninsula. Bang (1976) noted, however, that the concepts of upwelling systems in general are so pervaded by upwelling that the requirements of large scale sinking have received scant attention. Sinking has been noted near the front (Bang, 1973; Andrews, 1974; Andrews & Hutchings, 1980; Nelson, in press) and can occur there both as a slow diffuse process and in fairly dramatic overturning events (Bang, 1973). In addition, Bang (1976) suggested that parcels of cold water could break away from the cell and be dissipated by mixing and sinking. Both he and Nelson (in press) have drawn attention to the dissipative rôle which Langmuir circulations could play in the system.

The surface current patterns inferred from drift card recoveries (Duncan & Nell, 1969; Shannon *et al.*, 1983) have been substantiated by direct measurements using drogues and current meters. The earliest direct measurements of currents west of the Cape Peninsula (excluding nearshore pollution-related work) were made by Duncan (1966) using Pisa tubes (jelly bottles). His study showed the existence of three mesoscale cyclonic cells at a depth of 100–200 m in water of about 10 °C. The main eddy was situated off the northern part of the Peninsula and there was a suggestion of an equatorward jet of 27 $cm \cdot s^{-1}$ about 30 km offshore with southerly flow further east and a retroflection zone near 34°10′ S (Slangkop). His surface drifts under northwesterly winds were equatorwards (>50 $cm \cdot s^{-1}$) over the shelf break west of Cape Point. The existence of a strong shelf-edge jet (Fig. 38) 60 km west of the Peninsula at 34° S, having velocities of about 60 $cm \cdot s^{-1}$ at the surface and 120 $cm \cdot s^{-1}$ at 150 m, was demonstrated by Bang & Andrews (1974). Their work showed poleward flow east and west of the jet with a poleward undercurrent of about 40 $cm \cdot s^{-1}$ near the bottom on the shoreward side of the jet. Bang (1974) suggested that the jet could be expected to be closer inshore off Cape Point, over the Cape Point Valley. The findings of Bang & Andrews (1974) and Bang (1974) were substantially confirmed by Shannon, Nelson & Jury (1981) using a profiling current meter during CUEX. (Nelson, in press, has discussed the CUEX results in more detail.) Except close inshore where shear was evident (noted also by Boyd, 1982), currents were generally barotropic. The jet (about 90 cm·s^{-1}) was most pronounced between 34°30′ S and 34°10′ S just west of the 230 m isobath, and there was evidence of a convergence zone around 34°10′ S, *i.e.* in a similar position to the area of retroflection noted by Duncan (1966). The results of an intensive buoy and drogue tracking experiment undertaken in the upper 50 m during CUEX have been reported by Nelson (in press) and again confirmed the existence of an equatorward jet of about 50 cm·s^{-1} over the shelf break. His study also showed the cyclonic eddy and the convergence zone near 34°10′ S. A buoy with a drogue set at 10 m was released southeast of Cape Point by Shelton & Hutchings (1982) and showed rapid movement (55 cm·s^{-1}) around the Cape along the oceanic front, accelerating to 70 cm·s^{-1} west of the Cape Peninsula. Other studies involving the tracking of drifters by ships and satellites have confirmed the existence of topographically controlled equatorward flow in the southern Benguela near the shelf break (Harris & Shannon,

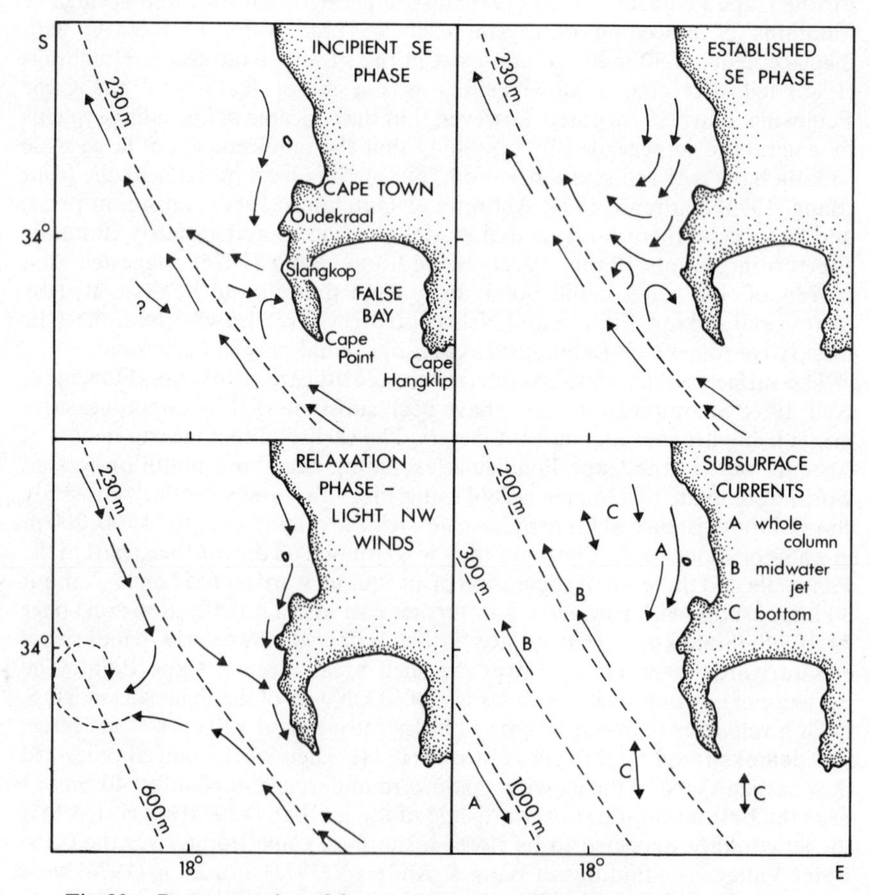

Fig. 39.—Representation of the currents under different winds and subsurface flow around the Cape Peninsula (from Nelson, in press).

1979; Nelson & Hutchings, 1983; Brown & Hutchings, 1984) and variable currents closer inshore (Van Ieperen, 1971; Boyd, 1982; Brown & Hutchings, 1984). The two mode numerical model of Van Foreest & Brundrit (1982) predicted the equatorward jet west of the Cape Peninsula in a similar position to that measured by Bang & Andrews (1974) and Shannon *et al.* (1981), as it did the poleward flow west of the jet. Their model also predicated that velocities would decrease with depth contrary to the observations of Bang & Andrews (1974) but in accordance with Shannon *et al.* (1981).

The only published information from moored current meters is that of Shannon *et al.* (1981) and Nelson (in press) who had two meters positioned within 50 m of the bottom in 253 m of water near the head of the Cape Point Valley during CUEX. The measured currents showed abrupt changes which correlated with local winds, the response time being a few hours, but evidently caused by pressure adjustments occurring on a scale many times larger than the Cape Peninsula upwell cell. Since the CUEX work, several moorings have been positioned in the vicinity of the Cape Peninsula. The results of this work are largely unpublished, but Nelson (1983) shows the existence of a southward coastal current and north–south direction switching over periods of typically six days, presumably as a barotropic response to atmospheric forcing. This appears to be a characteristic of the whole shelf zone as far as Cape Columbine (G. Nelson, pers. comm.).

The diagrammatic representation of near-surface and deep currents around the Cape Peninsula given by Nelson (in press) is shown in Figure 39, and illustrates a very plausible mechanism for the generation of frontal eddies as suggested by the author.

The dynamics of False Bay which forms the eastern seaboard of the Cape Peninsula does not appear to have any substantial influence on the upwelling system and, therefore, will not be discussed here. Readers are, however, referred to the following papers dealing with the Bay: Atkins (1970), Cram (1970), Harris (1978), Shannon, Walters & Moldan (1983), Jury (1984, in press), Van Foreest & Jury (in press).

THE CAPE COLUMBINE UPWELLING AREA

Upwelling in the Cape Columbine area, as off the Cape Peninsula, is controlled to a large extent by the bathymetry (see p. 174) and the influence of the orography on the wind field. Cape Columbine itself is a headland jutting seawards at 33° S consisting of low smooth hills rising to about 250 m (Jury, in press). The main shelf break is convoluted by a major submarine valley, namely the Cape Canyon (see Fig. 4, p. 113 and 41, see p. 174), the axis of which runs approximately from north to south. West of Cape Columbine the 300 m isobath is only about 30 km offshore, and this means that deep water is available close to the coast. Between 33° S and 32° S the shelf broadens substantially with the outer shelf break (400 m) running in a northwesterly direction, and the shallower isobaths tending to curve eastwards and northwards around the Columbine peninsula.

The surface wind vectors, stress and curl in the area between 31° S and 33° S, averaged over half degree rectangles of latitude and longitude, and described by Kamstra (in press) showed distinct seasonal variation. During summer the wind vectors indicated cyclonic curvature around Cape Columbine and into

St Helena Bay, with a zone of strong divergence evident from the wind stress curl north of 31° S. The pronounced curving of the wind has also been shown by the records from coastal weather stations (Hutchings, Nelson, Horstman & Tarr, 1983) and from aircraft drift measurements (Jury, 1984, in press). The last author also showed a distinct wake in the lee of the Columbine peninsula during a period of shallow southerly winds (inversion between 60 m and 350 m). Kamstra's (in press) winter wind data showed a rather confused pattern, which could have been expected as the area between 31° S and 33° S is intermediate between the zone of winter westerlies in the south and the perennial southerly winds north of 31° S. The two case studies of Jury (in press) showed the surface expressions of upwelling characteristic of periods of shallow and deep (inversion between 350 m and 500 m) southerly winds. In the shallow wind case the lowest sea surface temperatures occurred immediately west of Cape Columbine, approximately coinciding with a sausage-shaped zone of strongly negative wind stress curl, *i.e.* the classic cape effect off this headland. Reduced cyclonic curvature of the wind occurred under deep southerly wind conditions, with no significant headland wake, although a small convergence zone was evident in St Helena Bay (the latter could have implications for the retention of pollutants in the bay—see Shannon *et al.*, 1983). Evident in this case study was a cool tongue extending north from the Columbine peninsula and Ekman upwelling along the coast further north and east. The growth and decay of the Columbine upwelling tongue and coastal upwelling during a summer wind cycle was illustrated by Shannon, Schlittenhardt & Mostert (1984)—see Figure 40. The curving of the narrow tongue around the peninsula is evident—a feature often noted in the satellite infrared and ocean colour images of the region (Shannon, Walters & Mostert, in press). Several satellite images have indicated a subsequent anticyclonic curving of the tongue north of 32° S—*i.e.* an inverted 'S' configuration—suggesting topographic control. At times the base of the tongue or plume was around 33°20′ S (Shannon & Anderson, 1982).

The water masses present in the area have been described by Clowes (1954), Buys (1957, 1959), Duncan (1964), De Decker (1970), Bailey & Chapman (in press), and others. The properties of the surface layer (upper 20–30 m) are modified locally by the inflow from the Berg river during winter and in summer by insolation (Clowes, 1954), although recent studies have shown that the latter cannot account for the rapid warming of the water noted on occasion (G. Hughes, pers. comm.). Obviously advection, cross frontal mixing, and the entrainment of water in the area must be important considerations. According to Clowes (1954) a subsurface salinity maximum was common in the area during his study (1951 and 1952) at a depth of about 30 m, most marked during summer. The author attributed this to a subsurface current, although it may have been characteristic of advanced upwelling (Shannon, 1966). No estimates have been made of the rate of upwelling between 31° S and 33° S, although from the rapid response of the surface expression thereof (*e.g.* Fig. 40) it is probably of the same order as off the Cape Peninsula *viz.* a maximum of around 20 m·day^{-1}. The upwelled water originates from depths of 100–300 m (Clowes, 1954). An eight-year time series of monthly measurements in the upper 50 m at two stations, one just north of Cape Columbine, the other 120 km further west, showed distinct bimodality in the uplift of near-surface water (30–50 m), with maxima during October–November and March–April (Mostert, 1970).

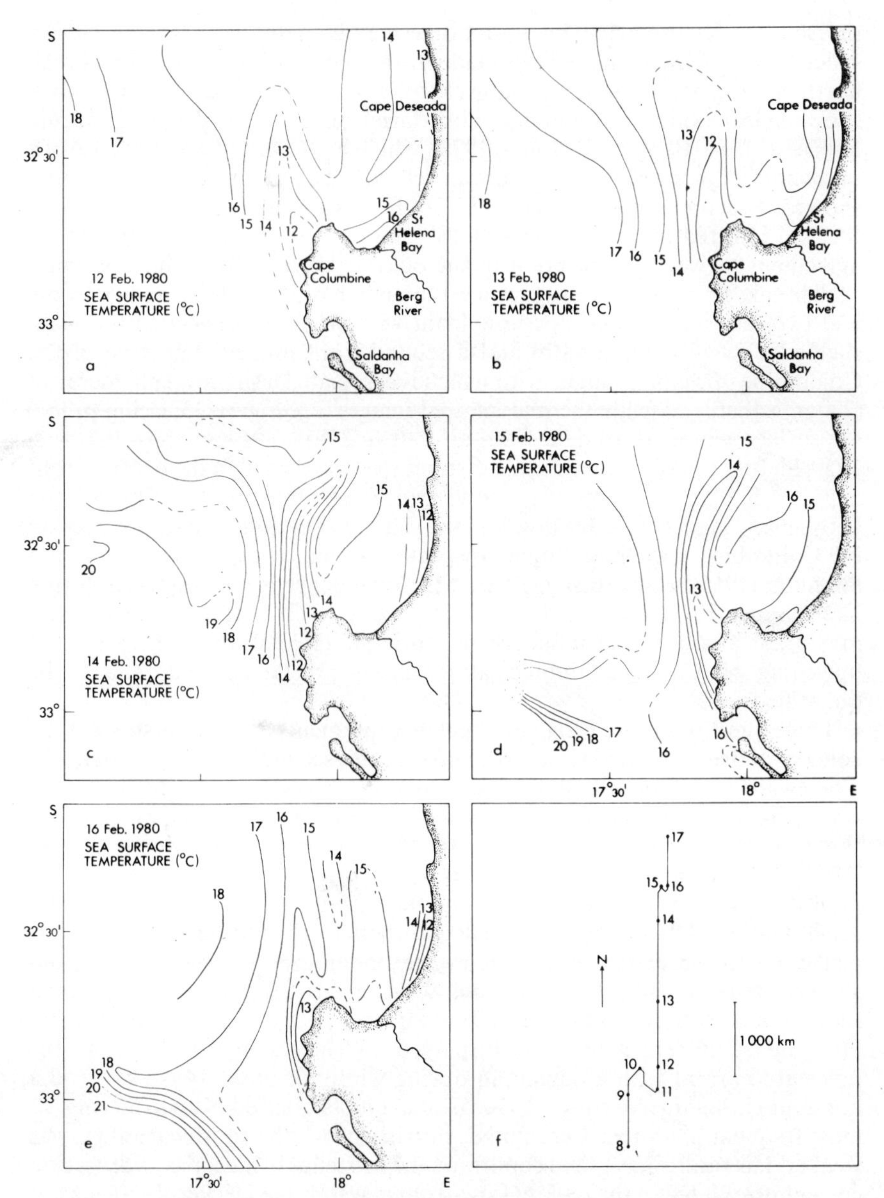

Fig. 40.—Changes in the Columbine upwelling tongue in response to local winds (after Shannon *et al.*, 1984).

The subsurface characteristics of the frontal systems in the Cape Columbine–St Helena Bay area have not been studied in any detail. The figures in Clowes (1950), Buys (1957, 1959), and Duncan (1964), however, showed the existence of a front in the upper 100 m around longitude 17°30′ E, *viz.* in the vicinity of the upper shelf break (20 km to 30 km offshore from Cape Columbine and further offshore north of 33° S) which suggest the existence of the shallow jet predicted by Van Foreest & Brundrit (1982). Few closely spaced measurements have been made in the region of the outer shelf break north of 33° S, except those of Bang (1976) along a line at 32°30′ S. His work showed clear evidence of a major subsurface front near the shelf edge during February 1966 (Bang, 1971; Jones, 1971), which he suggested was evidence of a baroclinic jet with an estimated surface velocity of about 60 cm·s^{-1}. His figure showed a second shallower jet at 17°30′ E. Also significant is that satellite thermal infrared images of the area (Shannon *et al.*, in press) showed close agreement between the position of the oceanic thermal front and the outer shelf break. Thus, there is a suggestion that, north of 33° S, there are two fronts and two jets, *viz.* the main oceanic front with which is associated the main shelf-edge jet (400 m isobath) and a second front immediately west of the Columbine upwelling tongue with which is associated a shallow ephemeral jet (200 m isobath). Satellite thermal infrared images (*e.g.* Figure 35, facing p. 158) sea-surface temperature data from aircraft (*e.g.* Fig. 40) and ships (Clowes, 1950; Shannon, 1966) often show a distinct westward bend in the oceanic front near 33° S. Thus, it seems that a divergence zone may exist at this latitude between the main shelf-edge flow and the shallower cyclonic curving jet west of the Columbine upwelling tongue. Inspection of figures in Clowes (1954) and Shannon (1966) seems to suggest that the divergence is most marked during late summer. From published diagrams it seems likely that the Cape Canyon may exert a substantial influence on both the circulation in the area and upwelling off Cape Columbine, although its precise rôle has still to be quantified.

Three questions arise. First, do direct measurements of currents support the concept of the postulated Columbine divergence; secondly, is there evidence of a poleward undercurrent; and thirdly, does sinking occur at the fronts? With respect to the last question the answer is probably yes. Clowes (1954) has shown strong evidence for this in some of his sections, while Shannon *et al.* (1983) showed a zone of high radiance in NIMBUS-7 CZCS band 520 nm immediately east of the oceanic thermal front, indicating the possible accumulation of floatables (oils, particulates) there, suggesting surface convergence. As to the existence of a poleward undercurrent, Holden (1984) has shown that perennial net southward flow occurred at depths below 80 m over the 180-m isobath west of Cape Columbine over periods longer than 10 days. His moored current meter data displayed a clear seasonal response in the poleward current with a maximum during winter (average 14 cm·s^{-1} and a minimum in summer (6 cm·s^{-1}). Occasional bursts of speed of up to 50 cm·s^{-1}, most frequent in winter, were noted. Further north the southward flow was weaker. His results have thus confirmed the existence of a deep compensation current postulated in the region. Low-oxygen water (De Decker, 1970) may at times be associated with a poleward coastal undercurrent. From data from the moored current meters and from a current profiling line west of Cape Columbine, Holden (1983) suggested that the interface between the surface and deeper currents sloped downwards away from the coast and that this interface tended to rise in winter, and occasionally during summer, to form a broad countercurrent at the surface, rather analogous with the California situation. This concept was to some extent supported by the intensive current profiling cruises undertaken within 20 km (180-m isobath) of the Columbine peninsula during September 1974 and June 1975 (NRIO, 1976). While the latter cruise showed southward or southwestward flow (10–60 $cm{\cdot}s^{-1}$) throughout the water column under northerly wind conditions, the currents during September 1974 (southerly winds) were, however, equatorward and barotropic, around 50 $cm{\cdot}s^{-1}$, showing no evidence of any poleward subsurface current. A drogue tracking experiment in November 1975 (NRIO, 1976) under strong southerly winds showed surface flow (typically >50 $cm{\cdot}s^{-1}$) directed to the north with the strongest current or jet (up to 135 $cm{\cdot}s^{-1}$) over the 100-m isobath *i.e.* in or immediately west of the Columbine upwelling tongue. A drogue released off the Cape Peninsula in recently upwelled water (Barlow, 1984) approximately followed the 150-m isobath northwards, and accelerated dramatically past the Columbine peninsula. The ship's drift measurements by Clowes (1954) east of 17°30′ E showed a similar equatorward set of about 50 $cm{\cdot}s^{-1}$ off the Columbine peninsula. Thus, there seems to be ample evidence of accelerated northward flow within 30 km of this peninsula which is probably strongest near the surface during summer, in particular during strong southerly wind events which favour the rapid development of the Columbine upwelling tongue. This is the Columbine jet. West of the 17°30′ E meridian the situation appears, however, to be somewhat different. The paths followed by three satellite tracked drifters during late summer to winter (Harris & Shannon, 1979; Lutjeharms & Valentine, 1981; Nelson & Hutchings, 1983) while also exhibiting accelerations west of Cape Columbine tended to diverge offshore in the region between 33° S (Cape Columbine) and 32° S, approximately following the run of the outer shelf break. Ship's drift observations (Clowes, 1954) showed similar westward or northwestward sets in these latitudes west of the 17°30′ meridian. Furthermore, the one current profiling line during June 1975 (NRIO, 1976) which extended 50 km offshore showed marked westward flow at the outer stations in sharp contrast to the currents within 20 km of the coast. Thus, there seem to exist two distinct branches of the current with a divide or divergence zone, the Columbine Divergence or Divide, near the 17°30′ E meridian. Whether the outer branch exists at subsurface depths is not certain, although the evident topographic steering of the satellite drifters and the shelf-edge jet inferred by Bang (1976) at 32°30′ S seems to suggest that it does. It is also not certain whether the two branches of the current coexist (Fig. 39, see p. 168, suggests that they do) or whether the current flips between the northward and westward configurations, with a preponderance for the former during spring and early summer and the latter from late summer to winter. If in fact the westward branch only forms in response to sea level adjustment during wind reversals then in fact the Columbine Divide could be associated with a confluence and retroflection of the currents rather than divergence. In any event the two branches do provide alternative mechanisms for the northward transport of material either into the St Helena Bay area or into the northern Benguela. The currents in the coastal area north of St Helena Bay are generally sluggish (Clowes, 1954; Holden, in press) and a cyclonic cell having mesoscale dimensions has been suggested (Duncan & Nell, 1969; Holden, in press). The residence time of water in this area is probably substantial. The northern boundary of this cell (30° S) is the perennial Namaqua upwelling area discussed earlier. The influence of the Namaqua region on the St Helena Bay area would appear to be largely restricted to the subsurface countercurrent, although a surface countercurrent may form in the absence of upwelling.

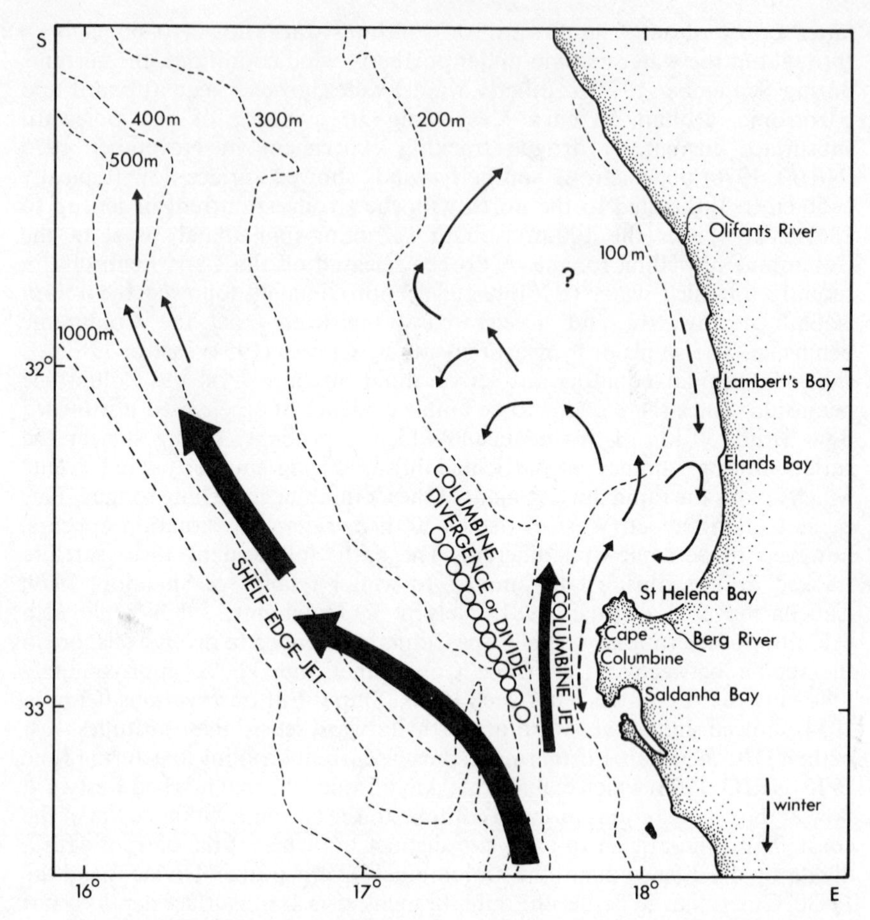

Fig. 41.—Schematic representation of currents in the Cape Columbine–St Helena Bay area.

The main currents in the region between 31° S and 33° S are shown schematically in Figure 41.

A CONCEPTUAL MODEL OF THE BENGUELA

The distinguishing features of the Benguela are highlighted in Figure 42 which is perhaps more of a conceptual image rather than a model. In it I have attempted to paint a cohesive picture of the system as a whole, and it is hoped that it may prove useful to workers in other disciplines and to those not familiar with the Benguela. While it does represent my interpretation of the system, it should, however, be viewed by readers in conjunction with the appropriate text and not in isolation.

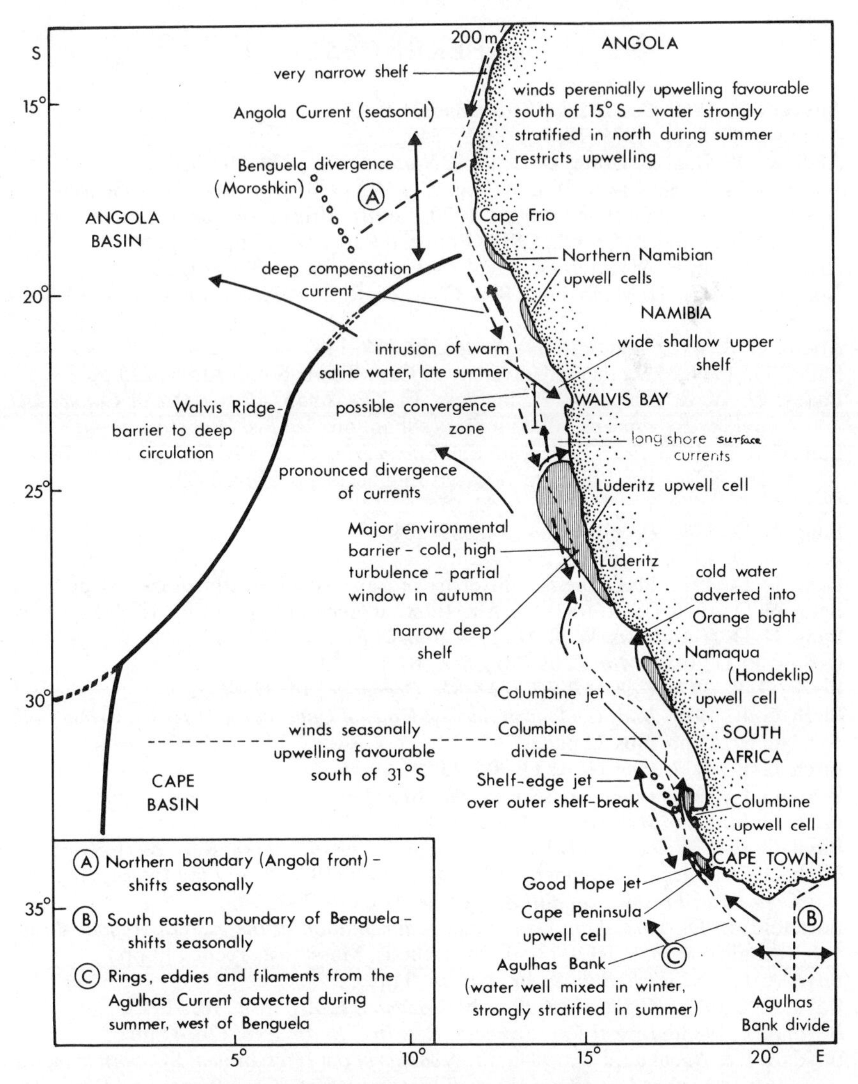

Fig. 42.—A conceptual model of the Benguela system.

ACKNOWLEDGEMENTS

I am indeed appreciative of the comments, encouragement, and assistance given to me by many of my colleagues while preparing this review. In particular I wish to thank Geoff Brundrit, Grev Nelson, Alan Boyd, John Rogers, Mike Bremner, Tony van Dalsen, Christine Illert, and Mariana van Niekerk. Many of the concepts embodied in this article crystallized as a result of the weekly seminars of the Benguela Ecology Programme, which also provided the stimulus for attempting this synthesis in the first place.

REFERENCES

Albrecht, F., 1960. *Ber. Dtsch. Wetterdienstes*, **9,** 1–19.

Andrews, W. R. H., 1974. *Téthys*, **6,** 327–340.

Andrews, W. R. H. & Cram, D. L., 1969. *Nature, Lond.*, **224,** 902–904.

Andrews, W. R. H., Cram, D. L. & Visser, G. A., 1970. *Proc. Symp. Oceanography in South Africa*, Durban, August 1970, South African National Committee for Oceanographic Research, CSIR, Pretoria, Paper B3, 13 pp. (unpubl.).

Andrews, W. R. H. & Hutchings, L., 1980. *Prog. Oceanogr.*, **9,** 1–81.

Apel, J. R., Byrne, H. M., Proni, J. R. & Charnell, R. L., 1975. *J. geophys. Res.*, **80,** 865–881.

Atkins, G. R., 1970. *Trans. R. Soc. S. Afr.*, **39,** 139–148.

Bailey, G., 1979. M.Sc. thesis. University of Cape Town, South Africa, 225 pp.

Bailey, G. W. & Chapman, P., in press. In *The South African Ocean Colour and Upwelling Experiment*, edited by L. V. Shannon, *Sea Fish. Res. Inst. S. Afr.*

Bain, C. A. R., 1983. *Proc. Eighteenth Cstl Engineering Conf. ASCE*, Cape Town, South Africa, November 1982, Am. Soc. cstl Engineers, pp. 2390–2402.

Bang, N. D., 1970. *S. Afr. geogr. J.*, **52,** 67–76.

Bang, N. D., 1971. *Deep-Sea Res.*, **18,** 209–224.

Bang, N. D., 1973. *Tellus*, **15,** 256–265.

Bang, N. D., 1974. Ph.D. thesis, University of Cape Town, South Africa, 181 pp.

Bang, N. D., 1976. CSIR/NRIO, S. Afr., Internal General Report SEA IR 7616, 14 pp.

Bang, N. D. & Andrews, W. R. H., 1974. *J. mar. Res.*, **32,** 405–417.

Barlow, R. G., 1984. *Mar. Ecol. Prog. Ser.*, **16,** 121–126.

Berrit, G. R., 1976. *Cah. O.R.S.T.O.M. Sér. Océanogr.*, **14,** 273–278.

Birch, G. F., 1975. *Joint Geological Survey, Univ. of Cape Town Mar. Sci. Group Bull.* No. 6, 142 pp. plus 22 plates.

Birch, G. F., 1977. *Mar. Geol.*, **23,** 305–337.

Birch, G. F., 1978. *Trans. geol. Soc. S. Afr.*, **81,** 23–34.

Birch, G. F., 1979. *Mar. Geol.*, **29,** 313–334.

Birch, G. F. & Rogers, J., 1973. *S. Afr. Shipping News Fish. Ind. Rev.*, **28**(7), 56–65.

Birch, G. F., Rogers, J., Bremner, J. M. & Moir, G. J., 1976. *Proc. First Interdisciplinary Conf. Mar. Freshw. Res. Sth Afr.*, Fiche 20A, C1–D12.

Bornhold, B. D., 1973. *Late Quaternary sedimentation in the eastern Angola Basin.* Technical Rept, WHOI-73-80 (Ph.D. thesis, Mass. Inst. Tech., 213 pp.).

Boyd, A. J., 1982. *Fish. Bull. S. Afr.*, No. 16, 9 pp.

Boyd, A. J., 1983a. *Colln scient. Pap. int. Commn S.E. Atl. Fish.*, **10,** 41–73.

Boyd, A. J., 1983b. *Investl Rep. Sea Fish. Res. Inst. S. Afr.*, No. 126, 47 pp.

Boyd, A. J. & Agenbag, J. J., 1984a. *Proceedings of the International Symposium on the Most Important Upwelling Areas off Western Africa* (*Cape Blanco and Benguela*), Barcelona, Spain, November, 1983. *Invest. pesq.* in press.

Boyd, A. J. & Agenbag, J. J., 1984b. *S. Afr. J. Sci.*, **80,** 77–79.

Boyd, A. J. & Cruickshank, R. A., 1983. *S. Afr. J. Sci.*, **79,** 150–151.

Boyd, A. J., Potgieter, E. & Buys, M. E. L., 1983. *Sea Fish. Res. Inst., S. Afr.* (unpubl. manus.), 11 pp.

Boyd, A. J. & Thomas, R. M. 1984. *Tropical Ocean—Atmosphere Newsletter*, No. 27, Sept. 1984, pp. 16–17.

Branch, G. M., 1984. *S. Afr. J. Sci.*, **80,** 61–65.

Bremner, J. M., 1978. *Tenth Int. Congr. on Sedimentology*, Jersualem, July 1978, Abstr., **1,** 87–88.

Bremner, J. M., 1980a. *Mar. Geol.*, **34,** M67–M76.

Bremner, J. M., 1980b. *J. geol. Soc. Lond.*, **137,** 773–786.

Bremner, J. M., 1981. *Geo-Mar. Lett.*, **1,** 91–96.

Bremner, J. M., 1983. In, *Coastal Upwelling: Its Sedimentary Record, Part B:*

Sedimentary Records of Ancient Coastal Upwelling, edited by J. Thiede & E. Suess, Plenum Press, New York, pp. 73–103.
Brown, P. C. & Hutchings, L., 1984. *Invest. pesq.* (in press).
Brundrit, G. B., 1981. *Trans. R. Soc. S. Afr.*, **44** (Part 3), 309–313.
Brundrit, G. B., de Cuevas, B. & Shipley, A. M., 1984. *S. Afr. J. Sci.*, **80,** 80–82.
Brundrit, G. B., Shipley, A. M., de Cuevas, B. & Brundrit, S., 1983. *Proc. 5th Nat. Oceanogr. Symp.*, S288, Grahamstown, S. Afr., January 1983 (paper B41).
Buys, M. E. L., 1957. *Investl Rep. Div. Fish. S. Afr.*, No. 27, 114 pp.
Buys, M. E. L., 1959. *Investl Rep. Div. Fish. S. Afr.*, No. 37, 176 pp.
Calvert, S. E. & Price, N. B., 1971a. *ICSU/SCOR Inst. Geol. Sci.*, Rep. No. 70/16, 171–185.
Calvert, S. E. & Price, N. B., 1971b. *Deep-Sea Res.*, **18,** 505–523.
Carmack, E. C. & Aagaard, K., 1977. *Estuar. cstl mar. Sci.*, **5,** 135–142.
Chapman, P. & Shannon, L. V., 1985. *Oceanogr. Mar. Biol. Ann. Rev.*, *Vol. 23*, pp. 183–251.
Christensen, M. S., 1980. *S. Afr. J. Sci.*, **76,** 541–546.
Clowes, A. J., 1950. *Investl Rep. Mar. Biol. Survey, S. Afr.*, No. 12, 42 pp. plus 20 pp. charts.
Clowes, A. J., 1954. *Investl Rep. Div. Fish. S. Afr.*, No. 16, 47 pp.
Coetzee, J. A., 1978. In, *Antarctic Glacial History and World Palaeoenvironments*, edited by E. M. van Zinderen Bakker, Balkema, Rotterdam, pp. 115–127.
Coetzee, J. A., Scholtz, A. & Deacon, H. J., 1983. In, *Fynbos Palaeoecology: A Preliminary Synthesis*, edited by H. J. Deacon, Q. B. Hendey & J. J. N. Lambrechts, S. Afr. Nat. Scientific Programmes Rep. No. 75, CSIR, Pretoria, pp. 156–173.
Connary, S. D. & Ewing, M., 1974. *J. geophys. Res.* **79,** 463–469.
Copenhagen, W. J., 1953. *Investl Rep. Div. Fish. S. Afr.*, No. 14, 32 pp.
Cram, D. L., 1970. *Trans. R. Soc. S. Afr.*, **39**(2), 129–137.
Crawford, R. J. M., Shelton, P. A. & Hutchings, L., 1984. *Proc. Expert consultation to examine changes in abundance and species composition of neritic fish stocks*, San Jóse, Costa Rica, April 1983, FAO Fish Rept, No. 291, part 2, pp. 407–448.
Currie, R., 1953. *Nature, Lond.*, **171,** 497–500.
Darbyshire, J., 1964. *Deep-Sea Res.*, **11,** 781–815.
Darbyshire, M., 1963. *Deep-Sea Res.*, **10,** 623–632.
Darbyshire, M., 1966. *Deep-Sea Res.*, **13,** 57–81.
Deacon, G. E. R., 1937. *'Discovery' Rep.*, **15,** 1–123.
Deacon, H. J., 1983. In, *Fynbos Palaeoecology: A Preliminary Synthesis*, edited by H. J. Deacon, Q. B. Hendey & J. J. N. Lambrechts. S. Afr. Nat. Sci. Programmes Rep. No. 75, CSIR, Pretoria, pp. 1–20.
De Decker, A. H. B., 1970. *Investl Rep. Div. Sea Fish. S. Afr.*, No. 84, 24 pp.
Defant, A. 1936. *Landerkdl. Forsch.*, Festchr. N. Krebs, Stuttgart, pp. 52–66.
Defant, A., 1961. *Physical Oceanography*, *Vol. 1*, Pergamon Press, London, 729 pp.
Defant, A. & Wüst, G., 1938. *Wiss. Ergebn. Dt. Atlant. Exp. 'Meteor'*, **6**(2:3), 105–181.
De Wet, L. W., 1979. M.Sc. thesis, University of Pretoria, South Africa, 79 pp.
Dias, C. A., 1983a. *Colln scient. Pap. int. Commn S.E. Atl. Fish.*, **10,** 103–116.
Dias, C. A., 1983b. *Colln scient. Pap. int. Commn S.E. Atl. Fish.*, **10,** 99–102.
Diester-Haass, L. & Schrader, H. J., 1979. *Mar. Geol.*, **29,** 39–53.
Dietrich, G., 1935. *Veröff. Inst. Meereskunde, Berlin, N. F., Reihe A, Heft* 27, 1–79.
Dingle, R. V. & Hendey, Q. B., 1984. *Mar. Geol.*, **56,** 13–26.
Dingle, R. V., Moir, G. J., Bremner, J. M. & Rogers, J., 1977. *Dept Mines S. Afr., Mar. Geosci. Ser. 1*, map—compiled by S. M. L. Sayers.
Dingle, R. V., Siesser, W. G. & Newton, A. R., 1983. *Mesozoic and Tertiary Geology of Southern Africa*, Balkema, Rotterdam, 375 pp.
Duffy, D. C., Berruti, A., Randall, R. M. & Cooper, J., 1984. *S. Afr. J. Sci.*, **80,** 65–69.
Duing, W., Ostapoff, F. & Merle, J., 1980. Editors, *Physical Oceanography of the Tropical Atlantic during GATE*, University of Miami, Florida, 117 pp.

Duncan, C. P., 1964. *Investl Rep. Div. Sea Fish. S. Afr.*, No. 50, 15 pp.
Duncan, C. P., 1965. *J. mar. Res.*, **23,** 350–354.
Duncan, C. P., 1966. *J. mar. Res.*, **24,** 124–130.
Duncan, C. P., 1968. *J. geophys. Res.*, **73,** 531–534.
Duncan, C. P. & Nell, J. H., 1969. *Investl Rep. Div. Sea Fish. S. Afr.*, No. 76, 19 pp.
Du Plessis, E., 1967. *Investl Rep. mar. Res. Lab. S.W.A.*, No. 13, 16 pp. plus 19 pp. figures and tables.
Dyer, T. G. J. & Tyson, P. D., 1975. *S. Afr. geogr. J.*, **75,** 104–110.
Filippov, E. A. & Kolesnikov, G. I., 1971. *Trudy Atlantniro*, **33,** 43–49.
Fuglister, F. C., 1960. *Atlantic Ocean Atlas*, Woods Hole Oceanographic Institution, Woods Hole, 209 pp.
Gillooly, J. F. & Walker, N. D., 1984. *S. Afr. J. Sci.*, **80,** 97–100.
Gründlingh, M. L. & Lutjeharms, J. R. E., 1979. *S. Afr. J. Sci.*, **75,** 269–270.
Gunn, B. W., 1977. *The Dynamics of Two Cape Embayments*, University of Cape Town (unpubl. manus.), 269 pp.
Hagen, E., 1979. *Geod. Geoph. Veröff.*, Berlin, Reihe 4, Heft 29, 1–71.
Hagen, E., 1981. In, *Coastal and Estuarine Sciences. I. Coastal Upwelling*, edited by F. A. Richards, American Geophysical Union, Washington D.C., pp. 72–78.
Hagen, E., 1984. *Proc. Internat. Symp. on the most important Upwelling Areas off Western Africa* (*Cape Blanco and Benguela*), Barcelona, Spain, November, 1983. *Invest. pesq.* in press.
Hagen, E., Schemainda, R., Michelsen, N., Postel, L., Schulz, S. & Below, M., 1981. *Geod. Geoph. Veröff.*, Berlin, Reihe 4, Heft 36, 1–99.
Harris, T. F. W., 1978. *S. Afr. Natl Sci. Programmes*, Rep. No. 30, CSIR, Pretoria, 103 pp.
Harris, T. F. W. & Shannon, L. V., 1979. *S. Afr. J. Sci.*, **75,** 316–317.
Hart, T. J. & Currie, R. I., 1960. *'Discovery' Rep.*, **31,** 123–298.
Hendey, Q. B., 1983. In, *Fynbos Palaeoecology: A Preliminary Synthesis*, edited by H. J. Deacon, Q. B. Hendey & J. J. N. Lambrechts, S. Afr. Nat. Sci. Programmes Rep. No. 75, CSIR, Pretoria, pp. 100–115.
Henry, A. E., 1975. *Investl Rep. Div. Sea Fish. S. Afr.* No. 95, 66 pp.
Hirst, A. C. & Hastenrath, S., 1983. *J. phys. Oceanogr.*, **13,** 1146–1157.
Holden, C. J., 1984. Poster paper presented at *Internat. Symp. on the Most Important Upwelling Areas off Western Africa* (*Cape Blanco and Benguela*), Barcelona, Spain, November, 1983. *Invest. pesq.*, in press.
Holden, C. J., in press. In, *South African Ocean Colour and Upwelling Experiment*, edited by L. V. Shannon, *Sea Fish. Res. Inst. S. Afr.* (in press).
Hutchings, L., Holden, C. J. & Mitchell-Innes, B., 1984. *S. Afr. J. Sci.*, **80,** 83–89.
Hutchings, L., Nelson, G., Horstman, D. A. & Tarr, R., 1983. In, *Sandy Beaches as Ecosystems*, edited by A. MacLachlan & T. Erasmus, Junk Publishers, Netherlands, pp. 481–500.
Isaac, W. E., 1937. *Geogr. Rev.*, **27,** 651–664.
Jackson, S. P., 1947. *S. Afr. geogr. J.*, **29,** 1–15.
Jones, P. G. W., 1971. *Deep-Sea Res.*, **18,** 193–208.
Jury, M. R., 1980. M.Sc. thesis, University of Cape Town, South Africa, 131 pp.
Jury, M. R., 1981. *Trans. R. Soc. S. Afr.*, **44** (Part 3), 299–302.
Jury, M. R., 1984. Ph.D. thesis, University of Cape Town, South Africa, 161 pp.
Jury, M. R., in press. In, *South African Ocean Colour and Upwelling Experiment*, edited by L. V. Shannon, *Sea Fish. Res. Inst. S. Afr.* (in press).
Kamstra, F., in press. In, *South African Ocean Colour and Upwelling Experiment*, edited by L. V. Shannon, *Sea Fish. Res. Inst. S. Afr.* (in press).
Kennett, J. P., 1978. *Mar. Micropaleontol.*, **3,** 301–345.
Kennett, J. P., 1982. *Marine Geology*. Prentice-Hall, New Jersey, 813 pp.
Latun, V. S., 1962. *Bull. Acad. Sci. USSR geophys. Ser.* (translated by American Geophysical Union), **9,** 770–773.

Lutjeharms, J. R. E., 1977. *Tellus*, **29**, 375–381.
Lutjeharms, J. R. E., 1981a. In, *Oceanography from Space*, edited by J. F. R. Gower, Marine Science Vol. 13, Plenum Press, New York, pp. 195–199.
Lutjeharms, J. R. E., 1981b. *Deep-Sea Res.*, **28A**, 1289–1302.
Lutjeharms, J. R. E., 1981c. *CSIR Research Report, S. Afr.*, No. 384, 39 pp.
Lutjeharms, J. R. E. & Heydorn, A. E. F., 1981. *Deep-Sea Res.*, **28A**, 631–636.
Lutjeharms, J. R. E. & Valentine, H. R., 1981. *Adv. Space Res.*, **1**, 211–223.
Lutjeharms, J. R. E., van Ballegooyen, R. C. & Valentine, H. R., 1981. CSIR Report T/SEA 8104, 121 pp.
Martin, A. K., 1981. *S. Afr. J. Sci.*, **77**, 547–554.
Mazeika, P., 1968. *Am. Geogr. Soc., Serial Atlas of Mar. Envir.*, Folio 16.
Meyer, H. F., 1923. *Veröff. Inst. Meereskunde*, Berlin, N. F., Reihe A, **11**, 1–35.
Meyers, P. A., Brassell, S. C., Huc, A. Y., Barron, E. J., Boyce, R. E., Dean, W. E., Hay, W. W., Keating, B. H., McNulty, C. L., Nohara, M., Schallreuter, R. E., Sibuet, J.-C., Steinmetz, J. C., Stow, D. & Stradner, H., 1983. In, *Coastal Upwelling. Its Sedimentary Record, Part B. Sedimentary Records of Ancient Coastal Upwelling*, edited by J. Thiede & E. Suess, Plenum Press, New York, pp. 453–466.
Moore, D. W., Hisard, P., McCreary, J. P., Merle, J., O'Brien, J. J., Picaut, J., Verstraete, J. & Wunch, C., 1978. *Geophys. Res. Lett.*, **5**, 637–640.
Moroshkin, K. V., Bubnov, V. A. & Bulatov, R. P., 1970. *Oceanology*, **10**, 27–34.
Mostert, S. A., 1970. Physical and chemical changes in the sea water off St Helena Bay during the years 1961–1968. *Sea Fish. Res. Inst., S. Afr.* (unpubl. manus.), 13 pp. plus tables and figures.
Muhry, A., 1862. *Klimatologische Ubersicht der Erde*. Leipzig (quoted in Clowes, 1950).
Nelson, G., 1981. *Trans. R. Soc. S. Afr.*, **44** (Part 3), 303–308.
Nelson, G., 1983. *S. Afr. J. Sci.*, **79**, 147 only.
Nelson, G., in press. In, *South African Ocean Colour and Upwelling Experiment*, edited by L. V. Shannon, *Sea Fish. Res. Inst. S. Afr.* (in press).
Nelson, G. & Hutchings, L., 1983. *Progr. Oceanogr.*, **12**, 333–356.
Nelson, G., Kamstra, F. & Walker, N. D., 1983. *Proc. Symp. Atmos. Sci. South Africa*, S331, Pretoria, S. Afr., October, 1983 (Abstract, p. 34).
Nelson, G. & Shannon, L. V., 1983. *S. Afr. J. Sci.*, **79**, 147 only.
Nelson, G. & Walker, N. D., 1984. *S. Afr. J. Sci.*, **80**, 90–93.
Nguyen Ngoc Anh & Gill, A. E., 1981. *Q. Jl. Roy. met. Soc.*, **107**, 521–530.
NRIO, 1976. *CSIR Report C/SEA 7611, Stellenbosch, S. Afr.*, 15 pp.
Olson, S. L., 1983. *S. Afr. J. Sci.*, **79**, 399–402.
Orren, M. J., 1963. *Investl Rep. Div. Sea Fish. S. Afr.*, No. 45, 61 pp.
Orren, M. J., 1966. *Investl Rep. Div. Sea Fish. S. Afr.*, No. 55, 36 pp.
Orren, M. J. & Shannon, L. V., 1967. *Deep-Sea Res.*, **14**, 279 only.
O'Toole, M. J., 1980. *Investl Rep. Sea Fish. Inst. S. Afr.*, No. 121, 25 pp.
Parrish, R. H., Bakun, A., Husby, D. M. & Nelson, C. S., 1983. In, *Proc. Expert consultation to examine changes in abundance and species of neritic fish resources*, San Jóse, Costa Rica, April 1983, FAO Fish Rept, No. 291, part 3, edited by G. D. Sharp & J. Csirke, pp. 731–777.
Pearce, A. F., 1977. *Prof. Res. Ser. No. 1, NRIO, CSIR, S. Afr.*, 220 pp.
Petermann, A., 1865. *Petermann's Mitteilungen*, **11**, 146–160 (quoted in Clowes, 1950).
Picaut, J., 1981. In, *Recent Progress in Equatorial Oceanography* (*A report of the final meeting of SCOR Working Group 47 in Venice, Italy, April 1981*), edited by J. P. McCreary, D. W. Moore & J. M. Witte, Nova University/NYIT Press, pp. 271–281.
Pieterse, F. & van der Post, D. C., 1967. *Investl Rep. mar. Res. Lab. S.W. Afr.*, No. 14, 125 pp.
Pugh, J., 1982. B.Sc. (Hons) project, University of Cape Town, South Africa, 10 pp.
Rabinowitz, P., Shackleton, L. Y. & Brenner, C., 1980. *Lamont-Doherty Geol. Obs./Univ. Cape Town, Chart 526A*.

Rawson, H. E., 1908. *Q. Jl Roy. met. Soc.*, **34,** 165–185.
Rennell, J., 1832. Charts of the prevalent currents in the Atlantic Ocean, engraved by J. and C. Walker, London.
Robson, S., 1983. *J. Micropalaeontol.*, **2,** 31–38.
Rogers, J., 1977. *Joint Geological Survey/University of Cape Town Marine Science Group Bulletin* No. 7, 162 pp.
Ross, J. C., 1847. *Voyage of Discovery and Research in the Southern and Antarctic Regions during the Years 1839–43.* John Murray, London (as cited by Hart & Currie, 1960).
Schell, I. I., 1968. *Dt. hydrogr. Z.*, **21,** 109–117.
Schell, I. I., 1970. *J. geophys. Res.*, **75,** 5225–5241.
Schultz, S., Schemainda, R. & Nehring, D., 1979. *Geod. Geophys. Veröff.*, Reihe IV, **28,** 1–7 and 43 pp. data tables.
Schulze, G., 1984. *S. Afr. J. Sci.*, **80,** 94–97.
Schumann, E. H. & Beekman, L. J., 1984. *Trans. R. Soc. S. Afr.*, **45,** 191–203.
Schumann, E. H. & Perrins, L.-A., 1983. *Proc. Eighteenth cstl Engineering Conf. ASCE*, Cape Town, South Africa, November, 1982, Am. Soc. cstl Engineers, pp. 2562–2580.
Schumann, E. H., Perrins, L.-A. & Hunter, I. T., 1982. *S. Afr. J. Sci.*, **78,** 238–242.
Scrutton, R. A. & Dingle, R. V., 1975. *Trans. geol. Soc. S. Afr.*, **77,** 253–260.
Shackleton, L. Y., 1982. *Univ. Cape Town, Chart 126A.*
Shannon, L. V., 1966. *Investl Rep. Div. Sea Fish. S. Afr.*, No. 58, 22 pp., plus 30 pp. figures.
Shannon, L. V., 1970. *Fish. Bull. S. Afr.*, No. 6, 27–33.
Shannon, L. V., 1977. In, *SANCOR Physical and Chemical Oceanography*. Rep. on a seminar held at the University of Cape Town, 6–7th Sept. 1977, pp. 21–22.
Shannon, L. V., 1983. *Tropical Ocean—Atmosphere Newsletter*, **22,** 8–9.
Shannon, L. V. & Anderson, F. P., 1982. *S. Afr. J. Photogrammetry, Remote Sensing and Cartography*, **13,** 153–169.
Shannon, L. V., Brundrit, G. B., Crawford, R. J. M., Payne, A. I. L. & Shelton, P. A., 1984. *Proc. Int. Symp. on the Most Important Upwelling Areas off Western Africa* (*Cape Blanco and Benguela*), Barcelona, Spain, November, 1983. *Invest. pesq.* (in press).
Shannon, L. V. & Chapman, P., 1983a. *S. Afr. J. mar. Sci.*, **1,** 231–244.
Shannon, L. V. & Chapman, P., 1983b. *S. Afr. J. Sci.*, **79,** 454–458.
Shannon, L. V., Chapman, P., Eagle, G. A. & McClurg, T. P., 1983. *Oil Petrochem. Pollut.*, **1,** 243–259.
Shannon, L. V., Cherry, R. D. & Orren, M. J., 1970. *Geochim. Cosmochim. Acta*, **34,** 701–711.
Shannon, L. V., Crawford, R. J. M. & Duffy, D. C., 1984. *S. Afr. J. Sci.*, **80,** 51–60.
Shannon, L. V., Hutchings, L., Bailey, G. W. & Shelton, P. A., 1984. *S. Afr. J. mar. Sci.*, **2,** 109–130.
Shannon, L. V. & Lutjeharms, J. R. E., 1983. *Proc. of Earth Data Information Systems Symp.*, Pretoria, S. Africa, September, 1983, edited by C. Martin, S. Afr. Soc. Photogrammetry, Remote Sensing and Cartography, 14 pp.
Shannon, L. V., Mostert, S. A., Walters, N. M. & Anderson, F. P., 1983. *J. Plankton Res.*, **5,** 565–583.
Shannon, L. V., Nelson, G. & Jury, M. R., 1981. In, *Coastal Upwelling*, edited by F. A. Richards, Coastal and Estuarine Sciences 1, AGU, Washington, D.C., pp. 146–159.
Shannon, L. V., Schlittenhardt, P. & Mostert, S. A., 1984. *J. geophys. Res.*, **89,** 4968–4976.
Shannon, L. V. & Stander, G. H., 1977. *Trans. R. Soc. S. Afr.*, **42,** 441–459.
Shannon, L. V., Stander, G. H. & Campbell, J. A., 1973. *Investl Rep. Sea Fish. Brch S. Afr.*, No. 108, 31 pp.

Shannon, L. V. & van Rijswijck, M. 1969. *Investl Rep. Div. Sea Fish. S. Afr.*, No. 70, 19 pp.
Shannon, L. V., Walters, N. M. & Moldan, A. G. S., 1983. *S. Afr. J. mar. Sci.*, **1,** 111–122.
Shannon, L. V., Walters, N. M. & Mostert, S. A., in press. In, *South African Ocean Colour and Upwelling Experiment*, edited by L. V. Shannon, *Sea Fish. Res. Inst. S. Afr.* (in press).
Shannon, L. V., Walters, N. M., Mostert, S. A. & Anderson, F. P., 1983. *J. Plankton Res.*, **5,** 565–583.
Shelton, P. A., 1984. *S. Afr. J. Sci.*, **80,** 69–71.
Shelton, P. A. & Hutchings, L. 1982. *J. Cons. perm. int. Explor. Mer*, **40,** 185–198.
Shelton, P. A. & Kriel, F., 1980. *Fish. Bull. S. Afr.*, **13,** 107–109.
Siesser, W. G., 1978. In, *Antarctic Glacial History and World Palaeoenvironments*, edited by E. M. van Zinderen Bakker, Balkema, Rotterdam, pp. 105–113.
Siesser, W. G., 1980. *Science*, **208,** 283–285.
Siesser, W. G. & Dingle, R. V., 1981. *J. Geol.*, **89,** 83–96.
Siesser, W. G., Scrutton, R. A. & Simpson, E. S. W., 1974. In, *The Geology of Continental Margins*, edited by C. A. Burke & C. L. Drake, Springer-Verlag, New York, pp. 641–654.
Stander, G. H., 1958. *Investl Rep. Div. Fish. S. Afr.*, No. 35, 40 pp.
Stander, G. H., 1962. *Investl Rep. mar. Res. Lab. S.W. Afr.*, No. 5, 63 pp.
Stander, G. H., 1963. *Investl Rep. mar. Res. Lab. S.W. Afr.*, No. 9, 57 pp.
Stander, G. H., 1964. *Investl Rep. mar. Res. Lab. S.W. Afr.*, No. 12, 43 pp., plus 77 pp. figures.
Stander, G. H. & De Decker, A. H. B., 1969. *Investl Rep. Div. Sea Fish. S. Afr.*, No. 81, 46 pp.
Stander, G. H., Shannon, L. V. & Campbell, J. A., 1969. *J. mar. Res.*, **27,** 293–300.
Stetsjuk, G. A., 1983. *Colln scient. Pap. int. Commn S.E. Atl. Fish*, **10,** 173–178.
Strogalev, V. D., 1983. *Colln scient. Pap. int. Commn S.E. Atl. Fish.*, **10,** 179–182.
Summerhayes, C. P., Bornhold, B. D. & Embley, R. W., 1979. *Mar. Geol.*, **31,** 265–277.
Sverdrup, H. U., Johnson, M. W. & Fleming, R. H., 1942. *The Oceans, their Physics, Chemistry and General Biology*, Prentice-Hall, New York, 1087 pp.
Taljaard, J. J., Schmidt, W. & van Loon, H., 1961. *Notos* (S. Afr. Weather Bureau, Pretoria), **10,** 25–58.
Tankard, A. J. & Rogers, J., 1978. *J. Biogeogr.* **5,** 319–337.
Taunton-Clark, J., in press. In, *South African Ocean Colour and Upwelling Experiment*, edited by L. V. Shannon, *Sea Fish. Res. Inst. S. Afr.* (in press).
Tripp, R. T., 1967. *An atlas of coastal surface drifts, Cape Town to Durban*, University of Cape Town, 10 pp. (unpubl.).
Tromp, B. B. S., Lazarus, B. I. & Horstman, D. A., 1975. *Gross features of the S.W. Cape Coastal waters.* Sea Fish. Res. Inst. S. Afr. (unpubl. rep.), 43 pp.
Tyson, P. D., 1981. *J. Climatology*, **1,** 115–130.
Van Andel, T. H. & Calvert, S. E., 1971. *J. Geol.*, **79,** 585–602.
Van Foreest, D. & Brundrit, G. B., 1982. *Progr. Oceanogr.*, **11,** 329–392.
Van Foreest, D. & Jury, M. R., in press. *S. Afr. J. Sci.* (in press).
Van Foreest, D., Shillington, F. A. & Legeckis, R., in press. *Continental Shelf Res.* (in press).
Van Ieperen, M. P., 1971. *Hydrology of Table Bay.* Dept Oceanogr., Univ. Cape Town, Int. Rept, 48 pp.
Van Loon, H., 1972a. In, *Meteorology of the Southern Hemisphere*, edited by C. W. Newton, American Meteorological Society, Boston, Meteorological Monographs 13(35), pp. 59–86.
Van Loon, H., 1972b. In, *Meteorology of the Southern Hemisphere*, edited by C. W. Newton, American Meteorological Society, Boston, Meteorological Monographs 13(35), pp. 101–111.
Van Zinderen Bakker, E. M., 1975. *J. Biogeogr.*, **2,** 65–73.

Van Zinderen Bakker, E. M., 1976. *Palaeoecol. Africa*, **9,** 160–202.
Van Zinderen Bakker, E. M., 1978. In, *Antarctic Glacial History and World Palaeoenvironments*, edited by E. M. van Zinderen Bakker, Balkema, Rotterdam, pp. 129–135.
Van Zinderen Bakker, E. M., 1982. In, *Palaeoecology of Africa and the Surrounding Islands, Vol. 15*, edited by J. C. Vogel, E. A. Voigt & T. C. Partridge, Balkema, Rotterdam, pp. 77–99.
Visser, G. A., 1969a. *Investl Rep. Div. Sea Fish. S. Afr.*, No. 75, 26 pp.
Visser, G. A., 1969b. *Investl Rep. Div. Sea Fish. S. Afr.*, No. 77, 23 pp.
Walker, N. D., Taunton-Clark, J. & Pugh, J., 1984. *S. Afr. J. Sci.*, **80,** 72–77.
Walter, H., 1937. In, *Jahrbücher für Wissenschaftliche Botanik*, edited by N. Pringsheim, Verlag von Gebrüder Borntraeger, Leipzig, pp. 58–222.
Ward, J. D., Seely, M. K. & Lancaster, N., 1983. *S. Afr. J. Sci.*, **79,** 175–183.
Welsh, J. G., 1964. *Deep-Sea Res.*, **11,** 43–52.
Welsh, J. G. & Visser, G. A., 1970. *Investl Rep. Div. Sea Fish. S. Afr.*, No. 83, 23 pp.
Wooster, W. S., 1973. *Proc. South African Nat. Oceanogr. Symp.*, Cape Town, S. Afr., August, 1973, edited by H. G. v. D. Boonstra, Sea Fisheries Branch of Department of Industries, Cape Town, p. 1 (abstract).
Wooster, W. S. & Reid, J. L., 1963. In, *The Sea, Vol. 2*, edited by M. N. Hill, Wiley Interscience, New York, pp. 253–276.
Yelizarov, A. A., 1967. *Oceanology*, **7,** 344–347.

Oceanogr. Mar. Biol. Ann. Rev., 1985, **23**, 183–251
Margaret Barnes, Ed.
Aberdeen University Press

THE BENGUELA ECOSYSTEM PART II. CHEMISTRY AND RELATED PROCESSES

P. CHAPMAN and L. V. SHANNON
Sea Fisheries Research Institute, Private Bag X2, Rogge Bay 8012, Cape Town, South Africa

INTRODUCTION

This review is the second in a tetralogy on the Benguela and deals with the chemistry of the system and related processes. It follows on logically from the discussion of the physical oceanography (Part I, Shannon, 1985) and has been structured so as to set the scene for the review of the biology of the Benguela ecosystem, which is in preparation. The present paper, the first to attempt a synthesis of information on the marine chemistry of the region, has been subdivided into four broad sections dealing with chemical inputs, the formation and advection of oligoxic water, nutrients and minor elements. As in the case of Part I the list of references is fairly comprehensive, although it is probable that we have missed some of the less readily available works *e.g.* reports in Russian published in the U.S.S.R. and possibly some recent papers by scientists in the German Democratic Republic. Nevertheless, the papers cited will serve to lead to other literature on the Benguela and the chemistry of comparable upwelling systems. In general we have excluded from our discussion publications dealing with the chemistry of the nearshore zone as these relate mainly to very localized pollution problems, or to the study of kelp beds and the intertidal region and, as such, fall outside the scope of this review.

Prior to 1925 very little was known about the chemistry of the South East Atlantic Ocean and adjacent areas. Following the era of the great expeditions *e.g.* by the ships VALDIVIA, DISCOVERY, METEOR, a picture of the macroscale distribution of elements of major importance such as oxygen, phosphorus, and silicate began, however, to emerge (Schott, 1902; Wattenberg, 1928, 1938, 1939; Hentschel & Wattenberg, 1930; Clowes, 1938) which, together with the progress in physical oceanography, resulted in an understanding of the large scale processes. Although the rôle of upwelling in the supply of nutrients in the Benguela was appreciated at an early stage (*e.g.* Copenhagen, 1934), it was not until the publication of the definitive text of Hart & Currie (1960) that the nutrient chemistry was placed in a proper biological and physical perspective. These authors also provided an excellent summary of the early published work. Building on the foundations laid by Hart & Currie (1960), were subsequent investigations by *inter alia* De Decker (1970), Calvert & Price

(1971), and Jones (1971), while more recently Andrews & Hutchings (1980) and Olivieri (1983a) have provided a good understanding of the distribution, supply, and utilization of nutrients on the mesoscale. With the improved speed and precision of analyses afforded by the advances in instrumentation during the 1960s, there was a dramatic expansion in the collection of data on the major and minor elements in the Benguela system. Sadly, however, publications have generally not kept pace and relatively few really good papers on the chemistry of the Benguela system have appeared during recent years. Accordingly we have tried, where possible, to provide our own interpretation of the processes in order to place the published data and papers into a system perspective.

TERRIGINOUS INPUTS

Although much of the coastline adjacent to the Benguela system is arid and virtually unpopulated, there is still some input of material either *via* rivers or *via* aeolian dust. There is, however, little information on the effects of either of these sources.

The mean volumes of run-off, together with estimates of their chemical composition, are given in Table I for the four major rivers that flow into the Benguela system. Because of poor accessibility to the Kunene, data are taken

TABLE I

Run-off and chemical data for rivers entering the Benguela system: data from Directorate of Water Affairs, Pretoria, and Department of Water Affairs, Windhoek; chemical data in mg·l^{-1} ± 2SD, except for Kunene River where very few data are available

	River			
	Kunene	Orange	Olifants	Berg
Mean run-off $m^3 \cdot yr^{-1} \times 10^6$	5200	7300	708	557
Peak flow	Jan.–Apr.	Oct.–Apr.	May–Oct.	May–Oct.
Ca	9	24±10	1±2	10±12
Mg	3	10±4	2±1	9±10
K	3	1·9±1·2	0·7±0·4	3·0±2·0
Na	4	18±16	10±4	47±50
Cl	3	14±20	18±8	85±96
F	0·1	0·3±0·2	0·1±0·2	0·2±0·2
SiO_2	12	7·6±3·0	1·5±1·0	1·9±1·6
SO_4^{2-}	3	18±24	4±4	15±18
NH_4^+		0·07±0·18	0·08±0·28	0·06±0·38
$NO_3^- + NO_2^-$	"trace"	0·17±0·66	0·04±0·10	0·43±1·22
PO_4^{3-}		0·03±0·06	0·01±0·01	0·03±0·24

at the Ruacana dam, some 320 km from the sea, but it is thought that side contributions from the lower catchment only compensate for main channel losses, and that the figures are, therefore, reasonably accurate (A. D. Hattle, pers. comm.). Both the Orange and Kunene are typical of subtropical rivers in having highly erratic flows, with one or more floods per year, which are responsible for the majority of the run-off and sediment transport.

The Orange River is by far the best studied of the four rivers, and data on flow go back to at least 1928. A very detailed description of sediment transport in the river is given by Rogers (1977). Although the major portion of the river's length is through arid and semi-arid areas, the occasional rare storm can have a catastrophic effect. De Villiers & Söhnge (1959) witnessed such a storm in 1943 on the lower Orange at Vioolsdrif, and commented that ". . . restricted areas may receive more rain in one hour than they usually receive in a year or more. . . . The storm approached from the northwest and was heralded by a wind of gale force that swept sand and dust to a height judged to be at least 2000 feet. This dust-storm swirled and eddied down all the tributary gorges into the Orange River and beyond to the sandy plains, and was followed by a deluge of rain swept along almost horizontally by the wind. So much rain fell that the shallow drainage-channels on the level portions of the Neint Nababeep Plateau were unable to accommodate the waters, which fell as roaring cataracts over the edge of the escarpment and formed a magnificent waterfall, miles long and as much as 700 feet high. All the tributaries to the Orange River were raging torrents, and water flowed where it had not been known to flow for years, transporting enormous quantities of sand and rubble. . . . Old alluvial fans were deeply incised . . . and a new fan, containing gigantic boulders the size of a room, was deposited at the edge of the old."

Within the last hundred years there have been major changes in the flow characteristics of the Orange, partly as a result of changes in agricultural practice, which have led to a decline in the rate of erosion in the upper catchment, the origin of most of the sediment load (Rooseboom & Maas, 1974), and partly through changes in the extant material (Rooseboom & Harmse, 1979). The construction of dams which have acted as sediment traps since the 1960s has also had its effect. Thus, while flows between 1939–1960 averaged 9300×10^6 $m^3 \cdot yr^{-1}$, since that date the mean annual flow has been only 5500×10^6 m^3. A similar decline in sediment transport since the 1930s from 90×10^6 $tonnes \cdot yr^{-1}$ to about 60×10^6 tonnes by 1965 is also evident in the figures (data from Directorate of Water Affairs). Rooseboom (pers. comm.) considers that since the construction of the Hendrik Verwoerd Dam (1970), the sediment load of the lower Orange has probably decreased to $<10 \times 10^6$ $tonnes \cdot yr^{-1}$.

Whether this change in flow rates and sediment loading has had any effect on the Benguela system as a whole is, however, very doubtful. Data on the chemistry of the Orange River suggest that elemental concentrations are very similar to world figures (see Burton, 1976; Liss, 1976) with the exception of Na, Ca, Mg and Cl, which are about double the average. The Olifants River appears to be low in almost all constituents, while the Berg River is particularly enriched in Na and Cl. Nutrients, with the exception of silica, which appears low, are within the normal ranges for other rivers. Similar data have been obtained by Eagle & Bartlett (1984) following survey trips to the estuaries of these rivers.

The mean annual flow of 7300×10^9 m^3 for the Orange equates to 20×10^9 $m^3 \cdot day^{-1}$. This is the same volume as is upwelled in a strip of water 100 km × 20 km if the upwelling rate is 10 $m \cdot day^{-1}$, a reasonable assumption (Shannon, 1985). Since the region round the mouth of the Orange is not one of the major upwelling areas, there is an important local effect near the estuary, and river water may be detected up to between 30–80 km offshore (Hart & Currie, 1960; Stander, 1964; Jones, 1971). The last author in fact found 35·8 μM silicate in the surface water at his station nearest the mouth, but $<0{\cdot}1$ μM at the next offshore station. There might also be effects on the dune formation in the Namib Sand Sea, the source of which appears to be coarse sediment transported northwards from the Orange mouth (Rogers, 1977), and on trace metal geochemistry to the south; Birch (1975) showed that high Fe and particularly Mn concentrations supposedly derived from the Orange were found between Spencer Bay (26° S) and St Helena Bay (33° S). In the latter case, there may well be additional inputs from the other two rivers.

The Berg and the Olifants seem to have equally little effect further south; although there is, particularly in winter, some decrease in surface salinity, this appears to affect only the surface 5–10 m (see Table VII, p. 232; Buys, 1959; Orren, 1969), and then only an area within 10 km of the Berg River mouth. Nutrients may, however, be more affected. Despite the low values (in riverine terms) seen in the inflowing water, since the St Helena Bay area appears to be a semi-closed system (Bailey & Chapman, in press), nutrient build-up in the area is possible, and sedimentary inputs could then become important. Dissolved nutrients in the river water are of the same order of magnitude as in the bottom water in the Bay (Eagle & Bartlett, 1984). Further data on the effects of river input to the Benguela system may be found in the section on particulate matter on page 227.

Copenhagen (1953), Hart & Currie (1960), and others have commented on the input of wind-blown dust and sand into the sea off Namibia. In the second case the authors noted that the ship was covered in a fine layer of dust following an offshore wind. Aeolian transport of dust offshore to a distance of 150 km between 18° S and 30° S during a katabatic ("Berg") wind event was demonstrated by Shannon & Anderson (1982). This was shown in Figure 10 in Part I (Shannon, 1985). Although it is difficult to quantify this input of terriginous material to the Benguela, the authors considered that this single event might add as much solid material as the annual supply from the Orange River ($\pm 50 \times 10^6$ tonnes). If this is so, then the aeolian transport of particulates has to be considered in any mass balance estimates.

DISSOLVED OXYGEN

One of the major features of the Benguela region is the occurrence of large areas throughout which very low oxygen concentrations are found. South Atlantic Central water commonly contains between 4·8–5·2 $ml \cdot l^{-1}$ dissolved oxygen (Shannon, 1966) and is about 80–85% saturated. In contrast, the shelf waters along the west coast of southern Africa frequently contain lower levels, and anoxic water has been reported from the area near Walvis Bay (Copenhagen, 1953; Hart & Currie, 1960; Stander, 1964; Pieterse & Van der Post, 1967; Eisma, 1969). Incursions of this low-oxygen water have been implicated in kills of both fish and other organisms (Copenhagen, 1953; Brongersma-Sanders, 1957; Hart & Currie, 1960; De Decker, 1970; and others). Subsurface oxygen concentrations anywhere between these limits may be found on the shelf following upwelling, although the lowest concentrations are usually confined to particular areas (see below), while in the surface layer where phytoplankton production is high, levels naturally increase and supersaturation is common. Pieterse & Van der Post (1967) found values of up to 185% saturation immediately off Walvis Bay in December 1965 following a bloom of *Peredinium triquetrum*, and Bailey (1979) has recorded 180% saturation at North Rocks (24°30′ S). Further south, Chapman (1983a) has reported values of up to 178% saturation in St Helena Bay, and Olivieri (1983a) found that surface oxygen concentrations increased from <90% to >140% saturation in 56 h off the Cape Peninsula. (This latter increase followed a mixed bloom, whose dominant species were *Chaetoceros compressus* and *Skeletonema costatum*.) Such high levels are only associated with quiescent well-stratified conditions, when an oxycline is normally found associated with the thermocline (Bailey, 1979; Boyd, 1983a; Chapman, 1983a; Olivieri, 1983a). In such cases, the rate of change of oxygen with depth is commonly 0·4 ml·m^{-1} or greater (De Decker, 1970; Bailey & Chapman, in press).

A discussion of oxygen depletion in the Benguela is complicated by the variety of terms that have been used by various authors to refer to different oxygen concentrations. Wattenberg (1938), for example, defined any value of less than 70% saturation (about 4·2 ml·l^{-1} at 9 °C) as "oxygen-depleted". This criterion was also used by Bubnov (1972), who also spoke of it as part of the "oxygen minimum layer". De Decker (1970), on the other hand, referred to "oxygen-poor" water, but termed anything containing less than 2 ml·l^{-1} oxygen "oligoxic". For the purposes of this review oxygen concentrations of less than 2 ml·l^{-1} will be termed "oxygen-deficient", while levels above 2 but less than 5 ml·l^{-1} will be termed "oxygen-depleted". This is in line with international usage (Deuser, 1975; Bailey, Beyers & Lipschitz, 1985), although strictly speaking any water containing less than the saturation level could be considered as depleted.

THE FORMATION AND ADVECTION OF LOW OXYGEN WATER

The first comprehensive account of the macroscale distribution of dissolved oxygen in the South Atlantic Ocean was provided by Wattenberg (1938, 1939) and was based on the measurements made during the 1925–1927 expedition of the METEOR. His work demonstrated the existence of a wedge-shaped tongue of oxygen-deficient water with a core depth of 300–400 m, which extended zonally from its base between the equator and latitude 20° S at the African continent across the tropical South Atlantic (Fig. 1). Lowest concentrations (<0·5 ml·l^{-1}) were recorded east of the 0° meridian at about 15° S. The existence of an oxygen-deficient layer overlying the continental shelf from 10° S to Walvis Bay was demonstrated by Van Goethem (1951) and Copenhagen (1953), while subsequent studies by Hart & Currie (1960), Stander (1964), De Decker (1970), Calvert & Price (1971) and others showed that oxygen-depleted subsurface water was characteristic of much of the Namibian shelf. Hart & Currie (1960) suggested that this oxygen-depleted water might be transported

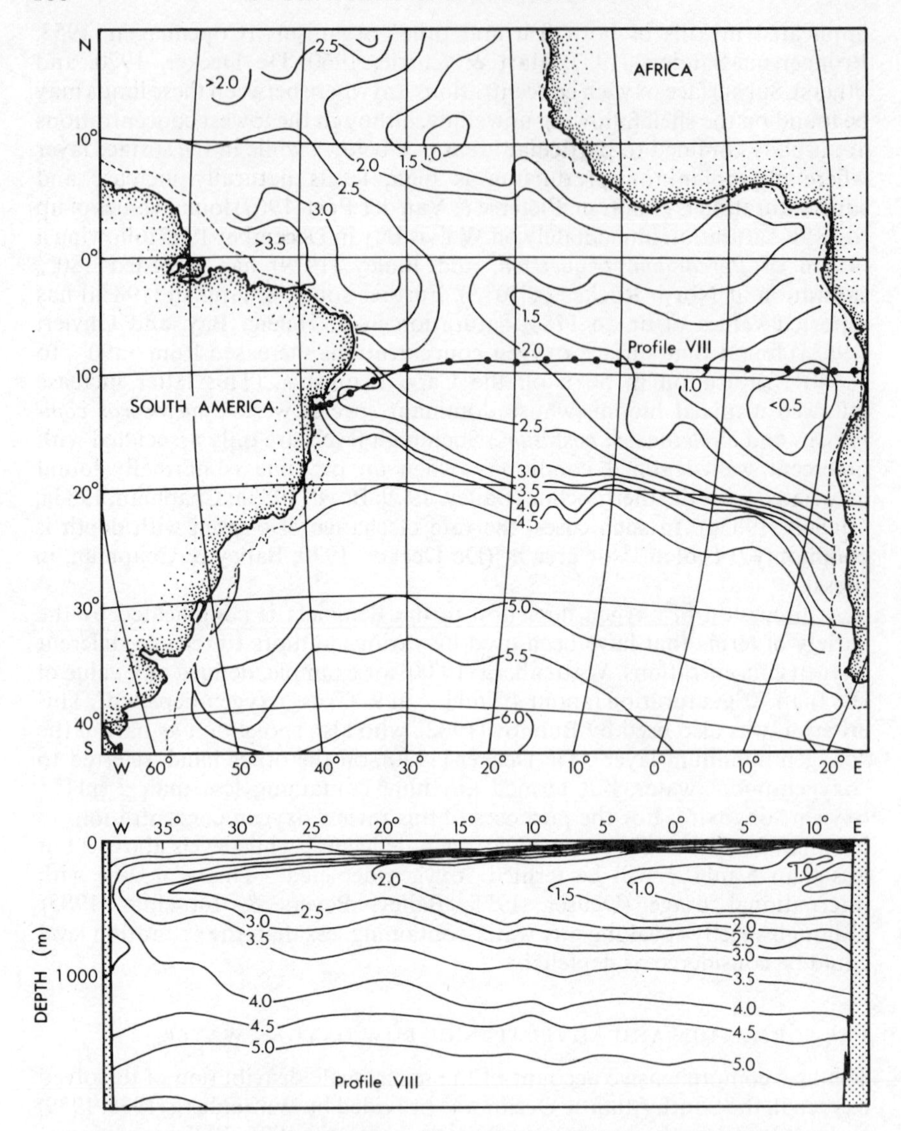

Fig. 1.—Horizontal distribution of dissolved oxygen in the South Atlantic at 400 m and zonal transect showing vertical distribution in the top 2000 m near 9° S: values in ml·l^{-1}; (after Wattenberg, 1939).

southwards from the tropical South East Atlantic along the edge of the shelf in a deep compensation current (200–400 m). This water could be elevated onto the shelf during upwelling whereupon it could be further depleted in oxygen by the local decomposition of organic matter.

This hypothesis was endorsed by Stander (1964), who noted a characteristic subsurface wedge of oxygen-depleted water near the northern Namibian coast in sharp contrast to the well-aerated water to the south and west (Fig. 2). This wedge was most conspicuous north of 25° S but extended at times as far south as the Orange River (29° S) with the oxygen minimum ($<0\cdot5$ ml·l^{-1} at 17° S, but increasing southwards) lying at a depth of about 300 m. In De Decker's paper (1970) his Figure 12 showed an obvious boundary between 24° S and 26° S. This was marked by a change in the depth and extent of the oxygen-deficient mass, although De Decker assumed that the water mass was the same and that it was uplifted onto the shelf further south. This point will be discussed later. The meridional distribution of oxygen off southern Africa described by Parrish, Bakun, Husby & Nelson (1984) also showed a distinct closing of the isopleths where their transect intersected the southwestern boundary of Stander's wedge.

Stander (1964) showed clear evidence of poleward flow immediately west of the shelf break between 200 m and 400 m north of 25° S (see Part I, Shannon, 1985), which he viewed as the mechanism producing the southward elongation of the tropical oxygen-depleted zone. He noted further that the anoxic conditions on the shelf were most pronounced during summer following periods of calms. Figure 3 (from Stander, 1964) which illustrates the vertical distribution of oxygen along four transects during April 1959 shows both the oxygen-poor core off the shelf at about 300 m and the virtually anoxic shelf water.

As the poleward undercurrent is implicit in the explanations of Hart & Currie (1960) and of Stander (1964) for the existence of oxygen-deficient water off the shelf off Namibia, it would be appropriate at this stage to comment on the tropical oxygen minimum layer. Waters with the lowest oxygen content in

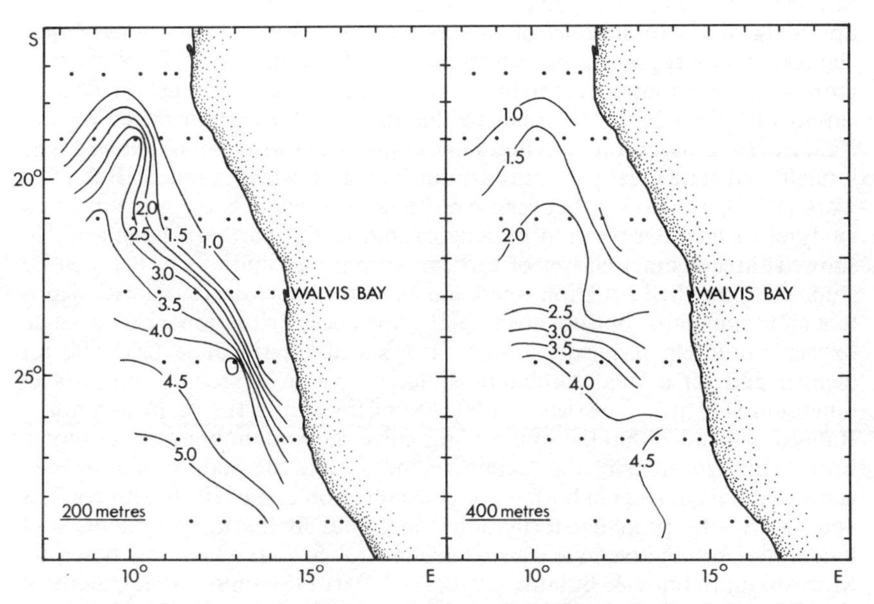

Fig. 2.—Horizontal distribution of dissolved oxygen (ml·l^{-1}) off Namibia during January, 1960 (after Stander, 1964).

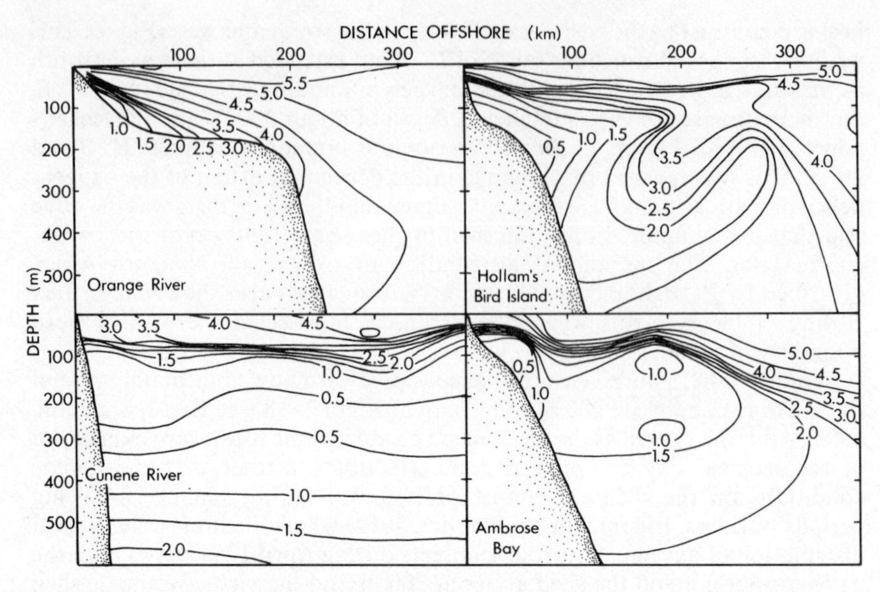

Fig. 3.—Vertical distribution of dissolved oxygen (ml·l^{-1}) off Namibia during April, 1959 (after Stander, 1964).

the Atlantic form in the tropical South East Atlantic (Bubnov, 1972) and it has been suggested that the oxygen minimum layer throughout the Atlantic could result from the horizontal advection of low oxygen water produced in or near upwelling areas. In support of this, Menzel & Ryther (1968) showed that changes in the oxygen minimum between 8° S: 30° W and 36° S: 52° W could be more easily attributed to mixing of water masses along constant density surfaces (sigma-*t* 27·0–27·2) than to the oxidation of carbon in the region. Bubnov (1972) has, however, cited some of his earlier work which showed that it might originate locally in areas distant from an upwelling source. Bubnov's work (1972), carried out between April and June 1968, is the most detailed study of oxygen distribution off Angola and in the northern Benguela. He showed three distinct classes of vertical oxygen distributions in the area he studied: (a) the shelf situation where the oxygen content decreases with depth to a minimum in the bottom layer; (b) the open ocean situation where a single oxygen minimum (typically 1 ml·l^{-1}) exists at intermediate (300–600 m) depths; and (c) a dual minimum situation where a second (subsurface) minimum (1–2 ml·l^{-1}) exists at 100–200 m, overlying the main minimum. Bubnov concluded that the source of the subsurface minimum was the anoxic water which forms over the Namibian shelf, while the main minimum was generated west of Angola both in the coastal region east of 10° E between 7° S and 18° S—the principal formation zone—and in the large cyclonic and moderately productive gyre centred around 13° S: 4° E which was noted by Moroshkin, Bubnov & Bulatov (1970)—see Part I (Shannon, 1985). Some of the diagrams in Stander (1964) and Figure 4, however, suggest that processes over the lower shelf and near the lower shelf break off Namibia may contribute

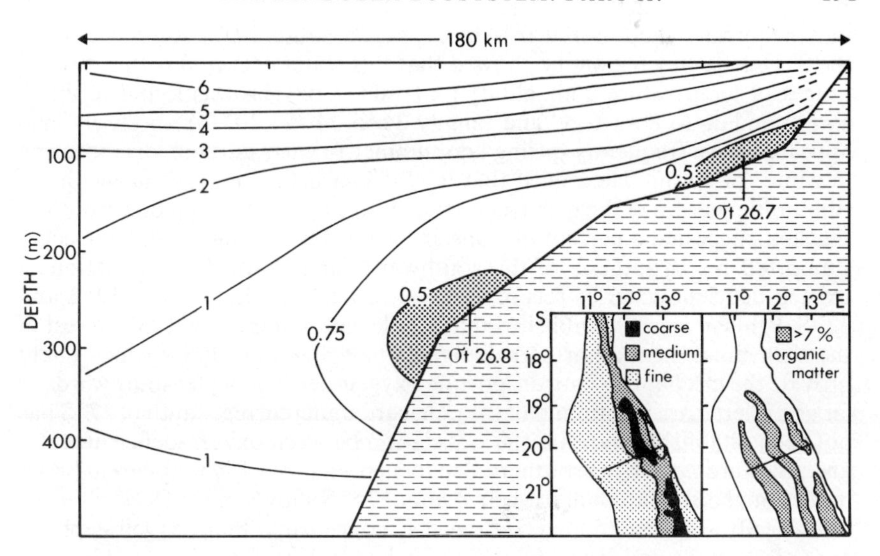

Fig. 4.—Vertical distribution of dissolved oxygen (ml·l^{-1}) along a transect near 20° S from 17th–25th October, 1976, showing discrete zones of oxygen-deficient water over the upper and lower shelves: the transect is superimposed on the inset diagrams showing sediment texture and organic matter based on the work of Bremner (1981, 1983); oxygen data from Schultz *et al.* (1979); diagram by courtesy of Dr D. van Foreest.

to the formation of the main oxygen minimum as well *i.e.* there seem to be three areas where the main minimum could be generated not just two.

Visser (1970) has reviewed the rôles of both circulation (*e.g.*, Wüst, 1935) and biochemical mechanisms (*e.g.*, Seiwell, 1937, 1938) for the formation of the oxygen minimum layer, and while the current evidence is overwhelmingly in favour of the latter it is none the less evident that advection and convection do play important rôles. It is perhaps significant that the eddy observed by Moroshkin *et al.* (1970) in the upper 300 m coincided with the oxygen-deficient (<0·5 ml·l^{-1}) pool recorded by Wattenburg (1938) in his 400-m distribution map, and that the main oxygen-poor zone is confined to the area north of the Walvis Ridge. (This ridge effectively restricts meridional circulation in the deep layers and affects upper-layer circulation patterns—see Part I.) If Bubnov (1972) is correct in his view that the coastal area off Angola is the principal region in which the main oxygen minimum is formed, then following Hart & Currie (1960) and Stander (1964) it must be concluded that processes off Angola are important for the northern Benguela region. The work of Van Goethem (1951) supports this—he found that oxygen concentrations from 10° S to 22° S on the shelf were generally less than 2 ml·l^{-1} at the bottom, but that the concentrations in the northern part of this area were variable at different periods of the year.

De Decker (1970) suggested that the oxygen-deficient water which occurs on the shelf as far south as Cape Town might be associated with the deep compensation current, postulated by Hart & Currie (1960), which he viewed as moving upwards and shorewards on its passage south. In the south, mainly in the St Helena Bay region, he showed that this water occurred within narrow ranges of temperature and salinity (<1 ml·l^{-1} oxygen was found in water corresponding to 8·9–9·6 °C and salinity 34·65–34·75 × 10^{-3}, respectively) and appeared most frequently during late summer to early autumn, disappearing in winter or spring. He argued that its *T-S* characteristics together with the ribbon-like nature of the occurrence and shifts in position from one month to the next suggested a current of a specific water type on the shelf. De Decker supported his hypothesis of the southwards and upwards propagation of oxygen-deficient water by sets of seasonal vertical profiles between 19° S and 32° S. While it is well established that a poleward undercurrent exists west of the shelf break in the northern Benguela (see Shannon, 1985) which could provide the mechanism for transporting oxygen-deficient water southwards as far as Lüderitz, the continuity of the poleward undercurrent south of 27° S has not been established, nor has the association between oxygen-deficient water and poleward flow in the southern Benguela been proved. The apparent break in the zonal distribution of oxygen at about 25° S shown in De Decker's work has already been mentioned. Indeed it is conceivable that oxygen-depleted conditions in the undercurrent could even be associated with reduced flow. In this respect Bailey (1979) only found evidence of low oxygen in the deep compensation current off Lüderitz during one (May 1976) of his five surveys, and he concluded that this low oxygen manifestation was a special case of the poleward undercurrent.

The concept of a current of oxygen-depleted water flowing southwards along the shelf from the area near Walvis Bay (23° S) was supported by Andrews & Hutchings (1980) and, like De Decker, these authors recorded oxygen-deficient water within fairly narrow temperature ranges near Cape Town, with some exceptions. They noted that oxygen-depleted water in the southern Benguela seldom reached the surface, and they considered it as underlying the upwelling water. (Somewhat curiously their low oxygen water was slightly warmer and less dense than their defined "upwelling" water!), although along the whole coast north of Cape Columbine (33° S) Nelson & Hutchings (1983) have stated that oxygen-depleted water is brought to the surface during upwelling, after which it rapidly gains oxygen by diffusion and phytoplankton growth. This is borne out by data in Hart & Currie (1960) and Bailey (1979), who show that the 2·0 ml·l^{-1} oxygen isopleth cuts the surface near the shore. Andrews & Hutchings classified two types of oxygen-depleted water off the Cape Peninsula using N:Si ratios *viz.* "local" (high ratio) and "distant" (low ratio), which they explained in terms of nitrogen and silica regeneration rates. Off Lüderitz (Bailey, 1979) and St Helena Bay (Bailey & Chapman, in press) no such distinction is apparent.

If indeed the oxygen-deficient water in the southern Benguela does originate off Namibia or Angola, then it might be expected to have similar *T-S* and density characteristics. Examination of the available evidence (unfortunately few authors have quoted sigma-*t*) shows that this is not the case. Temperature envelopes of oxygen in Stander (1964) show a progressive reduction in the temperature of oxygen-deficient water from >10 °C at 19° S to 10 °C at 21° S and 23° S, 9 °C at 25° S and <9 °C at 27° S, south of which it disappears. De Decker (1970) showed that at 19° S it occurred over a wide range of temperature and salinity (<9 °C to >14 °C; $<34·6$ to $>35·4$ × 10^{-3}), at 22° S

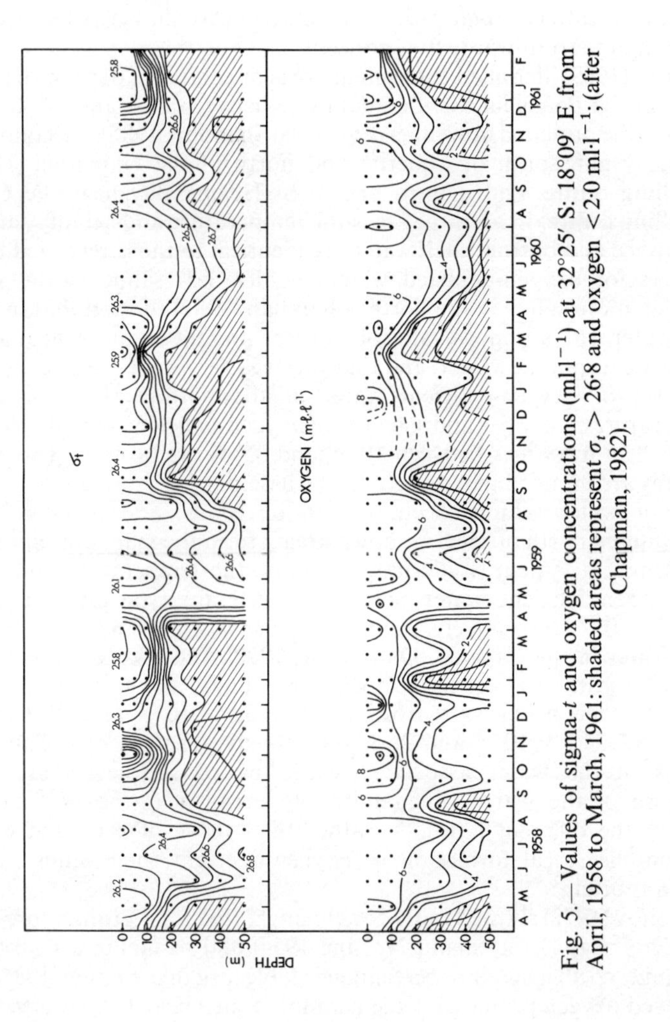

Fig. 5.—Values of sigma-*t* and oxygen concentrations (ml·l^{-1}) at 32°25′ S: 18°09′ E from April, 1958 to March, 1961: shaded areas represent $\sigma_t > 26{\cdot}8$ and oxygen $< 2{\cdot}0$ ml·l^{-1}; (after Chapman, 1982).

typically >11 °C and $>35{\cdot}0\times10^{-3}$ with the scatter narrowing progressively to 9 °C, $34{\cdot}7\times10^{-3}$ at 32° S. Near Lüderitz, Bailey (1979) found that the shelf low-oxygen water had temperature and salinity ranges of 10·5° to 12·5 °C and 34·9 to $35{\cdot}1\times10^{-3}$, respectively. The above imply a change in sigma-*t* between northern Namibia and the Cape of from 26·4–27·0 to 26·8–26·9 for the oxygen-deficient water which seems rather improbable, as one would expect an increase in scatter *en route* rather than the decrease shown by De Decker's data assuming the southwards flow concept.

Visser (1969) identified three main areas or centres of oxygen-depleted shelf water in the Benguela, *viz.* the main one between 2° S and 24° S which lies north of the principal upwelling centre (see Shannon, 1985), a second one in the Orange bight, south of Lüderitz and north of the Namaqua (Hondeklip) upwelling centre, and a third one in St Helena Bay near the Columbine upwelling centre. In all three areas the temperature and salinity ranges of the subsurface and bottom shelf water are identical to those recorded by various workers for oxygen-depleted water—colder, less saline in the south and warmer, more saline in the north. Chapman (1982) showed that in St Helena Bay, water with a sigma-*t* value of 26·8 was present on the shelf at all times of the year except mid winter, and that this was very closely associated with the presence of oxygen-deficient water at the bottom (Fig. 5). The mean temperatures, salinities and sigma-*ts* of oxygen-deficient water both off and on the shelf in five areas between 10° S and 35° S are given in Table II. Three features are immediately evident from these data. First, the off-shelf oxygen-deficient water seldom occurs south of 30° S, and is most frequently encountered north of 25° S *i.e.* downstream from the main upwelling centre in the Benguela system (Lüderitz). Secondly, the temperature and salinity of the oxygen-deficient water north of 25° S are typically $>10{\cdot}5$ °C and $>34{\cdot}9\times10^{-3}$, respectively, *i.e.* appreciably warmer and more saline than such water in the Cape (*cf.* De Decker, 1970; Andrews & Hutchings, 1980). The third feature is that the temperature, salinity, and sigma-*t* of the oxygen-deficient shelf water vary from 15 °C, $35{\cdot}6\times10^{-3}$ and 26·3 off Angola to 9·5 °C, 34.7×10^{-3} and 26·8, respectively, off the Cape. Thus, the properties of the water associated with low oxygen levels are the same as those of the water which normally exists near the bottom over the shelf in the different latitudes, rather than those of a unique water type. This implies local formation of oxygen-deficient shelf water rather than a distant origin.

Bubnov (1972) showed fairly convincingly that for a primary production of 3 g $C{\cdot}m^{-2}{\cdot}day^{-1}$ (Steemann Neilson, 1954) and assuming a realistic vertical exchange coefficient and percentage stable organic matter, the calculated dissolved oxygen profile over the Namibian shelf near 19° S was very close to the observed distribution. Bailey (1979) subsequently showed that this was also the case near Lüderitz.

Chapman (1983a), citing figures by Henry (1979) which ranged between 0·48–11·74 g $C{\cdot}m^{-2}{\cdot}day^{-1}$ for the production rate near St Helena Bay, assumed a mean summer production level of 3900–5200 mg $C{\cdot}m^{-2}{\cdot}day^{-1}$, and calculated that in the southern area oxygen in the surface layer could be consumed in a period of 17–23 days. This period would be shorter if total consumption is not required (Bubnov, 1972), or longer if advection or zooplankton grazing are considered. Despite the very high production that

TABLE II

Characteristics of oxygen-deficient water along the west coast of Africa between 10° S and 35° S: ± SD and number of samples given in parentheses

Area	Shelf (water depth <300 m)						Off-shelf (water depth >300 m)					
	O_2 <1 ml·l^{-1}			O_2 <2 ml·l^{-1}			O_2 <1 ml·l^{-1}			O_2 <2 ml·l^{-1}		
	Temp. °C	Sal. ×10³	σ_t	Temp. °C	Sal. ×10³	σ_t	Temp. °C	Sal. ×10³	σ_t	Temp. °C	Sal. ×10³	σ_t
Angola, 10° S–15° S, coast –0° E	15·19 (2·13) (21)	35·60 (0·27) (19)	26·28 (0·22) (19)	16·30 (1·50) (119)	35·65 (0·21) (113)	26·17 (0·29) (113)	10·98 (2·85) (657)	35·05 (0·32) (596)	26·79 (0·30) (596)	12·11 (3·27) (1916)	35·20 (0·34) (1762)	26·65 (0·34) (1762)
South Angola/North Namibia, 15° S–20° S, coast –3° E	13·77 (1·22) (134)	35·35 (0·22) (133)	26·52 (0·18) (128)	14·32 (1·42) (307)	35·38 (0·20) (305)	26·43 (0·21) (298)	11·11 (2·14) (517)	35·03 (0·50) (491)	26·77 (0·41) (491)	11·26 (2·82) (994)	35·06 (0·44) (945)	26·74 (0·38) (943)
Central Namibia, 20° S–25° S, coast –5° E	12·38 (0·74) (653)	35·16 (0·14) (513)	26·65 (0·11) (510)	12·48 (0·81) (1001)	35·16 (0·14) (791)	26·63 (0·12) (787)	10·63 (1·39) (67)	34·94 (0·17) (70)	26·81 (0·13) (67)	10·28 (2·04) (228)	34·90 (0·21) (228)	26·83 (0·22) (224)
South Namibia/Orange, 25° S–30° S, coast –7° E	10·81 (0·98) (38)	34·91 (0·11) (40)	26·75 (0·15) (38)	10·54 (1·05) (100)	34·89 (0·11) (102)	26·78 (0·14) (99)	9·55 (0·47) (7)	34·83 (0·04) (7)	26·91 (0·05) (7)	8·88 (1·24) (24)	34·78 (0·12) (24)	26·98 (0·10) (24)
Cape, 30° S–35° S, coast –10° E	9·47 (0·67) (308)	34·72 (0·05) (363)	26·84 (0·11) (307)	9·55 (0·96) (932)	34·73 (0·08) (1053)	26·83 (0·14) (923)	13·90 — (1)	35·37 — (1)	26·51 — (1)	11·91 (2·55) (6)	35·06 (0·26) (6)	26·64 (0·33) (6)

occurs above the thermocline, little of the excess oxygen will diffuse through to the deeper layers, being more likely to escape to the atmosphere (Bailey & Chapman, in press). Since currents in this area are sluggish, the gyre in this region (see Shannon, 1985) is likely to stabilize the oxygen-deficient water mass at the bottom. Further evidence for a local source is provided from sedimentation rates in St Helena Bay, where Bailey (1983) found sedimentary fluxes of organic carbon and nitrogen of up to 4000 and 600 $mg \cdot m^{-2} \cdot day^{-1}$, respectively.

These examples would seem to suggest that, even in the absence of oxygen-depleted source water, the formation of oxygen-deficient water on the shelf north of 33° S can be explained biochemically. Close inshore, however, levels of dissolved oxygen rise again as a result of entrainment of air by breaking waves (Copenhagen, 1953; Bailey, 1979). It is conceivable, however, that the subsurface oxygen minimum observed by Andrews & Hutchings (1980) south of 33° S may have had a "distant" origin, *viz.* in St Helena Bay, and could have been transported southwards around Cape Columbine in an inshore undercurrent predicted by Van Foreest & Brundrit (1982) (see Shannon, 1985).

A feature of the shelf-edge oxygen-minimum layer off Namibia which requires further attention is the possible rôle played by the Antarctic Intermediate Water in determining its position. Several authors, *inter alia* Calvert & Price (1971) and Jones (1971), have shown that oxygen-depleted water occurs immediately above the Antarctic Intermediate core, and it would be tempting to speculate as to what extent the latter constrains this water. The density of the upper parts of the Antarctic Intermediate layer off Namibia corresponding to a temperature and salinity of 6 °C and $34{\cdot}5 \times 10^{-3}$ respectively is sigma-*t* 27·2, a value very close to the buoyancy of particulates, calculated by Karl, La Rock, Morse & Sturges (1976) at 27·2–27·3 for the North Atlantic.

A feature of the oxygen-deficient Namibian shelf water is that it can exist as discrete pools or tongues over both the upper and lower shelf. This is shown particularly well in Figure 4 near latitude 20° S. The separation evident over the upper shelf break corresponds to a region where the sediments are coarser and organically poorer. Similarly, the inshore areas corresponding to oxygen enrichment through breaking waves referred to above also correspond to coarser sediments with less organic matter (Bremner, 1981, 1983). Whether this implies that the water column oxygen is being depleted by the carbon in the sediments or whether the upper shelf break is a zone of higher turbulence (or jet as suggested by Shannon, 1985) is a matter for speculation. Nevertheless, the distribution of the oxygen-deficient shelf water in relation to the mid-shelf coarse sediment ribbon and the rôle of the sediments in the system warrant further study.

From the foregoing it appears that in the Benguela system there are two distinct bodies of oxygen-deficient water. The first type occurs at a depth of about 300 m usually west of the shelf break (but at times spilling up onto the shelf) as a wedge which tapers southwards, usually no further than 25° S, but occasionally as far as 29° S. This oxygen minimum is often, but not always, associated with a poleward undercurrent and has its origin in the tropical South East Atlantic (largely off Angola). It appears to be further depleted of oxygen at the lower shelf or shelf edge off Namibia on its way south, and is often upwelled onto the shelf (Hart & Currie, 1960; Stander, 1964).

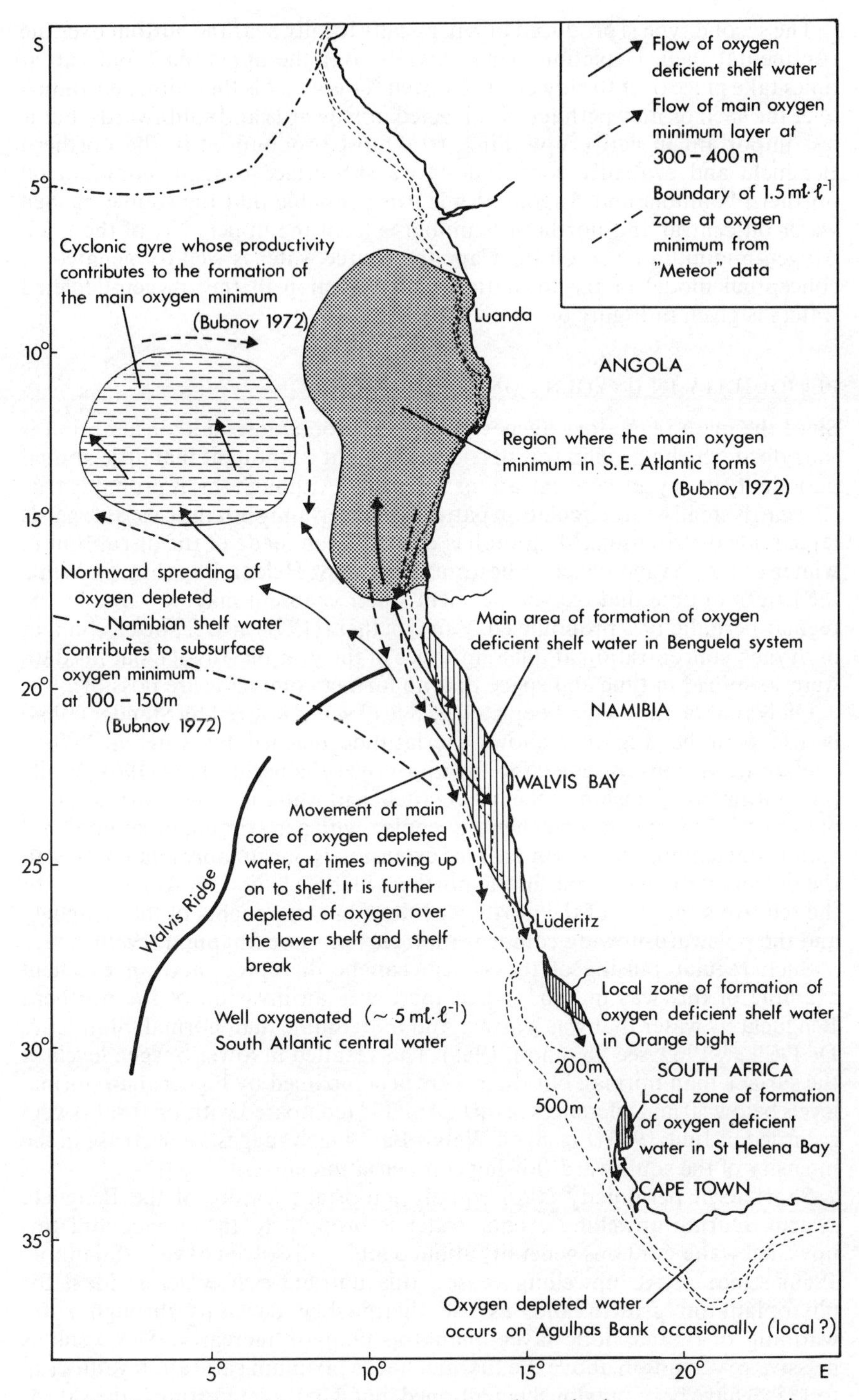

Fig. 6.—Conceptual model showing areas where the low oxygen water is formed in the South East Atlantic and its inferred movement.

The second type is produced biochemically locally near the bottom over the continental shelf. Depletion occurs mainly over the upper shelf, but can at times take place over the lower shelf as well. This water is then either entrained over the shelf or may perhaps be advected northwards and southwards, but is less important in active upwelling. It is most pronounced in the northern Benguela and evidently contributes to a subsurface oxygen minimum off northern Namibia and Angola. While it is probable that the source of shelf water off central and northern Namibia is from the upper part of the main oxygen-minimum layer, off the Cape the source water is well oxygenated. A conceptual model of the formation and advection of the oxygen-depleted waters is given in Figure 6.

PERIODICITY IN OXYGEN CONCENTRATIONS

Since the main regulatory agency of oxygen production and consumption is the rate at which upwelling events occur, it might be thought that little annual periodicity in oxygen concentrations will result. Although this is probably true as regards small-scale circulation patterns, it is certainly not the case as regards large scale distribution. Mention has already been made of the disruption in winter of the oxygen-deficient bottom water in St Helena Bay (Fig. 5). While the length of time that oxygen-deficient water is absent may be variable, the regular periodicity is pronounced. Van Goethem (1951) also reported changes in oxygen concentration at different times of the year off Angola, but his data were so spread in time and space that no further comments are possible.

Off Namibia, the oxygen-depleted wedge of water noticed by Stander (1964) is also periodic. Figure 7 shows the latitude reached by water at 200 m containing various oxygen concentrations over the period 1959–1961. While the southward extension of the oxygen-deficient water is rather variable, that of the 4 $ml \cdot l^{-1}$ isopleth is much more regular, with a maximum in summer and spring and a minimum in winter. This extension is presumably related to both the productivity in summer in the northern Benguela and off Angola, and to the relative strengths of the northward-flowing components of the Benguela and the poleward-flowing compensation currents (see Shannon, 1985).

Such regular pulsing of the system can be disrupted, and an excellent example of this was in 1963, when there was an invasion of the northern Benguela by water that was warmer and more saline than normal (Stander & De Decker, 1969; see Shannon, 1985). This resulted in lower oxygen levels at the surface than normal, but these were accompanied by higher than normal levels below 50 m. Differences of up to 4 $ml \cdot l^{-1}$ (compared with 1960 data) were recorded in June 1963 (Fig. 8) off Walvis Bay, which suggests a decrease in the intensity of the southward-flowing compensation current.

Short term periodicity is an equally important feature of the Benguela system. During upwelling events, water is brought to the surface and this upwelled water contains generally about 5 $ml \cdot l^{-1}$ dissolved oxygen (Shannon, 1966). Once active upwelling ceases, this nutrient-rich water is ideal for phytoplankton growth and, as the thermocline develops through solar warming of the euphotic layer, plankton biomass increases. The result is massive oxygenation above the thermocline. Maximum saturation values (up to 185%) have been previously mentioned, but what is interesting is the rate at which such levels are reached. Recently, staff at the Sea Fisheries Research

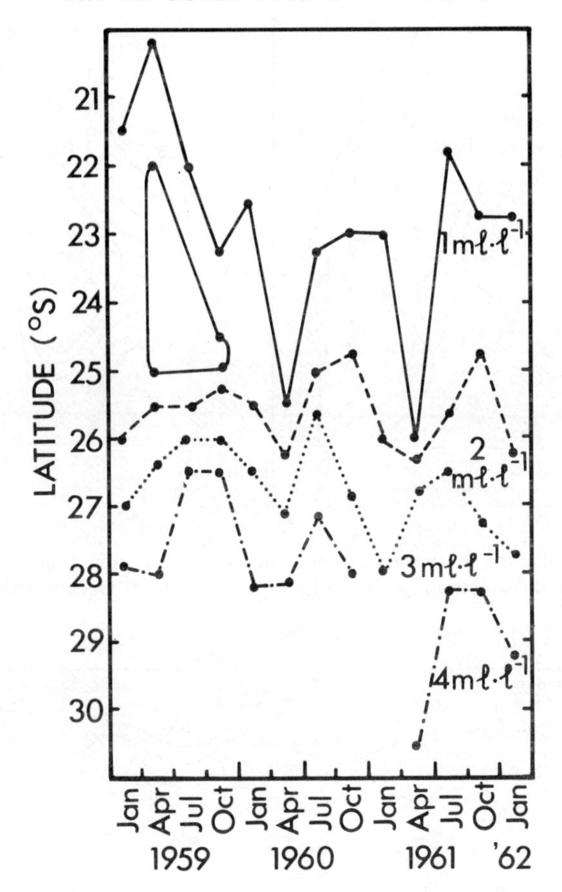

Fig. 7.—Southward extension of oxygen-depleted water at 200 m between January, 1959 and January, 1962: redrawn from data (ml·l^{-1}) in Sea Fisheries Annual Reports.

Institute have been monitoring single parcels of upwelled water over several days, using subsurface drogues to mark the track of the water mass. Sampling is carried out several times per day (Barlow, 1982, 1984a,b; Olivieri, 1983a). While temperature changes in the water during one such event between Cape Town and Cape Columbine were of the order of 2 °C at the surface, from 10–12 °C, the oxygen saturation increased from less than 90% (5·0–5·5 ml·l^{-1}) to over 140% (8·0–8·5 ml·l^{-1}), as shown in Figure 9. In the same time period, the 1% incident light level decreased, from between 50–60 m to 12 m, and the mean chlorophyll *a* concentration in the euphotic zone increased from 0·2 to 18 mg·m^{-3} (Olivieri, 1983a). The time scale of 5–6 days is typical of such upwelling events in the southern part of the area (Nelson & Hutchings, 1983).

Even more rapid increases have been noticed in particularly favourable environments. For example, Pieterse & Van der Post (1967) showed that in Walvis Bay, where admittedly the water was less than 10 m deep, local production could result in changes in oxygen level of nearly 8 ml·l^{-1}

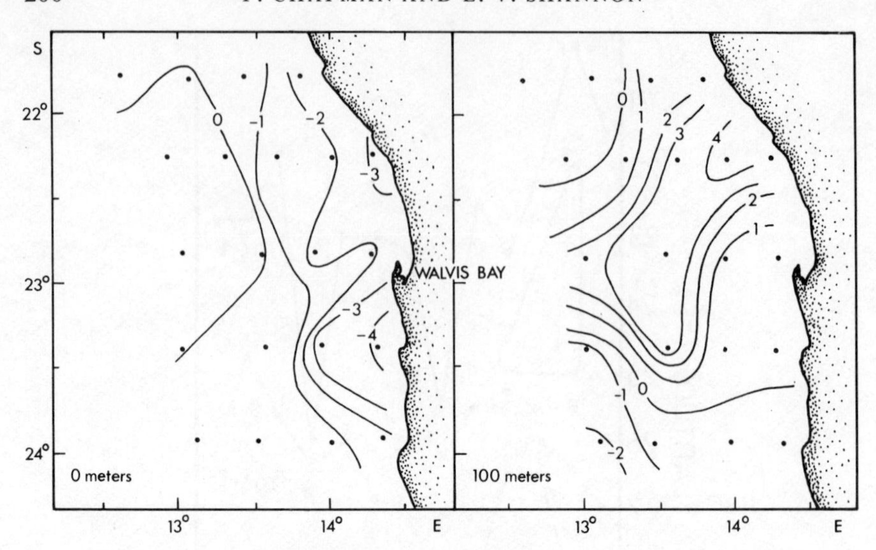

Fig. 8.—Difference between oxygen values (ml·l^{-1}) in June, 1960 and 1963 at 0 m and 100 m off Namibia (after Stander & De Decker, 1969).

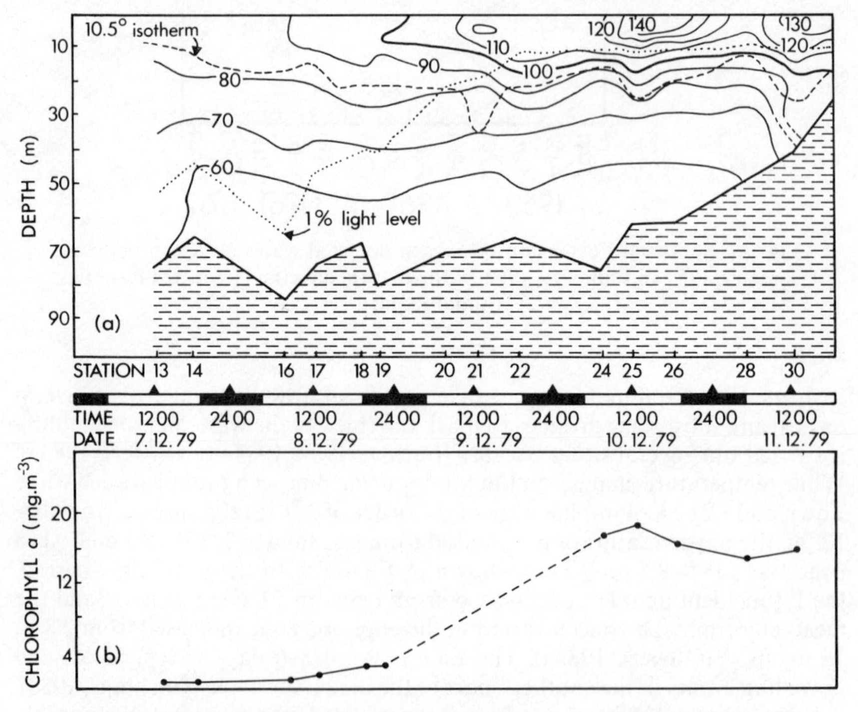

Fig. 9.—Changes in per cent oxygen saturation in an upwelled cell during December, 1979 (a): also shown are the 1% light level depth, the 10·5 °C isotherm, and (b) mean integrated chlorophyll *a* concentration (mg·m^{-3}) within the euphotic zone; (after Olivieri, 1983a).

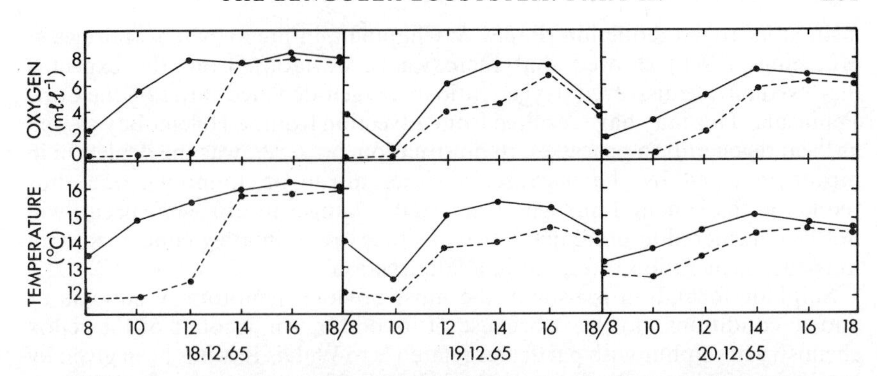

Fig. 10.—Changes in oxygen concentration (ml·l^{-1}) and temperature in Walvis Bay between 18th–20th December, 1965: solid line denotes surface values; broken line shows bottom values (± 7 m); (after Pieterse & Van der Post, 1967).

throughout the day (Fig. 10). This was accompanied by a 4 degree rise in temperature. During the night, however, respiration, water renewal and settling of decomposing organisms caused a drop in the oxygen concentration to virtually zero again. These sudden fluctuations also had a marked effect on nutrient concentrations in the water column—at the same station the phosphate concentration increased from 0·7–0·8 μM on 14th December 1965 to 3·3 μM for the period 18th–20th December, showing the importance of decomposition in such a restricted zone.

THE EFFECT OF OXYGEN DEPLETION ON REDOX COUPLES

As might be expected, such extreme changes in oxygen concentration can have a large effect on the dissolved concentrations and speciation of other elements. Breck (1974) has discussed the general effects of changes in oxygen tension and hence pE on speciation in the sea, and has differentiated between anoxic and suboxic regions. In the former, nitrate is reduced to elemental nitrogen gas, sulphate to HS^-, and ultimately CO_2 and HCO_3^- to hydrocarbons, while the last are characterized by the intermittent appearance of nitrite.

Much of the Benguela system can, therefore, be classed as permanently suboxic. The nitrogen distribution has been studied by several workers (Calvert & Price, 1971; Jones, 1971; Bailey, 1979; Andrews & Hutchings, 1980; Boyd, 1983a; Bailey & Chapman, in press), and it is clear that nitrogen reduction occurs intermittently at the bottom on the shelf in oxygen-deficient water. Calvert & Price (1971) reported that off Cape Cross (22° S), more than half the nitrate in the water column had been reduced, and a nitrite maximum (3 μM) was found at the bottom. This occurred under conditions of pronounced stratification, and was also reported (without the nitrite build up) by Boyd (1983a) for the same region and by Bailey (1979) near Possession Island (27° S). In the last instance, the water was anoxic and nitrate and nitrite were almost completely reduced to ammonia or elemental nitrogen.

South of the Orange River, the St Helena Bay area has also been shown to be a site of nitrogen reduction (Bailey & Chapman, in press), while Andrews & Hutchings (1980) showed that there was a deviation from the expected dissolved nitrate:dissolved oxygen ratio in oxygen-depleted water off the Cape Peninsula. This may have resulted from advection from St Helena Bay as well as local regeneration processes. In no situation, however, was the depletion in nitrate matched by the increase in either nitrite or ammonia, and thus reduction to elemental nitrogen is inferred. Changes in nitrogen species will not be considered in more detail here, as they are intimately concerned with nutrient regeneration rates, but in a later section.

Sulphide formation is usually the most obvious symptom of suboxic or anoxic conditions, not least because of its odour. An account of the redox chemistry of sulphur with particular reference to Walvis Bay has been given by Boulégue (1974) and Boulégue & Denis (1983). The occurrence of sulphide in the Benguela, first reported by Marchand (1928), but not positively identified until later (Copenhagen, 1934), has been blamed for the periodic mortalities of fish and other organisms near Walvis Bay, records of which go back at least to 1828 (Copenhagen, 1953). These occurrences are reported only in late summer near Walvis Bay, when the northerly winds are at a maximum and southerly winds at a minimum, *i.e.* quiescent upwelling conditions, and although localized they may be related to the southwards movement of the oxygen-deficient wedge of water described by Stander (1964). The most graphic effects of sulphide generation include the penetration of sulphurous fumes up to 60 km inland, metalwork and white paint on buildings being turned black, the appearance and disappearance of mud islands (the cause of which still remains to be found), and the apparent "boiling" of the sea surface (Copenhagen, 1953; Hart & Currie, 1960).

The main area within which such eruptions are found is the so-called "azoic zone" which lies off the coast between the Kunene River (17° S) and about 25° S (Hart & Currie, 1960). This corresponds closely to the inshore patches of fine sediment shown by Bremner (1981—see Fig. 14) and Copenhagen (1934) found up to 0·1% sulphide in the sediment near Walvis Bay. Bailey (1979) reported that sulphide was present in surface sediments further south, with levels decreasing from a maximum of 65 $mg{\cdot}l^{-1}$ in pore water at North Rocks (24° S) to about 6 $mg{\cdot}l^{-1}$ at Chamais Bay (28° S). Although Bailey found no free H_2S in the water column, Copenhagen (1953) reported a value of 5·88 $ml{\cdot}l^{-1}$ one metre above the bottom at a station 10 km west of Pelican Point, Walvis Bay, in May 1950, while in September of that year at William Scoresby Station 1076, in virtually the same position, he recorded oxygen and H_2S coexisting 5 m above the sediment–water interface. From differences between the results of oxygen samples analysed immediately and after storage, Hart & Currie (1960) reported that the water from Walvis Bay itself contained about 0·8 $ml{\cdot}l^{-1}$ H_2S, which compares with the maximum value of 1·12 $ml{\cdot}l^{-1}$ recorded by Pieterse & Van der Post (1967). These last authors, and Van Goethem (1951) stated, however, that free hydrogen sulphide was present in several samples. Pieterse & Van der Post reported that the maximum thickness of the totally anoxic layer was about 5 m, following an intense bloom of *Peredinium triquetum*—in deeper waters outside the Bay itself it would be expected that the anoxic layer could thicken, particularly during well-stratified conditions.

South of the Orange River, the only report of free hydrogen sulphide is that of Copenhagen (1953), who mentions its occurrence in the muds near

Lamberts Bay. Muds taken in St Helena Bay (Sea Fisheries Research Institute, unpubl. data) also smell of sulphide but the area is not anoxic, since various animal species are found therein. Oxygen data reported by Bailey & Chapman (in press) suggest, however, that the coexistance of oxygen and hydrogen sulphide is possible in this region.

The high quantities of sulphide in the sediment in the northern Benguela are apparently produced both by anaerobic biological breakdown of organic substances and by the reduction of sulphate by bacteria (*Desulphovibrio* sp., Stander, 1964). Reduction of sulphate in this way leads to the production of pyrite in the sediment near Lüderitz (Siesser & Rogers, 1976; Rogers, 1977). Rogers (1977) has suggested that the limiting agent in pyrite formation is the supply of iron, which is probably derived from either the Kunene or Orange Rivers. Free sulphur is also a feature of such sediments, and Boulégue & Denis (1983) have discussed its formation through bacterial (Thiobacteriaceae) oxidation of H_2S. They also found various sulphur-containing chemical species in the sediment, which could affect metal solubilities.

After nitrate and sulphate have been reduced, the next step in the redox cycle in the sea is the reduction of CO_2 and HCO_3^- to hydrocarbons (Breck, 1974). While this is unlikely in any but the most stagnant basins where all the sulphate may be reduced, bacteria are certainly capable of producing hydrocarbons in anoxic sediments. Scranton & Farrington (1977) have looked at methane production across the shelf to the west of Walvis Bay. The measured concentrations were all much higher than predicted from solubility and atmospheric methane concentration data, ranging from a factor of 2 to 440 times the equilibrium concentration (Fig. 11). Off the shelf, however, deep samples were depleted in methane. Somewhat surprisingly, higher levels were

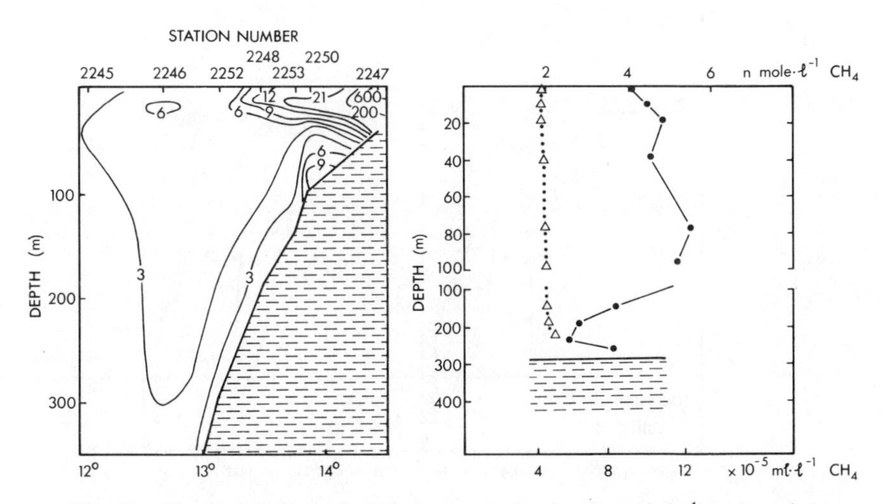

Fig. 11.—Vertical section of methane concentration (n·mole·l^{-1}) across the shelf at Walvis Bay (left) and (right) vertical profile at Station Atlantis 2252 (solid line) compared with predicted values (dotted line) (after Scranton & Farrington, 1977).

found near the pycnocline rather than at the bottom. Scranton & Farrington explained the increase at the bottom on the basis of methane supply from the sediments, but had also to assume *in situ* production at the thermocline.

The three examples above show that considerable changes in pE must be occurring in the bottom waters and surface sediments near Walvis Bay. While no studies have been made on metal speciation in this area (*cf.* comparable work in the Black Sea by Spencer & Brewer, 1971, and Saanich Inlet by Emerson, Cranston & Liss, 1979), iodine has been studied in both the water column, near St Helena Bay, by Chapman (1983a), and in the sediments off South West Africa by Price & Calvert (1973, 1977). In their later paper, the sedimentary distribution was compared with that of bromine. The iodine in surface sediment samples was found to vary from less than 100 $\mu g \cdot g^{-1}$ (by dry weight) at the inner edge of the Namibian shelf to a maximum of 1990 $\mu g \cdot g^{-1}$ off the shelf edge at 25°11′ S: 13°18′ E (Price & Calvert, 1973). This value is the highest reported in the literature. The increase was not regular, but was shown to be related to both the organic carbon content and the oxidizing character of the sediment, which varies considerably between organic-rich muds and carbonate-rich sands (Bremner, 1983). Thus the lowest I:C ratios were found inshore, while the highest were at the shelf edge (Fig. 12), and followed closely

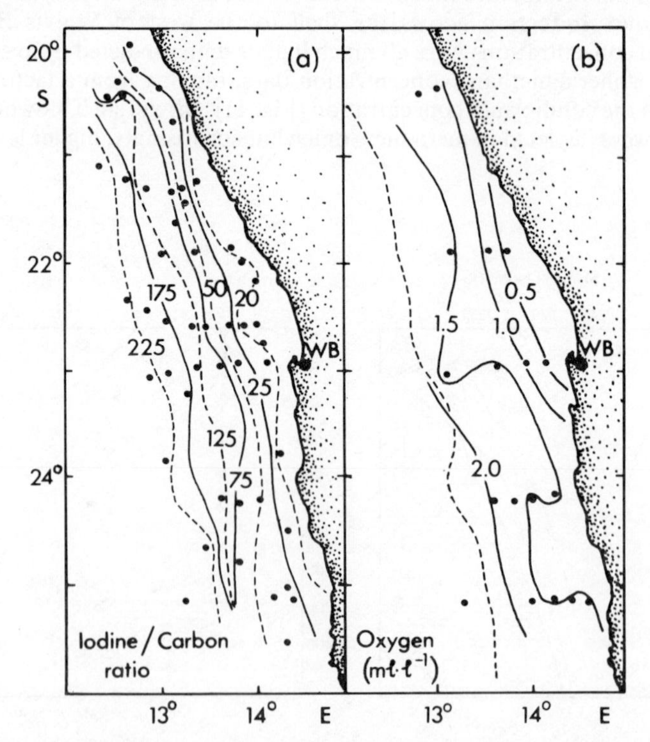

Fig. 12.—Distribution of I:C ratios ($\times 10^4$) in surface sediments of the Namibian shelf (a) and distribution of dissolved oxygen ($ml \cdot l^{-1}$) in bottom waters in the same area (b): dashed line in (b) denotes the shelf edge ($\pm$500 m); (after Price & Calvert, 1973).

the distribution of bottom-water dissolved oxygen concentrations. This distribution is similar to that seen for other recent shallow oxidized (I:C 250–350 $\times 10^{-4}$) and reduced (I:C 20–30 $\times 10^{-4}$) sediments around the world. In the deep oceans diagenesis in oxidized sediments and variations in sediment supply cause a loss of iodine from the surface layers and I:C ratios of 20–30 $\times 10^{-4}$ are found (Price & Calvert, 1973, and references therein). While oxidized sediments on the shelf had an iodine content that decreased with depth, those laid down under reducing conditions had a lower iodine content (100–200 $\mu g \cdot g^{-1}$, which remained virtually constant down the core (Price & Calvert, 1977).

Surprisingly, bromine behaved differently to iodine in these sediments, being correlated directly, at the surface, with the organic carbon content (Price & Calvert, 1977). The relationship was equivalent to an increase in bromine by 60 $\mu g \cdot g^{-1}$ (dry weight) per 1% carbon content. This difference is thought to be due to variations between the mechanisms of incorporation of the two elements into biological material. Iodine may be adsorbed to dead sestonic material, possibly as iodoamides (Harvey, 1980), and would be held below the surface in anoxic conditions which do not favour organic degradation. While the method of incorporation of bromine into sediments is unknown, it is certainly different from iodine (Harvey, 1980), and may be related to uptake by living plankton (Saenko, Kravtsova, Ivanenko & Sheludko, 1978).

No studies have been made on dissolved iodine concentrations off Namibia, but Chapman (1983a) has investigated the distribution of the element in St Helena Bay, another area with reducing sediments and substantial organic carbon deposits (Birch, 1977). This area varies between being well-oxygenated and possibly even anoxic (De Decker, 1970; this paper, above). Specific total iodine (concentration compared with unit salinity) remained almost constant throughout the area at 0·012–0·014 $\mu M \cdot g^{-1}$, apart from an anomaly which will be discussed below. Specific iodate, however, declined both across the shelf and down the water column (Fig. 13). At the surface and on the outer edge of the lines, the specific iodate content was 0·008–0·009 $\mu M \cdot g^{-1}$, while at the innermost stations and near the bottom it decreased to 0·006 or less. There was a similar decrease from the northern lines sampled (32° S) to those in the southern end of the Bay, which contained the lowest concentrations of dissolved oxygen. The 0·008 specific iodate isopleth was very close to the 10° isotherm, which followed closely both the 26·8σ_t boundary and the 2 $ml \cdot l^{-1}$ dissolved oxygen isopleth. These waters also contained enriched nitrate, phosphate and silicate concentrations.

Chapman (1983a) has suggested that phytoplankton uptake at the surface may account for the difference between specific total iodine and iodate concentrations above the thermocline, but that below this layer iodate is exchanged for iodide during breakdown of organic matter. A large anomaly in the total iodine content, but with normal iodate concentrations, was found at mid-depth on the last lines sampled, total iodine concentrations of up to 6·6 μM being found. This suggests that iodine is taken up by sinking particles in oxygenated water, as suggested by Price & Calvert (1973, 1977), but then released at the sediment surface as the dissolved oxygen level drops. Phytoplankton decomposition could not have produced enough iodine to account for this anomaly. Presumably such a cycle of adsorption and release will recur several times a year between upwelling events in this area, in contrast

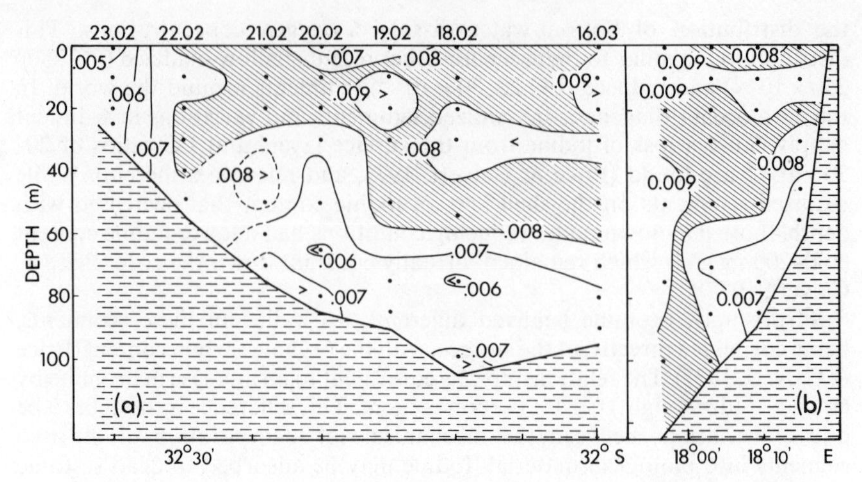

Fig. 13.—Distribution of specific total iodine along (a) 18°05′ E, (b) 32°20′ S: isopleths in μ·mol·g^{-1}; shaded area contains concentrations between 0·008–0·009 μ·mol·g^{-1}; (after Chapman, 1983a).

to the more stable regime off Walvis Bay where oxygen levels remain low throughout the year.

NUTRIENTS AND PARTICULATE MATTER

The first data on nutrient distributions in the Benguela System were obtained during the expeditions by DISCOVERY (silicate and phosphate—Clowes, 1938) and METEOR (phosphate—Wattenberg, 1928). The stations from the former expedition were too far offshore to show any of the typical features of the region, while those from the latter were too far apart to show the fine structure. Nitrate and silicate analyses were also made during the 1948 Belgian expedition (Van Goethem, 1951), but problems with the analytical methods used precludes much comment on their results, other than that nitrate maxima were associated with oxygen minima. It was not until the cruises of the WILLIAM SCORESBY in 1950 (Hart & Currie, 1960) that any meaningful work was done in the area. Unfortunately, phosphate was the only nutrient analysed, but the two surveys, carried out in March and September 1950, gave good coverage of the system in periods of both quiescence and active upwelling. Phosphate distribution was found to be similar on both surveys, with low values at the surface and offshore, and higher values in the recently upwelled water and at the bottom on the shelf. These concentrations of up to 3·0 μM were much higher than the concentrations found off the shelf at the same depth, and rivalled the values found at the nutrient maximum in the Antarctic Intermediate water (about 800 m).

Since that time, studies on nutrients have been carried out by, *inter alia*, Calvert & Price (1971), Jones (1971), Bishop, Ketten & Edmond (1978), Bailey (1979), Schultz, Schemainda & Nehring (1979), Andrews & Hutchings (1980),

Boyd (1983b), Olivieri (1983a), and Bailey & Chapman (in press), and a much clearer view of both the general features of the system and the short-term periodicity has emerged.

The general features of nutrient distribution in the Benguela resemble closely those of other upwelling regions. The upwelling water is enriched with nutrients relative to the surface layers and during active upwelling this water reaches the surface near the shore. Following the establishment of the thermocline, phytoplankton production will consume the nutrients in the upper layers, leaving them very much depleted, while nutrient re-enrichment occurs below the thermocline as the phytoplankton decay. Typical concentrations of nutrients off Namibia are shown in Figure 18 (see p. 214; Bailey, 1979) for both active upwelling and post upwelling situations. Horizontal distributions during upwelling have their highest concentrations at the sites of the active upwelling cells described in Shannon (1985; see Fig. 19, p. 218). Similar vertical and horizontal sections are shown by Calvert & Price (1971) and Jones (1971) from Cape Town north to Dune Point (21° S). Offshore, nutrient concentrations have been found to decrease. Andrews & Hutchings (1980) showed clearly for the Cape Peninsula upwelling plume that the sharpest drop in nutrient concentrations occurred at the outer edge of the plume where chlorophyll *a* concentrations were at a maximum. This appears to be related to the edge of the thermal front (Shannon & Anderson, 1982).

THE STABILITY OF THE SYSTEM AND ITS EFFECT ON NUTRIENT CONCENTRATIONS

One of the problems of discussing the nutrient status of the Benguela as a whole is that chemistry is very much ‘site-specific’. It is, therefore, less easy to generalize about rates of supply or removal, and the published data should only really be used as a guide to the instantaneous conditions prevailing at the time of collection. One generalization that can be made, however, is that the importance of nutrient regeneration in the supply process increases northwards. This is clear from the changes in the concentration of nutrients in the bottom water over the shelf, which have been discussed by Bailey (in press). Near the Cape of Good Hope, the shelf is very narrow, and there is very little settlement of decomposing material. Further north, in St Helena Bay, the gyral system on the wider shelf (Shannon, 1985) tends to keep material in the area and a build up of nutrients in bottom waters occurs (Bailey & Chapman, in press). The process occurs to its maximum extent north of Sylvia Hill (25° S), where the wide shelf allows the settlement of vast amounts of material, as evidenced by the high proportions of biogenic material in the sediment (Rogers, 1977; Bremner, 1983; see Fig. 14).

It is also expected that nutrient concentrations, like oxygen, will depend on whether or not upwelling is occurring, rather than on seasonal distribution patterns. This is particularly the case off Namibia, where Stander (1962) and Pieterse & Van der Post (1967) have shown that in Walvis Bay the highest phosphate concentrations coincide with lowest temperatures. Further south in St Helena Bay the situation is more confused as some input of nutrients would be expected from the Berg and Olifants Rivers during the winter, and possibly even from the Orange River during the summer (Birch, 1975). Clowes (1954) showed the possibility of phosphate input from this source, but his data

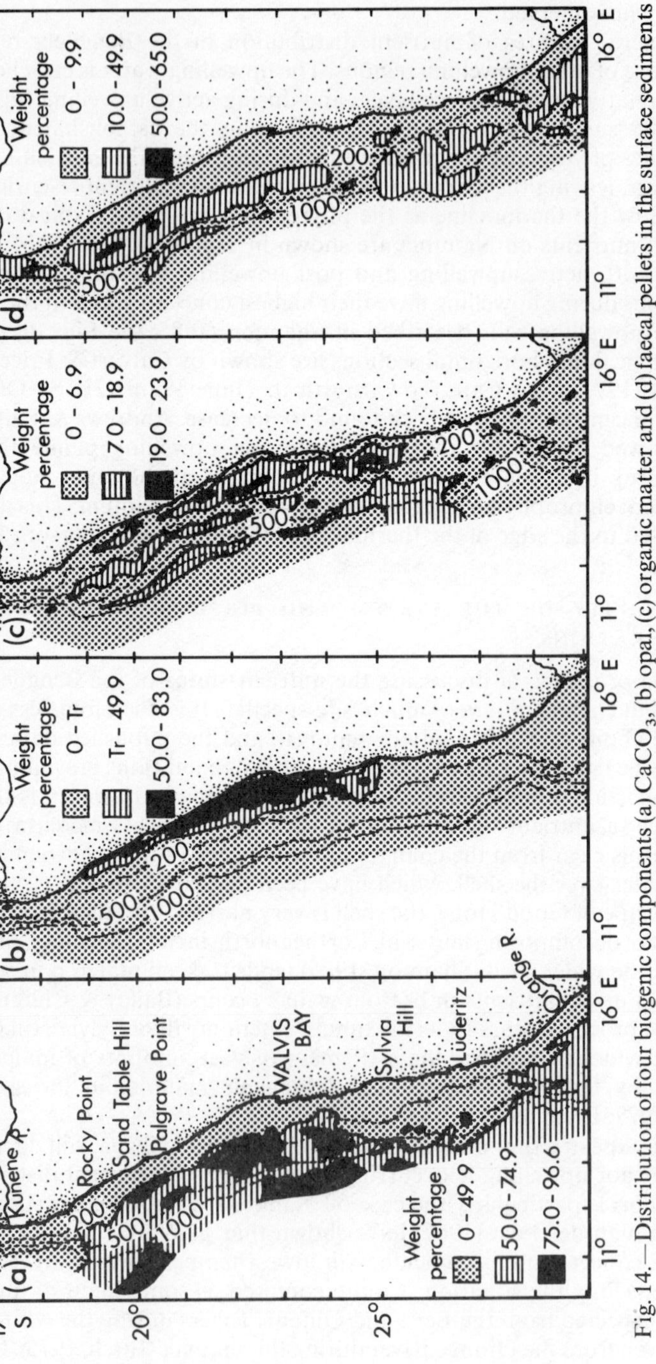

Fig. 14.—Distribution of four biogenic components (a) $CaCO_3$, (b) opal, (c) organic matter and (d) faecal pellets in the surface sediments of the Namibian continental margin derived from the analysis of 729 samples; isobaths in m; (after Bremner, 1983).

were averaged over each season and were not conclusive. Bailey (1979), however, has reported increased silicate concentrations at the surface emanating from the Berg River.

Near the Cape of Good Hope, Andrews & Hutchings (1980) found that the influx of oxygen-depleted water at the bottom correlated well with higher nutrient concentrations. Surface values were highest during active upwelling in summer, dropping dramatically as phytoplankton blooms developed. In winter, despite warmer water temperatures at the surface, nutrient concentrations were higher than in such post-bloom periods.

The stability of the system will also affect the magnitude of the changes in nutrient concentrations. In the south, the steep topography, the close presence of the coastal jet, and the periodic weather changes mean that the system is continually being pulsed on a time scale of about one week. Where the shelf is wide, near Walvis Bay, quasi steady-state conditions prevail instead, and both production (Steemann Neilson, 1954; Steemann Neilson & Jensen, 1957) and settlement rates (Bremner, 1983) are, therefore, greater. The area north of Cape Columbine can be considered intermediate between the two, but the gyral system in the area will favour settlement of detritus and hence nutrient build up.

These features have been studied by various workers. The changes over a five-day period in an upwelling plume off the Cape Peninsula are shown in Figure 15. These can be compared with the concomitant changes in dissolved oxygen shown in Figure 9 (see p. 200, data from Olivieri, 1983a). From these figures it is clear that primary productivity was responsible for the observed nutrient decrease, with nitrate being reduced to very low levels ($< 1\ \mu$M) before silicate. Phosphate showed no signs of becoming limiting. Other workers (Andrews & Hutchings, 1980; Barlow, 1980, 1982; Olivieri, 1983b; Bailey & Chapman, in press) have also reported similar rates of change and that nitrate appears to be the limiting nutrient.

Further north, off Namibia, conditions are more stable. Although Hart & Currie (1960) and Bailey (1979) reported that upwelling was strong off Lüderitz, north of Walvis Bay there may be little change. Hart & Currie in fact found little difference in phosphate distribution during their two surveys in March and October, despite strong upwelling on the second cruise. Boyd (1983a) found little vertical change over the period of a fortnight in either isotherms or nutrient isopleths along a line off 22° S, and data from the cruise of the A. VON HUMBOLDT in November 1976 (Schultz *et al.*, 1979) are similar. Figure 16 shows the distribution of sigma-*t*, phosphate, nitrate, and nitrite during one transect of their line which was occupied seven days running. This can be compared with Figure 4 (see p. 191), which shows the mean oxygen concentration over this period. A close study of their data showed that what little nutrient uptake took place occurred at Stations B, C, and to a lesser extent D on their line (Stations 260–262 in Fig. 16). There was very little change in the nutrient status of the inshore station (Station 259) despite an increase in chlorophyll *a* concentration from less than 1 to 11 $mg{\cdot}m^{-3}$ in four days, after which it declined and remained approximately constant at about 6 $mg{\cdot}m^{-3}$. This set of data is interesting in that it appears to show either subsurface upwelling at the shelf break or the passage of an internal shelf wave. Spectral analysis (Postel, 1982) appears to confirm the latter hypothesis, with the period being about five days.

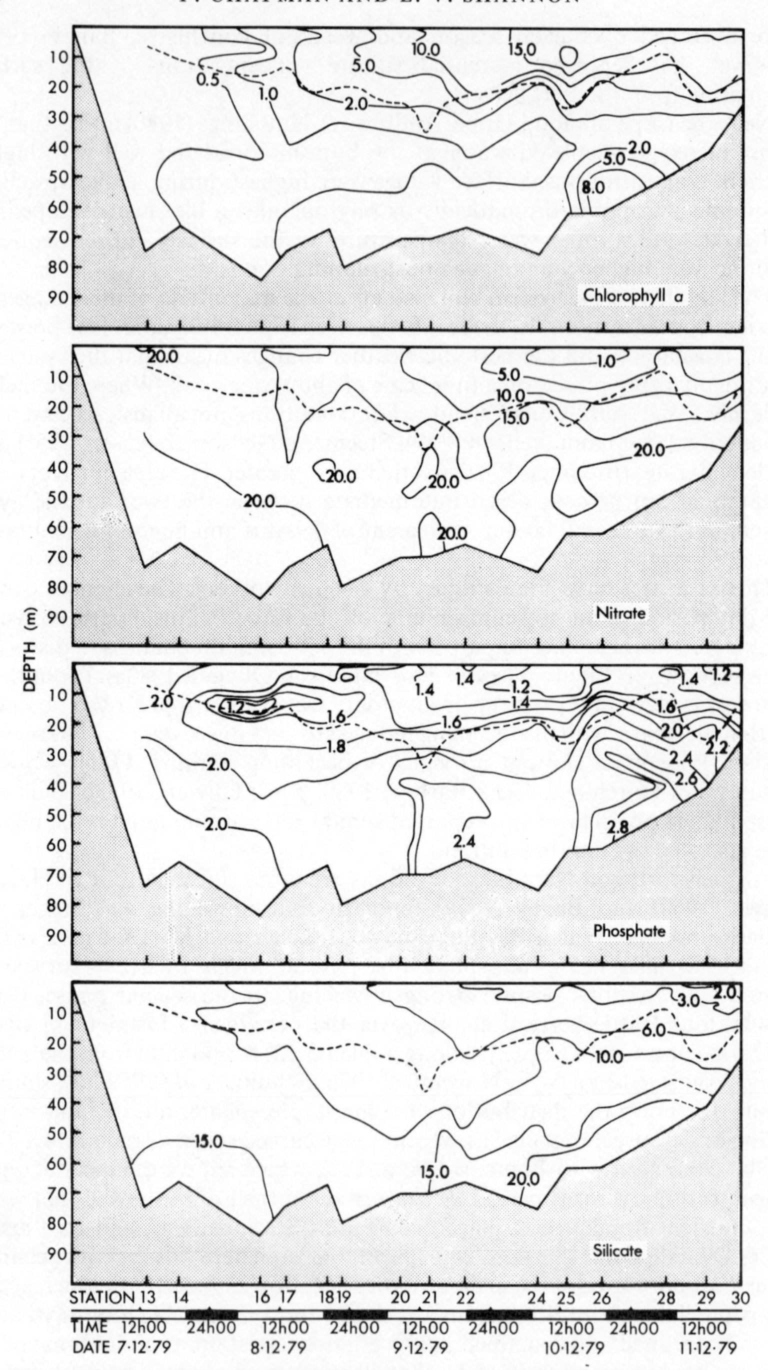

Fig. 15.—Changes in chlorophyll *a*, nitrate, phosphate, and silicate during a bloom of phytoplankton following upwelling off the Cape Peninsula: all figures in μM except chlorophyll *a* ($mg \cdot m^{-3}$); dashed line marks the position of the 10·5 °C isotherm; (after Olivieri, 1983a).

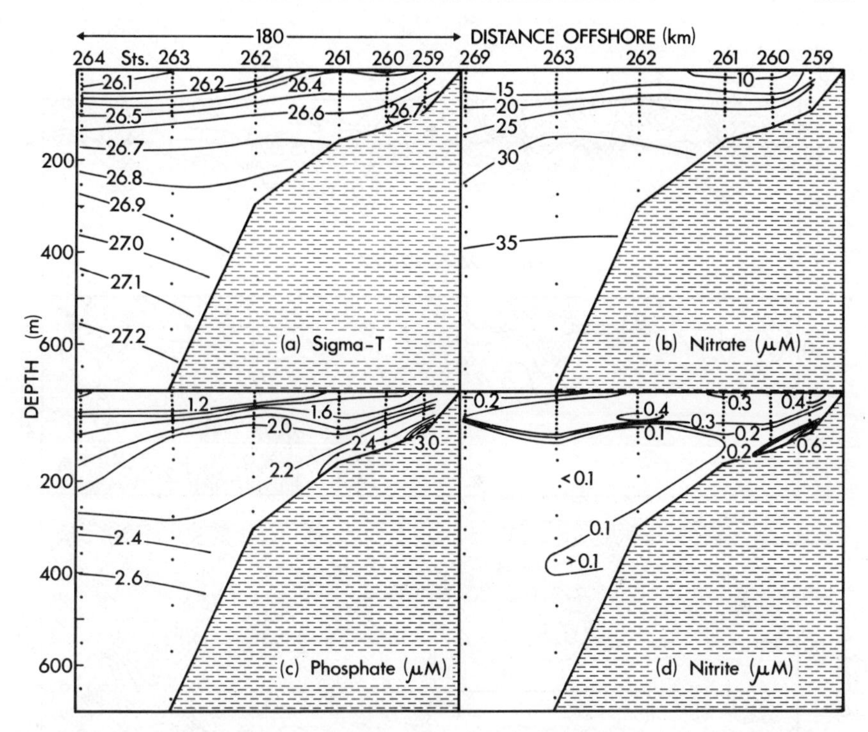

Fig. 16.—Distribution of sigma-t, phosphate, nitrate, and nitrite off Dune Point (20° S) on 17th–18th October, 1976: nutrient results in μM; data from Schultz *et al.* (1979).

The stations were occupied in the same order each day, from the coast outwards, and the depths of particular phosphate, nitrate and sigma-t isopleths on each day are shown in Table III. Winds of greater than 5 $m \cdot s^{-1}$ can be taken as favourable for upwelling (Shannon, 1985), and upwelling 'bursts' could be distinguished on the 20th–21st, 22nd, and 24th October, 1976. The steady rise in all three isopleths between 18th–21st October, followed by a relaxation, is marked.

This cruise also sampled one station, on the upper shelf at 22°16′ S: 14°05′ E in a depth of 81 m, over a period of 15 days. Temperature, salinity, dissolved oxygen, phosphate, and chlorophyll *a* were measured every three hours, while nitrate and nitrite were measured four times daily. The 15-day time series has been plotted at six hourly intervals in Figure 17, and shows several interesting features. First, there was a large degree of coherence between changes in the various components, and concentrations in the deeper layers fluctuated in sympathy with those in the upper 20 m. Secondly, a distinct short-term periodicity is discernible which might be related to internal tides. The amplitude appears to be about 10 m in the vicinity of the pycnocline, but perhaps 20 m below this depth. The uniform density (sigma-t 26·76–26·77) of the lower 40 m suggests the existence of a well-mixed, relatively thick, bottom layer. Conceivably, internal waves may play a rôle in its formation, in which

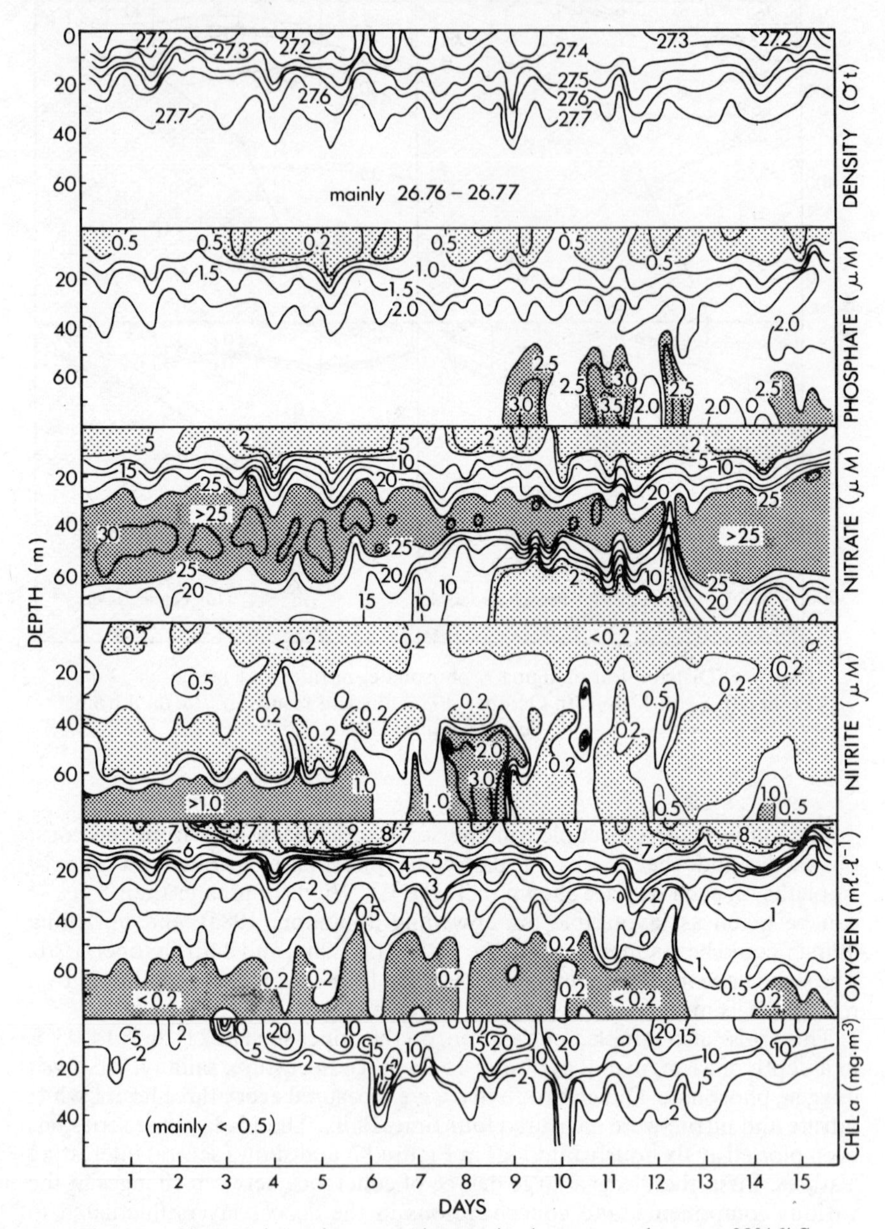

Fig. 17.—Changes in various constituents in the water column at 22°16′ S: 14°05′ E near Henties Bay between 2nd and 17th November, 1976: drawn from data in Schultz *et al.* (1979).

TABLE III

*Changes in the depth (m) of the 26·7 sigma-*t, *2·2 μM phosphate, and 30 μM nitrate isopleths at Station D (20°17′ S: 12°15′ E) over the period 18th–24th October, 1976: also given are the mean wind speeds in m·s^{-1}; data from Schultz, Schemainda & Nehring (1979)*

Oct. 1976	Sigma-*t*	Phosphate	Nitrate	Wind speed (offshore)
18	160	215	170	4
19	160	225	140	5
20	150	175	135	5–8
21	135	150	100	8–4
22	160	140	115	2–7
23	140	120	125	3–5
24	175	175	140	5–10

case they may also cause a stirring up of the surface sediments, resulting in drastic changes in the chemistry of the bottom water.

Both nitrate and phosphate were substantially depleted in the top 10–15 m of the water column (silicate was not, unfortunately, analysed). A well-defined nitrate maximum in mid-water was present throughout virtually the whole of the survey period, lower values in the bottom 20 m indicating progressive nitrate reduction after Day 5. A substantial perturbation evidently occurred during Days 8 and 9, as shown by the spike in the sigma-*t* trace, the appearance of high ($>2{\cdot}5$ μM) phosphate levels, and the decrease in nitrate and increase in nitrite, followed by the disappearance of both, in the bottom layer. Whether the changes shown in Figure 17 were due to internal tides, to shelf waves (as suggested by Postel, 1982), to local nutrient re-cycling rates or to the advection of another water mass through the system can be debated. Full moon occurred on 7th November, 1976, and consequently, if the waves evident in Figure 17 were tidal, then they would have been at a maximum around this time. We feel that it is reasonable to speculate that the dramatic changes in nutrients were related to such internal tidal action, which by stirring up the sediment layer caused an injection of both phosphate and sulphide into the bottom-water layer, resulting in the changes in chemistry mentioned above. If such features were regular, they would have large effects on redox processes and could help to maintain the thick oxygen-deficient layer at the bottom.

SUPPLY AND REMOVAL PROCESSES

There has been some disagreement as to the concentrations of nutrient elements in, and hence the source of, the upwelled water off South Africa. Bailey (1979) has discussed this, and has shown that one must be far enough offshore to remove the problem of interference caused by local shelf regeneration. From Figure 18 it is clear that phosphate concentrations in

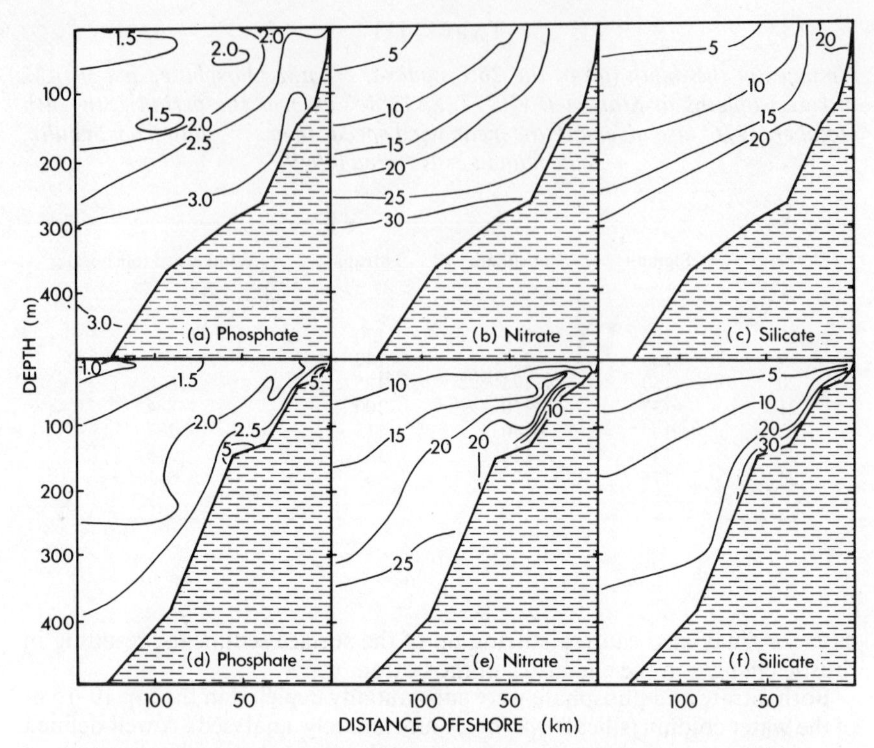

Fig. 18.—Vertical nutrient distributions at (a–c) Marshall Rocks (26°20′ S) during upwelling conditions, November, 1976; (d–f) North Rocks (24°30′ S) during quiescent conditions, February, 1977: all figures in μM; (after Bailey, 1979).

particular have been enriched by the passage of the upwelling water over shelf sediments containing very high levels of phosphorus—the maximum value found in bottom water at North Rocks during quiescent conditions amounted to 6·25 μM, and Boyd (1983a) recorded levels of up to 8 μM at 22° S. Silicate may also be enriched relative to the upwelling source water, and Bailey reports a maximum of 50 μM silicate in the surface layer, which cannot possibly arise from water upwelled from 200–350 m (Hart & Currie, 1960; Calvert & Price, 1971; Jones, 1971; Bailey, 1979). Table IV lists the variations in upwelling water: offshore water, specified as 9 °C off the Cape of Good Hope by Andrews & Hutchings (1980; off Namibia it is about 2–3 °C warmer); water on the shelf; and oceanic (offshore) surface water found by various investigators. These may be compared with values for 'true' South Atlantic Central water of 10–18 μM nitrate, 0·8–1·5 μM phosphate, and 6–15 μM silicate (Jones, 1971; Henry, 1975). Care should, however, be taken with the figures in Table IV, which are taken from vertical sections, since Andrews & Hutchings (1980) defined upwelling water as water that upwells into the photic zone, and such water frequently does not reach the surface. In the Cape, freshly upwelled summer water is unusual in having an ice blue to turquoise colour, implying little

TABLE IV

*Nutrient concentrations (μM) in (a) offshore upwelling, (b) shelf, and (c) oceanic surface waters at different points along the coast in the Benguela system: *, offshore stations still over shelf*

Reference		Nitrate	Phosphate	Silicate
Cape Peninsula				
Andrews & Hutchings (1980)	(a)	20	1·5	16
	(b)	23	1·5	19
	(c)	<1	0·5	≈5
St Helena Bay–Orange River				
Jones (1971)	(a)	15	1·5	15
	(b)	20–25	2·2	20–30
	(c)	<2	<0·2	<2
Bailey & Chapman (in press)	(a)	20–25	2	10–20
	(b)	25	2·5–3	30–>40
	(c)	<1	0·6–1	8
Namibia				
Calvert & Price (1971)	(a)	15–25	1·5	5–10
	(b)	25–30	2–2·5	20–>30
	(c)	<5	<1	<5
Bishop, Ketten & Edmond (1978)	(a)	—	1·2	6–7
	(b)	—	>2·5?	>20?
	(c)	—	<0·25	<1
Bailey (1979)	(a)	15–20	2–2·5	≈10
	(b)	10–30	2–3	30–>40
	(c)	<5	—	<5
Schultz, Schemainda & Nehring (1979)	(a)	20	1·5–2	—
	(b)	25–30	2·5	—
	(c)	10	≈1	—
Boyd (1983a)	(a)*	20–25	2–3	>20
	(b)	5–15	3–6	30–>50
	(c)	<5	<2	<1

recycling of nutrients in this area. While data from Calvert & Price (1971), Bishop *et al.* (1978) and Bailey (1979) all indicate that phosphate enrichment increases northwards in upwelled water, nitrate and silicate data are unclear in this respect. Intuitively, the relative widths of the shelf at different latitudes suggest that nutrient cycling should increase equatorwards.

The supply and removal of each nutrient will first be considered in terms of what is known from other areas, and these will be contrasted with observed changes in different sectors of the Benguela system.

Phosphate

South Atlantic Central water contains about 0·8–1·5 μM phosphate (Jones, 1971; Henry, 1975). Higher values are, however, found all along the coast, reaching maxima of 6–8 μM off Namibia (Bailey, 1979; Boyd, 1983a), and these must result from *in situ* regeneration processes. Grill & Richards (1964) and Ketchum & Corwin (1965) studied the rate of release of phosphate from decomposing phytoplankton, and found that the initial rate was high enough to allow 50% of the total to be regenerated within 24 h. This initial release occurred without loss of nitrogen, and only later were N and P released simultaneously at a constant ratio (Grill & Richards, 1964). This suggests that in shelf regions, phosphate should be released in the more inshore waters, with nitrate being released further offshore, assuming a general offshore movement of the water on the shelf.

Sedimentary processes can undoubtedly play a rôle in phosphate regeneration and/or removal. While Grill & Richards (1964) assumed that bacteria would be the main route by which phosphate is taken up from the water column, Suess (1981) has also suggested adsorption by detrital matter, such as fish debris. Once in the sediment, hydroxyapatite (Sillén, 1961), carbonate apatite (Hallberg, Bagander, Engvall & Schippel, 1972; Birch, 1975), or Fe–Mn complexes (Suess, 1976) can control the phosphate concentration in sea water, depending on the local pH levels. Pore water levels of phosphate in Namibia are high (up to 26 μM; Baturin, 1972; Rogers, 1977; Senin, 1977), as are concentrations of phosphorite rocks (see *e.g.* Baturin, 1971b,c, 1974; Baturin, Merkulova & Chalov, 1972; Summerhayes, Birch, Rogers & Dingle, 1973; Romankevitch & Baturin, 1974; Bremner, 1978c, 1980; Price & Calvert, 1978; Birch, 1979; Brongersma-Sanders, 1983), although there appears to be an inverse correlation between them (Baturin, 1972). Although many of the phosphorites off Namibia have been shown to be relict deposits (Bremner, 1983), it is believed that they are still forming in certain areas (Veeh, Calvert & Price, 1974). Bailey (1979) found that high organic nitrogen levels in sediments correlated well with Rogers' (1977) areas of high sediment phosphorus, supporting the idea of present-day formation. While this review is not the place for an account of the mechanisms of phosphorite formation, it seems clear that both pH and E_h will be important controlling factors, and the changes in oxygen concentration and hence pE observed in the water overlying the sediment will, therefore, affect the rates of supply or removal of phosphorus by this facies.

Taking the area to the north of Sylvia Hill, there are extensive patches of the sediment that contain both pelletal phosphate and authigenic phosphate, the latter occurring inshore (Birch, Thomson, McArthur & Burnett, 1983). Elsewhere, phosphate rocks occur outside the shelf break (Birch, 1975), but phosphate in unconsolidated sediments is between 1–10% in St Helena Bay and along the coast to north of the Olifants River (Birch, 1975). This would, therefore, account for the high phosphate concentrations (2–4 μM) found in bottom waters off Namibia (Hart & Currie, 1960; Calvert & Price, 1971; Bailey, 1979) and St Helena Bay (Jones, 1971; Bailey & Chapman, in press). Near Cape Point, phosphate levels are lower (Jones, 1971; Andrews & Hutchings, 1980) at 1·5–1·8 μM as a result of the narrow shelf and stronger currents close inshore. Clowes (1954) suggested that increased phosphate concentrations near St Helena Bay stemmed from the southward-flowing inshore current, which transported phosphate from Namibian waters. A more likely distant source is the Orange River, which has been shown to affect the sediments to at least 32° S (Birch, 1975), although Bailey & Chapman (in press) have argued in favour of a local source.

Bailey (1979) and Bailey & Chapman (in press) showed that off Namibia and St Helena Bay, upwelling rates of about 1 and 2 $m \cdot day^{-1}$, respectively, would be enough to supply the necessary phosphate requirements for phytoplankton growth. Off Cape Point, Andrews & Hutchings (1980) calculated a maximum potential production rate of 37 $g\,C \cdot m^{-2} \cdot day^{-1}$, which seems incredibly high. If the phosphate concentration in the water is 1·5 μM, this would require an upwelling rate of 20 $m \cdot day^{-1}$ to replenish the stock. A more realistic figure of 3·1 $g\,C \cdot m^{-2} \cdot day^{-1}$ (Brown & Hutchings, in press) requires an upwelling rate of only 2 $m \cdot day^{-1}$. Thus, it is clear that phosphate is unlikely to ever become limiting in the Benguela system.

Nitrate

While nitrate is the predominant form of reactive nitrogen in the ocean in well-oxygenated water below the euphotic zone, in shallow inshore waters, and particularly in upwelling areas, other reduced forms of nitrogen can occur. In the Benguela, these include particularly nitrite and ammonia, but presumably nitrous oxide will also be formed, although no analyses for this compound have been made in this area.

Regeneration of nitrogen is not as simple a process as that for phosphate. Nitrate itself is of course only produced during regeneration under oxygenated conditions, and when anoxic conditions result, ammonia is produced according to Equations 1 and 2, depending on whether nitrate or sulphate is the oxidizing species (Brewer & Murray, 1973).

$$(CH_2O)_{106}(NH_3)_{16}(H_3PO_4)+84{\cdot}8HNO_3 = 106CO_2+42{\cdot}4N_2+16NH_3+H_3PO_4+148H_2O \quad (1)$$

$$(CH_2O)_{106}(NH_3)_{16}(H_3PO_4)+53SO_4^{2-} = 106CO_2+53S^{2-}+16NH_3+H_3PO_4+106H_2O \quad (2)$$

This is in contrast to the normal Equation 3 for organic decomposition under aerobic conditions

$$(CH_2O)_{106}(NH_3)_{16}(H_3PO_4)+138O_2 = 106CO_2+16HNO_3+H_3PO_4+122H_2O \quad (3)$$

Equation 1 also shows that under conditions where nitrate is used as the oxidant, free molecular nitrogen can be produced. This can normally be considered as lost to the system, as nitrogen-fixing organisms are generally tropical species.

Bacteria, however, may also cause partial oxidation or reduction of the dissolved nitrogen species, and Spencer (1975) gives a good account of the various reactions that can occur. Much organic nitrogen is released as ammonia, which may be either taken up by phytoplankton directly or converted to nitrite and nitrate under aerobic conditions. Conversely, nitrate may be reduced to nitrite, by either phytoplankton or heterotrophic bacteria using nitrate as an electron acceptor during respiration, after which it is excreted in this form, or it may be reduced further to nitrous oxide and elemental nitrogen by anaerobic bacteria. Nitrate reduction is, however, inhibited by oxygen concentrations above 0·15 $ml \cdot l^{-1}$ (Skerman & MacRae,

1959). Nitrous oxide will be produced during nitrification in the water column or from the sediment under oxygenated conditions (Cohen & Gordon, 1979; Seitzinger, Pilson & Nixon, 1983) but lost under anoxic conditions. Concentrations of this species are, however, very small in comparison with other combined nitrogen compounds, being about 10^{-3} times that of nitrate in oxygenated areas.

Because of the large changes in oxygen concentration in the waters of the Benguela, it is clear that the capacity exists for all the above processes to occur. Nitrate occurs generally at the bottom in concentrations greater than in the upwelling source water (Fig. 18). This is clearly related to aerobic decomposition of organic matter, and accounts for the increased concentrations of nitrate found at the surface during upwelling events (Fig. 19).

Reduced forms of nitrogen are also found in the system. Figure 20 shows the nitrate, nitrite and ammonia levels found by Bailey & Chapman (in press) off St Helena Bay during February 1979. These data are from the only case in which the three species have been measured simultaneously in the southern Benguela. Off Namibia, simultaneous data exist in Schultz, Schemainda & Nehring (1979), although the ammonia data are from the surface layer only. Figure 20 shows the typical distributions of the three species. The high nitrate levels at the bottom are clearly visible, as is the depletion to less than 1 μM in the surface layer. Nitrite and ammonia, on the other hand, were concentrated around the thermocline at about 20 m depth. Both species were distributed in patches, although closer-spaced sampling might have shown a more continuous layer of enhancement. Ammonia levels were enhanced inshore, and a patch containing greater than 4 μM was found in the bottom water, associated with low nitrite ($<0{\cdot}05$ μM) levels. Although the levels of ammonia from St Helena Bay would normally be considered very high, we believe them to be real, as such concentrations are consistently found there. Elsewhere, other Sea Fisheries data show levels of less than 1 μM using the same analytical technique. Such high levels might denote either sediment release of ammonia or the onset of nitrate reduction, since a low nitrate concentration (<20 μM)

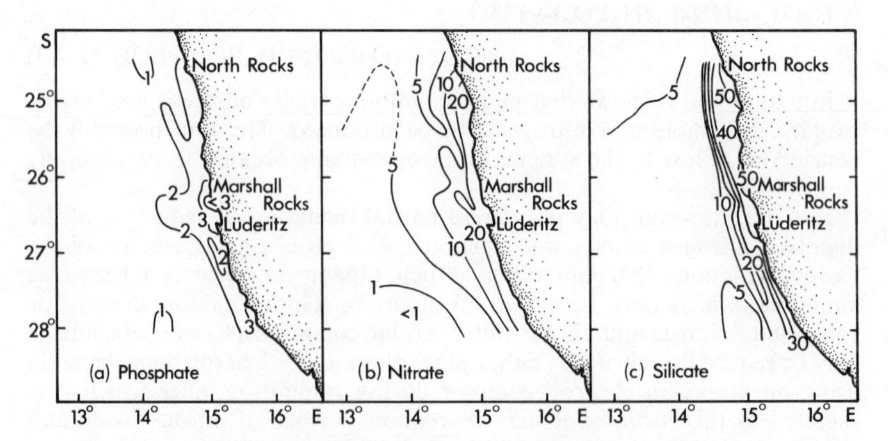

Fig. 19.—Surface nutrient distributions during upwelling conditions near the Lüderitz upwelling centre, August, 1976: all figures in μM; (after Bailey, 1979).

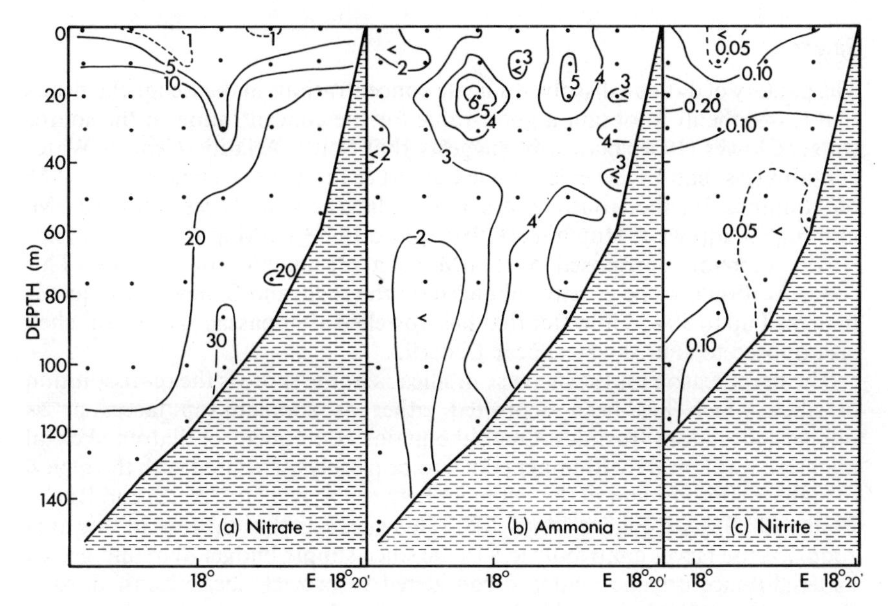

Fig. 20.—Nitrate, ammonia, and nitrite concentrations (μM) along 32°15′ S during February, 1979 (after Bailey & Chapman, in press).

was found in the same area just above the sediment, and plots of apparent oxygen utilization (AOU) against nitrate concentration showed a distinct departure from linearity at AOU values greater than 4 $ml \cdot l^{-1}$. The maximum discrepancy recorded was about 8 μM nitrate, which was not made up by nitrite or ammonia concentrations (Bailey & Chapman, in press).

Further north, off Namibia, such processes are more intense. Bailey (1979) has shown that more than 50% of the expected nitrate can be lost by nitrate reduction (see Fig. 18), and similar effects have been demonstrated by Calvert & Price (1971), Andrews & Hutchings (1980), and Boyd (1983a,b). Calvert & Price in fact found enhanced nitrite concentrations at the bottom compared with mid-water values. On their Cape Cross line (22° S), maximum nitrite levels were $>3{\cdot}0$ μM. This line was the only one on which a thermocline had definitely been established. Similar levels were found by Schultz *et al.* (1979), who also found lower levels of nitrite off Dune Point (20° S) (see Figs 16, 17). It seems, therefore, that the process is widespread throughout the Benguela system, and that the area is a sink for nitrogen. In this regard, it is interesting to look at C:N ratios in the sediments. Surface sediments in the Chamais Bay–North Rocks area, where much sedimentation of biological material occurs, have ratios of about 10:1 (Bailey, 1979). This is higher than would be expected from the settlement of fresh plankton (about 6–7:1 on the basis of Equation 3 and data from Chester & Stoner, 1974c; Bishop, Ketten & Edmond, 1978; Bailey, 1983, in press). There would appear to be little loss of nitrogen by absorption by zooplankton, since faecal pellet deposits are scarce inshore, where Bailey's samples were taken (Bremner, 1983).

Silicate

The paucity of data on dissolved silicate concentrations in the Benguela means that it is difficult to obtain a good value for the concentration in the source water. Clowes (1938, plate viii) suggests that South Atlantic Central Water contains less than 10 μM silica, while Calvert & Price (1971) suggest 10–15 μM off Namibia. Further south, Jones' (1971) diagrams imply less than 10 μM, although Andrews & Hutchings (1980) quote $15{\cdot}7 \pm 6$ μM at the Cape. Bailey (1979), therefore, proposed 5–15 μM as a reasonable compromise. This concentration may be greatly enhanced on the shelf, and Bailey (1979) quotes figures of up to 50 μM in water that has upwelled after passage across the shelf (as opposed to source water) near Lüderitz.

The main reason for the increase in silica near the shore is the re-dissolution of silica which has been deposited either as zooplankton faeces or as undigested diatom frustules. Opal, the major component of diatom skeletal material, was found by Bremner (1983) to be closely associated with the region of high organic content in the sediment between Sylvia Hill (25° S) and Rocky Point (19° S)—see Figure 14 (p. 208). This suggests that inshore in this area zooplankton faeces contribute less to the silica supply budget than undigested material. Since organic matter is considered to protect silica against dissolution (Spencer, 1975), it could also account for the apparent silica limitation found offshore in Namibia (Carmack & Aagaard, 1977; Bailey, 1979; Boyd, 1983a).

If, however, silica is protected in this way from re-solution, then the high silicate levels inshore pose a problem. Possibly the low pE found in the inshore sediments causes more rapid decomposition of the organic coating and loss of the stabilizing chelator Fe (III), as has been suggested for the Black Sea by Grasshoff (1975). The Black Sea is, however, permanently anoxic, and other than in the immediate vicinity of Walvis Bay, this process may not be as important in the Benguela. Bailey (1979) did, however, report high silicate concentrations in conjunction with oxygen-deficient conditions off Easter Point (25°20′ S). A more likely explanation is an alternative source of silica in the inshore region, and terrigenous material is the natural alternative source.

As has been stated above, there is very little riverine silica input along the Namibian coast. Aeolian dust has been observed offshore (Chester & Stoner, 1972; Chester, Griffiths & Stoner, 1978; Shannon, 1985, and his Fig. 10), particularly close to the coast, and the coastal desert regime makes this an attractive source. Rogers (1977) has described in detail sediment transport from the sea north of the Orange River mouth to the Namib Sand Sea and back again to the coast to the north of Lüderitz, and Bailey (1979) has suggested that diatom frustules must accompany this. Certainly, the dissolved silicate distribution tends to contradict the statement of Chester *et al.* (1978) that aeolian transport is relatively unimportant in the South Atlantic, although their samples were taken away from the coast.

Silica-rich sediments also occur south of the Orange River, and are found as far south as Cape Point (Birch, 1977). Size fraction analysis has shown them to be mainly terrigenous in origin, although in St Helena Bay they are diluted with muddy, organic-rich sediment. Since, however, the conditions are more oxidizing to the south of the Orange River than to the north, except during the summer in St Helena Bay, there will be less re-solution of the silica from this source, which appears anyway to be less susceptible to dissolution than biogenic silica (Nelson & Goering, 1977). This is reflected in the lower silica concentrations found inshore south of the Orange River—Bailey & Chapman (in press) report maximum values for dissolved silicate of about 40 μM, while Andrews & Hutchings (1980) found less than 20 μM off the Cape Peninsula.

Removal of silica would appear to be dependent on zooplankton grazing activity, particularly off Namibia. Willey (1978) has suggested that the ratio of clays to biogenic material can control the competing adsorption and desorption reactions that occur in sediments. Off Namibia, the predominance of biogenic over terrigenous material is so pronounced (see Fig. 14, p. 208) that adsorption is unlikely. Thus, the situation arises that silicate is taken up by phytoplankton inshore and redeposited either locally or offshore (after consumption by zooplankton). This leads to the depletion of silicate in offshore surface water, as mentioned above. Such depletion is, however, only a temporary phenomenon, as Bremner (1983) has calculated the silica budget for Sylvia Hill, and considers the supply of silicate in upwelling water to be about four times the demand as evidenced from the size of the silica-rich deposit on the shelf. Working from known phytoplankton production rates, Bailey (1979) and Bailey & Chapman (in press) have calculated required upwelling rates of 2–3 $m \cdot day^{-1}$ to supply the necessary silicate inshore in both the Lüderitz and St Helena Bay regions, which are well within known rates of upwelling in these areas.

NUTRIENT LIMITATION

There is some disagreement about nutrient limitation in upwelling areas. While nitrate is generally considered to be the major limiting nutrient in the sea, Dugdale & Goering (1970) showed that silica could become limiting in the Peru upwelling system. Bailey (1979) and Boyd (1983a) have both presented evidence for silica limitation in the northern Benguela, where silicate concentrations were <1 μM while nitrate concentrations were greater than 5 μM. Carmack & Aagaard (1977), using the data of Calvert & Price (1971), calculated the volumes of water occupied by particular nutrient concentration ranges. They showed, on the basis of T-S characteristics, that about 50% of the water on the shelf could be related to water derived from upwelling, and that while the major class in the nitrate-phosphate correlation diagram was derived from bottom water on the shelf, the phosphate-silicate diagram was biased towards low values, as found in surface waters. They suggested that this showed that silicate would be the limiting nutrient. Unfortunately, it is not possible to state this conclusively, since in none of these three papers are there any data on phytoplankton studies. In the southern area, near the Cape Peninsula and St Helena Bay, published data (Andrews & Hutchings, 1980; Barlow, 1980; Bailey & Chapman, in press) indicate that nitrate is the limiting nutrient. Despite the periodic influx of Agulhas Bank water and occasionally possibly also of Agulhas water, which contains <1 μM nitrate and 3–5 μM silicate (P. Chapman, unpubl. data), into the southern Benguela, one would expect intuitively that because of the large amounts of biogenic silica on the shelf off Namibia, which are not found near the Cape, that silica would be limiting in the south and nitrogen in the north.

This paradox may perhaps be resolved by considering the major removal process—uptake by phytoplankton—and the relative stabilities of the two parts of the system. Although diatoms are considered to be the major contributors to the phytoplankton in the Benguela (Hart & Currie, 1960; Kollmer, 1963; Kruger, 1980; Olivieri, 1983a,b), it is clear that dinoflagellates, which do not require silicate, are also frequently found; indeed, the red tides that have been reported all along the coast are generally caused by dinoflagellates (Pieterse & Van der Post, 1967; Horstman, 1981). Changes in the relative numbers of diatoms and dinoflagellates could thus affect the relative nitrate:silicate balance and determine which should be limiting. There is also the problem of species dominance. In Olivieri's (1983a) study, the dominant species were *Skeletonema costatum*, *Chaetoceros compressus*, and *C. debilis*. Changes in cell size, and hence surface to volume ratio, division rate, and nutrient uptake rate are perhaps all important in determining dominance, and have been discussed by Olivieri (1983a), who found that at the end of the study period the cells were larger. It may be that larger diatom cells develop as a result of silicate limitation (smaller surface to volume ratio), while small cells occur when nitrate is limiting, and thus that the apparent importance of nitrate in the southern area may be an artefact of the small number of cruises so far undertaken.

The stabilities of the two parts of the system may also affect nutrient limitation. In the south, the "pulsing" of upwelling referred to above means that there is less chance of establishing maximum numbers of zooplankton. This does not occur in the north, where quasi steady-state conditions prevail. If we can assume that the majority of silica is removed from the surface layer by faecal pellet deposition, and that these sink faster than decomposing phytoplankton cells, which seems very reasonable, then more rapid silicate depletion will occur off Namibia than off the Cape Peninsula, where recycling occurs more in the water column. The more rapid currents in the south will also tend to remove slowly-sinking material from the area by advection, so that recycling is less important here than further north.

INTER-NUTRIENT RELATIONSHIPS

The fact that plankton take up nutrients in fairly constant ratios has been long known (Spencer, 1975). From Equation 3 (p. 217) it is possible to calculate the ratio of oxygen, nitrogen, and phosphorus that will be taken up or released under normal oxidative conditions as $\Delta O:\Delta N:\Delta P = -276:16:1$ (the Redfield ratio). This ratio has been found to approximate to reality by many workers, and $\Delta N:\Delta P$ ratios generally range from 12–17:1 in either offshore waters or well-nourished cultures (Spencer, 1975). In inshore waters, however, lower results are often found, particularly when near-limiting conditions apply. Deviations from the standard 16:1, $\Delta N:\Delta P$, ratio can also be expected under conditions of anoxia. Equation 1 (p. 217) shows that in such circumstances, ammonia is produced, but nitrate is used up, and that $\Delta NO_3^-:\Delta NH_3$ is about $-5{\cdot}3:1$. Since ammonia is frequently omitted when calculating $\Delta N:\Delta P$ ratios, it is to be expected that the overall $\Delta N:\Delta P$ ratio will drop dramatically in such areas. Changes in Si:P and Si:N ratios are generally more varied than $\Delta N:\Delta P$ ratios. This arises from the variable use of silica by phytoplankton (Spencer, 1975). Although in certain areas the $\Delta Si:\Delta N$ ratio approximates to 1:1, both higher (Grill & Richards, 1964) and lower (Schott & Ehrhardt, 1969) ratios have been recorded.

The available data on Δ nutrient ratios for the Benguela system (from Calvert & Price, 1971; Andrews & Hutchings, 1980; Bailey & Chapman, in press) are shown in Table V. All relationships have been calculated from regression equations with the exception of the "a" and "b" sets of Andrews & Hutchings. These were calculated by integrating the total concentrations of nutrients in the 0 to 50-m and 50 to 100-m layers, respectively, and then obtaining the inter-element ratio. Since phosphate is always in large excess in the Cape Peninsula, these ratios would be expected to be lower than the ratios obtained from regression equations. A similar result was obtained by Bailey & Chapman (in press), who found that although the slope of the N:P graphs remained reasonably constant at between 16–25:1, the intercept on the phosphate axis decreased from 1·2 μM to 0·6 μM over a four-day period. The ΔN:ΔP ratios for these two sets of data are in good agreement. Deviation from the expected 16:1 Redfield ratio for ΔN:ΔP may well relate to the preferential initial release of phosphate from decomposing phytoplankton reported by Grill & Richards (1964). Calvert & Price (1971), however, found rather lower ΔN:ΔP ratios, particularly at the inshore stations on their Walvis Bay line. They suggested that this was probably related to loss of nitrate *via* reduction, which seems reasonable, as the ratio increased towards the outermost station on the line, where it was 14:1.

TABLE V

Nutrient relationships for the Benguela system, normalized to phosphate: all figures are atomic ratios; a, calculated from ratios of total nutrient contents in 0 to 50-m layer; b, calculated from ratios of total nutrient contents in 50 to 100-m layer; c, calculated from regression equations using 1971–1972 data; d, calculated from regression equations using 1972–1973 data; e, mean values across shelf; f, inshore samples, Walvis Bay; g, other inshore stations; h, calculated from regression against apparent oxygen utilization (AOU), not O_2

Reference and area	Phosphate	Nitrate	Silicate	Oxygen	Notes
Cape Peninsula					
Andrews & Hutchings	1	9·3	11·3	—	a
(1980)	1	11·7	10·8	—	b
	1	19·7	20·0	−227	c
	1	23·3	17·7	−293	d
Namibia					
Calvert & Price (1971)	1	12 (<16)	10–14	—	e
	1	7·2	25	—	f
	1	—	29–95	—	g
St Helena Bay					
Bailey & Chapman (in press)	1	16–25	20–34	−420	h

ΔSi:ΔP ratios were rather more variable. Andrews & Hutchings (1980) again recorded differences in the ratios depending on whether regression equations or total quantities were used. The main points of interest are the very low ΔSi:ΔP ratios found by Calvert & Price (1971) offshore, and the very high (<95) ratios inshore. Grill & Richards (1964) showed that the ΔSi:ΔP ratio changed with time from low to high values, and suggested that a ratio of 23 was reasonable for much of the time cells were decomposing. Calvert & Price suggested that their ratios meant that water on the outermost stations of their lines was the most recently upwelled, and that nutrient regeneration would therefore be at an earlier stage than closer inshore. In view of the mechanism of upwelling in this area described by Shannon (1985), this seems highly unlikely, and we prefer to explain the high inshore levels as resulting from the exceptionally high silica deposits in the region (see Fig. 14, p. 208). Low offshore ratios would result from the low dissolved silicate levels in the 'oceanic' surface water, since silica regeneration is confined to bottom layers in contrast to phosphate regeneration (see above).

Relationships between silica and dissolved oxygen concentrations are reported to vary considerably. While Andrews & Hutchings (1980) found a parabolic relationship between the two variables, Bailey & Chapman (in press) found no simple relationship at all. This they explained on the basis of the different silica concentrations on the bottom in the two areas, which are related to current patterns. A similar explanation can be used to account for the discrepancy in ΔO values given in Table V. In St Helena Bay, sluggish currents allow a larger build-up of organic matter at the bottom, and this has a disproportionate effect on oxygen levels since it decays slowly. Further south, much of the decomposing material may be advected out of the area.

Andrews & Hutchings (1980) compared the direct element ratios with oxygen concentrations in an attempt to separate uptake from regeneration. Their arbitrary cut-off limit between the two processes was 4 $ml \cdot l^{-1}$ oxygen. As the oxygen content increased from 4 to 7 $ml \cdot l^{-1}$, so both N:P and N:Si ratios decreased, the former from 10–20 to less than 1, the latter from 1–3 to zero. Thus, nitrate was found to decline faster than either phosphate or silicate and was taken to be the prime limiting nutrient in the area at that time. At oxygen concentrations below 4 $ml \cdot l^{-1}$ the N:P ratio remained virtually constant at between 10–20, while the N:Si ratio declined more sharply from 1·5 to 0·5 (Fig. 21). This suggests either that silicate is regenerated faster than nitrate under conditions of oxygen-depletion, which seems unlikely, or, more probably, provides further evidence for nitrate reduction to ammonia or elemental nitrogen.

These authors also postulated the separation of oxygen-depleted water into two masses, one in mid-water with a high N:Si ratio which they considered "local" in origin, and the bottom water with low N:Si ratio which they considered "distant". Bailey (1979) could, however, find no such relationship off Lüderitz. If nitrate is indeed regenerated faster in mid-water than silicate, it would be expected that silica enrichment would occur on the bottom as the silica-rich diatom remains settle out, and we believe this is a more reasonable explanation for this phenomenon.

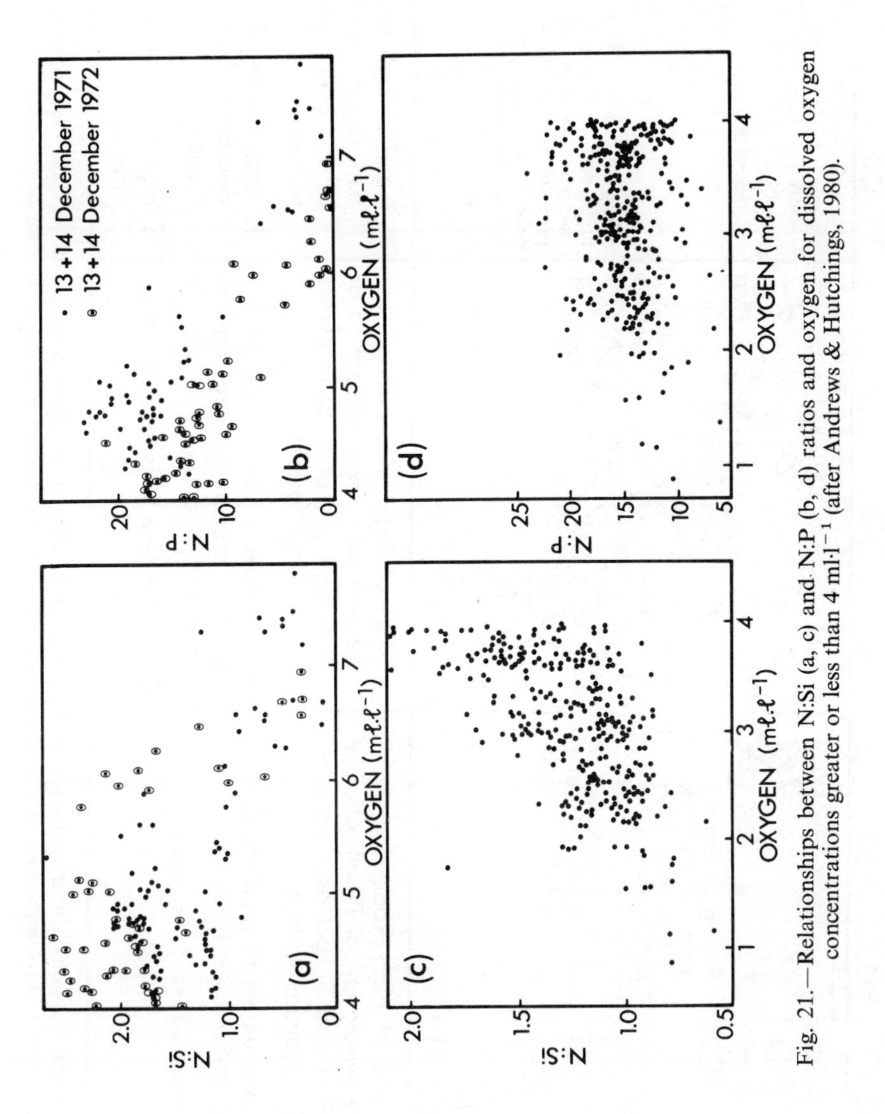

Fig. 21.—Relationships between N:Si (a, c) and N:P (b, d) ratios and oxygen for dissolved oxygen concentrations greater or less than 4 ml·l^{-1} (after Andrews & Hutchings, 1980).

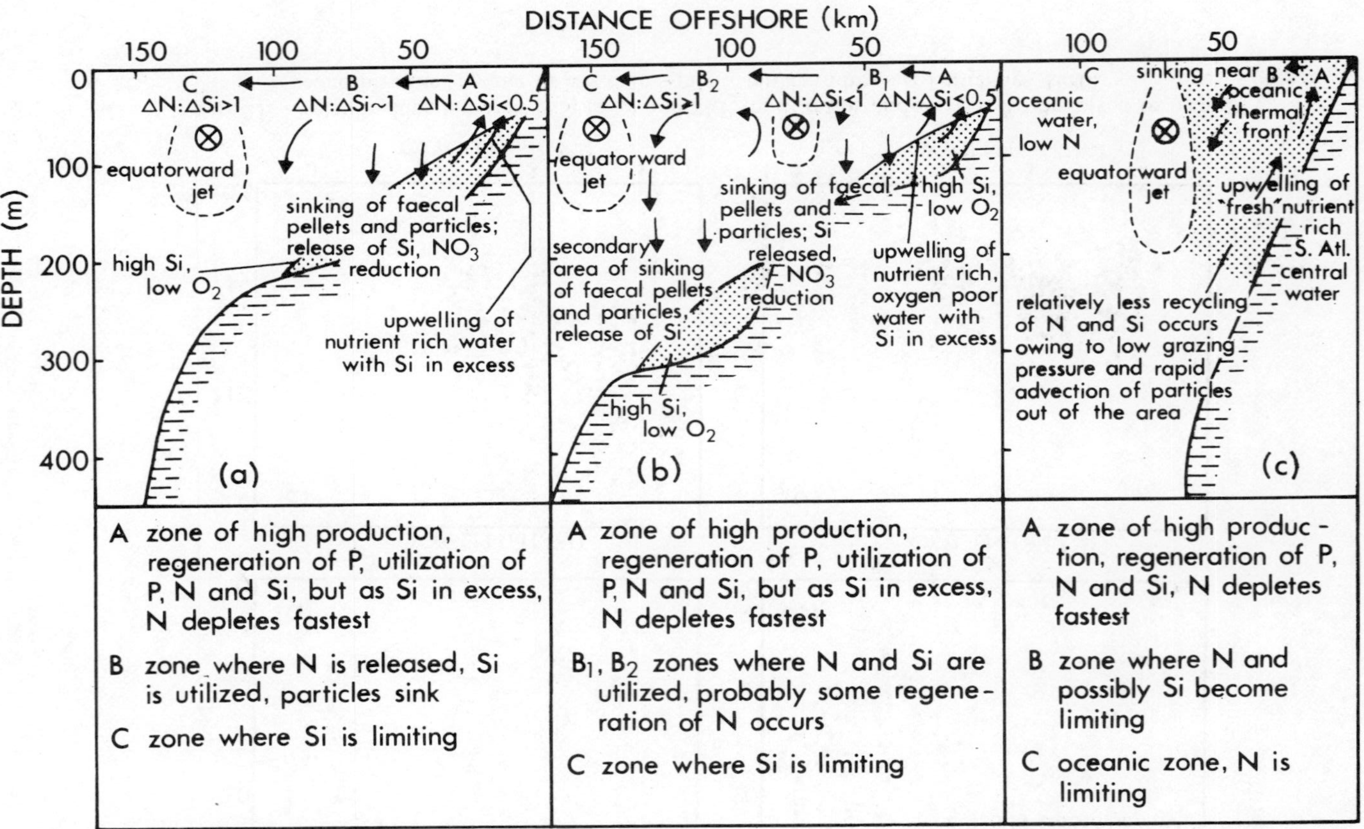

Fig. 22.—Conceptual diagrams of nutrient cycling off Namibia (a, b) and in the southern Benguela off the Cape Peninsula (c), based on interpretation of the published data; in the absence of more information, they should be considered speculative.

THE CONCEPTUAL MODEL OF NUTRIENT CYCLING

The processes described above have led to the conceptual diagram shown in Figure 22. Where there is only one shelf break (Fig. 22a), we consider that phosphorus and nitrogen are recycled relatively fast in the inshore zone leading to a pool of oxygen-depleted water and the prospect of nitrate reduction. Nitrate is liable to be the limiting nutrient, which leads to low N:Si ratios inshore, particularly if dinoflagellate blooms occur. Faecal pellets and dead plankton are redeposited inshore and also exported as the surface water moves offshore. Since nitrogen is released faster than silicate from decomposing phytoplankton, the N:Si ratio offshore changes, increasing to greater than 1 and leading to silicate depletion. This situation is also found between the mouth of the Orange River and St Helena Bay. Where a double shelf break occurs (Fig. 22b), the situation is similar, but two regions of oxygen depletion or deficiency can occur (see Fig. 4, p. 191). The boundary of the silicate-limited area is then moved further offshore.

South of Cape Columbine, however, the topography is totally different with a very narrow shelf (Fig. 22c). The pronounced thermal front and shelf-edge jet act as a barrier to offshore transport, creating an intense zone of convergence between the cold upwelled inshore water and the the warm nutrient-poor oceanic surface water. Nutrient regeneration thus occurs inshore of the frontal zone within the water column. Both nitrate and silicate could be limiting here, as a result of the rapid water movements in the area, and the advection into the system of lowish-nutrient Agulhas Bank water. Such a situation may also exist north of Cape Frio, where the topography is similar, although we feel that the stronger thermocline in this region may cause nutrient recycling to take place within the euphotic zone. Unfortunately, although data on this area have probably been published by Russian workers we have to date been unable to locate them.

PARTICULATE MATTER

While particulate matter is obviously implicated in nutrient cycling in the Benguela, again very little work has been done. Jacobs & Ewing (1969) and Chester & Stoner (1972, 1974c) took a few samples in the south east Atlantic, but the data were obtained far offshore and averaged 51 and 140 mg·m^{-3}, respectively. This is rather lower than the 800 mg·m^{-3} reported by Gordeyev (1963), but Russian data are generally an order of magnitude higher than those reported by other workers. Chester & Stoner did, however, do carbon and nitrogen analyses on their samples, and showed that the C:N ratio was not significantly different to surface values from the rest of the world's oceans.

Emery, Milliman & Uchupi (1973) found that total suspended matter in the South Atlantic east of 20° W correlated well with the Forel colour of the water when both were plotted on log scales. They also showed a high coherency between organic matter and total suspended matter (Fig. 23), although the organic content dropped close to the coast and round the mouths of the major rivers where the terrigenous inputs were enhanced. In this respect J. Agenbag (pers. comm.) has noted, from a LANDSAT IV image, a substantial input of terrigenous material, presumably fine particulates, during a recent flood of the Kunene River. The material was transported northwards adjacent to the

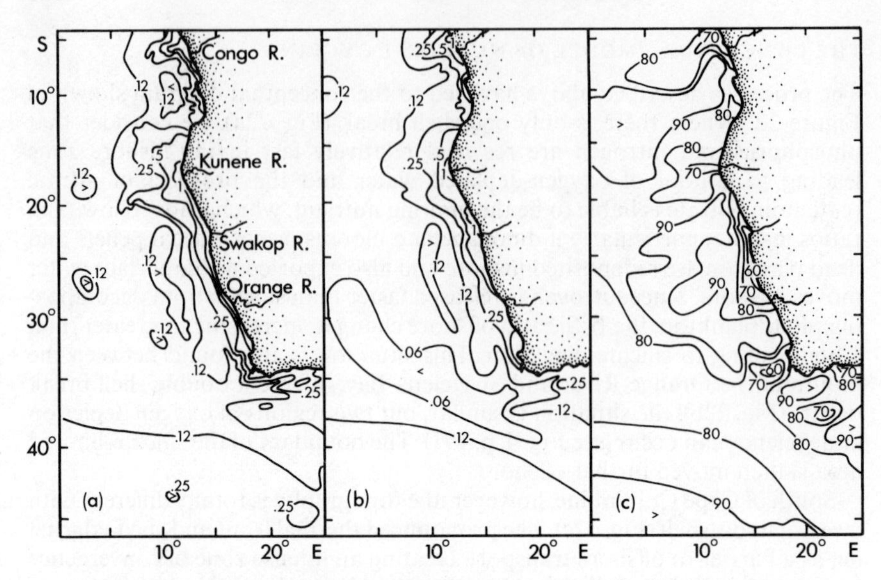

Fig. 23.—Loadings of (a) total particulate matter ($mg{\cdot}l^{-1}$), (b) organic matter ($mg{\cdot}l^{-1}$), and percentage organic matter (c) in suspended material in the south east Atlantic (after Emery *et al.*, 1973).

coast, resembling a wedge, with a sharp colour front on its western boundary. Total concentrations from the work of Emery *et al.* (1973) were much higher closer to the coast, and the figures agreed with those found by Chester & Stoner (1974c) along the South African south coast.

Fluxes of particulate matter in St Helena Bay have been described by Bailey (1983, in press). He found that the highest fluxes of particulate organic carbon and nitrogen, amounting to 4·2 and 0·65 $g{\cdot}m^{-2}{\cdot}day^{-1}$, respectively, occurred during the summer. This high organic carbon loading (20% of the total particulate matter) decreased (to 5%) during winter when fluxes of only 1·2 and 0·21 $g{\cdot}m^{-2}{\cdot}day^{-1}$ were measured. This decline was found to depend both on the life cycle of *Chaetoceros*, which blooms in spring and summer, and on the transport of inorganic matter by the Berg River during the winter.

Bishop, Ketten & Edmond (1978) also analysed the particulate matter distribution with depth along a line of stations near Walvis Bay. Using large volume filters they found that over 90% of the organic matter produced in the euphotic zone was recycled in the upper 400 m, with more than 99% being recycled in areas of low productivity. Particulate maxima for both the greater and less than 53 μm portions were found at or above the pycnocline. These increased inshore to >100 $mg{\cdot}m^{-3}$ for the >53 μm portion, and >500 $mg{\cdot}m^{-3}$ for the <53 μm portion. Particulate organic carbon, major ions (Ca, K, Mg), silica, and biological abundance data all followed a similar pattern, with C:N ratios being clustered between 7–8 for the larger size fraction and between 5–7 for particles of <53 μm. The data are in agreement with Hobson's (1971) values, for sample taken between the Congo and Orange Rivers. The maximum organic carbon loadings were 205–

260 mg·m^{-3} above 20 m at the inshore station, with diatoms forming by far the largest numbers of organisms. Hobson, in contrast, using bottle samples, found 30–90 mg·m^{-3} (mean 49 mg·m^{-3}) phytoplankton carbon and 10–60 (mean 25) mg·m^{-3} zooplankton carbon in the surface mixed layer at inshore stations apart from those off Cape Frio and south of Walvis Bay where the contribution from phytoplankton was an order of magnitude higher.

As a result of recycling in the surface layer, the percentage of organic matter in the particulates decreased to only 50–60% in deeper samples (Bishop *et al.*, 1978). The organic matter which survived to 400 m was mainly faecal material or foraminiferans, but such particles, although contributing the majority of the vertical flux, in fact made up less than 4% of the total suspended matter. Hobson (1971) found that less than 6% of particulate C was deeper than 100 m. The flux across the 400-m depth contour varied from between 84–91 mmol C·cm^{-2}·1000 yr^{-1} (10–11 g·m^{-2}·yr^{-1}) to 1–4 mmol (120–500 mg·m^{-2}·yr^{-1}) on going seawards, reflecting changes in productivity in the euphotic zone.

MINOR ELEMENTS

In contrast to oxygen and nutrient distributions in the Benguela, very little work has been done on other elements. Iodine and sulphur have been mentioned above, which leaves only the alkali and alkaline earth elements, and the so-called heavy metals. While chemical oceanographers are now changing to using molar concentrations for their data, since nearly all the previous work has been described in terms of ppm and μg·l^{-1}, this nomenclature will be retained as g·m^{-3} or mg·m^{-3}. This is necessary because of problems inherent in trying to redraw diagrams originally depicted in μg·l^{-1} where the data themselves are missing.

ALKALI ELEMENTS

Potassium, rubidium, and lithium concentrations have been analysed by Orren (1969, 1971). The results, from 198 samples around South Africa, nearly all from 20 m depth, showed only minor differences between the South Atlantic and South Indian Oceans (Table VI). Modal classes have been used, rather than means, because of skewness in the data. These modes compare to generally accepted marine abundances of 388 g·m^{-3} ($1{\cdot}02 \times 10^{-2}$ M) for K, 180 mg·m^{-3} ($2{\cdot}6 \times 10^{-5}$ M) for Li, and 120 mg·m^{-3} ($1{\cdot}4 \times 10^{-6}$ M) for Rb at a salinity of $35{\cdot}00 \times 10^{-3}$ (Brewer, 1975). The elements are generally considered to be conservative, and element:chlorinity ratios were close to other values in the literature (K 0·0190–0·0206 g·kg^{-1}; Li 8·47–9·65 μg·kg^{-1}; Rb $4{\cdot}99 \times 10^{-3}$–$5{\cdot}80 \times 10^{-3}$ μg·kg^{-1}). When the data were divided into inshore and offshore sets, however, it was found that water in the inshore regions subject to upwelling was lower in all three elements than elsewhere. Covariance between the three elements was found, with the Li:Rb ratio being about 1·7, K:Li 2200, and K:Rb 3500–4000, all on a mg/kg basis.

Depth profiles were obtained at Stations 133, 144, 211, and 212, of the "Circe" cruise of the ARGO in 1968 in the Cape and Angola Basins. At the three

TABLE VI

*Potassium, lithium, and rubidium data from the west and south coasts of South Africa: values for K in g·m^{-3}, Li and Rb in mg·m^{-3}; n, number of results; *, data imprecise because of analytical problems; data from Orren (1969)*

Area	Potassium			Lithium			Rubidium		
	Range	Mode	*n*	Range	Mode	*n*	Range	Mode	*n*
Offshore Atlantic	290–440*	370–390	41	140–200	165–175	41	85–130	100–110	41
Cape West coast (including upwelling)	270–430	390–410	34	125–205	180–185	34	55–150	100–105	34
Cape South coast (Agulhas)	310–430	400–410	51	155–205	180–185	51	75–135	105–110	51
Total	270–440	400–410	198	125–205	175–190	198	55–150	100–105	198

deeper, offshore, Stations (133, 29° S: 8° E; 144, 26° S: 9° E; 212, 18° S: 8° E), the three elements covaried and showed a distinct minimum below the core of the low salinity Antarctic Intermediate Water. This minimum was lower than expected from conservative theory, as levels were only about 50% of surface values despite a change in salinity of just over 1×10^{-3}. This minimum was not found at Station 211 (20° S: 11° E), where the water depth was only 1400 m, and the profiles were virtually constant with depth.

Detailed inshore surveys between Lamberts Bay and Dassen Island (32° S to 33°30′ S) and between 20° S to 25° S were made to investigate further the effect of upwelling on the alkaline metals. Off Namibia, there was strong upwelling (Orren, 1969; Calvert & Price, 1971) which decreased towards the north of the area. Remarkable changes in the alkali distributions were seen (Fig. 24). All five lines sampled showed a decrease in concentration inshore, and a zone of higher concentration about 35–70 km offshore. Orren (1969) has suggested that these relationships, together with similar effects observed for Cu, Fe and Mn (see below) show uptake of the elements by phytoplankton in this zone of intense production. He suggested that not only is the necessary amount required for metabolism ingested, but that adsorption to the surfaces of planktonic particles can occur. This material is then lost to the system as the organism decays, being regenerated at depth away from the area of uptake.

Similar changes in concentration with upwelling were found at a series of stations in the upwelling zone between Lamberts Bay and Dassen Island (Orren, 1969). The variations in concentration were, however, less than those observed off Namibia. Data suggested that upwelling (offshore) water was slightly depleted in alkalis (by 5–10%), and separated from the coast by a zone of regeneration. From data taken at the mouth of the Berg River, Orren concluded that riverine input had very little effect on elemental concentrations in the coastal water (Table VII). His figures showed that the dilution effect of the river water was restricted to the winter, and confined to the upper part of the water column. Data on salinity at a station in St Helena Bay some 12 km northwest of the Berg River collected between 1950–1957 (Buys, 1959) confirm this—of the 81 monthly values of surface salinity collected at this station, only three were less than $34{\cdot}00 \times 10^{-3}$, with a minimum of $32{\cdot}38 \times 10^{-3}$ in May 1954.

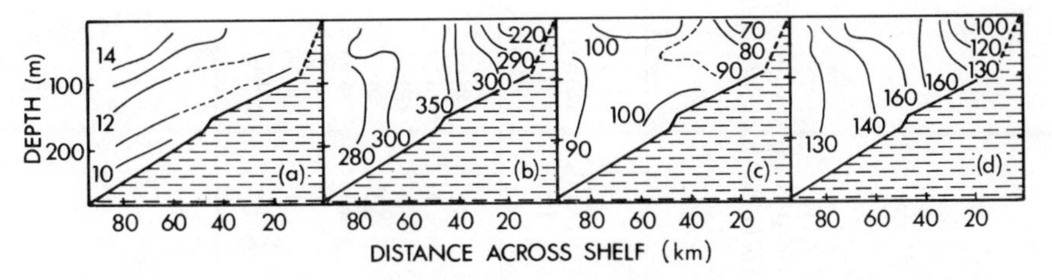

Fig. 24.—Distribution of temperature and alkali metals across the Namibian shelf at 24°15′ S in October, 1968, Argo Stations 160–163: values in $g{\cdot}m^{-3}$ (K), $mg{\cdot}m^{-3}$ (Rb, Li); (a) temperature; (b) potassium; (c) rubidium; (d) lithium; (after Orren, 1969).

TABLE VII

Effect of Berg River on concentrations of alkalis and Cu, Fe, and Mn in coastal waters of St Helena Bay: concentrations in g·m^{-3} (K), mg·m^{-3} (others): n.d., not detectable; Org., organic; Part., particulate; Sol., soluble; parentheses, figures less reliable because of analytical problems; data from Orren (1969)

Month	Depth (m)	Salinity × 10^{-3}	Cu			Fe		Mn		K	Rb	Li
			Org.	Part.	Sol.	Part.	Sol.	Part.	Sol.			
May	0	33·91										
	4	34·77	1·7		4·2		11·5		1·8	(355)	87	(159)
July	0	33·90										
	10	34·27	1·4	0·3	3·7	96	2·3	n.d.	2·3	413	107	163
November	0	34·95	n.d.	0·4	0·5	64	1·1	0·8	1·7	366	123	148
	8	34·76	n.d.	0·3	0·7	41	0·3	0·6	1·7	353	123	147
January	0	34·76	0·1	0·9	2·4	96	6·8	n.d.	1·4	362	97	174
	10	34·75	0·3	1·0	1·3	59	5·3	n.d.	1·4	362	97	174

BARIUM

Barium was studied in samples taken during the GEOSECS expedition throughout the Atlantic Ocean by Chan, Drummond, Edmond & Grant (1977), but of the 121 profiles, only eight (Stations 100–107) were taken in the eastern Atlantic between 35° S and 10° S, and only two were east of 10° E, off Cape Town. In all sections, barium was shown to covary with the silica and alkalinity distributions with a mid-water maximum in the Antarctic Intermediate Water. While surface values were about 5·5 mg Ba·m^{-3}, the mid-water maximum was about 11 mg·m^{-3} at 42° S (Station 93), 10 mg·m^{-3} at 23° S (Station 103), and 8·8 mg·m^{-3} at 12° S (Station 107) as the percentage of "pure" Antarctic Intermediate Water declined. Bottom water maxima also declined from 13·7–13·9 mg·m^{-3} in the Cape Basin to 9·5–12·4 mg·m^{-3} in the Angola Basin. Values from the eastern Atlantic were about 0·6–1·4 mg·m^{-3} higher than those from corresponding latitudes in the western Atlantic, despite the presence of the Walvis Ridge. Eastern Atlantic stations did, however, show an excess of barium against salinity in bottom water compared with western stations (Chan *et al.*, 1977). This coincided with silica enrichment, which the authors suggested arose through dissolution of biogenic remains in the Angola Basin. This seems reasonable when the oxygen distribution in the area is considered (see above). The excess amounted to 1 mg·m^{-3} at Station 107 compared with equatorial values. Phosphate and nitrate did not covary with Ba, which is probably, therefore, linked to a slowly dissolving phase such as carbonate, opal, or perhaps barite.

Barium in the sediment has been discussed by Calvert & Price (1983), who found that on the Namibian shelf there was, surprisingly, very little correlation with silica distribution. This is in contrast to data from Brongersma-Sanders, Stephan, Kwee & De Bruin (1980), and Brongersma-Sanders (1983), who found high values in diatomaceous ooze from the same area. Clearly, scope still exists for research into the geochemistry of this element, and the rôle of biological transport, in the Benguela.

HEAVY METALS

As has been stated above, there is little effect from river input on the Benguela system as a whole, except perhaps near the mouths of the Kunene and Orange Rivers. Since most metal input to the sea arises from riverine sources, it would be expected that heavy metal levels would also be low generally. Available data on metal concentrations in west coast rivers and estuaries, obtained from Drs P. Bartlett and G. Eagle of the National Research Institute for Oceanology, Stellenbosch, are shown in Table VIII. While the ranges for all the elements studied are wide, the geometric means are towards the lower limits of 'normal' estuarine values, particularly when compared with Northern Hemisphere estuaries (*cf.* Burton, 1976; de Groot, Salomons & Allersma, 1976). Sediment values, in particular, are an order of magnitude or more lower than common European concentrations. Any variations in the Benguela are, therefore, likely to result from oceanographic or biological activity.

Heavy metal concentrations in the waters of the Benguela system were first reported by Orren (1967) and Liesegang & Orren (1966). These authors studied the distribution of copper, iron, and manganese in samples from 20 m

TABLE VIII

Trace metal data for the Orange, Olifants and Berg Rivers: all data in $mg \cdot m^{-3}$ (water) or $\mu g \cdot g^{-1}$ (sediments); results are expressed as geometric means, with ranges in parentheses; n, number of results; n.d., not detectable; data from Drs P. Bartlett and G. Eagle, National Research Institute for Oceanology, Stellenbosch

River	Month	*n*	Sample (water/ sediment)	Cd	Co	Cu	Fe	Mn	Ni	Pb	Zn
Berg	Oct. 1975	13	sediment	0·14 (n.d.–1·73)	1·2 (0·3–4·0)	2·5 (0·5–30·3)	4007 (1530–13490)	32·8 (8·4–253)	3·2 (0·9–1601)	4·9 (2·4–16·1)	14·0 (2·3–101·1)
	Aug. 1976	17	sediment	0·23 (n.d.–1·30)	1·3 (0·5–6·2)	1·7 (0·1–45·9)	3751 (990–31170)	43·8 (8–498)	2·4 (0·5–21·5)	4·7 (1·8–91·1)	10·2 (2·6–203·5)
		12	water	0·07 (n.d.–0·17)		1·2 (0·7–3·9)	66 (29·7–120·0)	16·7 (9·8–25·2)	0·9 (0·5–1·2)	1·4 (0·5–2·0)	3·3 (1·9–9·5)
Olifants	Feb. 1976	11	sediment	0·11 (n.d.–0·49)	5·3 (0·6–16·7)	12·6 (0·6–49·4)	18058 (1560–73950)	276·7 (40·2–929)	9·8 (0·9–31·5)	8·6 (3·2–20·2)	26·7 (1·8–89·0)
	July 1977	11	sediment	0·10 (n.d.–0.26)	3·7 (0·3–15.0)	7·6 (n.d.–39·9)	9030 (710–26980)	167·6 (39·0–660)	6·8 (n.d.–28·9)	4·1 (1.6–16·3)	3·7 (0·3–15·0)
	Feb. 1980	28	water (filtered)	0·18 (0·04–0·42)		0·4 (n.d.–2·60)	5·0 (n.d.–212·0)	9·8 (n.d.–117·7)	0·2 (n.d.–1·3)	0·6 (0·3–3·2)	4·8 (0·7–80·6)
		27	sediment (suspended)	0·01 (n.d.–0·04)		0·4 (n.d.–2·0)	368·0 (23–2763)	13·8 (0·6–153·1	0·2 (n.d.–2·4)	0·4 (n.d.–1·4)	1·2 (0·3–5·3)
		29	sediment	0·02 (n.d.–0·11)		4·8 (n.d.–15·8)	8368 (209–22911)	216·1 (38·2–867)	4·8 (n.d.–13·4)	10·7 (1.2–124)	16·3 (1·3–57.5)
Orange	Jan. 1979	15	sediment	0·04 (0·02–0·11)	7·7 (5·3–12·0)	12·2 (7·8–115)	23893 (11140–419260)	132·1 (58–257)	15·3 (11·2–20·0)	6·0 (3·6–15·0)	25·7 (14·8–96)

depth during cruises off the South West Cape in April, July, and December 1964, and from surf zone samples near Cape Town. Analysis was by spectrophotometry. Very few samples were taken for Fe and Mn (December cruise only), and values for the two elements varied between <1 and 13 $mg{\cdot}m^{-3}$ for Fe (10 samples), and <1 to 6 $mg{\cdot}m^{-3}$ for Mn (7 samples). While it is difficult to draw conclusions from so few analyses, iron appeared to be more concentrated in upwelled and surf zone samples (9–13 $mg{\cdot}m^{-3}$) than in water sampled offshore. Manganese was not noticeably affected by hydrology, but surf zone samples increased after heavy rain. Such terrigenous input of these metals would be expected, particularly near urban areas.

Copper concentrations showed large variations, even between duplicate aliquots from the same sample bottle. Orren (1967) ascribed these to phytoplankton in the samples which could release organically-bound copper into the water, although as samples were frozen without acidification, there may have been some adsorption effects taking place. Orren (1967) gives more details on this. Deep-sea samples (April, 1964) had a bimodal distribution with modes near 4 and 9–10 $mg{\cdot}m^{-3}$. There was a distinct difference between water from the Agulhas Current and that from west and south of South Africa, with Agulhas Current water being depleted to between 2 and 5 $mg{\cdot}m^{-3}$. South and west of the Cape Peninsula, however, Cu concentrations were in the range 10–13 $mg{\cdot}m^{-3}$. Such differences were found in both July (winter) and in December (summer). In contrast, surf zone samples had concentrations of less than 5 $mg{\cdot}m^{-3}$. The higher concentrations near the Cape Peninsula were ascribed to upwelling of South Atlantic Central Water, which was found to have similar concentrations. On both sides of the Peninsula the concentration increased from winter to summer, from 9 to 17 $mg{\cdot}m^{-3}$ in False Bay and from 6 to 15 $mg{\cdot}m^{-3}$ to the west.

This work was later extended (Orren 1969, 1971, 1973) to encompass stations all round the coast of South Africa and South West Africa. Some 1700 samples were analysed for Cu, Fe, Mn, K, Li, and Rb by atomic absorption spectroscopy. Both 'free' and organically-complexed forms of the metals were analysed, and particulate analyses were made on some samples. Liesegang & Orren (1966) and Orren (1971) suggested that the Cu content could be used to trace the mixing of the 'low-in-Cu' Agulhas water into the Atlantic. Offshore west coast copper levels lay between 0–9 $mg{\cdot}m^{-3}$, with values clustering around 3. Elsewhere, levels were less than 5 $mg{\cdot}m^{-3}$, tending towards 0·5–2, particularly in the Agulhas water. Soluble Fe and Mn also showed lower levels on the south coast (5–7 and 0–1 $mg{\cdot}m^{-3}$, respectively), compared with ranges of 0–30 and 0–4, respectively, on the west coast. In deep stations in the Benguela area, maxima in soluble and particulate Fe and Cu occurred just below the Antarctic Intermediate Water layer. Copper levels were up to 12 (soluble) and 6 $mg{\cdot}m^{-3}$ (particulate), while iron concentrations were 5 and 50 $mg{\cdot}m^{-3}$, respectively. Such a maximum has also been noticed for iodine south of South Africa by Chapman (1983b), who related it to the oxygen minimum occurring there, and suggested that it was due to the regeneration of elements during decomposition of sedimenting material. Particulate Mn was very low at all stations, rarely exceeding 0·3 $mg{\cdot}m^{-3}$ and being undetectable ($<0{\cdot}1$ $mg{\cdot}m^{-3}$) at most.

In upwelling areas, the upwelling water was generally low in all metals, but was enriched in a regeneration zone about 75 km offshore where water was sinking (Orren, 1973, Fig. 25). Lower values occurred inshore at the surface as a result of planktonic uptake. This is in contrast to the results reported previously near Cape Point (Liesegang & Orren, 1966). There was again little influence from the Berg River (Table VII) except possibly for particulate Fe. As shown in Table VIII, however, Fe levels in the Berg are quite low, and a separate sample taken off Namibia showed a particulate Fe concentration of 109 mg·m^{-3}. Organic and particulate Cu concentrations were also low (0–1·5 and 0–2 mg·m^{-3}, respectively), and appeared to be independent of either water movement or phytoplanktonic activity. As a result of his analyses, Orren (1973) suggested that the Cape Canyon could affect upwelling processes by acting as a priming pump—see Part I (Shannon, 1985) for further comments on this topic. Off Namibia, conditions were similar to those off Cape Columbine.

Orren (1969, 1973) explained these results in terms of a variation to the model of "internal modification of concentration" of Schutz & Turekian (1965). This postulates that "trace elements in upwelling water are removed by the high level of activity characteristic of such zones. As organisms die organic debris sinks into the upwelling current and decomposes releasing its trace element content into the water which is then returned to the zone of high activity where further concentration takes place. The concentration increases until a steady state is reached. Thus a zone of relatively high trace element concentration can occur without necessarily implying a net supply or removal from the ocean as a whole." This is very similar to the circulation patterns described above for nutrients.

Off the west coast of southern Africa, Orren found that the following modifications were necessary. First, the process is not continuous, as implied by the steady-state hypothesis, since upwelling itself is discontinuous in this area (see Shannon, 1985), except perhaps along the Namibian coast in summer. Regeneration apparently also takes place in several different layers.

(a) Offshore surface layers in areas of strong water movement.

(b) A layer at about 50 m depth generally corresponding with the

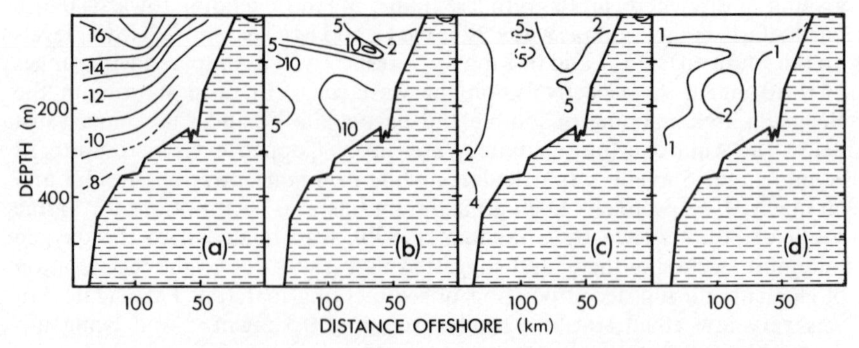

Fig. 25.—Temperature (a), dissolved copper (b), iron (c), and manganese (d) profiles across the shelf at 32°45′ S in July, 1968: all data in mg·m^{-3}; (after Orren, 1969).

thermocline (this can, of course, be much shallower—*e.g.* Bailey & Chapman (in press) found thermoclines off St Helena Bay at 10–15 m depth).

(c) A lower layer at 200–400 m. This layer would appear to correspond with water of σ_t 26·8 (see Shannon, 1985), which is the density level at which oxygen-deficient water is habitually found. Onshore, this layer may merge with (d).

(d) A near bottom layer, where sinking organic matter is broken down by benthic organisms and bacteria.

(e) Any boundary between different water masses.

Although many of Orren's data would now be considered too high (Wong *et al.*, 1983), their regularity suggests that qualitatively they are acceptable. During the Cape Upwelling Experiment, Orren & Cleverley (in prep.) measured dissolved heavy metals between Cape Point and Cape Columbine. Data on copper concentrations suggest that this area is enriched in the element, and confirm the circulation pattern suggested above. These data of Orren & Cleverley are rather lower than previous figures, with maxima of about 2 $mg{\cdot}m^{-3}$, but there were systematic differences between the upwelling water reaching the surface ($<1{\cdot}25$ $mg{\cdot}m^{-3}$) and that found in the water at 50–75 m (1·75 $mg{\cdot}m^{-3}$) postulated above as the site of initial decomposition.

Apart from the work of Orren, there has been little interest in the metal content of the water column in the Benguela system. Chester & Stoner (1974a,b, 1975) and Chester, Gardner, Riley & Stoner (1973) have taken a few samples in South African waters as part of a generalized survey of dissolved and particulate metals in the top 5 m of the world ocean. Since sampling was performed by pumping sea water into the collection system while the vessel was under way, it is not possible to specify sample positions, and the data should be considered as mean values over several hundred km. This is especially the case for the particulate fraction (specified as that retained by a 0·45 μm filter). Data on dissolved metals in their samples are given in Table IX, along with data from Fourie (1975) for a coastal site at Melkbosstrand, 20 km north of Cape Town. The agreement between the sets of data appears good, with the possible exception of Mn, which Orren (1967) suggested was enhanced near the shore. Fourie's samples would certainly have been classed as nearshore by Chester & Stoner (1974a), who divided their data into sets within 400 km of the coast and open ocean samples. It can be seen from Table IX that there is little sign of any metal pollution in the coastal waters (Chester & Stoner samples H, 37, and Fourie's samples), compared with open ocean South Atlantic Water. These results are lower than Liesegang & Orren's (1966) inshore results and are all at the low end of Orren's (1969) ranges, particularly those for Fe, although Orren's results were all positively skewed with 60% of Cu data being less than 3 $mg{\cdot}m^{-3}$, 65% of Fe data being less than 10 $mg{\cdot}m^{-3}$, and 88% of Mn results being below 2 $mg{\cdot}m^{-3}$. Chester & Stoner (1974a) also confirmed Orren's (1967, 1971) findings that Cu is lower in Agulhas Water than in the Atlantic, and found a similar distribution for Ni.

Data in Table IX compared with oceanic mean figures given by Brewer (1975) suggest that both the South Atlantic proper and the coastal region of southern Africa are low in most metals, particularly when compared with the coastal seas of the North Atlantic. This is not perhaps surprising given the low

TABLE IX

Concentrations of trace metals in surface sea water along the west coast of southern Africa: all concentrations in mg·m^{-3} except Hg μg·m^{-3}; data from (a) Chester & Stoner (1974a), (b) Chester et al. (1973), (c) Fourie (1975), and (d) Brewer (1975)

Source	Sample	Month	Position S:E	Fe	Mn	Cd	Zn	Cu	Ni	Hg	Cr	Co	Sb
a	E	April	15°26′:02°11′	1·0	0·36	0·17	1·2	1·5	0·8				
	F	April	20°05′:05°54′	1·1	0·19	0·10	1·2	2·6	0·8				
	39	July	26°18′:11°15′	1·1	0·18	0·05	0·6	0·5	0·7				
	40	July	23°40′:09°00′	1·4	0·24	0·06	1·9	1·3	0·9				
	41	July	21°18′:07°03′	1·1	0·23	0·06	2·0	0·4	1·1				
	42	July	16°17′:03°07′	3·1	0·29	0·05	1·1	1·2	2·7				
	G	April	29°32′:13°52′	0·8	0·29	0·13	3·7	0·4	1·1				
	H	April	33°55′:18°01′	1·1	0·39	0·16	12·6	1·1	5·0				
	37	July	34°01′:17°58′	1·1	0·21	0·05	1·0	0·5	3·9				
	38	July	31°32′:15°41′	0·8	0·18	0·06	1·2	0·3	2·7				
b	3	April	11°36′:00°47′							43			
	4	April	29°32′:13°52′							22			
	I	July	21°18′:07°03′							48			
	J	July	13°34′:00°49′							52			
c		"Surfzone"	33°45′:18°25′	1·7	0·7		1·2				0·08	0·02	0·7
				±0·3	±0·2		±0·5				±0·07	±0·01	±0·7
		"offshore"		3·99	0·7		3·2				0·06	0·06	0·3
				±2·9	±0·2		±2·1				±0·04	±0·05	±0·2
d			Oceanic mean	2	0·2	0·1	4·9	0·5	1·7	30	0·3	0·05	0·24

levels in local river waters (Table VIII), but even in industrialized regions such as Saldanha Bay or Table Bay levels remain low (see Table X).

The only other available data on offshore metal concentrations in this region refer to arsenic. Burton, Maher & Statham (1983) studied vertical distributions of the element at six stations in the Cape Basin. They found very little difference between samples taken above 110 m (mean 1·49, range 1·28–1·72 $mg \cdot m^{-3}$) and those taken below this depth (mean 1·58, range 1·28–2·10 $mg \cdot m^{-3}$), and no significant correlation with phosphate. This suggests that the transport of arsenic to deeper waters by biogenic material is only of minor importance, and that cycling between oxidation states may be a greater influence on its distribution in the euphotic zone. No speciation studies were, however, reported for these samples.

Data on particulate matter also reflect the lack of metal input. Chester *et al.* (1972) and Chester & Stoner (1972, 1974b, 1975) have investigated both the concentration of suspended particulate matter off the west coast and the concentration of aeolian dust in the atmosphere (Table XI). They found that dust in the atmosphere (particle size $>0{\cdot}5\ \mu m$) was less than 1 $\mu g \cdot m^{-3}$, and that this contained crustal abundances of Fe, Mn, Co, Ga, Cr, and Sr. They concluded that concentrations of Sn, Zn, and Pb were enriched, even in open ocean waters of the southeastern Atlantic, and suggested that these were anthropogenically produced. This seems unlikely in this area, and their results may reflect naturally higher mineral concentrations along the western coastal strip. Cu, Ni, V, and Ba were also found to be enriched, but this may have been related to a marine source (Chester & Stoner, 1972), since Cu and Ni were both found to be affected by planktonic production (Orren, 1969; Chester & Stoner, 1974a). Clearly, Chester & Stoner did not sample any of the intense dust clouds (see above, p. 186) which are found during periods of northeastern winds between Kleinsee and the Kunene River extending up to 150 km out to sea (Copenhagen, 1934; Shannon & Anderson, 1982; see Fig. 10 in Shannon, 1985).

Most of the work reported by Chester & Stoner (1972, 1975) would also appear to be outside the main production area of the Benguela system, since they reported (1972) only 67–448 $mg \cdot m^{-3}$ suspended matter (mean 150) for 23 samples from the South Atlantic, and 84–1530 $mg \cdot m^{-3}$ (mean 456) for 11 coastal samples. This contrasts with 3–5 $g \cdot m^{-3}$ over the Namibian shelf (Bailey, 1979), and even higher concentrations in areas of preferential settlement or high production, and is a factor of 10–30 lower than found by Fourie (1975) close to the coast north of Cape Town, although Chester & Stoner worked in April and July, when production would be expected to be at a minimum. All elements, with the exception of Co, were depleted in surface particulates relative to their concentrations in dust particles (Table XI), and, apart from V, fell in the ranges for marine plankton (Eisler, 1981).

Sediment data for the region are again sparse, with most information being available only from the carbon-rich muddy anoxic sediments off Namibia. In the southern part of the Benguela system, the only data relate to pollution surveys carried out in areas of sand-sized or larger particles, such as Hout Bay (Fricke *et al.*, 1979) or Green Point (Table Bay, Orren *et al.*, 1979, 1981; Eagle, Bartlett & Long, 1982). In these cases, only the acid-leachable fractions of the metals were studied, leaching being carried out with a 4:1 nitric:perchloric acid mixture. Levels were found to be low (Table X), even close to industrial

TABLE X

Trace metals in Table Bay: all concentrations in mg·m^{-3} (water) or μg·g^{-1} (sediment); results are geometric means, with the range in parentheses; n.d., not detectable; n, number of results; from Eagle, Bartlett & Long (1982)

Month 1980	*n*	Cd	Cu	Fe	Mn	Ni	Pb	Zn
Water								
July	51	0·04 (0·02–0·10)	0·3 (0·04–0·9)	2·0 (n.d.–40·8)	0·5 (0·1–1·9)	0·2 (n.d.–0·9)	1·3 (n.d.–3·5)	0·4 (n.d.–28·5)
Sept.	81	0·03 (n.d.–0·10)	0·2 (n.d.–1·4)	0·4 (n.d.–8·4)	0·3 (0·02–1·6)	0·2 (n.d.–0·8)	0·9 (0·1–5·0)	0·8 (n.d.–7·70)
Sediment								
Sept.	39	0·03 (n.d.–0·27)	1·4 (n.d.–64)	1423 (30–8700)	13·3 (3·4–102)	0·2 (n.d.–7·2)	4·6 (0·7–68)	5·6 (0·6–119)

TABLE XI

Concentrations of metals in dust and surface water particulates collected near the west coast of South Africa and in the South East Atlantic: data from Chester & Stoner (1974b, 1975); all concentrations in $\mu g \cdot g^{-1}$, except Fe in %; plankton data from Eisler (1981)

Dust samples	Mn	Cu	Co	Ga	V	Ba	Pb	Zn	Ni	Cr	Sr	Sn	Fe
No. 13	900	96	8	30	123	137	135	660	63	71	67	9	5·1
14	1500	320	5	12	126	57	185	215	105	50	52	25	6·4
17	720	160	5	17	340	480	450	290	189	61	116	20	6·2
59	1080	146	11	21	176	670	955	515	121	106	118	91	5·6
61	810	230	4	7	99	400	490	330	180	147	33	20	6·5
Oceanic particulates	85	52	16	3	69	72	72	260					
Marine plankton (dry mass)	20–200	2–60	1–15	5	1–5	35–300	10–80	30–1000	1–40	0·4–10	5–100	17–100	<0·1

outfalls, although up to an order of magnitude higher for Cd, Cu, Pb, and Zn than unpolluted south coast beaches (Orren *et al.*, 1979).

Further north, the metal content of the organic-rich sediments has been investigated by, among others, Calvert & Price (1970, 1983), Calvert & Morris (1977), and Brongersma-Sanders *et al.* (1980). Cores taken during the "Circe" cruise of the ARGO in 1968 from both sandy and muddy substrata were analysed for Cu, Ni, Pb, and Zn by Calvert & Price (1970). The muddy sediments were found to correspond with high carbon concentrations (see also Bremner, 1978a, 1980, 1983; Brongersma-Sanders *et al.*, 1980), and all four elements were associated with areas of increased sediment carbon, especially Cu and Ni (Fig. 26). These high concentrations were confirmed by Bremner (1978a), who also reported high Cd levels in the sediments from the same area. Uranium (Baturin, 1971a; Veeh, Calvert & Price, 1974) and molybdenum (Calvert & Morris, 1977) were also high in these sediments. The area of maximum metal concentration was the intermediate zone away from the coast at the boundary between predominantly diatomaceous and predominantly calcareous sediments, shorewards of the area of highest faecal pellet concentration (Bremner, 1983). This was not the area of lowest dissolved oxygen concentrations at the time of the survey (Price & Calvert, 1973), but formed the outer edge of the azoic zone originally mentioned by Von Bonde (1928) and Marchand (1928).

While the levels of metals are high compared with other marine sediments, they are not particularly high when compared with values in bitumenous shales (Wedepohl, 1960; Brongersma-Sanders *et al.*, 1980). Calvert & Price (1970) suggested that the metals were introduced to the sediments by sinking dead phytoplankton—certainly the amounts of biogenic sediment on the Namibian shelf are very large, with up to 96·6% calcium carbonate, 88% opal, and 24% organic carbon being present (see Fig. 14, p. 208; Bremner, 1983). Bremner (1978a) considered that this relationship was based not only on the differences in dissolution rates between diatom frustules and organic remains, but also on the affinity of these elements for sulphur. The reducing environment would cause concentration of the elements as insoluble sulphides, in similar fashion to the sediments of the Black Sea, where up to 1562 and 3133 $\mu g \cdot g^{-1}$ of Cu and Ni have been reported by Volkov & Fomina (1972). While in most cases diagenesis will affect the metal–organic associations after sedimentation, the extreme examples for each element given by Calvert & Price (1970) have been shown to be similar to those found in particulate matter off the upwelling zones of northwestern Africa (Chester, Griffiths & Stoner, 1978). The similarity between the positions of zones of high Cu, Ni, and organic carbon again bears out the conclusions of Orren (1969, 1973) and Chester & Stoner (1974a). Bremner (1978a) also examined the distribution of Ca, Sr, Co, Rb, Mg, Mn, Fe, Al and K in the surface sediments between 17° and 25° S. The first three elements were found to be associated with the calcareous sediments near the shelf break off Walvis Bay, and were found seawards of the zone of highest trace metal concentrations reported above. The remainder were all associated with sediments of mainly terrigenous origin, and their concentrations increased northwards and offshore from Walvis Bay. The highest concentrations in all cases occurred north of 19° S, where the influence of the Kunene River becomes more important and the northward flow of the Benguela Current is opposed by southward-moving water of the Angola Current.

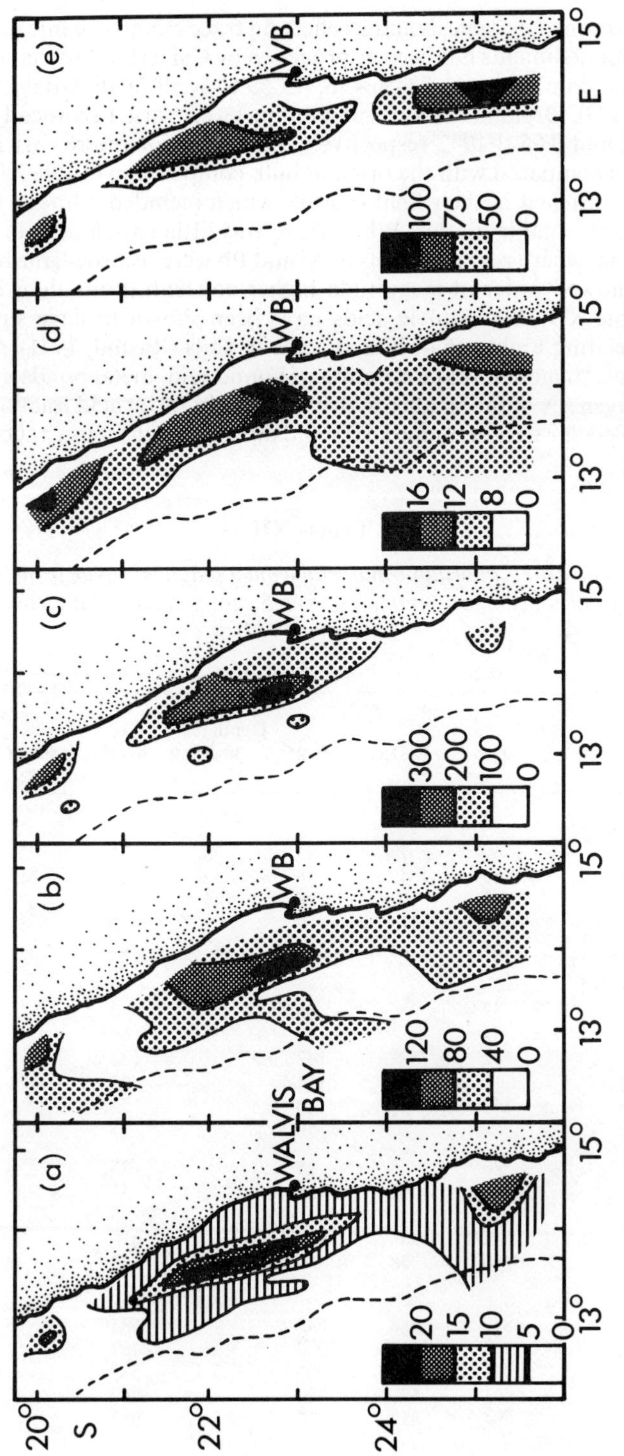

Fig. 26.—Distribution of (a) organic carbon, (b) copper, (c) nickel, (d) lead, and (e) zinc in Namibian shelf sediments: all values $\mu g \cdot g^{-1}$ except organic carbon (%) and on a salt-free basis (after Calvert & Price, 1970).

The question of diagenesis and its effect on trace metal–organic interactions in Namibian sediments has been approached by Calvert & Morris (1977), who examined a core taken at 22°56·4′ S: 13°59·6′ E off Walvis Bay (Morris & Calvert, 1977). Organic carbon and sulphur levels varied erratically between 8.1–15.1% and 0·65–1·19%, respectively. Lipid fractions were enriched in Ni, Pb, and Zn compared with the original bulk composition (Table XII), and Ni was again enriched in their final residue, which included refractory material and resistant organic matter. While there was little enrichment in the fulvic acid fraction, where very low levels of Ni and Pb were recorded, the humic acid fraction showed order-of-magnitude higher concentrations than bulk sediment (Table XII). Since fulvic acids have been shown to have significantly higher chelating ability than humic acids in soils (Rashid, 1971), Calvert & Morris have suggested that formation of humic acid–metal bonds must occur during diagenesis of the smaller molecular weight fulvic acid fraction into the higher molecular weight humic acids, and discounted the importance of pyrite

TABLE XII

Metal concentrations in organic material from a sediment taken from the shelf off Walvis Bay: all concentrations in $\mu g{\cdot}g^{-1}$; n.d., not detected; data from Calvert & Morris (1977)

	Depth (cm)						
	0–5	10–15	20–25	30–35	40–45	50–55	60–65
Bulk sediment							
Cu	69	79	83	73	72	75	74
Mo	59	42	36	37	42	51	38
Ni	115	107	86	145	107	167	85
Pb	4	16	10	8	5	5	5
Zn	61	64	136	112	64	38	91
Organic extract I							
Cu	52		46		40		43
Mo	64		60		66		83
Ni	142		142		122		180
Pb	114		114		95		128
Zn	243		116		79		78
Fulvic acid							
Cu	26	23	25		25		20
Mo	102	41	36		41		56
Ni	n.d.	n.d.	n.d.		5		n.d.
Pb	n.d.	n.d.	n.d.		n.d.		4
Zn	160	58	69		62		50
Humic acid							
Cu	877	686	683		533		797
Mo	2368	1168	1000		437		1525
Ni	333	409	500		533		491
Pb	491	671	283		267		306
Zn	894	817	817		933		712

in metal cycling reactions in this area. This view was subsequently modified (Calvert & Price, 1983) but the part played by sulphides in metal cycling is still unclear.

Much more work has been done on the trace metal geochemistry of certain components of the sediment, in particular the phosphorite deposits, and on the Walvis Ridge. It is not the intention of this review to consider these here, but the interested reader is referred to the work of Baturin (1971b,c, 1974), Baturin, Merkulova & Chalov (1972), Bremner (1978b, 1980), Bremner & Willis (1975), Price & Calvert (1978) and Calvert & Price (1983), among others.

Transuranic elements have been analysed by Baturin (1971a), Veeh, Calvert & Price (1974) and Birch *et al.* (1983), mainly as a means of dating the sediments. Veeh *et al.* (1974) suggested that U is being enriched in the surface sediment at present, since the ^{234}U:^{238}U ratios in their sediments were very similar to the 1·15 found in sea water. They showed that U deposition was related to the phosphate content of the sediment, suggesting that it was incorporated into apatite following reduction of U^{6+} to U^{4+} in anoxic pore waters. This work suggested a sedimentation rate of up to 1 mm·yr^{-1} on the Namibian shelf, which agrees with the work of Bremner (1983).

^{210}Po and ^{210}Pb concentrations in surface waters from 11 stations around the Cape of Good Hope and from 18 stations in the southeastern Atlantic have been investigated by Shannon, Cherry & Orren (1970). ^{210}Po concentrations ranged from 8 ± 2 to 41 ± 4 pCi·m^{-3}, while ^{210}Pb varied between less than 10 (detection limit) to 135 ± 16 pCi·m^{-3}. There was good agreement between the ^{210}Po concentrations and the inferred water movement around Cape Point, water of Agulhas Current origin showing higher ^{210}Po but lower ^{210}Pb contents. The existence of a gyre at 36° S: 15° E was clearly shown (Fig. 27), although there was little difference in vertical distribution at a station at 32°40′ S: 14°33′ E in the top 600 m of the water column. Plankton data showed enrichment factors of about 870 (Pb) and 20 000 (Po) for wet zooplankton and 890 and 5400 for phytoplankton, respectively, and such data were used to show that surface water in the Benguela system (*i.e.* upwelled South Atlantic Central Water) had low ^{210}Po levels, presumably through isolation from atmospheric fall-out (Shannon, 1973). An alternative possible explanation is that the much greater production in the Benguela region strips out the radionuclides to a greater degree than in the relatively impoverished Agulhas Current. ^{228}Th (Cherry, Shannon & Gericke, 1969) and ^{226}Ra (Shannon, 1972a) were also shown to vary in a similar fashion in zooplankton samples. Near the Walvis Ridge, Cherry *et al.* (1969) found ^{228}Th at a concentration of 27×10^{-13} g·m^{-3}, compared with a mean content (Shannon, 1972b) of about 10^{-14} g·m^{-3}. This may have resulted from upwelling over the ridge, as suggested by Shannon & Van Rijswyk (1969). Shannon (1972b) has calculated a removal time for Th of something less than eight years for the highly productive waters of the Benguela system, rather shorter than other authors, and has suggested that biological removal of the element is significant. While Th is again depleted in the Benguela relative to Agulhas Current Water (based on phytoplankton and zooplankton concentrations), it is higher in upwelled water than in the surface waters near the Subtropical Convergence where such water is formed. This may result from diffusion of ^{232}Th from the sediments into the overlying water (Moore, 1969). With the advent of nuclear power in South Africa as the Koeberg Nuclear Power

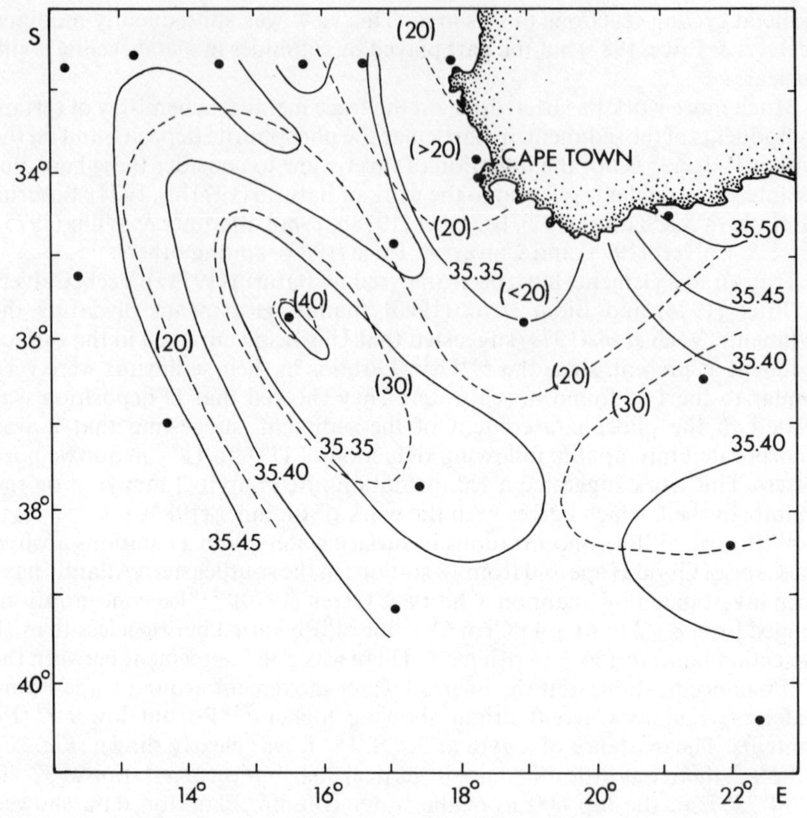

Fig. 27.—Distribution of ^{210}Po (pCi·m^{-3}—dashed lines) and salinity (solid lines) around the south west Cape in March 1969 (after Shannon *et al.*, 1970).

Station comes on stream in 1984, it is expected that interest in radionuclides in the area will increase.

ACKNOWLEDGEMENTS

The authors are appreciative of the comments of colleagues. In particular we wish to acknowledge the many discussions that we had with Geoff Bailey and Grev Nelson and the assistance received during the preparation of the manuscript from Dan Hunter, Anastasia Polito, Heather Sessions, Tony van Dalsen, Christine Illert, and Mariana van Niekerk.

REFERENCES

Andrews, W. R. H. & Hutchings, L., 1980. *Progr. Oceanogr.*, **9**, 1–81.
Bailey, G. W., 1979. M.Sc. thesis, University of Cape Town, South Africa, 225 pp.
Bailey, G. W., 1983. *S. Afr. J. Sci.*, **79**, 145–146.

Bailey, G. W., in press. *Invest. Pesq.* (in press).
Bailey, G. W., Beyers, C. de B. & Lipschitz, S. R., 1985. *S. Afr. J. Mar. Sci.*, **3,** in press.
Bailey, G. W. & Chapman, P., in press. In, *The South African Ocean Colour and Upwelling Experiment*, edited by L. V. Shannon, Sea Fish. Res. Inst. S. Afr. (in press).
Barlow, R. G., 1980. *J. exp. mar. Biol. Ecol.*, **45,** 83–93.
Barlow, R. G., 1982. *J. exp. mar. Biol. Ecol.*, **63,** 239–248.
Barlow, R. G., 1984a. *J. Plankton Res.*, **6,** 385–397.
Barlow, R. G., 1984b. *Mar. Ecol. Progr. Ser.*, **16,** 121–126.
Baturin, G. N., 1971a. *Dokl. Akad. Nauk S.S.S.R.*, **198,** 224–226.
Baturin, G. N., 1971b. *Nature, Lond.* (*Phys. Sci.*), **232,** 61–62.
Baturin, G. N., 1971c. *Oceanology*, **11,** 373–376.
Baturin, G. N., 1972. *Oceanology*, **12,** 849–855.
Baturin, G. N., 1974. *Oceanology*, **14,** 856–860.
Baturin, G. N., Merkulova, K. I. & Chalov, P. I., 1972. *Mar. Geol.*, **13,** 37–41.
Birch, G. F., 1975. Ph.D. thesis, University of Cape Town, South Africa, 210 pp. (Also published as *Joint Geol. Surv./U.C.T. Mar. Geosci. Gp. Bull.* No. 6, 142 pp.)
Birch, G. F., 1977. *Mar. Geol.*, **23,** 305–337.
Birch, G. F., 1979. *J. sediment. Petrol.*, **49,** 93–110.
Birch, G. F., Thomson, J., McArthur, J. M. & Burnett, W. C., 1983. *Nature, Lond.*, **302,** 601–603.
Bishop, J. K. B., Ketten, D. R. & Edmond, J. M., 1978. *Deep-Sea Res.*, **25,** 1121–1161.
Boulégue, J., 1974. *C. R. Acad. Sci. Paris*, **278D,** 2723–2726.
Boulégue, J. & Denis, J., 1983. In, *Coastal Upwelling, Its Sediment Record. Part A. Responses of the Sedimentary Regime to Present Coastal Upwelling*, edited by J. Thiede & E. Suess, Plenum Press, New York, pp. 439–454.
Boyd, A. J., 1983a. *Investl Rep. Sea Fish. Res. Inst. S. Afr.*, No. 126, 47 pp.
Boyd, A. J., 1983b. *Coll. scient. pap. int. Commn SE Atl. Fish.*, **10,** 41–73.
Breck, W. G., 1974. In, *The Sea, Vol. 5*, edited by E. D. Goldberg, John Wiley, New York, pp. 153–179.
Bremner, J. M., 1978a. Ph.D. thesis, University of Cape Town, South Africa, 300 pp.
Bremner, J. M., 1978b. In, *10th Int. Congr. on Sedimentology, Jerusalem*, 9th–14th July, 1978, Abstracts, **1,** 87–88.
Bremner, J. M., 1978c. In, *Proterozoic–Cambrian Phosphorites*, edited by P. J. Cook & J. H. Shergold, Australian National University Press, Canberra, pp. 56–58.
Bremner, J. M., 1980. *J. geol. Soc. Lond.*, **137,** 773–786.
Bremner, J. M., 1981. *Geo-marine Lett.*, **1,** 91–96.
Bremner, J. M., 1983. In, *Coastal Upwelling, Its Sediment Record. Part B. Sedimentary Records of Ancient Coastal Upwelling*, edited by J. Thiede & E. Suess, Plenum Press, New York, pp. 73–103.
Bremner, J. M. & Willis, J. P., 1975. *Proc. Electron Micros. Soc. S. Afr.*, **5,** 127–128.
Brewer, P. G., 1975. In, *Chemical Oceanography, Vol. 2*, edited by J. P. Riley & G. Skirrow, Academic Press, London, 2nd edition, pp. 415–496.
Brewer, P. G. & Murray, J. W., 1973. *Deep-Sea Res.*, **20,** 803–818.
Brongersma-Sanders, M., 1957. In, *Treatise on Marine Ecology and Palaeoecology, Vol. 2*, edited by J. W. Hedgpeth, *Geol. Soc. Am. Mem.*, No. 67, 941–1010.
Brongersma-Sanders, M., 1983. In, *Coastal Upwelling, Its Sediment Record, Part A. Responses of the Sedimentary Regime to Present Coastal Upwelling*, edited by J. Thiede & E. Suess, Plenum Press, New York, pp. 421–437.
Brongersma-Sanders, M., Stephan, K. M., Kwee, T. G. & De Bruin, M., 1980. *Mar. Geol.*, **37,** 91–132.
Brown, P. C. & Hutchings, L., in press. *Invest. Pesq.* (in press).
Bubnov, V. A., 1972. *Oceanology*, **12,** 193–201.
Burton, J. D., 1976. In, *Estuarine Chemistry*, edited by J. D. Burton & P. S. Liss, Academic Press, London, pp. 1–36.

Burton, J. D., Maher, W. A. & Statham, P. J., 1983. In, *Trace Metals in Sea Water*, edited by C. S. Wong, E. Boyle, K. W. Bruland, J. D. Burton & E. D. Goldberg, Plenum Press, New York, pp. 415–426.

Buys, M. E. L., 1959. *Investl Rep. Div. Fish. S. Afr.*, No. 37, 176 pp.

Calvert, S. E. & Morris, R. J., 1977. In, *A Voyage of Discovery*, edited by M. V. Angel, Pergamon Press, Oxford, pp. 667–680.

Calvert, S. E. & Price, N. B., 1970. *Nature, Lond.*, **227,** 593–595.

Calvert, S. E. & Price, N. B., 1971. *Deep-Sea Res.*, **18,** 505–523.

Calvert, S. E. & Price, N. B., 1983. In, *Coastal Upwelling, Its Sedimentary Record. Part A. Responses of the Sedimentary Regime to Present Coastal Upwelling*, edited by J. Thiede & E. Suess, Plenum Press, New York, pp. 337–375.

Carmack, E. C. & Aagaard, K., 1977. *Estuar. cstl mar. Sci.*, **8,** 135–142.

Chan, L. H., Drummond, D., Edmond, J. M. & Grant, B., 1977. *Deep-Sea Res.*, **24,** 613–650.

Chapman, P., 1982. Ph.D. thesis, University of Wales, 309 pp.

Chapman, P., 1983a. *Deep-Sea Res.*, **30,** 1247–1259.

Chapman, P., 1983b. *5th S. Afr. Natl Oceanogr. Symp., Grahamstown.* Abstract D1 (unpubl.).

Cherry, R. D., Shannon, L. V. & Gericke, I., 1969. *Earth Planet. Sci. Lett.*, **6,** 451–456.

Chester, R., Elderfield, H., Griffin, J. J., Johnson, L. R. & Padgham, R. C., 1972. *Mar. Geol.*, **13,** 91–105.

Chester, R., Gardner, D., Riley, J. P. & Stoner, J., 1973. *Mar. Pollut. Bull.*, **4,** 28–29.

Chester, R., Griffiths, A. & Stoner, J. H., 1978. *Nature, Lond.*, **275,** 308–309.

Chester, R. & Stoner, J. H., 1972. *Nature, Lond.*, **240,** 552–553.

Chester, R. & Stoner, J. H., 1974a. *Mar. Chem.*, **2,** 17–32.

Chester, R. & Stoner, J. H., 1974b. *Mar. Chem.*, **2,** 157–188.

Chester, R. & Stoner, J. H., 1974c. *Mar. Chem.*, **2,** 263–275.

Chester, R. & Stoner, J. H., 1975. *Nature, Lond.*, **255,** 50–51.

Clowes, A. J., 1938. *'Discovery' Rep.*, **19,** 1–120.

Clowes, A. J., 1954. *Investl Rep. Div. Fish. S. Afr.*, No. 16, 47 pp.

Cohen, Y. & Gordon, L. I., 1979. *J. Geophys. Res.*, **84,** 347–353.

Copenhagen, W. J., 1934. *Investl Rep. Fish. Mar. Biol. Surv. Div. Un S. Afr.*, No. 3, 11 pp.

Copenhagen, W. J., 1953. *Investl Rep. Div. Fish. S. Afr.*, No. 14, 35 pp.

De Decker, A. H. B., 1970. *Investl Rep. Div. Sea Fish. S. Afr.*, No. 84, 24 pp.

De Groot, A. J., Salomons, W. & Allersma, E., 1976. In, *Estuarine Chemistry*, edited by J. D. Burton & P. S. Liss, Academic Press, London, pp. 131–157.

Deuser, W. G., 1975. In, *Chemical Oceanography, Vol. 3*, edited by J. P. Riley & G. Skirrow, Academic Press, London, 2nd edition, pp. 1–37.

De Villiers, J. & Söhnge, P. G., 1959. *Mem. geol. Surv. S. Afr.*, No. 48, 295 pp.

Dugdale, R. C. & Goering, J. J., 1970. *Anton Bruun Rep.* (*Texas A. and M. University*), **4,** 5.3–5.8.

Eagle, G. A. & Bartlett, P. D., 1984. *Council for Scientific and Industrial Research, S. Afr.*, Report T/SEA 8307, 46 pp.

Eagle, G. A., Bartlett, P. D. & Long, M. V., 1982. *Council for Scientific and Industrial Research, S. Afr.*, Report 571, 79 pp.

Eisler, R., 1981. *Trace Metal Concentrations in Marine Organisms.* Pergamon Press, Oxford, 687 pp.

Eisma, D., 1969. *Rep. Ned. Inst. Onderzoek. Zee*, 34 pp.

Emerson, S., Cranston, R. E. & Liss, P. S., 1979. *Deep-Sea Res.*, **26,** 859–878.

Emery, K. O., Milliman, J. D. & Uchupi, E., 1973. *J. sediment. Petrol.*, **43,** 822–837.

Fourie, H. O., 1975. *S. Afr. J. Sci.*, **71,** 151–154.

Fricke, A. H., Eagle, G. A., Gledhill, W. J., Greenwood, P. J. & Orren, M. J., 1979. *S. Afr. J. Sci.*, **75,** 459–461.

Gordeyev, Ye. I., 1963. *Dokl. Akad. Nauk S.S.S.R.*, **149,** 181–184.

Grasshoff, K., 1975. In, *Chemical Oceanography, Vol. 2*, edited by J. P. Riley & G. Skirrow, Academic Press, London, 2nd edition, pp. 455–597.

Grill, E. V. & Richards, F. A., 1964. *J. mar. Res.*, **22,** 51–69.

Hallberg, R. O., Bagander, L. E., Engvall, A. G. & Schippel, F. A., 1972. *Ambio*, **1,** 71–73.

Hart, T. J. & Currie, R. I., 1960. *'Discovery' Rep.*, **31,** 123–298.

Harvey, G. R., 1980. *Mar. Chem.*, **8,** 327–332.

Henry, A. E., 1975. *Investl Rep. Div. Sea Fish. S. Afr.*, No. 95, 66 pp.

Henry, J. L., 1979. *4th S. Afr. Natl Oceanogr. Symp., Cape Town*, Abstract (unpubl.).

Hentschel, E. & Wattenberg, H., 1930. *Annln Hydrogr. Berl.*, **58,** 273–277.

Hobson, L. A., 1971. *Invest. pesq.*, **35,** 195–208.

Horstman, D. A., 1981. *Fish. Bull. S. Afr.*, **15,** 71–88.

Jacobs, M. B. & Ewing, M., 1969. *Science*, **163,** 380–383.

Jones, P. G. W., 1971. *Deep-Sea Res.*, **18,** 193–208.

Karl, D. M., La Rock, P. A., Morse, J. W. & Sturges, W., 1976. *Deep-Sea Res.*, **23,** 81–88.

Ketchum, B. H. & Corwin, N., 1965. *Limnol. Oceanogr.*, **10** (Suppl.), R148–R161.

Kollmer, W. E., 1963. *Investl Rep. Admin. S.W. Afr. Mar. Res. Lab.*, No. 8, 78 pp.

Kruger, I., 1980. *Fish. Bull. S. Afr.*, **13,** 31–53.

Liesegang, E. C. & Orren, M. J., 1966. *Nature, Lond.*, **211,** 1166–1167.

Liss, P. S., 1976. In, *Estuarine Chemistry*, edited by J. D. Burton & P. S. Liss, Academic Press, London, pp. 93–130.

Marchand, J. M., 1928. *Un. S. Afr. Fish. Mar. Biol. Survey, Dept Mines and Industries, Spec. Rep.*, No. 5, 11 pp.

Menzel, D. W. & Ryther, J. H., 1968. *Deep-Sea Res.*, **15,** 327–337.

Moore, W. S., 1969. *Earth Planet. Sci. Lett.*, **6,** 437–446.

Moroshkin, K. V., Bubnov, V. A. & Bulatov, R. P., 1970. *Oceanology*, **10,** 27–34.

Morris, R. J. & Calvert, S. E., 1977. In, *A Voyage of Discovery*, edited by M. V. Angel, Pergamon Press, Oxford, pp. 647–665.

Nelson, D. M. & Goering, J. J., 1977. *Deep-Sea Res.*, **24,** 65–73.

Nelson, G. & Hutchings, L., 1983. *Progr. Oceanogr.*, **12,** 333–356.

Olivieri, E. T., 1983a. *S. Afr. J. mar. Sci.*, **1,** 77–109.

Olivieri, E. T., 1983b. *S. Afr. J. mar. Sci.*, **1,** 199–229.

Orren, M. J., 1967. *Investl Rep. Div. Sea Fish. S. Afr.*, No. 59, 40 pp.

Orren, M. J., 1969. Ph.D. thesis, University of Cape Town, South Africa, 135 pp. + 50 figs.

Orren, M. J., 1971. In, *The Ocean World*, edited by M. Uda (Proc. Joint Oceanogr. Assembly, Tokyo, 1970), Japan Soc. for Promotion of Science, Tokyo, pp. 187–189.

Orren, M. J., 1973. *Proc. Symp. Hydrogeochem. Biogeochem., Tokyo, 1970*, edited by E. Ingerson, Clark Co., Washington, D.C., **2,** 544–555.

Orren, M. J., Eagle, G. A., Fricke, A. H., Greenwood, P. J., Hennig, H. F.-K. O. & Bartlett, P. D., 1981. *S. Afr. J. Sci.*, **77,** 183–188.

Orren, M. J., Fricke, A. H., Eagle, G. A., Greenwood, P. J. & Gledhill, W. J., 1979. *S. Afr. J. Sci.*, **75,** 456–459.

Parrish, R. H., Bakun, A., Husby, D. M. & Nelson, C. S., 1984. In, *Proc. Expert consultation to examine changes in abundance and species composition of neritic fish stocks*, San Jóse, Costa Rica, April, 1983, FAO Fish Rept, No. 291, part 3, pp. 731–777.

Pieterse, F. & Van der Post, D. C., 1967. *Investl Rep. Admin. S.W. Afr. Mar. Res. Lab.*, No. 14, 125 pp.

Postel, L., 1982. *Rapp. P.-v. Réun. Cons. perm. int. Explor. Mer*, **180,** 274–279.

Price, N. B. & Calvert, S. E., 1973. *Geochim. Cosmochim. Acta*, **37,** 2149–2158.

Price, N. B. & Calvert, S. E., 1977. *Geochim. Cosmochim. Acta*, **41,** 1769–1775.
Price, N. B. & Calvert, S. E., 1978. *Chem. Geol.*, **23,** 151–170.
Rashid, M. A., 1971. *Soil Sci.*, **111,** 298–305.
Rogers, J., 1977. Ph.D. thesis, University of Cape Town, South Africa, 212 pp. (Also published as *Joint Geol. Surv./U.C.T. Mar. Geosci. Gp. Bull.*, No. 7, 162 pp.)
Romankevich, Y. A. & Baturin, G. N., 1974. *Oceanology*, **14,** 529–533.
Rooseboom, A. & Harmse, H. J. v. M., 1979. In, *The Hydrology of Areas of Low Precipitation*, I.A.H.S. Publ., **128,** pp. 459–479.
Rooseboom, A. & Maas, N. F., 1974. *Tech. Rep. Dept Wat. Affairs, S. Afr.*, No. 59, 48 pp.
Saenko, G. N., Kravtsova, Y. Y., Ivanenko, V. V. & Sheludko, S. I., 1978. *Mar. Biol.*, **47,** 243–250.
Schott, F. & Ehrhardt, M., 1969. *Kiel. Meeresforsch.*, **25,** 272–278.
Schott, G., 1902. *Wiss. Ergebn. dt. Tiefsee-Exped. 'Valdivia'*, **1,** 1–403.
Schultz, S., Schemainda, R. & Nehring, D., 1979. *Geodat. Geophys. Veröff.*, Reihe iv, **28,** i–vii + 1–43.
Schutz, D. F. & Turekian, K. K., 1965. *Geochim. Cosmochim. Acta*, **29,** 259–313.
Scranton, M. I. & Farrington, J. W., 1977. *J. geophys. Res.*, **82,** 4947–4953.
Seitzinger, S. P., Pilson, M. E. Q. & Nixon, S. W., 1983. *Science*, **222,** 1244–1246.
Seiwell, H. R., 1937. *Pap. phys. Oceanogr. Meteor.*, **5**(3), 1–24.
Seiwell, H. R., 1938. *Pap. phys. Oceanogr. Meteor.*, **6**(1), 1–60.
Senin, Yu. M., 1977. *Oceanology*, **16,** 586–590.
Shannon, L. V., 1966. *Investl Rep. Div. Sea Fish. S. Afr.*, No. 58, 22 pp. plus 30 pp. figs.
Shannon, L. V., 1972a. *Investl Rep. Div. Sea Fish. S. Afr.*, No. 98, 80 pp.
Shannon, L. V., 1972b. *Investl Rep. Div. Sea Fish. S. Afr.*, No. 99, 20 pp.
Shannon, L. V., 1973. *Investl Rep. Div. Sea Fish. S. Afr.*, No. 100, 34 pp.
Shannon, L. V., 1985. *Oceanogr. Mar. Biol. Ann. Rev.*, **23,** 105–182.
Shannon, L. V. & Anderson, F. P., 1982. *S. Afr. J. Photogrammetry, Remote Sensing, and Cartography*, **13,** 153–169.
Shannon, L. V., Cherry, R. D. & Orren, M. J., 1970. *Geochim. Cosmochim. Acta*, **34,** 701–711.
Shannon, L. V. & Von Rijswyk, M., 1969. *Investl Rep. Div. Sea Fish. S. Afr.*, No. 70, 19 pp.
Siesser, W. G. & Rogers, J., 1976. *Sedimentology*, **23,** 567–578.
Sillén, L.-G., 1961. In, *Oceanography*, edited by M. Sears, Am. Acad. Adv. Sci., No. 67, pp. 549–581.
Skerman, V. B. D. & MacRae, J. C., 1957. *Can. J. Microbiol.*, **3,** 505–526.
Spencer, C. P., 1975. In, *Chemical Oceanography, Vol. 2*, edited by J. P. Riley & G. Skirrow, Academic Press, London, 2nd edition, pp. 245–300.
Spencer, D. W. & Brewer, P. G., 1971. *J. geophys. Res.*, **76,** 5877–5892.
Stander, G. H., 1962. *Investl Rep. Admin. S.W. Afr. Mar. Res. Lab.*, No. 5, 63 pp.
Stander, G. H., 1964. *Investl Rep. Admin. S.W. Afr. Mar. Res. Lab.*, No. 12, 43 pp. plus 77 pages of figures.
Stander, G. H. & De Decker, A. H. B., 1969. *Investl Rep. Div. Sea Fish. S. Afr.*, No. 81, 46 pp.
Steemann Neilsen, E., 1954. *J. Cons. perm. int. Explor. Mer*, **19,** 309–328.
Steemann Neilsen, E. & Jensen, E. A., 1957. *Galathea Rep.*, **1,** 49–136.
Suess, E., 1976. In, *The Benthic Boundary Layer*, edited by I. N. McCave, Plenum Press, New York, pp. 57–80.
Suess, E., 1981. *Geochim. Cosmochim. Acta*, **45,** 577–588.
Summerhayes, C. P., Birch, G. F., Rogers, J. & Dingle, R. V., 1973. *Nature, Lond.*, **243,** 509–511.
Van Foreest, D. & Brundrit, G. B., 1982. *Progr. Oceanogr.*, **11,** 329–392.
Van Goethem, C., 1951. *Rés. Scient. Exped. Océanogr. Belge Eaux côt. afr. Atl. Sud (1948–49)*, **2,** 1–151.
Veeh, H. H., Calvert, S. E. & Price, N. B., 1974. *Mar. Chem.*, **2,** 189–202.

Visser, G. A., 1969. *Investl Rep. Div. Sea Fish. S. Afr.*, No. 75, 26 pp.
Visser, G. A., 1970. *Fish. Bull. S. Afr.*, **6,** 10–22.
Volkov, I. I. & Fomina, L. S., 1972. Litologiya i Poleznye Iskopaemye, No. 2, 18–24 (English translation).
Von Bonde, C., 1928. *Un. S. Afr. Fish. Mar. Biol. Survey, Dept Mines Industries, Ann. Rep.*, No. 6, 18–21.
Wattenberg, H., 1928. *Rapp. P.-v. Réun. Cons. perm. int. Explor. Mer*, **53,** 90–94.
Wattenberg, H., 1938. *Wiss. Ergebn. dt. Atlant. Exped.* '*Meteor*', *1925–1927*, **9,** 1–132.
Wattenberg, H., 1939. *Wiss. Ergebn. dt. Atlant. Exped.* '*Meteor*', *1925–1927*, **9** (Atlas), plates I–LXXII.
Wedepohl, K. H., 1960. *Geochim. Cosmochim. Acta*, **18,** 200–231.
Willey, J. D., 1978. *Mar. Chem.*, **7,** 53–65.
Wong, C. S., Boyle, E., Bruland, K. W., Burton, J. D. & Goldberg, E. D., 1983. Editors, *Trace Metals in Sea Water*, Plenum Press, New York, 920 pp.
Wüst, G., 1935. *Wiss. Ergebn. dt. Atlant. Exped.* '*Meteor*', *1925–1927*, **6,** 109–288.

Oceanogr. Mar. Biol. Ann. Rev., 1985, **23**, 253–312
Margaret Barnes, Ed.
Aberdeen University Press

DIVERSITY AND STRUCTURE IN AQUATIC ECOSYSTEMS

SERGE FRONTIER
Laboratoire d'Ecologie Numérique, Université des Sciences et Techniques de Lille, F-59655 Villeneuve d'Ascq, France

INTRODUCTION

A community which involves many species is usually considered as more evolved than a community which involves only a few. This paradigm arose from the observation that species appear progressively when a new environment is invaded; it is in terms of this hypothesis that ecosystems are compared with regard to their species diversity. The quantity gives us an image simplified overall of the ecological complexity, since it is based on an inventory of organisms, instead of an inventory of interactions which constitute the system. It gives, however, any image, and not a groundless one seeing that the more numerous the categories of organisms, the more numerous are the communications.

I propose here a critical review of the different methods of defining and describing the ecological diversity. I state again the problem of the functional significance of the diversity, taking into account some recent modellings. I shall emphasize some ambiguities arising from a too cartesianist approach of complex systems. I shall also evoke some problems of sampling methodology and finally problems arising when using models of distributions, particularly in the light of the recent theory of fractals: the ecological diversity seems to be a facet of the fractal organization of biomass.

I shall, therefore, (i) characterize, with a functional viewpoint, the variations of structure in ecosystems, particularly as they result from a maturation in natural conditions or develop under the effect of disturbances; (ii) introduce the notion of observation scale in modelling an ecosystem; (iii) explain some problems of sampling.

WHY AND HOW ECOLOGICAL DIVERSITY HAS TO BE MEASURED

INTERACTION SYSTEMS AND OBSERVATION SCALES

The studying or managing of ecosystems has already demonstrated that biomass is not a product but a system constituted of elements in perpetual interaction, and in which the heterogeneity is a functional character and not a mere constraint of sampling. The bewildering complexity of each real system, and the fact that all systems are heterogeneous at each scale of observation, always result in a major difficulty of the analysis of ecosystems. Numerous approaches have been suggested in order to introduce into the concept of ecosystem this 'necessary heterogeneity'. A homogeneous ecosystem could not operate. Describing the reality at different scales of observation (taking into account simultaneously the complexity and the unity at each scale) is a challenge for the ecologist at all stages of his investigation. A purely cartesianist approach fails, because it consists of dividing the object into more and more tiny parts, until we 'understand' it. By dividing the object, even though abstractedly, we destroy it. Any object depending on a system is altogether (i) organized by itself as a system (or subsystem) at an inferior level of perception, and (ii) a part of a more complex system, defined at a superior level of observation; this duality is to be observed at each observation scale.

Before continuing, let me define what I mean by a "system". Thom (1966, in Thom, 1980) ascertains that "all is a system" and, finally, defines a system as "the content of any domain of space-time". This definition is, however, of such a generality that it becomes of little use, and I prefer to use my previous attempt of definition of systems in general and particularly ecosystems (Frontier, 1977): a system is a set of parts which are interconnected between themselves by way of reciprocal actions, in such a manner that it results in new global properties and new global functioning in respect to those of the elements. The latter properties are not sufficient to explain the properties of the system, although they are determinant, because it is necessary to add to them the properties of the set of interactions.

The macroscopical level which emerges from a system of interactions is quite as real as the constituent parts, in the same way that a house is as real as its bricks, the organism as real as its cells, and the population as real as the individuals which constitute it; each level of integration has its genuine properties. Moreover, all the existing systems show a hierarchical organization of their interaction network. There results a wide variety of pertinent observation scales of which the constituent phenomena can be very different, but in which can be recognized the same laws of organization, of interactions, and of circulation of the information. I shall call "laws of systems" these laws of organization, common for the various levels; they may be realized in different spatio-temporal designs, and under different physico-chemical constraints: the surface tension, the Brownian motion, the various photoperiods acting at well-defined levels, and not at others.

While the human scales of observation *a priori* seem to vary continuously, the organization of ecosystems includes some scales more significant than others; the organization is "intermittent". The levels of organization are more clearly disjointed into those of the organites, of the cells, of the organs, of the organisms, of the populations and of the ecosystems, than at intermediate levels (which also exist from the viewpoint of the observer). Since in the last part of this paper I relate the diversity to the notion of fractal organization of the living systems (Mandelbrot, 1977, 1982; Frontier & Legendre, in prep.), I now show that the 'selfsimilarity' (i.e., the fact that all a subsystem is to some degree a miniature model of the more comprehensive system), and the intermittent character of the levels of integration, are fractal properties.

From my viewpoint, an ecosystem is defined as a system of interactions. It is, of course, developed in a spatio-temporal frame, but this frame does not participate in the definition. No size is to be assigned to it, because it contains short-term and long-term, local and long-distant interactions. At each scale of observation in time or space, it can be described as interactions between populations, and between the populations and the physical environment. We arbitrarily can study as an ecosystem either a pond, or one of its parts (for example the plankton, the bottom...), or a pond plus the river which crosses it, or the pond plus the watershed, or a whole region. The action of living populations on the physical environment (which can act as far as the entire creation of an environment, as for example in the humus) is a part of the ecosystem. The equation "ecosystem = biocoenosis + biotope" is to be abandoned, the "+" needing to be replaced by another mathematical (still not imagined) sign which should indicate the integration and the emergence of global properties derived from an association.

On the contrary, the simple duality macroscopical/microscopical, which indicates nothing but a shift of the human scale of observation (such as the duality linking molecular cinetics and the laws of perfect gases), does not involve the existence of a system. But as soon as the set interferes in the behaviour of the parts, there is a system.

In reverse, the set acts upon the parts in a system; the properties of an element (or of a subsystem) are not the same if included in a system or if isolated (assuming that it can exist isolated, which is not the case for living systems). We approach here one of the most important characters of the ecosystem concept: autecology alone cannot explain the general behaviour of the ecosystem because we have also to consider the properties of the interactions. In addition, autecology cannot explain the ecological behaviour of a single species, because the whole ecosystem acts on this species. In this reciprocal communication between the system and an element or between the system and a subsystem, we do not know what is the cause and what is the effect. A pure analytical approach lapses and a synthetic one is necessary, consisting in (i) assuming that each macroscopical level exists with autonomy (such a consideration coincides with the "rehabilitation of the macroscopical" from Prigogine, 1980) and (ii) trying to understand the communication between one observation level and another. The second objective is now more a dream than a way of approaching the phenomena, probably because of the lack of an adequate intellectual tool. Perhaps it is the most important scientific problem of our time.

In fact, the classical statistical approach is to be excluded; we have not to deal with a global sight of a set of innumerable microscopical phenomena, as in statistical thermodynamics, but with some global properties resulting from an organized network, the detailed analysis of which escapes us because of its complexity. What results is not a simple 'statistical' phenomenon, because the network is very different from a sum of independent actions. We can, however, consider what we could call "statistical cybernetics" (perhaps initiated by the information theory), which should be able to give us an account of some of these global properties of the "managing the internal information by the system", particularly by the ecosystem. This managing differs from a statistical summing of the information, and statistical cybernetics ought to give us some global conditions for it. I shall specify, in this paper, what is thought at the present time about this managing—without giving the key of the whole systemics but only a way of investigation. The theory of dissipative systems (Morowitz, 1968; Glansdorff & Prigogine, 1971; Prigogine, 1980) and the catastrophe theory (Thom, 1972, 1980; Jones, 1977) are other aspects of a synthetic vision; it will be necessary to try a synthesis between them, perhaps introducing new mathematical concepts, such as fractals and fuzzy sets (Blanchard, 1984; Dujet, 1984) able to provide an adequate language for the description of realities which are not univocal nor sharply bounded. Endeavours in this way are actually being undertaken and the actual state of advance is given by Frontier & Legendre, in prep.

SIGNIFICANCE OF DIVERSITY AND SOME SAMPLING PROBLEMS

All the interactions constituting an ecosystem are exchanges of energy, of matter, and of information. Exchanges of energy and matter could be gathered into the same category since the flow of matter across a trophic net is, in fact, a flow of chemical energy, and since the biomass is nothing but a retardment of the energy (Margalef, 1979, 1980a,b). Two reasons should, however, lead us to consider separately the energy flow and the matter flow. First, the matter is recycled (at very different scales of time and space) in contrast to the energy, which is definitively dissipated from its assimilation by the ecosystem, up to its loss in a thermic form. This recycling is one aspect of the thermodynamics of dissipative systems, according to which each open flow of energy is necessarily associated with at least one cycle of matter (Morowitz, 1968).

The second reason for considering separately the circulation of matter through the ecosystem, is that it is not indifferent, such an energy flow being realized by a transfer of nitrogen or carbon, or by an exchange of potassium or calcium. It concerns here the qualitative aspect of energy-matter flow; the importance, and even the direction of the energy transfer may vary following the kind of material vehicle. In fact, an action happens only when the 'good molecule' arrives on the 'good site', which introduces the concept of 'qualitative' (or 'significant' or 'semantic') information, independently of any quantification of this information.

The information is the third 'product' to be transferred or exchanged in a system; it is the realization of a circumstance which allows, or modulates, an energy or matter transfer. A trajectory of the energy-matter corresponds, qualitatively speaking, to a chain of circumstances, *i.e.* to a succession of 0 ('do not pass') and 1 ('pass'), which is precisely a circulation of the information.

The species diversity corresponds well to this qualitative insight on the ecosystems; the shifts of matter and energy across distinct species are not equivalent. Nevertheless, with the definition of the Shannon-Wiener diversity index (Shannon & Weaver, 1963) which expresses a quantity of information, this information gets reduced to a mathematical definition clearly independent of its qualitative contents. Here an ambiguity exists, particularly emphasized by Thom (1980), but it can be shown that the qualitative aspect is not so excluded as it appears in such a mathematical formula.

Remember that the quantification of information, the object of the information theory, is obtained starting from the *a priori* probability of an issue. The more improbable is an issue, the more important is the quantity of information provided by that issue. For some mathematical reasons, the "quantity of information" (I) is defined as the logarithm of the inverse of the probability (P); logarithms to basis 2

$$I = \log_2(1/P) = -\log_2 P.$$

If, in a plurispecific community, the proportions of species are known, the attribution of an individual to a species called i, whose proportion in the community is p_i, provides a quantity of information equal to $-\log p_i$. The diversity index (H) is the average quantity of information for the set of species, *i.e.*:

$$H = -\sum (p_i \cdot \log_2 p_i)$$

(sum extended to the S species).

When we have only a sample to get acquainted with the community, and if the species in the sample are in proportions f_i, the quantity

$$H' = -\sum (f_i \cdot \log_2 f_i)$$

(sum extended to the species present in the sample) represents the diversity index of the sample, which is a biased estimator of the diversity index of the community as proved by Pielou (1975). The smaller the sample, the lower the diversity index we expect, because the community is only partially represented in the sample; the sample contains fewer species than the complete community, and with frequencies more or less different. Bowman, Hutcheson, Odum & Shenton (1969) calculate the variance and the bias of the estimator H', but only under the hypothesis of aleatory and independent distributions of individuals in space, which never happens. The bias and the variance are, in fact, much magnified by the heterogeneity of the spatial distributions. Individuals of the same species are most frequently over-dispersed, *i.e.* show a tendency to gathering, in a manner that is partially independent from one species to another; it follows that samples may be dominated sometimes by some species, sometimes by others, which results in a sample diversity index always lower than the diversity index of the community. When we accumulate a set of samples from the same community, we obtain a description which better approaches the reality.

Remember here the method of "diversity spectra" of Margalef (1956, 1967, 1970, 1980b): accumulating progressively the samples starting from a point and going away from it, we obtain an increase of the diversity index (diversity of the first sample; then diversity of the two first together; then of the three first, *etc.*). The pattern of increase varies following the types of ecosystems and the direction of going away from a point. Sometimes, the diversity increases very sharply, then remains stable ("rectangular" spectrum), which signifies that the ecosystem is relatively homogeneous and the complete diversity is realized over a little space. In other cases, the diversity increases slowly, and eventually reaches a high value, that is the spectrum is "obliquous" and indicates a high spatial heterogeneity of the ecosystem. There is, between the two situations, a change of the spatial scale of the diversity. Ultimately, a shift from one community to another may be revealed in the diversity spectrum by an abrupt increase of the index. It is surprising that this method, which seemed promising, has been rarely applied in the description of ecosystems, even by its author. The reason is probably the amount of the necessary sampling and the time to spend in analysing the samples, as well as the difficulty of maintaining the sampling within a 'homogeneous' community, and the definition itself of this 'homogeneity'. It is difficult to distinguish an obliquous increase of the curve, practically always more or less irregular, from a succession of steps indicating successive changes of communities. We recognize the problem of the observation scales of a system; it is arbitrary to describe a spatially heterogeneous community as a mosaic of distinct sub-communities or as a single system showing various possible aspects.

I have studied elsewhere (Frontier, 1983) some problems of sampling the ecological diversity. The background noise observed in a compact survey of the diversity index, for example in time series including one measurement each day in the same place (Fig. 1, from Dessier, 1979), has an amplitude of one unity of diversity, for a maximum value of 4·5. This amplitude of the aleatory fluctuations is about ten times greater than the amplitude calculated following the formula of Bowman *et al.* (1969), based on an aleatory distribution of the organisms.

One of the main causes of the sampling variability of diversity index is the contribution of the rare species. The contribution of a species of frequency f_i in the diversity index is the quantity $(-f_i \cdot \log f_i)$. Now the graph of the function $y = -x \cdot \log x$, with $x \in (0,1)$, indicates a maximum of about 0·531 for $x = 0{\cdot}368$, and shows that the increase is very sharp starting from 0, since the curve is tangent at 0 to the vertical axis (Fig. 2). The contribution of a species attains half of the maximum contribution for $f_i = 0{\cdot}0687$, and a quarter part for $f_i = 0{\cdot}0249$ (2·49% of the sample). These rare species, although often qualitatively important in the analysis of samples, are precisely the most badly

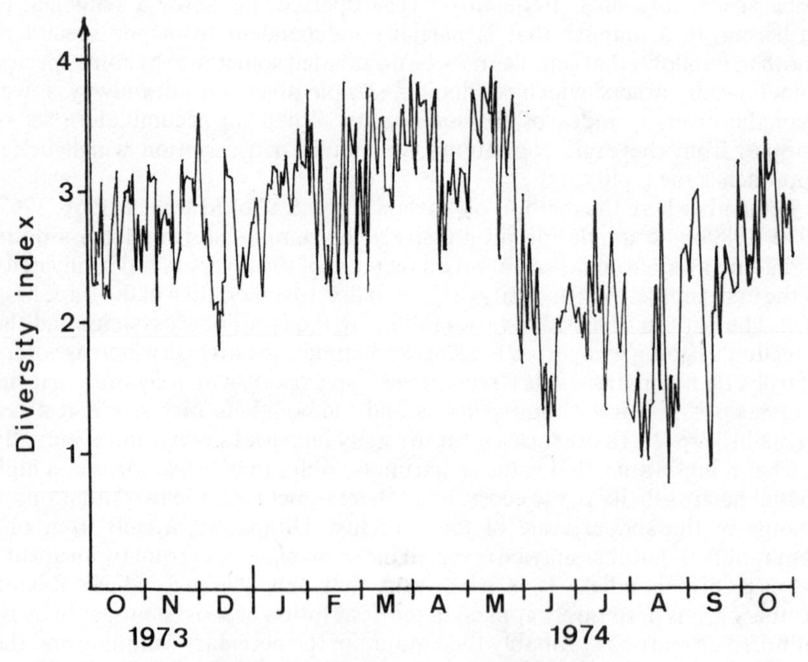

Fig. 1.—Daily variations of the diversity index of planktonic copepods off Pointe-Noire, Congo (from Dessier, 1979).

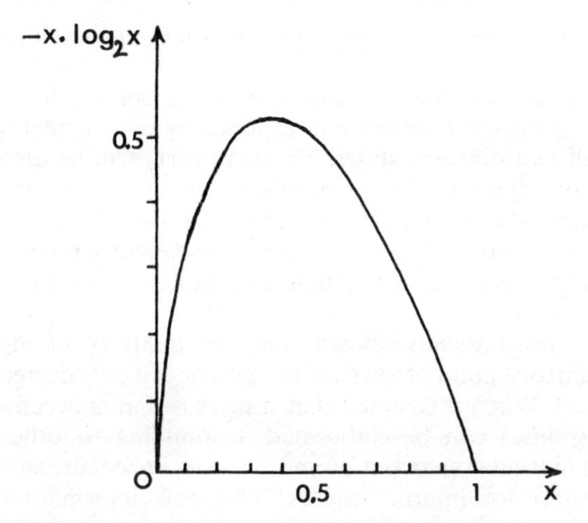

Fig. 2.—Graph of the function $-x \cdot \log_2 x$ against x.

sampled, with an important variance sampling. Ultimately, rare species are often numerous. Hence, an important fraction of the diversity index is affected by a great variability. I shall later propose a method to avoid this inconvenience (see p. 306).

CIRCULATION OF INFORMATION IN A COMPLEX SYSTEM

I shall now try to explain why the notion of information viewed as a 'product' which can circulate, is pertinent to the probabilistic quantification stated above.

When an event starts or evolves in a way that depends on its environment, we say that it is controlled by that environment. Similarly, a subsystem in the midst of a system can be controlled by another subsystem; it 'responds' taking one of different possible states. Each of these can, then, constitute a signal or an information which in turn controls a third subsystem, and so on. In such a manner, a transfer of the initially introduced information will take place.

What is too rarely emphasized is that, even though a quantity of information is transferred, this quantity has not a value for itself but is relative to the set of environmental states that the receiving system can discern. In fact, events are 'rare' or 'frequent' depending on the fineness with which the receiving system can distinguish between them. If one colour (for example a certain gradation of red) cannot be discerned from other gradations by an organism, the information it brings to this organism is not different from the information brought by any red, and the quantity of information carried by the event will be low, including when the particular gradation is a rare one. On the contrary, if an organism is able to distinguish the gradation and if that gradation rarely occurs, then the quantity of information will be high. It is hence necessary, in order to calculate quantities of information, to know precisely the set of discernible states, depending on the organisms or subsystems receiving the information. Afterwards, these discernible states being known, we have to be informed on their probabilities in a given environment.

Furthermore, we can consider that the subsystem discerns between two states of its environment only when it responds by two distinct behaviours or by taking itself two different states. These, in turn, will be discerned or not discerned by an adjacent subsystem, and so on. Two responses are, in this context, considered as different if they are received as different by at least one subsystem. The quantity of information is consequently relative not only to the capacities of a receiving subsystem, but also to a chain of transfers in a system of interactions.

It can be mathematically shown that the quantity of information so transmitted from one point of a system to another can only decrease or, at best, remain constant. When it is stated that it increases, it is because some more diversified responses can be elaborated, responding to other signals not emitted by the first subsystem but by another one, or because some discernible states can appear for internal causes. This new information, that is independent of the examined information transfer, constitutes, in respect of this transfer, any background noise. We agree with Thom (1980) that the theory of information is, in fact, a theory of the communication of information. The set of discernible states, and of the probabilities p_i of each one, constitute the "information channel"; the "capacity" of this channel is the quantity $(-\sum p_i \cdot \log p_i)$, also called "neguentropy" of the set of discernible states in conformity with the formalism of the statistical thermodynamics (in which the considered discernible states belong to the molecular or atomic levels). The law of no-increase of the information is then formally identical to the law of no-decrease of entropy.

In an analysis from 1973, rewritten in 1980, Thom (1980) emphasizes the ambiguity of the term "information" in the sense used in semantics. In the theory of information, we limit ourselves to the restricted sense of pattern and transmission of a pattern, in the same way that a programme is transmitted. Each combination of signals, coming from the environment of a receiving subsystem, has the meaning of a programme for this subsystem, which receives what it is able to discern. Moreover, the significance (*i.e.* the semantic contents) of the programme depends on the receiver. The quantity of transferred information and the significance of this information are then linked, in respect to the receiving subsystem, and I think it is not necessary to consider, following Thom, that there are two completely independent notions unluckily gathered under the same word. The fundamental phenomenon that we have to consider in our analysis of ecosystems, is the transmission of some quantity of information that each subsystem "semantically" receives and repercusses. It remains that the capacity of information (which is, at each time of the transfer, linked to the calculable diversity of discernible states and possible responses) can only decrease or remain constant, what imposes various constraints to the structure of system. Each structure is characterized by a maximum capacity which limits its performances.

The paradox in the calculation of a diversity index as a capacity of information surely comes from the fact that this index is calculated starting from a set of species discerned by man and whose accuracy can vary depending on what each specialist acknowledges as a species, or a subspecies, or a set of species. The value of the diversity index varies following this, and the "receiving subsystem" is the zoologist! Now it is very rare to encounter a subsystem having exactly the same set of discernible states as determined by the zoologist who studies it; for example, the range of preys of a given predator rarely coincides with a taxonomical collection established by man.

It is obviously impossible to catalogue the discernible states of each element of the ecosystem, so much more so that each element is also a subsystem, and that the overall hierarchical organization states some problems of observation scale. We have then to try to quantify a 'total information' that is, practically, the information originating from all the states discernible by man. We justify this method by two hypothesis: (1) the states discernible by man are also discernible by at least one element of the ecosystem, and (2) the quantity of information initiated by this distinction of the states may be transferred to at least one other point of the ecosystem (and, practically, gradually along a trajectory, *i.e.* along a chain of interactions, since at each point a fluctuation is discernible only if it provokes a response discernible in another point). What concerns ecological diversity, is that the hypothesis justifying its calculation admits that fluctuations concerning each species (as recognized by man) have any importance at one point of the ecosystem—which seems pertinent with the definition of ecosystems.

SOME OTHER CONSIDERATIONS

The same arguments also justify the calculation of a diversity based on taxonomic units other than species. As usual in all hierarchical classification, a species itself is heterogeneous and constituted from subspecies, ecophasis, age and size classes, whose rôles in the ecosystem are different. Moreover, superspecific units also may have a significance, *i.e.* correspond to certain types of action in the ecosystem: for example, they are trophic classes, taxonomic classes such as families or genus having a homogeneous biology *etc.* They may be then divided into sub-categories such as species. A description is always made at the human level and, because of this, is relative to certain pre-occupations or certain constraints.

It is pertinent to take an interest in the composition of a community, its families or even orders, because each taxonomic category may represent a type of solution to the problem of occupation of the ecological space; many entomological analyses are then based upon a distribution of the principal groups and no species. Moreover, the diversity inside a group is also pertinent. Remember too that we can study the genetic diversity of a species or the biochemical diversity in a bacterial community. The pre-eminence (sometimes dogmatic) generally given to the description by species is only traditional. Many zoologists and botanists consider that it cannot be matter of ecology before having previously taken the list of species through meticulous determinations, because "it is necessary to know about what we talk", they say. It is the inheritance of a great desire to make an inventory of all the living species. Now the analysis of ecosystems imposes another suppleness, stating that diversity concerns each hierarchical level and that the true ecological diversity is a diversity of rôles in ecosystem.

I shall expose in the second part of this review a better method of analysing ecological diversity: the analysis of the distribution of individuals among species. I shall then introduce another argument to justify the arbitrary choice of the level of description, by showing that the fundamental laws of distribution are of the same kind from one level to another, with variations expressing the degree of integration of the different hierarchical levels. The arbitrary character of the exclusive reference to species level appears, too, in the cases when the receiving subsystem is more difficult to localize and, however, the diversity acts. For example, it is classically said that the species diversity allows the permanence of a resource, realizing a homeostasy (which consumes information). The substitution, casual or in the frame of an ecological succession, of some species by others in a set of preys, guarantees a permanence of the disposable biomass for a predator; the latter profits from the diversity of preys, while it does not perceive it because all the preys are equivalent to food for it. The receiver of information, in this set of interactions and retroactions, is the controlled quantity, that is, the total biomass of potential preys. This biomass is also controlled by the biomass of predators. If there are various predator species, their diversity also guarantees the permanence of the control system, because one predator species may replace another. On the whole, it is the total diversity—preys plus predators—which guarantees the permanence of the whole system.[1] Some sets of species can here behave as single species for such a receiver of information. Conversely, some subspecific categories such as development stages, ecophasis *etc.*, may play as diversified rôles as species. The only reference to 'true' species paralyses the analysis, and is only justified by the facility of separating categories (when this facility exists, which often does not, particularly in cases when the distinction of species is fuzzy).

It is the diversity of possible energy flows which ought to be reached, that is, a diversity of rôles inside the system. A place might be occupied by different taxonomic or ontogenic categories, which only partially coincide with the division into species. The distribution of biomass among species is frequently studied for convenience, and probably reflects as best as it might the ecological diversification, but we have to emphasize some attempts of analysing the ecosystem through other categories, especially in entomology.

In addition, the diversity should be studied at different scales of space and time. In fact, the system of interactions is included in a spatio-temporal frame (see the analysis of Margalef, 1979). Some interactions have a local and immediate effect, but the local variations will have some long-term and long-distance consequences. The 'managing of information' will be different on

[1]A law of instability of too complex systems was stated by May (1971, 1972, 1973) and seems verified by the low connectance (*i.e.* rate of actual interactions in respect to the total number of mathematically possible links) in true ecosystems. Practically, a given species is actually linked by direct interactions with only few other species in respect to all coexisting species, and indifferent to others. The complete set of species may, however, be considered linked through what mathematicians call a "transitive closing": A interacts with B and B with C, then A with C (but with a time delay, and with a loss of information). The real ecosystems are complex because they differ from aleatory assemblages, such as those modelled by May. We suggest that the very significance of the hierarchical organization is to make compatible a maximization of the connectance with a stability of the system: each subsystem has to manage a reduced number of functional links, and is linked to other subsystems by equally few interactions. Thus, the stability–complexity (or diversity) problem could be solved from the evidence that systems formed by numerous randomly distributed interactions are unstable, whereas true ecosystems may be complex and stable because they are hierarchical.

these different scales and, as a matter of fact, we verify the independence of the diversity over a reduced space-and-time domain (practically represented by the diversity measured in a single sample) and the diversity over a larger domain which includes heterogeneities and gradients of the ecological factors. In the second case, diversity is stated cumulating various samples taken over the domain, with a sampling pattern which, as well as possible, eliminates any bias.

The same remarks also are valid in the description of diversity through not a simple numerical index but a distribution of the species frequencies, as shown below. A more suggestive survey of the ecological changes than with the mere diversity index that is the distribution of individuals among species, allows us to reveal (under some hypotheses) the inaccurate samples, and to determine the scales at which the analysis of diversity is pertinent or not. It seems that the scales of observation are not of equal value: as said above, there is an 'intermittency' in the organization of ecosystems into subsystems, instead a perfect continuity among the scales of integration. A system of short-term interactions may exist locally, and systems may interact over long distance and time lags; on these two spatio-temporal scales, some statistical structures may be described in the community. On intermediates scales, on the contrary, a sampling pattern does nothing but overlap various communities, without expressing a statistical structure, for we have a mosaic and not a mixing.

A simple numerical index, that it is always possible to calculate, rarely allows us to judge the pertinence of an analysis in respect to the observation scale, or to the division into taxonomic classes. A finer analysis of diversity is hence necessary. We attempt it by studying the species distributions through rank-frequency diagrams.

DISTRIBUTION OF INDIVIDUALS AMONG SPECIES: AN EMPIRICAL STUDY

GENERAL CONSIDERATIONS

Studying frequency distributions is in no way recent. The collectors always have verified that in a sample, whatever its importance, there is a set of frequent species and various rare species, with all degrees of abundance and rarity. Regularities are to be sought in the progressive shifting from the most abundant species to the rarest (see *e.g.* Williams, 1964). This regularity is mathematically expressed by a statistical distribution.

Giving a statistical distribution means giving a set of values that may be taken by a random variate and the set of probabilities of each of these in the population (that generally cannot be entirely known), or their frequencies in a sample, or probabilities in a theoretical population given by a model. If the variate is continuous (for example, the size or weight of organisms), giving the distribution means giving the probability or frequency of each interval of values. In practice, it is sufficient to divide the interval of variation into a number of classes, and to indicate the probability or frequency of each one. In this case, the distribution may be represented by a histogram in which the probability of an interval coincides with the surface of the rectangle rising above the class interval (or by the height of the rectangle when all the classes have the same amplitude). If the number of classes increases, with a diminution of the amplitude of each one, the histogram tends towards a limit-curve, that sometimes has an equation. In fact, the histogram is built from a finite sample, it is not possible to increase indefinitely the number of classes, and we try to fit the histogram to a curve with an expressible equation which describes it approximately (see elementary books on statistics concerning this).

In biology, the number of observations is often not very great and the histogram shows an irregular shape which makes it difficult or impossible to fit a continuous probability equation. The continuous distribution is more easily represented by using the cumulated distribution, which consists of calculating, for each value, the frequency with which the variate is inferior or equal to this value (or retrocumulate distribution, with the frequency of values superior or equal). The cumulated distribution can be easily described starting from the limits of classes already fixed in the histogram, following an arrangement such as:

Classes of the random variate X	Frequencies of classes	Cumulated frequencies	Retrocumulated frequencies
X_0 to X_1	f_1	f_1	$f_1+f_2+\cdots+f_n=1$
X_1 to X_2	f_2	f_1+f_2	$f_2+f_3+\cdots+f_n$
X_2 to X_3	f_3	$f_1+f_2+f_3$	$f_3+\cdots+f_n$
$\vdots$	$\vdots$		$\vdots$
X_{n-1} to X_n	f_n	$f_1+f_2+\cdots+f_n=1$	f_n

Figure 3 gives an example of this double representation, from the same set of classes. We see that the cumulative representation gives an easier way of fitting a continuous distribution, since it is here easier to trace a theoretical or empirical regular curve very close to the observed points.

Another advantage of the cumulative (or retrocumulative) curves is that we can use all the data, and not only the bounds of classes, because we can observe the number of results inferior or equal to a given value, including whether this value lies inside a class. Hence, the fitting is based on all the observed values.

Now we rank the values in increasing order, saying that a result has the rank r and the value X_r, is equivalent of saying that there are r results which are inferior or equal to X_r. If S is the total number of data, r/S is the cumulative frequency of the value X_r (or, if the results are ranked in decreasing order, r/S is the retrocumulate frequency of X_r). A rank-frequency curve, often utilized to represent a distribution of individuals among species in a sample or population, is obtained putting the rank on the ordinate and the frequency on the abscissa: it is a retrocumulative curve of a statistical distribution. But a rank-frequency curve is generally represented with the rank on the abscissa, that is, the two axes are inversed (Fig. 4).

For mathematical reasons that I shall show below, I have chosen to represent frequencies and ranks in logarithms. A log-normal distribution then corresponds to a convex curve, and not an S-shaped one as when the ranks were on an arithmetic scale. One of the consequences of this log-log

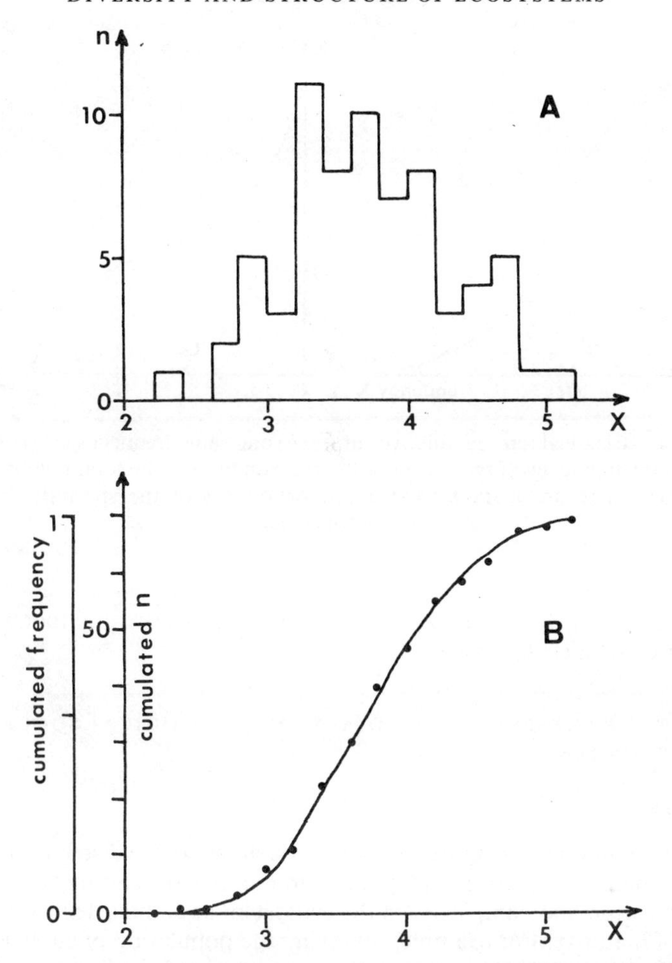

Fig. 3.—Fitting a distribution from a sample: A, the original histogram; B, the cumulative curve (only the bounds of classes are plotted); the cumulative numbers (N) are also the ranks of data after being ranked in increasing order; the irregularities of the histogram are much less apparent on the cumulative curve, to which it is easier to fit a smoothed curve, than to the upper histogram; (from Frontier, 1981).

representation, is that it is possible to represent absolute frequencies (*i.e.* numbers of individuals of the species) or relative ones (*i.e.* percentages of the various species), because the difference between the two curves is only a vertical translation and the shape of the curve remains constant (if $f_r = N_r/N$, then $\log f_r = \log N_r - \log N$, $\log N$ being the same for all species of the same sample).

The retrocumulative distributions of species, or rank-frequency diagrams, are used in two ways: (1) empirically, to describe samples or populations which are to be compared on this basis, and (2) with reference to models of

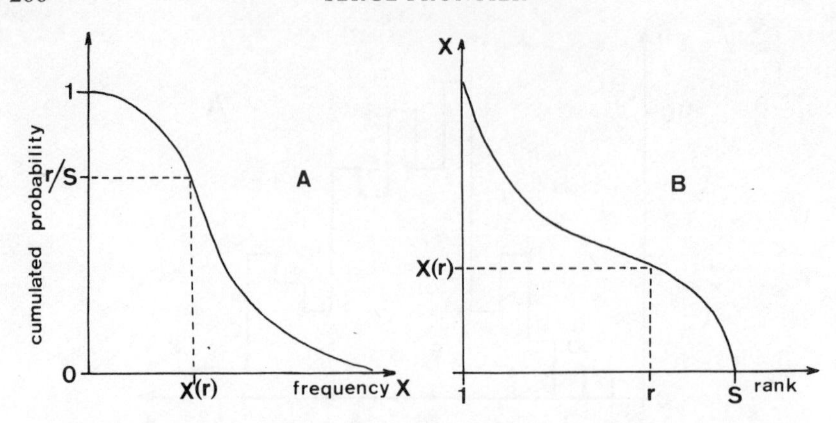

Fig. 4.—Classical retrocumulative curves (A) and rank–frequency curves (B): X_r is the frequency of the rth species in the sample; S is the total number of species; there are r species (or a proportion r/S of the species) whose abundance is $\geqslant X_r$.

distributions, which may be explicative or only descriptive and that we try to fit to observed distributions.

EMPIRICAL RANK-FREQUENCY DIAGRAMS. ECOLOGICAL SUCCESSIONS AND DISTURBANCES

An example

When the sampling is accurate, characteristic shapes of the distribution curves can be recognized in numerous types of communities, depending on the degree of organization and complexity of the ecosystem and, thus, depending on its degree of maturity. For example, surveying the population of euphausiids of the Pacific Equatorial Current from the east (submitted to an upwelling) to the far west through a regular set of plankton samples, an increase in species diversity was found (Roger, 1974). Simultaneously, the shape of the rank-frequency curves changes following a recognizable pattern (Frontier, 1969): they are concave on their left side, at the beginning of the westwards drift, then become progressively convex in the western part of the transect (Fig. 5).

The first curve shapes express the predominance of one or two species; this is a characteristic feature of juvenile ecosystems. Then, an increase of the species diversity is realized through a more even distribution of the species, but without attaining an even distribution (with a maximum diversity of $\log_2 18 = 4{\cdot}17$, since there are 18 species), nor a MacArthur broken stick model (diversity for 18 species: 2·84): the maximum actual diversity is 2·42. I have called the first shape "stage 1" and the second "stage 2" (Frontier, 1976a, 1977). Curves showing an approximate rectilinearity, as intermediate between stages 1 and 2, were called "stage 1′". Finally, we always observe to the right of the curves a "tail" of distribution owing to the rarest species; the curve falls abruptly, following the same shape along the whole succession, whereas the number of species increases; it is an invariant feature.

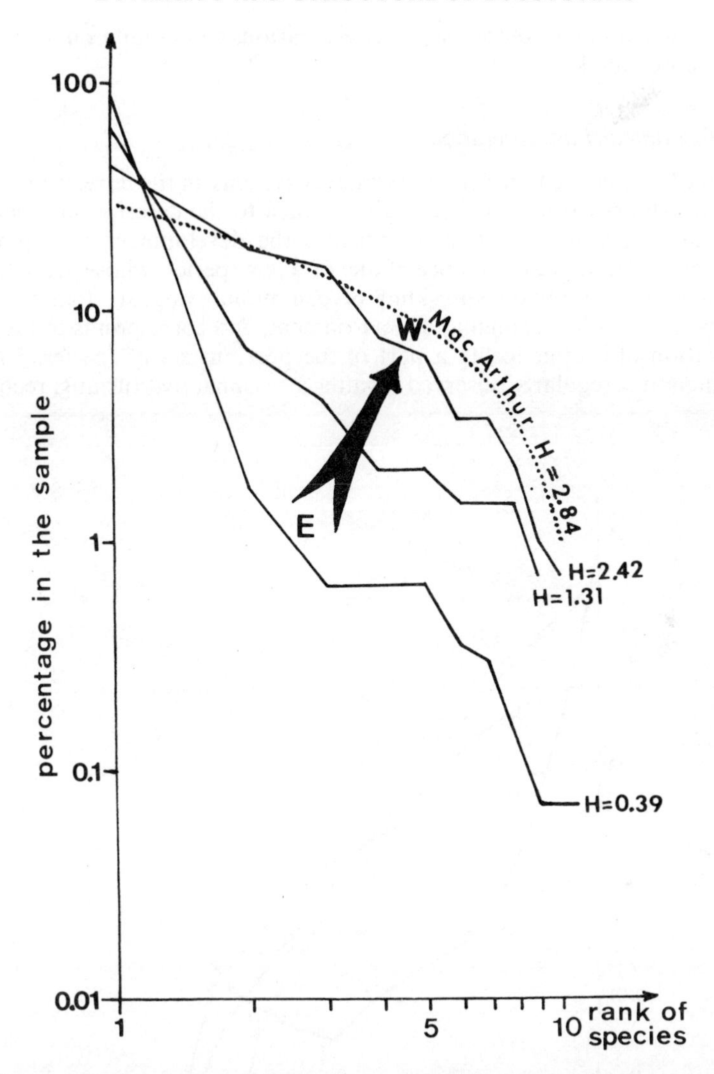

Fig. 5.—Rank-frequency diagrams for euphausiids sampled along a transect from east to west in the Pacific Equatorial Current: note an ecological succession from a juvenile community in the upwelling, towards a progressive maturation along the drift; simultaneously the diversity index increases (H passes from 0·39 to 2·42); dotted line is the MacArthur distribution for the maximum number of species, *i.e.* 18; rarest species are not represented here.

The rank-frequency curves permit us to follow the shift of the quantitative structure of the species percentages, during the evolution of the community from pioneer to mature (climax) stages. Moreover, it is possible to indicate the names of species on the graph, facing the points: the diagram then indicates not only the quantitative changes, but also the species which participate in this change. So we have a particularly synthetic tool for describing the evolution of a community, and the comparison between various communities may be very clearly represented.

Generalization and interpretation

The same features are found in numerous ecosystems, in the terrestrial as well in the aquatic environment. Let us then return to the description of stages, shown in Figure 6. The stage 1 indicates the development of a pioneer community, with a predominance of one or a few species, whose growth and multiplication are rapid. This kind of community appears following all disturbances or a fast changing of environment, and corresponds to the new colonization of it after losing a part of the previous set of species. Such a phenomenon is regularly observed (1) after a seasonal overturning: reconsti-

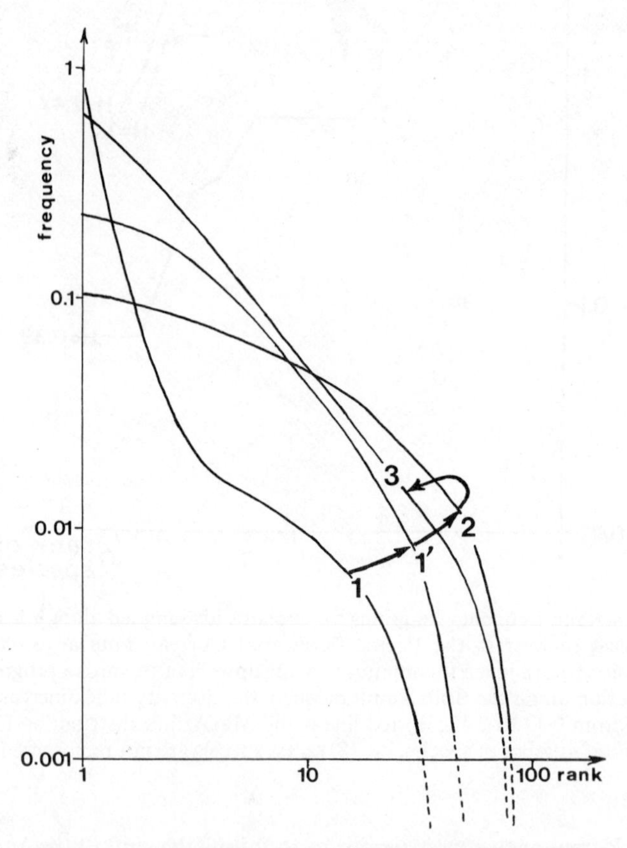

Fig. 6.—Diagrammatic representation of an ecological succession by the rank-frequency method (ranks and frequencies are in log): 1, first stage, juvenile ecosystem, pioneer community; 2, stage of maximum diversity (1′, intermediate stage); 3, re-arrangement of species frequencies resulting in a slight decrease or diversity, and perhaps the beginning of the aging of the ecosystem; a semantic problem is, where is the "climax"?

tution of the plankton of a lake after winter (Fig. 7, from Devaux, 1980); (2) after an intensive nutritional importation, *e.g.* upwellings, terrigenous deposits, eutrophication *etc.*; (3) after an accidental breakdown of the substratum, *e.g.* collapse of a cliff, forest fire, immersion of an area *etc.*; or (4) after a human disturbance, *e.g.* destruction of the environment, pollution *etc.* (Fig. 8, from Hily, 1983).

In the case of recolonization of a solid substratum, an increase of diversity in the very first stage immediately following the ending of the stress, sometimes

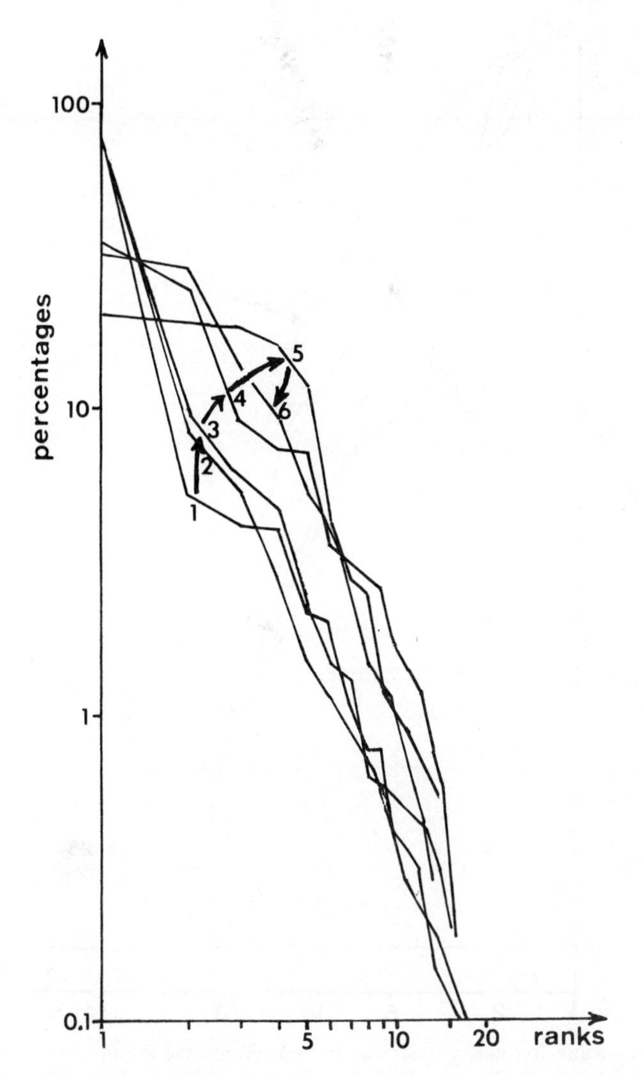

Fig. 7.—Seasonal variations of the rank-frequency curves in the phytoplankton of Lake Pavin, France: Stage 1, 1 is 11th April, 2 is 23rd April, 3 is 7th May; Stage 2, 4 is 22nd May, 5 is 5th June; Stage 3, 6 is 23rd June; diversity index increases from 1·45 to 2·97; (from Devaux, 1980).

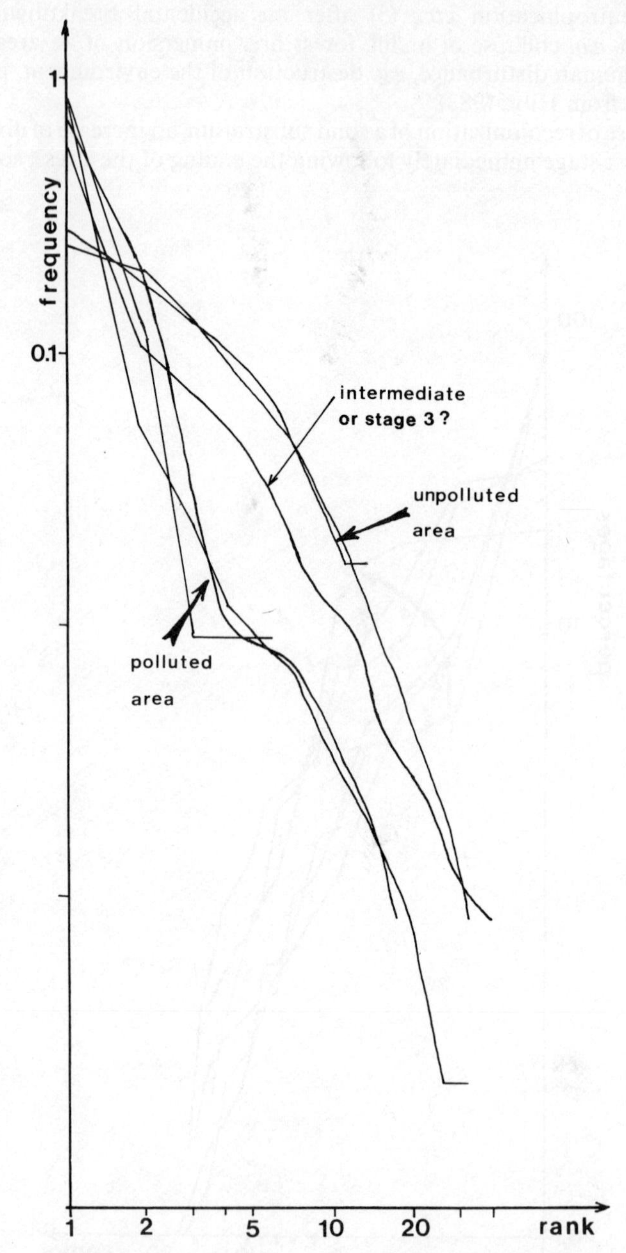

Fig. 8.—Rank-frequency diagrams in benthic marine fauna (Bay of Brest) along a gradient of pollution of the sediment (from Hily, 1983).

takes place for a short time. For example, a rocky shore left open after a storm (or experimentally denuded) first collects a wide variety of benthic invertebrates, whose larvae were in the plankton and settled there. After a time, however, a few species impose their dominance through their high growth rate and their tolerance of more drastic conditions, such as exposures to waves, and then the diversity rapidly falls. We have also observed the phenomenon in the first settlement of animals in marine sediments where the benthos has been destroyed (pollution, dredging of harbour regions, estuaries) before the colonization by a few pioneer species. In these cases, the rank-frequency diagram on a log-log scale appears irregular, but coarsely rectilineal with a slope equal to 1 (Hily, 1984; Diaz-Castañeda, 1984; Prygiel, 1983) as seen in Figure 9.

With respect to the previous diversified state of the community, the stage 1 represents a regressive phenomenon for the structure of an ecosystem (sometimes accompanied by an increase of biomass, but that biomass is controlled by external more than by internal factors). The rapid development

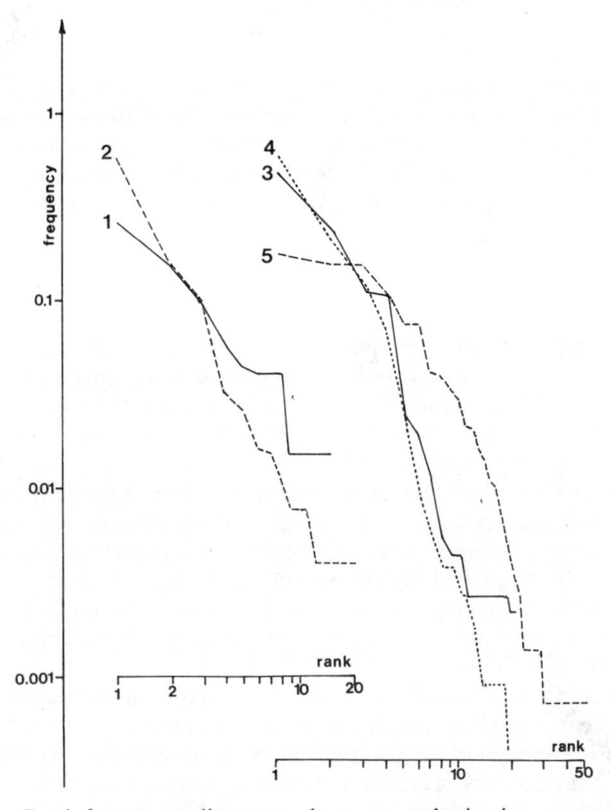

Fig. 9.—Rank-frequency diagrams along a recolonization process of a dredges area: 1, May, 1978; 2, September, 1978; 3, January, 1979; 4, May, 1979; 5, September, 1979; (from Hily, 1984).

of pioneer species occurs under conditions of lowered reciprocal demographic control of these species; there only exists an intensive competition for space and for the principal resource. After a time, some higher trophic levels and interactions appear, initiating a new integrated system. The term "juvenile" as applied to the pioneer community hence refers to a "normal" subsequent evolution, which will be realized provided the motive of the disturbance has disappeared. If the disturbance is permanent, then the system is maintained in its "juvenile" stage. Thus a stage 1 can be observed permanent in the immediate surroundings of an upwelling for as long a time as the upwelling is maintained, or in an estuary, and the evolution towards a stage 2 only occurs in the drift of the surface waters towards the open sea.

The trend towards realizing stage 2 is spontaneous, and indicates the appearance of a more and more complex network of interactions as new species appear. These are generally species with lower growth and reproduction rates and, above all, which demand previous changes of the ecosystem: for example, some species develop only below the cover of pioneer species, and trophic levels appear successively, giving higher and higher trophic and size pyramids. The appearance of larger and smaller sizes can be observed simultaneously, that is, an increase of the diversity of sizes, besides the diversity of species. It would be interesting to undertake a precise study of the variations of the diversity of sizes, parallel to that of the species diversity, along the ecological successions.

Returning from a stage 2 to a stage 1 is sometimes, but rarely, progressive; one can then say that the system fulfil a "rejuvenation". In most of the observed cases, the transition is a catastrophic one. In contrast, the evolution from stage 1 to stage 2 is always slow. Finally, the choice of the qualifications "juvenile" and "mature" traduces the fact that a complex structure is slow to acquire, and fast to lose.

A limit for diversity

The ecological diversity increases between stage 1 and stage 2, but it is observed that the diversity index never attains the maximum value as calculated with the Shannon-Wiener formula, equal to the logarithm to base 2 of the number of species. This theoretical maximum would be reached if all species were equally represented. In true communities, the diversity index rarely exceeds 4·5, including in extreme circumstances corresponding to a very convex rank-frequency curve; the right part of the diagram still describes the set of rare species. This observation strengthens the theory of May (1971, 1972, 1973) that a too diversified system becomes unstable and ought to disappear, for it is unable to manage its internal information when the interactions become too numerous. For this reason, the diversity always remains far from the theoretical maximum.

May's models were based on casual assemblages of elements—which are not ecosystems—and on an *a priori* non-hierarchical structure. A true ecosystem is elaborated in the outline of a progressive evolution of the structures, the latter being selected following the resilience they guarantee to the system. This is a truism, since a system which does not have sufficient resilience will disappear. The progressive appearance of species, which are maintained if they integrate themselves to the previous system, or get eliminated if they do not integrate, is a regularly observed phenomenon. Moreover, the fact that the control of artificially introduced species is progressively established suggests that control mechanisms take place as soon as they become necessary. An artificially introduced species may have two possible fates; either it disappears quickly (if not artificially maintained), or it first invades the biotope because of the lack of demographic control of this new species in the previous community, then it regresses (sometimes slowly) until a definitive level, because some control mechanisms have emerged. A re-adjustment of the ecosystem happened. For example, the brown alga *Codium bursa* accidentally introduced in the 1960s on to the French coast by a Japanese ship first invaded the rocky shores of the Channel and the Atlantic, up to an elimination of other species of algae. A regression was later initiated and led to the integration of the new species into the littoral flora, as observed today. It is a pity that, owing to the slowness of the re-adjustment of ecosystem, the mechanism of the appearance of the new control was never minutely studied. Recently, a *Sargassum* species was also introduced with oyster spat from Japan, and its population is actually exploding; some algologists are studying this demographic phenomenon (Delepine, pers. comm.).

The necessary limitation of complexity of an ecosystem in order to persist leads me to return to the important notion of hierarchical structure. A hierarchical system of interactions is built in such a manner that a given subsystem has to manage directly only a few interactions inside itself and, on the other hand, forms with some adjacent subsystems a rather simple new system. Such a hierarchical organization involves the permanent presence of a lot of rare species which occupy key places in the structure: species belonging to high trophic levels, large-sized organisms, *etc.* Moreover, some reference to the spatio-temporal scale is necessary; for example, a parasite which develops in a large host can, in some cases, multiply in great numbers in a single host organism, adopting an *r*-strategy during the colonization of its own biotope. But its total biomass is as low as the biomass of a predator in respect to that of the prey. With respect to the complete ecosystem, the parasite represents only a tiny fraction of the biomass, whereas it plays a determinant rôle in the control of demographies. Furthermore, this example states the problem (that will be evoked later) of the double possible representation of ecological diversity; the abundance of species can be expressed either as number of individuals or as biomass.

In conclusion, the high (although limited) diversity obtained at the end of the ecological succession can be interpreted as the acquisition of a hierarchical structure, through the progressive appearance of species which integrate themselves as soon as they appear. Each appearance of a new species depends on the previous state of the ecosystem; this inference is fundamental and will be the basis of a model of distributions that I shall preconize (see p. 288): the Zipf-Mandelbrot model.

STAGE 3

When appearing at the end of an ecological succession, the stage 3 indicates some decrease of diversity, following the maximum value realized at the stage 2; nevertheless, there is no returning to the simple structure observed at the beginning of the succession. Hence it is not a question of rejuvenation, but rather of a re-adjustment of the diversity, which was too high to permit a long-term stability of the system.

The rank-frequency diagram then becomes more rectilinear, at least over a part of its course excluding the tail of distribution due to rare species; this tail is often more distinct at this stage than in the previous ones. An inflection generally subsists in the left extremity of the curve, as a reminder of the previous convexity. Such distributions have been observed in the later stages of some ecological successions, for example in the phytoplankton of a lake (Devaux, 1976, 1977, 1980). More often this type of diagram is obtained through accumulating samples taken in the same area or along a gradient, as observed in plankton: chaetognaths sampled at 11 stations along the continental slope (Frontier & Bour, 1976 and here Fig. 10) or phytoplankton of a lake, when adding samples taken at various depths (Berthon, 1983, reprinted in Frontier, 1983 and here Fig. 11).

The last example allows us to ascertain that the planktonic community is distinct in its detail at different depths, where it becomes adapted (the species are not the same), but the quantitative structure of the community (*i.e.* distribution of frequent species) remains comparable. In contrast, the structure is different at another observation scale; under the circumstances we can

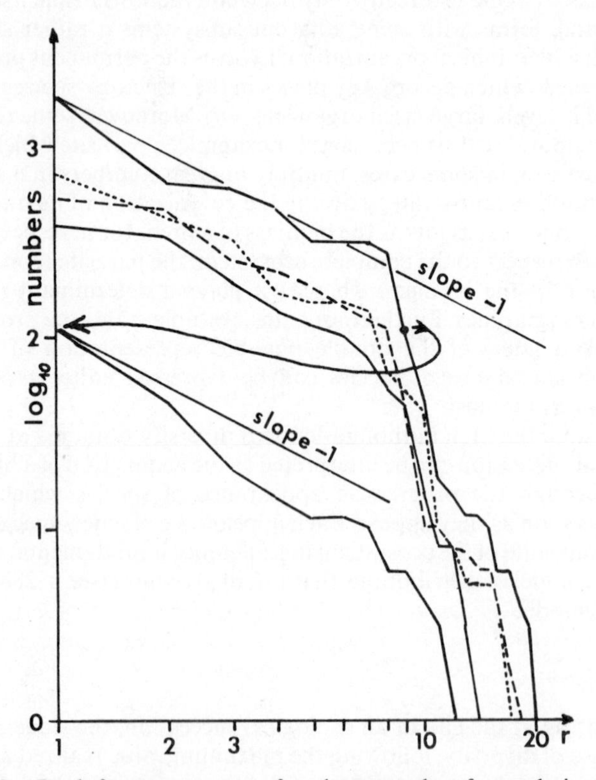

Fig. 10.—Rank-frequency curves for chaetognaths of cumulative samples from the continental slope off Madagascar (from Frontier & Bour, 1976).

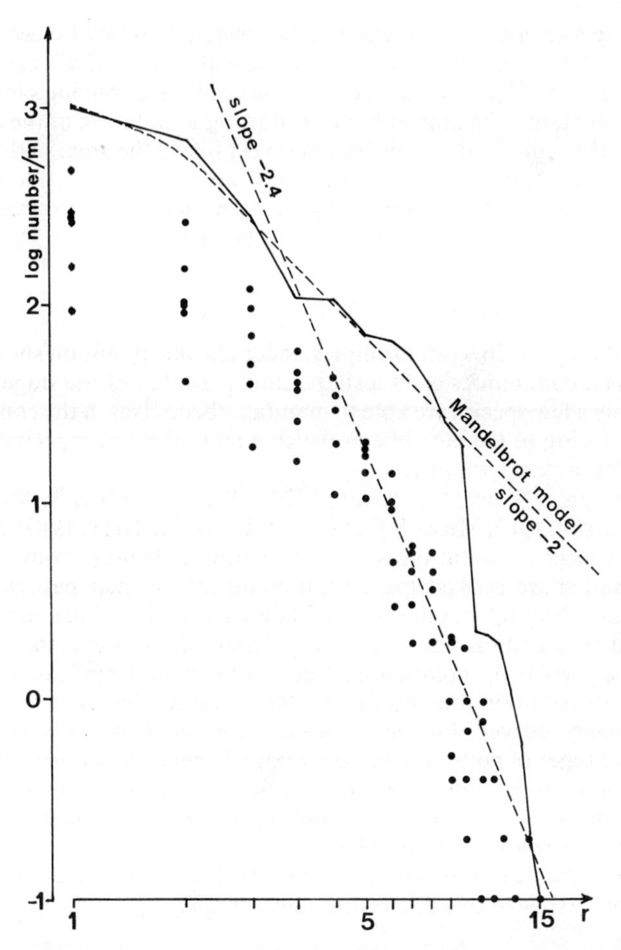

Fig. 11.—Rank-frequency diagrams in phytoplankton of Lake Dayet-er-Roumi (Morocco) at five different depths and for the total: a stage 3 seems to be realized for the cumulative sample; the left part of the distribution agrees with a Mandelbrot model; (from Berthon, 1983).

obtain an idea of the distribution at the scale of the whole water column, mixing a lot of samples scattered in the water column. One may suggest that different phytoplankton communities have interactive dynamics owing to the vertical migrations not of phytoplankton, but of the zooplankton which exploits it, and which transports organic and inorganic material between the various depths. So the water layer is organized as a whole better than at each single depth. It follows that species of the water mass may show a genuine distribution, different from those at individual depths. It can be seen on the upper curve of Figure 10 (see also Fig. 11), that the breakpoint between frequent and rare species is much more distinct than on the curves of the individual samples.

The emergence of a curve of type 3, sometimes, at the end of successions was interpreted by Travers (1971) as a "shift in the scale of diversity". The two ways of obtaining a stage 3 (at the end of certain successions, or mixing samples from a sufficient area) may be unified by the following hypothesis: at the end of the succession, the ecological mosaic becomes very fine in the area, so that a single sample becomes representative of a greater area, whereas at the preceding stages, the community had to be sampled over a large area to give the same diversity. This hypothesis has, of course, to be verified.

EFFECTS OF POLLUTION

As was said on page 269, our examples under chronic pollution show that the structure of a community gives a shape similar to that of the stage 1 (Fig. 8) because only a few species are able to maintain themselves in the conditions of stress and, owing to the lack of competition and other interspecific controls, these species multiply rapidly.

Gray and his collaborators (Gray, 1979, 1980, 1981; Gray & Mirza, 1979; Gray & Pearson, 1982; Mirza & Gray, 1981; Ugland & Gray, 1982) describe in detail the effects of pollution on the structure of benthic communities in estuaries and shore ecosystems. I shall comment on their papers later (see p. 284) because they refer exclusively to lognormal distribution and affirm that, under fugitive pollution, the community departs from lognormal, but if the disturbance persists the community "returns to equilibrium" and fits again a lognormal distribution, but with a greater variance (that is, a more vertical rank-frequency curve). Further analyses are necessary, diversifying the biotope and types of pollution, but the rank-frequency or species distribution method seems here a really useful tool to detect a disturbance of ecosystems due to pollution, in cases where the faunistical changes are difficult to ascertain because of the variability of samples.

In the terrestrial environment it seems that other disturbances, such as over-exploitation, also lead to a regression towards stage 1.

FACIES AND THE AGING OF ECOSYSTEMS

The stage 3 is not to be confused with a regression of the whole structure, that sometimes appears at the end of an ecological succession, when one or a few species are in exceptionally favourable conditions and escape the demographic control performed by the other species. This species then develops, invading the biotope and almost eliminating the other species. This senescent stage is very different from the initial stage of the succession: the principal species are not the pioneer species of the initiating community, but generally species of large size, slow growing but long-living, and robust. Good examples are found in the terrestrial environment: a high tree forest eliminates a greater part of the lower strata, and the diversity falls as well for the vegetation as for insects.

In the aquatic ecosystems, the phenomenon is less frequent, except for the marine benthos, in which many facies have been described and probably correspond to the local escape of a species from the overall control. For example, on exposed rocky shores, *Fucus* can develop almost exclusively in belts parallel to the coast, and eliminates the rest of the community; all species can be seen, however, under the alga, but stunted and scarce. A community of few species (fishes, epiphytes), adapted to this shady place, develop and the diversity, as in high-tree forests, falls. The rank-frequency curve is of type 1. When a facies is destroyed (through aging and death of trees, or through experimental denudation of rocks), what regenerates is not directly the facies, but a pioneer community similar to that of the first stages of the colonization of the biotope. The facies can settle again after a long time.

It is sometimes difficult to distinguish an initiating senescent facies and a stage 3; perhaps it is the same thing. For example, Devaux, Millerioux & Amblard (1983) observed that the decrease of diversity in lake phytoplankton was linked to the increase of two big diatoms (*Cyclotella comta* and *Synedra acus*), eaten with difficulty by the herbivorous crustaceans present in the last stage of the ecological succession, and so escaping the demographic control. We have here a beginning of facies through aging of the community.

CURVES WITH STEPS

Tracing rank-frequency diagrams does not always give regular curves and does not always agree with one of the types of curves described above. When we observe one or some steps in the curve, we often discover that the sampling was bad: either it has been selective (what is described is then an artefact) or it has overlapped two or more distinct communities but in too little number to reach a superior statistical level.

The first case concerns essentially trapping or similar methods. Sampling through trapping will most often give an extremely biased picture of the community, for it is based on any ethological selection such as attraction by light *etc.* The pictures of communities arising from a fishery are affected by this difficulty since the proportions of different landed species not only depend on their presence and abundance in the surveyed area, but also on fishing efforts, which are principally directed by the commercial value of the fishes, by the ease of capturing them in great quantities, *etc.* It is a selected sampling. We give in the Figure 12 a rank-frequency diagram obtained in a village fishery in Senegal, Africa.

The overlap of communities is illustrated with a set of Isaac–Kidd midwater trawl samples of myctophids, in the tropical Pacific. Taking into account all the samples together, we have a curve with two main steps, the first including the four first species, and the second species five to nine. We verify that, among the four first species, two are dominant in the daytime and two at night; the species of the second group also permute. Then, tracing two curves, one for the days and another for the nights, we obtain two more regularly shaped graphs, each of these following approximately a MacArthur distribution (Fig. 13, from Frontier, 1969).

All the methods of sampling communities give a figure more or less biased because they are selective: many species escape or avoid a plankton net in various proportions; but it can be expected that the plankton net simulates rather well the utilization of plankton by predators, since the catch by the predators is based upon the filtration of a range of particle sizes. We do not wonder in observing an integrated composition of the sampled plankton, which ought to be near the composition of the natural resource. In contrast, light sampling rarely provides an interpretable distribution.

Steps in the curve can be obtained at intermediate stages of the ecological

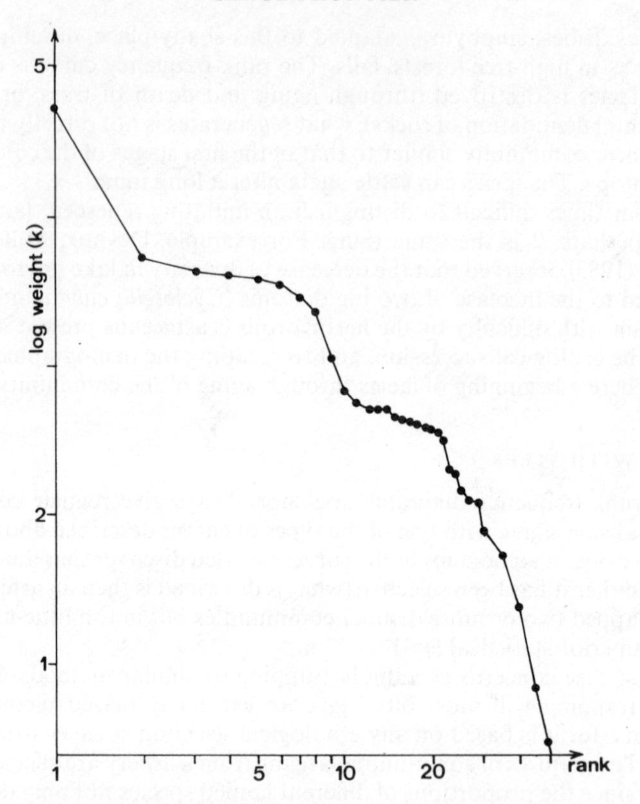

Fig. 12.—Rank-frequency diagram for the total catch of a non-commercial fishery at Dakar, Senegal in two weeks at one landing beach.

succession. They indicate then a transitory stage in a re-adjustment of the abundances of species: some species are decreasing in abundance, while others are increasing, for example between the stage 1 and the stage 2.

Finally, steps can also appear arising from aleatory fluctuations of abundance, resulting in a difference between the distribution of species in the sample and in the population. There is actually no way of testing whether an observed step in a sample distribution is statistically significant or not, without use of a model of distribution selected as a working hypothesis. I shall now expose the principal models of distributions used in ecology.

MODELS FOR SPECIES DISTRIBUTIONS

Various models of distributions have been proposed for ecology, each of these corresponding to an argument which describes a plausible way of partitioning the ecological space amongst the existing species. The models most often used are the Motomura model (Motomura, 1932, 1947, re-explained by Inagaki, 1967) or geometric series (Whittaker, 1972); the Preston or lognormal model (Preston, 1948, 1962; Daget, 1976); and the MacArthur or broken stick model

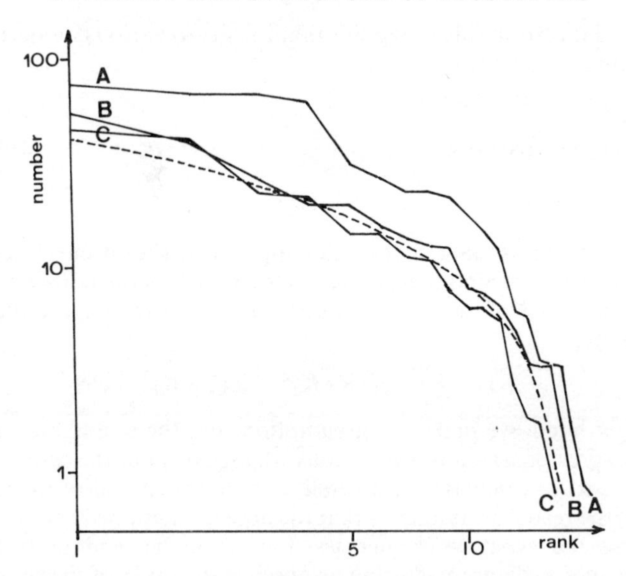

Fig. 13.—Rank-frequency diagrams for a set of myctophids samples (Isaac-Kidd midwater trawl) along the Pacific Equatorial Current: A, all samples together; B, night samples; C, day samples; broken line is the MacArthur model for 17 species; (from Frontier, 1969).

(MacArthur, 1957, 1960). A fourth model has been designed (Frontier, 1976a, 1977) derived from the Zipf-Mandelbrot model for linguistics and economy (Zipf, 1949, 2nd edition, 1965; Mandelbrot, 1953, 1977, 1982).

The various models essentially differ by the method of generating the probabilities of appearance of the species, depending on the argument adopted for modelling the partitioning of resources amongst species. In fact, it is extremely probable that no model is perfectly or exclusively realized in nature, and that the real partitioning of the ecological space arises in different, and probably composite, ways which vary according to the conditions. Each single model is an over-simplified image. It is interesting to note that each proposed model was justified by the authors by means of well-adjusted real examples. When an ecologist discovers a series of species which fits with one of the models, he publishes a paper; that explains why each model has been proved many times.

One of the interests of models is that they emphasize the consequences of a set of hypotheses, which are the bases of the model. I give below a review of the above-mentioned models and the underlying hypotheses. It appears that all the hypotheses are verified in some circumstances, so that the underlying patterns are probably actually present in the settlement of biomass in the ecological space. A single pattern ought to be occasionally recognized; it is more probable that various patterns simultaneously exist in the same community or (as we shall see) that intermediate patterns exist, making precarious the fitting of an empirical distribution to a model. Nevertheless, in all cases where a model of mechanism seems prominent, it is worthwhile noting it, and it is from this viewpoint that the fits to various models found in the literature are to be considered.

THE CLASSICAL MODELS

The Motomura model

The basic statement is that a first species appropriates for itself a fraction K of the resource, then a second species takes a fraction K of the remainder, and so on. The proportions of the total resource which are successively allocated to species are then:

$$K, \quad K(1-K), \quad K(1-K)^2, \quad K(1-K)^3, \quad etc.$$

On the other hand, we make the assumption that the abundance of species (generally expressed as a number of individuals, except in the cases where the biomass is used to calculate the diversity) are proportional to the respective fractions of the resource. It follows that the above-mentioned geometric series also expresses the respective frequencies of species in the community. If we plot the rank-frequency diagram, putting on abscissa the ranks of species without a logarithmic transformation and on the ordinates the logarithm of abundances:

$$y_r = \log K + (r-1)\cdot\log(1-K),$$

then we obtain a straight line.

Some rectilinear (or almost rectilinear) alignments were found in community samples, for example Lecal (1965) for planktonic coccolithophorids and Daget (1976) for molluscs.

Although it was often criticized for the silliness of the underlying hypotheses (particularly, that the proportion K is supposed the same for all species), the Motomura model is one of those which introduces a succession of the species, in conformity with ecological evidence. K can be modified in the course of the succession; we can then obtain a series of linear segments with various slopes (*e.g.* Binet & Dessier, 1972, for planktonic copepods off Pointe-Noire, Congo; Fig. 14). If K is changing from one species to another Motomura's law, of course, does not fit. In terrestrial vegetation, Motomura's distributions have been described at the beginning of ecological successions, then progressively passing to a lognormal model (Horn, 1975).

The MacArthur model or "broken stick"

The notion of ecological succession is here suppressed: the resource is divided at random and, what is very important in the building of the model, all together amongst the species. The fractions of allocated resource are then very diverse; the diversity index is high, generally higher than it will be in any true sample. The mathematical model is built following the classical problem of "broken stick", that is, the resource is admitted to be divided as a "stick" of given length divided into S segments by $(S-1)$ randomly distributed points over the total length. The S segments are then ranked in decreasing order, and the expected distribution, represented by a theoretical rank-length curve, can be calculated. The theory (Barton & David, 1956, 1959) shows that, if the

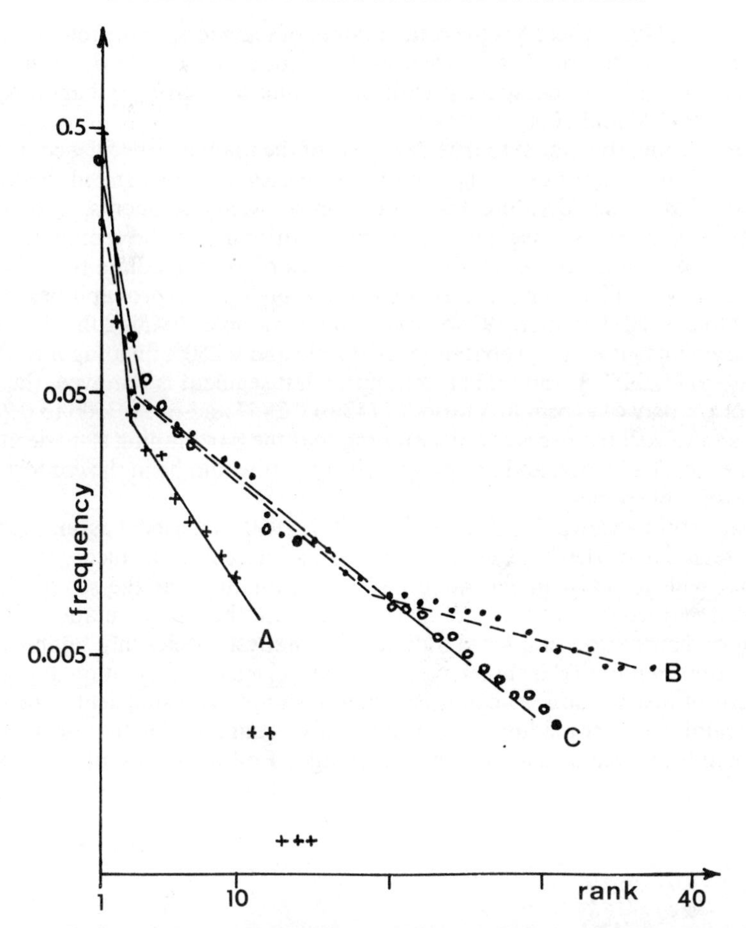

Fig. 14.—Rank-frequency diagrams (without logarithmic transformation of the ranks) for copepods of three plankton samples off Pointe-Noire, Congo: note that some rectilinear segments evoke different Motomura models, the constant of the model being variable along the series of species; A, surface sample; B, 15 m-sample; C, vertical 200 to 0 m-sample; (from Binet & Dessier, 1972).

length of the "stick" is equal to 1 (which allows comparisons of the lengths of segments to probabilities, for their sum is 1), the expected lengths of segments are given by:

$$L(r) = \frac{1}{S} \cdot \sum_{i=1}^{i=r} \frac{1}{S+1-i} = P(r),$$

where $P(r)$ is the probability of the species of rank r, if the species are ranked in increasing order of frequencies, or

$$P(r) = \frac{1}{S} \cdot \sum_{i=1}^{i=S-r} \frac{1}{r+i}$$

(Frontier, 1976b), where S is the total number of segments, or the total number of species in the sample. Figure 15 gives the shape of the distribution (on a log-log scale in order to compare it with other rank-frequency of diagrams) for $S = 10$, 20, 100 and 1000.

When fitting the model to true data, one of the main inconveniences is that the distribution depends on the number of species. It can be stated (Frontier, 1976b) that, when dividing the "stick" into twenty segments, and when excepting the greatest segment, then the partitioning of the remainder into nineteen does not reproduce the probabilities of an immediate partitioning into nineteen. That can be seen in a table giving the probabilities, from $S = 2$ to $S = 20$ (Frontier, 1976b). As a simple example, dividing the stick into two segments gives the probabilities of 0·7500 and 0·2500; dividing into three gives 0·6111, 0·2778, and 0·1111, but, if the first segment is removed, the two others are between them in a ratio 0·7143 to 0·2857 and not 0·7500 to 0·2500. This shows well the necessity of assuming that the partitioning is made at the same time, at a glance, and not progressively as it would be in the context of a succession of species.

MacArthur's viewpoint (MacArthur, 1957, 1960) consists of assuming that each segment of the broken stick represents an ecological niche, that two species will not exist in the same niche and, finally, that the set of niches exhaustively covers all the ecological space. Another interpretation of this random distribution of a set of individuals amongst species in a sample (and not in the community) is that it expresses only a lack of information about the pattern of distribution, as can arise when a sample is insufficient—the only constraint being the additive and exhaustive character of the "segments", which indeed is realized for species percentages. Following this interpretation,

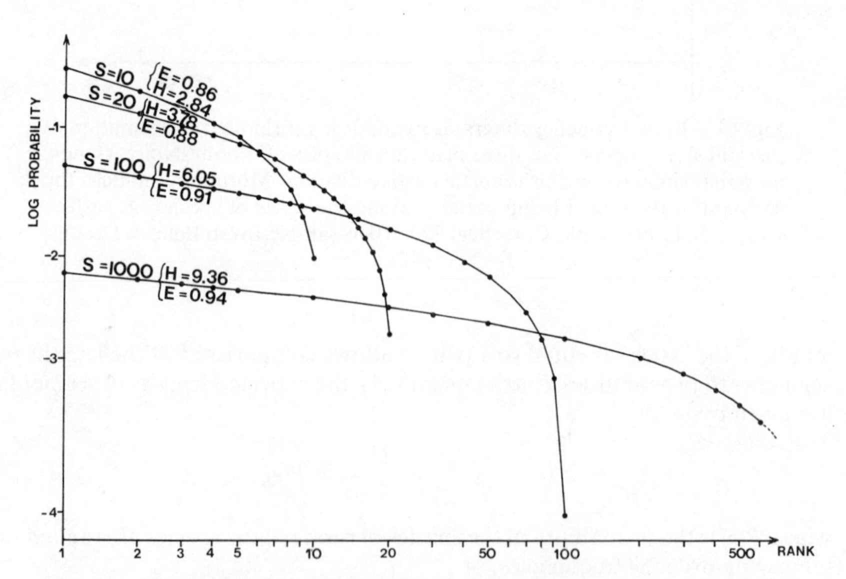

Fig. 15.—MacArthur or "broken stick" model represented on a log-log scale: S, total number of species, or of segments of the "stick"; H, diversity and E, evenness, both calculated from the model.

when the MacArthur model fits, we have to infer a casual distribution of species in the sample. In fact, King (1964) remarks that MacArthur's model never fitted except for series of data containing a few individuals and few species and hence was little representative of the population. We shall show that when a sample contains many individuals and many species, the model often fits for the distribution of rare species among themselves, whereas the other species show a very different distribution (see Fig. 26, p. 303). The rare species correspond to the situation described by King since they are wrongly sampled; the sampling variance of the frequency of a rare species is very great, and many of these rare species of the population are absent in a sample. Their fitting to a MacArthur model denotes an absence of singularity in their actual distribution.

In addition, the MacArthur distribution can be derived from completely different hypotheses. For example, Cohen (1968) demonstrates that if n independent aleatory variates $X_1, X_2, \ldots, X_n$ have the same distribution, that is

$$P(r)\{X_i \leqslant x\} = 1 - e^{-\lambda x},$$

it follows that the quantities

$$f_i = \frac{X_i}{\sum_i X_i}$$

are expected to be ordered as in the MacArthur model. We insist that the correct fitting of a set of data to a theoretical distribution, which starts from some hypotheses, never constitutes a proof of that hypotheses, because frequently different hypotheses may result in the same distribution, as was demonstrated for example by Irwin (1941, in Anscombe, 1950) for the binomial negative distribution (which has also been proposed as a model for species distribution, Brian, 1953).

The diversity index calculated following the model is high and attains values which are never observed in natural communities. In the latter, the Shannon index rarely exceeds 4·5. Figure 16 shows the increase of this theoretical index when the number of species, S, increases. We verify that the diversity index H is almost linear with respect to log S, and that it is >5 for $S > 30$; $H = 6{\cdot}05$ for hundred species and $= 9{\cdot}36$ for thousand. It should be remembered that the maximum diversity index (obtained with all species equally abundant) is $\log_2 S$. The evenness $E(= H/\log_2 S)$ increases very slowly, passing from 0·86 for ten species to 0·88 for twenty, 0·91 for hundred and 0·94 for thousand. It follows that the diversity index is almost proportional to log S in this model.

In addition, it can be observed that MacArthur distributions are not very far from lognormal for small or medium number of species. Figure 17 shows some MacArthur distributions plotted on a log-probit paper (a lognormal distributions would then result in a straight line), S varying from 5 to 100. The alignment is not far from being rectilinear for $S < 30$. It follows that the choice between fitting an observed distribution to a MacArthur, or to a lognormal distribution is sometimes hazardous, as can be seen from Figure 18 where a distribution built as lognormal through a simulation with 27 species is compared with a MacArthur for the same number of species. Using statistical tests of departure, it is often found that there is no significant difference

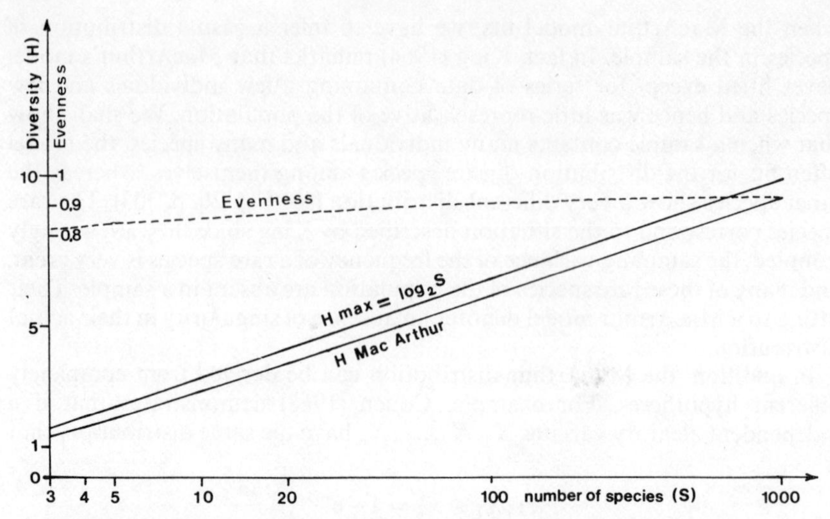

Fig. 16.—Increasing diversity index and evenness when increasing the number of species S, following a MacArthur model.

between the lognormal and MacArthur hypotheses, and the random irregularities of an empirical distribution from a true sample make the choice of model difficult; the "best fitting" is widely fortuitous when determined on the basis of a minimum squared distance.

Preston or lognormal model

Here also, no reference is made to an ecological succession and to the way of constituting the community. We only assume that the variate "number of

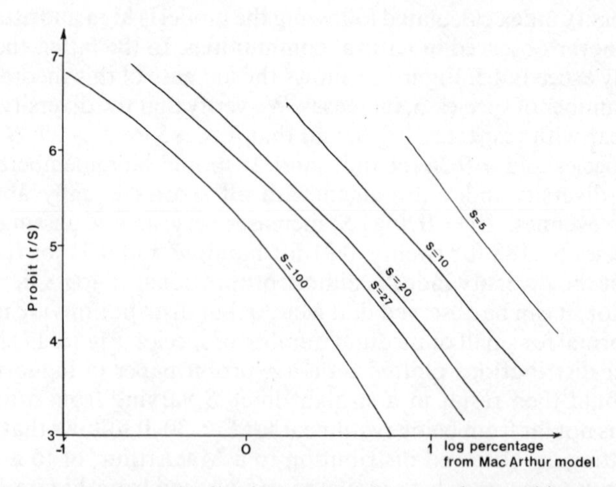

Fig. 17.—MacArthur distributions plotted on a log-probit paper: the distribution approaches a lognormal for low numbers of species S; the cumulative probabilities are the quantities r/S, where r is the rank of the species following the decreasing order of probabilities.

species whose abundance is included inside the interval $(X, X+\mathrm{d}X)$" has a lognormal distribution. The justification of the lognormality generally consists in saying that the action of environmental factors on the biological variates is a multiplicative one; then, the logarithms of the biological variates, and not the variates themselves, are linked in a linear way to the environmental factors. If these factors are numerous and additive in their effects (a classical hypothesis in statistics) then, owing to the central limit theorem, the logarithms of the biological variates will be normally distributed.

Particularly in what concerns the populations of a pioneer community, each monospecific population is characterized by a growth curve having an equation $N(t) = N_0 \cdot e^{rt}$ where the value of the growth rate r depends on the species, on the environmental conditions, on the concurrence, and on the predation. In particular, r is the sum of a positive natality rate and a negative mortality rate, each one of these being the sum of various terms depending on various factors. If these numerous factors are independent, and if their effects are additive, then the global effect on r is normally distributed and N is lognormally distributed.

Other patterns of actions lead a lognormal distribution of abundances to be expected; it is enough that we imagine a sum of effects whose actions are multiplicative. We shall develop here a very general pattern which is comparable to another pattern, and which leads to another distribution—the Mandelbrot distribution, that we shall examine further.

Let us suppose that the appearance of one individual of a given species depends on the realization of a number of previous conditions arising in the environment. No details are given about the nature of these previous conditions, which indeed can include the preliminary presence of individuals of the same species (at least as genitors), or numerous factors of the environment. We only make the assumption that each previous condition can arise under the form of various alternatives, and each of these possesses a probability. We assume that the numerous species differ by the combination of these alternatives; each combination allows the presence of one well-determined species, as follows.

Previous conditions	Alternatives
A	$a_1, a_2, a_3, \ldots$
B	$b_1, b_2, b_3, \ldots$
C	$c_1, \ldots$
$\vdots$	
Z	$z_1, z_2, \ldots$

Combinations	*Species*
$a_1 b_1 c_1 \ldots z_1$	S_1
$a_1 b_1 c_2 \ldots z_2$	S_2
$\vdots$	
$a_8 b_1 c_3 \ldots z_4$	S_i
$\vdots$	
$a_8 b_4 c_4 \ldots z_1$	S_j
$\vdots$	

This way of generating probabilities is represented in Figure 21 (p. 295). If the successive alternatives are independent (which is a very rough approximation), the probability of a combination equals the product of the probabilities of all the alternatives that compound it; then the logarithm of the probability of the combination is the sum of the logarithms of the realized alternatives. Once more, owing to the central limit theorem, this sum will be normally distributed; then the probabilities of the combinations (and hence of the species) show a lognormal distribution.

The convergence towards a lognormal is fast. We have verified it by simulating a simple situation with three previous conditions, each one having three alternatives whose probabilities are arbitrarily fixed. We have $3^3 = 27$ combinations, that is 27 possible species, according to the model. The table of probabilities is as follows.

Previous conditions	Probabilities of alternatives 1	2	3	
A	1/7	2/7	4/7	$\Sigma = 1$
B	1/6	2/6	3/6	$\Sigma = 1$
C	1/10	2/10	7/10	$\Sigma = 1$

Combinations	Probabilities	Ranks
1 1 1	1/7·1/6·1/10 = 1/420	27
1 1 2	1/7·1/6·2/10 = 2/420	24
1 1 3	1/7·1/6·7/10 = 7/420	16
1 2 1	2/420	26
1 2 2	4/420	19
1 2 3	14/420	9
1 3 1	3/420	23
1 3 2	6/420	18
1 3 3	21/420	7
2 1 1	2/420	25
2 1 2	4/420	20
2 1 3	14/420	10
2 2 1	4/420	21
2 2 2	8/420	14
2 2 3	28/420	5
2 3 1	6/420	17
2 3 2	12/420	11
2 3 3	42/420	3
3 1 1	4/420	22
3 1 2	8/420	13
3 1 3	28/420	4
3 2 1	8/420	15
3 2 2	16/420	8
3 2 3	56/420	2
3 3 1	12/420	12
3 3 2	24/420	6
3 3 3	84/420	1

Figure 18 represents the so obtained rank-frequency diagram. Figure 18A, is log–probit, showing that the lognormality is rather well verified, despite the

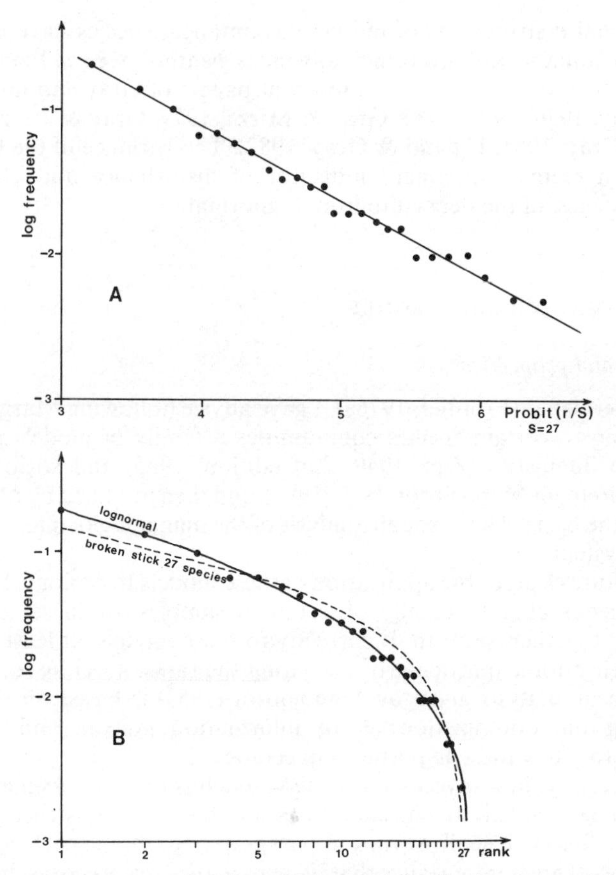

Fig. 18.—Simulated sample fitting a lognormal distribution (27 species): A, on a probit scale; B, on a log-log scale and comparison with MacArthur model for 27 species.

small number of previous conditions and small number of alternatives of each one, and despite the presence of tied ranks. Figure 18B, is log–log, as in our rank–frequency diagrams, showing that the distribution is not very far from a MacArthur one with 27 species.

The variance of a "canonical lognormal distribution" (Preston, 1962) is always higher than in a MacArthur distribution, that is, the diversity is lower (because the lower is the variance, the more similar between themselves are the species frequencies, hence the higher the diversity index). The following shows the difference.

		Number of species 10	15	20	36	100
Standard deviation of $\log_2$	MacArthur	1·50	1·57	1·61	1·67	1·76
	Canonical, Preston	2·45	2·69	2·85	3·19	3·69
	Ratio between the two	0·61	0·58	0·56	0·52	0·48

Lognormal distributions of individuals amongst species have often been reported in aquatic ecology, principally in the benthos, see *e.g.* Patrick (1972), Hicks (1980) and, above all, the important papers of Gray and his collaborators (Gray, 1979, 1980, 1982; Gray & Mirza, 1979; Gray & Pearson, 1982; Mirza & Gray, 1981; Ugland & Gray, 1982). The variance of the lognormal distribution seems a pertinent indicator of disturbance and also, in the transient stages, of the departure from lognormal.

THE ZIPF-MANDELBROT MODEL

Definition and properties

In an earlier paper (Frontier, 1976a), I gave advice (following Margalef, 1957) on applying to certain species communities a family of models previously applied to linguistics (Zipf, 1949, 2nd edition, 1965) and socio-economic studies (Pareto, in Mandelbrot, 1977, 1982), and then modified by Mandelbrot (1953) on the basis of a thorough analysis of the managing of information in a complex system.

Two features lead to the application of these models to ecology. First, some rank-frequency curves actually observed in samples, or in set of samples considered together, seem to fit correctly to these models, at least in a great part of their outline (particularly, excluding the rarest species); secondly, the mathematical analysis given by Mandelbrot (1953) is based on hypotheses concerning the optimization of an information system, and it seemed appealing to apply these hypotheses to ecology.

More recently, new investigations of Mandelbrot (1977, 1982) allow these models to be related to considerations on the "fractal structure" of an ecosystem. Below I shall devote a paragraph to this aspect, and to the perspectives of great importance that it opens in the area of ecosystem analysis, especially the successional aspects.

The original Zipf model was:

$$f_r = f_0 \cdot r^{-\gamma}$$

or

$$\log f_r = f_0 - \gamma \cdot \log r$$

where f_r is the frequency of the rth species after ranking the species in decreasing order of their frequency. The frequency is thus linked with the rank through a linear relation on a log-log basis. We have $f_1 = f_0$. The parameter γ is the slope of the straight line in the log-log graph, and is directly linked with the diversity of the sample—more precisely, with the evenness, since a low value of γ signifies a low decrease of the species abundances (hence, a more even distribution of individuals amongst species), and a high γ, a rapid decrease (hence, a more heterogeneous distribution). The relation $\Sigma f_r = 1$ has to be verified.

Mandelbrot (1953), as a result of an analysis of information and a detailed examination of experimental curves, recommends the model:

$$f_r = f_0 \cdot (r + \beta)^{-\gamma}$$

or

$$\log f_r = \log f_0 - \gamma \cdot \log(r+\beta) = \alpha - \gamma \cdot \log(r+\beta).$$

He obtains, in log-log coordinates, a curve asymptotic to the Zipf straight line, the latter being identical to the curve of Mandelbrot for $\beta = 0$; γ is then the slope of the asymptote.

In the Mandelbrot's theory, β is always $\geqslant 0$; the curve then lies below the asymptote and moves away from the asymptote in its left part, so more β is higher (Fig. 19). But negatives values of β can be introduced: the curve then lies above the asymptote and ascends rapidly along the axis of ordinates. In fact, β can only be between 0 and -1; it cannot be $\leqslant -1$ since, in this case, $(r+\beta)$ becomes negative for the first ranks, and the logarithms cannot be calculated. When β tends towards -1, the curve tends towards a limit position including a first point at $+\infty$.

The parameter $\alpha = \log f_0$ corresponds to the constraint $\sum_\infty f_r = 1$ and does not influence the shape of the curve.

Moreover, Mandelbrot demonstrated that a $\gamma < 1$ leads to a divergence of the sum $\sum_\infty f_r$, which then cannot tend towards 1; $\gamma = 1$ is the minimum value consistent with a frequency distribution model, and corresponds to a curve the

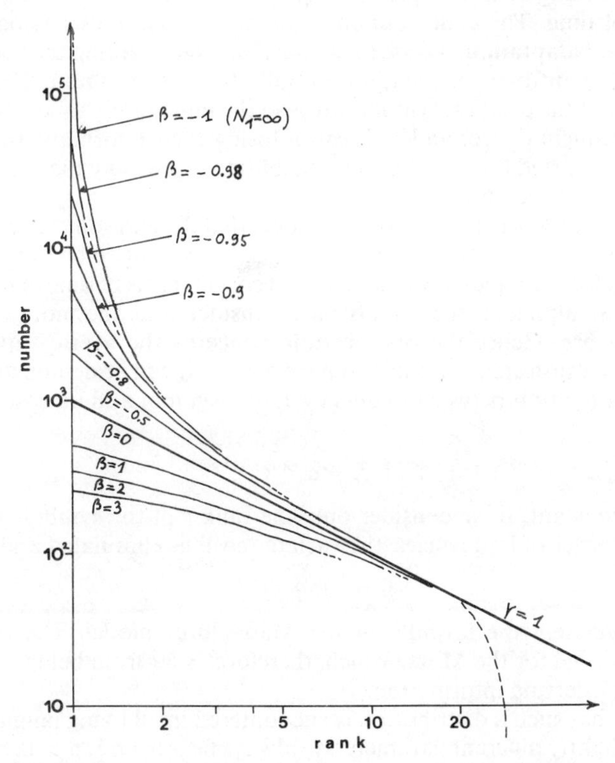

Fig. 19.—Curves for $N_r = N_0(r+\beta)^{-\gamma}$ on a log-log scale, with $N_0 = 1000$, $\gamma = 1$ and various values of β; broken line, distribution of the rare species (see comment p. 302).

least inclined as possible, and hence to a maximum diversity index (but always inferior to the maximal value $\log_2 S$, and even inferior to the MacArthur distribution index). In the observed distributions, the slopes are generally >1. In extremely evolved information transfer systems, such as natural languages, a γ included between 1 and 1·2 is found. In ecology, we find more often slopes between 2 and 3 (for all that these slopes can be easily estimated, which is not always true).

First interpretation by Mandelbrot (1953) and the influence of the observation scale

Before proposing an ecological interpretation, let us examine the logical justification of the Mandelbrot model in the area of the transmission of information. Mandelbrot (1953), recapitulated by Brillouin (1959), developed an analysis of the adaptation of a code to an information channel in the case of sequential messages such as those constituting a discourse, either in a living language or an artificial coding (Morse alphabet or Esperanto). One of the aims of the analysis was to explain the very regular rank-frequency distributions, fitting the model, amongst the words of a language.

The capacity of an information channel may be defined, in the case of sequential messages, as the maximum quantity of information it can transmit in a unit of time. The code contains a number of symbols having different "costs". The "adaptation of code to channel" signifies making the code so that the average quantity of information per unit of cost is maximum. For this, it is necessary that the costliest symbols are also the most rarely used, but without having a frequency zero, under threat of losing their informative value. The optimum is realized for a particular distribution—precisely that of the model calculated by Mandelbrot.

The notion of "cost" of a symbol in a code, that depends on the reasoning of Mandelbrot, has not to be more precise for the reasoning to be worthwhile. In the first treated examples, it was a matter of cost in terms of time of transfer; so in the Morse alphabet, the symbols are considered as the more costly, the longer they are. Hence the optimization concerns the average quantity of information transferred per unit of time in Morse. It will be demonstrated that the optimal relation between frequency f_r of a symbol and its cost t_r has the form:

$$f_r = e^{-k \cdot t_r}$$

k being a constant. If we consider only the rank r of the symbols ranked in decreasing order of frequencies, the variate "cost" is eliminated and we find:

$$f_r = f_0(r+\beta)^{-\gamma}$$

which is precisely the formula of the Mandelbrot model. The relation is wrongly realized for the Morse which, therefore, is far from being an optimal code for transferring information.

The fact that such a distribution is encountered in all living languages, and with only slightly different parameters (γ always near from 1; $\beta > 0$, but low) is to be considered. The classification of words following their frequency by no means corresponds to their lengths. Hence it is not the length of a word which, in the case of natural languages, constitutes its principal "cost", but rather any psychophysiological cost, perhaps linked to the apprehending of signals and of their significance by the brain. We actually cannot talk more precisely about this cost. The fact remains, that millenaries of evolution of this complex system today result in an optimal distribution. This is not the case of some new artificial languages such as Esperanto or Volapuk, which do not show a Mandelbrot distribution. On the other hand, the language of a child shows a Mandelbrot distribution with γ higher than by adults (1·6); the Latin language has a low γ (0·6), perhaps a β slightly negative, and a tail of rarer words (Mandelbrot, 1953).

Another interesting remark is that the law is verified in a 'natural' sub-vocabulary of a language, such as the list of surnames, or in a subset of economics such as the scales of incomes in a socio-economic category, in comparison with the distribution observed in the whole population (Mandelbrot, 1977). But the model does not fit well a more 'artificial' subset, such as the set of Latin citations in an English text. The remarks agree with the fact that, in ecology, we find distributions which confirm Mandelbrot distribution, restricting the distribution to a taxonomical category (*e.g.*, copepods, fishes) or to an ecological category, within a range of sizes (*e.g.* plankton, endobenthos, meiobenthos), that are subsystems of the community. Assuming that these categories correspond more or less to a similar way of life and insertion in the ecosystem, it will be admitted that this subsystem exerts an internal management of some ecological information. In socio-economics, many examples of Mandelbrot distributions in subsystems are given by Petruscewicz (1972).

Margalef (1957) was the first to propose fitting the Mandelbrot model to a collection of organisms; he fitted a numeration of tintinnids from the Mediterranean plankton to a Mandelbrot distribution with $\gamma = 4{\cdot}5$ and $\beta = 8{\cdot}4$ (but taking into account the tail of rarest species, which I do not recommend: see below).

I have observed possible fits to a Mandelbrot distribution in marine and lake plankton, for defined states of maturity of the ecosystem and defined scales of observation. Travers (1971), Devaux (1976, 1980), and Devaux, Millerioux & Amblard (1983) found log–log diagrams which were noticeably rectilinear, at the end of the ecological succession in phytoplankton (see Fig. 28, p. 306). Frontier & Bour (1976) described alignments, with a slope (-1), for chaetognaths of a set of samples considered together (see Fig. 10, p. 274) and Frontier (1977) with a slope (-2) for pteropods of an annual cycle off Madagascar, and also in a pteropod sediment at the bottom of the same stations (Fig. 20). In the two examples, β was very low and positive. The rarer species were excluded from the fit; the Mandelbrot model of distribution exclusively concerns the most frequent species, the remainder fitting another model (for example, the MacArthur model). In Figures 10, 20, and 28, the breakpoint initiating the different distribution of the rare species is evident.

When a straight line is not observed in a sample, it is often obtained by adding a number of samples from a same area, or from one station sampled at adequate intervals of time. Summing the number of individuals of a set of samples (taking care not to sum the frequencies), the various species of a set of samples have a distribution completely different from the distribution at a single station. We noted above that the diversity index depends on the scale of time and space at which we try to determine it; we see here that the remarks can

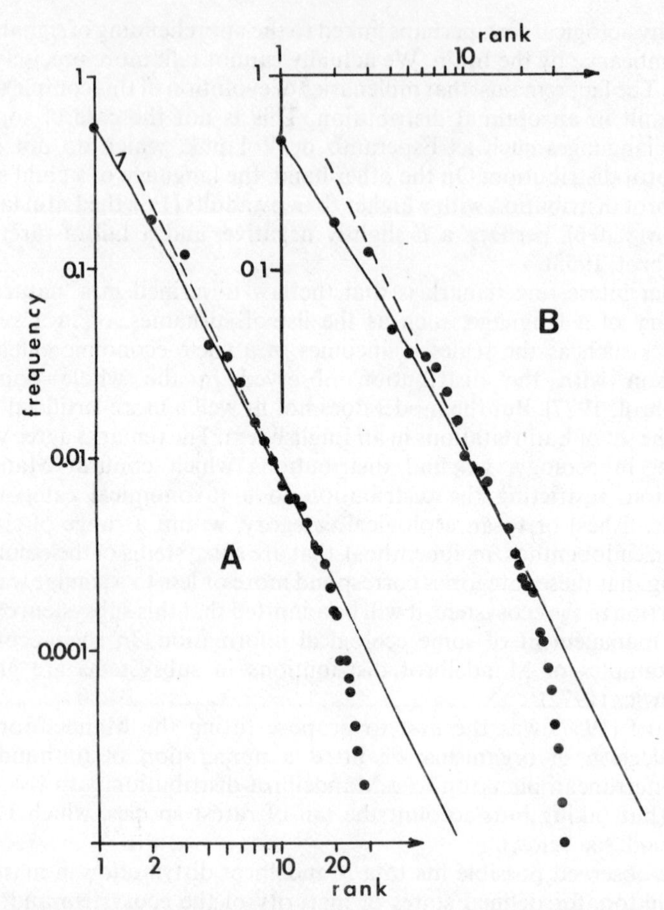

Fig. 20.—Rank-frequency diagrams for pteropods in the plankton off Madagascar: A, five offshore stations, all surface samples of an annual cycle considered together; B, census of pteropods in a sediment, 600 m deep below the same stations.

be extended to the distribution of species. We can cumulate the sample either from a relatively homogeneous area (*e.g.* Fig. 10, see p. 274) or along a gradient (*e.g.* Fig. 11, see p. 275), or along a lapse of time (*e.g.* Fig. 20). Similar features have been found in the terrestrial environment (Leprêtre, in prep.). Generally speaking, our "stage 3" (see above) is clearly related to Mandelbrot model.

It will be objected that it would not seem justified to mix samples taken at different points of a heterogeneous environment or ecosystem. It has been demonstrated that an ecosystem, that is a system of numerous short-term and long-term interactions, may be defined on different scales of space and time. The circulation of energy, matter, and information indeed can be understood on these different scales, and stochastic distributions may be manifest on a large time-and-space scale if the sampling gives a figure of the relative abundances of species on this scale. On the contrary, a single sample made in a well-delimited environment sometimes gives a wrong idea of the organization of the system, for it is too small and too local. In the example given in Figure 11 (from Berthon, 1983), already commented on (see p. 274) in the scope of diversity index, we obtain a Mandelbrot model for the 11 first species of phytoplankton cumulating the samples taken at different depths in a lake. The phytoplankton populations at various depths form a real system, for they are interconnected through the action of zooplankton, which regularly migrates across the water layers.

Sampling along a time lag, and then cumulating the samples in order to obtain a Mandelbrot distribution, seems to be less easy to justify. But an ecosystem is an evolving system; some species are successively replaced by others, and successive developmental stages of the same species play different rôles in the system. No species is independent of another species arising in the environment at another time. For example phytoplankton, at a given time, is controlled by a herbivorous population, but that herbivorous population could have been reduced earlier by a carnivore (which then disappeared), resulting in the non-destruction of the phytoplankton population. In this way, the carnivore acts upon the phytoplankton across time; taking into account the whole ecosystem in a season, a managing on the ecological information took place with time, and the pertinent description of the ecosystem as a whole includes taking into account its elements at different times. The changes, in an ecosystem, are an aspect of its diversity.

Returning to the model and its possible interpretation in ecology, the 'cost' of a species in an ecosystem is indeed equivocal. It can be an energetic cost; obviously, a predator costs more primary production than a herbivore, so it ought to be rarer. The necessary information and the necessary time-lag are concerned in the appearance of a species; each species requires a great number of previous conditions in its physical and biological environment. The more numerous the necessary previous conditions, the more information and time costs to the species. A banal species is a species whose requirements are few; on the contrary, a specialized species, or a species which appears late in the ecological succession, is a species requiring a great quantity of previous information and time-lag. Probably all these forms of 'cost' (energy, time, internal information of the ecosystem) are linked and combined, constituting all together the 'ecological cost' of the species. Despite the heterogeneity of the mechanisms involved in the cost, at certain scales of observation we find a Mandelbrot distribution, whose equation was obtained by eliminating the variate 'cost'. Since the cost is mathematically eliminated, the nature of cost does not interfere, and only the expected relation between rank and frequency remains.

The maturation of an ecosystem is an increase of its internal information (Margalef, 1968). The probability of a species provides a rough estimation of the information it added to the ecosystem, and that information is, in the model, connected to the cost of the species for the ecosystem. A more recent analysis of Mandelbrot, applicable to ecology, allows us to show what could be the cost of a species; it is the analysis of the "lexicograph tree".

Second interpretation by Mandelbrot (1977, 1982)

Let us consider that a community involves, on the one hand, banal species which require only a few previous conditions for them to develop in the ecosystem and to multiply and, on the other hand, species showing various degrees of rarity and specialization, and so demanding what have been previously realized as a great number of very specific previous conditions. Among such conditions may be, in particular, the presence of some species of the first category. The more numerous the previous conditions that a species demands, the less frequent is this species in the samples. Although this analysis is imprecise (nothing being said about the nature and the probabilities of the previous conditions), it can be demonstrated that it results in a Mandelbrot distribution.

We shall detail the pattern of the appearance of species. Let A be the first circumstance, which can be realized under the alternatives $a_1, a_2, a_3, \ldots$ (as in Fig. 21, already commented on for the lognormal distribution), whose probabilities are $\Pr(a_1), \Pr(a_2), \Pr(a_3), \ldots$. It is not necessary to know precisely these probabilities, only that their sum is 1.

Assume that the first species, S_1, a generalist and banal one, can arise when the alternative a_1 is realized; its probability is then proportional to $\Pr(a_1)$. The only additional hypothesis is that $\Pr(a_1)$ is great with respect to the probabilities of the other alternatives for A. The second species demands what will be realized, for example, as the conditions a_2 and b_1, the probability of b_1 being greater than that of all the other alternatives of B. The probability of S_2 is hence proportional to the product $\Pr(a_2) \times \Pr(b_1)$, assuming, as in the lognormal model, that A and B are independent; and so on. The probabilities of species are, of course, less and less in this series.

The main difference from the pattern given on page 285 for the lognormal distribution is that an increasing number of combinations of alternatives do not emerge since a combination of the few first conditions results in the appearance of a species, thus concealing the remaining combinations having the same beginning. Mandelbrot expresses the pattern, saying that it produces a "barren trunk" in the arborization of the possible combinations, at the level of the appearance of a species. The arborization of possible cases was called a "lexicograph tree" because it was a matter, initially, of statistical linguistics. The author demonstrates that such a pattern of generating the probabilities of species unavoidably results in a distribution following the above given model:

$$P_r = P_0(r+\beta)^{-\gamma}.$$

Figure 21B symbolizes this pattern in comparison with that resulting in a lognormal distribution. Species that are less and less banal appear gradually as we progress in the tree. This pattern illustrates at once the evolving character of an ecosystem, and the above introduced notion of cost. Indeed, for an over-particular species to exist, it is necessary that a good number of previous conditions are realized, and this can need much time to happen, whether the conditions are successively bound together or fortuitous and independent. Moreover, it is not inconsistent to assume that some of the conditions are the presence of banal species previously present, such as prey for a predator or a host for a parasite. In this case, the presence of a previous species is simply re-injected into the model as a previous condition. The general rule of evolution of ecosystems follows immediately: a community, in this meaning of the phenomena, cannot but evolve from an initial state poorly differentiated, including essentially a few banal species which agree with the ambient conditions, and moves towards more and more complex and diversified states.

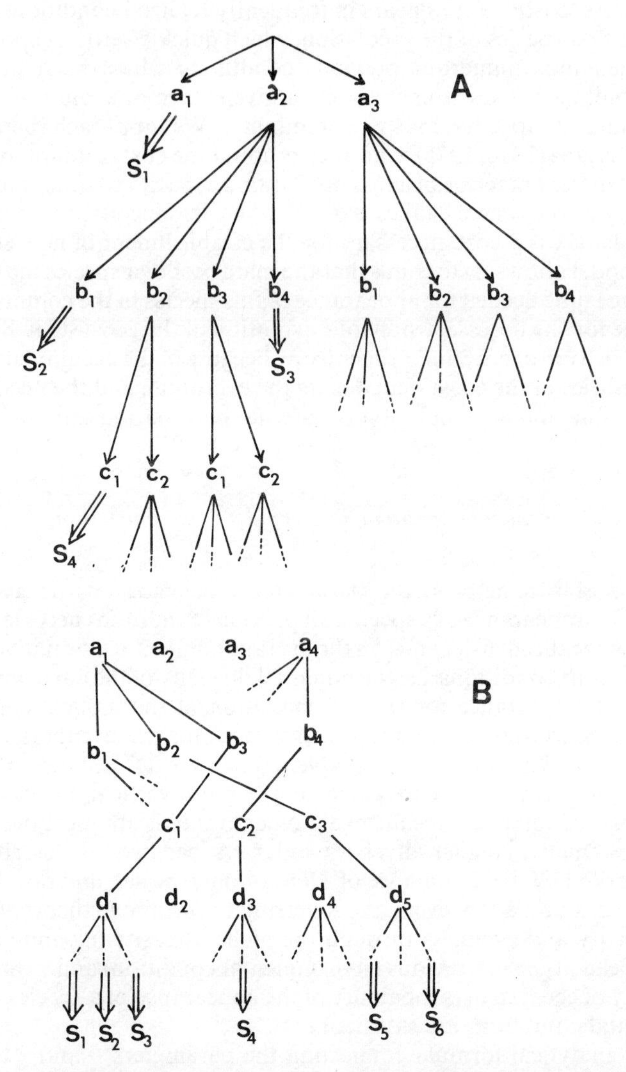

Fig. 21.—Diagrammatic representation of two models of generating probabilities of species, resulting in a Mandelbrot distribution (A) or in a Preston lognormal distribution (B): $a_1, a_2, a_3, \ldots$ are modalities of the first "previous condition", $b_1, b_2, \ldots$, of the second, etc.; $S_1, S_2, S_3, \ldots$ are the species resulting from combinations, for example in A, S_1 results from the ecological condition a_1, S_2 results from the conditions a_2, and b_1, S_4 from the conditions a_2, b_2, and c_1, etc.; it is to be seen that in B, all possible combinations can provide a species, whereas in A, only a few are functional.

Each stage of the community modifies the environment, and allows the appearance of new species.

A species is less 'costly' if it appears in frequently realized conditions, as is the case for the first species of the succession, which quickly settle. A species is costlier if it demands numerous previous conditions which need time to develop, especially a species which expects a given stage of evolution of the ecosystem before it appears in the community. We approach here the conception of Orians (1974, 1975), who speaks about the cost of adaptation to a specific environment in terms of a loss of adapting capacity to other possible environments; the ecosystem "takes a risk" when it depends on specialist species, and this risk is a cost, necessary for the establishment of new energy circuits. The model allows us to think that the main cost of a species may be a time: the average time needed for appearance of this species in the community, that is, the time for the necessary previous evolution of the ecosystem. Such a species is also costly in energy and neguentropy because of the accumulation of these two quantities in the ecosystem during the evolution, and the necessary flow of them. Time and accumulation of internal information are obviously linked.

Significance of the parameters of Mandelbrot model, and connection with diversity

In the analysis of Mandelbrot, the parameter γ depends on the average probability of the appearance of a species, all previous conditions necessary for this species being realized. β depends on the average number of alternatives per category of previous conditions, *i.e.* the potential diversity of the environment. This has a great importance for the interpretation of the model. The two parameters are conditioning the species diversity and the evenness of the community. A γ not far from 1, for example, gives a greater evenness than a higher γ; the latter gives a rank-frequency curve more vertical, hence a low evenness. A positive value of β results in a greater evenness amongst the 10 or 15 first species, then a higher diversity index. A negative β describes a community marked by the dominance of a few, or one, species, and provides a low diversity index and a low evenness. Referring to the above theory, β acts upon the diversity and evenness through the niche diversity (*i.e.* number of alternatives in each type of previous environmental condition) and γ through the previsibility of ecosystem (probability of the appearance of a species when its environmental conditions are satisfied).

There is no analytical formula connecting the parameters β and γ to the diversity index and the evenness, nevertheless we are aware, using the model, that there is a strong connection between the two sets of coefficients. Figure 22 represents the relation between the inverse of the slope, $1/\gamma$ and the evenness for different values of β; it is an almost rectilinear relation when β is positive or slightly negative. P. Legendre (pers. comm.) was already aware of a high statistical correlation between $1/\gamma$ and the evenness, for true samples fitted to the Mandelbrot model.

Mandelbrot (1977, 1982) demonstrates that the quantity $D = 1/\gamma$ is a fractal dimension; it is the dimension (< 1) of the set of species whose probabilities are generated as in the above model. The demonstration is rather arduous and brings in the Cantor's sets. Excluding the cases where β is not far from (-1), we

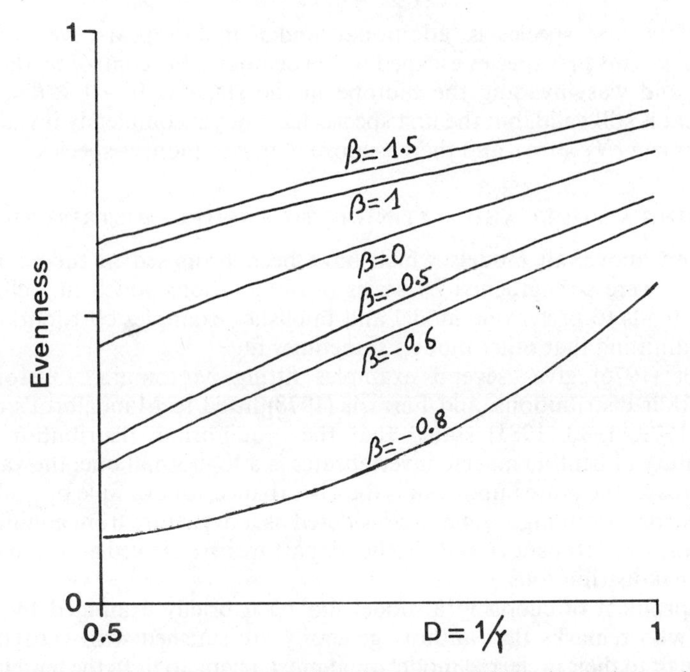

Fig. 22.—Relation between $D = 1/\gamma$ and evenness $H/H_{\max}$ for various values of β; the relation is almost linear, except for β not far from -1 (which is the uppermost negative value).

can say that the evenness represents, after a linear transformation, the fractal dimension. This conclusion is very appealing, knowing that the evenness measures the level of complexity and evolution of the community.

β may have another, purely empirical meaning. Remember that if $\beta = 0$, we have the Zipf formula:

$$f_r = f_0 \cdot r^{-\gamma}$$

which is represented, in log-log coordinates, by a straight line with slope $(-\gamma)$. When $\beta \neq 0$, we can assign to a species S_r a "fictitious rank":

$$\rho = r + \beta$$

hence we have

$$f_\rho = f_0 \cdot \rho^{-\gamma}$$

that is, the law of Zipf for the shifted ranks ρ. For example, if $\beta = 2$ we assign to the first species the "rank" 3, to the second the "rank" 4, *etc.* It results, with these shifted ranks, in a Zipf's straight line, but two first species are missing; they are recovered if the rank-frequency diagram evolves towards a curve with $\beta \simeq 0$, as observed in the stage 3 of the successions when stage 3 occurs.

On the other hand, let us suppose $\beta = -1$ (it is an extreme case resulting in an infinite N_1 number of individuals of the first species). When this first species is suppressed, *i.e.* when $\rho = r - 1$, we find a Zipf line starting from the second species; the first species is 'additional' and if it disappears, we return to Zipf's law. This first species escaped to the demographic control by the other species and was invading the biotope in the stage 1. If $-1 < \beta < 0$, the argument is still valid, but the first species has not yet completely invaded the environment ($N_1 \neq \infty$), and there is more than one pioneer species.

CHOOSING A MODEL AND FITTING IT TO A SAMPLE DISTRIBUTION

As stated above, all models which have been proposed in the ecological literature were strengthened by series of observations which fit well. Each worker tends to prefer one model and publishes examples corroborating it, while admitting that other models sometimes fit.

Daget (1976) gives several examples fitting Motomura, Preston and MacArthur distributions, and Ferraris (1978) fitted to Mandelbrot's model. Gray (1979, 1980, 1981) stated that the "equilibrium distribution" of a community of benthic marine invertebrates is a lognormal one, the variance being greater the more important is the disturbance, for example by pollution; the departure from lognormal is considered as a departure from equilibrium, and always as a transitory state. Other departures are viewed as a sum of two lognormal distributions.

The problem of choosing a model has been briefly examined by Daget (1976), who remarks that authors generally are satisfied with verifying the possible fit to their preferred model or, at most, trying to fit to the few classical models in order to see which fits the best, but rarely using a statistical test. The problem is more thoroughly analysed by Amanieu, Gonzalez & Guelorget (1981) and Lam Hoai, Amanieu & Lasserre (1984). The rank-frequency diagram being a cumulative distribution curve, the Kolmogorov-Smirvov test (*e.g.* Siegel, 1956) is convenient. But, as with all statistical tests, it only permits rejecting the hypothesis of a model when the experimental graph escapes the confidence area drawn from the Kolmogorov table. A single point of the experimental curve outside the confidence area is enough to reject the hypothesis, since the test is based on the maximum deviation observed between a theoretical and an empirical frequency. The test never permits affirmation of a model to be the right one.

A χ^2 test can also be used, but only on numbers of individuals, not on biomass. The quantity:

$$\sum_{r=1}^{r=S} \frac{(N_r - \hat{N}_r)^2}{\hat{N}}$$

theoretically permits rejecting of a model. Amanieu *et al.* (1981) remarked that the value of χ^2 considerably depends on the size of the sample, and consider this as an inconvenience of the test. It is not surprising, however, since in all tests the smaller the sample, the more difficult it is to reject a hypothesis; the χ^2 is by no way a 'distance' between a theoretical and observed distribution independent of the size of the sample. For the same shape of the frequency curves, the test leads to rejecting a model if the total number of organisms is great, not if it is small.

Another criticism of the method of testing is that the statistical tests, Kolmogorov and χ^2 tests, are not safe from rank changes of species in the sample, due to aleatory variations of abundances. When the ordering of the species is not the same in the sample and in the population, there is a bias in both tests. The ultimate criticism of the use of these tests is that they generally lead either to simultaneous rejection of all models, or acceptance of all. The usual models indeed are so designed that it is possible, by adequately choosing the values of parameters, to make them very similar to each other (see above the similarity found between MacArthur and lognormal distribution, Figs 17 and 18). Other examples concerning Mandelbrot and lognormal distribution are given below (Figs 23 and 24). In contrast, some observed distributions (particularly, curves with steps) are rejected from all models.

When a theoretical distribution can be represented by a straight line, it is generally easy to see whether an experimental alignment of points is rectilinear or not. This is the case when testing a Motomura model, linear in r and $\log N_r$, or a lognormal model, linear in probit of r/S and $\log N_r$. Nevertheless, in the latter case, there is a risk of taking into account only the species of a single sample, when the sample is small and contains few species. If the distribution of species is lognormal in the population, it will not be so in the sample because many species are missing; the lognormal distribution concerns the whole distribution of species and not a part of it. It is never certain that all species are recorded. Thus it follows that, if the cumulated probability at the level of a species of rank r is r/S, substituting the true frequencies of species by the frequencies in the sample, we make an error on each probability because the number of species in the sample is $< S$. For example, when a population with a lognormal distribution contains 100 species, if the 10th species has an abundance N_{10}, it is to say that there is 10% of the species which have an abundance $\leqslant N_{10}$. But if the sample contains only 50 species, this cumulated frequency or probability becomes 20%. Since the probit of 10% and the probit of 20% are quite different, the shape of the curve on a log-probit paper is no longer straight.

Preston (1948) speaks about a "truncated lognormal distribution" to take into account the absence of some rare species in the sample, and estimates the supposed total number of species using the symmetry of the theoretical curve. It is, of course, necessary to know the position of the mode and theoretically, if we have sampled a good number of species, the mode arises. In fact, there are two main difficulties. First, the mode is difficult to position precisely, for it corresponds to the inflexion point of the cumulated curve; the curve is approximately linear in the vicinity of its inflexion, and the true modal number (hence, the true total number of species) is thus very coarsely determined. Secondly, the distribution of the rare species often differs from that of the species of the left part of the distribution.

These inconveniences, however, have little importance when the distribution of the abundant and fairly abundant species is lognormal, including when the actual distribution distorts in its tail; the model suits the main part of the community. On the contrary, in the Mandelbrot distribution, which is not symmetrical, it is not a matter of correcting for rare species and the shape of the probit diagram differs when we take into account the number of species S' in the sample, or the number in the population $S > S'$. At best, we can compare Mandelbrot's model with the distribution of the sample without reference to a population.

When the Mandelbrot distribution is represented in a log-probit diagram, following the theoretical frequencies with a fixed $N_0 = 1000$, $\gamma = 2$ and various values of β, it is clear that in each curve with $\beta > 0$, a part of the curve is fairly linear, and would be confused with a lognormal when fitting a number of species from a true sample (Fig. 23). Figure 24 represents two rank-frequency diagrams in log-probit (the number of species S being that of the sample, and the percentage r/S transformed to probits) from a terrestrial environment. They agree with the Mandelbrot model with β low and high, γ being not far from 2 (data from Brunel, 1983); nevertheless in the upper curve, neglecting rhe numerous rare species and, in the lower curve, neglecting the rare species and also the species No. 1, the samples could be interpreted as lognormal or slightly departed from lognormal, because the main middle part of the curve is straight. It will be noted that in the Mandelbrot's model, the theoretical number of rare species is much greater than in a lognormal distribution.

Amanieu *et al.* (1981) propose a distance between two profiles, following the theoretical work of Gonzalez (1979) on this kind of data. It is the Hellinger

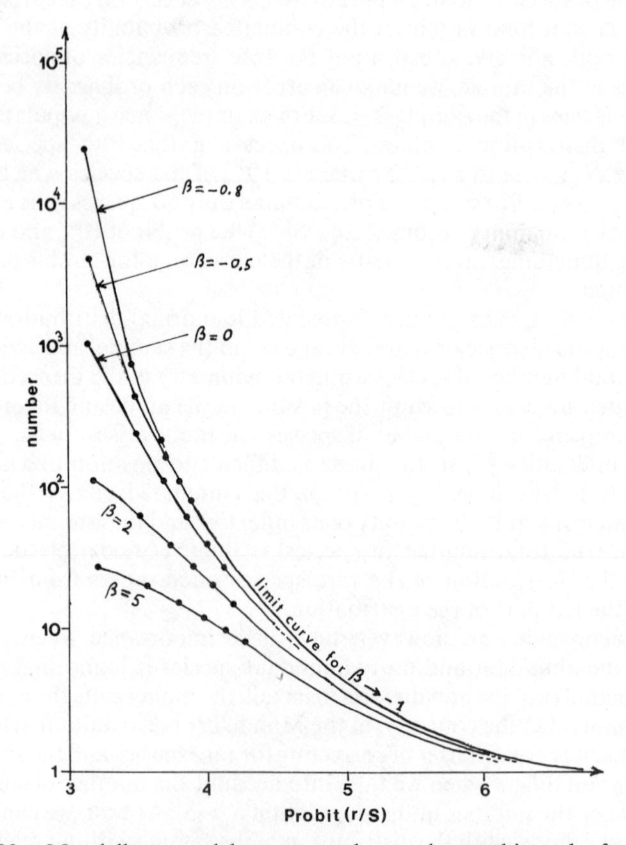

Fig. 23.—Mandelbrot model represented on a log-probit scale, for comparison with lognormal distributions, $N = N_0(r+\beta)^{-\gamma}$ with $N_0 = 1000$ and $\gamma = 2$: the calculation is made up to $N_r = 1$, resulting in a number of species varying between 27 and 33.

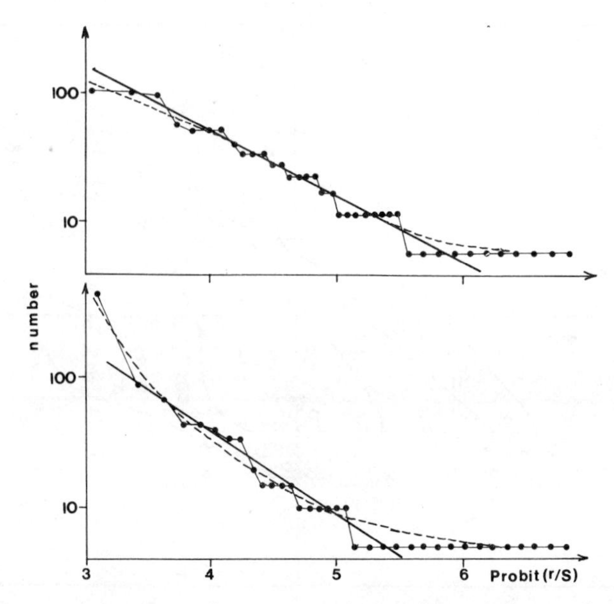

Fig. 24.—Rank-frequency diagrams on a log-probit scale for two families of insects in a terrestrial environment: broken lines, Mandelbrot model; straight lines, lognormal model; compare with the theoretical curves of Figure 23; (from Brunel, 1983).

distance:

$$D^2(i,j) = \sum_{r=1}^{r=S} (\sqrt{p_{r,i}} - \sqrt{p_{r,j}})^2$$

where $p_{r,i}$ and $p_{r,j}$ are the probabilities or frequencies of the species of rank r (the species can be different) in the profiles i and j. This distance possesses some interesting mathematical properties, and especially allows representations of distances between two curves in an euclidian space. The authors apply this metrics to a number of samples, and to their profiles after fitting three models: Motomura, Mandelbrot, and Preston. They then represent the sample points and the model points in the plane of the two first axes of a principal component analysis. They state that no one model agrees; the models seem to be nearer between themselves than for the raw data (Fig. 25). The authors also make, in the same paper, an interesting simulation of samples and find the same result and, moreover, a variation of the results with respect to the number of species actually recorded (some rare species can escape the simulated samples). Finally, the simulation being repeated 50 times starting from the same theoretical set of species frequencies, at no time did the simulated sample exactly reproduce the ordering of species in the population.

The least we can conclude is that our present catalogue of models of distributions is clearly insufficient. It is not sufficient only to observe that one of the four classical models is well realized in some circumstances and at some observation scales, when in the main cases none of these models is pertinent.

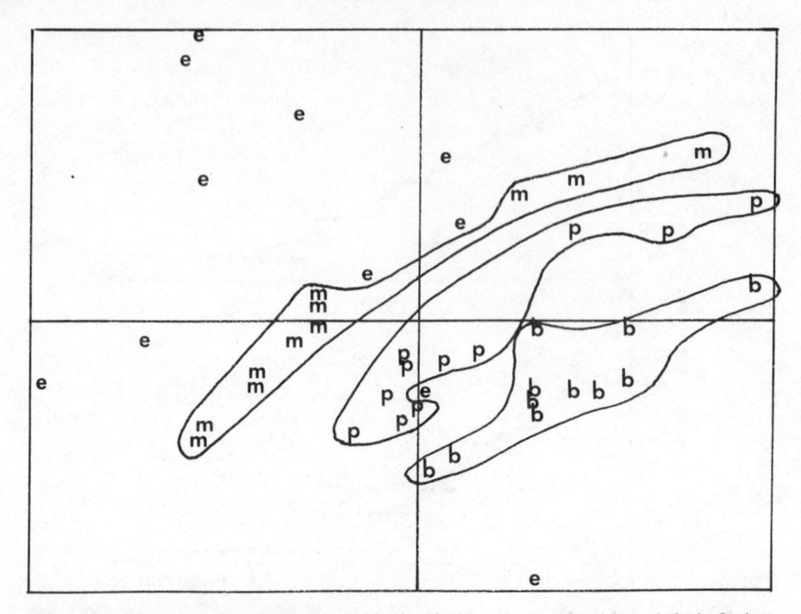

Fig. 25.—Principal component analysis of 11 true samples (e) and their fitting to Motomura (m), Preston (p) and Mandelbrot (b) models: the two first axis extract 84% of the inertia, and distances between points in the plane are the projection of distances, in a particular metrics; it can be noted that the adjustments to the three models form three parallel sets of points, whereas the non-transformed samples are far from the three models; (from Amanieu *et al.*, 1981).

CAN OBSERVATIONS FIT COMPOSITE OR INTERMEDIATE DISTRIBUTIONS?

Each model is based on hypotheses concerning the underlying patterns of the sharing of ecological space amongst species. It is obvious that all proposed hypotheses oversimplify the reality.

One of the main simplifications is that all species behave in the same way, particularly the rare and the dominant ones. In fact, it is sometimes found that a part of the sample distribution fits one law, and another part another law. For example, several Motomura distributions can follow one another, with a change in slope (see Fig. 14, p. 281). Other examples are given above: the distribution of the rarest species often follows another distribution, for example a MacArthur's, but this may be an effect of the sampling. It seems necessary to eliminate them before fitting. But this elimination is made by the computation when the models are fitted by means of a no linear least-squares method (Lam Hoai *et al.*, 1984) applied to the raw data (ranks and frequencies), and not to their logarithms, which are used in the graphs. In the non-transformed frequencies, the absolute differences (and their squares) between the experimental data and the model are much greater in the high frequencies than in the low ones; it follows, since the method minimizes the sum of squares of these differences, that the fitted model principally depends on the high frequencies, and almost never on the lower ones. As a result, plotting the fitted model on a log-log scale, it appears that the curve fits the frequent species very well and is independent of the rarest ones, as in Figure 26: the elimination of the rarest species occurred automatically. The curve obtained of course, does not fit the observed distribution on the whole, but only its left part, and the right part does not agree with a Mandelbrot model. We can see that, following the model, the rare species ought to be much more numerous than in the actual sample. A lot of rare species do not 'exist', either in the community, or in the sample owing to the sampling method.

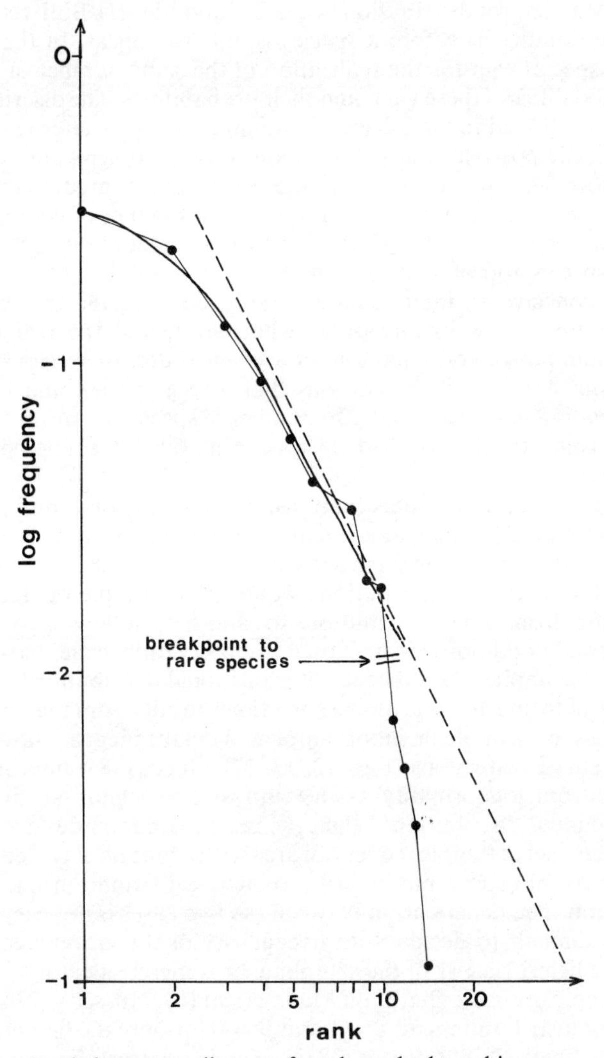

Fig. 26.—Rank-frequency diagram for phytoplankton biomasses over an annual cycle in Lake Aydat, France: fitting a Mandelbrot distribution after eliminating the rare species, that fit another distribution; $\gamma = 2$, $\beta = 1{\cdot}47$, $f_0 = 1{\cdot}778$; (data from Devaux *et al.*, 1983).

In addition, when assuming that the mechanism of ecological distribution is the same for a set of species, the invoked patterns are too simple. Such a criticism has often been made against modelling, but the fact that one or another is occasionally very well verified proves that the situation is not so entangled as originally thought (or quite the contrary, that it is entangled in such a manner that a statistical law emerges).

Nevertheless, composite situations may exist: either the pattern changes from the left to the right of the distribution, or it is intermediate between two simple models. For example, let us examine the two patterns of generating probabilities, the one resulting in a lognormal distribution (see p. 285), and the other in a Mandelbrot distribution (see p. 294 and Fig. 21). Both refer to the set of previous conditions before a species is able to appear. In the lognormal model, all species wait for the realization of the same number of conditions, only the modalities of these vary, and their probabilities. The discrimination of species is only linked to the different combinations of the alternatives. All the mathematically possible combinations are considered here, and no timing of them is specified. In contrast, in the Mandelbrot model, the previous conditions are awaited in a given order; a great number of combinations do not arise since, in the "tree" of the possible cases, a trunk becomes "barren" as soon as a species appears.

We can conceive an intermediate pattern. Between the realization of all possible combinations in any order whatsoever, and the realization of a restricted number of combinations in a given order, there can occur: (1) a combination of a few first conditions that give a species, and nevertheless the continuation of the tree; and (2) a number of species arising at the extremities of "barren" trunks, and others arising at the extremities of complete combinations.

We have to expect, in such situations, intermediate distributions between Mandelbrot and lognormal, or else a distribution not yet modelled. Perhaps a general situation is to be sought in a linear system of distributions, in which the simple so far described distributions would play the rôle of eigenfunctions. Many shifts from one distribution to another, following a change in environmental conditions or a change of observation scale, have also been described; this implies the existence of transitional distributions.

The lack of fitting any one model sometimes results from the mixing of two communities in a sample or set of samples. We have then, as shown in above examples, curves with steps (Figs 12, 13, 27). Gray (1981) interprets certain departures from lognormality as the sum of two lognormal distributions, giving a bimodal distribution. When the mixed communities are more than two, or when a set of samples covers an area with a mosaic of two communities, the rank-frequency curve may describe a concrete distribution at a higher level of integration. The demarcation between the two cases is not easy to see; it is sometimes difficult to decide if an irregularity of the graph signifies a step proving the heterogeneity of the sampling, or is merely aleatory.

The example of chaetognaths off Madagascar (Fig. 10, see p. 274) suggests a Mandelbrot distribution with $\gamma = 1$ and $\beta = 0$ for only the first eight species, then a sharp break-point towards rare species. In contrast, Figure 27 shows an alignment of the first six species, which seems to be a step and not a part of a Mandelbrot distribution since the slope is < 1, which is not consistent with the model (see p. 289). The nine following species as a whole fit a MacArthur

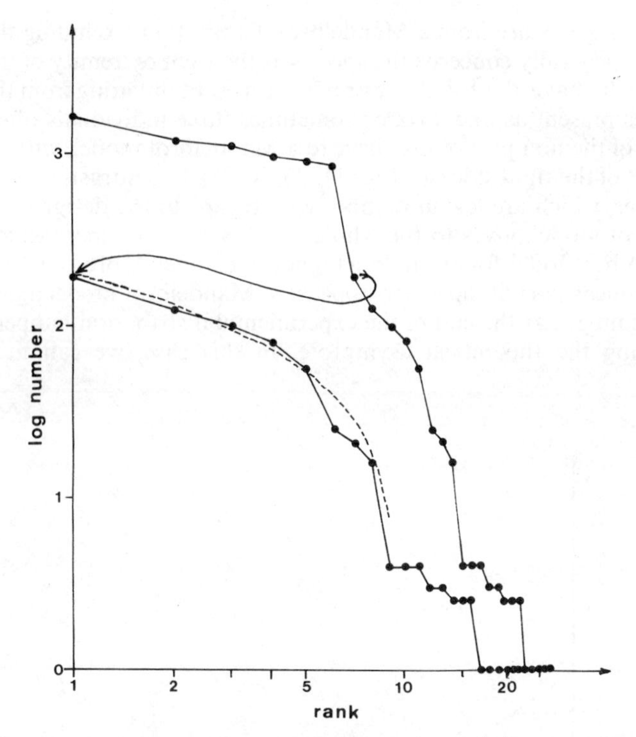

Fig. 27.—Rank-frequency diagrams for the total sample for a year in a pond in Morocco: the species 7 to 15 fit a MacArthur distribution (broken line); (data from Boutin *et al.*, 1982).

distribution; the remaining species do not follow any representable distribution. This example concerns the total sampling of zooplankton in a Moroccan pond during a year. The dominant species is not the same throughout the year and, mixing the samples taken each month, a quasi-equal frequency of the six dominant species is obtained, which explains the step (data from Boutin, Lesne & Thiery, 1982). On the other hand, we should compare the distribution with a Mandelbrot with $\gamma \simeq 2{\cdot}3$ (rather a good fit for species 7 to 27) and β high, which explains the curvature of the left part of the diagram, and aleatory fluctuations resulting in a flatness of this left part. We get here a clear idea of the complexity and the very hazardous character of the problem of fitting an experimental curve to a particular model. Before trying to choose and fit a model we have, of course, to examine the empirical curve, asking ourselves what is the representative character of the sample (that is, what represents a phenomenon) and what is aleatory.

RETURNING TO THE MANDELBROT MODEL

Despite the above criticism about the aim of seeking a model, cost what it will, I believe that the Mandelbrot model, so far little considered by the ecologists, has some general use.

The main departure from a Mandelbrot distribution, excluding the curves with steps, generally concerns the species at the right extremity of the graph, and to which I have alluded already several times. Eliminating from the census the species present as one, two, or sometimes three individuals often makes clear a fit of the first part of the curve to a Mandelbrot model, with a distinct linear part of the right side (see Figs 11, 20, 26, 28). In contrast, preserving the rare species, which are few in number with regard to the designed one from Mandelbrot model, gives to the whole graph a convex aspect which would justify another model, for example, a lognormal model. Moreover, in the cases where no linear part of the graph appears, a Mandelbrot model again can be used, assuming that the end of the experimental distribution happens before approaching the theoretical asymptote. In this case, we can fit either a

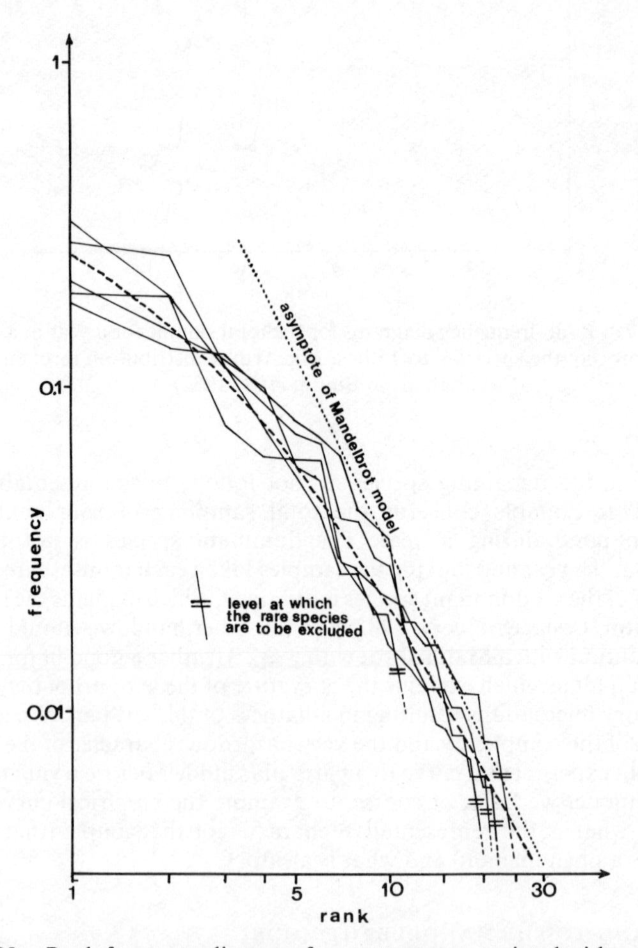

Fig. 28.—Rank-frequency diagrams for crustaceans associated with pocillo-porid corals from five Pacific atolls (the rarest species have been removed because they fit another distribution, see Fig. 19); broken line, Mandelbrot model with $\gamma = 2$, $\beta = 2{\cdot}55$, $p_0 = 3{\cdot}254$; (data from Odinetz, 1983).

lognormal or a Mandelbrot distribution. No statistical test will reject both models, which are a matter of choice depending on the hypotheses about the ecological pattern. The proximity between Mandelbrot and lognormal distributions was also shown by utilizing a probit transformation on ranks, as seen in Figure 24 (see p. 301).

Comparing the patterns of generating the species probabilities in the Mandelbrot and the lognormal model (Fig. 21, see p. 295), we can see that the second results in a limited number of species for a fixed number of previous conditions (for example, if there are four previous conditions and three alternatives of each, we have $3^4 = 81$ combinations, hence 81 possible species, among which only a few arise in the Mandelbrot model). In contrast, the Mandelbrot model tends to infinity: the asymptote has no limit. The fact that, in the true ecosystems, there are a limited (even if great) number of species implies that the Mandelbrot model no longer suits the rare species. Hence, the fractal model never fits entirely, but only fits over a range of abundance and rarity of species.

On page 259 I emphasized the inconvenience which arises from the rare species in the calculation of a diversity index; they provide a background noise which conceals the significant variations of the diversity of well-sampled species. A solution would be to calculate the index without these rare species and to take into account only those species having a frequency fairly well estimated by the sampling. (In all the cases, rarest species are omitted from the calculation because they are absent in the sample.) The evenness should also be so calculated. The problem is how to choose the limit between rare and frequent species, and it should be solved by examining the rank–frequency diagram on log-log scale. The breakpoint often clearly appears between the two kinds of species, referring to the Mandelbrot model. It would be interesting to calculate again some long series of diversity indices such as those of Dessier (1979) (Fig. 1, p. 258) and, after tracing the rank-frequency curves, to localize the rare species and to eliminate them, in order to see whether the background noise has been significantly reduced.

Another criticism of the Mandelbrot model, as above applied, concerns the cases where β is negative (more precisely $-1 < \beta < 0$), describing curves located above the asymptote and rising toward high values of abundance for a few first species. Negative values of this parameter do not appear in the analysis of Mandelbrot (1953), although certain figures indicate a curve rising slightly above the asymptote, and they cannot be interpreted in the model of generating probabilities (Mandelbrot, 1977) since β is linked with the average number of alternatives in every previous condition. I arbitrarily introduced this $\beta < 0$ in order to generalize the descriptive model to the "stage 1" curves, where few species attain high frequencies. It should also be noted that no other model describes a stage 1, which is always indicated as a departure from the model.

$\beta < 0$ cannot, however, describe all "stage 1" curves. Often an S-shaped rank–frequency curve shows an inflection point with a slope much weaker than the slope of the asymptote for other stages of the community. Sometimes the slope at the inflection is <1 and cannot agree with the model (see above). In the Mandelbrot model extended to $\beta < 0$, the inflection ought to coincide with the asymptote with a part of the curve below the asymptote due to rare species, as seen in Figure 19 (see p. 289). If the slope at the inflection point is assumed to be equal to the asymptotic slope, then it is often found too weak in experimental curves, and another interpretation has to be found.

Another interpretation is a Mandelbrot model with $\beta \geqslant 0$, but starting from the second or third species instead of the first of the sample. I said above (p. 297) that we can imagine a "fictive rank" $\rho = r+\beta$, and then the new ordering of species follows an exact Zipf model. The first species justifying $\beta < 0$, hence appears as an "extra-species", escaping the overall demographic control and no longer fitting to a Mandelbrot model; but a Mandelbrot distribution is still realized for the subsequent species.

OTHER METHODOLOGICAL PROBLEMS

The selection of the descriptive variate is a matter of researcher's initiative, as it is also for the diversity index and the rank-frequency diagrams. The main constraint is that the sampled organisms have sizes of the same order of magnitude, since we cannot calculate a diversity index for a number of copepods and baleen whales. Now most of the sample methods are size-selective and the size of organisms of a sample made with a trawl net, or a plankton net of a given mesh size, or a closing bottle, does not vary excessively.

In order to make the abundances of species more comparable, it was proposed to utilize the biomasses. Devaux (1980) stated that, in lake phyto-plankton, the biomasses (more precisely, the biovolumes calculated on the basis of number and dimensions of cells) often give rank-frequency diagrams smoother and easier to interpret than those with the mere number of cells (Fig. 26, see p. 303). In other cases, it is the opposite (Cuisinet & Moguedet, 1983, for marine benthos). It is possible that biomasses and numbers of individuals do not follow the same law. Two ecological significances of diversity can be considered; one taking into account, as the most important signal, the presence of individuals of species, emphasizing the predictability of the ecosystem; the other, considering the biomass (or the energy) and referring to the necessary variety of energy flows in an energetic network. The two viewpoints are worthwhile and depend on the aim of the scientist.

The choice of the target-group also depends on the aim of the investigation. It seems justified to apply the rank-frequency method to taxonomic groups, that are subsets of the community, as in statistical linguistics when applied to subvocabularies (p. 291). At any rate, it would be impossible to describe the whole ecosystem altogether because of the heterogeneity of sizes of organisms. We are constrained to describe only the structure of particular taxocoenoses. Structures of various taxocoenoses of the same ecosystem often coincide. The distributions and the evolution of distributions of copepods and chaetognaths of the same sample of plankton are alike. Margalef (1970) reported an exception: planktonic diatoms have a pattern different from that of the other main groups, the maximum diversity taking place when other groups are less diversified. The reason is that this group as a whole is adapted to an environment influenced by terrigenous material, and diversifies itself in it, whereas other groups are stressed there and regress in their structure.

Partitioning the biomass into taxonomic groups other than species should also be considered. It seems plausible, owing to the hierarchical organization of ecosystems, that this functional hierarchy is more or less reflected in the taxonomic hierarchy and also in the hierarchy of sizes, diets and other biological characters. In fact, it is necessary to investigate the diversity and structure between taxonomic or ecological sets of species, as well as within them.

CONCLUSIONS

Rank-frequency diagrams, which express the distribution of individuals amongst species in natural communities, are to be first considered as empirically describing the diversity in a better manner than simple diversity indices, since they contain more information than a mere number. The high random variability of the diversity index often conceals interesting and significant variations; on the contrary, the shape of a rank-frequency curve can be seen despite some random fluctuations of abundance of the individual species.

It has been shown that rank-frequency diagrams are appealing to follow ecological situations and their changes, such as natural successions (see Figs 6, 7), stresses by pollution (see Figs 8, 29) or by over-exploitation, senescence or

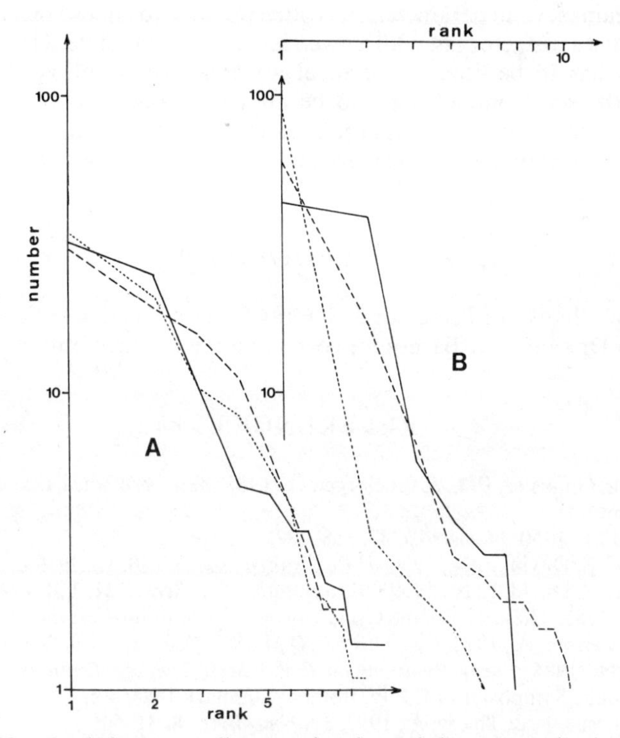

Fig. 29.—Rank-frequency diagrams for the *Ophelia* community (endobenthos in the Channel): A, offshore; B, in a polluted area (near Dunkerque); (from Prygiel, 1983).

contacts between ecosystems (Frontier, 1978, 1984). Characteristic deformations of the curves happen and may constitute a worthwhile test of the actual condition of an ecosystem, which cannot always be found in the mere census of species because it shows, in addition, changes in the reciprocal proportions of species.

The models of distributions hitherto proposed still are insufficient in number and complexity, and are too 'manichaean' because they involve a single hypothesis for each one. Nevertheless, it is of great interest to observe the evidence that sometimes one of these simple models agrees with a set of true data. At other times, it ought to be possible to combine them. We have not to be too amazed in discovering that some distributions fit well; the fact is that the underlying hypotheses of these distributions are very abstract and very general. The patterns of generating the probabilities of species are defined without any reference to the concrete mechanisms; they handle only a law of determining successive or simultaneous probabilities of elements of a set. Mandelbrot (1963) demonstrated that the distributions leading to his model are very robust with respect to the highly variable "filter function", that is, the function through which we are acquainted with these distributions thanks to a sampling or measuring procedure.

It follows from this that models cannot supply much information about the concrete mechanisms that determine the observed distribution. Many hypotheses lead to the same distribution of probabilities. It is well known that if a model contains several parameters, it is often possible to choose them in such a manner that a satisfactory fit with a given set of data is possible. The model is a tool which has to be used in a critical way and with caution. This being admitted, the set of models (still to be improved) ought to be a powerful method of analysing and comparing ecosystems, the final aim being an acknowledgement of the statistical organization of them in various conditions.

ACKNOWLEDGEMENTS

I address my thanks to Professor J. S. Gray for constructive criticism of this text and to Dr Margaret Barnes for correcting my English draft.

REFERENCES

Amanieu, M., Gonzalez, P. L. & Guélorget, O., 1981. *Acta oecologica, Oecol. gener.*, **2**, 265–286.

Anscombe, F. J., 1950. *Biometrika*, **37**, 358–382.

Barton, D. E. & David, F. N., 1956. *Jl Roy. statist. Soc. Ser. B*, **18**, 79–94.

Barton, D. E. & David, F. N., 1959. *Jl Roy. statist. Soc. Ser. B*, **21**, 190–194.

Berthon, C., 1983. Thèse Troisième Cycle Université Clermont-Ferrand II, 91 pp.

Binet, D. & Dessier, A., 1972. *Cah. O.R.S.T.O.M., Sér. Océanogr.*, **10**, 243–250.

Blanchard, N., 1984. *Fuzzy Relations on Fuzzy Sets: Two Applications to Ecology.* Communic. Symposium I.E.E.E., Bombay, January 1984, 4 pp.

Boutin, C., Lesne, L. & Thiery, A., 1982. *Ecol. mediterr.*, **8**, 31–56.

Bowman, K. O., Hutcheson, K., Odum, E. P. & Shenton, L. R., 1969. In, *International Symposium on Statistical Ecology*, Pennsylvania State University Press, Philadelphia, pp. 3–10.

Brian, M. V., 1953. *J. anim. Ecol.*, **22,** 57–64.
Brillouin, L., 1959. *La Science et la Théorie de L'information.* Masson, Paris, 302 pp.
Brunel, C., 1983. D.E.A. Université Lille I, 105 pp.
Cohen, J. E., 1968. *Am. Nat.*, **102,** 165–172.
Cuisinet, H. & Moguedet, P., 1983. D.E.A. Université Lille I, 195 pp.
Daget, J., 1976. *Les Modèles Mathématiques en Écologie.* Masson, Paris, 172 pp.
Dessier, A., 1979. Thèse Doct. Etat Université Paris VI, 274 pp.
Devaux, J., 1976. *C.-r. Acad. Sci. Paris, Sér. D,* **282,** 1499–1501.
Devaux, J., 1977. *Ann. Stat. biol. Besse-en-Chandesse,* **10** (suppl.), 1–185.
Devaux, J., 1980. *Acta oecologica, Oecol. gener.*, **1,** 11–26.
Devaux, J., Millerioux, G. & Amblard, C., 1983. In, *Stratégies D'échantillonnage en Écologie,* edited by S. Frontier, Masson, Paris, pp. 291–310.
Diaz-Castañeda, V., 1984. Thèse Troisième Cycle Université Marseille II, 180 pp.
Dujet, C., 1984. *Application of the Separating Power of a Fuzzy Set to Ecological Systems.* Communic. Sympos. I.E.E.E., Bombay, January 1984, 4 pp.
Ferraris, J., 1978. Mém. Ingénieur Université Montpellier II, 108 pp.
Frontier, S., 1969. *Doc. sci. Centre O.R.S.T.O.M. Nosy Be,* **7,** 33 pp.
Frontier, S., 1976a. *J. Rech. océanogr.*, **1,** 35–48.
Frontier, S., 1976b. *J. exp. mar. Biol. Ecol.*, **25,** 67–75.
Frontier, S., 1977. *Bull. Ecol.*, **8,** 445–464.
Frontier, S., 1978. *Ann. Inst. océanogr. Paris,* **54,** 95–106.
Frontier, S., 1981. *Méthode Statistique. Applications à la Biologie, la Médecine et l'Écologie.* Masson, Paris, 242 pp.
Frontier, S., 1983. Editor, *Strategies d'Échantillonnage en Écologie.* Masson, Paris, 578 pp.
Frontier, S., 1984. *Vous avez dit écosystème?* Unpubl. preprint, 52 pp.
Frontier, S. & Bour, W., 1976. *Cah. O.R.S.T.O.M., sér. Océanogr.*, **14,** 267–272.
Glansdorff, P. & Prigogine, I., 1971. *Structure, Stabilité et Fluctuations.* Masson, Paris, 288 pp.
Gonzalez, P. L., 1979. D.E.A. Université Montpellier II, 67 pp.
Gray, J. S., 1979. *Phil. Trans. R. Soc. Ser B,* **286,** 545–561.
Gray, J. S., 1980. *Rapp. P.-v. Réun. Cons. perm. int. Explor. Mer,* **179,** 188–193.
Gray, J. S., 1981. *Mar. Pollut. Bull.*, **12,** 173–176.
Gray, J. S. & Mirza, F. B., 1979. *Mar. Pollut. Bull.*, **10,** 142–146.
Gray, J. S. & Pearson, T. H., 1982. *Mar. Ecol. Prog. Ser.*, **9,** 111–119.
Hicks, G. R. F., 1980. *J. exp. mar. Biol. Ecol.*, **44,** 157–192.
Hily, C., 1983. *Ann. Inst. Océanogr. Paris,* **59,** 37–56.
Hily, C., 1984. Thèse Doct. Etat Université Brest, 2 volumes, 326 and 389 pp.
Horn, H. S., 1975. In, *Ecology and Evolution of Communities,* edited by M. L. Cody, Harvard University Press, London, pp. 196–211.
Inagaki, H., 1967. *Vie Milieu, Sér. B,* **18,** 153–166.
Jones, D. D., 1977. *Simulation,* **29,** 1–15.
King, S. E., 1964. *Ecology,* **45,** 716–727.
Lam Hoai, T., Amanieu, M. & Lasserre, P., 1984. *Acta oecologica, Oecol. gener.*, **5,** 159–167.
Lecal, J., 1965. *Vie Milieu, Sér. B,* **16,** 251–260.
MacArthur, R. H., 1957. *Proc. natn. Acad. Sci. U.S.A.*, **43,** 143–190.
MacArthur, R. H., 1960. *Am. Nat.*, **94,** 25–36.
Mandelbrot, B. B., 1953. *Publ. Inst. Stat. Univ. Paris,* **2,** 1–121.
Mandelbrot, B. B., 1963. *Intern. Econ. Rev.*, **4,** 111–115.
Mandelbrot, B. B., 1977. *Fractals. Form, Chance and Dimension.* W. H. Freeman and Co., San Francisco, 365 pp.
Mandelbrot, B. B., 1982. *The Fractal Geometry of Nature.* W. H. Freeman and Co., San Francisco, 468 pp.
Margalef, R., 1956. *Invest. pesq.*, **3,** 99–106.

Margalef, R., 1957. *Mem. real Acad. Ciencias Artes Barcelona*, **32,** 373–449.
Margalef, R., 1967. *Oceanogr. Mar. Biol. Ann. Rev.*, **5,** 257–289.
Margalef, R., 1968. *Perspectives in Ecological Theory.* University of Chicago Press, Chicago, 111 pp.
Margalef, R., 1970. In, *Seminario de Ecologica Matematica. Invest. pesq.*, **34,** 65–72.
Margalef, R., 1979. *Oikos*, **33,** 152–159.
Margalef, R., 1980a. *Ecologia.* Ediciones Omega, Barcelona, 951 pp.
Margalef, R., 1980b. *La Biosfera.* Ediciones Omega, Barcelona, 236 pp.
May, R. M., 1971. *Math. Biosci.*, **12,** 59–79.
May, R. M., 1972. *Nature* (*Lond.*), **238,** 413–414.
May, R. M., 1973. *Stability and Complexity in Model Ecosystems.* Princeton University Press, Princeton, N.J., 225 pp.
Mirza, F. B. & Gray, J. S., 1981. *J. exp. mar. Biol. Ecol.*, **54,** 181–207.
Morowitz, H. J., 1968. *Energy Flow in Biology.* Academic Press, New York, 179 pp.
Motomura, I., 1932. *Zool. Mag.* (*Tokyo*), **44,** 379–383.
Motomura, I., 1947. *Physiol. Ecol.*, **1,** 55–60.
Odinetz, O., 1983. Thèse Troisième Cycle Université Paris VI, 221 pp.
Orians, G. H., 1974. In, *Proc. 1st int. Congr. Ecology* (The Hague, Sept. 1974), edited by Centre for Agricultural Documentation, Wageningen, Netherlands, pp. 64–65.
Orians, G. H., 1975. In, *Unifying Concepts in Ecology*, edited by W. H. van Dobben & R. H. Lowe-McConnell, Junk B.V. Publishers, The Hague, pp. 139–150.
Patrick, R., 1972. *Trans. Conn. Acad. Arts Sci.*, **44,** 281–282.
Petruscewicz, M., 1972. *Informatique et Sciences humaines*, **14,** 1–20.
Pielou, E. C., 1975. *Ecological Diversity.* Wiley-Interscience, New York, 165 pp.
Preston, F. W., 1948. *Ecology*, **29,** 254–283.
Preston, F. W., 1962. *Ecology*, **39,** 185–215.
Prigogine, I., 1980. *Physique, Temps et Devenir.* Masson, Paris, 275 pp.
Prygiel, J., 1983. D.E.A. Université Lille I, France, 51 pp.
Roger, C., 1974. *Mém. O.R.S.T.O.M.*, **71,** 1–265.
Shannon, C. E. & Weaver, W., 1963. *The Mathematical Theory of Communication.* University of Illinois Press, Urbana, Illinois, 117 pp.
Siegel, S., 1956. *Nonparametric Statistics.* International Student Edition, McGraw-Hill, London, 312 pp.
Thom, R., 1972. *Stabilité Structurelle et Morphogénèse.* W. A. Benjamin, Paris, 315 pp.
Thom, R., 1980. *Modèles Mathématiques de la Morphogénèse*, Christian Bourgeois, Paris, see pp. 9–35.
Travers, M., 1971. *Mar. Biol.*, **8,** 171–178.
Ugland, K. I. & Gray, J. S., 1982. *Oikos*, **39,** 171–178.
Whittaker, R. H., 1972. *Taxon*, **21,** 213–251.
Williams, C. B., 1964. *Patterns in the Balance of Nature and Related Problems in Quantitative Ecology.* Academic Press, London, 324 pp.
Zipf, G. K., 1949. *Human Behaviour and the Principle of Least Effort.* Hafner, New York, 2nd edition 1965.

Oceanogr. Mar. Biol. Ann. Rev., 1985, **23**, 313–371
Margaret Barnes, Ed.
Aberdeen University Press

TRANSOCEANIC AND INTEROCEANIC DISPERSAL OF COASTAL MARINE ORGANISMS: THE BIOLOGY OF BALLAST WATER

JAMES T. CARLTON[1]
Department of Biology, Woods Hole Oceanographic Institution, Woods Hole, MA 02543, U.S.A.

INTRODUCTION

Between Ernst Haeckel's comment "Die weitaus wichtigsten von allen Ursachen, welche die wechselnde und ungleichmässige Vertheilung des Plankton in Meere bedingen, sind die Meeresströmungen" and summary in 1890 of the processes determining the distribution of plankton in the oceans and Charles Elton's implication in 1958 that other forces may be increasingly at work—"... we can discern the setting in of a very strong historical move, the interchange of the shore fauna of continents, and also sometimes the plankton of different seas. It is only an advance guard . . .", systematists and biogeographers began to seek other than natural explanations to account for the appearance of neritic planktonic organisms in coastal regions far from their presumed native shores. A phenomenon increasingly called into play to explain such appearances has been the transport of marine organisms in the sea-water ballast tanks of ocean-going vessels. Although the rôle of water ballast in trans- and interoceanic dispersal has been speculated upon and invoked for more than 75 years, there has been no previous review of this subject.

I review here the history of biogeographic and ecological interest in ballast water as a dispersal mechanism, the hypothesis of ballast-water transport, and the evidence for such transport. A model is presented in which the sequential events of ballast-water transport are broken into divisions capable of investigation. The potential significance of such transport—as the sole mediator of dispersal of certain neritic taxa in directions, across barriers, and over time scales distinct from that which could be achieved by natural current systems—is explored here.

[1]Present address: Williams College, Program in American Maritime Studies, Mystic Seaport, Mystic, CT 06355, U.S.A.

BALLAST-WATER TRANSPORT

Ballast-water transport is the movement of living organisms in the water ballast tanks of vessels. Such movement can be along coasts, into inland waterways and lakes, along and through canals, and across and between oceans. The water may be fresh, brackish, or marine. The organisms are those species in the water column. These include nekton, holoplankton, meroplankton, demersal zooplankton, and tychoplankton *i.e.* benthic species which are passively swept up into overlying waters by water turbulence, such as wave action and tidal currents (Gerlach, 1977; Thomas & Schafer, 1982; Palmer, 1984) or by the action of a vessel's propellers. Pleuston and neuston (Banse, 1975; Cheng, 1975; Champalbert, 1981) are probably not generally entrained into ballast systems, due to the subsurface intake of water into vessels (described below). Organisms entering ballast tanks either remain in suspension in the water or come to rest on the bottom or sides of the tanks. Fouling communities inside ballast tanks, from which larvae could be derived, are not known to exist.

The use of water as ballast overlapped and then succeeded the use of "dry" or "wet" ballast. This consisted of any heavy material, terrestrial or maritime, loaded and unloaded into and out of ships' holds (Dana, 1840; Lindroth, 1957; Carlton, 1979a). The rôle of such ballast in the dispersal of organisms is not reviewed here, but has been considered by Lindroth (1957) for some terrestrial species in the North Atlantic Ocean (and to a lesser extent in the Northwest Pacific Ocean), and has been mentioned in numerous botanical works since at least the nineteenth century. Beach sand and "shingle" ballast (round beach pebbles, cobbles and boulders), and concomitantly beach detritus (such as drift algae) were used for centuries as ballast. The world-wide movement of such materials may explain the widespread distribution of many species of strand animals (such as maritime insects and talitrid amphipods), as well as of various meiofaunal taxa, but has received only limited attention (Gerlach, 1977; Carlton 1979b).

Also not considered here is ballast water transported in seaplanes, about which little is known. Wirth (1947) speculated that the salt pond brine-fly (Diptera) *Ephydra gracilis* was carried from San Francisco Bay, California to Hawaii by seaplanes. Arnal (1961, p. 469) "favored" the hypothesis that transport of foraminiferans "in the bilge water of seaplanes" during World War II was a partial explanation for the similarity of the sarcodinid fauna of the Salton Sea, in southern California, and the Gulf of Mexico.

BALLAST-WATER SYSTEMS

The existence of tens of thousands of vessels with water-ballast holding capacity precludes more than a generic description of ballast-water systems. The following descriptions are based upon discussions with maritime industry personnel and upon my own experience with (and in) such systems.

Ballast tanks range in capacity from hundreds of gallons to hundreds of thousands of gallons of water. Their number and size vary with vessel type. The generally steel walls, floor, and ceiling of a tank may be uncoated (either originally or by erosion of former coatings) or coated with a wide variety of toxic or non-toxic compounds, ranging from metal-containing and epoxy-containing paints to lanolin-based anticorrosive films. Tank arrangement also varies with vessel type. Outer tank walls are generally the hull of the vessel and may be either above or below water line. The ceiling may be an internal deck floor or the main deck of the vessel, in the latter case receiving direct solar radiation. Access to tanks is through horizontal or vertical manhole covers. Ballast tanks are generally isolated from each other, and thus can be filled or emptied individually. Tanks are either set aside ("dedicated" or "designated") or designed solely ("segregated") to hold water. Sea water may also be carried as ballast in temporarily empty cargo tanks, including tanks which alternately hold petroleum products (when on the cargo leg) and sea water (when on the ballast leg). Such tanks contain varying amounts of petroleum products and are not the systems generally considered in this review.

Water is drawn, usually from well below the water line, into ballast tanks through one or more intake hatch covers. These covers consist of steel plates with numerous holes each about 1·0 to 1·5 cm in diameter. As noted by Springer & Gomon (1975), however, the diameter of such holes may increase through corrosion. Water may be pumped into the tanks, and thus pass through impeller blades, or may be drawn in through gravity flow, and thus avoid passage through the pumps. Tanks may range from "empty" (although some water may still be in the tank bottom) to almost full ("pressed"). The amount of water and length of time it is carried will depend upon the required stability to be achieved relative to cargo and fuel loads and weather conditions. Thus, ballast tanks may be filled with water both in port and at sea.

Discharge of ballast water may commence as soon as a vessel enters protected harbour or river waters and as progress is made towards port. By doing so, time is saved at dockside in deballasting while cargo is waiting to be loaded. Ballast water is frequently discharged at the dock as well. Although discharge of sea-water ballast in certain places may be illegal (as in the Panama Canal), as Dawson (1970) has noted, and as I can confirm from discussions with ships' engineers, deck officers, and other maritime industry personnel, sea-water ballast is frequently discharged in many technically restricted places, if the ships' personnel believe that no petroleum products discharge will be involved. Caution must, therefore, be exercised in assuming where and when ballast water is actually discharged.

The physical-chemical environment in a ballast tank may be characterized as follows. There is no light. Tank temperatures may either remain close to the original temperature of the ballasted water or, more commonly, mirror (with some lag time), within one or two degrees, the water or air temperature the vessel is in or passing through. Such variations depend upon the position and size of the ballast tank. Oxygen content may vary considerably, depending upon initial concentration, the amount of air space remaining in the tank after it is filled (the ullage, or the height of the space above the water surface), the size of the tank, and the nature of the tank walls (for example, whether heavily rusted or not). A wide range of oxygen values have been found in ballast tanks by Medcof (1975), Howarth (1981), and Carlton, Navarret & Mann (1982). Other variables dependent upon the location and time of ballasting may include water quality (extent of organic or inorganic pollutants), salinity, pH, and sediment load. Some of these (such as salinity) may remain stable during a given voyage, while others (such as temperature and oxygen) may change considerably.

Ballast water is not bilge water (Dawson, 1973; Springer & Gomon, 1975). Bilge water, which consists of rain water, sea water, waste water, and oil, seeping from decks, the engine room, pump room, and other machinery spaces, collects in the bilge, the lowest internal part of a vessel, adjacent to the keelson. The environment so created is generally considered abiotic, although I have found no studies on it. Most authors referring to "bilge" water appear to mean ballast water.

HISTORY OF BALLAST WATER

Although experiments with built-in water-ballast tanks in ships date from the mid-1840s (Bowen, 1932), it is generally considered that the use of water as ballast commenced a decade later with the introduction of built-in compartments in colliers trading between the Tyne and London, in order to reduce the time and expense in loading and unloading solid (dry) ballast (Pollock, 1905; Walton & King, 1926; Casson, 1964). The concept of water as ballast, although aided by the increasing popularity of iron as a shipbuilding material (Thayer, 1897), appears, however, to have been relatively slow in catching on; while Grantham (1868) devotes several pages to the design of double-bottom systems for water ballast, Meade (1869) makes no mention of water ballast or water-ballast tanks. By 1880, however, Hamersly's *Naval Encyclopedia* contained a definition of "tank-water ballast" (Hamersly, 1880), White (1882) wrote that "double-bottom compartments are commonly used for water ballast" and, perhaps most significantly, *Lloyd's Register*, from 1880 on (but not before) began noting types and capacities of water ballast tanks, illustrating various vessels of "cellular construction of double bottom for water ballast", as well as after peak, fore peak, and midship deep tanks, holding up to 500 long tons of water (approximately 130 000 U.S. gallons) (*Lloyd's Register*, 1880). By the 1890s, ballast water had come into extensive use (Little, 1890; White 1894; Frear, 1897) and in 1896 Walton wrote, "Probably most cargo steamers in these days are fitted with some means of carrying water as ballast. . . ." Submarines with ballast tanks did not come into conventional maritime use until World War I (Casson, 1964; Guthrie, 1970), well after other types of vessels had been regularly transporting ballast water. It may be assumed then, for the purposes of this review, that the regular use of sea water as ballast began in the late 1870s and early 1880s, or somewhat more than 100 years ago.

In the 1980s, "clean" sea-water ballast, in tanks segregated or dedicated solely for that purpose, is carried in a wide variety of merchant vessels. These include tankers (petroleum and chemical), bulk carriers (single-deck vessels with large hatchways and holds, varying in size from 17 000 to 70 000 deadweight tonnage, corresponding to the largest ships that can pass through the Panama Canal), ore carriers (often with ballast tanks extending for the full length of the hold space along each side), combination carriers (including the "OBO", ore-bulk-ore, carriers), general cargo liners, barge carriers, and container ships (Couper, 1983). Tankers, bulk carriers, and general cargo vessels make up over 83% by tonnage of the world merchant vessel fleet (Couper, 1983).

World patterns of ballast-water movement over the past 100 years have probably paralleled the changing patterns of world shipping routes since the late nineteenth century. These patterns, in turn, are complex, and "reflect the world distribution of resources, population, location of industries, the characteristics of markets, economic growth rates, political and military factors, as well as short-term meteorological conditions" (Couper, 1983) that change with time. Although the intensity, number, and direction of trade routes have changed many times, a clearer pattern emerges relative to the growth of maritime trade. From the beginning of World War I to the beginning of World War II tanker fleets grew from 1·4 million tons in 1913 to 11·6 million tons in 1938; from the end of World War II until 1979 the growth in seaborne trade averaged 8% per year, and from 1971 to 1981 alone the world fleet of merchant ships increased from 33 000 to 38 000 (Couper, 1983). In the 1980s, the major commodities transported are crude petroleum and petroleum products, iron ore, coal, grain, bauxite and alumina, and phosphate in bulk (Couper, 1983) all of which are generally carried in vessels with a ballast leg. While no records are kept in the maritime industry of the amount of ballast water which has been or is transported, the generally steady increase in total seaborne trade throughout most of the past century, suggests that the amount of sea-water ballast carried has concomitantly, also increased. This is due to both the greater number of vessels and to the greater amount of ballast water now carried in larger vessels. While there is doubtless more sea water transported around the world as ballast today than ever before, the changed conditions of such transport, in terms of water quality and of time elapsed between ports, in addition to the increased amounts carried, may have been important factors in the rôle played by ballast-water transport as a dispersal agent. These changes are discussed at the end of this review.

RECOGNITION OF BALLAST WATER AS A DISPERSAL MECHANISM

SURVIVAL OF PLANKTON THROUGH SHIPS' PUMPS

It was recognized relatively early that planktonic organisms would survive passage through ships' pumps. Indeed, several investigators at the turn of the century tapped the salt-water supply leading to bathrooms and bathtubs aboard ocean-going steamers to collect plankton. Herdman, Thompson & Scott (1897) described the composition of plankton samples pumped aboard a steamer into a deck tank as the vessel crossed the north Atlantic Ocean. Taxa passing through the pumps into and through the ship's sea-water system included numerous species of protists (diatoms, dinoflagellates, radiolarians, foraminiferans), and of adult and larval animals (hydromedusae, polychaetes, cladocerans, ostracods, barnacle cyprids, harpacticoid, cyclopoid, and calanoid copepods, hyperiid amphipods, mysids, isopods, shrimp and crab larvae, euphausiids, bivalve and gastropod larvae, pteropods, chaetognaths, and fish the last including eggs, embryos, and larvae). Giesbrecht (1897) described collections made by A. Krämer in a similar fashion in the Red Sea. In addition to some of the taxa listed above, he also noted tunicates and echinoderm larvae in pump samples. Herdman again used a ship's pump and sea-water system to collect plankton at a bathroom tap during a passage from the English Channel to India in 1901–1902 (Thompson & Scott, 1903).

Johnstone, Scott & Chadwick (1924, p. 177), noting that "on board ocean-going steamships there is always a tank kept full of seawater by a pump", suggested attaching a silk net to bathroom taps in connection to such tanks, as a general means by which "a series of plankton samples" could be "obtained in a very convenient way". Herdman *et al.* (1897) concluded "that little or no damage is received in passing through the pumps" by the plankton, and Giesbrecht similarly noted that damage to organisms passing through the pumps "ist ohne Gewicht". None of these workers, interestingly enough, commented on the possible presence of plankton in the ballast tanks aboard the same vessels they were using. Herdman's studies, especially of those samples collected near each port, may be taken as at least a conjectural description of the plankton that was entrained in the ballast tanks by these steamships.

SUGGESTIONS OF BALLAST-WATER TRANSPORT

A decade after the ship pump collections by Herdman and Krämer, the discovery in the North Sea in 1903 of the Asian diatom *Biddulphia sinensis* led Ostenfeld (1908) to suggest "no other explanation" for its appearance than transport by ship, "for instance attached to the outside or growing in the water of the hold"; its occurrence on ships' hulls appears unlikely, however, for this "truly planktonic" (Lebour, 1930; Hendey, 1964) species. For the next 50 years, however, the rôle of ballast water as a dispersal agent attracted little attention. Broch (1924) considered the possibility that the hydroid *Gonionemus vertens* was carried to Norway in ballast water, and Moll (1936) suggested transport in the wooden sea-water tanks (presumably short-lived phenomena about which there are few data) as a means of shipworm (Teredinidae) dispersal. Kofoid & Miller (1927, p. 309) suggested that the wood-boring isopod (gribble) *Limnoria* may be carried in the ballast water of steel-hulled ships: "It is conceivable that, in the intake of some tons of ballast water by a ship lying at a wharf, the piles of which are infected, many small organisms of all kinds, including *Limnoria* and other crustaceans, will be taken in. Rubbing and chafing of the ship against the wharf may remove small fragments of wood which carry *Limnoria*, and these may then be drawn in with the ballast water."

In the most often cited case of hypothesized ballast-water transport, Peters (1933) described in detail the evidence for such transport, before 1912, for the Chinese mitten crab *Eriocheir sinensis* from the deltas of Chinese or Korean rivers to Germany. In 1949 Rees & Cattley suggested that "ships returning from the Middle East during the War, with empty holds" may have introduced southern species of crustaceans "in water ballast" to Western Europe.

It was not, however, until the mid-1950s that sea-water ballast entered the literature as a more regularly invoked transoceanic or interoceanic dispersal mechanism. Barnard & Reish (1959), for example, suggested that ballast water "may play a role in the dispersal of estuarine organisms. It is possible that larval stages of marine invertebrates can be pumped into the holds of ships with seawater ballast. Any suspended material would settle to the bottom of the hold and serve as suitable substrate and food for some species of animals." None the less, while other workers in the 1950s suggested ballast-water transport for specific taxa (*e.g.*, Wolff, 1954; Holthuis & Gottlieb, 1955; Hynes, 1955; Nebolsina, 1959), 75% of the primary references in Table I date from

1960 or later. It is unfortunate, perhaps, that for the greater portion of the time during which ballast-water transport has existed as a potential dispersal agent, few workers examined it as a possibility, and none until 1970 (discussed below) directly examined ballast tanks for evidence of such dispersal.

Between 1968 and 1978, during debates on the potential biological and ecological changes that might result from the construction of a sea-level canal across the Isthmus of Panama, numerous opinions were offered on whether sea-water ballast was or was not a viable transport mechanism for marine organisms. The issue raised was whether or not ballast water, taken in and then discharged on opposite sides of the Panama Canal, had since its opening in 1914 led to a biotic interchange between the Atlantic and Pacific Ocean sides of Panama over some 60 years of such activity. That it had was championed (Sheffey, 1968, 1972, 1978; Voss, 1978), supported (Chesher, 1968), doubted (Rubinoff, 1970; Menzies, 1972; Beeton, 1978), or considered moot (Jones, 1972; Newman, 1972; Dawson, 1973; Challinor, 1978). A wide variety of versions of where and when ballast water was taken in and released in the Canal Zone region also appeared (Chesher, 1968; Sheffey, 1968, 1972, 1978; Rubinoff, 1970; Springer & Gomon, 1975; Beeton, 1978; Constant, 1978). Despite estimates of the volume or frequency of ballast-water transport through the Panama Canal (Chesher, 1968; Sheffey, 1978), no records are generally kept (Constant, 1978). Studies were called for (Rubinoff, 1970; Jones, 1972; Dawson, 1973; Jones & Dawson, 1973; Challinor, 1978) but none apparently were made. None the less, these discussions led to the elucidation of certain features about ballast-water transport which are incorporated in this review. The rôle of ballast-water transport through the Panama Canal is treated below under the geographic analysis of probable transport incidences.

SAMPLING OF BALLAST-WATER TANKS

Although Rubinoff (1970) refers to the "few ballast tanks" he sampled, apparently in the Panama Canal region (noting that the samples "contained no living plankton" but giving no further details) and Peters (1933) noted the presence of the mitten crab *Eriocheir* and the amphipod *Gammarus* in a steamer's ballast tanks in 1932 (a record based upon a vessel in Hamburg Harbour which had remained at dockside for an unspecified period and had taken on water while never leaving port), it remained for J. C. Medcof and E. A. Scribner to report in detail on perhaps the first samples collected from a ship's ballast tanks at the end of a voyage (Medcof, 1975). Medcof and Scribner sampled the ballast water "in two holds of a ship" in 1973, on a vessel that had completed a 14·5-day passage from Tagonoura, Japan to New South Wales, Australia (see Tables III and IV). The ballast water had been derived in part from Tagonoura Harbour and in part at sea during the first four days of the voyage; thus, the plankton in the holds was at least 10 days old. Plankton samples revealed "numerous, swimming organisms; crustaceans being the most conspicuous". The plankton included copepods, ostracods, amphipods, crustacean larvae, polychaete larvae, adult polychaetes (said to be a benthic species), and chaetognaths. Medcof and Scribner found the ballast water to have an average temperature of 25·7 °C, and the surface water in Twofold Bay, the point of discharge, to be 16·7 °C. The dramatic temperature change experienced by the plankton while being discharged "could well be directly or indirectly lethal to most or all of them" (Medcof, 1975), although Medcof noted that "at other seasons and in other harbours the discharge of ballast water might be less shocking to organisms it contains".

Since Medcof's report, several more extensive studies have been made in Australia, Canada, and the United States of America. New South Wales (N.S.W.) State Fisheries investigators have intercepted and sampled the ballast water in bulk cargo carriers coming from Japan to northwestern Western Australia (where sea-water ballast is exchanged for iron ore), to Gladstone, Queensland (where ballast is exchanged for coal), and to Eden, N.S.W. and to Triabunna, Tasmania (where it is exchanged for wood chips) (Williams, 1977, 1982 in Middleton, 1982; Williams, van der Wal & Story, 1978). After nine months of investigation between 1976 and 1977 it was reported that "living plankton had been obtained at each location on each sampling occasion" (Williams, 1977). The Environmental Protection Service, Environment Canada, commissioned a study (Howarth, 1981) to sample the ballast water of 55 merchant vessels from many countries in the summer and fall of 1980 at Montreal, as the vessels entered the Great Lakes through the St Lawrence Seaway. Identifiable living organisms were found in 51 of the 55 vessels. In 1980–1981, Carlton, Navarret & Mann (1982), Woods Hole Oceanographic Institution, conducted a series of coastal and transatlantic experiments in plankton survival in the sea-water ballast tanks of an oceanographic research vessel, examined the plankton surviving transport in several merchant vessels arriving in United States Atlantic coast ports, and conducted an experiment in plankton survival from New England to the Gulf of Mexico in the ballast tanks of an oil tanker. Other than these few ports in Australia and on the Atlantic coasts of the United States and Canada, plankton survival in sea-water ballast tanks remains uninvestigated in the rest of the world.

The studies in Australia (including those of Medcof and Scribner) and in the United States used vertical or horizontal hauls with plankton nets to sample ballast tanks. In the Canadian studies, tank surface water was sampled in half of the vessels by a "grab sample" with a bucket, and tank bottom water was sampled in the remaining vessels by filtering water pumped out of the bottom of the tanks through the vessels' discharge pumps. Some of the available results of these studies are incorporated in the tables and discussions herein.

THE HYPOTHESIS OF BALLAST-WATER TRANSPORT

I analyse here instances of introductions for which ballast water has been suggested as being either the only, or one of several, possible dispersal agents (Table I). I also analyse additional instances, largely from the literature, where ballast-water transport may have played a rôle but was not the mechanism of dispersal originally suggested. A few freshwater taxa are included, especially if these involve oligohaline forms, or are recorded from major maritime corridors, such as the St Lawrence Seaway to the Great Lakes or the Panama Canal. Of the almost 100 taxa in Tables I and II, none are known to have teleplanic larvae, and none (with the exception of the two anguillid eels in Table II) have ever been collected in oceanic or mid-ocean waters as either larvae or adults.

Table I is not necessarily a complete synopsis of all literature suggestions. Most references to "ballast water" or "bilge water" are buried in taxonomic or distributional studies, and are difficult to locate. The word "ballast" in a biological sense rarely appears in paper titles, key word indices, or computer literature searches. Table II, in turn, is intended only as a set of examples, selected from hundreds of potential cases, where ballast-water transport may function as an alternative mechanism of dispersal with explanatory power as great as or greater than the dispersal mechanisms previously proposed.

ALTERNATIVE DISPERSAL MECHANISMS

Alternative dispersal mechanisms, other than ballast-water transport, are indicated in Table I (as ADM) and in Table II (as DM), and are explained in the caption to Table I. For many taxa the available data are insufficient to resolve between ballast-water transport as the mechanism of dispersal and one or more of these alternative mechanisms. Distinguishing transport in ship-fouling communities from transport in ballast-water tanks may be particularly difficult. Ship fouling is of two kinds: external fouling on the hull (ADM-2 in Table I) and internal fouling in sea-water pipe systems (ADM-3, in part, in Table I).

The transport of living marine organisms (as fouling "assemblages" or "communities") on the hulls of ships has been a long-recognized phenomenon (Hentschel, 1923, 1924; Visscher, 1928; Woods Hole Oceanographic Institution, 1952; Allen, 1953; Skerman, 1960; Clapp & Kenk, 1963; Carlton, 1979a,b). As a dispersal mechanism of innumerable taxa it is centuries old. Ship fouling may be composed of both sedentary and sessile species (sponges, hydroids, anemones, serpulid and other polychaetes, mussels, oysters, tube-dwelling crustaceans, barnacles, bryozoans, ascidians, algae, and so forth) and, if the fouling accumulation is thick and stable enough, free-living invertebrates and fish. Free-living taxa may include polychaetes, peracarid crustaceans (amphipods, isopods, tanaids) and crabs (such as xanthids) and fish (such as gobies and blennies) with a strong propensity to hide deep within a fouling matrix. Bertelsen & Ussing (1936) record the transoceanic crossing of tropical crabs (grapsids) and fish (blennies) in a very heavily fouled (a "crust" up to 25 cm thick) vessel which had been in Bermuda "for a series of years" before being towed for 45 days to Copenhagen. Wooden vessels, in addition, until the late nineteenth century, by virtue of the often deep and extensive galleries, tunnels and holes in their keels and hulls created by shipworms and other boring organisms, provided a further means for mobile species to colonize hull spaces yet avoid being swept off at sea (Chilton, 1910; Carlton, 1979a). Presumably the probability of successful transport of free-living species decreases with increasing distance of travel (and thus a greater likelihood of being lost at sea).

Species that tend to feed only on the outside of a fouling community but not hide within it, such as active, swimming shrimp and fishes, are likely to be washed away from the hull of a vessel soon after it leaves the harbour (Newman, 1963). Peters (1933) thus argued that the mitten crab *Eriocheir sinensis* did not exhibit the necessary crevice-dwelling behaviour to be a member of a ship's fouling community: "... denn wir trauen diese nur solchen Formen zu, die von Natur mehr als die Wollhandkrabbe gewohnt sind in

TABLE I

*Literature suggestions of ballast-water transport (BWT) of marine and some freshwater organisms: a, not necessarily the native region of the species; b, does not include regions to which the species may have subsequently spread; c, date of first collection, not necessarily the date of introduction; d, ADM, alternative dispersal mechanism (see key, below) as suggested by the same worker(s) who suggested BWT; ADMs suggested herein or by others are under Reference/Notes column; e, the first cited reference is the work suggesting BWT and the ADMs, not necessarily the first report of the introduction; f, BWT likelihood, ballast-water transport likelihood, see text for full definitions; * = probable; + = possible; — = not likely; g, NE, not established (no reproducing populations known); alternative dispersal mechanisms—1 = by "ships", author specifies neither BWT nor ADM-2; 2 = fouling on or boring into ships' hulls; 3 = in sea-water pipe systems of ships; 4 = with oyster introductions; 5 = with aquatic plant introductions; 6 = with algae packed with fishery products (such as lobsters, bait worms); 7 = intentional release as private introduction, or as discarded specimens imported for public or private aquaria, or as discarded fish bait; 8 = natural dispersal by currents or on drifting objects (algae, wood); ∅ = naturally occurring in region and not previously reported or recognized*

Species	Introduced: From[a]	Introduced: To[b] (Date)[c]	ADM[d]	Reference[e]/Notes	BWT Likelihood[f]
PORIFERA					
Trochospongilla leidii	Eastern North America	Panama Canal: Gatun Locks (1930s–1970s)	7; ∅	Jones & Rützler, 1975	+
CNIDARIA					
Hydrozoa: Hydroida					
Maeotias inexpectata	Black Sea—Sea of Azov	Virginia: Chesapeake Bay (1968)	2	Calder & Burrell, 1969	+
Gonionemus vertens	Western North Atlantic	Norway: Oslofjord (1921)	—	Broch, 1924; Edwards (1976) suggests ADM-4; ADM-2 possible	+
Hydrozoa: Siphonophora					
Chelophyes contorta	Panama: Pacific side	Panama: Atlantic side (pre-1969)	3	Alvariño, 1974; A. Alvariño, pers. comm., 1982; NE[g]?	*

Chelophyes appendiculata	Panama: Atlantic side	Panama: Pacific side (pre-1969)	3	(as above)	*
Muggiaea atlantica	Panama: Pacific side	Panama: Atlantic side (pre-1969)	3	(as above)	*
Muggiaea kochi	Panama: Atlantic side	Panama: Pacific side (pre-1969)	3	(as above)	*
Lensia challengeri	Panama: Atlantic side	Panama: Pacific side (pre-1969)	3	(as above)	*
ROTIFERA					
Keratella cochlearis	Eastern North America?	Canada: Nova Scotia: open sea ≈150 km SE of Halifax (1978)	—	Coates *et al.*, 1982, in food bolus of pleustonic juvenile marine fish	*
ANNELIDA					
Polychaeta					
Nereidae					
Nereis acuminata	Western North Atlantic	California: Los Angeles—Long Beach Harbors (1954)	—	Barnard & Reish 1959; as swimming post-larvae which occur in plankton, there being no true planktonic stage (Reish, 1957); ADM-2 possible	+
Dorvilleidae					
Ophryotrocha labronica	Mediterranean Sea	California: Los Angeles Harbor (pre-1975)	—	Åkesson, 1977; ADM-2 possible	+
Spionidae					
Boccardiella ligerica	Western Europe	California: San Francisco Bay (pre-1954)	—	Carlton, 1979a; Light, 1978; ADM-2 possible	+
Polydora ligni	Western North Atlantic	Florida: Tampa Bay population only (1970s)	2, 4	Rice & Simon, 1980: local genetically distinct population	+
Serpulidae					
Ficopomatus enigmaticus	Western Europe	Netherlands (1967)	2, 8	ten Hove, 1974	+
Hydroides elegans	Western Europe	Netherlands (1972)	2, 8	ten Hove, 1974	+

TABLE I—*continued*

Species	Introduced From[a]	To[b] (Date)[c]	ADM[d]	Reference[e]/Notes	BWT Likelihood[f]
Janua brasiliensis	Japan	England: Portsmouth (1973–1975)	—	Knight-Jones *et al.* 1975: on alga *Sargassum*; consider and reject ADM-8; ADM-4 more likely (Farnham, 1980)	+
Pileolaria rosepigmentata	Japan	England: Portsmouth (1973–1975)	—	(as above)	+
MOLLUSCA					
Gastropoda					
Mesoclanculus ater	Western North Pacific (probably Japan)	Canada: British Columbia: Queen Charlotte Sound (1964)	—	Carlton, 1979a; Clarke, 1972 (who rejects ADM-4), NE?	*
Sabia conica	Western North and South Pacific (probably Japan)	Canada: British Columbia: Queen Charlotte Sound (1963); Vancouver Island (1940)	—	Carlton, 1979a; Cowan, 1974: ADM-1; NE?	*
Bivalvia					
Crassostrea gigas	Japan–China	New Zealand ($\approx$1970)	2, 7	Dinamani, 1971	+
Mya arenaria	Mediterranean Sea?	Black Sea (1960s)	—	Gomoiu, 1975; ADM-2 also possible (occurrence in fouling: Graham & Gay, 1945; WHOI, 1952, see p. 200; Merrill, 1959; J. T. Carlton, field obs.)	+
Ensis directus	Western North Atlantic	Germany and Denmark (1978)	—	Cosel *et al.* 1982: suggest ballast water was released off mouth of Elbe or Weser Rivers	

Theora fragilis	Western North and South Pacific	California: Anaheim Bay (1968), Newport Bay (1971), Los Angeles Harbor (1973)	—	Carlton, 1979a; Seapy, 1974: ADM-1	*
Teredinidae	—	—(around the world, 19th–20th centuries)	2 and log booms	Moll, 1936: wooden sea water tanks of ships	+
Laternula limicola	Western North Pacific: China, Japan, Korea	Oregon: Coos Bay (1963)	—	Carlton, 1979a; Keen, 1969	*
CRUSTACEA					
Copepoda: Cyclopoida					
Limnoithona sinensis	China	California: San Francisco Bay (1979)	—	Ferrari & Orsi, 1984: Sacramento/San Joaquin Estuary	*
Oithona davisae	Indo-West Pacific?	California: San Francisco Bay (1979; 1963?)	∅	Ferrari & Orsi, 1984: described as new species from San Francisco Bay, but taxon is of Indo-Pacific stock	*
Copepoda: Calanoida					
Sinocalanus doerii	China	California: San Francisco Bay (1978)	—	Orsi *et al.*, 1983: Sacramento/San Joaquin Estuary	*
Pseudodiaptomus marinus	Japan	Hawaii: Oahu (1964)	8	Jones, 1966: BWT “remote possibility”, but notes that life span of this neritic species is less than necessary residence time in oceanic currents; Grindley & Grice (1969) favour BWT	*
Pseudodiaptomus marinus	Japan	Indian Ocean: Mauritius (1964)	2	Grindley & Grice, 1969; Bowman (1978) suggests ADM-2 “unlikely” for congener *P. acutus* dispersal	*

TABLE I—*continued*

Species	Introduced From[a]	Introduced To[b] (Date)[c]	ADM[d]	Reference[e]/Notes	BWT Likelihood[f]
Acartia tonsa	Western North Atlantic	Denmark: Zuiderzee (1916)	2, Ø	Brylinski, 1981: demonstrates restricted distribution suggestive of introductions; Remy, 1927: ADM-2, 8, 9; Redeke, 1934: 9, but favoured introduction; Conover, 1957: data insufficient to determine status	+
Branchiopoda: Cladocera					
Penilia avirostris	Tropical to warm temperate regions	North Sea off Netherlands (1947)	—	Rees & Cattley, 1949; Cattley & Harding, 1949 (original report) suggest rôle of resting eggs in appearance of new populations; ADM-8?, -Ø?	+
Ostracoda					
"ostracods"	Indo-Pacific, via Panama and Suez Canals	Caribbean (after 1914)	2, 8	Teeter, 1973: on algae taken up into tanks during ballasting	+
Malacostraca					
Isopoda					
Cirolana arcuata	New Zealand or Chile	California: San Francisco Bay (1978)	2	Bowman *et al.*, 1981	+
Dynoides dentisinus	Japan, Korea	California: San Francisco Bay (1977)	2	Carlton, 1979a	+
Sphaeroma walkeri	Western South Pacific	California: San Diego Bay (1973)	2	Carlton, 1979a; Carlton & Iverson, 1981 (only ADM-2)	+

Limnoria	—	—(around the world, 19th–20th centuries)	2	Kofoid & Miller, 1927: in "small fragments of wood" rubbed off pilings by ships	+
Amphipoda					
Gammarus tigrinus	Western North Atlantic	Britain (1931)	—	Hynes, 1955; Bousfield, 1973	*
Gammarus tigrinus	Germany	Netherlands (1960–65)	7	Nijssen & Stock, 1966	+
Crangonyx pseudogracilis	England or Wales	England: E. Norfolk: Barton Broad (1940s?)	—	Hynes, 1955	*
Ampithoe longimana	Western North Atlantic	California: Newport Bay (1949); Morro Bay (1960)	2	Carlton, 1979a; Barnard & Reish, 1959 suggest ADM-6	+
Decapoda: Caridea					
Palaemon macrodactylus	Western North Pacific: Japan or Korea	California: San Francisco Bay (1957)	—	Hedgpeth, 1968 (and Dawson, 1973; Carlton, 1975); Newman, 1963 suggests ADM-3, considers and rejects ADM-2, -4	*
Palaemon macrodactylus	Western North Pacific or San Francisco Bay	California: Los Angeles Harbor (1962)	—	Carlton, 1975	*
Palaemon macrodactylus	San Francisco Bay?	California: Elkhorn Slough (1979)	7	Standing, 1981	+
Processa aequimana	"Middle East" (Red Sea and Suez Canal)	Southern North Sea (1946)	—	Rees & Cattley, 1949, as larvae. Williamson & Rochanaburanon (1979) describe this as *P. modica modica* n.sp., from western Europe	—
Decapoda: Brachyura					
Rhithropanopeus harrisii	Western North Atlantic	Netherlands (pre-1874)	—	Wolff, 1954: rejects ADM-2 for physiological reasons, which, however, is possible (see text for discussion)	+
Rhithropanopeus harrisii	Western North Atlantic	California: San Francisco Bay (1937)	2, 4	Carlton, 1979a	+

TABLE I—*continued*

Species	Introduced From[a]	Introduced To[b] (Date)[c]	ADM[d]	Reference[e]/Notes	BWT Likelihood[f]
Rhithropanopeus harrisii	Black Sea—Sea of Azov	Caspian Sea (1958)	—	Nebolsina, 1959; ADM-2 also possible	+
Neopanope sayi	Western North Atlantic	Wales: Swansea (1959?)	3	Naylor, 1960; ADM-2 also possible?	+
Neorhynchoplax alcocki	Iraq: near Al-Basrah, Persian Gulf	Panama Canal: East Pedro Miguel Locks (1969)	—	Dawson, 1973; Abele (1972) suggests ADM-5	+
Callinectes sapidus	Western North Atlantic	Netherlands (1932)	—	Wolff, 1954: considers and rejects ADM-2 (ADM-7 may explain single 1900 specimen from France); still regularly arriving in Netherlands ports: IJmuiden (1981), North Sea Canal (1982), Europort: Rotterdam (1983) (S. J. de Groot, pers. comm., 1983, 1984; Adema, 1982)	*
Callinectes sapidus	Western North Atlantic	Israel (1955)	—	Holthuis & Gottlieb, 1955, who consider and reject ADM-8	*
Callinectes sapidus	Western North Atlantic	Japan: Hamana Lake (1975)	—	Sakai, 1976a,b; suggests by submarine into Sea of Enshū-nada	*
Callinectes sapidus	Western North Atlantic	Germany: Elbe River (1964)	2, 7	Kühl, 1964	*
Eriocheir sinensis	Western Europe	Lake Erie (1965; 1973)	—	Nepszy & Leach, 1973; NE	*

Eriocheir sinensis	China or Korea	Germany: tributary of Weser River (1912)	—	Peters, 1933, who considers and rejects ADM-2, -7; as larvae taken up in ballast water "in den Mündungsgebieten chinesischer und koreanischer Flüsse"	*
CHORDATA					
Urochordata: Ascidiacea					
Styela clava	Western Europe	Denmark (1978)	2, 4	Christiansen & Thomsen, 1981, who favour ADM-4; BWT possible only over very short distances, as planktonic larval life generally less than 48 hours	+
Vertebrata: Osteichthys					
Anguillidae					
Anguilla anguilla	Western Europe	California: San Francisco Bay (1969)	—	Skinner, 1971, who considers and rejects ADM-7, -8; NE	*
Anguilla rostrata	Western North Atlantic	California: San Francisco Bay (1964)	—	Skinner, 1971, who considers and rejects ADM-7, -8; NE	*
Blenniidae					
Omobranchus ferox	Indo-West Pacific	Mozambique: Delagoa Bay (early 20th century?)	—	Springer & Gomon, 1975	*
Omobranchus punctatus	Indo-West Pacific	Venezuela and Trinidad (1930)	—	Lachner *et al.*, 1970; Springer & Gomon, 1975	*
Omobranchus punctatus	Indo-West Pacific	Mozambique: Delagoa Bay (early 20th century?)	—	Springer & Gomon, 1975	*
Omobranchus punctatus	Trinidad or Venezuela	Panama Canal and Atlantic Panama (1966)	—	Springer & Gomon, 1975	*
Eleotridae					
Prionobutis koilomatodon	Indo-West Pacific	Panama Canal (1972)	—	Dawson, 1973; considers and rejects ADM-4, -7; NE?	*

TABLE I—*continued*

Species	Introduced From[a]	To[b] (Date)[c]	ADM[d]	Reference[e]/Notes	BWT Likelihood[f]
Gobiidae					
Barbulifer ceuthoecus	Western Atlantic: Bahamas to Brazil	Panama: Pacific side: Panama Reef (1973)	—	McCosker & Dawson, 1975; NE?	*
Acanthogobius flavimanus	Japan	Australia: New South Wales: Sydney Harbour (1971)	2, 3	Hoese, 1973; Friese, 1973; Middleton, 1982	*
Acanthogobius flavimanus	Japan	San Francisco Bay (1963)	—	Lachner *et al.*, 1970; BWT also suggested by Dawson, 1973, and Springer & Gomon, 1975; Brittan *et al.*, 1963, 1970, suggest ADM-3, not BWT	*
Tridentiger trigonocephalus	Japan	Australia: New South Wales: Sydney Harbour (1973)	2, 3	Hoese, 1973; Friese, 1973	*
Tridentiger trigonocephalus	Japan	California: San Francisco Bay (1964)	—	Lachner *et al.*, 1970; Dawson, 1973; C. L. Hubbs & J. Prescott in Hubbs & Miller, 1965: ADM-4, as eggs	
Cyprinodontidae					
Lucania parva	Western North Atlantic	Oregon: Yaquina Bay (1958)	4	Hubbs & Miller, 1965, who, however, note no regular oyster shipments since the 1930s (last in 1943–44, if ADM-4, perhaps overlooked previously)	*
Lucania parva	Western North Atlantic	California: San Francisco Bay (1958)	4	Hubbs & Miller, 1965; however, commercial oyster importations ceased by the 1930s (small experimental plantings thereafter) (Carlton, 1979a)	*

Pleuronectidae					
Platichthys flesus	Western Europe	Lake Erie (1974, 1976)	—	Emery & Teleki, 1978; NE	*
Platichthys flesus	Western Europe	Lake Superior (1981)	—	Crossman, 1984; one specimen (179 mm TL, about two + years old), from Thunder Bay area off McNabb Point, mixed bottom, less than 31 m depth (H. D. Howell & E. J. Crossman, pers. comm.); NE	*
BACILLARIOPHYTA					
Biddulphia sinensis	Pacific or Indian Oceans or Red Sea	Western Europe: North Sea (1903)	2	Ostenfeld, 1908	*
Asterionella formosa	Eastern North America?	Canada: Nova Scotia: open sea ≈ 150 km SE of Halifax (1978)	—	Coates *et al.*, 1982: in food bolus of pleustonic juvenile marine fish	*
Melosira granulata	(as above)	(as above)	—	(as above)	*
CHRYSOPHYTA					
Dinobryon bavaricum	(as above)	(as above)	—	(as above)	*
Dinobryon cylindricum	(as above)	(as above)	—	(as above)	*
CHLOROPHYTA					
Codium fragile tomentosoides	Japan	Netherlands (1900)	4	Walford & Wicklund, 1973; ADM-2 generally more likely (Carlton & Scanlon, 1985)	+
PHAEOPHYTA					
Sargassum muticum	Japan	England: Portsmouth (1973–1975)	—	Knight-Jones *et al.*, 1975, who consider and reject ADM-8; ADM-4 more likely (Farnham, 1980)	+

Löchern and Gehäusen zu leben, wie zum Beispiel die *Menippe convexa.* Wir glauben nicht, dass die jugen Wollhandkrabben eine so weite Reise am freien Schiffsboden bei der starken Wasserbewegung überstehen würden."

The transport of living marine organisms as fouling in ships' sea-water pipe systems (pipes leading to and from tanks, bilges, fire hose, sanitary systems, and so forth) is a less well recognized and thus a poorly understood phenomenon. Newman (1963) has described one such fouled system, consisting of serpulid polychaete worms and barnacles, in the sea-water pipes of a naval vessel in 1954. The 4-cm thick fouling mass had reduced the bore of the pipes to 5–6·5 cm. While to my knowledge no free-living species have yet been recorded from such fouled systems, it would none the less appear that errant or semi-errant species could also live within such a matrix, as described above for external fouling, and thus avoid being carried along by the often constant and strong water flow in such pipes (Dawson, 1973). This flow would function to either sweep an organism out of the systems or crush it, by water pressure, against hatch cover gratings. Swimming, mobile species or non-crevice dwellers are probably less likely to survive in such pipes. For example, although Newman (1963) has suggested that the shrimp *Palaemon macrodactylus* could have been transported transoceanically from Asia to the United States in fouled sea-water pipe systems, it is a species which swims above or rests upon, rather than hiding within, fouling (J. T. Carlton, pers. obs., San Francisco Bay, California), and would thus probably be carried along by any strong water flow in a sea-water system.

The long overlap between fouling and ballast water as mechanisms of dispersal thus renders it difficult, if not impossible, to distinguish practically for certain taxa between juvenile or adult transport in fouling communities and larval transport in ballast water. Thus, hypothetically, barnacles may be transported as either adults in fouling on ships' hulls or as nauplii and cyprids in ballast water, hydroids as either polyps or as hydromedusae, scyphozoans as either scyphistomae or as ephyrae, serpulid polychaetes as either attached adults or as trochophores, and so forth.

Nevertheless, changes in the nature of ship fouling assemblages over the last century suggest that the problem of making a distinction between one or the other method of transport may be of greater importance to the analysis of dispersal events and introductions between the 1880s and the 1940s to 1950s than it is now. Since World War II, several factors appear to have provided mechanisms to reduce the importance of ship fouling as a major dispersal agent. The development of antifouling paints with remarkable efficacy, significantly increased ship speeds (and thus increased water motion across the hull) and decreased harbour residence times of modern merchant vessels (often only 24 hours or less) may have decreased the frequency of extensive colonization by fouling organisms on ships' hulls (Allen, 1953; Carlton, 1979a; Farnham, 1980). The frequency of extensive build-ups of fouling many centimetres thick, which would secondarily carry free-living species, on vessels regularly engaged in trans- or interoceanic commerce, is also less likely now than in the late nineteenth or early twentieth centuries. Fouled sea-water pipes are, in turn, presumably less frequent and indeed were considered uncommon in 1954, Newman (1965) noting that "such sea water systems are not ordinarily allowed to become fouled".

I thus judge that the successful transport of errant species in massive fouling on the hulls of ships has become increasingly less important since the nineteenth century. Conversely, prior to the late nineteenth century, when heavily fouled ships or heavily bored wooden vessels were more common, the survival of a fish or crab amidst the fouling interstices, or in shipworm borings or enlarged boring isopod galleries, was probably more likely. As the extent of ship fouling has decreased and presumably continues to decrease, ballast water may provide a continued and in some cases renewed mechanism for the transport, now as larvae, of both fouling and errant species that may have been previously dispersed more frequently in fouling communities. The same argument may be advanced for wood-boring species. With the decline of wooden vessels in the late nineteenth century, long-distance dispersal of shipworms and boring isopods (*Limnoria*) in ships' hulls virtually ceased. Continued transport of shipworms (those species with planktotrophic larvae) or of isopods (as free-swimming individuals) in ballast water, however, could hypothetically continue the long-distance dispersal by ships of such species.

It may be noted that a further complexity in distinguishing transport in ballast water from transport in fouling is the relative distance of the dispersal event. For many species transport by either mechanism is more likely to be successful over short distances than over long distances. Thus, the transport of an errant species in or on ships' hulls may be successful if between two relatively near coastal points, but may be unsuccessful in a transoceanic crossing. Similarly, the movement of heavily fouled local vessels from one end of a canal system (such as the Suez Canal or Panama Canal) to the other end, assuming that the taxa involved are able to withstand the canal's physical and chemical environment, may provide an effective means of dispersal of even relatively errant species. In contrast, taxa such as most bryozoans and ascidians, with planktonic larval stages lasting no more than a few hours, may be hypothetically transported between two nearby ports along a coast in ballast water as larvae, but would not probably survive as competent larvae in passages of several days or more. Distinguishing between natural larval dispersal by local coastal currents and ship transport (whether fouling or ballast water) along coastlines may also be difficult if both currents and shipping traffic are in the same direction.

LIKELIHOOD OF BALLAST-WATER TRANSPORT

With all of the above considerations in mind, it is thus necessary to assign a status of either "possible" or "probable" (and in one suggested case "unlikely") to all the taxa in Tables I and II relative to the likelihood that ballast water transport was the mechanism of dispersal. Future studies will doubtless change some of these assignments. Transport incidences are considered "possible" if alternative dispersal mechanisms could operate to transport the species and there is no clear evidence favouring any one mechanism. Incidences are considered "probable" if, given the biology and ecology of the taxon and the historical details of the dispersal incident, there is no clear evidence that a known alternative dispersal mechanism, other than ballast water, has a high likelihood to transport the species the distances involved. A largely conservative approach has been taken in this regard. I have assigned "possible" to several cases where ballast-water transport was perhaps the most harmonious explanation, but where I could not reasonably reject an

TABLE II

Examples suggested here of probable or possible ballast-water transport (BWT) of marine and some freshwater organisms: a, b, c, see notes to Table I; d, DM (Dispersal Mechanism): see key to numbers in Table I; e, BWT likelihood: ballast-water transport likelihood; see note to Table I; f, NE, not established (no reproducing populations known)

Species	Introduced From[a]	Introduced To[b] (Date)[c]	Dispersal mechanism originally suggested, if any (Reference/Notes)[d]	BWT Likelihood[e]
CNIDARIA				
Hydrozoa: Hydroida				
Blackfordia virginica	Black Sea	Virginia: Chesapeake Bay (1904)	DM-2 (Thiel, 1935); Mayer, 1910	+
Blackfordia virginica	Black Sea	India: Calcutta: Ganges River (1926)	DM-2 (Kramp, 1955)	+
Gonionemus vertens	Western Europe	Massachusetts Woods Hole (1894)	DM-4 (Edwards, 1976, but evidence presented for oyster transplantation not consistent in space (Maryland) or time (1881) with this record); DM-2 and BWT possible if free medusae or attached polyps on drift carried by local currents into harbours	+
Ostroumovia inkermanica	Black Sea	Netherlands: North Sea Canal (1959)	DM-2 (Saraber, 1962)	+
Ostroumovia inkermanica	Black Sea	India: Visagapatnam Channel (1926)	DM-2 (Saraber, 1962, from P. Kramp, *in litt.*)	+
Moerisia lyonsi	Eastern Mediterranean Sea: Egypt	Virginia: Chesapeake Bay inlets (1965)	DM-1 (Calder & Burrell, 1969); Calder & Burrell, 1967; DM-2 possible	+
Scyphozoa: Rhizostomeae				
Cotylorhizoides pacificus	Philippine Islands	Hawaii: Oahu Pearl Harbor (1941–1945)	DM-2 as scyphistomae (Cutress, 1961); BWT as ephyrae?; taxonomy uncertain: Cooke, 1984	+

Cassiopea medusa	Philippine Islands	(as above)	DM-2 as scyphistomae (Cutress, 1961); BWT as ephyrae?	+
Anomalorhiza shawi	Philippine Islands	Hawaii: Oahu: Kaneohe Bay (1983)	DM-2 as scyphistomae (Cooke, 1984); BWT as ephyrae?	+
ANNELIDA				
Polychaeta: Spionidae				
Boccardia proboscidea	Japan, or Eastern North Pacific	Australia: Victoria: Port Phillip Bay (1977)	Blake & Kudenov, 1978 (suggest introduction, no DM noted); BW release at Melbourne?; DM-2 possible	+
Pseudopolydora paucibranchiata	Japan, New Zealand, or Eastern North Pacific	Australia: New South Wales: Botany Bay (1973); Victoria: Hobsons Bay (1975)	Blake & Kudenov, 1978 (suggest introduction, no DM); BW release at Sydney (NSW) and Melbourne (Victoria)? DM-2 possible	+
MOLLUSCA				
Gastropoda				
Littorina meleagris	Tropical Western Atlantic	West Africa: Ghana (1946)	DM-1 (Rosewater & Vermeij, 1972); DM-2 unlikely for non-fouling intertidal species	*
Littorina scabra angulifera	Tropical Western Atlantic	Panama: Pacific side: Panama City (1933)	Bequaert, 1943 (suggests introduction, no DM); DM-2 unlikely for non-fouling intertidal species; appears not to have become established (Rosewater, 1980)	*
Littorina ziczac	Tropical Western Atlantic	Panama: Pacific side: Panama City (1914, 1933)	Bequaert, 1943 (suggests introduction, no DM); DM-2 unlikely for non-fouling intertidal species; "established" (Bequaert), not found there by J. Rosewater (pers. comm.); DM for 1914 record?	*
Bivalvia				
Notospisula trigonella	Eastern and southeastern Australia	Southwestern Australia: Swan River (1964)	DM-8 (Wilson & Kendrick, 1968, who note that the distances involved and current directions do not favour this method); BWT to Fremantle?	+

TABLE II—*continued*

Species	Introduced From[a]	Introduced To[b] (Date)[c]	Dispersal mechanism originally suggested, if any (Reference/Notes)[d]	BWT Likelihood[e]
Martesia cuneiformis	Western subtropical and tropical Atlantic	Panama: Pacific side: Balboa (1940s?)	Turner, 1955 (suggests introduction, no DM); DM-2 possible	+
Bankia gouldi	Western Atlantic: New Jersey to Brazil	Panama: Pacific side: Balboa (1930s?)	Clench & Turner, 1946; United States Navy, 1951 (Gulf of California populations of this species derived at least in part from pre-Canal introduction, as *B. mexicana* Bartsch, a synonym, catalogued at U.S. National Museum in 1907 (J. Rosewater, pers. comm.)); DM-2 also	+
Bankia destructa	Western Atlantic: Honduras to Brazil	(a) Panama: Puerto Armuelles (date?); (b) Mexico: Sinaloa: El Tanque Canal, Caimanero Lagoon (1978–1979)	DM-2 (Clench & Turner, 1946 (a) and Hendrickx, 1980 (b))	+
Bankia fimbriatula	Tropical Western Atlantic	Panama: Pacific side: Balboa (1930s?)	Clench & Turner, 1946 (= *B. canalis* Bartsch, 1944); DM-2 possible	+
Bankia cieba	Western Atlantic: Cuba to Colombia	(a) Panama: Pacific side: Balboa (1930s?) (b) Gulf of California (date?)	(a) Clench & Turner, 1946; (b) Turner, 1971; DM-2 possible	+
CRUSTACEA				
Copepoda: Cyclopoida				
Oithona oculata	Indo-Pacific	Caribbean: Puerto Rico (1957)	DM-8 (Sander & Moore, 1979 who, however, note that life span of this neritic species is less than necessary minimal residence time in oceanic currents)	*

Copepoda: Calanoida				
Acartia centrura	Indian Ocean, Red Sea, or Suez Canal	Eastern Mediterranean (1968)	DM-8 through Suez Canal implied (Berdugo, 1974)	+
Cirripedia				
Balanus subalbidus	Western Atlantic: Maryland to Gulf of Mexico	Massachusetts: Boston: Charles River (1972)	Henry & McLaughlin, 1975, who suggest neither introduction nor DM; DM-2 possible; NE[f]?	+
Balanus eburneus	Western Atlantic: Massachusetts to Brazil	Panama: Pacific side: Balboa (1964)	DM-1 (Newman, 1964); Spivey (1976) gives no further Panama Pacific records; Henry & McLaughlin (1975) note introductions to seaports on west coat of Mexico; DM-2 possible	+
Branchiopoda: Cladocera				
Ilyocryptus agilis	West, North, and central Europe	Chesapeake Bay: Potomac River (1974), James River (1977?)	Williams, 1978, who suggests neither introduction nor DM	*
Malacostraca				
Isopoda				
Munna reynoldsi	Southeast United States: Georgia and Louisiana	Panama Canal: Atlantic and Panama locks (early 1970s?)	Schultz, 1979, who implies dispersal but no mechanism noted; DM-2?	+
Amphipoda				
Crangonyx pseudogracilis	Western North Atlantic	Britain (about 1930)	Hynes, 1955, who suggests introduction but no transoceanic DM; Bousfield, 1973	*
Mysidacea				
Praunus flexuosus	Northwest Europe	Massachusetts: Cape Code: Barnstable Harbor (1960)	DM-2 (Wigley & Burns, 1971); Wigley, 1963; DM-2 unlikely in transoceanic dispersal	*
Mysidopsis almyra	Gulf of Mexico	Chesapeake Bay: Patapsco River (1980)	Grabe, 1981, who suggests neither introduction nor DM	*
Decapoda: Caridea				
Palaemon macrodactylus	Western North Pacific: Japan or Korea	Oregon: Yaquina Bay (1984)	J. Chapman, pers. comm. (1984)	*
Palaemon macrodactylus	(as above)	Australia (pre-1977)	Williams *et al.*, 1978 and Holthuis, 1980, who note introduction but no DM	*

TABLE II—*continued*

Species	Introduced From[a]	Introduced To[b] (Date)[c]	Dispersal mechanism originally suggested, if any (Reference/Notes)[d]	BWT Likelihood[e]
Decapoda: Brachyura				
Pyromaia tuberculata	California	Japan: Sagami Bay (1970)	DM-2 (Sakai, 1976a, as larvae in fouling, but unlikely in transoceanic dispersal)	*
Brachynotus sexdentatus	Mediterranean Sea	Britain: Wales: Swansea (1957)	DM-3 (Naylor, 1957a,b); DM-2?	+
Rhithropanopeus harrisii	Western North Atlantic	Panama Canal: Pedro Miguel Locks (1969)	Carlton, 1979a (L. Abele collector); DM-2?	+
Eurypanopeus dissimilis	Western Atlantic: Florida to Brazil	Panama Canal: Miraflores Third Lock (1971)	McCosker & Dawson, 1975, who imply introduction but suggest no DM; DM-2?	+
ECTOPROCTA				
Gymnolaemata				
Electra monostachys	Western Atlantic: Bay of Fundy to Brazil; western Europe	Panama: Pacific side: Panama Harbor (pre-1924)	DM-1 (Powell, 1971); DM-2 possible	+
CHORDATA				
Vertebrata: Osteichthys				
Blenniidae				
Lupinoblennius dispar	Western Atlantic: Caribbean	Panama Canal: Miraflores Third Lock (1967)	DM-2, as eggs (Dawson, 1970) or as adults (McCosker & Dawson, 1975)	+
Hypleurochilus aequipinnis	Eastern and Western Atlantic	Panama Canal: Miraflores Third Lock (1971)	DM-2 implied by McCosker & Dawson, 1975	+

Gobiidae				
Lophogobius cyprinoides	Western Atlantic: Caribbean	Panama Canal: Miraflores Third Lock (1967?)	DM-2 implied by McCosker & Dawson, 1975; natural dispersal through Panama Canal suggested by Rubinoff & Rubinoff 1968, 1969	+
Gobiosoma nudum	Tropical Western Pacific	Panama: Atlantic side: Galeta Reef (1962)	DM-2 as eggs (Rubinoff & Rubinoff, 1969); McCosker & Dawson, 1975; NE?	+
Tridentiger trigonocephalus	Japan	California: Los Angeles–Long Beach Harbors (1960)	DM-2 as eggs (C. L. Hubbs & J. Prescott, in Hubbs & Miller, 1965); DM-3, -7 (Haaker, 1979)	*
Acanthogobius flavimanus	Japan or San Francisco Bay	California: Los Angeles–Long Beach Harbors (1977)	DM-2, -3, -7 (Haaker, 1979)	*
Cyprinodontidae				
Cyprinodon variegatus	Western North Atlantic	Washington: Strait of Juan de Fuca: Sequim Bay (Dungeness River) (1949)	DM-4 rejected by Greenfield & Grinols, 1965, but implied as possible by Hubbs & Miller, 1965; DM-7 as aquarium discard (Greenfield & Grinols, 1965) considered "unlikely" by Hubbs & Miller, 1965; Greenfield & Grinols also suggest indirect transport with "unofficial transplantation" of fish or shellfish; NE	+
BACILLARIOPHYTA				
Pleurosigma planctonicum	Indian Ocean	Britain: English Channel (1966)	Boalch & Harbour 1977a,b, who imply introduction but no DM	*
Coscinodiscus wailesii	Western Atlantic or Pacific?	Britain: English Channel (1977)	Boalch & Harbour, 1977b ("cannot offer an explanation"); Robinson *et al.*, 1980 (both, as *C. nobilis*)	+

alternative mechanism based on the data at hand. In turn, by assignment of a species to a "probable" category, I do not eliminate entirely all alternatives, as none of the suggested cases in Tables I or II are "proof" of transport by ballast water of marine organisms.

It may be further noted that ballast-water transport may be "probable" for a species in one case but "possible" for the same species in another case. For example, the Asian brackish-water shrimp *Palaemon macrodactylys* (Tables I and II) has probably been carried to Australia and the Pacific coast of North America by ballast water, but its subsequent occurrence in a small port in California, where it is used for fish bait, may be due to its release by fishermen.

ANALYSIS OF TABLES I AND II

Seventy-two cases, involving 64 species, of suggested ballast-water transport are considered in Table I (counting *Janua* and *Pileolaria* as one "instance", or case; *Keratella* and four species of associated phytoplankton as one instance, and not counting teredinids, *Limnoria*, or ostracods, as these last three, in the form presented by the authors in question, represent no particular dispersal event). Forty-five instances, involving 40 species, of potential ballast-water transport cases are considered in Table II (instances where a dispersal mechanism other than ballast water was originally suggested; three species are newly suggested herein as possible introductions to the localities noted, *viz. Balanus subalbidus*, *Mysidopsis almyra*, and *Ilyocryptus agilis*). There are six species in common between Tables I and II, but different localities are involved. Of the taxa in Tables I and II, several fish species (*Platichthys flesus*, *Anguilla* spp., and *Cyprinodon variegatus*) and one crab (*Eriocheir sinensis*) are considered not have established reproducing populations in the localities noted. For a number of other species the status of the introduced populations is not reported in the literature or is not known.

Of the 72 cases in Table I, I consider one unlikely (*Processa*), 28 as "possible", and 43 as "probable" instances of ballast-water transport. Of the 45 cases in Table II, I consider 31 as "possible" instances and 14 as "probable".

Those species that are "possible" cases of ballast-water transport are largely those for which dispersal as fouling or boring organisms on or in ships has been or is possible, as discussed above. These include hydroids, scyphozoans, polychaetes, boring and epifaunal bivalves (and in one instance, an infaunal bivalve, *Mya arenaria*, known to occur in fouling communities), barnacles, ostracods, isopods, amphipods, xanthid crabs, bryozoans, ascidians, and algae. For the remaining "possible" cases, either too little is known about the species or the incidence cited to resolve clearly a dispersal mechanism (for example, the sponge *Trochospongilla*, the bivalve *Notospisula*, the copepod *Acartia tonsa*, the cladoceran *Penilia*, and the fish *Cyprinodon*) or dispersal mechanisms other than fouling are possible (for example, the copepod *Acartia centrura*, the shrimp *Palaemon macrodactylus* in California—Elkhorn Slough record—and the alga *Sargassum*). Blenniid and gobiid fishes are regarded as "possible" cases of ballast-water transport if relatively short distances of transport are involved (as through the Panama Canal, discussed below), but as "probable" cases if transoceanic or interoceanic dispersal is involved. The case reported by Bertelsen & Ussing (1936) of blennies carried on the bottom of a ship from Bermuda to Denmark, where unusual circumstances were involved, is an exception to this generalization.

The "probable" cases include the adult or larval stages of siphonophores, of littoral gastropods and infaunal bivalves not known to occur in fouling communities (and for which other mechanisms such as movements of commercial oysters do not appear to apply), cyclopoid and calanoid copepods, the cladoceran *Ilyocryptus*, the amphipods *Crangonyx* and *Gammarus*, mysids, the shrimp *Palaemon*, the crabs *Pyromaia* (Inachidae), *Callinectes* (Portunidae), and *Eriocheir* (Crapsidae), and gobiid, blenniid and eleotrid fishes. Of these probable instances, several cases require additional comment beyond the remarks in Tables I and II.

The dispersal of the calanoid copepod, *Pseudodiaptomus marinus*, has been discussed by Jones (1966), Grindley & Grice (1969) and Bowman (1978), and the dispersal of the cyclopoid copepod, *Oithona oculata*, by Sander & Moore (1979). Jones (1966) remarked that "it is not unreasonable to assume" that *Pseudodiaptomas* "could survive the salinity of the open ocean" and thus be transported by oceanic currents from Japan to Hawaii. Jones noted, however, that such oceanic transport would require *Pseudodiaptomus* to pass through several generations on the way to Hawaii. Sander & Moore (1979) in an extended account, suggested that the Indo-Pacific *Oithona oculata* was transported in oceanic currents to the Caribbean, around Africa and across the South Atlantic Ocean, and in doing so "must have managed to survive its full life cycle between major continents". They noted that a crossing of the Atlantic alone "would require a minimum period of one year" and that even if islands along the proposed route were colonized, "there are expanses of ocean, uninterrupted by island-masses that would require a minimum of six months at sea", a time period which would also appear to require *Oithona* to pass through several generations.

The probability of free-swimming neritic copepods being transported inter- or transoceanically on oceanic currents, especially if such transport exceeds the life span of an individual copepod, would appear to be low. I have found no data indicating that any strictly neritic species of copepod could or does pass through multiple generations while being transported on open ocean currents. Yeatman (1962) noted that "the most obvious method of dispersal of littoral harpacticoids and some cyclopoids is by clinging to marine algae drifting in the open ocean currents". In a limited series of stations across the North Atlantic Ocean from Europe to North America, Yeatman took plankton hauls "to determine if any littoral species may survive unsupported in the open ocean currents" and collected drifting algae. He found no littoral species in the plankton samples, and one cyclopoid species and six harpacticoid species on drifting holopelagic *Sargassum*. Both *Pseudodiaptomus marinus* and *Oithona oculata* are neritic taxa, not associated with drifting algae, and have not been recorded from oceanic waters. Dispersal on transoceanic currents appears to be most unlikely. Calanoid and cyclopoid copepods are, however, known to survive in ballast water (as discussed below), the routes indicated for both of the above species taking a matter of days if by ship.

The singular case of several freshwater taxa (the rotifer *Keratella* and four species of diatoms and chrysophytes, Table I, found as a food bolus in the stomach of a near-surface juvenile white hake (*Urophycis tenuis*) in the open ocean off Canada is an event regarded by Coates, Roff & Markle (1982) as linked to the discharge of these freshwater taxa from a ballast tank or "contaminated freshwater holding tank" of a passing vessel. The studies summarized by Howarth (1981) and discussed below indicate that such taxa are common in ballast-tank water. As discussed by Coates *et al.* (1982) it is difficult to identify another mechanism.

A temporal analysis, by date of first collection, of the 57 "probable" cases of ballast-water transport in Tables I and II is hindered by several factors. The date of first collection is not necessarily and indeed rarely is coincident with the date of introduction (Carlton, 1979a) and thus any detectable patterns in time may be accurate only within a period of several years at best. Such patterns may be clearest where recent prior investigations demonstrate the absence of the species in question. It is also difficult to separate completely the increased shipping activity over the twentieth century (as discussed earlier) from the increased number of investigators over the same period. Finally, elimination of the "possible" cases of ballast-water transport, due to difficulties in resolving the rôle of fouling as a dispersal mechanism in these instances, may tend to make ballast-water transport appear less frequent than it may have been, especially in the years following World War II.

Given these restrictions, an analysis of the time sequence involved reveals the following patterns. Pre-World War II records of hypothesized ballast-water transport are limited. This may reflect differences in maritime traffic during this time (fewer and slower ships which may have resulted in fewer successful ballast-water dispersal incidences, as discussed below), combined with the factors listed above and a general lack of awareness by investigators that such a dispersal mechanism existed. The earliest probable occurrences are the appearance of the Asian diatom, *Biddulphia sinensis*, in 1903 in the North Sea and of the Asian crab, *Eriocheir sinensis*, in 1912 in Germany. These occurrences suggest the possible onset of ballast-water traffic in the first decade of this century from Japan, Korea and China to the North Sea region of western Europe. In the early 1930s the amphipod, *Gammarus tigrinus*, and the blue crab, *Callinectes sapidus*, both North American species, first appeared in Europe. The blenny, *Omobranchus*, was evidently also being transported around the Pacific, Indian, and Atlantic Oceans at this time. The first probable case of a ballast-water introduction in the Northeastern Pacific Ocean is about 1940 with the collection of the intertidal gastropod, *Sabia conica*, on Vancouver Island, perhaps introduced as larvae by Japanese vessels arriving on the ballast leg. Ballast-water introductions in the Northeastern Pacific Ocean are particularly difficult to detect due to the intensive introduction of oysters from Japan to that region over many years.

In contrast, between 1962 and 1984, only one year (1967) is without a "first collection" record. Probable instances of ballast-water transport become relatively common in the mid- to late 1970s. Of particular interest is the discovery in the eastern North Pacific Ocean within the three-year period 1962–1964 of a number of Japanese species regarded here as probable ballast-water introductions (the shrimp *Palaemon*, the fish *Tridentiger* and *Acanthogobius*, and the molluscs *Laternula*, *Sabia*, and *Mesoclanculus*). At the same time, the Japanese copepod, *Pseudodiaptomus marinus*, was first collected in both Hawaii and Mauritius.

A combined geographic and temporal analysis suggests that active regions of ballast water-mediated introductions in the last decade and a half may be

(1) northwestern Europe (the 1978 appearance of the American razor clam, *Ensis directus*, and the continual introduction of the American blue crab, *Callinectes*), (2) Australia (the appearance in the 1970s of various Japanese species, including the gobies *Acanthogobius* and *Tridentiger*, the shrimp *Palaemon*, and possibly the polychaetes *Boccardia* and *Pseudopolydora*), and (3) the Pacific coast of the United States (the appearance of the bivalve *Theora* and the goby *Acanthogobius* in southern California, and of the copepods *Limnoithona, Oithona*, and *Sinocalanus* since 1978 in San Francisco Bay, all of which are Asian species).

Whether ballast-water transport is an "active" mechanism of transport through the Panama Canal is not yet known. Possible or probable incidences of ballast-water dispersal through or into the Panama Canal listed in Tables I and II are as follows.

Possible

Trochospongilla leidii, freshwater sponge
Martesia cuneiformis, boring pholad clam
Bankia spp., shipworms
Balanus eburneus, barnacle
Munna reynoldsi, isopod
Rhithropanopeus, harrisii, mud crab
Eurypanopeus dissimilis, mud crab
Neorhynchoplax alcocki, freshwater crab
Electra monostachys, bryozoan
Lupinoblennius dispar, blenny
Hypleurochilus aequipinnis, blenny
Lophogobius cyprinoides, goby
Gobisoma nudum, goby

Probable

Chelophyes contorta and *C. appendiculata*, siphonophores
Muggiaea atlantica and *M. kochi*, siphonophores
Lensia challengeri, siphonophore
Littorina scabra angulifera, periwinkle
Littorina ziczac, periwinkle
Prionobutis koilomatodon, eleotrid fish
Barbulifer ceuthoecus, goby

The existence of various species of fouling and boring organisms (pholad clams, shipworms, barnacles, bryozoans, and others) normally regarded as native to one or the other ocean, on opposite sides of the Panama Canal, has attracted the attention of a number of workers. These records are generally regarded as representing euryhaline taxa that as adults may survive transport through the freshwater sections of the canal, presumably on or in local vessels that have been resident in Canal Zone waters for a sufficient time to accumulate a fouling or boring fauna (Hay & Gaines, in press). Several of these species have planktotrophic larvae hypothetically transportable in ballast water. The oviparous shipworm, *Bankia*, has planktotrophic larvae which may remain in the plankton for two or more weeks (Turner, 1966; Turner & Johnson, 1971). At least four species of this tropical western Atlantic genus are regarded as having been transported by shipping through the Panama Canal to the west coast of Panama and Mexico. Except for pre-1914 records of these species in Pacific waters, ballast-water transport of these shipworms can neither be rejected nor accepted as a dispersal mechanism given the present state of our knowledge. The potential rôle of ballast-water dispersal in the movement of the larvae of membranipore bryozoans is also of interest here.

Zimmer & Woollacott (1977) have noted that "the pelagic and planktotrophic cyphonautes larvae of *Membranipora* and *Electra* are among the most commonly recognized of any marine larva because of their regular presence and often incredible abundance in the plankton". While Powell (1971, p. 776) lists five species of *Membranipora* and *Electra* that may be introductions to the Pacific entrance of the Panama Canal, the data are insufficient to draw clear conclusions, and I have included only one species, *Electra monostachys*, in Table II (although *Membranipora savartii*, which occurs at either end of the Canal, may also be a candidate here).

Of the remaining "possible" cases, four species of blennies and gobies may have been transported either on or in vessel fouling or in ballast water. McCosker & Dawson (1975) have demonstrated that *Lupinoblennius*, *Hypleurochilus*, and *Lophogobius* are all euryhaline and could survive transport through the freshwater sections of the Canal, while Rubinoff & Rubinoff (1969) experimentally determined that the eggs of the stenohaline, *Gobiosoma*, can survive in freshwater portions of the Canal and still remain viable. It is of interest to note that a species of *Hypleurochilus* from Bermuda survived a slow passage in a dense fouling community on a ship's bottom across the Atlantic to Denmark (Bertelsen & Ussing, 1936).

It is perhaps surprising that, overall, so few species may have been transported through the Panama Canal by ballast water, given the volume of traffic over the past 70 years, and the relatively short distance (and thus time in transport) and presumably relatively stable temperature regimes involved (see discussion below). Indeed, Glynn (1982) has noted that increasing numbers of oil tankers (arriving in ballast to receive Alaskan oil from tankers too large to transit the Canal) are releasing up to 20 000–30 000 tons of clean ballast sea water each from the Caribbean into Parita Bay in the Gulf of Panama. It is of interest to observe that Chesher (1968) noted that the greatest traffic in ballasted vessels in the 1960s was in the opposite direction. If ballast-water transport has played a rôle, it may either be largely unrecognized (Clarke, 1969) or unrecognizable (Chesher, 1968), because of the lack of pre-canal, and for most of the Canal's history, post-canal, biological and distributional data, especially for such candidate taxa as hydromedusae, polychaetes, copepods, peracarid crustaceans, membranipore bryozoans, and others. Whether the motto on the official seal of the Panama Canal Zone ("A land divided; the oceans united") applies to some portions of the marine biota as well as to maritime transportation is not yet well understood.

EVIDENCE FOR BALLAST-WATER TRANSPORT

SAMPLES FROM BALLAST TANKS: ORGANISMS SURVIVING TRANSPORT

Twelve phyla representing a diverse array of neritic (marine and estuarine) protists (ciliate protozoans), invertebrates, and fish have been recorded as having survived transport in ballast water (Table III). Data sources for these records are given in Table IV (references 1–50). The major sources were briefly discussed in the literature survey at the beginning of this review.

Most commonly recorded are nematodes, rotifers, polychaetes, cladocerans, barnacle larvae (nauplii and cyprids) and harpacticoid, cyclopoid and

TABLE III

Protists, invertebrates, and fish surviving transport in ballast tanks: for data sources see Table IV

	Reference number, see Table IV
CILIOPHORA	
Unidentified	2G, 11, 18, 24, 34, 44
Gymnostomatia	44
Peritrichida	13, 44
Oligotrichida (Tintinnina)	9
Hypotrichida	2G, 23, 44
PLATYHELMINTHES	
Turbellaria (Acoela)	13
NEMATODA	
Unidentified	2G, 15, 18, 23, 24, 33, 34, 39, 42, 43
ROTIFERA	
Unidentified	13, 44
Keratella spp.	6, 9, 10, 15, 20, 27, 28, 29, 33
GASTROTRICHA	
Unidentified	44
ANNELIDA	
Polychaeta	
Adult benthic species, unidentified	50
Larvae, unidentified	3, 4, 23, 50
Spionidae (adults?)	30
Spionidae, larvae, unidentified	6, 40
Spionidae, *Polydora* sp., larvae	6
Capitellidae?, larvae	1
Oligochaeta	
Naididae	6
MOLLUSCA	
Gastropoda	
Unidentified veligers	2C, 6, 16, 22, 38, 42
Bivalvia	
Unidentified larvae	6, 11, 21, 22, 38
CRUSTACEA	
Unidentified crustacean larvae	50
Branchiopoda: Cladocera	
Unidentified	13
Acroperus sp.	42
Bosmina sp.	9, 10, 28, 29, 32, 33, 38, 43
Evadne sp.	2A, 23, 38
Podon sp.	2A
Ephippia	28, 33
Ostracoda	
Unidentified	7, 14, 16, 29, 30, 43, 50

TABLE III—*continued*

	Reference number, see Table IV
Cirripedia	
Nauplii, unidentified	38
Cyprids, unidentified	11, 23
Semibalanus balanoides, nauplii and cyprids	3, 4
Elminius modestus, nauplii	14, 16, 26, 27, 30, 32, 33
Elminius modestus, cyprids	27
Verruca sp., nauplii	42
Copepoda	
"several" planktonic species	50
"20 identified copepod species" (at least six of which were Japanese)	48
Copepoda: Harpacticoida	
Unidentified	1, 2F, 3, 5, 9, 11, 17, 23, 25, 35, 38, 42
Euterpina sp.	15, 21, 31, 37
Euterpina acutifrons	14
Macrosetella sp.	8
Parathalestris croni	26
Copepoda: Cyclopoida	
Unidentified	6, 7, 10, 13, 14, 19, 22, 23, 25, 38, 41
Corycaeus sp.	8, 15, 32, 37, 42
Corycaeus anglicus	20, 22, 29
Oithona similis	2F
Oithona spp.	8
Copepoda: Calanoida	
Unidentified	2D, 3, 6, 7, 9, 10, 11, 12, 14, 15, 16, 17, 19, 20, 22, 23, 26, 27, 28, 29, 30, 31, 32, 33, 35, 36, 37, 38, 39, 40, 41, 42, 43, 45
Nauplii	2B, 6, 9, 10, 14, 15, 17, 19, 22, 26, 27, 28, 29, 30, 32, 33, 35, 36, 37, 38, 39, 41, 43, 45
Acartia sp.	21
Acartia clausii	2D, 3, 4
Acartia longiremis	2A
Centropages sp.	2E
Centropages hamatus	3, 4
Eurytemora sp.	3, 4
Metridia lucens	3
Pseudocalanus sp.	3
Temora longicornis	2A, 3, 4
Tortanus discaudatus	3
Malacostraca	
Mysidacea	
Mesopodopsis slabberi	14
Neomysis spp.	13, 17, 19, 41
Unidentified	47, 48
Amphipoda: Gammaridae	
Unidentified	48, 50
Melitidae	13
Amphipoda: Caprellidae	
Caprella sp.	42
Isopoda	
Asellus sp.	42
Idotea baltica (juvenile)	28

TABLE III—*continued*

	Reference number, see Table IV
Malacostraca: Decapoda	
Penaeidea	
Unidentified nauplii	26
Anomura	
Porcellana sp., zoea	7
Porcellana longicornis, zoea	26
UNIRAMIA (Insecta)	
Diptera: Chironomidae	
Orthocladinii, larvae	15
CHELICERATA	
Acariformes	
"Hydracarina", unidentified	23
CHAETOGNATHA	
Unidentified	50
Sagitta sp.	14, 22
CHORDATA	
Vertebrata: Osteichthyes	
"a number of exotic fishes"	46
Blenniidae	
Petroscirtes breviceps	49
Bothiidae	
Pseudorhombus arsius	47
Clupeidae	
Hyperlophus vittatus	47
Gasterosteidae	
Gasterosteus aculeatus	13
Gobiidae	
Favonigobius sp.	47
Scorpaenidae	
Centropogon australis	47

calanoid copepods. Less common but with multiple records are ciliates, gastropod and bivalve larvae, ostracods, mysids, gammarid amphipods and fish. Rare are turbellarians, gastrotrichs, oligochaetes, isopods, decapod larvae, insects and mites. These "relative frequencies" are based upon limited records, and will doubtless change with additional data. In addition to these taxa, numerous species of freshwater and marine phytoplankton have been recorded by Howarth (1981) and have been observed by myself. These include diatoms, dinoflagellates, euglenophyceaens, chrysophyceans, desmids, and cryptomonads.

Fifty per cent of the records are based upon ballast water originating from the northeastern Atlantic Ocean, 20% are from the Mediterranean and Black Sea, 18% are from the northwestern Atlantic Ocean, 6% are from the Persian

Gulf–Indian Ocean–Australia region, 4% are from Japan, and one vessel (2%) is from the southwestern Atlantic Ocean (Brazil).

The days-in-transit of the ballast water and plankton range from 4 to 95. While the days-in-transit are not listed for individual vessels by Howarth (1981) (data source B in Table IV), the maximum length of time in transit for any ship was 18 days (Howarth, 1981). Most of the longer voyage records are based upon the studies of Carlton, Navarret & Mann (1982). Their studies examined the quantitative patterns of plankton survival in the ballast tanks (14 624 gallons of water per tank) of the oceanographic research vessel KNORR (Woods Hole Oceanographic Institution) on coastal cruises of nine and 15 days duration (ballast water from Woods Hole, Massachusetts, with no port stops prior to arrival back at Woods Hole, references 3, 4, 5, Table IV), and on longer voyages where sea-water ballast was carried from Iceland to St John's, Newfoundland (30 days, reference 11, Table IV), and from St John's to Woods Hole (31 days, reference 1). In all cases the original ballast water was retained and none was added at sea or at other ports.

Taxa surviving the 30-day passage from Iceland to St John's, during which time the initial tank-water temperatures of 10·5 °C fell to a low of 2·8 °C before returning to 12·5 °C upon arrival in St John's, were polychaete larvae, harpacticoid copepods, calanoid copepod nauplii and adults and, in rare numbers, bivalve larvae and barnacle cyprids (Carlton *et al.*, 1982, who detailed the quantitative results of these studies). Taxa surviving the 31-day passage from St John's to Woods Hole, during which time tank-water temperatures, initially at 10·5 °C rose to 25·0 °C (when the KNORR sailed to latitudes southeast of Bermuda) before returning to 13·1 °C, were capitellid polychaete larvae and harpacticoid copepods. A long-term experiment utilizing the same ballast water in one tank examined the quantitative survival of plankton ballasted in Scotland after 31 days transit to Iceland (reference 23, Table IV), after 64 days transit to Newfoundland (reference 24), and after 95 days transit to Woods Hole (reference 25). Taxa surviving included polychaete larvae, cladocerans, calanoid copepods, barnacle cyprids, and mites after 31 days, while cyclopoid and harpacticoid copepods survived 95 days to Woods Hole (Carlton *et al.*, 1982). Another experiment involved the *en route* and *in situ* sampling of plankton in the forepeak ballast tank (461 286 gallons of water) and in smaller ballast tanks of an oil tanker from Portsmouth, New Hampshire (north of Boston, Massachusetts), the source of the water, to Corpus Christi, Texas, a voyage of nine days. Differential patterns of plankton survival were documented as water temperatures steadily rose from 7·5 °C to 22·0 °C by the end of the voyage. Taxa surviving transport to the south Atlantic coast (Georgia–Florida region), with water temperatures rising to 13·8–20·0 °C, included gastropod larvae, the calanoids *Acartia* and *Centropages*, the cyclopoid *Oithona*, and harpacticoid copepods (Carlton *et al.*, 1982). Taxa surviving into the Gulf of Mexico (water temperatures of 20–22 °C) included *Oithona* and harpacticoids, as well as nematodes and protozoans. During the course of the voyage, additional water was ballasted into a cargo tank off the southeastern coast of Florida and again 24 hours later off the Dry Tortugas. Taxa surviving the subsequent four-day voyage in this tank across the Gulf of Mexico to Corpus Christi (and thus through the Loop Current) included the cyclopoids *Oithona* and *Corycaeus* and the harpacticoid *Macrosetella*. The quantitative results of all of these experiments aboard the

KNORR and the oil tanker, and the detailed temperature histories of the tank water, are presented by Carlton *et al.* (1982). The species surviving in these experiments are incorporated here in Tables III and IV.

Also incorporated in Tables III and IV are data on various planktonic and nektonic species collected alive in the ballast tanks of several merchant vessels arriving on the Atlantic coast of the United States. One of these cases is detailed here as an example of a ballast-water dispersal event. These observations are based upon field work by myself and A. Navarret in July 1981. A bulk cargo vessel was met as it arrived at dock in Wilmington, Delaware (on the lower Delaware River just above the head of Delaware Bay) at the completion of a 12-day transit from Brake, on the Weser River, Germany, where the ballast water had been taken on by gravity flow. An upper wing port ballast tank, with a deck manhole cover, was sampled. The top of the tank was the deck of the ship, and the walls of the tank were above the waterline. The steel tank, with uncoated bare walls, 5·8 m in depth, was filled within 0·5 m of the tank ceiling. The water was mesohaline ($S = 5^{\circ}/_{\circ\circ}$), warm (24 °C) and oxygen was 71·0% saturation (about 5·8°/₀₀). As cargo was loaded for Japan, the vessel released this water at Wilmington, where July salinities and water temperatures average about 1°/₀₀ and 25·0 °C. Living plankton recovered by vertical net hauls in the tank included juvenile and adult mysids (*Neomysis* sp.), melitid gammarid amphipods, unidentified cladocerans, and unidentified cylopoid copepods. In addition, flagellates, peritrich protozoans, rotifers, dinoflagellates (*Ceratium*) and various diatoms and algal filaments were collected. Also collected was a threespine stickleback fish *Gasterosteus aculeatus* (2·5 cm total length), which was observed to be swimming actively in the tank. Over a 30-minute period due, apparently, to their being attracted by the light entering through the now-open deck manhole cover more were found. Five of seven collected fish survived transport back to Woods Hole, Massachusetts, where they lived (in water of the same temperature and salinity as the ballast tank) and consumed brine shrimp (*Artemia*) for two weeks. The data of Jordan & Garside (1972) suggest that these fish may have been near their upper lethal temperature limits. Jordan & Garside found that *Gasterosteus aculeatus* acclimated at 10 °C in 0°/₀₀ salinity have an upper lethal temperature, at a test salinity of 5°/₀₀, of about 26·5 °C; fish acclimated to 20 °C under the same test conditions have a limit only slightly higher, of about 27·3 °C. In concert with these experimentally determined temperature limits, *Gasterosteus aculeatus* reaches its southern limits on the Atlantic coast in the Delaware–Chesapeake Bay region (Burgess & Lee, 1980).

The discovery of the Eurasian stock of the stickleback *Gasterosteus* confirms that living fish can survive a transoceanic voyage in a ballast tank. Whether such events are common is not known. It may be predicted, however, that *Gasterosteus*, among other brackish-water fish, will be found to be a member of a "ballast-water biota" that may be making regular transatlantic crossings. The discovery of live mysids (and the other reports of living mysids in ballast tanks, Table III) lends some support to the concept that the common western European mysid, *Praunus flexuosus*, which was first collected in Massachusetts in 1960, may have been a ballast-water introduction.

TABLE IV

Data sources for Table III: a, vessel type—BC, bulk carrier; CS, container ship; MS, merchant ship (not specified); PCT, petroleum cargo tanker; RV, oceanographic research vessel; b, References—A, Carlton, Navarret & Mann, 1982; B, Howarth, 1981; C, Medcof, 1975; D, Williams, 1977, 1982 (in Middleton, 1982), Williams, van der Wal & Story, 1978; E, Middleton, 1982; F, Springer & Gomon, 1975; G, Wheeler, 1958; H, J. T. Carlton & A. Navarret, unpubl. obs., 1981; c, sample examined 18 days later, thus 35 days old

Reference number, given in Table III	Ballast water from	Ballast water sampled at	Vessel type[a]	Year	Days in transit	References[b]
1	Newfoundland: St John's	Massachusetts: Woods Hole	RV	1981	31	A
2	New Hampshire: Portsmouth	Texas: Corpus Christi	PCT	1981	9	A
	[Same vessel (2) sampled daily while en route to Corpus Christi: Reference (Days in Transit): 2A(2), 2B(4), 2C(5), 2D(6), 2E(7), 2F(8), 2G(9)]					
3	Massachusetts: Woods Hole	Massachusetts: Woods Hole	RV	1981	9	A
4	Massachusetts: Woods Hole	Massachusetts: Woods Hole	RV	1981	15	A
5	Massachusetts: Woods Hole	Bahamas: Freeport	RV	1981	15	A
6	Maryland: Baltimore	Montreal	MS	1980	—	B
7	Virginia: Norfolk	Montreal	MS	1980	—	B
8	Southeast Florida and Gulf of Mexico near Dry Tortugas	Texas: Corpus Christi	PCT	1981	4	A
9	Caribbean Sea, near Cuba	Montreal	MS	1980	—	B
10	Brazil: Santos	Montreal	MS	1980	—	B
11	Iceland: Reykjavik	Newfoundland: St John's	RV	1981	30	A
12	Finland and Baltic Sea	Montreal	MS	1980	—	B
13	Germany: Weser River: Brake	Delaware: Wilmington	BC	1981	12	H
14	North Sea	Montreal	MS	1980	—	B
15	Netherlands: Rotterdam	Montreal	MS	1980	—	B
16	Netherlands: Delfzijl	Montreal	MS	1980	—	B
17	Netherlands: Delfzijl	Montreal	MS	1980	—	B
18	Belgium: Ghent	Virginia: Norfolk	BC	1981	13	H
19	Belgium: Antwerp	Montreal	MS	1980	—	B

20	England: Immingham and Atlantic Ocean	Montreal	MS	1980	—	B
21	England: Felixstowe	Boston	CS	1981	9	H
22	France: Dunkirk; North Sea	Montreal	MS	1980	—	B
23	Scotland: Firth of Clyde; Greenock	Iceland: Reykjavik	RV	1981	31	A
24	Scotland: Greenock	Newfoundland: St John's	RV	1981	64	A
25	Scotland: Greenock	Massachusetts: Woods Hole	RV	1981	95	A
26	Irish Sea	Montreal	MS	1980	—	B
27	England: Birkenhead	Montreal	MS	1980	—	B
28	Britain: Atlantic coast	Montreal	MS	1980	—	B
29	English Channel	Montreal	MS	1980	—	B
30	France: Caen	Montreal	MS	1980	—	B
31	Spain: Bilbao	Montreal	MS	1980	—	B
32	Portugal: Lisbon	Montreal	MS	1980	—	B
33	Spain: Atlantic coast	Montreal	MS	1980	—	B
34	Azores	Massachusetts: Woods Hole	RV	1981	20	H
35	Algeria: Mostaganem	Montreal	MS	1980	—	B
36	Spain: Tarragona	Montreal	MS	1980	—	B
37	France: near Marseilles	Montreal	MS	1980	—	B
38	Mediterranean Sea	Montreal	MS	1980	—	B
39	Italy: Venice	Montreal	MS	1980	—	B
40	Italy: Adriatic Sea	Montreal	MS	1980	—	B
41	Adriatic Sea	Montreal	MS	1980	—	B
42	Adriatic Sea	Montreal	MS	1980	—	B
43	Near mouths of Nile River	Montreal	MS	1980	—	B
44	Black Sea	Boston	BC	1981	17[c]	H
45	Black Sea	Montreal	MS	1980	—	B
46	Persian Gulf or Singapore	England: Grimsby	Dredge Carrier	1954	—	G
47	Indian Ocean: Western Australia: Port Hedland	Australia: New South Wales: Newcastle	BC	1980/81?	—	E
48	Northwest Australia ("a continuous flow of seawater in the ship's hold")	Papua New Guinea: Port Moresby	MS	1972	—	F
49	Japan	Australia	BC	1976+	—	D
50	Japan: Tagonoura, "and partly at sea during the first four days of the passage"	Australia: New South Wales: Eden: Twofold Bay	MS	1973	14·5	C

SAMPLES FROM BALLAST TANKS: ADDITIONAL SPECIES ENTRAINED ALIVE

In addition to those species shown in Table III as having survived in ballast tanks for a specified voyage, a number of other species are known to have been pumped into or gravitated into ballast tanks alive which did not subsequently survive that voyage that followed. If such taxa can survive the ballasting process, it may be that under different transit conditions ballast water would act as a transport mechanism. These species are listed in Table V. The data are from my previously unpublished observations based upon ballast-water samples taken from ballast tanks shortly after each vessel was ballasted. These pre-transit samples were later compared to post-transit samples; the quantitative results of those species surviving the voyages are given by Carlton *et al.* (1982). I list here the taxa found alive, swimming or moving, in ballast tanks at the time of ballasting but not subsequently recovered. Other species did survive these same voyages (Tables III, IV); thus, the tank and transit conditions led to selective mortalities among the ballasted plankton.

Several records are of interest here relative to previously published suggestions of ballast-water transport (Table III), specifically for certain taxa for which there has been no previous evidence that they occurred in ballast tanks. These include siphonophores (which Alvariño, 1974, suggested may be moved through the Panama Canal by ballast water), hydromedusae, caridean (shrimp) larvae and brachyuran (crab) larvae (all suggested by various workers to have been candidates for ballast-water transport), and the boring isopod *Limnoria* (free-swimming; Kofoid & Miller, 1927, having suggested that this species could be taken up into ballast tanks in bits of wood).

BALLAST-WATER BIOLOGY: A SUMMARY

A comparison of Tables I, II, III, and IV suggests a high level of correspondence with some interesting gaps. Almost all higher taxonomic groups which have been hypothesized as being transported in ballast water (Tables I and II) have been found alive in ballast tanks (Table III). The correspondence is, in some cases, to the familial and generic levels. The study of ballast-water biota is, however, too recent to have generated sufficiently large data sets to yield correspondences at the specific level between hypothesized dispersal and actual presence in ballast tanks. An examination of Table I suggests the prediction that the first specific link could be the discovery of larvae of the blue crab, *Callinectes sapidus*, in ballast water in vessels returning from the mid-Atlantic United States coast (such as Chesapeake Bay) to western Europe (particularly vessels entering the North Sea Canal in the Netherlands, for example).

Of the remaining hypothesized incidents, there are no records of sponge, scyphozoan, or ascidian larvae from ballast tanks. Hydromedusae, siphonophores, shrimp, crab, and bryozoan larvae are known to have survived ballasting but not transport (Table V).

Conversely, several taxa are known to have survived transport but have not, to my knowledge, been suggested to have been dispersed by ballast water. These include mostly protists and small invertebrates (ciliate protozoans,

TABLE V

Zooplankton entrained alive in ballast tanks other than those taxa in Table III: for data sources see Table IV

	Reference number see Table IV
COELENTERATA (CNIDARIA)	
Hydrozoa	
Hydroida: hydromedusae, unidentified	8
Siphonophora, unidentified	8
PLATYHELMINTHES	
Turbellaria (Acoela)	8
ANNELIDA	
Polychaeta	
Terebellidae (?) larvae (tubiculous)	8
CRUSTACEA	
Cladocera	
unidentified species	8
Copepoda: Harpacticoida	
Tisbe sp.	8
Copepoda: Calanoida	
Acartia longiremis	2
Isopoda	
Limnoria lignorum	4
Amphipoda: Hyperiidae	
Unidentified (two species)	8
Amphipoda: Gammaridae	
Ischyroceridae	23
Malacostraca: Decapoda	
Caridea, unidentified larvae	8
Brachyura, unidentified megalopae and zoea	23
ECTOPROCTA	
Unidentified cyphonautes larvae	2
ECHINODERMATA	
Bipinnaria larvae	8
CHAETOGNATHA	
Unidentified species	8
CHORDATA	
Urochordata: Thaliacea	
Unidentified doliolids	8
Urochordata: Larvacea	
Oikopleura sp.	8
Fritillaria sp.	8
Vertebrata: Osteichthyes	
Eggs and larvae, unidentified	8

turbellarians, nematodes, gastrotrichs, and harpacticoid copepods), all of which are characterized by high numbers of "cosmopolitan" species. There are also records of living anomuran larvae and chaetognaths in ballast tanks (Tables III, V). I suggest here (Table II) that ballast water may have played a rôle in the dispersal of certain barnacles and mysids, taxa which have been recovered alive, post-voyage, in ballast tanks.

A DESCRIPTIVE MODEL OF BALLAST-WATER TRANSPORT

I present here a description of the probable sequence of events in time and space of the dispersal of marine organisms by ships' ballast water, in terms of a series of steps each capable of investigation (Fig. 1).

The transport of an organism in ballast water would appear to be governed by a series of increasingly smaller survival "filters" connected by "bottlenecks". The pictorial model (Fig. 1) represents a composite sequence of steps rather than any one dispersal event. I have shortened the length of the "filter" at each subsequent step, on the assumption that the number of species is reduced sequentially. As the processes between each step (filter) are complex, multiparametric, and poorly understood, I have maintained the same bottleneck width through the sequence. From the naturally-occurring donor area planktonic-nektonic community adjacent to the vessel (Stage I), a subset of species is drawn (gravitated) or pumped into the ballast tank. This subset (Stage II) is composed of those species in the water column at the time of ballasting, and is assumed to change with tidal cycles, as well as over longer day-month-year time scales. Thus, larvae of some local species that are reproducing may not be present at shipside at the time of ballasting; nektonic species may not be present in the water column next to the ship's intake during ballasting, and so forth. Active avoidance by some species of tank pumps or gravity intake pipes may be an additional factor, as would the effect of the pumping and screen system on the size range of species effectively ballasted. The initial ship ballast-tank assemblage (II) is thus developed through

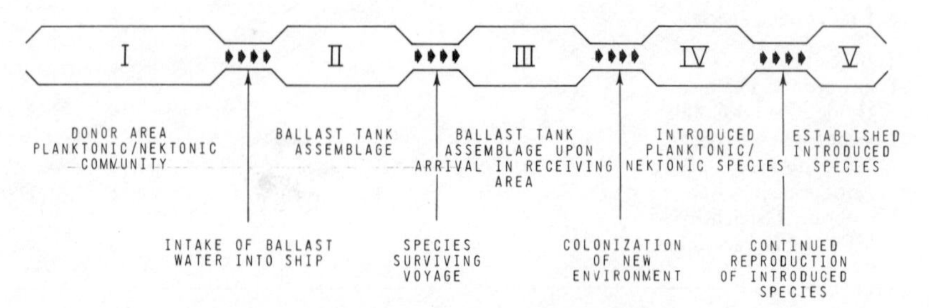

Fig. 1.—Sequence of events in the dispersal and introduction of exotic invertebrates by ships' ballast water.

selective processes. The third stage consists of the assemblage of species that have survived the voyage. Taxa that may be favoured here are discussed below. Some species, although surviving, may no longer be competent, being rendered incapable of completing development or of producing viable gametes due to sustained physiological stress. The ballast-water biota that arrives in the new region (Stage III) may now be released into an area with a suite of novel environmental conditions, including physical-chemical regimes, predators, and competitors, not previously experienced by the species involved. Those colonizing individuals able to survive initially (Stage IV) must, if sexual reproduction is necessary and if ovigerous females are not among the population initially released, remain sufficiently concentrated or localized to mate. A species must then maintain sustained reproducing populations to become established (Stage V) and disperse. Discussed below are certain aspects and conditions of each of these five stages.

STAGE I: AVAILABLE PLANKTON-NEKTON FOR INTAKE INTO BALLAST-WATER TANKS

As noted at the beginning of this review, any organism in the water column is a potential candidate for entrainment into water tanks. It is necessary, therefore, to avoid assumptions about the availability of a given taxon in the water column based upon its presumed biology or distributional ecology. For example, Wolff (1977) argued that the ballast-water transport of the horseshoe crab *Limulus polyphemus* from North America to northern Europe (where specimens have been found regularly since the late 1960s) was "highly improbable", because "*Limulus* larvae, although able to swim, do not usually occur pelagically, in particular not off main harbours where ships take in ballast water". *Limulus* larvae, however, do occur in the plankton, and Rudloe (1979) has described newly-emerged larvae as "swimming vigorously and continuously for hours"; indeed, *Limulus* larvae were first described from the plankton (Packard, 1872). *Limulus* larvae thus could be carried by tidal or longshore currents into harbour areas. Wolff (1977) has argued convincingly in this case, however, that the horseshoe crabs in question are released by east European fishermen returning from the North American Atlantic coast.

As another example, Edwards (1976) believed that the microhabitat of the hydrozoan, *Gonionemus*, was unlikely to make it a candidate for ship-transport, since "because of the shallow inshore weedy habitat favoured by this species, it is unlikely to settle on ships' bottoms" (or thus be taken up in ballast tanks). Indeed, the medusa of *Gonionemus* "for the great part of its life . . . moves around on algae in shallow water by help of the 'suckers' on the tentacles" (Tambs-Lyche, 1964), with occasional water-column feeding forays. None the less, medusae, eggs, or even polyp frustules (Edwards, 1976) of *Gonionemus* could be carried either freely, or adhering to loose algae or eelgrass, into a harbour region where the medusae (or eggs) might adhere to a hull, or where they might be drawn into a ballast tank.

Shallow-water plankton tows in harbours and estuaries frequently reveal a wide variety of benthic taxa (nematodes, benthic turbellarians, benthic harpacticoid copepods, pycnogonids, hydroid colonies, and others) alive, apparently undamaged, and in suspension in the water column. Also common in such tows are fragments of eelgrass and algae, upon which are numerous epiphytic invertebrates and plants (J. T. Carlton, pers. obs. and field collections, southern New England; see also Appendix in Gerlach, 1977). The donor area planktonic assemblage available for intake into a vessel may thus be fairly broad in its composition, and assumptions about what is and what is not available for ballast-water tranport must be carefully framed.

The behaviour of certain species around ships at dockside may, interestingly enough, influence their likelihood of being taken into ballast tanks. Thus, it has been suggested that certain fish, such as blennies, may have a great propensity than other fish to enter ballast tanks by their "general inclination . . . to inhabit tubelike holes" (Springer & Gomon, 1975). They may thus enter the holes of the hatch covers of the intake pipes to the tank while the ship is in port, and "then be sucked into the ballast tanks when the pumps were started" (Springer & Gomon, 1975). This hole-entering behaviour of blennies and gobies is well known, and has been documented, for example, for *Tridentiger obscurus*, a congener of *Tridentiger trigonocephalus* (see Tables I and II) by Umezawa & Yoshida (1981). The survival of adult or larval fish entering ballast tanks by pumping or by gravitation filling may rest, in part, of course, upon any initial injuries that may be sustained, a phenomenon amenable to experimentation (for example, Hickey, 1979). It is of interest that most of the fish in Tables I and II are either blennies or gobies or gobioid (the eleotrid *Prionobutis*) in nature.

The only studies comparing the planktonic assemblage in the water at shipside during ballasting to the plankton actually ballasted into tanks are those of Carlton *et al.* (1982). These studies show that not all taxa adjacent to the ship, at the same depth as water intake, are necessarily taken into the tank. Conversely, and as might be expected from the patchiness of plankton, several taxa may appear in the ballast tank which do not appear in shipside samples taken at the same time and depth as water intake. Such studies are limited, however, and far more needs to be done relative to the availability–intake sequence described here.

STAGE II: INITIAL BALLAST-TANK ASSEMBLAGE AND SURVIVAL DURING TRANSPORT

Limited data indicate that a number of species may be taken up and enter ballast-water tanks alive but that these may not subsequently survive the voyage that follows (see Table V). Little is known about the precise conditions that may mediate this selection. The highly variable nature of the physical-chemical environment of ballast tanks, discussed at the beginning of this review, may vary by ship and by voyage. The only study of *in situ en route* survival in a ballast tank is that noted earlier of plankton samples taken in tanks semi-daily aboard an oil tanker from New England to the Gulf of Mexico. The results of this study were briefly referred to earlier and are described in Carlton *et al.* (1982). Certain taxa surviving various legs of this voyage have been incorporated into Table III.

Certain "pre-adaptations" of various taxa would appear to make survival during transport in ballast-water tanks more likely for some groups than for others. These may include taxa with cysts or "resting" stages or eggs (such as dinoflagellate cysts, cladoceran ephippia, and the diapause eggs of calanoid copepods), and taxa with non-feeding larval stages (such as barnacle cyprids).

In a broad sense, taxa with life-cycle stages capable of utilizing fixed and/or decreasing food resources may have a higher likelihood of long-distance ballast-tank survival. These include, for example, larvae with large protein-lipid-carbohydrate reserves, and larvae capable of delaying metamorphosis in the absence of a suitable settling site.

More specifically, however, a planktonic organism may have to endure conditions of no light, reduced or non-existent food supplies, and changing temperature and oxygen regimes, sustained, in terms of a transoceanic crossing, for a period of one to two weeks or more. The probabilities of survival may perhaps thus be assessed indirectly from an examination of the results of experimental work on the physiological effects of manipulated light, temperature, and food regimes on planktonic invertebrate larvae and copepods. I briefly review several examples of these here, as illustrative of the application of such studies to a broader understanding of the potentially predictable nature of ballast-water transport survivors.

Light

The biology of organisms, that normally experience natural photoperiods, under a regime of sustained darkness (as in a ballast tank) has rarely been studied, although numerous classical experimental studies are available that have manipulated light to various degrees to study their effects on marine organisms. Several examples are considered here of the potential effects of continuous dark on taxa that may be entrained in and transported by ballast tanks.

Experimental studies on the production of subitaneous ("hatching without delay") and diapause ("resting") eggs in calanoid copepods have elucidated the rôle of photoperiod in controlling the life cycle (Marcus, 1979, 1980; Grice & Marcus, 1981). Marcus (1980) found that individuals of the calanoid *Labidocera aestiva* reared under a photoperiod regime of 18 h of light (L) and six hours of dark (D) produced subitaneous eggs; copepods developing under a short-day regime of 8L:16D produced mostly diapause eggs. Marcus did not investigate the effects of continuous dark on egg production. It may thus be predicted that under sustained darkness, if survival conditions are otherwise optimal, dormant-egg producing copepods are more likely to produce diapause than subitaneous eggs, which may, in turn, facilitate dispersal in ballast tanks.

The absence of light in ballast tanks may further be critical to visual predators, although I find no reference to this in discussions involving ballast-water transport of coastal fishes. Both larval and adult fish may, however, be able to obtain food in the dark. Studies by Blaxter (1968) and Blaxter & Staines (1971) have shown, for example, that herring and pilchard larvae are entirely visual feeders, that plaice larvae could take food in the dark as they reached metamorphosis but not before (suggesting that dark feeding may be possible by virtue of another sense, chemical or mechanical, developed in older fish), and that sole larvae could take food in the dark from an early stage. The discovery of small adult sticklebacks (*Gasterosteus*) in ship's water ballasted in Germany and carried to the United States demonstrates that the dispersal of visual predators in the absence of light, for at least 12 days, can occur. Other visual predators transportable in ballast water may include predatory copepods. Gophen & Harris (1981) have shown that for the cyclopoid *Corycaeus anglicus* (a species recorded from ballast tanks; Table III) brine shrimp (*Artemia*) nauplii are ingested at much higher rates (up to 15 times) in light than in dark regimes.

Temperature

Ballast-tank temperatures may increase, decrease, remain steady, or become cyclic. A vast literature addresses the effects of temperature on the biology of marine organisms. Reviewed here are several examples pertinent to the relationships of temperature, season, tolerance, and reproductive biology in planktonic copepods.

At food satiation, temperature appears to be a critical controlling factor in copepod egg production (Corkett & McLaren, 1969; McLaren & Corkett, 1981). The influence of temperature upon fecundity has been examined by Landry (1975a,b) and Uye (1981) for *Acartia*, by McLaren (1963, 1965) and Corkett & McLaren (1969, 1978) for *Pseudocalanus*, and by Corkett & Zillioux (1975) for *Acartia*, *Pseudocalanus*, and *Temora*. Uye (1981), for example, found that daily egg production was a function of temperature, increasing with increasing temperature up to about 20 °C, and decreasing thereafter.

Of particular interest is the relationship between temperature tolerance and season, and between temperature, egg production, and season. Bradley (1975) noted significant differences in temperature tolerances between March and August populations of *Eurytemora affinis*, finding the latter to be more tolerant. Halcrow (1963) found that seasonally adapted *Calanus finmarchicus* would not readily acclimate to temperatures outside its seasonal range, but would within the tolerance range of the species. Landry (1975b) found that, at 20 °C, *Acartia* collected in colder (12 °C) water in the spring produced offspring that developed faster at every stage than offspring of copepods collected in warmer fall waters (18 °C). Hart & McLaren (1978) found that *Pseudocalanus* embryos in the spring take longer to develop than those from warmer summer and autumn waters, and Uye (1981) established that there were significant differences in egg production between more-prolific summer females and winter females.

The existence of seasonally-dependent temperature tolerances and egg-production rates by a number of species of copepods has important implications for ballast-water biology, suggesting that populations of the same species ballasted at different seasons will have different response regimes to changing temperatures during transport.

Food availability

Food availability in a ballast tank may be visualized as a fixed and/or declining resource. An exception may be continual bacterial production, which could support populations of, for example, harpacticoid copepods for several months (Hardy, 1978). Two sets of experimental studies are of interest relative to food availability: first, those studies concerned with the physiological and ecological effects of starvation on larval development and plankton survival, and secondly, those concerned with the effects of food limitation on copepod egg production.

Bayne (1976) reviewed feeding and starvation studies of larval bivalve molluscs. Bayne (1965) found that larvae of the mussel, *Mytilus edulis*, could survive for up to 26 days without food, and that for at least 10 days after the onset of starvation, *Mytilus* larvae could resume feeding and subsequent growth was normal. Millar & Scott (1967) found that larvae of the oyster, *Ostrea edulis*, could survive for up to eight days without food and grow normally when feeding resumed. Survival potential of bivalve larvae is related to successful storage and utilization of lipid reserves (Holland & Spencer, 1973). Other invertebrate larvae appear to have similar adaptations to periods of starvation (Holland, 1978).

Anger & Dawirs (1981) determined that for the spider crab, *Hyas araneus*, no larval stage reared with a continuing lack of food had sufficient energy reserves to complete moulting to the next instar, and that in general starvation resistance decreased as temperature increased. Anger & Dawirs considered two categories of sublethal limits to starvation: (1) a "point-of-reserve-saturation" (the minimum time in which sufficient reserves are accumulated during the early phase of a developmental stage to ensure successful completion of the moult to the next instar without food; even short periods of initial feeding are sufficient for successful development, and (2) a "point-of-no-return" (beyond which larvae are unable to recover from starvation and cannot moult even if re-fed). Cronin & Forward (1980) found that starved larvae of the crab, *Rhithropanopeus harrisii*, swam slower than well-fed larvae from the same brood (of potential importance, for example, in food capture), but they suggest that this is a result of energy conservation mechanisms rather than deterioration, as late-stage larvae can survive up to three days of starvation with "good recovery" on the resumption of feeding. Lang & Marcy (1982) experimentally determined that at 15 °C partially starved nauplii of the barnacle, *Balanus improvisus*, retained the ability to complete nauplius development, although the rate was slower and mortalities were higher compared with controls. The boring isopod, *Limnoria*, which has been suggested to be transported by ballast water (Table I) and which has been found to be taken up into a ballast tank (Table V), can survive starved, without wood, for periods of up to 11 weeks (Kühne, 1973) and still be capable of re-burrowing into wood after at least two weeks of starvation (J. T. Carlton & S. Gallager, unpubl. experiments).

The effect of starvation regimes upon copepod survival has been established for a number of species. Ikeda (1971) kept *Calanus* alive without food for 36 days at 7–12 °C with a mortality rate twice that of fed controls; starved *Calanus* fed on their own faeces and after about 15 days of starvation fed on the dead bodies of other *Calanus*. A high utilization of lipids during starvation was observed. Conover (1964) demonstrated that *Calanus* could survive up to 80 days of starvation. Mayzaud (1976) studied the physiological effects of starvation on *Calanus finmarchicus* and *Acartia clausii* (as well as upon the chaetognath *Sagitta elegans*). *Calanus* survived the longest (more than 30 days) relying principally on protein and lipid reserves. Dagg (1977) found that with total starvation, *Centropages typicus* survived six days, *Acartia tonsa* ten days, *Pseudocalanus minutus*, 15 to 20 days, and *Calanus finmarchicus* more than 20 days.

The biology and physiology of starvation in larval fishes have been investigated by a number of workers (Ehrlich, 1974a,b; Blaxter & Ehrlich,

1974; O'Connell, 1976). May (1971) studied the effects of delayed initial feeding on newly-hatched larvae of the grunion *Leuresthes*; 80% of starved larvae began feeding when food was subsequently offered, regardless of how long initial feeding was delayed, and at least 40% of the larvae that were alive when food was offered survived the 20-day experiment.

Numerous studies have addressed the relationship of food supply and limitation to starvation and egg production copepods (*e.g.*, Corkett & McLaren, 1969; Ikeda, 1971; Mayzaud, 1976; Paffenhofer & Harris, 1976; Dagg, 1977; Landry, 1978; Parrish & Wilson, 1978; Checkley, 1980a,b; Uye, 1981). These data sets are of interest relative to the dispersal in ballast tanks of copepods and to their subsequent reproductive capabilities upon release into open waters. Dagg (1977) found that egg production by *Acartia tonsa* and *Centropages typicus* declined rapidly during brief periods of starvation. Parrish & Wilson (1978) and Uye (1981) obtained similar results for *Acartia tonsa* and *A. clausii*, respectively. Egg production by *Pseudocalanus minutus*, however, was not impaired by periods of starvation (Dagg, 1977). Dagg suggested that the reproductive behaviours of *Calanus* and *Pseudocalanus* are linked to their capability of being able to withstand longer periods without food. Critical to successful post-ballast transport survival is the recovery capability of species that have experienced low or discontinuous food abundances. Runge (1980) found that starved *Calanus pacificus* fed at rates up to three times higher than previously fed animals, even at low food concentrations; such rates may further be influenced by temperature (Anraku, 1964). Corkett & McLaren (1969) demonstrated that *Pseudocalanus* continued to reproduce during a two-week starvation period, although at a somewhat reduced level and, after this two-week period and the resumption of feeding, egg production did not differ significantly from the number of eggs produced after two weeks by continuously fed females. Parrish & Wilson (1978) noted that for *Acartia tonsa*, after food was restored to starving females, the rate of recovery of egg production was approximately 25% that of the drop-off rate following starvation.

In summary, certain larval and adult invertebrates have evolved a number of mechanisms to compensate naturally for low food concentration conditions. These starvation and post-starvation recovery adaptations may lend a selective advantage to some taxa (such as *Pseudocalanus* and *Calanus*) for survival in ballast tanks. Indeed, Dagg's "starvation survival sequence" may reflect an approximate order of the likelihood of copepod species survival in a ballast tank over time.

STAGE III: BALLAST-TANK BIOTA UPON ARRIVAL

The taxa known to have survived the sequence of events described so far are given in Table III (see p. 345) and were discussed earlier. A description of the general nature of a "ballast-water biota" remains elusive, however, pending more data.

STAGES IV AND V: RELEASE, SURVIVAL, CONTINUED REPRODUCTION, AND ESTABLISHMENT

The events in Stages IV and V leading to the successful establishment of an introduced species are, in broad terms, not unique to dispersal by ballast water but pertain to almost any colonization event by any mechanism. As with all dispersal mechanisms, natural and man-induced, the likelihood of successful colonization is partially attendant upon the nature of the inoculating population and partially upon the nature of the environment being colonized. While it is possible to document the survival of a wide variety of organisms in ballast tanks, and that they are released into regions where they are not native and do not occur, subsequent documentation of the fate of such organisms is an elusive task (but perhaps no more so than the documentation of the general fate of any one native larval cohort dispersed by currents through an estuary or along the coast). In essence, the appearance of reproducing populations of exotic species in a new environment (Tables I, II) is the only direct measure that a transport event has been successful (even if such appearances cannot be precisely tied to a given mechanism or date of introduction).

SURVIVAL POTENTIAL AND CHANGES OVER TIME

A step-by-step analysis of the sequence of events in time and space mediating dispersal of a planktonic or nektonic organism in sea-water ballast tanks suggests that a complex set of variables are at work. If broken into stages, however, these events and variables become accessible for sampling and experimentation. A model of ballast transport survival potential presented by Carlton *et al.* (1982) stresses temperature as an example of how one major variable may act throughout the system. Long voyages over extreme, one-way temperature changes are likely to result in low survival; short voyages (for example, of 72 h or less) with moderate temperature changes are likely to yield higher survival among the tank biota given, in both cases, that other conditions are suitable for existence. Similar arguments may be elucidated for other variables or for a set of variables acting together. The overall survival potential for organisms in ballast tanks has changed over the last half century, relative to the *in situ* conditions of transport, and to the time elapsed between ports.

Due to an increasing number of both national and international conventions focusing upon oil pollution of the oceans, a constantly greater number of vessels are transporting ballast water in tanks which hold only sea water, as opposed to decades of older practices where tanks were more regularly used alternately for sea-water ballast and, for example, for carrying petroleum products. These practices suggest that the amount of "clean" sea water transported has increased and will continue to do so.

Of equal importance is that the time elapsed between ballast-water uptake and release has decreased dramatically due to significantly increased vessel speeds. As Dawson (1973) has shown, the average new vessel speed doubled between World War II and the 1970s. In turn, time in port has been reduced to days (Couper, 1983) and even hours, such that once ballast water is taken on, departure is virtually imminent.

Two historical events that may have further played a rôle in increasing the potential of ballast water as a dispersal mechanism were the opening of the Suez Canal in 1869 and the Panama Canal in 1914. By vastly shortening the distances travelled between ocean basins, transit time was decreased, and the probabilities for survival for ballast-dwelling, as well as for fouling and boring, species probably increased greatly. Necessary minimal residence time in ballast tanks for organisms moved from the Atlantic to the Pacific was thus reduced from weeks (by going around Cape Horn) to hours (by going through the Panama Canal). Por's (1978) comment that "Passive transport on ships or in ballast tanks has no relevance to the problem of migrations. This is a process enhanced by human navigation, with or without canals" is of interest in this regard, as canals have doubtless enhanced the successful ship-mediated transport of innumerable taxa by vastly decreasing the distances (and therefore times) involved.

Finally, steady amelioration, starting in the 1960s, of harbour and estuarine pollution has been documented in many coastal areas of the world, and changes in the diversity and abundance of the biota recorded as a result. These generally "cleaner" harbour and port conditions may have similarly led to changes in the diversity and overall composition of ballast-water biota being transported.

IMPLICATIONS AND POTENTIAL SIGNIFICANCE OF BALLAST-WATER TRANSPORT

The potential significance of sea-water ballast dispersal may lie in its relevance to problems of ecology, genetics, systematics, and biogeography. These topics are addressed here in terms of potential mixing patterns in time and space relative to (1) transoceanic and interoceanic dispersal and (2) coastal dispersal.

TRANSOCEANIC AND INTEROCEANIC DISPERSAL OF NERITIC TAXA

In its most obvious form, sea-water ballast may act as a mechanism by which strictly neritic phytoplankton, holoplankton, and benthic invertebrates (lacking teleplanic larvae) can be transported across ocean basins or between oceans. There exists little opportunity for such taxa (other than long-term rafting on floating substrata) to be so transported, and ballast-water dispersal thus may act to transcend natural dispersal barriers separating shelf-dwelling species around the world.

Transport *via* ballast water may thus lead (a) to genetic mixing of stocks widely dispersed but generally held to be isolated from each other and (b) to the introduction of new species.

With regard to the potential mixing of genomes of widely occurring taxa, it is of interest to note the remarks of Barnes & Barnes (1976) that since the planktonic life of the barnacle *Balanus balanoides* is only three to four weeks long (in boreal regions), ". . . it seems unlikely that there could be any . . . exchange between the major populations in the eastern and western Atlantic and none between the Atlantic and Pacific." The survival of both nauplii and cyprids of this barnacle in ballast tanks (Table III) over time periods comparable to transatlantic ship crossings suggests that a potential for such exchange may exist.

The discovery of the stickleback *Gasterosteus aculeatus* transported from the Weser River, Germany to Delaware provides another example of the potential mixing of Atlantic North American and Atlantic European populations of this species, regarded as non-interbreeding (Burgess & Lee, 1980). If new genomes were to be introduced and remain as localized populations, it may be predicted that these would, at least for a time, be detectable. Thus, Rice & Simon (1980) have suggested that a genetically anomalous population of the polychaete *Polydora ligni* within Tampa Bay, Florida, may have been introduced in ballast water, among other means.

Entirely new taxa may be introduced to a region by sea-water ballast, and both possible and probable cases have been detailed in Tables I and II. The problems attendant upon the introduction of exotic species have been reviewed by many workers (see Lachner, Robins & Courtenay, 1970).

COASTAL DISPERSAL OF NERITIC SPECIES

Sea-water ballast transport may play a rôle along coastlines in enhancing the dispersal potential of species (a) with very short planktonic lives (as noted earlier), (b) generally restricted to estuaries, and (c) in directions opposite to those of the prevailing coastal currents.

Retention in estuaries of the larvae of benthic invertebrates is well known (Carriker, 1967; Wood & Hargis, 1971; Cronin & Forward, 1979). A number of species of estuarine invertebrates have apparently evolved mechanisms that permit their planktonic larvae to reduce the "risk" of being carried out to sea. These mechanisms include active vertical migration timed to patterns of tidal ebb-and-flood. The result is that planktonic larval populations are retained in estuaries near parent populations, for example, the xanthid mud crab, *Rhithropanopeus harrisii* (Cronin & Forward, 1979; Forward & Cronin, 1980). Other taxa (such as barnacle cyprids) may be retained in estuaries by physical processes (Wolf, 1974). In either case, little natural opportunity for long-distance dispersal out of the estuary to distant estuaries appears to exist for certain species in ecological time. Indeed, for some species adapted at all life stages to estuarine conditions, such dispersal through open waters of higher salinity and lower temperatures may not normally be physiologically possible, and transport within a ballast tank would provide a unique means of dispersing such species while retaining the appropriate environmental conditions during the dispersal event.

Patterns of ship traffic along coasts indicate that there is considerable movement of ballast water in directions opposite to those of coastal currents. Ballast water is transported both to the north and to the south of Cape Cod in New England, for example, and McAlice's (1981) study of the postglacial history of the calanoid copepod *Acartia tonsa* in this region is of interest in this regard. McAlice has suggested that this copepod, which ranges south to Florida, once occurred continuously during a warm hypsithermal period from Cape Cod to Northumberland Strait in the Gulf of St Lawrence. Now, however, in this same northern region, *Acartia tonsa* is represented by apparently separated populations, a distribution particularly marked by a 670 kilometre disjunction between populations in the Bay of Fundy and on Cape Breton Island. On the basis of (a) the presence of warm-water relict species in the seasonally warm estuaries in and between the Gulf of St Lawrence and Cape Cod, (b) the reproductive biology and temperature requirements of *Acartia tonsa*, (c) postglacial oceanographic changes, and (d) present patterns of net surface circulation (including generally southern- and western-flowing currents in the region under consideration), McAlice (1981) concluded that the disjunct populations of *Acartia* north of Cape Cod "are not consonant with a modern, northward range extension" (as had been suggested by others) but are rather relict northern populations. "The existing populations of *A. tonsa* north of Cape Cod are now isolated from the main east coast stocks of the species, and probably from one another as well. Certainly there is now no communication between the populations in the Gulf of Maine and those in the Gulf of St. Lawrence.... The Gulf of Maine populations of *A. tonsa* have been isolated from those in the Gulf of St. Lawrence for 2000 to 5000 years. They have thus been separated for at least 6000, and perhaps as many as 20,000, generations" (McAlice, 1981).

Complicating McAlice's interpretation is, then, the likelihood that *Acartia* has been transported from waters south of Cape Cod into the Gulf of Maine (and possibly into the Gulf of St Lawrence) in sea-water ballast many times (calanoid copepods, including other species of *Acartia*, are taken up in ballast water both north and south of Cape Cod and can survive transport in ballast tanks (Tables III, IV)). Water ballast is also regularly transported from the Gulf of Maine to the Gulf of St Lawrence, in, for example, bulk carriers arriving in ballast from the United States at the Madeleine Islands for salt. Indeed, this activity raised questions about the potential dispersal from Maine to Canada, via sea-water ballast, of red-tide causing dinoflagellates (G. Turner, Fisheries and Oceans Canada, pers. comm., 1983). McAlice's isolation model could perhaps be tested by means of electrophoretic genetics and cross-breeding experiments. The existence of ballast water as a potential transport mechanism renders strictly geological-historical interpretations for the origin of *Acartia* populations north of Cape Cod moot.

Finally, it may be noted that, despite the hundreds of cases of the relatively sudden appearance of dinoflagellates and other organisms causing discoloured water in coastal locations, in regions where they were previously unknown (for example, LoCicero, 1975; Taylor & Seliger, 1979), there appear to have been relatively few speculations about the rôle of passive transport in serving to inoculate new localities. Anderson & Wall (1978) speculated that benthic cyst transport of dinoflagellates could be effected by sediment transfer (dredging and dumping operations) or by the transplantation of commercial shellfish stocks. Ballast water may also play a rôle in certain instances. Living dinoflagellates, as noted earlier, are known from ballast-water samples.

BALLAST-WATER DISPERSAL: A PERSPECTIVE

For an increasing number of taxa, in groups as widely ranging as diatoms, hydromedusae, polychaetes, copepods, mysids, and barnacles, ballast-water transport in modern times becomes an alternative hypothesis to explain their patterns of distribution in estuarine and shelf waters around the world. For a large suite of neritic taxa, it is difficult to determine the rôle ballast-water dispersal may have played in the twentieth century in creating the modern distributions of many "cosmopolitan" species. The implications of this relative to conventional interpretations of the biogeography of many species, in terms of their origin and maintenance over time, are extensive. Particularly perplexing may be the possibility that extensive blending of the plankton of coastal waters of the world by the transport of sea-water ballast may have commenced many years prior to the onset of most biological surveys, such that many workers would come to believe in a "natural cosmopolitanism" of certain taxonomic groups. That there is no uniformity of opinion as to the magnitude of this phenomenon is demonstrated by contrasting the remarks made by various workers during the Panama sea-level canal debates (reviewed above) on the unlikelihood of ballast water acting as a dispersal mechanism, with the comment made by Springer & Gomon (1975) that, as a result of ballast-water movement, "one must generally hold in suspect the naturalness of the apparent distributions of small, benthic shore fishes in the Indo-west Pacific".

It appears probable that ballast dispersal incidents will increase with increased maritime traffic, faster ship speeds, and the movement of "cleaner" sea-water ballast. The extent to which ships' ballast waters, perhaps the only major synanthropic dispersal mechanism for marine organisms operating in the world today, have acted, are acting, and will act as international biotic conveyor belts remains a largely unexplored field in coastal biogeography.

ACKNOWLEDGEMENTS

I thank B. Åkesson, A. Alvariño, J. Chapman, L. Coates, E. J. Crossman, S. J. de Groot, M. E. Hay, H. D. Howell, J. Rosewater, and G. E. Turner for providing valuable information. I am particularly indebted to P. Heinermann and J. Schormann for providing information on the Canadian studies. The studies aboard the KNORR and the oil tanker MARTHA R. INGRAM were made possible by the cooperation of numerous enthusiastic colleagues at the Woods Hole Oceanographic Institution (particularly R. S. Edwards, Marine Superintendent, and his staff) and at Ingram Tankships, Inc. (particularly R. B. Ramsey and Captain L. Palmer and their staffs), acknowledged in Carlton *et al.* (1982), and to whom thanks are again extended. Mr P. W. Wilson provided an interesting sample of Black Sea ballast water from a vessel arriving at Boston. The United States Lines, particularly Captain W. McManus and his staff, and Getty Marine Corporation, particularly J. B. Lawson and C. Sayre, cooperated by sampling ballast tanks on vessels under their supervision with interest and unlimited cooperation. It is a pleasure to further acknowledge A. Navarret and R. Mann, Department of Biology, WHOI, who worked with me on many aspects of the studies described herein. D. Carlton provided unflagging support in uncountable ways. This work is based in part upon studies supported by the National Science Foundation under Grant No. DAR-8008450.

REFERENCES

Abele, L., 1972. *Crustaceana*, **23**, 209–218.
Adema, J. P. H. M., 1982. *Het Zeepaard*, **42**, 6–9.
Åkesson, B., 1977. *Mikro. Meeresbod.*, **61**, 11–18.
Allen, F. E., 1953. *Aust. J. mar. freshwat. Res.*, **4**, 307–316.
Alvariño, A., 1974. *Fishery Bull. N.O.A.A.*, **72**, 527–546.
Anderson, D. M. & Wall, D., 1978. *J. Phycol.*, **14**, 224–234.
Anger, K. & Dawirs, R. R., 1981. *Helgoländer Meeresunters.*, **34**, 287–311.

Anraku, M., 1964. *Limnol. Oceanogr.*, **9,** 195–206.
Arnal, R. E., 1961. *Bull. geol. Soc. Am.*, **72,** 427–478.
Banse, K., 1975. *Int. Revue ges. Hydrobiol.*, **60,** 439–447.
Barnard, J. L. & Reish, D. J., 1959. *Occ. Pap. Allan Hancock*, No. 21, 1–12.
Barnes, H. & Barnes, M. 1976. *J. exp. mar. Biol. Ecol.*, **24,** 251–269.
Bayne, B. L., 1965. *Ophelia*, **2,** 1–47.
Bayne, B. L., 1976. In, *Marine Mussels: Their Ecology and Physiology*, edited by B. L. Bayne, Cambridge University Press, Cambridge, pp. 81–120.
Beeton, A. M., 1978. In, *Sea-level Canal Studies*, Hearings, House of Representatives, 95th Congress, Ser. No. 95-51, pp. 224–234.
Bequaert, J., 1943. *Johnsonia*, **7,** 1–27.
Berdugo, V., 1974. *Rapp. P.-v. Réun. commn int. Explor. scient. Mer Méditer.*, **22,** 85–86.
Bertelsen, E. & Ussing, H. 1936. *Vidensk. Meddr dansk naturh. Foren.*, **100,** 237–245.
Blake, J. A. & Kudenov, J. D., 1978. *Mem. natl Mus. Victoria*, **39,** 171–280.
Blaxter, J. H. S., 1968. *J. exp. mar. Biol. Ecol.*, **2,** 293–307.
Blaxter, J. H. S. & Ehrlich, K. F., 1974. In, *The Early Life History of Fish*, edited by J. H. S. Blaxter, Springer-Verlag, New York, pp. 575–588.
Blaxter, J. H. S. & Staines, M. E., 1971. In, *Fourth European Marine Biology Symposium*, edited by D. J. Crisp, Cambridge University Press, Cambridge, pp. 467–485.
Boalch, G. T. & Harbour, D. S., 1977a. *Nova Hedwigia*, **54,** 275–280.
Boalch, G. T. & Harbour, D. S., 1977b. *Nature, Lond.*, **269,** 687–688.
Bousfield, E. L., 1973. *Shallow Water Gammaridean Amphipoda of New England.* Cornell University Press, Ithaca, New York, 212 pp.
Bowen, F. C., 1932. *Ships We See.* Sampson Low, Marston & Co., London, 311 pp.
Bowman, T. E., 1978. *Crustaceana*, **33,** 249–252.
Bowman, T. E., Bruce, N. L. & Standing, J. D., 1981. *J. crust. Biol.*, **1,** 545–557.
Bradley, B. P., 1975. *Biol. Bull. mar. biol. Lab., Woods Hole*, **148,** 26–34.
Brittan, M. R., Albrecht, A. B. & Hopkirk, J. D., 1963. *Calif. Fish Game*, **49,** 302–304.
Brittan, M. R., Hopkirk, J. D., Conners, J. D. & Martin, M., 1970. *Proc. Calif. Acad. Sci.*, (4), **38,** 207–214.
Broch, H., 1924. In, *Handbuch der Zoologie, Vol. 1*, edited by W. Kükenthal & T. Krumbach, Walter de Gruyter, Berlin, pp. 459–484.
Brylinski, J. M., 1981. *J. Plankton Res.*, **3,** 255–260.
Burgess, G. H. & Lee, D. S., 1980. In, *Atlas of North American Freshwater Fishes*, edited by D. S. Lee *et al.*, North Carolina State Museum, Raleigh, North Carolina, p. 563 only.
Calder, D. R. & Burrell, Jr, V. G., 1967. *Am. Midl. Nat.*, **78,** 540–541.
Calder, D. R. & Burrell, Jr, V. G., 1969. *Nature, Lond.*, **222,** 694–695.
Carlton, J. T., 1975. In, *Light's Manual: Intertidal Invertebrates of the Central California Coast*, edited by R. I. Smith & J. T. Carlton, University of California Press, Berkeley, 3rd edition, pp. 17–25.
Carlton, J. T., 1979a. Doctoral dissertation, University of California, Davis, 904 pp.
Carlton, J. T., 1979b. In, *San Francisco Bay: the Urbanized Estuary*, edited by T. J. Conomos, California Academy of Sciences, San Francisco, pp. 427–444.
Carlton, J. T. & Iverson, E. W., 1981. *J. nat. Hist.*, **15,** 31–48.
Carlton, J. T., Navarret, A. & Mann, R., 1982. Final Proj. Rept, Natl Sci. Fdn, Div. Applied Sci., DAR-8008450, Dept Biol., Woods Hole Ocean. Inst., Woods Hole, Massachusetts, 161 pp.
Carlton, J. T. & Scanlon, J., 1985. *Bot. Mar.*, **25,** in press.
Carriker, M. R., 1967. In, *Estuaries*, edited by G. Lauff, American Association for the Advancement of Science, Washington, D.C., pp. 442–487.
Casson, L., 1964. *Illustrated History of Ships and Boats.* Doubleday & Co., New York, 272 pp.
Cattley, J. G. & Harding, J. P., 1949. *Nature, Lond.*, **164,** 238–239.

Challinor, D., 1978. In, *Sea-level Canal Studies*, Hearings, House of Representatives, 95th Congress, Ser. no. 95-51, p. 248 only.
Champalbert, G., 1981. *Oceanis*, **7,** 131–147.
Checkley, D. M., 1980a. *Limnol. Oceanogr.*, **25,** 430–446.
Checkley, D. M., 1980b. *Limnol. Oceanogr.*, **25,** 991–998.
Cheng, L., 1975. *Oceanogr. Mar. Biol. Ann. Rev.*, **13,** 181–212.
Chesher, R. H., 1968. *Limnol. Oceanogr.*, **13,** 387–388.
Chilton, C., 1910. *Trans. N.Z. Inst.*, **43,** 131–133.
Christiansen, J. & Thomsen, J. C., 1981. *Steenstrupia*, **7,** 15–24.
Christiansen, M. E., 1969. *Mar. Inverts. Scandinavia*, No. 2, 143 pp.
Clapp, W. F. & Kenk, R., 1963. *Marine Borers, an Annotated Bibliography*. Department of the Navy, Office of Naval Research, Washington, D.C., 1136 pp.
Clarke, A. H., 1969. *Ann. Rept Am. Malacol. Union*, 1969, 51–52.
Clarke, A. H., 1972. *Can. Fld Nat.*, **86,** 165–166.
Clench, W. J. & Turner, R. D., 1946. *Johnsonia*, **2,** 1–28.
Coates, L. J., Roff, J. C. & Markle, D. F., 1982. *Environ. Biol. Fish.*, **7,** 69–72.
Conover, R. J., 1957. *Ann. Mag. nat. Hist.*, Ser. 12, **10,** 63–67.
Conover, R. J., 1964. *Occ. Publ. Univ. Rhode Island*, **2,** 81–91.
Constant, T. M., 1978. In, *Sea-level Canal Studies*, Hearings, House of Representatives, 95th Congress, Ser. no. 95-51, p. 197 only.
Cooke, W. J., 1984. *Proc. biol. Soc. Wash.*, **97,** 583–588.
Corkett, C. J. & McLaren, I. A., 1969. *J. exp. mar. Biol. Ecol.*, **3,** 90–105.
Corkett, C. J. & McLaren, I. A., 1978. *Adv. mar. Biol.*, **15,** 2–231.
Corkett, C. J. & Zillioux, E. J., 1975. *Bull. Plankton Soc. Jap.*, **21,** 77–85.
Cosel, R., Dörjes, J. & Mühlenhardt-Siegel, U., 1982. *Senckenbergiana marit.*, **14,** 147–173.
Couper, A., 1983. Editor, *The Times Atlas of the Oceans*. Van Nostrand Reinhold Co., New York, 272 pp.
Cowan, I. McT., 1974. *Veliger*, **16,** 290 only.
Cronin, T. W. & Forward, Jr, R. B., 1979. *Science*, **205,** 1020–1021.
Cronin, T. W. & Forward, Jr, R. B., 1980. *Biol. Bull. mar. biol. Lab., Woods Hole*, **158,** 283–294.
Crossman, E. J., 1984. In, *Distribution, Biology, and Management of Exotic Fishes*, edited by W. Courtenay & J. R. Stauffer, Johns Hopkins University Press, Baltimore, pp. 78–101.
Cutress, C. E., 1961. *Pacif. Sci.*, **15,** 547–552.
Dagg, M., 1977. *Limnol. Oceanogr.*, **22,** 99–107.
Dana, R. H., 1840. *Two Years Before the Mast and Twenty-four Years After*. Harper & Brothers, New York, 405 pp.
Dawson, C. E., 1970. *Proc. Biol. Soc. Wash.*, **83,** 273–286.
Dawson, C. E., 1973. *Copeia*, 1973, 141–144.
Dinamani, P., 1971. *N.Z. Jl mar. freshwat. Res.*, **5,** 352–357.
Edwards, C., 1976. *Adv. mar. Biol.*, **14,** 251–284.
Ehrlich, K. F., 1974a. *Mar. Biol.*, **24,** 39–48.
Ehrlich, K. F., 1974b. In, *The Early Life History of Fish*, edited by J. H. S. Blaxter, Springer-Verlag, New York, pp. 301–322.
Elton, C. E., 1958. *The Ecology of Invasions by Animals and Plants*. John Wiley & Sons, New York, 181 pp.
Emery, A. R. & Teleki, G., 1978. *Can. Fld Nat.*, **92,** 89–91.
Farnham, W. F., 1980. In, *The Shore Environment, Volume 2, Ecosystems*, edited by J. H. Price, D. E. G. Irvine & W. F. Farnham, Academic Press, London, pp. 875–914.
Ferrari, F. D. & Orsi, J., 1984. *J. crust. Biol.*, **4,** 106–126.
Forward, Jr, R. B. & Cronin, T. W., 1980. *Biol. Bull. mar. biol. Lab., Woods Hole*, **158,** 295–303.
Frear, H. P., 1897. *Trans. Soc. Naval Architects mar. Engrs*, **5,** 151–161, 171–174.

Friese, U. E., 1973. *Koolewong*, **2,** 5–7.
Gerlach, S. A., 1977. *Mikro. Meeresbod.*, **61,** 89–103.
Giesbrecht, W., 1897. *Zool. Jahrb.*, **9,** 315–328.
Glynn, P. W., 1982. *Adv. mar. Biol.*, **19,** 91–132.
Gomoiu, M.-T., 1975. *Cercetari marine* (*IRCM*), **8,** 105–117.
Gophen, M. & Harris, R. P., 1981. *J. mar. biol. Ass. U.K.*, **61,** 391–399.
Grabe, S. A., 1981. *Proc. Biol. Soc. Wash.*, **94,** 863–865.
Graham, H. W. & Gay, H., 1945. *Ecology*, **26,** 375–386.
Grantham, J., 1868. *Iron Ship-building with Practical Illustrations*, 5th edition. Virtue & Co., London, 321 pp.
Greenfield, D. W. & Grinols, R. B., 1965. *Copeia*, 1965, 115–116.
Grice, G. D. & Marcus, N. H., 1981. *Oceanogr. Mar. Biol. Ann. Rev.*, **19,** 125–140.
Grindley, J. R. & Grice, G. D., 1969. *Crustaceana*, **16,** 125–134.
Guthrie, J., 1970. *Bizarre Ships of the Nineteenth Century*. Hutchinson & Co., Publ., London, 212 pp.
Haaker, P. L., 1979. *Bull. So. Calif. Acad. Sci.*, **78,** 56–61.
Haeckel, E., 1890. *Jena Z. Naturw.*, **18,** 232–336.
Halcrow, K., 1963. *Limnol. Oceanogr.*, **8,** 1–8.
Hamersly, L. R., 1880. *A Naval Encyclopaedia*. L. R. Hamersly & Co., Philadelphia, 1017 pp.
Hardy, B. L. S., 1978. *J. exp. mar. Biol. Ecol.*, **34,** 143–149.
Hart, R. C. & McLaren, I. A., 1978. *Mar. Biol.*, **45,** 23–30.
Hay, M. E. & Gaines, S. D., in press. *Biotropica*, **15,** in press.
Hedgpeth, J. W., 1968. In, *Between Pacific Tides*, edited by E. F. Ricketts, J. Calvin & J. W. Hedgpeth, Stanford University Press, Stanford, pp. 376–380.
Hendey, N. I., 1964. *An Introductory Account of the Smaller Algae of British Coastal Waters. Part V. Bacillariophyceae* (*Diatoms*). Her Majesty's Stationery Office, London, 317 pp.
Hendrickx, M. E., 1980. *Veliger*, **23,** 93–94.
Henry, D. P. & McLaughlin, P. A., 1975. *Zool. Verh.*, No. 141, 254 pp.
Hentschel, E., 1923. *Int. Revue ges. Hydrobiol.*, **11,** 238–264.
Hentschel, E., 1924. *Mitt. zool. StInst. Hamb.*, **41,** 1–51.
Herdman, W. A., Thompson, I. C. & Scott, A., 1897. *Trans. Lpool biol. Soc.*, **12,** 33–90.
Hickey, G. M., 1979. *J. exp. mar. Biol. Ecol.*, **37,** 1–17.
Hoese, D. F., 1973. *Koolewong*, **2,** 3–5.
Holland, D. L., 1978. In, *Biochemical and Biophysical Perspectives in Marine Biology*, edited by D. C. Malins & J. R. Sargent, Academic Press, New York, pp. 85–123.
Holland, D. L. & Spencer, B. E., 1973. *J. mar. biol. Ass. U.K.*, **53,** 287–298.
Holthuis, L. B., 1980. *F.A.O. Fish. Syn.*, No. 125, 271 pp.
Holthuis, L. B. & Gottlieb, E., 1955. *Bull. Res. Counc. Israel*, **5B,** 154–156.
Howarth, R. S., 1981. *Presence and implication of foreign organisms in ship ballast waters discharged into the Great Lakes*. Rept to Environmental Protection Service, Environment Canada, Ottawa. Bio-Environ. Services, Georgetown, Ontario, Vol. 1, 1–97, Vol. 2, unpaginated.
Hubbs, C. L. & Miller, R. R., 1965. *Miscell. Publ. Mus. Zool. Univ. Mich.*, No. 127, 104 pp.
Hynes, H. B. N., 1955. *Verh. int. Verein. theor. angew. Limnol.*, **12,** 620–628.
Ikeda, T., 1971. *Bull. Fac. Fish Hokkaido Univ.*, **21,** 280–298.
Johnstone, J., Scott, A. & Chadwick, H. C., 1924. *The Marine Plankton*. University Press of Liverpool Ltd, Liverpool, 194 pp.
Jones, E. C., 1966. *Crustaceana*, **10,** 316–317.
Jones, M. L., 1972. *Bull. biol. Soc. Wash.*, **2,** vii–viii.
Jones, M. L. & Dawson, C. E., 1973. *Mar. Biol.*, **21,** 86–90.
Jones, M. L. & Rützler, K., 1975. *Mar. Biol.*, **33,** 57–66.

Jordan, C. M. & Garside, E. T., 1972. *Can. J. Zool.*, **50,** 1405–1411.
Keen, A. M., 1969. *Veliger*, **11,** 439 only.
Knight-Jones, P., Knight-Jones, E. W., Thorp, C. H. & Gray, P. W. G., 1975. *Zool. Scripta*, **4,** 145–149.
Kofoid, C. A. & Miller, R. C., 1927. In, *Marine Borers and Their Relation to Marine Construction on the Pacific Coast*, edited by C. L. Hill & C. A. Kofoid, San Francisco Bay Marine Piling Committee, San Francisco, pp. 188–343.
Kramp, P. L., 1955. *Rec. Ind. Mus.*, **53,** 339–376.
Kühl, H., 1964. *Arch. FischWiss.*, **15,** 225–227.
Kühne, H., 1973. In, *Proc. Third int. Cong. mar. Corr. Fouling*, edited by R. F. Acker *et al.*, Northwestern University Press, Evanston, Illinois, pp. 814–821.
Lachner, E. A., Robins, C. R. & Courtenay, Jr, W. R., 1970. *Smithson. Contr. Zool.*, No. 59, 29 pp.
Landry, M. R., 1975a. *Limnol. Oceanogr.*, **20,** 434–440.
Landry, M. R., 1975b. *Limnol. Oceanogr.*, **20,** 854–857.
Landry, M. R., 1978. *Int. Revue ges. Hydrobiol.*, **63,** 77–119.
Lang, W. H. & Marcy, M., 1982. *J. exp. mar. Biol. Ecol.*, **60,** 63–70.
Lebour, M. V., 1930. *The Planktonic Diatoms of Northern Seas.* Ray Society, London, 244 pp.
Light, W. J., 1978. *Invertebrates of the San Francisco Bay Estuarine System. Spionidae, Polychaeta, Annelida.* Boxwood Press, Pacific Grove, California, 211 pp.
Lindroth, C. H., 1957. *The Faunal Connections between Europe and North America.* John Wiley & Sons, New York, 344 pp.
Little, G. H., 1890. *The Marine Transport of Petroleum.* E. & F. N. Spon, London, 251 pp.
Lloyd's Register, 1880. *Lloyd's Register of British and Foreign Shipping from 1st July 1880 to the 30th June 1881.* Wyman & Sons, Printers, London (no pagination).
LoCicero, V. R., 1975. Editor, *Proc. First int. Conf. on Toxic Dinoflagellate Blooms.* Massachusetts Science Technology Foundation, Wakefield, Massachusetts, 541 pp.
Marcus, N. H., 1979. *Biol. Bull. mar. biol. Lab., Woods Hole*, **157,** 297–305.
Marcus, N. H., 1980. *Biol. Bull. mar. biol. Lab., Woods Hole*, **159,** 311–318.
May, R. C., 1971. *Fishery Bull. N.O.A.A.*, **69,** 411–425.
Mayer, A. G., 1910. *Medusae of the World. Volume II. The Hydromedusae*, Carnegie Institute Washington, Publ. No. 109, pp. 231–498.
Mayzaud, P., 1976. *Mar. Biol.*, **37,** 47–58.
McAlice, B. J., 1981. *Mar. Biol.*, **64,** 267–272.
McCosker, J. E. & Dawson, C. E., 1975. *Mar. Biol.*, **30,** 343–351.
McLaren, I. A., 1963. *J. Fish. Res. Bd Can.*, **20,** 685–727.
McLaren, I. A., 1965. *Limnol. Oceanogr.*, **10,** 528–538.
McLaren, I. A. & Corkett, C. J., 1981. *Can. J. Fish. Aquat. Sci.*, **38,** 77–83.
Meade, R. W., 1869. *A Treatise on Naval Architecture and Ship-building.* J. B. Lippincott & Co., Philadelphia, 496 pp.
Medcof, J. C., 1975. *Proc. natl Shellfish. Ass.*, **65,** 54–55.
Menzies, R. J., 1972. *Mar. Biol.*, **17,** 149–157.
Merrill, A. S., 1959. *Nautilus*, **73,** 39–43.
Middleton, M. J., 1982. *J. Fish Biol.*, **21,** 513–523.
Millar, R. H. & Scott, J. M., 1967. *J. mar. biol. Ass. U.K.*, **47,** 475–484.
Moll, F., 1936. *Mitt. Ges. Vorratsch.*, **12,** 3–4.
Naylor, E., 1957a. *Nature, Lond.*, **180,** 616–617.
Naylor, E., 1957b. *Ann. Mag. Nat. Hist.*, Ser. 12, **10,** 521–523.
Naylor, E., 1960. *Nature, Lond.*, **187,** 256–257.
Nebolsina, T. K., 1959. *Prirorda*, **6,** 116–117.
Nepszy, S. J. & Leach, J. H., 1973. *J. Fish. Res. Bd Can.*, **30,** 1909–1910.
Newman, W. A., 1963. *Crustaceana*, **5,** 119–132.

Newman, W. A., 1964. Addendum, p. 144, in Tetsuo, M., Shane, G. & Newman, W., *Crustaceana*, **7,** 141–145.

Newman, W. A., 1972. *Bull. biol. Soc. Wash.*, **2,** 247–259.

Nijssen, H. & Stock, J. H., 1966. *Beaufortia*, **13,** 197–206.

O'Connell, C. P., 1976. *J. exp. mar. Biol. Ecol.*, **25,** 285–312.

Orsi, J. J., Bowman, T. E., Marelli, D. C. & Hutchinson, A., 1983. *J. Plankton Res.*, **5,** 357–375.

Ostenfeld, C. H., 1908. *Meddr Kommn Havunders., Ser. Plankton*, **1,** No. 6, 44 pp.

Packard, A. S., 1872. *Mem. Bost. Soc. nat. Hist.*, **2,** 155–202.

Paffenhofer, G.-A. & Harris, R. P., 1976. *J. mar. biol. Ass. U.K.*, **56,** 327–344.

Palmer, M. A., 1984. *Mar. Behav. Physiol.*, **10,** 235–253.

Parrish, K. K. & Wilson, D. F., 1978. *Mar. Biol.*, **46,** 65–81.

Peters, N., 1933. *Zool. Anz.*, **104,** 59–156.

Pollock, D., 1905. *The Shipbuilding Industry. Its History, Practice, Science and Finance.* Methuen & Co., London, 199 pp.

Por, F. D., 1978. *Lessepsian migration.* Springer-Verlag, Berlin, 228 pp.

Powell, N. A., 1971. *Bull. mar. Sci.*, **21,** 766–778.

Redeke, H. C., 1934. *J. Cons. perm. int. Explor. Mer*, **9,** 39–45.

Rees, C. B. & Cattley, J. G., 1949. *Nature, Lond.*, **149,** 367 only.

Reish, D. J., 1957. *Pacif. Sci.*, **11,** 216–228.

Remy, P., 1927. *Ann. Biol. Lacustre*, **15,** 169–186.

Rice, S. A. & Simon, J. L., 1980. *Ophelia*, **19,** 79–115.

Robinson, G. A., Budd, T. D., John, A. W. G. & Reid, P. C., 1980. *J. mar. biol. Ass. U.K.*, **60,** 675–680.

Rosewater, J., 1980. *Nautilus*, **94,** 158–162.

Rosewater, J. & Vermeij, G. J., 1972. *Nautilus*, **86,** 67–69.

Rubinoff, I., 1970. *Biol. Conserv.*, **3,** 33–36.

Rubinoff, R. W. & Rubinoff, I., 1968. *Nature, Lond.*, **217,** 476–478.

Rubinoff, R. W. & Rubinoff, I., 1969. *Copeia*, 1969, 395–397.

Rudloe, A., 1979. *Biol. Bull. mar. biol. Lab., Woods Hole*, **157,** 494–505.

Runge, J. A., 1980. *Limnol. Oceanogr.*, **25,** 134–145.

Sakai, T., 1976a. *Crabs of Japan and the Adjacent Seas.* Kodansha Ltd, Tokyo, 773 pp.

Sakai, T., 1976b. *Res. Crustacea*, **7,** 29–40.

Sander, F. & Moore, E., 1979. *Crustaceana*, **36,** 215–224.

Saraber, J. G. A. M., 1962. *Beaufortia*, **9,** 117–120.

Schultz, G. A., 1979. *Proc. biol. Soc. Wash.*, **92,** 577–579.

Seapy, R. R., 1974. *Veliger*, **16,** 385–387.

Sheffey, J. P., 1968. *Science*, **162,** 1329 only.

Sheffey, J. P., 1972. *Bull. biol. Soc. Wash.*, **2,** 31–39.

Sheffey, J. P., 1978. In, *Sea-level Canal Studies*, Hearings, House of Representatives, 95th Congress, Ser. no. 95-51, pp. 304–305 and pp. 317–318.

Skerman, T. M., 1960. *N.Z. Jl Sci.*, **3,** 620–648.

Skinner, J. E., 1971. *Calif. Fish Game*, **57,** 76–79.

Spivey, H. R., 1976. *Corr. Mar.-Fouling*, **1,** 43–50.

Springer, V. G. & Gomon, M. F., 1975. *Smithson. Contrib. Zool.*, No. 177.

Standing, J. D., 1981. *Proc. biol. Soc. Wash.*, **94,** 774–786.

Tambs-Lyche, H., 1964. *Sarsia*, **15,** 1–8.

Taylor, D. L. & Seliger, H. H., 1979. Editors, *Toxic Dinoflagellate Blooms.* Elsevier/North-Holland, New York, 505 pp.

Teeter, J. W., 1973. *Ohio Jl Sci.*, **73,** 46–54.

ten Hove, H. A., 1974. *Bull. Zool. Mus. Univ. Amsterdam*, **4,** 45–49.

Thayer, I. E., 1897. *Trans. Soc. Naval Architects mar. Engrs*, **5,** 163–169.

Thiel, M. E., 1935. *Zool. Anz.*, **111,** 161–174.

Thomas, F. C. & Schafer, C. T., 1982. *J. Foram. Res.*, **12,** 24–38.

Thompson, I. C. & Scott, A., 1903. *Ceylon Pearl Oyster Fish., Supp. Rept*, **7,** 227–307.

Turner, R. D., 1955. *Johnsonia*, **3,** 65–160.

Turner, R. D., 1966. *A Survey and Illustrated Catalogue of the Teredinidae (Mollusca: Bivalvia)*. Museum of Comparative Zoology, Harvard University, Cambridge, Massachusetts, 265 pp.

Turner, R. D., 1971. In, *Marine Borers, Fungi and Fouling Organisms of Wood*, edited by E. B. G. Jones & S. K. Eltringham, OECD, Paris, pp. 17–64.

Turner, R. D. & Johnson, A. C., 1971. In, *Marine Borers, Fungi and Fouling Organisms of Wood*, edited by E. B. G. Jones & S. K. Eltringham, OECD, Paris, pp. 259–301.

Umezawa, S.-I. & Yoshida, M., 1981. *Zool. Mag. (Dobut. Zasshi)*, **90,** 338–350.

United States Navy, 1951. *Report on the Marine Borers and Fouling Organisms in 56 Important Harbors and Tabular Summaries of Marine Borer Data from 160 Widespread Localities*. Department of the Navy, Bureau of Yards and Docks, Washington, D.C., NAVDOCKS TP-Re-1, 327 pp.

Uye, S., 1981. *J. exp. mar. Biol. Ecol.*, **50,** 255–271.

Visscher, J. P., 1928. *Bull. Bur. Fish.;* **43,** 193–252.

Voss, G. L., 1978. *Sea Frontiers*, **24,** 206–213.

Walford, L. & Wicklund, R., 1973. *F.A.O. Fish. Tech. Pap.*, No. 121, 49 pp.

Walton, T., 1896. *Know Your Own Ship: a Simple Explanation of the Stability, Tonnage, and Freeboard of Ships*. Charles Griffin & Co. Ltd, London, 242 pp.

Walton, T. & King, J., 1926. *Steel Ships: their Construction and Maintenance*. Charles Griffin & Co. Ltd, London, Part I, 70 pp., Part II, 270 pp.

Wheeler, A. C., 1958. *Proc. Zool. Soc. Lond.*, **130,** 253–256.

White, W. H., 1882. *A Manual of Naval Architecture*. John Murray, London, 2nd edition, 672 pp.

White, W. H., 1894. *A Manual of Naval Architecture*. John Murray, London, 3rd edition, 729 pp.

Wigley, R. L., 1963. *Crustaceana*, **6,** 158 only.

Wigley, R. L. & Burns, B. R. 1971. *Fishery Bull. N.O.A.A.*, **69,** 717–746.

Williams, J. L., 1978. *Proc. biol. Soc. Wash.*, **91,** 666–680.

Williams, R. J., 1977. *Aust. mar. Sci. Bull.*, No. 60, 9 only.

Williams, R. J., van der Wal, E. J. & Story, J., 1978. *Aust. mar. Sci. Bull.*, No. 61, 12 only.

Williamson, D. I. & Rochanaburanon, T., 1979. *J. nat. Hist.*, **13,** 11–33.

Wilson, B. R. & Kendrick, G. W., 1968. *J. malac. Soc. Aust.*, **11,** 25–31.

Wirth, W. W., 1947. *Proc. Hawaiian entomol. Soc.*, **13,** 141–142.

Wolf, P., 1974. *Thalassia jugosl.*, **10,** 415–424.

Wolff, T., 1954. *Nature, Lond.*, **174,** 188–189.

Wolff, T., 1977. *Vidensk. Meddr dansk naturh. Foren.*, **140,** 39–52.

Wood, L. & Hargis, Jr, W. J., 1971. In, *Fourth European Marine Biology Symposium*, edited by D. J. Crisp, Cambridge University Press, Cambridge, pp. 29–44.

Woods Hole Oceanographic Institution, 1952. *Marine Fouling and its Prevention*. United States Naval Institute, Annapolis, Maryland, 388 pp.

Yeatman, H. C., 1962. *Crustaceana*, **4,** 253–272.

Zimmer, R. L. & Woollacott, R. M., 1977. In, *Biology of Bryozoans*, edited by R. M. Woollacott & R. L. Zimmer, Academic Press, New York, pp. 57–89.

Oceanogr. Mar. Biol. Ann. Rev., 1985, **23**, 373–398
Margaret Barnes, Ed.
Aberdeen University Press

REPRODUCTIVE PATTERNS OF MARINE INVERTEBRATES

JOHN GRAHAME
Department of Pure and Applied Zoology, University of Leeds, Leeds LS2 9JT, England

and

GEORGE M. BRANCH
Zoology Department, University of Cape Town, Rondebosch 7700, South Africa

INTRODUCTION

The life history of an organism is the schedule of events from birth, through growth to reproduction and eventual death. An individual which dies before reproducing has no fitness. Of two which do reproduce, if one can enhance reproductive success, it also enhances its own fitness, and it may be supposed that life history attributes have been subject to selection and represent adaptive features. That this is so was recognized by Fisher (1958) among others. This review will concentrate on reproduction in marine invertebrates.

A problem is that life history features—such as when to breed first, or how much resource to put into breeding—must be the outcome of compromises between many potentially conflicting demands. They will, then, represent optimal solutions, not maximal ones. If all the factors involved in producing the solution have not been accounted for, the investigator risks committing 'adaptationist' excesses of the kind criticized by Gould & Lewontin (1979). There is also the danger of overlooking phenotypic effects. For instance, Fletcher (1984) has shown that the amount of environmentally determined variation in the reproductive output of the limpet *Cellana tramoserica* exceeds interspecific differences previously described for four species of limpet, considered to be adaptive responses to age-specific mortality in the four species (Parry, 1982a,b).

Research into the life histories of marine invertebrates parallels work on other organisms in the broad fields of studies on demography and resource allocation. Much of the literature, however, concerns a question unique to the sea: why have planktonic larvae—what are the costs and benefits?

MODE OF DEVELOPMENT

CLASSIFICATION OF MODES

Thorson (1946) summarized the work up to that time, with his own very considerable contribution, in a monograph on the larvae of the Øresund. He suggested a classification of larval types which, with modification, is still in use. The following larval types were described.

(1) Long-life planktotrophic—those larvae swimming in the plankton and feeding on other planktonic organisms, usually phytoplankton. Duration of larval life is from 1 week to 2 or 3 months. Some 63% of Øresund taxa showed this larval type, examples being *Littorina littorea*, *Polydora ciliata*, and *Asterias rubens*. The advantage of this mode lies in the dispersal gained from the long planktonic sojourn, but set against this is the disadvantage of high mortality in the plankton, causing great "waste" of larvae. Animals in this category produce large numbers of small eggs, each being "cheap" because of its size and small provision of yolk.

(2) Short-life planktotrophic—discriminated chiefly on the basis that their size and organization change hardly, if at all, in the course of the week or less spent in the plankton. The designation "planktotrophic" seems to have been on the criterion that the gut was developed and apparently functional when the larva hatched, even though some might settle without feeding. Only 6% of taxa were in this category, an example being *Spirorbis borealis*.

(3) Lecithotrophic pelagic larvae—in this category were 15% of echinoderms, 7% of polychaetes and lamellibranchs, and most chitons. The larva is large and provided with much yolk, hatching from a large yolky egg. The yolk provides all the energy needed by the larva until metamorphosis into a juvenile. In the starfish *Solaster endeca* 3000 eggs were spawned, floating at the surface to hatch after 6 days, the larva being pelagic for about 12 days. This type of development was seen by Thorson (1946) as a "typical compromise". He considered it to be absent from polar waters and the deep sea.

To this scheme must be added the animals which show 'direct development' in an encapsulated egg from which a juvenile eventually hatches, and those which show forms of brooding. This habit complicates the classification: it may precede a period in which larvae swim for a long time in the plankton, as in *Littorina angulifera* (Lebour, 1945) and *Balanus balanoides* (Thorson, 1946). Both of these animals may be said to have long-life planktotrophic larvae. *Spirorbis borealis* follows brooding with a short planktonic phase (Thorson, 1946). In *Littorina neglecta* and *Littorina rudis*, brooding is followed by the emergence of juveniles from the brood pouch of the female (Ellis, 1983), also many echinoderms brood their young to the juvenile stage (Himmelman *et al.*, 1982).

Mileikovsky (1971) reviewed larval types afresh and concluded that Thorson's classification largely held good. As with any attempt at simplification of this kind, there are cases which do not fit the finished scheme. The use of terms may reflect the interest of an individual worker. Thus Scheltema (1978), writing about dispersal, differentiates between pelagic and non-pelagic development. If resource use is of interest, the stress is on planktotrophy (when the developing larva finds much of the food needed during the egg to juvenile phase) *versus* lecithotrophy (when the parent must provide all the food needed). 'Lecithotrophic' may mean 'non-feeding pelagic' or 'non-pelagic' and will be qualified as such in the account which follows.

Some animals may have free, water-borne developmental stages which are not planktonic, but rather inhabit the boundary layer just above the bottom—demersal larvae. For instance, Pearse (1969) has reported a demersal larva for the Antarctic seastar *Odontaster validus*. The rate of development appears to be slower than would be expected on the grounds of temperature alone, although no explanation was offered for this. The embryos and larvae spend almost all their time near the bottom, where they are considered to feed on bottom material after the bipinnaria stage has been reached. Pearse (1969) suggested that demersal development in this species is an adaptation to avoid exposure to fluctuating surface salinities, and that in this respect it may be analogous to the brooding habit in other Antarctic echinoderms. Demersal larvae may occur among several deep-sea invertebrates (see below), and they are now recognized as being an important larval type, but the extent of their occurrence is poorly known.

MIXED LIFE HISTORIES

If a period of development in a benthic egg capsule (or indeed in a parental brood chamber) precedes a pelagic phase, the life history may be said to be "mixed" (Pechenik, 1979). This habit has been adopted particularly by gastropods, spionid polychaetes, most nemerteans and some turbellarians. It has been suggested that the advantage of this is in protecting very early developmental stages from predation, and Pechenik (1979) argued that the floating egg capsules of some littorinids are consistent with this hypothesis, in that they protect the eggs from benthic predators. This presumes that surface waters are safer than waters near the bottom. Thorson (1946) and Fish (1979) have noted that the egg capsules of *Littorina littorea* are especially abundant in surface waters after stormy weather, and Fish (1979) concluded that in calm water large numbers will sink to the bottom and develop there. This is somewhat inconsistent with the buoyancy required by Pechenik's (1979) arguments.

Caswell (1981) commented that the use by Pechenik (1979) of the probability of offspring survival as a measure of fitness is not adequate, and proposed instead the use of number of surviving offspring—*i.e.* the product of survival probability and fecundity. Crucial conditions in these models are the relative severity of capsule *versus* larval mortality, and the relationship between fecundity and the duration of the encapsulated phase—increased duration of encapsulation will reduce fecundity, given constant resources (see below).

The evolution of brooding has been considered by Strathmann & Strathmann (1982). Brooding is frequently associated with small body size. This may be because of an allometry of gamete production and ability to brood, where capacity for egg production increases more rapidly with increasing size than ability to brood. If so, brooding would be an evolutionary bottleneck at larger body sizes and would be disadvantageous. Brooding might be a 'conservative' response in small and short-lived animals unable to afford the risks of poor recruitment associated with reduced brood care. On the other hand, for large and long-lived animals, the advantages of dispersal could outweigh the risks involved. Strathmann & Strathmann (1982) pointed out that the hypotheses they advanced are not mutually exclusive, and indeed that present knowledge does not allow them to be adequately tested. There is a need for more information on the demography and energetics of brooding and related non-brooding species.

Grant (1983) developed a theoretical consideration of brood protection. Important features in his model were the reduction in predation gained by protecting the eggs, and egg size. While Perron & Carrier (1981) stated that egg size distributions were unimodal, Grant (1983) disagreed, and suggested that their analysis confounded information from different faunistic regions. His model predicted extremes of developmental type to be the evolutionarily stable behaviour (see also Vance, 1973; Christiansen & Fenchel, 1979). This prediction was considered to hold for bivalves, but not for the genus *Conus*, and Grant (1983) called for further field studies of species with mixed development. The biological basis on which further progress may rest must come from studies such as those by Spight (1975, 1976) on the ecology of egg and juvenile stages, paying particular attention to the costs of eggs and capsules, and the degree of mortality. It would be particularly useful to have more data for Antarctic invertebrates, where protected development is frequent and is probably related to the paucity of food for larval stages (Picken, 1980a).

MODE OF DEVELOPMENT IN THE DEEP SEA

Thorson (1946, 1950) stated that species with pelagic larvae decline in numerical importance towards the poles (especially in the south) and in deeper water. He considered that pelagic development is virtually absent among deep-sea forms (Thorson, 1950). It was reasoned that this is due to lack of a food supply for planktotrophic larvae in the deep sea, as they were thought to be unlikely to ascend to the rich surface plankton. If food resources for the adults are sparse, spawning may be asynchronous and perhaps opportunistic—occurring upon chance encounter (Pain, Tyler & Gage, 1982). The shallow-water infaunal starfish *Ctenodiscus crispatus* seems to parallel this behaviour: its food resources are said to be very constant, and it also indulges in chance encounter spawning (Schick, Taylor & Lamb, 1981).

It is now clear, however, that planktonic development may be used by at least a minority of deep-sea animals. Tyler, Grant, Pain & Gage (1982) have found that at ≈2000 m in the Rockall Trough, the echinoderms *Ophiura ljungmani*, *Plutonaster bifrons*, *Echinus affinis*, *Ophiocten gracilis* and *Dytaster insignis* produce a large number of small eggs giving rise (presumably) to planktotrophic larvae. Breeding in these species is synchronized and seasonal, spawning occurring in January to February. How this synchrony is achieved is not known; the authors point out that even at this depth seasonal 'signals' arrive in the form of variation in particle flux from the surface, and of current variations. The effect is to place larvae in the water column at approximately the time of the spring bloom, when downward flux of organic particles is greatest. It is not clear where the preferred depth is for these larvae, nor on what exactly they may feed. Discussing *Ophiura ljungmani*, Tyler & Gage (1980) question whether larvae can migrate to the surface, 2–3 km above the parent stock, although this is known to occur in some deep-sea animals (see below). Nevertheless, Tyler *et al.* (1982) distinguish between seasonal reproduction involving small eggs and planktotrophic larvae in such echinoderms as *O. ljungmani*, and non-seasonal (continuous?) production of fewer, larger eggs by such animals as *Bathybiaster vexillifer* and *Psilaster andromeda*. Here, development is believed to be demersal, as the animals do not brood (Tyler & Pain, 1982).

Rokop (1977) has reported seasonal reproduction in two species which occur in bathyal and shallow waters—the brachiopod *Frieleia halli* and scaphopod *Cadulus californicus*. *Frieleia halli* spawns small eggs (≈ 100 μm) in January through April, and these are believed to give rise to planktotrophic larvae feeding on organic matter originating in the surface production, enhanced at this time of year by upwelling. Again, the larvae are of unknown distribution in the water column. *Cadulus californicus* has larger eggs (≈ 240 μm) which, it is inferred, are lecithotrophic (pelagic or demersal), and it spawns in July to October. Rokop (1977) suggested that this seasonal variability may simply be retained by deep populations of these species as a remnant of a shallow water habit. Tyler & Gage (1980) echo this suggestion concerning *Ophiura ljungmani*, which they say may possibly be a recent invader in the

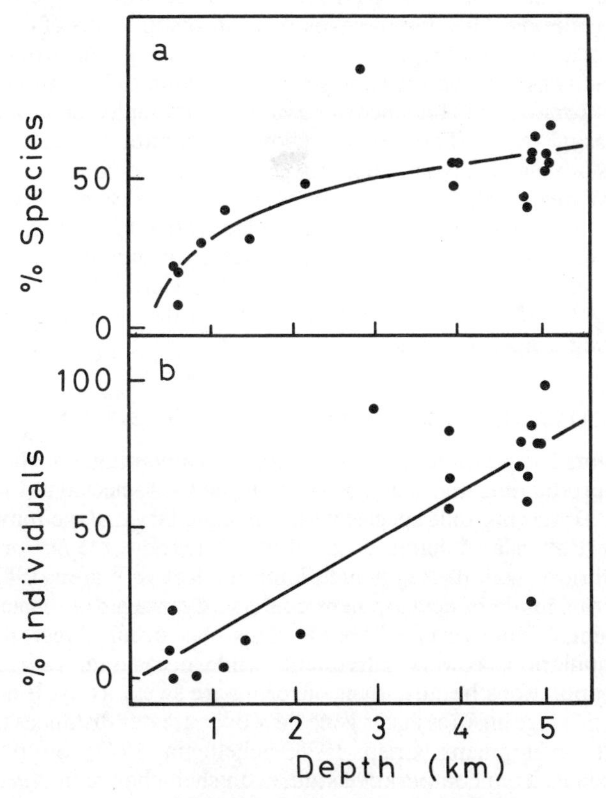

Fig. 1.—Percentage of all prosobranch species (a) and individuals (b) with planktotrophic development on the Gay Head–Bermuda transect: regression equations are, a, $Y = -88{\cdot}69 + 17{\cdot}20 \ln X, r = 0{\cdot}80, P < 0{\cdot}001$, b, $Y = -1{\cdot}51 + 0{\cdot}02X, r = 0{\cdot}83, P < 0{\cdot}001$; redrawn from Rex & Warén (1982).

bathyal habitat. They suggested that the rarity of planktotrophy in deep-sea forms may result from its irreversible loss in taxa spreading through the deep ocean from polar seas, where planktotrophy is rare. If *Frieleia halli* and *Ophiura ljungmani* have invaded the bathyal from shallower water at lower latitudes, they may have retained a planktotrophic larval stage under circumstances of reduced competition for particulate food. Lutz, Jablonski, Rhoads & Turner (1980) have analysed the shell of an undescribed mytilid bivalve from the hydrothermal vent community of the Galapagos Rift ($\approx$2500 m). On the basis of the larval shell, visible on the adult, they concluded that the animal has a planktotrophic larva with a long range dispersal ability. Hydrothermal vents are believed to be transient islands in the otherwise very stable deep sea, so the specialized biota associated with them are presumably dependent upon a relatively very precarious resource, and it might be expected that there would be a high premium on dispersal.

Similar arguments are used by Rex & Warén (1982) in discussing the occurrence of planktotrophic larvae among meso- and neogastropods in the deep sea. Analysing material from the western North Atlantic, and again using the form of the larval shell to assign species to lecitho- or planktotrophic modes, they found that planktotrophy is absent from slope and abyssal archaeogastropods. This may be explained on the grounds of constraints to adaptation: archaeogastropods all seem to lack planktotrophic larvae, although many may feed on the bottom (Strathmann, 1978). Among the meso- and neogastropods, the incidence of larval planktotrophy, however, increases with increasing depth. This trend is shown in Figure 1, where data for all gastropods are plotted against depth.

Rex & Warén (1982), noting the predominance of carnivores in the taxa concerned, interpreted this to mean that planktotrophic larvae are favoured among these animals in abyssal environments, presumably because of the advantages of dispersal to patchy food resources. Certainly, their findings are contrary to the statements of Thorson (1950) on the virtual absence of these larvae among deep-sea invertebrates.

THE SELECTIVE ADVANTAGES OF PLANKTONIC LARVAE

Many authors have commented on the supposed advantages of one or another mode of reproduction, and a summary of the views of selected authors is given in Table I. Observing that species with a pelagic larval stage may be absent from such tiny island habitats as Rockall (Moore, 1977) or from the continental slope with its fragmented habitats (Rex & Warén, 1982), we infer that while this mode of development confers advantages of dispersal, it may exclude animals from certain habitats. This may occur if recruitment from outside populations is very infrequent, while occasional vagrants cannot establish populations because their offspring are swept away. It is also likely that a pelagic stage ensures faster gene flow over greater distances than occurs with direct development (Crisp, 1978; Scheltema, 1978). Support for this inference comes from comparative studies on shell change in *Nucella lapillus* (direct development) and *Littorina littorea* (pelagic development) after the arrival of predator (the crab *Carcinus maenas*). Vermeij (1982a,b) found evidence of a response in *Nucella lapillus* populations north of Cape Cod, in which the shells became thicker and stronger after the spread of *Carcinus*

TABLE I

Summary of comments on mode of development

Thorson, 1946	Advantage of long-life planktotrophic lies in wide dispersal. Short-life planktotrophic seeks dispersal and not planktonic food.
Crisp, 1974, 1976	Advantage of pelagic larva lies in dispersal, achieving great range, maximizing gene flow, and producing even population density across habitats.
Clark & Goetzfried, 1978	In nudibranchs, lecithotrophic and direct development are more common in the tropics than in temperate waters—probably as a result of stable food supplies reducing the need for dispersal.
Clark, *et al.*, 1979	Pelagic development achieves dispersal, non-pelagic the local exploitation of resources.
Palmer & Strathmann, 1981	Dispersal of pelagic larvae is excessive for best advantage, pelagic habit may be a concomitant of small egg size adopted for its own sake, assuming that planktotrophic larvae confer some other advantage.
Todd & Doyle, 1981	Pelagic development bridges time between optimal spawning and settlement times.
Chia, 1974	Economy of egg production when resources are scarce will favour non-pelagic development, especially among small animals.
Vance, 1973	Only extremes of egg size confer maximum reproductive efficiency. Non-pelagic development favoured by high pelagic mortality, long development time, lack of planktonic food.
Christiansen & Fenchel, 1979	Only extremes will be found. Pelagic development disadvantageous at low temperatures because of long development times and lack of food at high latitudes. Direct development favoured when animals are small.

maenas into the area. *Littorina littorea* did not show this local evolutionary response, presumably because the interchange of larvae prevented response to relatively local selection pressures. Similarly, Currey & Hughes (1982) showed that individuals of *Nucella lapillus* from sheltered shores (where predation is high) have stronger shells than those from exposed sites, while *Littorina littorea* has shells of similar strength at these contrasting sites. Currey & Hughes ascribe the lack of differences between littorine populations to greater gene flow associated with pelagic larval development. They also demonstrated that increased shell strength in *Nucella lapillus* occurs at a cost, body mass (and hence, presumably, reproductive output) being smaller in strong-shelled animals. Amongst European *Littorina*, only two (*L. littorea* and *L. neritoides*) have a pelagic larva. Of those with direct development, the 'flat periwinkles' comprise two species (*L. obtusata* and *L. mariae*) and the 'rough periwinkles' four (*L. rudis, L. nigrolineata, L. arcana,* and *L. neglecta*)—see Raffaelli (1982) for a discussion. Evidently, the stocks with reduced gene flow have shown greater speciation. Johnson & Black (1982) suggested that while planktonic dispersal may cause uniformity on a large scale, it can give rise to fine-scale genetic patchiness where 'patches' on the shore can be distinguished over a distance of 50 m. This was said to arise capriciously as a result of settlement of larvae from entirely different parental stocks in adjacent portions of the shore.

If widespread dispersal is achieved by planktonic larvae, it may be intuitively sensible that the advantages of such dispersal underly the prevalence of planktonic larvae in most seas. Using a simulation model, Crisp (1974, 1976) has shown that different degrees of density-independent mortality across a series of habitat patches leads to local, and ultimately global extinction, in the absence of effective dispersal. Unpredictable density-independent mortality in habitat patches would then select for larval dispersal. Obrebski (1979) considered that scarcity of patches for colonization would select for larval dispersal, and that those invertebrates with abundant habitat patches might be expected to lose larval dispersal stages. Rex & Warén (1982) used similar arguments to explain the widespread occurrence of planktotrophic larvae in deep-sea carnivorous gastropods. Ayal & Safriel (1982) have shown that analysis of population growth parameters for the genus *Cerithium* provides evidence that the planktonic larva is of significance in maximizing dispersal.

Todd & Doyle (1981) have presented a variation on the dispersal arguments. They estimated that among nudibranchs the duration of the interval from egg to juvenile in the field is $\approx$55 days in *Adalaria proxima* and $\approx$71 days in *Cadlina laevis* (both of which are lecithotrophic), but $\approx$108 days in *Onchidoris bilamellata* (planktotrophic). This happens (at least partly) because of the longer developmental time taken by small eggs. Todd & Doyle (1981) suggest that the planktotrophic mode employed by *O. muricata* may be a device to bridge the time between optimal spawning of the adults (mid-January) and optimal settlement of the larvae, which, they argue, must arrive on the shore at about the time of the peak settlement of *Balanus balanoides*, occurring in May.

Strathmann (1974) suggested that pelagic larval phases of several weeks were a result of selection for spreading sibling larvae—*i.e.* achieving great separation not only between parents and offspring but also between individual offspring of a parent. Palmer & Strathmann (1981) advanced the idea that the scale of dispersal actually achieved by a great many invertebrates with planktotrophic larvae is greater than optimal on the basis of 'spreading' hypotheses. While stressing that adequate demographic data were not available for fully testing their hypothesis, Palmer & Strathmann (1981) used what was known about rates of spreading and the extent of environmental vicissitudes to suggest that there is no advantage to spreading beyond 40 km (on the open coast) or 20 km (in enclosed waters) on the coast of Washington State, U.S.A. This spatial scale corresponds to a planktonic life of a little more than 14 days (see Table 1 in Crisp, 1978). Palmer & Strathmann (1981) speculated that the advantages of planktotrophy over lecithotrophy must lie in other life history aspects, such as the ability to produce a greater number of smaller eggs, with dispersal occurring as a fortuitous by-product. From studies on the barnacles *B. glandula* and *B. cariosus*, Strathmann, Branscomb & Vedder (1981) have concluded that the loss of fine tuning in settlement responses is a real cost of wide larval dispersal.

Where the same species uses more than one mode of reproduction, it might be expected that close study should reveal the advantages of the different modes. There are likely candidates among the nudibranchs (Clark, Busacca & Stirts, 1979; Eyster, 1979), prosobranchs (Gallardo, 1977), and echinoderms (Scheibling & Lawrence, 1982). Gallardo (1977) and Scheibling & Lawrence (1982) believed that very closely related sibling species were involved in the cases studied by them. Only in the example of *Elysia cauze*, reported by Clark *et al.* (1979), is there good evidence of adaptive features. The development of *E. cauze* varies seasonally, with early generations of the animal producing planktotrophic veligers, later ones more advanced veligers which settle almost at once, and later still the development becomes direct. The eggs are deposited in a coil, provided with an extra-embryonic yolk string which runs the length of the coil. The adults feed obligately on the alga *Caulerpa racemosa*, which is not available in the area (southern Florida) in winter, and the change in developmental mode seems to be associated with tracking this seasonal food resource. Early egg masses have more closely packed egg capsules, implying that there is less yolk resource per embryo in the mass. These embryos hatch as veligers. Very late in the season, fewer eggs per mass are produced, and the embryo is encapsulated until after complete metamorphosis has occurred. There is an intermediate summer type in which the larva hatches as a very late veliger. This interesting case seems to be an instance of an animal which uses a dispersive mode to colonize a food resource as it becomes available, and non-dispersive modes to maximize use of the established resource. Where the initial colonists come from is unknown.

McKillup & Butler (1979) found that the gastropod *Nassarius pauperatus* produces more eggs when food is short than when it is abundant. They interpreted this behaviour as an iteroparous animal shifting towards semelparity when future adult survival is threatened, and pointed out the advantages of dispersal when habitats are patchy and liable to deterioration.

With all these contrasting views, the significance of pelagic development remains problematic. It is clearly of advantage in dispersal, and in some animals this seems to be reason enough for its existence. Paradoxically (because it was thought that pelagic larvae were virtually absent from the deep sea—*e.g.*, Thorson, 1950) this is especially apparent for some deep-sea animals (Lutz *et al.*, 1980; Rex & Warén, 1982). In the case of teleplanic larvae (Scheltema, 1978; Kempf, 1981; Pechenik, Scheltema & Eyster, 1984) there are metabolic adaptations for a very long larval life. Whether such species are exploiting long range dispersal, or simply possess larvae which are able to cope with a long and unpredictable planktonic phase, is a conundrum. Palmer & Strathmann (1981) have not found advantages in excessive dispersal, but admit that their approach was (of necessity) limited. Species with wide dispersal may have great "geological longevity" (Scheltema, 1978; Jablonski & Lutz, 1983), and this may partly underly their success.

RESOURCES AND REPRODUCTIVE MODE

Clearly it is cheaper to produce a small egg than a large one. When the costs of accessory capsular material are included, capsules and protective layers being a feature especially of larger eggs, the unit cost of large eggs rises still further. Perron (1981) has shown that in *Conus* species there is a positive correlation between puncture resistance of the capsule and duration of encapsulated development. He recorded that for *C. pennaceus* the cost of producing an egg capsule more protective than those of the other *Conus* species he studied was equivalent to an estimated reduction in fecundity of 34% per capsule. Among ten species, as the duration of development in capsules increased, so did the proportion of energy allocated to capsules (Fig. 2), indicating that there is a shift in the allocation of the total resource available for reproduction towards increased protection of the eggs. In other prosobranchs, capsules may offer limited protection against osmotic shock, by delaying exposure of the embryos to dilution of their immediate environment (Pechenik, 1982, 1983).

As for overall cost to a parent, taking number and size of eggs into account, Vance (1973) speculated that non-pelagic development is more expensive than pelagic. Crisp (1974), on the contrary, suggested that pelagic development is more expensive. Chia (1974) expressed the same view, and proposed that "the main selective force for shifting from feeding to non-feeding larvae is the limitation of energy for gamete production". Empirical studies have produced conflicting answers: Menge (1974) stated that non-pelagic development is more expensive than pelagic in two starfish, while Grahame (1977) and Todd (1979) stated the reverse for two prosobranchs and two opisthobranchs respectively. These studies used ratio indices to judge the cost of reproduction—*i.e.* the magnitude of gonad (or spawn output) as a fraction of body mass (or joules). Tinkle & Hadley (1975) have argued that this is not an adequate measure of reproductive effort, which should be expressed as a portion of the total energy budget. With this in mind, Grahame (1982) re-examined the question of reproductive effort in the prosobranchs *Lacuna*

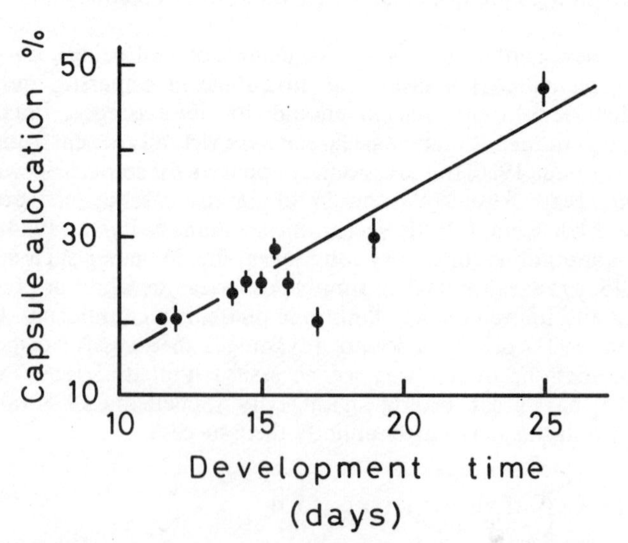

Fig. 2.—The relationship between the duration of development inside egg capsules and the percentage of total female reproductive output (as energy equivalents) allocated to capsules, in 10 species of *Conus* from Hawaii: means and standard deviations are shown; the regression equation is, $Y = 0{\cdot}821 + 1{\cdot}727X$, $r = 0{\cdot}883$, $P < 0{\cdot}001$; redrawn from Perron (1981).

pallidula (direct development) and *L. vincta* (brief encapsulated embryonic development followed by long-life planktotrophic development). It was found that when energy allocations to growth and respiration, as well as to egg production, were accounted for, *L. vincta* put only 4% more of its energy budget into eggs than did *L. pallidula*. As *L. vincta* has overall a higher energy flow than *L. pallidula*, the simple ratio index of reproductive effort is misleading—the 'effort' of *L. vincta* is higher all round. Therefore, there is as yet no consensus as to whether lecithotrophy or planktotrophy (if either) is more expensive to the parent, and indeed the work on *Lacuna* suggests that there may be very little difference. Kolding & Fenchel (1981) cite an unpublished theoretical study indicating that there is one evolutionarily stable value of reproductive effort, which species presumably will tend to adopt in a very wide range of circumstances (see also Christiansen & Fenchel, 1979).

It has been suggested that parental size may be important in governing reproductive mode (other than in the case of brooding already discussed). For example, among nudibranchs, in the case of a small species (*Onchidoris muricata*, which is planktotrophic) and a larger one (*Adalaria proxima*, which is lecithotrophic with a short pelagic life, if any), Todd (1979) argued that *Onchidoris muricata* may be too small to sustain the production of lecithotrophic eggs. Chia (1974) argued the opposite as a general rule: that smaller animals produce larger eggs on the whole, and that their development then tends to be lecithotrophic and brood protecting.

There are several other cases of closely related coexisting species which differ in size and have different modes of reproduction. For instance, *Asterina gibbosa* produces about 1000 benthic eggs with direct development, while *A. phylactica* (one fifth the size) produces only 60 eggs which it broods (Emson & Crump, 1979). *Alcyonidium digitatum* is larger than *A. hibernicum* and is planktotrophic compared with brooding (Hartnoll, 1977). *Crepidula fornicata* is planktotrophic while a smaller congener, *C. convexa*, broods its eggs (Hoagland, 1978). Similarly *Pisaster ochraceus*, a very large starfish, has planktotrophic larvae while *Leptasterias hexactis*, one-twentieth the size, broods its eggs (Menge, 1975). Menge argued that because *Pisaster ochraceus* is large it can produce a vast number of eggs—about 40×10^6 per year—and thus replace itself despite massive mortality of the planktotrophic larvae; *Leptasterias hexactis*, being smaller, could only produce 2·5% of this number of eggs if it were to be planktotrophic, and Menge estimates it would have to live 1626 years to produce enough eggs to maintain itself.

In all these cases the smaller of the pairs of species undertakes the "safer" method of development. While this is interesting, it raises another problem. If, for example, we consider *L. hexactis* "too small" for planktotrophy, then why is it that *Lacuna vincta*, a fraction of its size, manages to survive with its planktotrophic development?

Underwood (1979) has suggested that there may be two size thresholds operating in prosobranchs. The largest animals have available lecithotrophy or planktotrophy, or may brood, intermediate-sized ones are constrained to planktotrophy or brooding because they cannot provision enough lecithotrophic eggs, while the smallest species revert to lecithotrophy combined with cryptic habits. Investigation of this hypothesis requires studies of the energetics and demography of the smallest species, which have been largely neglected (but see Wigham, 1975; Underwood & McFadyen, 1983).

LIFE HISTORY PATTERNS

THE BACKGROUND

Life history patterns are often referred to as "strategies". A strategy is "the art of projecting and directing the larger military movements and operations of a campaign" (*Oxford English Dictionary*). If we take the view that survival and progeny-leaving are the outcome of a series of features (morphological, physiological and behavioural), then these features can be seen as a "strategy" assembled by natural selection and ensuring survival. It is implicit in this that the features under consideration are adaptive—*i.e.* that in the course of evolution they have been selected on account of advantages conferred.

It is inappropriate to review here the theoretical work on life histories, for there are several excellent reviews, of which Stearns' (1976) paper is one. Briefly, there are two main schools of thought: one concerning *r*- and *K*-selection, and the other what Stearns (1976) has called "bet-hedging".

In *r*- and *K*-selection, species are seen as lying somewhere on a continuum between extreme *r*- (with features such as early first reproduction, small and numerous eggs, high reproductive effort, semelparity, no parental care) and extreme *K*- (with features generally the opposite of *r*-). The choice of terms (from the logistic population growth equation) reflects the belief that selection might operate differently in expanding populations (*r*-) from those at or near carrying capacity (*K*-). Attractive as these ideas are to many ecologists, there are severe problems in applying them to real populations, causing others to reject them (Ebert, 1982; Stearns, 1983; Bergmans, 1984).

In bet-hedging, the crucial factor is the predictability of mortality at particular stages in the life of the animal. Unpredictable juvenile mortality favours reduced reproductive effort, smaller clutches and longer life of the adult—especially repeated breeding, or iteroparity. Unpredictable or heavy adult mortality favours early, substantial investment in reproduction, large clutches and once-only breeding, or semelparity.

As an example, Frank (1975) showed that at high latitudes the intertidal winkle *Tegula funebralis* is slower growing, has a lower mortality, is longer lived and reaches twice the size compared with individuals at lower latitudes. He associated these features with irregular recruitment at higher latitudes, compensation being gained by longer adult life. Longevity is presumably increased by slower growth and (possibly) lower reproductive output. When Frank transplanted winkles from Oregon (43·3° N) to California (36·5° N) they grew slightly faster than in Oregon but failed to match the high growth rates of the Californian residents. Genetic differences between the populations may thus put constraints on growth rates, and may have been favoured by natural selection because of the differences in the predictability of recruitment at different latitudes.

The notion of a trade-off embodies the intuitively sensible idea that if allocation to one function increases allocations to others (or, perhaps, one other) are likely to decrease in compensation. Likely trade-offs in life histories have received recent attention from Sibly & Calow (1983). A trade-off between egg size and number is common in Crustacea (Hines, 1982).

We have already considered the possible adaptiveness, and the relative costs and benefits, of reproductive mode in marine invertebrates. We now proceed to examine some other aspects of their life history patterns.

THE COST OF REPRODUCTION

Reproduction takes resources which might be devoted to other activities, and Calow (1979) states that in general there is a negative correlation between reproductive investment and future growth, survival and reproductive activity. Crisp & Patel (1961) measured the effect of reproduction on growth of the barnacle *Elminius modestus*, an obligate cross-fertilizing hermaphrodite. They grew the animal on plates, some arranged so that they could fertilize one another and reproduce, others isolated so that they could not do so. Non-reproducers attained a larger size than did reproducers (Fig. 3). As the difference in body weights between the two groups was found to be almost exactly equal to the weight of the egg mass, it was suggested that energy was channelled into reproduction instead of growth. Barnes (1962) argued that the decline of growth in reproducing cirripedes was due to the suspension of feeding during reproduction. Also it is clear that there must be a 'programmed' decrease in growth even in the absence of reproduction (Fig. 3). In any event, reproduction clearly costs *Elminius* something in terms of body growth.

Reproduction may be risky: for example, Menge (1974) has shown that the starfish *Leptasterias hexactis* broods smaller numbers of eggs on wave-exposed shores than on sheltered ones. Brooding animals have fewer tube feet attached to the substratum than ones without eggs, and seem to be at greater risk of being washed away.

Intuitively one expects reproduction to have a cost, and there is some evidence that this is so. As well as the direct evidence just referred to, there are

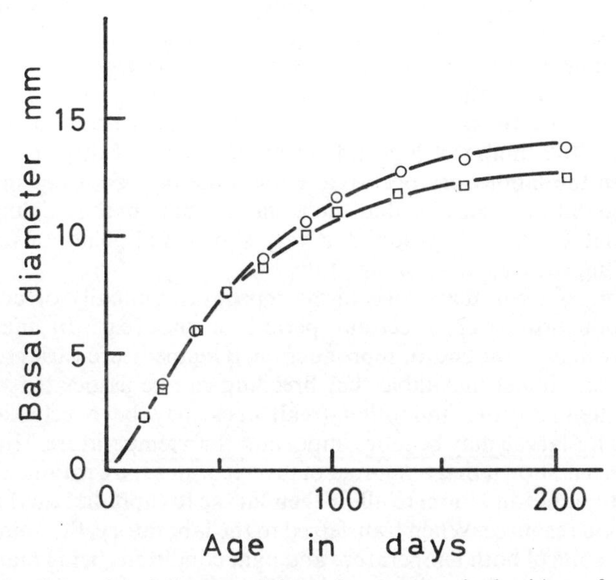

Fig. 3.—Mean basal diameters (along the rostro-carinal axis) as a function of time for non-breeding (○) and breeding (□) *Elminius modestus* growing on experimental panels: the difference in growth rate after ≈50 days was found to be significant; redrawn from Crisp & Patel (1961).

many instances where animals are known to put additional resources to reproduction when food is abundant—for example in barnacles (Wu, Levings & Randall, 1977), limpets (Fletcher, 1984) and in copepods (Checkley, 1980a,b; Cahoon, 1981; Uye, 1981; Kimmerer, 1984). Care must be taken in generalizing, however, for the prosobranch *Nassarius pauperatus* evidently increases fecundity when short of food, as discussed above (McKillup & Butler, 1979). This would be in accord with life history theory (Stearns, 1976) as a response to unfavourable conditions for future adult survival. Again, however, a contrary result is reported for two other prosobranchs by Spight & Emlen (1976): *Thais lamellosa* and *T. emarginata* both increased fecundity with increasing food supply.

If there are costs to reproduction, then selection may operate to minimize them, and also to maximize the expected return. There is a considerable literature exploring how these outcomes may be obtained and this will now be considered.

TIMING OF REPRODUCTION

In the short term, it may for example be critical to coincide breeding with particular phases of the tidal cycle. The pulmonate *Melampus bidentatus* inhabits the higher levels of salt marshes and is thus only covered by the tide during high spring tides. As it has a free-swimming veliger larva, both hatching of the eggs and settlement of the larvae are synchronized with spring tides (Russell-Hunter, Apley & Hunter, 1972). The synchrony persists as a rhythm in the laboratory under constant conditions.

Littorina littorea appears to have a lunar-tidal rhythm of spawn release (Grahame, 1975; Fish, 1979), presumably enhancing the dispersal of egg capsules released when tidal flow is greatest. The animal does not stop spawning on neap tides, rather, it increases output on springs. *L. melanostoma* restricts egg release to the days of spring tides around the full moon (Berry & Chew, 1973). Borkowski (1971) has suggested the presence of a "short term memory" for the timing of high tide in several tropical littorinids, because spawning in the laboratory occurred at the time of high water in the field, although he did not consider that there was a lunar rhythm. Schmitt (1979) reported that *L. planaxis* produces a very short-lived gelatinous egg mass, synchronizing with the time of high tide.

In the longer term, many organisms reproduce annually or concentrate their reproduction over a certain period of the year. In many cases, temperature may be the cue for reproduction. If temperature varies seasonally, it is, however, almost inevitable that breeding can be associated with some change of temperature, and other (real) cues may be overlooked. Food availability for larvae may be more important than temperature. Himmelman (1975, 1979) has shown that a number of invertebrates spawn at the time of the spring phytoplankton bloom to allow their larvae to capitalize on this rich but transient food resource. When transferred to the laboratory, the animals failed to spawn in spite of both temperature and light conditions being manipulated. Only when phytoplankton was introduced did spawning immediately follow. The use of temperature as opposed to larval food as a cue for breeding has been studied in barnacles by Hines (1978; and see also Page, 1984). Hines (1978) analysed natural populations in comparison with those living in the warm water of a nearby discharge canal of a power plant. The latter situation experienced the same amount of food, but a temperature 5 to 10 °C higher. In the case of *Chthamalus fissus*, breeding was spread over summer and very similar in both populations, suggesting that temperature was unimportant for this species, and laboratory experiments confirmed that food ration alone determined breeding. In *Tetraclita squamosa*, the breeding cycle was, however, closely linked to temperature. Normally occurring in summer, it was shifted by six months in the hot water outfall, thus taking place at the same temperature as in the normal population, but in winter.

If food abundance is important for the larvae, the possibility of competition exists. Reese (1968) described the reproductive cycles of three sympatric species of hermit crabs. The commonest (*Calcinus laevimanus*) had a prolonged summer breeding period, peaking between April and August. *C. latens* had a short peak in February and a lesser, second peak in August. *Clibanarius zebra* peaked only in August. Reese (1968) suggested that the less common species have had their reproductive cycles displaced and compressed to avoid larval competition with the most common species, although there is no direct evidence that competition does occur. It has also been suggested that as brittle stars cannot feed while breeding, some benthic gastropods time their breeding so that their larvae settle while these predators are not feeding (see Underwood, 1979). In the tropics, many invertebrates may breed continuously (Goodbody, 1965; Pearse & Phillips, 1968). At the same time, some seasons may be better than others for survival of juveniles (Goodbody & Gibson, 1974).

A variety of factors, then, may determine what time of the year an animal should reproduce. At times these factors may conflict, and it is not surprising that no single general cue for reproduction has emerged.

SEMELPARITY *versus* ITEROPARITY

Some organisms reproduce repeatedly during their lives and are termed iteroparous. Others reproduce only once and then die. They are termed semelparous, or more descriptively, 'big bang' reproducers. Many polychaete worms fit this description almost literally, packing their bodies with gonads and rupturing the body wall to release the gametes. A spectacular example of the semelparous pattern is seen in the opisthobranch *Onchidoris muricata*, where the digestive gland atrophies as spawning proceeds (Todd, 1978).

Among semelparous animals, the timing of the reproductive episode, and the circumstances under which it might occur earlier or later, are of interest. McLaren (1966) has drawn attention to the biennial reproductive cycle of the chaetognath *Sagitta elegans* in the Arctic, and contrasted it with the five or six generations per year in warmer waters. In the latter situation, food is available for long periods of time and the animal grows fast and reproduces at a small size. In the Arctic, food is available for newly-hatched chaetognaths for only a short period during the spring plankton bloom. An annual cycle would perhaps make sense, why be biennial? Cold waters slow down growth and result in a much larger size for *S. elegans*. In the Arctic, the average one-year old could produce about 138 eggs, but by postponing reproduction for another year a larger size is reached and egg production is about 543 eggs. The disadvantage of this is that the animal may not live that long. McLaren (1966), however, has suggested that the biennial cycle is in fact an optimal solution to the highly seasonal environment as it maximizes r—the intrinsic rate of natural increase.

In the supralittoral isopod *Ligia oceanica*, Willows (1984) has shown that small slow-growing females apparently delay reproduction until their second year, thus increasing their fitness over the expected level which would be attained if they were to breed as soon as they were able.

With iteroparity, there comes the question of how great the reproductive effort should be at successive breeding episodes. It is common to find that the relative amount of resource put into reproduction increases with age: an example is in species of *Conus* (Perron, 1982). This pattern is clearly revealed when the reproductive effort is quantified in terms of the amount of energy used for reproduction relative to that for other elements of production (Fig. 4). Simple ratios of gonadal to somatic weight (*e.g.*, Fletcher, 1984) are useful as comparative indices but fail to detect changes in reproductive effort with age. This is one reason why Tinkle & Hadley (1975) advocate energy as the currency for measuring reproductive effort.

If adequate data on the breeding and demography of a population are available, it is possible to estimate not only current reproductive effort (RE) but also what is called "residual reproductive value" (RRV—see Williams, 1966). This is an expression of the expected future offspring of the individual. It might be expected that if RRV is high, current RE will be low, with RE rising as RRV declines. Vahl (1981) has examined this proposition in the case of the scallop *Chlamys islandica* and found no evidence of the expected relationship between RE and RRV.

REPRODUCTION AND RISK

As was pointed out above, high unpredictability of mortality among adults (compared with juveniles) is expected to be correlated with early, urgent reproduction. Low unpredictability of mortality among adults is expected to

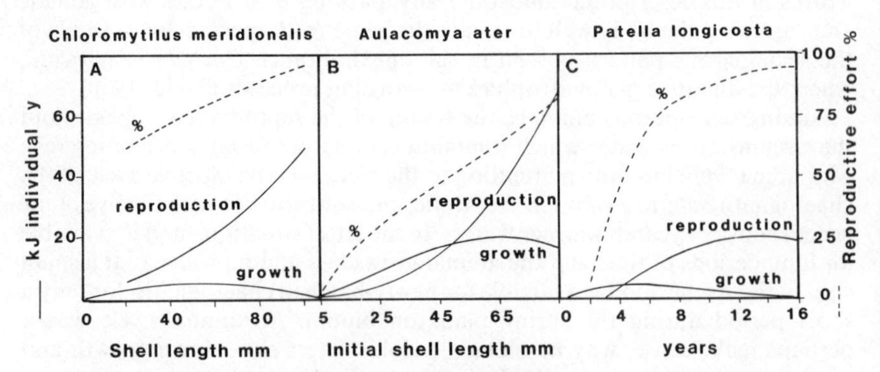

Fig. 4.—The relative allocation of energy (kJ·ind.$^{-1}$·yr^{-1}) to growth and reproduction, as related to age or size of (A) *Choromytilus meridionalis* (Griffiths, 1981), (B) *Aulacomya ater* (Griffiths & King, 1979), and (C) *Patella longicosta* (Branch, 1981).

be correlated with later and lower-effort reproduction. There is evidence for this sort of relationship in copepods (Lonsdale, 1981), among limpets (Parry, 1982a,b), and cone shells (Perron, 1982), although Fletcher (1984) could find no intraspecific correlation between reproductive effort and adult mortality in the limpet *Cellana tramoserica.* In amphipods (*Gammarus* spp.) winter-breeding forms have fewer and larger eggs than summer-breeding ones, and this is believed to be an adaptation countering larger juvenile mortality resulting from a low growth rate at low temperatures (Kolding & Fenchel, 1981). There were no apparent differences in 'effort' between the two groups, and these authors reject considerations of *r*- and *K*-selection as being futile for explaining their data. Similarly, comparing epi- and infaunal gammarids, Van Dolah & Bird (1980) found that the relative clutch volume was not different between the two groups, but epifaunal species produced more, smaller eggs. It was suggested that this was an adaptation to relatively greater adult mortality risk in epifaunal species.

Hines (1978) has compared the reproductive output of three barnacles, to which Hurley's (1973) data on a fourth can be added. *Balanus pacificus* occurs subtidally on ephemeral substrata such as small cobbles, surfaces that are periodically sand-scoured, and arthropod skeletons. These are short-lived and unstable substrata. Three intertidal species, *Chthamalus fissus*, *Balanus glandula* and *Tetraclita squamosa*, live in the high, mid and low shore, respectively. Figure 5 shows that the four species can be ranked in terms of a variety of life history characters. *Balanus pacificus* clearly makes a priority of early breeding, and breeds all year. For the other three species, their rank coincides with their shore position, with greater emphasis on reproduction high on the shore, and greater emphasis on somatic functions low on the shore. Thus, where the substratum for the adult is impermanent and colonization of new ones is particularly vital, reproduction is of paramount importance. Species on a more permanent substratum decrease their reproductive allocation along with decreasing physical hazards (from desiccation) and increasing biological hazards (from predation).

Choat & Black (1979) have investigated the life histories of the limpets *Acmaea insessa* and *A. digitalis. A. insessa* lives on the kelp *Egregia laevigata* and must mature and reproduce within a year, before the death of the alga. Up to 70% of the body weight of this limpet may be gonad, while in *Acmaea digitalis*, living on permanent hard surfaces, the gonad may be up to 30% of the body weight. Clearly the limpet on the ephemeral substratum makes a higher priority of reproduction and seems to have a higher reproductive effort. Whether these findings on barnacles and limpets may be better reconciled with *r*- and *K*-selection, or with bet-hedging, or indeed with some combination of the two, is not clear.

Perhaps it is a mistake to polarize bet-hedging and *r*- and *K*-selection. Different species may have their reproductive patterns moulded by different conditions, and either or both theories may be applicable under different circumstances. The balance between adult and juvenile mortality does seem important in relation to reproductive patterns (bet-hedging) but another factor is the degree to which adult mortality is determined by reproduction (intrinsic mortality) or by agencies unrelated to reproduction (extrinsic mortality) (*r*-, *K*-selection). Parry (1982a), for example, has attempted to unravel these effects for four limpets. If extrinsic mortality is high, then a high

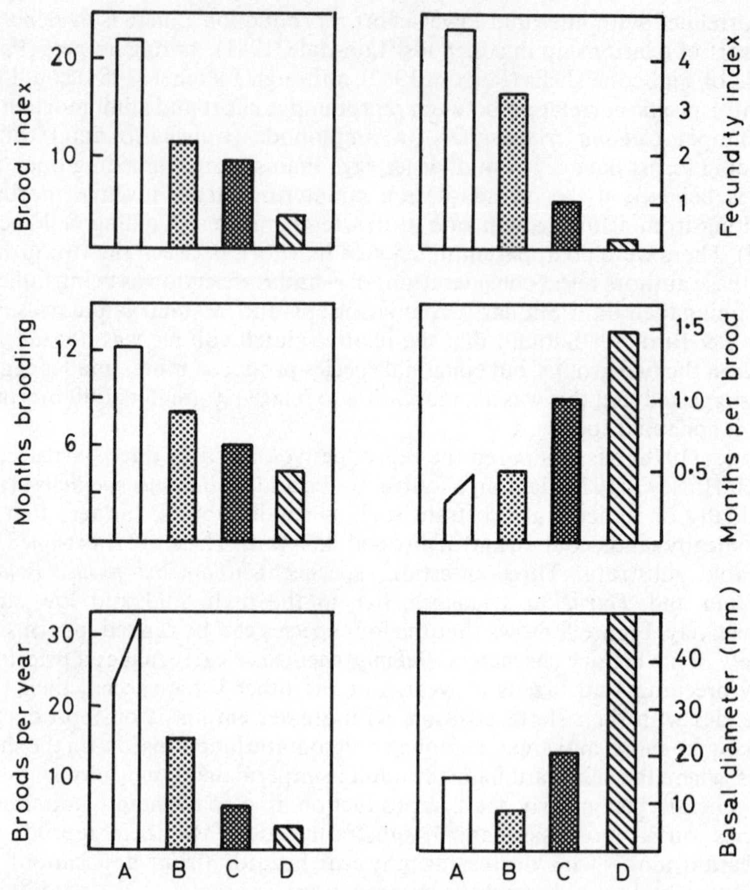

Fig. 5.—Six indices of reproductive performance in the barnacles *Balanus pacificus* (A), *Chthamalus fissus* (B), *Balanus glandula* (C), and *Tetraclita squamosa* (D): "brood index" is the total annual brood weight as a percentage of body weight; "fecundity index" is the number of eggs ($\times 10^4$) per mg body weight; drawn from data in Hurley (1973) and Hines (1978).

reproductive effort can be anticipated. There is little point in withholding energy from reproduction if the adult is not likely to survive long enough to reproduce at a later stage (Fig. 6A). Species living in transient habitats conform to this pattern—*e.g. Balanus pacificus* which lives on ephemeral substrata (Hurley, 1973) and *Tigriopus brevicornis* which occupies ultra-high intertidal pools (Harris, 1973). On the other hand, if there is a low extrinsic mortality, then investment in reproduction will depend first, on how much reproduction costs the parent, and secondly, on the balance between juvenile and adult mortality. Taking the case where intrinsic mortality is potentially high (reproduction carries a high cost) we can expect that reproductive effort will be much lower in animals with a potentially long life. It will be lowest if juvenile mortality is high relative to adult mortality, for an increase in reproduction

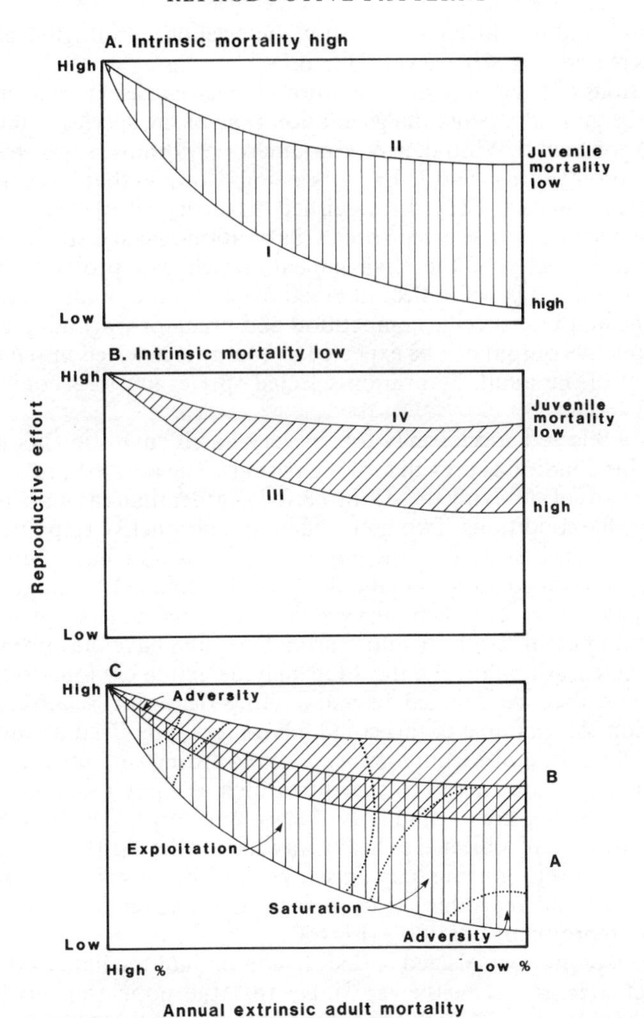

Fig. 6.—Hypothetical influence of extrinsic mortality on reproductive effort: A, when intrinsic mortality due to reproduction is high, and juvenile mortality (relative to that of the adult) varies from being high (curve I) to low (curve II); B, when intrinsic mortality is low and juvenile mortality ranges from high (curve III) to low (curve IV); C, superimposition of these circumstances showing the most likely reproductive patterns of adversity-, exploitation-, and saturation-selected species.

will reduce adult survival while reaping dubious returns because the survival of the offspring will be doubtful (Fig. 6A, curve I). If on the other hand juvenile mortality is low relative to that of the adult, a higher reproductive output will be desirable (Fig. 6A, curve II).

If reproduction costs the adult little, reproductive effort will again be increased, and although relationships corresponding to curves I and II still exist for high and low juvenile mortality, the level of reproductive effort can now be increased (Fig. 6B, curves III and IV).

Fluctuations of environmental conditions (which cause extrinsic mortality) must also be gauged against the generation time of the species—'stability' is relative to longevity. Whittaker & Goodman (1979) have suggested that a simple dichotomy into *r*- and *K*-habitats is simplistic and that three situations should be recognized. They distinguished adversity-selected, exploitation-selected, and saturation-selected species. Saturation-selected species live near the carrying capacity of the environment, which will probably be fairly constant. Adaptations will reflect interactions with other animals, including intraspecific and interspecific competition and predator-prey adaptations. A low reproductive output can be expected as energy is diverted into increasing the survival of the adult. Saturation-selected species are thus equivalent to *K*-selected species.

Adversity-selected species may live in very harsh environments in which unfavourable conditions prevail for long periods. The greatest pressure exists to evolve means of surviving these long periods, rather than capitalizing on the rare favourable conditions. Two quite different reproductive responses can be expected. The species may be strictly semelparous and have outbursts of reproduction when conditions make this possible, coupled with a tolerant or dormant phase that can then survive the following adverse period. Such species would put all resources into reproduction and have the appearance of being *r*-selected. Alternatively, the adult may be extremely long-lived. Most energy would then be devoted to maintaining the adult, relatively less to reproduction. An example of this could be the very long-lived urchins which Ebert (1982) has describe from areas exposed to strong surf, for they channel proportionately more energy to structures such as thick test walls than do urchins from calm areas. Similarly Lewis & Storey (1984) noted that *Echinometra lucunter* subjected to heavy wave action develop thick tests (at the apparent cost of reproducing only once a year) while those in sheltered areas have thin tests and reproduce twice a year. Thus, adversity may select for extremes of reproductive patterns (Fig. 6C).

Thirdly, exploitation-selected species live in an intermediate environment which fluctuates and is neither so stable that the population lives at the carrying capacity of the environment, nor so adverse that all resources need to be devoted to survival. Populations will fluctuate as conditions change. Under favourable circumstances *r*-selected characteristics will be advantageous, but the species will need to be flexible and reduce reproductive effort if resources are short (*e.g.* Fletcher, 1984).

These three possibilities are summarized by superimposing them on the hypothetical graphs relating mortality to reproductive effort (Fig. 6C).

Hart & Begon (1982) have examined reproduction in populations of *Littorina rudis* living in two habitats, boulders and crevices, argued to be respectively *r*- and *K*-selecting environments. They found that in the boulder population, reproduction was delayed and of lower 'effort' (*K*-characters) but offspring were smaller and more numerous (*r*-character). It was concluded that while the number and size of offspring reflected the '*r*-ness' or '*K*-ness' of that habitat, other attributes were a solution to the particular demands of the winkles' habitat. This arose because of selection for larger adult size and thicker shells among the boulders than in the crevice habitat, which when combined with selection for many, small young produced an outcome confounding *r*- and *K*-characters. Hart & Begon (1982) concluded that in this case at least, the life histories can only be understood by reference to the details of the ecology of the species: "general theories cannot provide short cuts". Furthermore, it is worth noting that Faller-Fritsch (1977), also working on *L. rudis* from exposed shores and sheltered boulder beaches, produced different results. While he too found larger sized winkles and delayed reproduction on boulder beaches, he recorded that "embryo production involves fewer numbers but greater individual size" than on exposed shores. Thus, even working on the same species under comparable conditions, contrasting results have been obtained.

CONCLUSIONS AND PROSPECTS

At the outset, it was stated that the attributes of an organism which make up its life history may usually be seen to be adaptive. It can be argued that ageing is not (Kirkwood & Holliday, 1979), but this aspect has not concerned us here. There is much evidence that the life history traits which have concerned us—such as egg size and number, time of breeding, degree of protection of the young—have been (and are) under selective pressure, and have evolved. Life history is not a unitary character, but composed of several traits. These evolve neither completely in concert, nor completely independently (Trendall, 1982; Rose, 1983). That evolutionary change in life history traits of modern animals is possible has been shown by experiment—for example, with *Gammarus lawrencianus* (Doyle & Hunte, 1981). In this case, the intrinsic rate of natural increase (*r*) was increased by artificial selection in the laboratory over 26 generations, as a consequence of changes in survival, age at maturity, and fecundity.

The cost of reproduction—either direct, in terms of resources used, or indirect, in terms of additional risks taken—is a central point in life history research. There is disagreement in the literature on how great such costs are, or even if they exist. Bell (1984) considered that in five freshwater species he studied there was no evidence of the expected negative correlation between present and future fecundity, and this particular cost hypothesis was falsified. All the animals in his experiments reproduced asexually; the use of such a special group of animals may mean that the results are not generally applicable. Another facet is that reproduction and somatic growth are often considered antagonistic (*e.g.* Ebert, 1982; Calow, 1983). But when closely related species are compared, rate of growth and reproductive effort may even be positively correlated (as, for example, in Fig. 7A). It seems that some species have a characteristically high turnover while others do not. Newell & Branch (1980) have contrasted these extremes as being "exploiters" and "conservers" in relation to the rate with which they use resources. Even within species reproductive effort and growth rate are correlated, as Fletcher (1984) has shown for different populations of the limpet *Cellana tramoserica*. Both reproductive effort and growth rate do seem inversely correlated with longevity (*e.g.* Fig. 7B,C), but whether high reproductive effort and rapid growth are responsible for low longevity or are selected in species with short

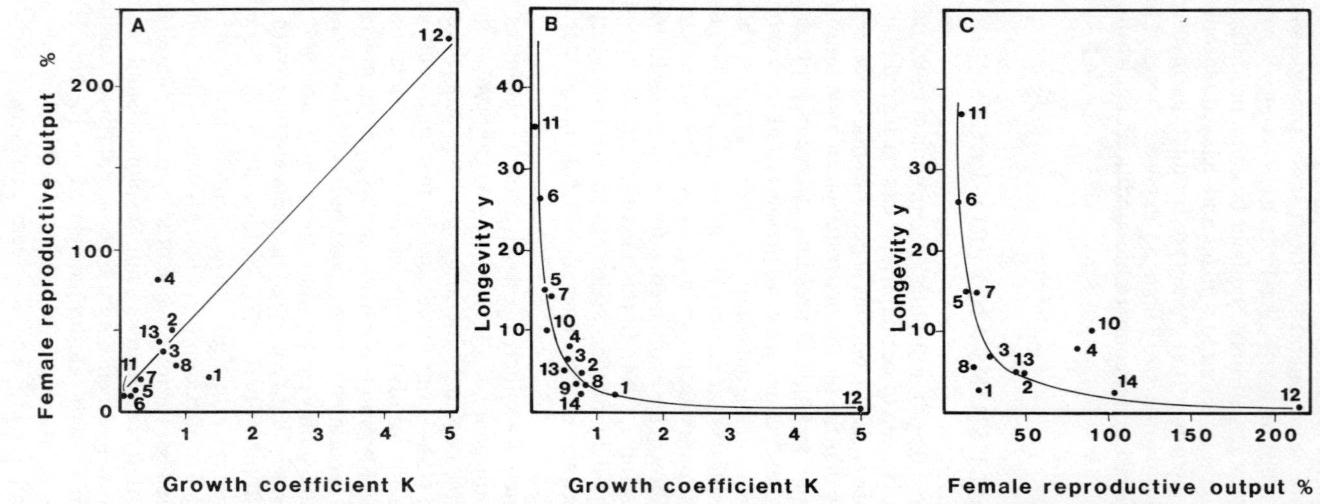

Fig. 7.—The relationships between (A) growth rate (measured as the growth coefficient K) and reproductive output (ratio of gametic output to somatic weight $\times$ 100), (B) growth rate and longevity, and (C) reproductive output and longevity, for 14 patellacean limpets: 1, *Patella barbara*; 2 and 3, *P. granularis*; 4, *P. granatina*; 5 and 6, *P. cochlear*; 7, *P. longicosta*; 8, *P. oculus*; 9, *P. compressa*; 10, *Nacella delesserti*; 11, *N. concinna*; 12, *Acmaea insessa*; 13, *A. digitalis*; 14, *Cellana tramoserica*; for A: $y = 14{\cdot}16 + 42{\cdot}78\,x$; $r^2 = 0{\cdot}80$; for B: $y = 1{\cdot}86\,x^{-1{\cdot}05}$; $r^2 = 0{\cdot}92$; for C: $y = 243\,x^{-1{\cdot}05}$; $r^2 = 0{\cdot}68$; 1–9 from Branch (1981); 10 from W. O. Blankley & G. M. Branch (in prep.); 11 from Picken (1980b); 12 and 13 from Choat & Black (1979); 14 from Parry (1982a,b).

life expectancies is often difficult to distinguish. Certainly, more work on the nature and magnitude of reproductive costs is needed.

In this context, it should be noted that very seldom when authors write of "reproductive effort" has this been quantified as rigorously as it should be (Tinkle & Hadley, 1975; Grahame, 1982). There is evidence that if the energy budget of the animal is not taken into account, the data on reproductive effort can be misleading (Grahame, 1977, 1982). We endorse the call by Calow (1983) for more life-time energy budget studies.

The advantages and the costs of any particular life history feature will, it seems, be different in different species, and the priority to be attached to a feature in understanding different life histories will vary. It seems to be futile to seek wide, all-embracing explanations. This is true when considering the advantages of the planktonic larva, and the life history attributes dealt with above (pp. 378–381). General theoretical schemes, such as *r*- and *K*-selection and bet-hedging, are useful as tools for thought; we cannot expect organisms to conform faithfully to one or the other, or indeed to any such scheme. This point has been made by several authors, for example, Hart & Begon (1982).

It is not sufficiently appreciated that there are limits to adaptation (Gould & Lewontin, 1979). Belief in the perfectibility of organisms under natural selection has sometimes led to error. Woodward (1909) cited evidence given by Tattersall on the breeding of the high shore winkle *Littorina neritoides*, which was said to be like *L. rudis* in being viviparous—"a remarkable instance this of adaptation to suit environmental conditions". Lebour (1935) corrected this, pointing out that *L. neritoides* has planktonic egss and larvae, and concluded that it was not a fixed rule for very high shore molluscs to have abandoned the veliger stage. Lebour (1935) reported that Tattersall was unable by then to say how he had drawn the conclusions he had done. Despite the correction, the original error still appears in an ecology text of two decades ago (Hesse, Allee & Schmidt, 1962). More recently, Menge (1975) suggested that brooding in the starfish *Leptasterias hexactis* was "a coadaptive consequence of competition-induced small size", whereas in fact brooding is probably a fixed trait in the genus *Leptasterias* (Himmelman *et al.*, 1982). Hines (1982) has examined the relationship between brood size and body size in a wide range of brachyuran crabs. He concluded that body size was the principal determinant of reproductive output, and challenged "the common theoretical assumptions that life-history traits are free to evolve under purely demographic forces". These cautionary tales remind us that while devising ingenious adaptive explanations for observed features, we must bear in mind that natural selection works with what is available to do only the best necessary job.

ACKNOWLEDGEMENTS

The second author's contribution was partly funded by a grant from the South African National Committee for Oceanographic Research and was completed at the Friday Harbor Laboratories, U.S.A.

REFERENCES

Ayal, Y. & Safriel, U. N., 1982. *Am. Nat.*, **119,** 391–401.
Barnes, H., 1962. *Limnol. Oceanogr.*, **7,** 462–473.

Bell, G., 1984. *Evolution*, **38,** 314–326.
Bergmans, M., 1984. *Evolution*, **38,** 292–299.
Berry, A. J. & Chew, E., 1973. *J. Zool.*, **171,** 333–344.
Borkowski, T. V., 1971. *Bull. mar. Sci.*, **21,** 826–840.
Branch, G. M., 1981. *Oceanogr. Mar. Biol. Ann. Rev.*, **19,** 235–380.
Cahoon, L. B., 1981. *Deep-Sea Res.*, **28A,** 1215–1221.
Calow, P., 1979. *Biol. Rev.*, **54,** 23–40.
Calow, P., 1983. *Biol. J. Linn. Soc.*, **20,** 153–165.
Caswell, H., 1981. *Am. Nat.*, **117,** 529–536.
Checkley, Jr., D. M., 1980a. *Limnol. Oceanogr.*, **25,** 430–446.
Checkley, Jr., D. M., 1980b. *Limnol. Oceanogr.*, **25,** 991–998.
Chia, F. S., 1974. *Thalassia jugosl.*, **10,** 121–130.
Choat, J. H. & Black, R., 1979. *J. exp. mar. Biol. Ecol.*, **41,** 25–50.
Christiansen, F. B. & Fenchel, T., 1979. *Theor. Pop. Biol.*, **16,** 267–282.
Clark, K. B. & Goetzfried, A., 1978. *J. moll. Stud.*, **44,** 283–294.
Clark, K. B., Busacca, M. & Stirts, H., 1979. In, *Reproductive Ecology of Marine Invertebrates*, edited by S. Stancyk, University of South Carolina Press, South Carolina, pp. 11–24.
Crisp, D. J., 1974. *Thalassia jugosl.*, **10,** 103–120.
Crisp, D. J., 1976. In, *Perspectives in Experimental Biology, Vol. 1, Zoology*, edited by P. Spencer-Davies, Pergamon Press, Oxford, pp. 145–155.
Crisp, D. J., 1978. In, *Marine Organisms; Genetics, Ecology, and Evolution*, edited by B. Battaglia & J. A. Beardmore, Plenum Press, New York, pp. 257–273.
Crisp, D. J. & Patel, B., 1961. *Limnol. Oceanogr.*, **6,** 105–115.
Currey, J. D. & Hughes, R. N., 1982. *J. anim. Ecol.*, **51,** 47–56.
Doyle, R. W. & Hunte, W., 1981. *Can. J. Fish. Aquat. Sci.*, **38,** 1120–1127.
Ebert, T. A., 1982. *Ecol. Monogr.*, **52,** 353–394.
Ellis, C. J. H., 1983. *J. moll. Stud.*, **49,** 98–106.
Emson, R. H. & Crump, R. G., 1979. *J. mar. biol. Ass. U.K.*, **59,** 77–94.
Eyster, L. S., 1979. *Mar. Biol.*, **51,** 133–140.
Faller-Fritsch, R. J., 1977. In, *Biology of Benthic Organisms*, edited by B. F. Keegan, P. O'Céidigh & P. J. S. Boaden, Pergamon Press, Oxford, pp. 225–231.
Fish, J. D., 1979. *J. moll. Stud.*, **45,** 172–177.
Fisher, R. A., 1958. *The Genetical Theory of Natural Selection.* Dover Publications, Inc., New York, 291 pp.
Fletcher, W. J., 1984. *Oecologia (Berl.)*, **61,** 259–264.
Frank, P. W., 1975. *Mar. Biol.*, **31,** 181–192.
Gallardo, C. S., 1977. *Mar. Biol.*, **39,** 241–251.
Goodbody, I., 1965. *Ecology*, **46,** 195–197.
Goodbody, I. & Gibson, J., 1974. *Biol. Bull. mar. biol. Lab., Woods Hole*, **146,** 217–237.
Gould, S. J. & Lewontin, R. C., 1979. *Proc. R. Soc. Ser. B*, **205,** 581–598.
Grahame, J., 1975. *J. exp. mar. Biol. Ecol.*, **18,** 185–196.
Grahame, J., 1977. *Mar. Biol.*, **40,** 217–224.
Grahame, J., 1982. *Int. J. Invert. Reprod.*, **5,** 91–99.
Grant, A., 1983. *Am. Nat.*, **122,** 549–555.
Griffiths, C. L. & King, J. A., 1979. *Mar. Biol.*, **53,** 217–222.
Griffiths, R. J., 1981. *J. exp. mar. Biol. Ecol.*, **52,** 231–241.
Harris, R. B., 1973. *J. mar. biol. Ass. U.K.*, **35,** 785–800.
Hart, A. & Begon, M., 1982. *Oecologia (Berl.)*, **52,** 37–42.
Hartnoll, R., 1977. In, *Biology of Benthic Organisms*, edited by B. F. Keegan, P. O'Céidigh & P. J. S. Boaden, Pergamon Press, Oxford, pp. 321–328.
Hesse, R., Allee, W. C. & Schmidt, K. P., 1962. *Ecological Animal Geography.* John Wiley, New York, 715 pp.
Himmelman, J. H., 1975. *J. exp. mar. Biol. Ecol.*, **20,** 199–214.
Himmelman, J. H., 1979. *Mar. Biol.*, **50,** 215–225.

Himmelman, J. H., Lavergne, Y., Cardinal, A., Martel, G. & Jalbert, P., 1982. *Mar. Biol.*, **68,** 235–240.
Hines, A. H., 1978. *Biol. Bull. mar. biol. Lab., Woods Hole*, **154,** 262–281.
Hines, A. H., 1982. *Mar. Biol.*, **69,** 309–320.
Hoagland, K. E., 1978. *Malacologia*, **17,** 365–391.
Hurley, A. C., 1983. *Limnol. Oceanogr.*, **18,** 386–393.
Jablonski, D. & Lutz, R. A., 1983. *Biol. Rev.*, **58,** 21–89.
Johnson, M. S. & Black, R., 1982. *Mar. Biol.*, **70,** 157–164.
Kempf, F. C., 1981. *Mar. Ecol. Prog. Ser.*, **6,** 61–65.
Kimmerer, W. J., 1984. *Mar. Biol.*, **78,** 165–169.
Kirkwood, T. B. L. & Holliday, R., 1979. *Proc. R. Soc. Ser. B*, **205,** 531–546.
Kolding, S. & Fenchel, T. M., 1981. *Oikos*, **37,** 167–172.
Lebour, M. V., 1935. *J. mar. biol. Ass. U.K.*, **20,** 373–378.
Lebour, M. V., 1945. *Proc. zool. Soc. Lond.*, **25,** 462–489.
Lewis, J. B. & Storey, G. S., 1984. *Mar. Ecol. Progr. Ser.*, **15,** 207–211.
Lonsdale, D. J., 1981. *Mar. Ecol. Progr. Ser.*, **5,** 333–340.
Lutz, R. A., Jablonski, D., Rhoads, D. C. & Turner, R. D., 1980. *Mar. Biol.*, **57,** 127–133.
McKillup, S. C. & Butler, A. J., 1979. *Oecologia (Berl.)*, **43,** 221–231.
McLaren, I. A., 1966. *Ecology*, **47,** 852–855.
Menge, B. A., 1974. *Ecology*, **55,** 84–93.
Menge, B. A., 1975. *Mar. Biol.*, **31,** 87–100.
Mileikovsky, S. A., 1971. *Mar. Biol.*, **10,** 193–213.
Moore, P. G., 1977. *J. mar. biol. Ass. U.K.*, **57,** 191–200.
Newell, R. C. & Branch, G. M., 1980. *Adv. Mar. Biol.*, **17,** 329–396.
Obrebski, S., 1979. *Mar. Ecol. Prog. Ser.*, **1,** 293–300.
Page, H. M., 1984. *J. exp. mar. Biol. Ecol.*, **74,** 259–272.
Pain, S. L., Tyler, P. A. & Gage, J. D., 1982. *Mar. Biol.*, **70,** 41–50.
Palmer, A. R. & Strathmann, R. R., 1981. *Oecologia (Berl.)*, **48,** 308–318.
Parry, G. D., 1982a. *Ecol. Monogr.*, **52,** 65–91.
Parry, G. D., 1982b. *Mar. Biol.*, **67,** 267–282.
Pearse, J. S., 1969. *Mar. Biol.*, **3,** 110–116.
Pearse, J. S. & Phillips, B. F., 1968. *Aust. J. mar. Freshwat. Res.*, **19,** 161–172.
Pechenik, J. A., 1979. *Am. Nat.*, **114,** 859–870.
Pechenik, J. A., 1982. *J. exp. mar. Biol. Ecol.*, **63,** 195–208.
Pechenik, J. A., 1983. *J. exp. mar. Biol. Ecol.*, **71,** 165–179.
Pechenik, J. A., Scheltema, R. S. & Eyster, L. S., 1984. *Science*, **224,** 1097–1099.
Perron, F. E., 1981. *Am. Nat.*, **118,** 110–118.
Perron, F. E., 1982. *Mar. Biol.*, **68,** 161–167.
Perron, F. E. & Carrier, R. H., 1981. *Am. Nat.*, **118,** 749–755.
Picken, G. B., 1980a. *Biol. J. Linn. Soc.*, **14,** 67–75.
Picken, G. B., 1980b. *J. exp. mar. Biol. Ecol.*, **42,** 71–85.
Raffaelli, D., 1982. *J. moll. Stud.*, **48,** 342–354.
Reese, E. S., 1968. *J. exp. mar. Biol. Ecol.*, **2,** 308–318.
Rex, M. A. & Warén, A., 1982. *Deep-Sea Res.*, **29,** 171–184.
Rokop, R. J., 1977. *Mar. Biol.*, **43,** 237–246.
Rose, M. R., 1983. *Am. Zool.*, **23,** 15–23.
Russell-Hunter, W. D., Apley, M. L. & Hunter, R. D., 1972. *Biol. Bull. mar. biol. Lab., Woods Hole*, **143,** 623–656.
Scheibling, R. E. & Lawrence, J. M., 1982. *Mar. Biol.*, **70,** 51–62.
Scheltema, R. S., 1978. In, *Marine Organisms; Genetics, Ecology and Evolution*, edited by B. Battaglia & J. A. Beardmore, Plenum Press, New York, pp. 303–322.
Schick, J. M., Taylor, W. F. & Lamb, A. N., 1981. *Mar. Biol.*, **63,** 51–66.
Schmitt, R. J., 1979. *Mar. Biol.*, **50,** 359–366.
Sibly, R. & Calow, P., 1983. *J. theor. Biol.*, **102,** 527–547.
Spight, T. M., 1975. *Oecologia (Berl.)*, **21,** 1–16.

Spight, T. M., 1976. *Biol. Bull. mar. biol. Lab., Woods Hole*, **150,** 491–499.
Spight, T. M. & Emlen, J., 1976. *Ecology*, **57,** 1162–1178.
Stearns, S. C., 1976. *Q. Rev. Biol.*, **51,** 3–47.
Stearns, S. C., 1983. *Evolution*, **37,** 601–617.
Strathmann, R., 1974. *Am. Nat.*, **108,** 29–44.
Strathmann, R. R., 1978. *Evolution*, **32,** 894–906.
Strathmann, R. R., Branscomb, E. S. & Vedder, K., 1981. *Oecologia* (*Berl.*), **48,** 13–18.
Strathmann, R. R. & Strathmann, M. F., 1982. *Am. Nat.*, **119,** 91–101.
Thorson, G., 1946. *Meddr Kommn dansk Fisk. Havunders.* (*Ser. Plankton*), **4,** No. 1, 523 pp.
Thorson, G., 1950. *Biol. Rev.*, **25,** 1–45.
Tinkle, D. W. & Hadley, N. F., 1975. *Ecology*, **56,** 427–434.
Todd, C. D., 1978. *J. moll. Stud.*, **44,** 190–199.
Todd, C. D., 1979. *Mar. Biol.*, **53,** 57–68.
Todd, C. D. & Doyle, R. W., 1981. *Mar. Ecol. Prog. Ser.*, **4,** 75–83.
Trendall, J. T., 1982. *Am. Nat.*, **119,** 774–783.
Tyler, P. A. & Gage, J. D., 1980. *Oceanologica Acta*, **3,** 177–185.
Tyler, P. A., Grant, A., Pain, S. L. & Gage, J. D., 1982. *Nature, Lond.*, **300,** 747–750.
Tyler, P. A. & Pain, S. L., 1982. *J. mar. biol. Ass. U.K.*, **62,** 869–887.
Underwood, A. J., 1979. *Adv. Mar. Biol.*, **16,** 111–210.
Underwood, A. J. & McFadyen, K. E., 1983. *J. exp. mar. Biol. Ecol.*, **66,** 169–197.
Uye, S.-I., 1981. *J. exp. mar. Biol. Ecol.*, **50,** 255–271.
Vahl, O., 1981. *Oecologia* (*Berl.*), **51,** 53–56.
Van Dolah, R. F. & Bird, E., 1980. *Estuar. cstl mar. Sci.*, **11,** 593–604.
Vance, R. R., 1973. *Am. Nat.*, **107,** 339–352.
Vermeij, G. J., 1982a. *Evolution*, **36,** 561–580.
Vermeij, G. J., 1982b. *Nature, Lond.*, **299,** 349–350.
Whittaker, R. H. & Goodman, D., 1979. *Am. Nat.*, **113,** 185–200.
Wigham, G. D., 1975. *J. mar. biol. Ass. U.K.*, **55,** 45–68.
Williams, G. C., 1966. *Am. Nat.*, **100,** 687–690.
Willows, R. I., 1984. Ph.D. thesis, University of Leeds, 184 pp.
Woodward, B. B., 1909. *Proc. Malac. Soc.*, **8,** 272–286.
Wu, R. S. S., Levings, C. D. & Randall, D. J., 1977. *Can. J. Zool.*, **55,** 643–647.

Oceanogr. Mar. Biol. Ann. Rev., 1985, **23**, 399–489
Margaret Barnes, Ed.
Aberdeen University Press

THE ECOLOGY OF MARINE NEMATODES

CARLO HEIP, MAGDA VINCX and GUIDO VRANKEN
Marine Biology Section, Zoology Institute, State University of Ghent, Ghent, Belgium

INTRODUCTION

Nematodes are probably the most abundant metazoans in the biosphere and of very great importance to man. As parasites of man they are responsible for disease in hundreds of millions of people and an estimated eight billion nematodes are enjoying food, warmth and shelter in human intestines today; *Ascaris lumbricoides* is, after the viruses responsible for diarrhoeal disease, the second most infectious organism in the world, about a quarter of the world's population being infected (Ash, Crompton & Keymer, 1984).

Plant-parasitic nematodes cause considerable damage to crops and many species are vectors of soil-borne viruses. Today, plant nematology is one of the most vital fields in agricultural research and is studied in many university departments over the world, yet fifty years ago only a few pioneers were engaged in the study of plant-parasitic nematodes and were convinced of the economic importance of these worms.

Whereas the importance of parasitic nematodes has now been recognized for many decades, this is not the case for the free-living species, especially those of aquatic environments. They remain relatively unstudied, despite the fact that they are extremely abundant, often numbering millions per m^2 in soils and sediments, and that they occur in a range of habitats which is unsurpassed by any other metazoan group, being absent only from the oceanic plankton.

Free-living nematodes are small and inconspicuous and have very rarely attracted amateur naturalists. They tend to live in environments such as intertidal muds that are not particularly appealing to many people. Many species are difficult to maintain in the laboratory and the taxonomic literature has been very scattered until quite recently, making determination of nematodes a frightening affair and putting ecologists in the embarrassing position of studying animals they could not even name. Yet, other fields have recently been attracted to nematodes, and species such as the small soil-dwelling *Caenorhabditis* have become very popular model organisms in molecular biology, genetics (Brenner, 1974), and even gerontology (Vanfleteren, 1978), where methods have been developed that greatly surpass the degree of sophistication reached in most ecological work. The lack of interest for a group that is as dominant in sediments as copepods are in the plankton seems no longer justified on methodological grounds.

It is also unjustified on ecological grounds. Nematodes are important in the ecology of the seas and they are interesting. Of course, they are small or even very small (down to 100 μm adult size) and structurally simple, basically consisting of two concentric tubes. In general most species are slender, with only a few tenths of μm diameter, but a considerable variation in habitus exists. As Platt & Warwick (1980) observed, not all nematodes look alike, many species of fine sediments are short while in coarse sands species are either very small or very elongate and thin. They are inconspicuous; many older papers do not even mention the occurrence of nematodes in sediments as nearly all specimens disappeared through the sieves that were used. Even today, when sieves of around 50 μm mesh size are universally used, estimates of nematode abundance are biased since many of the small species and juveniles of the bigger ones will be lost while processing the sample.

Despite their similar basic morphology nematodes occupy very different rôles and trophic positions in sediments. Many species feed on bacteria, on algae or on both, they eat detritus and possibly dissolved organic matter, and a considerable number are predators, feeding on other nematodes, oligochaetes, polychaetes, *etc.* This diveristy in feeding is reflected in species diversity; the number of nematode species in most habitats is much higher than that of any other metazoan group. As an example, a recent review counted a total of 735 nematode species in the North Sea (Heip, Herman & Vincx, 1983) and it is not uncommon to obtain 50 species or even more in a single 10 cm^2 core, often many congenerous, a situation that should be appealing to the theoretical ecologist trying to explain community structure in terms of predation or competition.

That nematodes are important in the energy flow through sediments has often been inferred from their numerical abundance. Yet, reliable data are scarce. In this paper we shall try to evaluate the existing literature, both the rather extensive literature describing the main structural features of nematode communities in the sea, and the far fewer papers on functional relationships and their possible use in pollution studies.

METHODOLOGY

General problems concerning benthic sampling have been treated in a number of publications (Holme & McIntyre, 1971; Elliott, 1971). For meiofauna the handbook edited by Hulings & Gray (1971) is still useful and it is in the process of revision.

SAMPLING

The most efficient sampling of nematodes from sediments is by hand-coring. The diameter of the cores used depends on the problem, but in general cores with an internal diameter of 3–4 cm are most efficient. In deeper water the different types of box-corers are to be preferred to grabs. An ideal corer should penetrate the sediment without a shock wave and as slowly as possible. We refer to the IBP-Handbook on Marine Benthos for further discussion.

THE NUMBER OF SAMPLES

The number of samples and sample size required depends on the problem being studied. Variability in nematode density appears to be mostly on a scale of a few centimetres (see p. 434) and again on a scale of many km or even more, depending on substratum heterogeneity. Intermediate scales have surprisingly low variability. If the problem consists of studying small-scale horizontal distribution, what we might call the intrinsic spatial pattern of nematode populations, obviously the largest amount of the smallest possible samples is required, with the important restriction that sample size must be large compared with the individuals being sampled. When the problem consists, as is usually the case, in obtaining an estimate of nematode density, there may be two alternative solutions. The first is to destroy all small-scale variability in the sample and take a sample as big as possible, mix it thoroughly and, if necessary, take subsamples. We have tested this many times in our laboratory and found it satisfactory. If the intrinsic pattern is not destroyed, aggregation will require that the sample size is as small as possible and the number of samples taken as large as possible. This is so because most statistical analyses are robust when they are based on a large number of error degrees of freedom (Green, 1979). When mean density is estimated from simple random sampling the error degrees of freedom is $n-1$, with n the number of samples. If sampling is done t times at each of s stations, the error degrees of freedom is $st(n-1)$. It therefore pays to sample more stations more frequently: as an example, the error degrees of freedom are the same when five stations are sampled twice (*e.g.* in a year) with three replicates or when one station is sampled once with 21 replicates. Green (1979) proposes three replicates per treatment combination (station-time) as a good round number in the absence of other information.

If a relationship between the variance s^2 and the mean $\bar{x}$ such as Taylor's power law exist ($s^2 = a\bar{x}^b$), some powerful generalizations can be made (Green, 1979). The required precision, as a percentage of the mean, can then be estimated as:

$$D^2 = an^{1-b}T_n^{b-2}$$

in which T_n is the cumulative number of individuals in sample n, *i.e.* the number of individuals in samples 1 plus 2 plus . . . plus n. When, as is often the case, $b = 2$, $D^2 = an$, and $n = a/D^2$; a can be evaluated from the fact that, if $b = 2$, $s^2 = a\bar{x}^2$ and $a = v^2$, in which v is the coefficient of variation $s/\bar{x}$. Taking the value obtained by Delmotte (1975) in a study of 49 samples covering a 7×7 grid of $0{\cdot}4 \times 0{\cdot}4$ m one gets $a = 0{\cdot}15$. If the desired precision is 10% of the mean (confidence intervals approximately 20% of the mean), the number of samples required is 15. For a precision of 20%, four samples are required; three replicates give a precision of 22%, two replicates 27%.

EXTRACTION TECHNIQUES

The extraction of nematodes from sediments is easy when the sediment is a sand with low amounts of detritus or silt-clay. Simple decantation on a sieve is often satisfactory, although more elaborate apparatus has been developed (Hulings & Gray, 1971). The maximum sieve size to be used is 50 μm, although juveniles and even adults of many species will not be retained by such a sieve and sieves of 20 μm have been used. Such small mesh sizes can only be used for very clean sands, and a 38 μm standard sieve seems a good compromise.

The extraction from muds or detritus (after the sand has been removed by decantation or other methods) is done most efficiently by using Ludox, although sugar can be used instead (Heip, Smol & Hautekiet, 1974). The method developed in our laboratory consists in the following procedure:

1. Rinse the sample thoroughly with tap water, to prevent flocculation of Ludox, over a sieve of 38 μm.
2. Bring the sample from the sieve in a centrifugation tube as large as available.
3. Add a small amount of kaolin, enough to cover the sample completely.
4. First centrifugation at 1800 *g* in water for 10 min.
5. Remove the supernatant.
6. Add Ludox HS 40‰, which is half diluted, to at least five times the volume of sediment and again centrifuge at 1800 *g* for 10 minutes. The supernatant is passed through a sieve of 38 μm. The centrifugation is repeated three times.

It has been shown by Heip (1974) that the number of nematodes collected after successive centrifugations is a constant proportion. When N_1 and N_2 are the numbers found in the first and second supernatant, then the total number in the sample is given by

$$N_t = \frac{(N_1^2/N_2)-1}{(N_1/N_2)-1}.$$

When the procedure can be standardized and the constant factor $a = N_1/N_2$ has been shown to be indeed constant, even one centrifugation permits a good estimate of the total number in the sample since:

$$N_t = \frac{aN_1-1}{a-1}.$$

FIXATION

Samples have to be fixed with 4–7% neutralized formalin. Both cold and warm fixation (70 °C) have been used. Many workers in the field fix marine nematodes with formalin in tap water.

MICROSCOPICAL EXAMINATION AND DETERMINATION

After fixation, animals must be transferred to anhydrous glycerol. Specimens are transferred from formalin to glycerol through a series of ethanol-glycerol solutions to prevent the animals from collapsing (Seinhorst, 1959; De Grisse, unpubl.).

When in glycerol, animals may be mounted on glass slides. Many nematologists use Cobb-slides (Cobb, 1917) which permit examination from both sides as the animals are mounted between two cover glasses together with glass rods that prevent flattening of the nematode. If permanent slides are to be made, the cover glass may be sealed with Glyceel, Clearseal or Bioseal.

In toto preparations are usually satisfactory for species identification. A good quality microscope with a 100 × oil immersion lens is required and interference-contrast equipment is useful, especially for microscope photography.

Comprehensive guides for identification of nematodes have been available for ten years; the most important are: the Bremerhaven *Checklist of Aquatic Nematodes I & II* (Gerlach & Riemann, 1973/1974); *An Illustrated Guide to Marine Nematodes* (Tarjan, 1980); and *Free-living Marine Nematodes. I. British Enoplids* (Platt & Warwick, 1983) (Parts II and III are in preparation). A review of the important systematic literature is given in Heip, Vincx, Smol & Vranken (1982).

ISOLATION TECHNIQUES AND MAINTENANCE IN THE LABORATORY

The different techniques used in the cultivation of marine nematodes have been reviewed by Kinne (1977). Pioneering work in the field of cultivation is that of Chitwood & Murphy (1964) and von Thun (1966, 1968). These early workers started to use agar as a maintenance medium. Von Thun (1966, 1968) developed and successfully used a modified Killian agar for the cultivation of six species (listed in Table VIII, p. 444). Von Thun's medium was used later by Gerlach & Schrage (1971, 1972) for life-cycle studies.

The use of fungal mats has been popular for some time in Florida (Meyers, Feder & Tsue, 1963, 1964; Hopper & Meyers, 1966a,b, 1967a; Meyers & Hopper, 1966, 1967). These mats were used either for trapping or for culturing. Two fungi (*Dendryphiella arenaria* and *Halosphaeria mediosetigera*) were particularly appropriate, both forming a compact mat that proved to be a good substratum on which to culture both nematodes and food organisms.

With the publications by Tietjen *et al.* (1970) and Lee *et al.* (1970) the first successful attempts to culture both the nematodes and their food in controlled conditions were made. Especially Erdschreiber-3 medium proved a good basis for culturing several chromadorids. Techniques were presented for aseptical working with nematodes and for establishing monoxenic (see Dougherty, 1960, for terminology) cultures of *Rhabditis marina** with the bacterial strain *Pseudomonas* sp. From such monoxenic cultures, Tietjen & Lee (1975) produced an axenic medium for *Rhabditis marina*, based on Grace's insect medium, with marine salt mixture and sheep blood.

The same techniques were used for many studies on life-cycles of marine nematodes (Tietjen & Lee, 1972, 1973, 1977a,b), on studies of feeding behaviour using tracer techniques, and on studies of trophic interactions (Alongi & Tietjen, 1980). When algae were used as food the nematodes were cultured in Erdschreiber medium, with bacteria as food the basic medium was autoclaved sea water with cereal.

Hopper, Fell & Cefalu (1973) and Warwick (1981a) used corn meal agar. Heip, Smol & Absillis (1978) used bacto-agar (DIFCO) and Vranken, Thielemans, Heip & Vandycke (1981), Geraert, Reuse, Van Brussel & Vranken (1981) and Vranken, Vincx & Thielemans (1982) cultured on bacto-agar (DIFCO) enriched with Vlasblom-medium (containing glycine) and silicate. Romeyn, Bouwman & Admiraal (1983) and Bouwman (1983) used a very similar medium for *Eudiplogaster pararmatus* and several Aufwuchs species, among which was the Ghent stock of *Monhystera microphthalma*.

Several other species were established on agar as substrate. Trotter &

*Andrássy (1983) transferred *Rhabditis marina* to *Pellioditus marina* (Bastian, 1865) n. comb. Andrássy, 1983.

Webster (1984) did feeding experiments with three species using bacto-agar (DIFCO) with bacteria and diatoms, grown separately on standard media, as substance. Jensen (1982) cultured *Chromadorita tenuis* on a brackish water agar-bottom enriched with a modified Erdschreiber medium after Hällfors. Findlay (1982a) and Findlay & Tenore (1982) established the nematode, *Diplolaimella chitwoodi*, in monoxenic cultures, maintained on Gerber's mixed cereal (Pablum).

Gnotobiotic culturing methods were developed by Vranken, Van Brussel, Vanderhaeghen & Heip (1984a), *i.e.* a completely chemically-defined medium based on artificial sea water enriched with amino acids, modified Provasoli-Walne nutrient medium and a sterol mixture was established and the nematodes were cultured monoxenically on an *Alteromonas haloplanktis* strain.

Most species cultured up to now are members of Aufwuchs communities and thrive well on agar. Other species, such as the typical mud-dwelling *Sabatieria* and *Daptonema*, or typical sand-dwellers from the open sea have not yet been cultured on agar. The methods existing at present do not permit permanent cultivation of many of the most important marine species.

NEMATODE ASSOCIATIONS

ESTUARIES AND BRACKISH WATER

It has been known for many decades that benthic communities in brackish water have fewer species than either marine or freshwater communities (Remane, 1933). This is also true for nematodes. A species-salinity curve was constructed by Gerlach (1953, 1954) from comparable sediments in brackish waters along the German coast. This curve (Fig. 1) shows a minimum number of species between 3–7‰ *S*. Freshwater species penetrate into brackish water to a maximum of 10‰ *S* and marine species can invade, in relatively high densities, the oligohaline area to 0·5‰ *S*. Bilio (1966) listed 60, 59, 2 and 4 meiofauna species (with nematodes dominant) in eu-, poly-, meso- and oligohaline water, respectively.

Brackish-water nematodes have been divided into six groups according to salinity by Gerlach (1953) and this was later followed and adapted by Bilio (1966), Skoolmun & Gerlach (1971), Warwick (1971), Brenning (1973), and Van Damme *et al.* (1980). However, such groupings appear to be artificial and differ from place to place; different environmental factors may interact, the most important being type of sediment. Warwick (1971) even found an increasing number of species in the Exe estuary, U.K., with decreasing salinity but increasing grain size.

A list of the dominant brackish-water and marine nematodes invading brackish water is given in Table I. From the 155 species listed, only 18 are restricted to brackish water. The salinity boundaries are found in the systematic literature not reviewed here.

Whether true brackish-water nematode species exist is still a matter of debate. Meyl (1954, 1955) described nematode assemblages from inland saline waters in Braunschweig, West Germany, and found no marine species although the salinity was high enough. Paetzold (1955, 1958) described similar assemblages in saline waters near Aseleben, West Germany. The species that occur in these habitats are well adapted, not only to low or high salinities, but especially to fluctuations. During dry weather, when salinity may become very high, some species survived 123°/₀₀ *S* (*Monhystera multisetosa*, *Theristus flevensis*, *Diplolaimelloides oschei*, *Tripyloides marinus*, *Paracyatholaimus intermedius*, and *Oncholaimus oxyuris*) whereas *Chromadora nudicapitata* was found surviving in up to 84·5°/₀₀ *S*.

Several estuaries around the North Sea have been well investigated. Riemann (1966) described the nematode communities along the Elbe in Germany and tried to classify them according to the Venice system of salinity. The polyhaline region of the Elbe is distinguished from the mesohaline zone by the presence of Desmodoridae. A series of species extend from the polyhaline zone into a salinity of 10°/₀₀ so that the boundary between α- and β-mesohaline zones can be found biologically. These species, however, have different salinity ranges in other estuaries. In the tidal freshwater region of the Elbe, the most abundant species is *Daptonema setosum*, whereas in the upper brackish-water regions an assemblage with *Axonolaimus spinosus* and *Theristus meyli* is characteristic. In areas with intermediate salinities the distribution of species is regulated by the variable degrees of sand movement.

Warwick (1971) described six different habitats along the Exe estuary by a combination of salinity, grain size, and degree of water retention. In muddy sediments, nematodes are small, with short setae. They are mainly deposit-

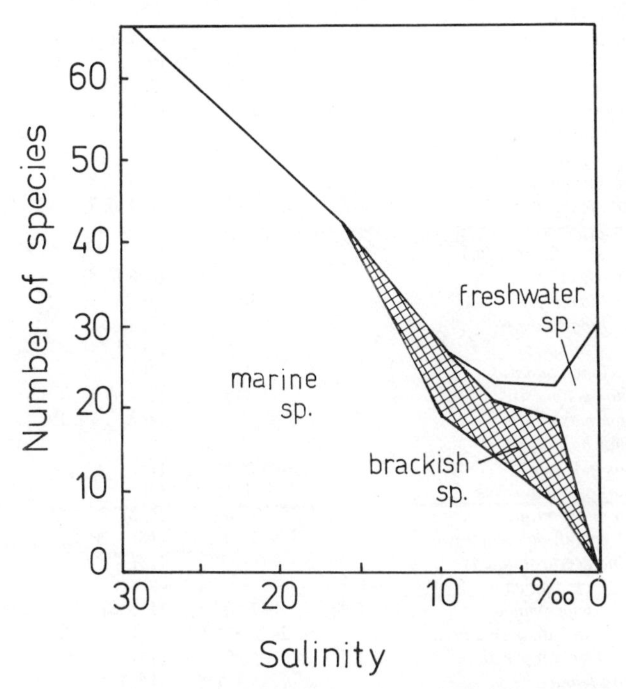

Fig. 1.—Species-salinity curve from comparable sediments in brackish waters along the German Bight (after Gerlach, 1954).

TABLE I

Salinity tolerances of brackish-water nematodes: except when otherwise mentioned, the tolerance is from sea water (35°/₀₀ S) at the upper level to the noted salinity value on the lower level; freshwater species are not included; reference numbers are given in parentheses. 1. Gerlach (1953): Kiel Bay, North Sea coast (W. Germany). 2. Meyl (1954, 1955): salt marsh (W. Germany). 3. Capstick (1959): Blyth estuary (U.K.). 4. Riemann (1966): Elbe estuary (W. Germany). 5. Bilio (1966): North Sea. 6. Skoolmun & Gerlach (1971): Weser estuary (W. Germany). 7. Warwick (1971): Exe estuary (U.K.). 8. Brenning (1973): Baltic coast (E. Germany). 9. Lasserre, Renaud-Mornant & Castel (1975): saline reservoirs (France). 10. Riemann (1975): Columbia (S. America). 11. Elmgren (1978): Baltic Sea. 12. Möller, Brenning & Arlt (1976): Baltic Sea (E. Germany). 13. Warwick & Price (1979): Lynher estuary (U.K.). 14. Van Damme et al. *(1980): Western Scheldt estuary (The Netherlands). 15. Jensen (1981a): Baltic Sea (Finland). 16. Bouwman (1983): Wadden Sea–Ems estuary (The Netherlands). 17. Schiemer, Jensen & Riemann (1983): N. Baltic Sea. 18. Jensen (1984): Baltic Sea (Finland). 19. Smol (unpubl.): Dievengat (Belgium)*

Species	°/₀₀ Salinity	Reference
Adoncholaimus fuscus	0·5	(3, 4, 7)
„ *thalassophygas*	0·5	(1, 2, 3, 4, 5, 6, 7, 8, 12, 16, 17, 18, 19)
Aegialoalaimus elegans	26·5	(13)
Anoplostoma viviparum	0·5	(1, 3, 5, 6, 7, 8, 12, 13, 16, 18, 19)
Antomicron elegans	2·1	(3, 14)
Aponema torosus	26·0	(13)
Ascolaimus elongatus	0·5	(1, 5, 6, 7, 8, 14, 17, 19)
Atrochromadora microlaima	0·5	(7, 13, 16)
Axonolaimus paraspinosus	0·9	(1, 3, 7, 8, 13)
„ *spinosus*	0·5	(1, 4, 6, 7, 8, 12, 16, 17, 18)
„ *typicus*	24·5	(3, 5)
„ sp. (aff. *spinosus*)	0·5	(18)
Bathylaimus assimilis	5·3	(1, 6, 7, 8)
„ *filicaudatus*	30·5	(3)
„ *longisetosus*	0·5	(1, 8)
„ *stenolaimus*	5·3	(7)
„ *tenuicaudatus*	0·5	(8)
Calomicrolaimus honestus	0·5	(1, 7, 8, 11, 13)
Calyptronema maxweberi	0·5	(4, 5, 7, 14, 19)
Camacolaimus barbatus	5·3	(7)
Campylaimus inaequalis	26·5	(13)
Choanolaimus psammophilus	0·5	(4)
Chromadora nudicapitata	17·0	(1, 8)
„ sp. (aff. *nudicapitata*)	17·0	(18)
Chromadorina erythropthalma	11·0 → 1·0	(18)
„ *germanica*	9·0	(9)
„ *microlaima*	24·0	(1, 3, 8)
„ sp. (aff. *germanica*)	24·0 → 9·0	(18)
„ sp. (aff. *viridis*)	5·0	(17)
Chromadorita fennica	15·0 → 4·0	(17, 18)
„ *guidoschneideri*	25·0 → 2·0	(1, 4, 5, 8, 18)
„ *nana*	2·0	(12)

Chomardorita tentabunda	26·0 → 0·5	(1, 4, 6, 8, 13, 17)
„ *tenuis*	25·0 → 4·0	(1, 8, 18)
Cyatholaimus punctatus	10·0	(1, 8)
Daptonema leptogastrelloides	0·5	(17)
„ *normandicum*	5·3	(1, 6, 7, 8, 13)
„ *oxycerca*	0·9	(1, 7, 8, 12, 13, 18)
„ *procerum*	26·0 → 11·0	(8, 13, 19)
„ *setosum*	0·5	(1, 2, 3, 4, 5, 6, 7, 8, 9, 12, 13, 14, 16, 17, 18, 19)
„ *tenuispiculum*	30·5	(3)
„ *trabeculosum*	26·0 → 0·5	(1, 4, 6, 8, 18, 19)
„ sp. (aff. *biggi*)	0·5	(17)
Desmodora communis	5·3	(7)
Desmolaimus fennicus	0·9	(7)
„ *zeelandicus*	0·5	(1, 8, 11, 12, 18, 19)
Desmoscolex falcatus	26·0	(13)
Dichromadora cephalata	0·5	(1, 7, 8, 13, 19)
„ *geophila*	0·5	(1, 2, 5, 8, 16, 17, 18)
„ *hyalocheile*	18·0	(1, 8)
Diplolaimelloides altherri	22·0	(2)
„ *islandicus*	0·5	(5)
„ *oschei*	27·5	(2)
Eleutherolaimus stenosoma	0·5	(1, 5, 6, 8, 11, 13, 16, 19)
„ sp. (aff. *stenosoma*)	0·5	(17)
Enoploides caspersi	18·0 → 0·5	(4)
„ *labiatus*	10·0	(1, 8, 16)
„ *spiculohamatus*	5·3	(6, 7)
Enoplolaimus balgensis	18·0 → 0·5	(1, 8, 17)
„ *litoralis*	2·1	(14)
„ *propinquus*	2·1	(1, 6, 8, 14)
„ *vulgaris*	0·5	(4)
Enoplus brevis	12·0	(1, 3, 5, 6, 8, 18)
„ *schulzi*	5·3	(7)
Eurystomina terricola	5·3	(7)
Halalaimus gracilis	7·5	(3, 5, 6, 8, 13, 14)
Halichoanolaimus robustus	11·0	(3, 8, 19)
Hypodontolaimus balticus	0·5	(1, 3, 5, 8, 13, 18, 19)
„ *geophilus*	0·5	(4, 7)
„ *inaequalis*	15·0	(16)
„ *setosus*	0·5	(6, 11, 16, 17)
Leptolaimus elegans	0·5	(8, 11, 17)
„ *limicolus*	11·0	(13, 19)
„ *papilliger*	0·5	(1, 5, 7, 8, 10, 11, 12, 14, 16, 17, 19)
„ *setiger*	22·0	(3)
Metachromadora remanei	5·3	(7)
„ *suecica*	12·0	(1, 6, 8)
„ *vivipara*	0·9	(3, 7, 13)
Metalinhomoeus filiformis	11·0	(19)
Metaparoncholaimus sp. (aff. *campylocercus*)	0·5	(4)
Microlaimus globiceps	0·5	(1, 2, 4, 5, 8, 11, 16, 17, 18, 19)
„ *marinus*	2·1	(14)
„ *robustidens*	5·4	(7)
Molgolaimus demani	24·5	(3, 13)
Monhystera disjuncta	0·5	(18)
„ *microphthalma*	25·0 → 11·0	(5, 19)
„ *parva*	0·5	(5, 8, 12, 13, 19)
„ *paramacramphis*	22·0	(2)
„ sp. (aff. *filiformis*)	0·5	(18)
„ sp. (aff. *parasimplex*)	3·5 → 2·0	(17)

TABLE I—*continued*

Species	‰ Salinity	Reference
Monhystrella parelegantula	11·0	(2, 19)
Monoposthia costata	25·0	(3)
„ *mirabilis*	24·0	(1, 8, 16)
Nemanema cylindraticaudatum	10·0	(1, 8)
Neochromadora izhorica	18·0 → 0·5	(1, 4, 8, 17, 18)
„ *poecilosoma*	8·6	(1, 8, 13, 16, 18)
„ *poecilosomoides*	18·0	(9)
„ *trichophora*	24·0	(16)
Odontophora armata	12·0	(3, 6)
„ *rectangula*	24·0	(16)
„ *setosa*	5·0	(1, 7, 8, 13, 16)
Oncholaimus brachycercus	5·0	(1, 6, 7, 8)
„ *conicauda*	18·0 → 10·0	(1, 8, 17)
„ *oxyuris*	0·5	(1, 2, 5, 8, 19)
Oxystomina elongata	11·0	(3, 13, 19)
„ *unguiculata*	31·0	(3)
Paracanthonchus caecus	0·5	(1, 3, 5, 8, 11, 16, 18, 19)
„ *bothnicus*	18·0 → 0·5	(17, 18)
„ *elongatus*	25·0	(3)
„ *sabulicolus*	24·0	(16)
„ *tyrrhenicus*	5·0	(7)
„ sp. (aff. *bothnicus*)	18·0 → 0·5	(18)
Paracyatholaimus intermedius	30·0 → 0·5	(1, 2, 4, 5, 6, 7, 8, 12)
„ *proximus*	0·5	(1, 4, 5, 6, 8, 16)
„ *ternus*	0·5	(10)
„ *truncatus*	0·5	(10)
Paralinhomoeus lepturus	11·0	(3, 19)
Pomponema sedecima	24·0	(16)
Praeacanthonchus punctatus	0·9	(7, 13)
Prochromadorella longicaudata	24·0	(15)
Ptycholaimellus ponticus	0·5	(1, 3, 5, 6, 7, 8, 9, 13, 18)
Quadricoma scanica	26·0	(13)
Sabatieria longispinosa	24·0	(16)
„ *pulchra*	0·5	(11, 13, 15, 16, 18, 19)
„ *punctata*	0·5	(1, 8)
„ *vulgaris*	0·9	(6, 7, 15)
Setoplectus riemanni	0·5	(10)
Sigmophoranema rufum	18·0	(1, 8)
Southernia zosterae	26·0	(1, 8, 13)
Sphaerolaimus balticus	22·0	(3, 8, 12, 16, 18)
„ *gracilis*	0·5	(3, 4, 5, 17, 19)
„ *hirsutus*	0·9	(3, 6, 7, 13)
Spilophorella paradoxa	2·1	(3, 14)
Spirinia parasitifera	22·0	(3)
Syringolaimus striaticaudatus	10·0	(5)
Terschellingia communis	0·9	(7, 13)
„ *longicaudata*	0·9	(7, 9, 13)
Thalassoalaimus tardus	22·0	(3, 13)
Theristus acer	5·3	(1, 3, 5, 7, 8, 12, 13, 18, 19)
„ *blandicor*	7·5	(6, 14)
„ *ensifer*	18·0	(5)
„ *flevensis*	0·5	(1, 2, 4, 5, 8, 10, 12, 17, 18)
„ *metaflevensis*	0·5	(10)
„ *meyli*	0·5	(4)

Theristus pertenuis	11·0	(5, 6, 8, 19)
„ *scanicus*	0·5	(4, 6, 17)
Trefusia conica	0·5	(10)
„ *longicauda*	5·3	(6, 7)
Trichotheristus mirabilis	2·1	(14)
Tripyla cornuta	18·0 → 0·5	(1, 4, 8)
Tripyloides amazonicus	0·5	(10)
„ *gracilis*	5·3	(7, 13)
„ *marinus*	0·5	(1, 3, 4, 5, 6, 8, 12, 14, 17)
„ sp. (aff. *marinus*)	0·5	(18)
Tubolaimoides tenuicaudatus	12·0	(6)
Viscosia rustica	24·0	(16)
„ *viscosa*	0·9	(1, 3, 6, 7, 8, 13, 14, 16, 19)

feeders. In sandy bottoms, species are longer with long setae and heavily ornamented cuticles. Epistratum-feeders and predators are numerous. Gourbault (1981) examined the nematodes in silt deposits along the Morlaix river bed (France). Species composition is primarily influenced by grain-size characteristics and secondarily by the range of the estuarine conditions (*e.g.* salinity ranges between 20–34°/₀₀). Most of the species are true marine, polyhaline species (and are, therefore, not included in Table I); the Comesomatidae (28%) and Linhomoeidae (27%) are the most abundant families.

In estuaries with semidiurnal tides, the interstitial salinity is rather constant and nematodes can burrow down, whereas in areas where salinity changes are irregular and unpredictable the only possible adaptation is a broad physiological tolerance. Brackish-water pools or salt marshes will have a lower number of species than estuaries. Bouwman (1983) thinks that the number of nematode species in an estuary is probably close to 200 (compare with 735 for the North Sea). The maximum number of species in inland brackish waters is about 50 (Gerlach, 1953; Meyl, 1954, 1955; Brenning, 1973; Smol, unpubl.).

In his study of the Ems estuary and the Wadden Sea (The Netherlands) Bouwman (1983) examined the occurrence of associations in relation to the conditions offered by particular habitats. Sublittoral marine and estuarine muds are dominated by members of the same genera (*e.g. Sabatieria, Spirinia, Terschellingia, Odontophora,* and *Desmolaimus*). Tidal flats differ from subtidal estuarine sediments in having a dense stock of microphytobenthos at the surface. A considerable part of the nematode community from the upper sediment layers in the Ems estuary is made up by diatom-feeders. Most of these species are not found in sublittoral marine sediments (*Atrochromadora microlaima, Dichromadora geophila, Hypodontolaimus balticus, Ptycholaimellus ponticus, Chromadorita guidoschneideri, Chromadora nudicapitata, Eudiplogaster pararmatus, Daptonema* aff. *normandicum, D. oxycerca, D. setosum, D. trabeculosum, D. xyaliforme, Paracyatholaimus proximus,* and *Praecanthonchus punctatus*). Bouwman thinks that the preference of these species for diatoms, as food and their tolerance to reduce salinities are two of the main differences between nematodes from sublittoral sediments and those from tidal estuarine sediments. The amount of diatom-feeders in sublittoral sediments, however, may be very important; Tietjen (1969) found epigrowth-feeders (mainly Desmodoridae) dominant with maximum abundance in spring and summer in four subtidal stations in two estuaries in New England. The distribution of marine nematodes in near-coast fresh water (0·5‰ *S*) was examined by Riemann (1975). Contrary to observations from the macrofauna, the abundance of the meiofauna and especially nematodes was not influenced by the latitudinal position of the habitats.

Table II summarizes data on density and biomass of nematode assemblages in brackish-water habitats. Most of the studies are from intertidal or shallow subtidal stations in estuaries or from salt marshes in the United States of America, where *Spartina alterniflora* is abundant.

Especially sheltered, muddy regions have an extremely high abundance of meiofauna, with nematodes always the dominant taxon. 16 300 ind.·10 cm^{-2} were counted in a *Spartina* salt marsh in Georgia (Teal & Wieser, 1966) and a record figure of 22 860 ind.·10 cm^{-2} was found on a mud-flat in the Lynher estuary, U.K. (Warwick & Price, 1979). Van Damme *et al.* (1980) studied the meiofauna of the Western Scheldt estuary in The Netherlands along five transects over a one-year period. The highest densities of nematodes (up to 17 500 ind.·10 cm^{-2}) were found in summer in sediments with a small grain size ($\pm$0·105 mm) and a relatively high organic content. The decrease of annual mean density from the sea to the inner part of this polluted estuary is shown in Figure 2. Biomass values follow the same trend. Diversity is relatively low in the eu- to polyhaline zones ($H = 2{\cdot}27 - 2{\cdot}44$), reaches a peak in the poly- to mesohaline zones ($H = 3{\cdot}01$) and declines to $H = 1{\cdot}63$ in the meso- to oligohaline zones (all values in bits/ind.). The relatively low diversity in the sandy sediment at the mouth of the estuary may be explained through high turbulence and periodical re-working of the sediment, a phenomenon that also occurs in the nearby Eastern Scheldt mouth (Heip *et al.*, 1979). A decrease of diversity with increasing environmental fluctuations has also been observed by Ott (1972b) on an intertidal sand-flat.

Schiemer, Jensen & Riemann (1983) examined the subtidal (>80 m depth) nematodes from the northern part of the Baltic (Bothnian Bay, salinity only 2–3·5‰), where the meiofauna is the dominant constituent of the benthic fauna in density as well as in biomass (Elmgren, 1978). The composition of the fauna is similar to the situation along the German coast described by Gerlach (1953). Diversity is relatively low, with 5–18 species present, $H = 1{\cdot}9$–$3{\cdot}3$ bits/ind. The area is quite uniform in species composition and similarity between stations is high.

The abundance of epiphytic nematodes in brackish water has been discussed by Jensen (1984) who compared benthic and epiphytic nematodes from several brackish-water areas in Finland and other European brackish waters. In winter, when the submerged vegetation in this area is destroyed and incorporated in the sediments as detritus, epiphytic nematodes stay within the sediment. In spring, the epiphytes leave the bottom and colonize the newly grown submerged macrophytes. The abundance of Chromadoridae appears to be related to three conditions: salinity, explaining large-scale horizontal distribution, and substratum and oxygen, which determine vertical distribution. *Chromadorita tenuis* inhabits all types of submerged vegetation; its salinity preference is 4–25‰. The four other *Chromadorita* species are benthic and have different types of locomotion: *C. tentabunda* jumps from sand grain

TABLE II

Nematode densities (ind.·10 cm^{-2}), biomass (mg dry wt·10 cm^{-2}) from estuarine and brackish-water areas: salinity values ‰, sediment characteristics and species composition (Sp. comp., — not given in text, + given in text) also included

Reference	Locality	Sediment	Salinity	Density	Biomass	Sp. comp.
Rees, 1940	Bristol Channel (U.K.)	mudflat	brackish	1000–10 000	—	—
Smidt, 1951	Danish Wadden Sea	mud/sand	?	223	0·03–0·43	—
Capstick, 1959	Blyth estuary (U.K.)	mud/sand	31–33	625–2210	—	+
			25–32	750–1880	—	+
			22–30	228–715	—	+
Teal & Wieser, 1966	Georgia (U.S.A.)	mud/sand	?	46–16 300	0·05–2·43	+
Muus, 1967	Danish estuaries lagoons	?	5–18	300–1400	—	—
Tietjen, 1969	New England estuaries (U.S.A.)	mud/sand	26–31	1000–4811	3·3–8·0	+
Skoolmun & Gerlach, 1971	Weser estuary (W. Germany)	sand	12–26	8–107	—	+
Nixon & Oviatt, 1973	New England (U.S.A.)	mud	20–25	1600–10 000	0·2–1·0	—
Lasserre *et al.*, 1975	Arcachon (France)	mud	3–40	200–12 600	0·1–10·4	+
Möller *et al.*, 1976	Baltic Sea (E. Germany)	sand	5–6	245–423	—	+
Arlt, 1977	Greifswalder Bodden (E. Germany)	sand–sandy mud	5–9	105–1559	—	—
Saad & Arlt, 1977	Tigris & Euphrates estuary (Arabian Gulf)	mud	fresh	71	—	—
			mixed	24	—	—
			marine	636	—	—
Dye & Furstenberg, 1978	Swartskop estuary (S. Africa)	sand	0–43	960–1380	0·1–0·4	—
Warwick & Price, 1979	Lynher estuary (U.K.)	mud	26	800–22 860	1·4–3·4	+
Van Damme *et al.*, 1980	Western Scheldt estuary (The Netherlands)	mud/sand	2–32	160–17 500	0·03–4·58	+
Coull & Wells, 1981	Wellington estuary (New Zealand)	mud	5–34	22–444	—	—
Sikora & Sikora, 1982	S. Carolina (U.S.A.)	mud	high-intertidal	4400	—	—
			mid-intertidal	3600	—	—
			low-intertidal	2300	—	—
			subtidal	1900	—	—
Bouwman, 1983	Ems estuary (The Netherlands)	mud/sand	3–32	40–10 000	—	+
Montagna *et al.*, 1983	S. Carolina (U.S.A.)	mud/sand	22–36	270–304	—	—
Schiemer *et al.*, 1983	N. Baltic Sea	mud/sand	2–3·5	?	?	+
Ellison, 1984	Cornwall (U.K.)	mud	33–35	1100	—	—
Jensen, 1984	Baltic Sea (Finland)	mud	18–30	1200–1500	—	+

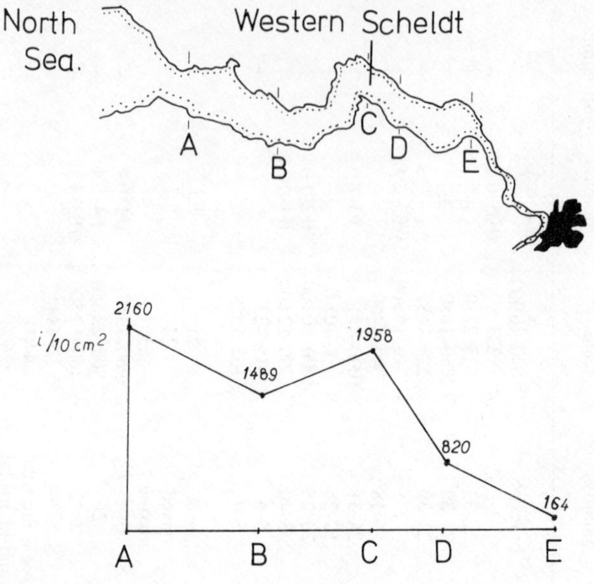

Fig. 2.—Annual mean density of nematodes communities in the Western Scheldt estuary, along 5 transects A–E (after Van Damme *et al.*, 1980).

to sand grain (*e.g.* Riemann, 1980) whereas *C. fennica*, an oligo- to mesohaline species, glides through the sediments like a snake. *C. leuckarti* is a freshwater species and *C. guidoschneideri* occurs mostly in sandy bottoms and does not enter oligohaline waters.

MARINE SANDY BEACHES

The intertidal environment of marine beaches has a particular brackish-water fauna which lives in the coastal subsoil water. This environment is a transition zone between sublittoral, truely marine bottoms and the continental subterranean waters with their phreatic freshwater fauna.

The different types of sandy beaches have been classified by McLachlan (1980) in relation to exposure. He distinguishes four types.

(1) Very sheltered beaches, with virtually no wave action, reduced layers close to the surface, abundant macrofauna burrows; this type of beach is to some extent modified and stabilized by the fauna.
(2) Sheltered beaches, with little wave action; the reduced layer is present and there are some macrofaunal burrows.
(3) Exposed beaches subject to moderate to heavy wave action; the reduced layer, if present, is deep and there are usually no macrofaunal burrows.
(4) Very exposed beaches subject to heavy wave action; there is no reduced layer and macrofauna consists solely of burrowers.

Table III summarizes the references on marine sandy beaches. When we compare density values in this table with those from muddy substrata (Table

TABLE III

Nematode densities (ind.·10 cm^{-2}) and biomass (mg dry wt·10 cm^{-2}) from marine sandy beaches: habitat characteristics (e.g. grain size), distribution depth (cm) and species composition (Sp. comp., – not given in text, + given in text) also included

Reference	Locality	Habitat	Density	Biomass	Depth	Sp. comp.
Smidt, 1951	Danish Wadden Sea	'hard' sand	10–1050	–	?	–
		'soft' sand	31–367	–	?	–
Renaud-Debyser & Salvat, 1963	Atlantic coast	279–343 μm	83–591	–	20	–
	Channel (France)	174–211 μm	19–389	–	20	–
Jansson, 1967, 1968	Baltic Sea	180–320 μm	1–169	–	11	–
	North Sea					
Renaud-Mornant & Serène, 1967	Malaya	exposed	8170	–	20	–
McIntyre, 1968	S.E. India	'soft bottom'	2233	–	?	–
Boaden & Platt, 1971	N. Ireland	205–228 μm	570	–	12	+
Gray & Rieger, 1971	Yorkshire (U.K.)	sheltered	827	–	10	+
		exposed	38–71	–	10	+
Harris, 1972	Cornwall (U.K.)	exposed	109–1062	–	50	–
Ott, 1972a	N. Carolina (U.S.A.)	fine sand	500–1100	–	30	+
McIntyre & Murison, 1973	W. Coast Scotland	210–279	371–1449	–	40	+
Platt, 1977a	N. Ireland	sheltered; 125–196 μm	200–3300	–	5 (95% of the fauna)	+
Munro *et al.*, 1978	W. Coast Scotland	sand: 208 μm	168–1905	–	–	–
	S.W. Coast India	sand: 175 μm	127–524	–	105	–
Fricke *et al.*, 1981	S. Africa	196–337 μm	±5000	–	30	+
McLachlan *et al.*, 1981	S. Africa	–	7000	0·8	?	–
Sharma & Webster, 1983	W. Canada	234 μm	230–670	–	6	+
		813 μm	36–160	–	6	+

II), it can easily be seen that densities in silty or fine sand (1000–5000 ind. $\cdot 10\ cm^{-2}$) are lower than in muds and that the lowest densities of nematodes are found in very exposed beaches, down to 100 ind.$\cdot 10\ cm^{-2}$. Along the beach, the highest densities are often found near M.T.L. which, on an average beach, is the place where the water table comes closest to the surface (McIntyre, 1969; Ganapati & Rao, 1968; Ott, 1972b; McIntyre & Murison, 1973). McLachlan, Wooldridge & Dye (1981) observed a shift in dominance between nematodes and harpacticoids when grain size increases, the percentage of both groups being correlated with the grain size. The shift occurs at about 330 μm, in coarser beaches harpacticoids dominate. An extreme example is a very coarse (1200 μm) tideless beach in Sweden, where only one nematode species was collected (Jansson, 1968).

Although nematodes tend to increase in density in finer sediments, diversity is higher in coarser sediments. In sediments finer than about 120 μm a true interstitial fauna is lacking and a poorer burrowing fauna remains. A true interstitial fauna exists in sediments between 125 and 500 μm grain size (Wieser, 1959; McIntyre & Murison, 1973), and sorting of the sediment is another important factor determining the available interstitial space.

Nematodes, like most other meiofaunal taxa, may penetrate very deep into coarse sandy beaches (Table III). Meiofaunal zonation is essentially three-dimensional here (McLachlan, 1977). Factors responsible for this distribution are desiccation during low tide and dissolved oxygen in the interstitial water. McLachlan (1980) made a scheme for an average beach in South Africa (Fig. 3) and recognized four strata in the sediment.

(1) A dry sand stratum near the top of the shore, wetted only at high tide; meiofauna consists mainly of nematodes.

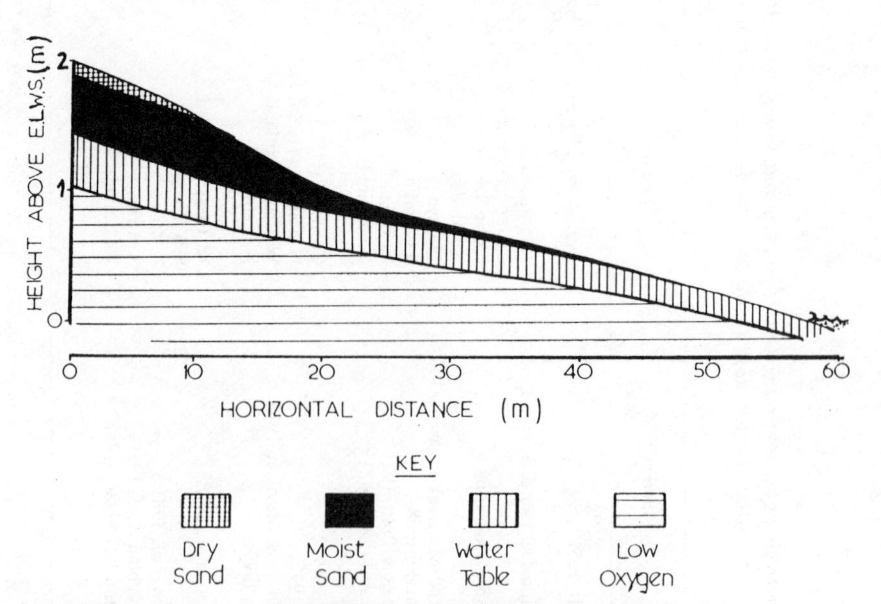

Fig. 3.—Diagrammatic representation of meiofauna strata on East Cape high energy beaches (after McLachlan, 1980).

(2) A moist sand stratum, under the dry sand and down to the permanent water table; a numerous and diverse meiofauna exists, dominated by crustaceans.
(3) The water table stratum, where lower oxygen tensions prevail; meiofauna occurs in moderate numbers and nematodes and crustaceans dominate.
(4) A low oxygen stratum, under the water table stratum, with low numbers of meiofauna and nematodes dominant.

This zonation is not static and shows tidal rhythms with the macrofauna moving upshore and the meiofauna towards the surface at higher tidal levels when the tide comes in (McLachlan, Winter & Botha, 1977).

The distribution of the meiofauna from such a model beach in South Africa is shown in Figure 4 (after McLachlan *et al.*, 1981). Meiofauna is abundant up to a considerable depth into the sand, penetrating below the permanent water

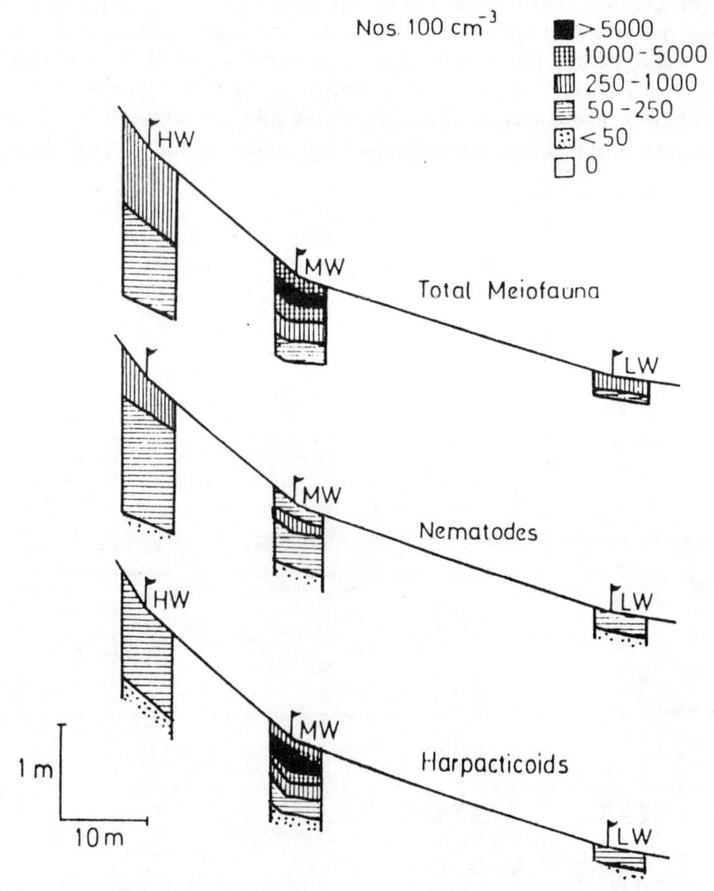

Fig. 4.—Distribution of total meiofauna and dominant taxa on Struisbaai beach (after McLachlan, Wooldridge & Dye, 1981).

table. Maximum densities occur just below the surface near mid-water and numbers generally drop more rapidly towards low water than towards high water. The dry sand stratum, the upper layer above high water neaps (HWN) harbours mostly small nematodes, adapted to live in water films around the sand grains. In the moist sand stratum, the fauna is more diverse, with harpacticoids extremely abundant. In the water table stratum larger nematodes become important and abundance drops. Below this level, nematodes are the dominant group.

In these studies from South Africa, nematode species composition is only partially known. Fricke & Fleming (1983) found the following genera: *Desmodora, Paramonhystera, Oncholaimellus, Bathylaimus, Trissonchulus*, and *Nudora.*

An extensive study of species composition from an intertidal sandy beach on the island of Sylt (North Sea) over one year has been made by Blome (1983). Densities vary between 67 and 366 ind.·100 cm^{-3}. The beach profile was divided in different regions. Dominant species which show a preference for a particular region are noted in Figure 5. The middle to upper regions of the beach slope show the greatest density and the damp sand zone is preferred. *Leptolaimus ampullaceus* and *Enoploides spiculohamatus* are important here, and also on the sand-flat before the beach slope. Over the whole shore, the dominant species is *Viscosia franzii* (27·4%), followed by *Microlaimus nanus* (8%), *Metepsilonema emersum* (6·4%), and *Paracanthonchus longus* (6·1%). The beach slope is poorer in nematode species than the sand-flat. The slope-living species

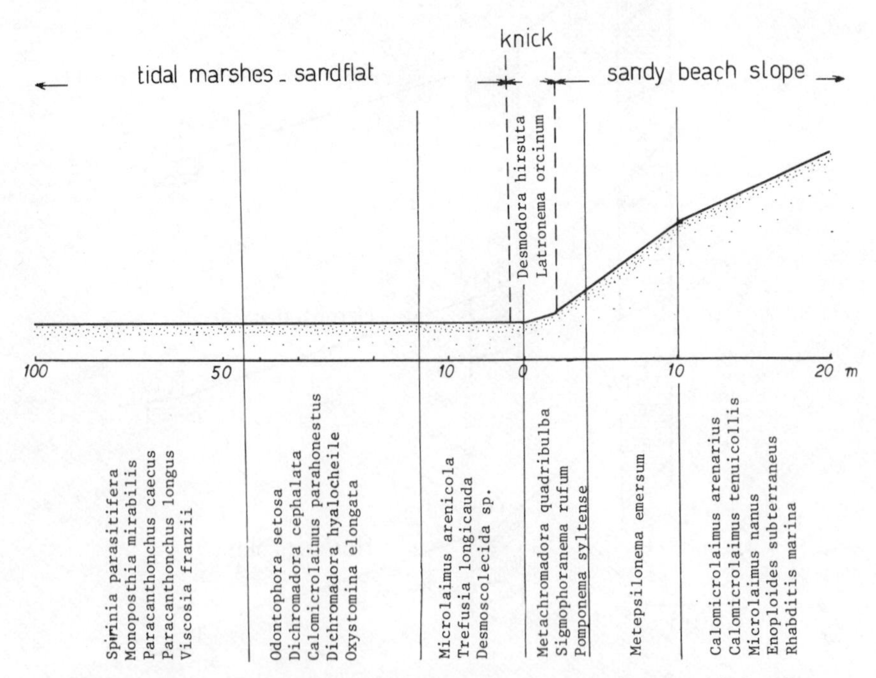

Fig. 5.—Species composition on a beach profile near the Isle of Sylt (after Blome, 1983).

show vertical migration, moving deeper into the sediment at lower temperatures. *Enoplolaimus litoralis* follows horizontal fluctuations of the high-water level over the year.

The intertidal meiofauna from exposed to sheltered beaches along the coast of North Carolina, U.S.A., close to *Spartina* marshes, has been studied by Ott (1972a). He distinguished four associations along a transect from high to low water.

(1) The upper centimetre of the high intertidal: the dominant species of this as well as the following association is *Metachromadora obesa*. Characteristic are *Microlaimus dimorphus* and *Desmolaimus zeelandicus*. Diversity is intermediate, 3·99 bits/ind.
(2) The upper centimetre from mid-tidal level to low tide: characteristic are *Cytolaimium exile*, *Terschellingia brevicauda*, *T. longicaudata*, *Eubostrichus parasitiferus*, *Gomphionema typicum*, *Synonchiella hopperi*, *Chromaspirina parma*, *Paradesmodora sinuosa*, *Camacolaimus prytherri*, *Pomponema hastatum*, *P. macrospirale*, *Xyzzors iubatus*, and *Nannolaimoides decoratus*. Diversity is high, 4·76 bits/ind.
(3) The deep fauna at high tidal levels: this fauna is dominated by a *Theristus* species and has the lowest diversity, 1·83 bits/ind.
(4) The deep fauna at lower tidal levels: the largest faunal unit; characteristic are *Axonolaimus paraponticus*, *Ptycholaimellus pandispiculatus* and *Chromadorina minor*. Diversity is highest here, 5·11 bits/ind.

These faunal assemblages are affected by temperature, salinity, pore water content and redox potential and variations in these factors (Ott, 1972b). The deeper layers of the sediment are not inhabited by a depauperate surface fauna but by a fauna of its own, containing the overall dominant *Metachromadora obesa*, penetrating often deepest into the sediment, and species such as *Cytolaimium exile*, confined to the deeper layers and which is tolerant of low oxygen levels.

In these beaches closely related species, mainly of the genera *Ptycholaimellus*, *Microlaimus*, *Theristus*, *Pomponema*, and *Terschellingia*, co-occur (Ott, 1972a). Within the first four genera, differences in the size of trophic structures (*e.g.* size of teeth in the mouth cavity) exist, so that food partitioning seems to be important. The distribution of the two *Terschellingia* species suggests actual competition for the same resources. *T. brevicauda* is limited to the mid-tidal region and out-competes in this area the closely related species *T. longicaudata*, a species that occurs under a wider range of conditions.

Hulings & Gray (1976) examined the interstitial sandy beach meiofauna from Mediterranean coasts. They found that meiofaunal abundance is controlled by waves, tide, and current action, which also control sorting in tidal beaches. In most atidal beaches, biological interactions such as competition and predation, were thought to control meiofaunal abundance, although no evidence for this was given.

A sheltered intertidal, fine-particle sand-flat in Northern Ireland was examined at three sites by Platt (1977a). The number of nematodes was generally lower than in intertidal muds, but higher than in coarse beaches (*cf.* Table III). Densities are influenced by sediment composition and the amount and nature of the available food. The number of species recorded ranges from 51 to 71 per station. Important species of this sand-flat are: *Spirinia parasitifera, Neochromadora poecilosoma, Daptonema setosum, Microlaimus zosterae, Theristus pertenuis, Chromadorita tentabunda, Monoposthia mirabilis, Pomponema sedecima,* and *Spirinia laevis.* This species assemblage is found in many shallow subtidal areas over the northern hemisphere.

EPIPHYTIC NEMATODES

When macrophytes are present in the littoral zone, an enormous increase in food availability, habitat complexity, and shelter for the fauna is created. The structure and shape of the plants is important as are the presence of epiphytes (hydrozoans, bryozoans, small algae *etc.*), sediment and detritus caught by the plant.

In a comparison between phytal nematodes of Chile and the Plymouth region of England, Wieser (1951, 1959) could detect some common patterns. The same species appear to be common on different species of algae (Table IV) but dominance differs. The size of the nematodes is correlated with the general shape of the algae and the amount of detritus caught in the rhizoids. In foliaceaous algae from exposed coasts, small chromadorids dominate (62–68% in the Plymouth area, 68% in Chile). The most abundant species are closely related: *Chromadora nudicapitata* in England, *Chromadorina laeta* in Chile, and the fauna is in general more diverse than in other types of algae. In tuft-like algae from exposed coasts, a slight dominance of the large Enoploidea was found, although Chromadoroidea remained important. *Enoplus communis* has a dominance of 48% in the Plymouth area, *Oncholaimus longus* has the same dominance in Chile. Shrub-like algae have a similar fauna. In encrusting algae, small chromadorids are dominant whereas in the holdfasts of large brown algae, the enoploids are most abundant; the holes within the holdfasts provide shelter for large animals, even in an exposed environment.

In sheltered areas, the algae accumulate sediment and detritus and these

TABLE IV

Relative abundance (%) of dominant nematode species on different algae (after Wieser, 1951)

	Gelidium	*Ceramium*	*Fucus*	*Gigartina*	*Nitophyllum*
Anticoma limalis	7·0	0·2	9·2	1·2	2·0
Enoplus communis	28·7	10·0	27·6	20·4	11·3
Dolicholaimus marioni	7·6	—	0·2	1·2	—
Paracanthonchus caecus	9·5	—	0·2	1·5	—
Desmodora serpentulus	11·0	2·5	0·7	3·3	0·5
Chromadora nudicapitata	3·8	56·4	17·2	35·5	18·2
Chromadorina germanica	0·2	20·7	25·9	13·6	—
Prochromadorella paramucrodonta	—	—	0·5	0·3	7·8
Neochromadora poecilosomoides	—	—	+	+	24·0

plants have a richer nematode fauna. *Theristus* species are dominant, a consequence of the presence of sediment. In an earlier paper, Wieser (1954) used dominance of the family Monhysteridae as an indicator of the degree of sedimentation in littoral areas.

Nematodes living in the soft sediments in sea-grass beds (*Spartina, Thalassia, Zostera, Posidonia*) are not really epiphytic (Hopper & Meyers, 1967b). These habitats contain large amounts of detritus derived from the plants and contain a typical fauna. *Metoncholaimus scissus, Daptonema fistulatum, Spirinia parasitifera*, and *Gomphionema typicum* are the dominant benthic species in *Thalassia*-beds, whereas *Oncholaimus dujardinii, Chromadora macrolaimoides, Paracanthonchus platypus*, and *Chromadorina epidemos* were dominant inhabitants of the epiphytic algae on the *Thalassia* plants (Hopper & Meyers, 1967b) and did not occur except sporadically in the sediments.

The nematodes inhabiting holdfasts of kelp (*Laminaria*) are mostly epigrowth-feeders and omnivores (Moore, 1971). Species with short setae, a body length above 1·5 mm, smooth cuticle and visual perception mechanisms are dominant. Most species are large enoplids, such as *Enoplus communis, Anticoma acuminata, Thoracostoma coronatum, Phanoderma albidum*, and *Pontonema vulgaris*. Chromadoroidea, Axonolaimidae and Monhysteroidea are poorly represented. Nematodes with strongly ornamented cuticles are present in very finely branched algae (as in the interstitial environment).

There exists a large resemblance between nematodes from holdfasts of kelp in the North Sea and littoral nematodes from Chile. The large omnivorous nematodes exploit the niches normally occupied by deposit-feeders in exposed environments, whereas in more sheltered conditions the smaller deposit-feeders are able to exist in the holdfasts.

Warwick (1977) found that the nematode fauna of finer, softer weeds is different from that of coarse weeds, stiff in texture. In these stiff weeds the dominant species were *Enoplus communis, Oncholaimus dujardinii, Thoracostoma coronatum*, and *Anticoma acuminata*, whereas in fine weeds the dominant species are *Oncholaimus dujardinii, Theristus acer, Symplocostoma tenuicolle*, and *Enoplus communis*. The coarse rigid weeds have a much higher percentage of large nematodes (>6 mm length) and of species with visual mechanisms, particularly true ocelli. Predators and omnivores (Group 2B) and nematodes with smooth cuticles predominate in all weeds (Figs 6 and 7).

The nematodes from the thalli of the brown algae *Sargassum confusum* have been studied by Kito (1980) in northern Japan. Nematodes were second in abundance after harpacticoids. Four families accounted for 98% of all nematodes: Monhysteridae (60·2%), Chromadoridae (26·2%), Axonolaimidae (9·5%) and Cyatholaimidae (2·1%). Four dominant species accounted for 80–90%: *Monhystera refringens, Chromadora nudicapitata, C. heterostomata*, and *Araeolaimus elegans*.

Two pairs of closely related species occurred: *Monhystera refringens* and *M. disjuncta* and *Chromadora nudicapitata* and *C. heterostomata*. The two monhysterids had differences in food preference: *M. disjuncta* takes fine, soft material while *M. refringens* also eats pennate diatoms. The two *Chromadora* species have asynchronous reproductive activity. It is also interesting to note that one of the species in each species pair has visual sensory means while the other has not (*Monhystera refringens* and *Chromadora nudicapitata* have).

The same two *Monhystera* species, *M. disjuncta* and *M. refringens*, together with *Prochromadorella neapolitana*, comprised 91–99% of the nematode fauna from the kelp *Macrocystis integrifolia* in western Canada (Trotter & Webster, 1983).

From these data it appears that nematode communities from marine algae are similar world-wide, and many species appear to be true cosmopolitans, occurring independently of the exact species of algae.

Submerged macrophytes from the Baltic (*e.g. Pomatogeton, Cladophora, Pilayella*) and European brackish waters also harbour a typical nematode assemblage. *Adoncholaimus thalassophygas, Axonolaimus spinosus, Chromadora* aff. *nudicapitata, Chromadorina* aff. *germanica, Chromadorita tenuis, Diplogaster rivalis* (freshwater), *Koernia fictor* (freshwater), *Prochromadora orleji* (freshwater), *Punctodora ratzeburgensis, Theristus acer*, and *Tripyloides* aff. *marinus* are the most important species, and occur independent of the precise nature of the plant (Jensen, 1984). These species are only rarely reported from the well-investigated Baltic and appear to inhabit the plants as long as possible. Most of these species were much more easily cultivated than benthic species (Jensen, 1982, 1984) and appear to be opportunists.

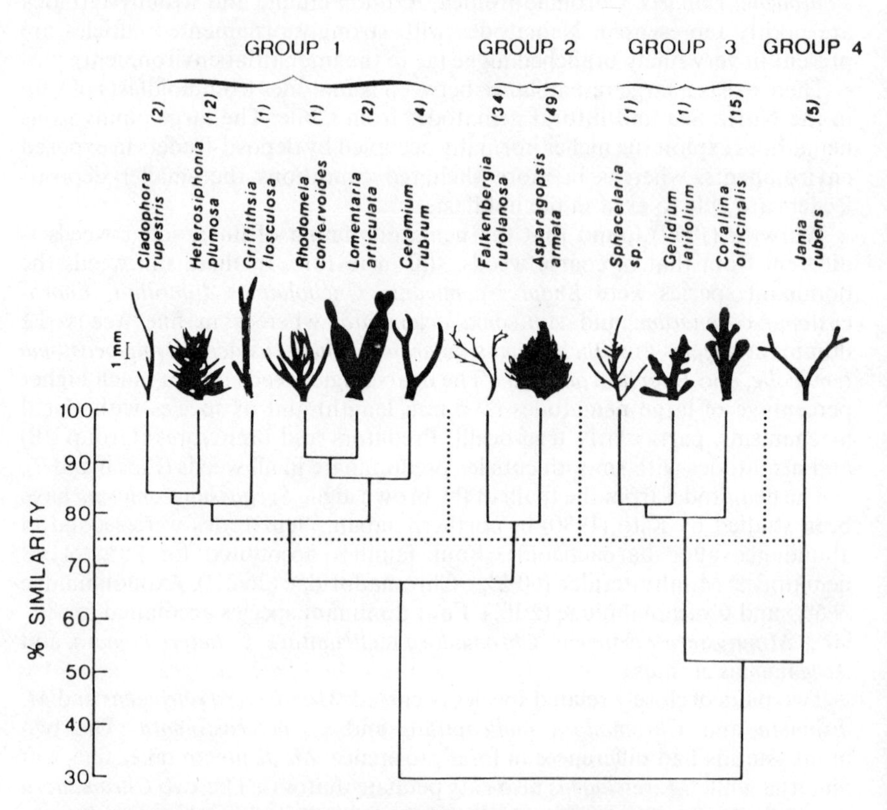

Fig. 6.—Dendrograms of faunal affinities between weed types, showing four weed groups defined at the 75% similarity level: number of samples used in analysis bracketed; after Warwick (1977).

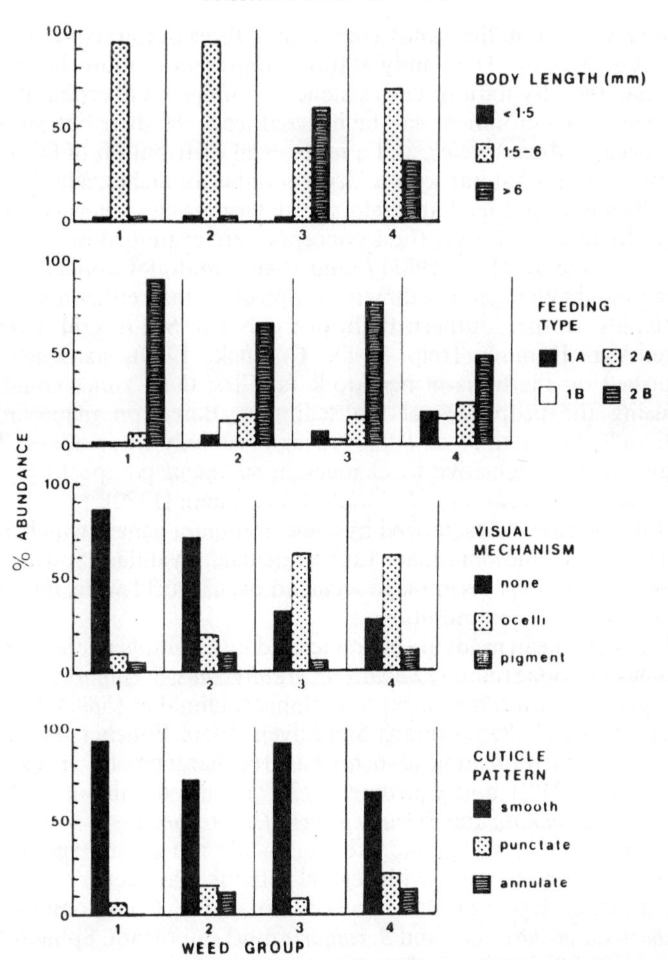

Fig. 7.—Distribution of physiognomic characters as defined by Wieser (1959), based on percentage of specimens in each weed group (after Warwick, 1977).

SHALLOW MARINE SUBTIDAL AREA

Most studies on nematodes of shallow subtidal environments deal with describing different communities occurring in different habitats, mostly characterized by sediment composition. McIntyre (1969) summarized the quantitative aspects of marine subtidal meiofauna (with nematodes, of course, very abundant); later, Heip, Herman & Vincx (1983) reviewed the state of knowledge of the subtidal meiofauna of the North Sea.

Wieser (1960a) examined the meiofauna from Buzzards Bay (U.S.A.), and was the first to deal with the community concept in marine nematodes. An *Odontophora–Leptonemella* community from sandy habitats and a *Terschellingia longicaudata–Trachydemus mainensis* (Kinorhynch) community from a silty habitat were described. In the 'silty community' one species dominated, whereas in the 'sandy community' three or four equally dominant species were present. The sandy stations represent a more heterogeneous habitat than the silty station. The existence of numerous microhabitats within a more general environment can be inferred from the distribution of closely related species. More species, and a more even distribution of feeding-types, will be present in a habitat with a larger number of niches; silt-clay content, sorting efficiency, and median grain size determine the heterogeneity of the sediment. In all later studies, these concepts were examined in more detail.

Herman, Vincx & Heip (1985) found that nematodes appear to be more sensitive to slight changes in sediment composition than either macrofauna or harpacticoids. In the Southern Bight of the North Sea (a grid described by Govaere, Van Damme, Heip & De Coninck, 1980), six zones can be distinguished on the basis of nematode families; these zones could also be found using the properties of the sediment. Based on macrofauna and harpacticoids, Govaere *et al.* (1980) recognized only three zones. That the meiofauna is more sensitive to changes in sediment composition than the macrofauna, is also found by Warwick & Buchanan (1970).

Coastal muds are characterized by a few dominant genera which all belong to the families Comesomatidae, Linhomoeidae, Xyalidae, Spiriniidae, and Sphaerolaimidae. This assemblage seems to occur world-wide, indicating the existence of parallel communities.

The Mediterranean muds are characterized by a homogeneous set of species with Comesomatidae (mainly *Sabatieria proabyssalis*, *S. vulgaris*, *S. granulosa*, and *Dorylaimopsis mediterranea*) and Sphaerolaimidae (*Sphaerolaimus dispar*) very abundant (Schuurmans Stekhoven, 1950; Boucher, 1972; Vitiello, 1974). The infralittoral area (2–6 m) has 16 characteristic species; mainly Linhomoeidae (31%) and Spiriniidae (29%) with dominant species *Terschellingia longicaudata* and *Spirinia parasitifera*, respectively. The first species is more abundant in more sandy regions while the second species is more abundant in more silty areas, with vegetal detritus. The deeper stations (25–90 m) have a higher degree of silt and are dominated by Comesomatidae (64%), with *Sabatieria proabyssalis* and *S. vulgaris* most important. Sphaerolaimidae are the second important family.

The muddy sediments along the Northumberland coast in the North Sea (U.K.) (Warwick & Buchanan, 1970, 1971) are characterized by *Dorylaimopsis punctata*, *Leptolaimus elegans*, and *Sabatieria celtica*. *Dorylaimopsis punctata* and *Sabatieria celtica* are also dominant in the mud samples of the Fladen Ground and Loch Nevis (McIntyre, 1961).

Sabatieria is the dominant genus in the muddy habitats of Liverpool Bay (Ward, 1973), with *Odontophora*, *Neochromadora*, and *Dichromadora* also important in the muddy sands and muddy coarse bottoms. The relative abundance of characteristic genera seems to be influenced by small differences in sediment granulometry.

The muds in the German Bight of the North Sea are characterized by *Sabatieria pulchra*, *Terschellingia longicaudata* and *Desmolaimus*, aff. *bulbulus* (Lorenzen, 1974; Juario, 1975), whereas in the Southern Bight of the North Sea, *Sabatieria breviseta*, *S. vulgaris*, *Daptonema tenuispiculum*, and *Monhystera disjuncta* are the characteristic species of the polluted muddy subtidal stations along the Belgian coast (Heip, Herman & Vincx, 1984).

North American subtidal muddy areas are generally characterized by the same genera (Tietjen, 1977). *Sabatieria pulchra*, *Terschellingia communis*, *Tripyloides gracilis*, and *Spirinia parasitifera* are the dominant species in the muds and muddy sands of Long Island Sound (U.S.A.). Those species showed very reduced abundances in sands, where *Tripyloides gracilis* was absent. *Sabatieria pulchra* is also dominant in medium sands which are contaminated by heavy metals (Tietjen, 1980a). The species which are normally very abundant in silt tolerate 'stressed' sands better than the 'typical' sand species.

With progressive increase in grain size (and decrease in silt-clay content), the numbers of Chromadoridae, Desmodoridae, Xyalidae, Axonolaimidae, and Enoplidae become increasingly abundant. Araeolaimida and Monhysterida are typical for fine sands with a small amount of silt.

The existence of parallel communities (at a species level) is not so obvious in sandy habitats as in silty ones. The fine and very fine sands along the Northumberland coast are characterized by *Odontophora longisetosa* (Warwick & Buchanan, 1970), *Mesacanthion* sp., and *Sabatieria hilarula*. In Liverpool Bay (Ward, 1973), where only genera are discussed, the sandy substrata are characterized by *Desmodora*, *Neochromadora*, *Microlaimus*, and *Richtersia*. In a fine sandy area of the German Bight of the North Sea, the following species have a relative abundance $>5\%$: *Sabatieria celtica*, *Daptonema leviculum*, and *Richtersia inaequalis*. The silty sands have high abundances of *Sabatieria pulchra*, *Molgolaimus turgofrons*, *Sabatieria celtica*, and *Longicyatholaimus complexus*. Lorenzen (1974) found also an increasing number of chromadorid species in the transition from mud to sand (*e.g.* *Prochromadorella ditlevseni*, *Dichromadora cucullata*). A very fine sand station (with $\pm 25\%$ silt) in the German Bight was studied in detail by Juario (1975). *Molgolaimus turgofrons* is the most abundant nematode, followed by *Sabatieria pulchra*, *Aponema torosus*, *Calomicrolaimus honestus*, *Marylinia complexa*, *Sabatieria celtica*, *Metalinhomoeus* aff. *typicus*, *Daptonema flagellicauda*, *D. longissimecaudatum*, *Prochromadorella ditlevseni*, and *Leptolaimus venustus*. These species make up for 75% of the total number of nematodes. Juario (1975) also compared different stations in the German Bight and distinguished the following nematode communities in sandy areas: a fine sand community dominated by *Sabatieria celtica*, *Metadesmolaimus heteroclitus*, and *Paracanthonchus caecus*; a silty sand community dominated by *Molgolaimus turgofrons*, *Sabatieria pulchra*, and *Aponema torosus*. In clean sand stations in the Southern Bight of the North Sea, mostly Chromadoridae such as *Chromadorita mucrocaudata*, *Neochromadora munita*, *N. paramunita*, *Ptycholaimellus ponticus*, and *Dichromadora cucullata*, and Desmodoridae species such as *Desmodora schulzi*, *Microlaimus* (several spp.), *Chromaspirina parapontica*, and *C. pellita* dominate the community (Jensen, 1974; Vincx, unpubl. results).

The sand community in Long Island Sound (Tietjen, 1977) is also characterized by low species dominance and high species diversity. It seems that more species are endemic to sands than to muds. *Theristus flevensis*, *Paralinhomoeus* sp., and *Theristus rusticus* were the most abundant species. The number of species per family increased in the progression muds → muddy sands → sands. In the Comesomatidae, for example, the number of species increased from one in muds to ten in medium-coarse sands, despite the fact that the relative abundance of the family decreased from an overwhelming 42% to only 10% in sands. The spatial distribution of species in the New York

Bight Apex (Tietjen, 1980a) agreed with previous findings. In silty sands especially *Sabatieria pulchra* and *Terschellingia* are dominant; in medium sands, the Comesomatidae were also important and the family is represented by more species.

Nichols (1980) examined the nematode community in sandy stations off the coast of Peru. The three species, *Tricoma* sp., *Latronema piraticum*, and *Choanolaimus* sp. that dominate the community, have a short, stout annulated body, especially adapted to the large interstitial spaces present. Ward (1975) found that the range of nematode lengths is greater in more heterogeneous sediments. Nematodes with very ornate cuticular ornamentation tended to be associated with coarser, silt-free sediments; it is suggested that this may be correlated both with their mode of locomotion and with the need for mechanical protection in unstable substrata.

A fine sand station in the Bay of Morlaix (English Channel) was examined by Boucher (1980a) during an annual cycle. The following species are important: *Richtersia kreisi*, *Calomicrolaimus monstrosus*, *Chromaspirina renaudae*, *Daptonema divertens*, *Actinonema celtica*, *Prochromadorella ditlevseni*, and *Dichromadora cucullata*.

Subtidal, very coarse sands, in a high energy environment were studied by Willems *et al.* (1982). The high abundance of Draconematoidea and Epsilonematidae is similar to the abundance of those taxa in beaches under strong hydrodynamical stress. No dominance of any feeding type indicates perhaps that this biotope is very heterogeneous and that nematodes occupy many niches. Wieser (1959) described genera confined to coarse sand; *e.g.* *Trefusia*, *Latronema*, *Campylaimus*, *Oxyonchus*, *Pomponema*, *Nudora*, *Bathylaimus*, and *Xyala*. The same genera are found in the fine sands of the sublittoral sandbank of the Belgian coast.

The occurrence of closely related nematode species with and without evidence of competition has been noted by Wieser (1960a), Ott (1972a), Boucher (1972, 1980a), and Juario (1975). It seems that microhabitats within a more general habitat will be occupied by closely related species or, conversely, if closely related species are found within one well-defined habitat, the segregation of the habitat into microhabitats is to be expected. Wieser (1960a) found two *Odontophora* species, *O. pugilator* and *O. papusi*, equally abundant in a sample area of 10 cm^2. The structure of the buccal plates of these two species is different which may reflect the adaptation to different food sources. Boucher (1972) proposed a competitive displacement of some species of *Sabatieria* in terms of vertical distribution. Juario (1975) found three very closely related dominant species in the upper 2 cm of sandy sediment: *Molgolaimus turgofrons*, *Aponema torosus*, and *Calomicrolaimus honestus*. These three species have the same structure of buccal cavity. It does not seem that these species are competing for limited food sources as they do coexist. Perhaps they actually feed on different kinds of food. Boucher (1980a) found five species of *Microlaimus s.l.* dominant in the same community.

Table V gives density and biomass values of some shallow subtidal areas. Muds and fine sands, with a large amount of silt, are the richest habitats in terms of nematode density; coarse sands and gravels have especially low density values, comparable with exposed sandy beaches. In most subtidal environments, nematodes represent 90–100% of the total meiofauna numbers and 50–100% of the total meiofauna biomass.

TABLE V

Nematode densities (ind.·10 cm^{-2}), biomass (mg dry wt·10 cm^{-2}) from shallow marine subtidal areas: sediment characteristics and species composition (Sp. comp., – not given in text, + given in text) also included

Reference	Locality	Sediment	Density	Biomass	Sp. comp
Wieser, 1960a	Buzzards Bay (U.S.A.)	silt–sand	1690–1860	0·1–0·6	+
McIntyre, 1964	N. North Sea	silt	1845	0·7–0·8	–
	W. Scotland coast	silt	853	0·7–1·6	–
Guille & Soyer, 1968	Mediterranean	silt	79	0·1–0·8	–
Stripp, 1969	German Bight of the North Sea	silt	788–884	0·8–3·1	–
Warwick & Buchanan, 1970	Northumberland coast (U.K.)	very fine sand	185	–	+
		fine sand	815	–	+
		silt	713	0·3–0·7	+
Soyer, 1971	Mediterranean	silt	400	1·0–1·4	–
Boucher, 1972	Mediterranean	silt	3665	3·8	+
McIntyre & Murison, 1973	Firemore Bay (U.K.)	fine sand	960–2765	0·5	–
Ward, 1973	Liverpool Bay (U.K.)	mud, sand	290–565	–	+
de Bovée & Soyer, 1974	Mediterranean	mud	4279	–	–
Lorenzen, 1974	German Bight of the North Sea	fine sand	530	–	+
		silty sand	2350	–	+
		silt	920	–	+
Juario, 1975	German Bight of the North Sea	very fine silty sand	3047–5261	0·6–1·3	+
Gray, 1976	River Tees estuary (U.K.)		393–1904	0·1–3·1	–
de Bovée & Soyer, 1977	Kerguelen Islands	mud/sand	138–3599	–	–
Tietjen, 1977	Long Island Sound (U.S.A.)	mud	530–2710	–	+
		muddy sand	560–1450	–	+
		fine sand	370–1650	–	+
		med. coarse sand	110–5010	–	+
Ito, 1978	Ishikari Bay (N. Japan)	sand	103–525	–	–
Heip *et al.*, 1979	North Sea	mud, sand	1650	1·0	–
Boucher, 1980a	Bay of Morlaix (France)	fine sand	1446–3432	–	+
Govaere *et al.*, 1980	North Sea	coast zone	1178	–	+
		transition zone	1423	–	+
		open sea zone	998	–	+
Nichols, 1980	E. Pacific (Peru)	sand	150–220	0·1–1·0	+
Tietjen, 1980a	New York Bight Apex (U.S.A.)	sand	221–1381	–	+
Willems *et al.*, 1982	North Sea	sand	58–1095	–	+

The correlation between sediment composition and trophic structure of the community has been studied by Wieser (1953, 1959), Hopper & Meyers (1967b), Tietjen (1969), Warwick & Buchanan (1970), Coull (1970), Boucher (1972, 1974, 1980a), Vitiello (1974), and Juario (1975). In general, they state that muddy sediments are dominated by non-selective deposit-feeders (50–60%) and that sandy bottoms are dominated by epistratum-feeders (50–60%); in most biotopes, selective deposit-feeders and omnivorous predators are numerically less important. A large number of selective deposit-feeders were found by Boucher (1980a) in sublittoral fine sand of the Bay of Morlaix (with important species of the Stilbonematidae) and by Willems *et al.* (1982) (mostly Epsilonematidae and Draconematoidea).

Warwick & Buchanan (1970), Heip & Decraemer (1974) and Juario (1975) found a correlation between diversity and sedimentological characteristics (silt-clay fraction and median grain size of the sand fraction). Tietjen (1977) found two basic faunistic units: a mud unit characterized by high species dominance, low species diversity (Shannon-Wiener diversity index, H', 1·00–2·00 bits/ind.) and low species endemism. The sand unit is characterized by low species dominance, high species diversity (H', 2·34–3·10 bits/ind.) and high species endemism. Same trends are found in medium and silty sands in the New York Bight Apex (Tietjen, 1980a). Willems *et al.* (1982) found very high diversities in the nematode communities in a sandbank in the Southern Bight of the North Sea (Brillouin diversity index, H, 3·30–4·60 bits/ind.); a high number of microhabitats may explain this high diversity.

DEEP SEA

Prior to 1960, no information was available on the abundance of meiofauna in the deep sea. Wieser (1960a), without having reliable data, postulated a decrease in meiofauna abundance with depth in a constant ratio to the already known decrease in macrofauna abundance. From that moment on, some effort has been made to gather quantitative data on the abundance of meiofauna in general and nematodes in particular. Table VI gives the density of nematode communities for the different deep-sea regions investigated. Densities range from a few to 1500 ind.·10 cm^{-2} and generally decrease with increasing depth. An extremely high density of 4000 ind.·10 cm^{-2} is found in the northeastern Atlantic near the Faeroes (Thiel, 1971).

Thiel (1979) discussed the abundance of meiofauna in the different deep-sea regions. In poor as well as in rich areas nematodes comprise between 85–95% of the meiofauna, when Foraminifera are excluded. Meiofauna densities are roughly three orders of magnitude higher and decrease slower with depth than macrofauna densities. The meiofauna gains in importance with increasing depth, community structure changes to smaller life forms (see also Thiel, 1975).

Contrary to Thiel's suggestions, Shirayama (1983) found in the western Pacific, that the rate of decrease in macrobenthic biomass was not significantly higher than that of meiobenthos. The size structure of the meiobenthic community (and nematode assemblages) examined by Shirayama (1983) was expressed by three indices: median size, sorting, and skewness. The correlation between median size of the nematodes and water depth was positive, which was thought to be related to the predominance of adults, a characteristic feature of deep-sea communities. The age structure of nematode communities

TABLE VI

Nematode densities (ind.·10 cm^{-2}), biomass (mg dwt·10 cm^{-2}) in deep sea communities, including species composition (Sp. comp., – not given in text, + given in text) and water depth (m)

Reference	Locality	Depth	Density	Biomass	Sp. comp.
Wigley & McIntyre, 1964	W. Atlantic, off New England	40–567	50–924	0·06–0·51	–
Thiel, 1971	N.E. Atlantic (Faroës)	290–2500	26–3999	0·08–1·78	–
Tietjen, 1971, 1976	W. Atlantic, N. Carolina (U.S.A.)	600–2500	32–140	–	+
Thiel, 1972a	N.E. Atlantic	250–2250	–	0·10–0·20	–
	North Sea (Spitsbergen)	750–3000	–	0·32–1·30	–
	W. Mediterranean	2500	600	–	–
Thiel, 1972b	N.E. Atlantic, Iberian deep sea	5271–5340	80–278	0·06–0·18	–
Dinet, 1973	S.E. Atlantic	1440–5170	294–961	–	–
Rachor, 1975	N.E. Atlantic	838–5510	6–187	0·001–0·023	–
Vitiello, 1976	W. Mediterranean	320–580	–	–	+
Coull *et al.*, 1977	W. Atlantic (N. Carolina, U.S.A.)	400	51–353	–	–
		800	280–942	–	–
		4000	4–60	–	–
Dinet & Vivier, 1977, 1979	Bay Biscayne	1920–4725	81–487	–	+
Vivier, 1978	W. Mediterranean	168–580	72–441	0·005–0·105	+
Dinet, 1979	N.E. Atlantic (Norwegian Sea)	2465–3709	9–1127	–	–
Shirayama, 1983	W. Pacific	2090–8260	31–1195	0·009–0·145	–
Soetaert, 1983	W. Mediterranean	175–1605	91–403	0·006–0·07	+
Rutgers *et al.*, 1984	Ibernian deep sea	4000–4800	101–989	0·008–0·028	–
Tietjen, 1984	Venezuela Basin	3517–5054	36–94	0·030–0·088	+

has only been studied by Vivier (1978) and Soetaert (1983) both in a Mediterranean transect from 168–580 m and from 175–1605 m, respectively. Vivier (1978) found between 43–65% juveniles and Soetaert (1983) found 48–61% juveniles in September. From these data it seems incorrect to assume that deep-sea nematode communities have a low frequency of juveniles, as hypothesized by Shirayama (1983).

The median size of nematodes is not controlled by the size of the interstitial space in the deeper clayey bottoms. In silty, shallow subtidal and intertidal environments, nematodes are longer than in coarse sandy sediments (*cf.* above). Soetaert (1983) found an opposite trend in the size structure of the nematode assemblage in relation to depth; nematodes are larger (5 times) in shallower areas (175 m) than in the deeper stations (from 305–1605 m). These last stations are more similar in species composition. On the basis of sediment granulometry, two station groups could be recognized, from 175–540 m and from 805–1605 m. It is obvious that sediment composition alone in this region is not responsible for the distribution of nematode species.

Only five studies from deep-sea environments are available in which species composition (mostly at generic level) and diversity of the nematode community are discussed: Tietjen (1976) in the Western Atlantic off North Carolina; Vivier (1978) in the Mediterranean off Marseilles; Dinet & Vivier (1979) from the Bay of Biscay (eastern Atlantic), Soetaert (1983) in the Mediterranean off Corsica, and Tietjen (1984) from the Venezuela Basin.

Tietjen (1976) recognized on a transect from 50–2500 m, four habitats. The upper sandy sediments (quartz-algal sands from 50–100 m and foraminiferan sands from 250–500 m) contain 35 and 17 species, respectively, mainly members of the Enoplidae, Ceramonematidae, Chromadoridae and Desmodoridae. The transition zone (500–800 m) consists of sandy silt with little species endemism. The lower part (800–2500 m) consists of clayey-silts with the highest number of stenotopic species, mainly members of the Leptosomatidae, Oxystominidae, Axonolaimidae, Leptolaimidae, Linhomoeidae, Siphonolaimidae, Sphaerolaimidae, and Comesomatidae.

Sabatieria stekhoveni, *S. conicauda*, *Sphaerolaimus uncinatus*, *Sabatieria pisinna*, *Acantholaimus spinicauda*, and *Sphaerolaimus paragracilis* are the dominant species of the nematode associations in the bathyal mud off Marseilles (Mediterranean), mostly members of the Chromadoridae, Comesomatidae, Sphaerolaimidae, and Monhysteridae (Vivier, 1978). Dinet & Vivier (1979) found the same families dominant in the Bay of Biscay, together with Oxystominidae, Desmoscolecidae, Microlaimidae, and Axonolaimidae. The following families are dominant in the deep sea off Corsica (Soetaert, 1983) (ordered in decreasing degree of abundance): Comesomatidae (mainly *Sabatieria*), Monhysteridae *s.l.*, Chromadoridae (mainly *Acantholaimus*, *Dichromadora*), Oxystominidae (mainly *Halalaimus* and *Oxystomina*), Selachinematidae (mainly *Richtersia*), and Leptolaimidae (mainly *Leptolaimus*) have an abundance of more than 5%. The 175 m sample is characterized by genera lacking in the deeper stations: *Odontophora*, *Spirinia*, and *Eubostrichus*.

The availability of food is very important for explaining the quantitative distribution of nematodes (and meiofauna in general) in the deep sea. Dinet (1979) found that proteins are correlated with meiofaunal densities while the organic matter is used by meiofauna; more precise analyses of the organic compounds in sediments are necessary for a better understanding of the energy pathways in the abyssal environment. The energy flow most probably goes through bacteria and different other microorganisms as recently pointed out by Burnett (1973) (*i.e.* amoeba, sarcodines, sporozoans). Organic matter is recycled in the food web and partitioned in surface waters. It reaches the deep-sea bottom mostly in a refractory stage and as small-sized particles. The decreasing density of organisms limits resources for predators; only 6–16% of omnivorous predators are found in the Mediterranean communities (Soetaert, 1983). Deposit-feeding is predominant (35–56%); species with very small buccal cavities (mainly selective deposit-feeders) are important in the deep-sea environment (16–35%), contrary to most shallow subtidal silty areas. The occurrence of deep-sea nematodes without mouth or gut (Hope, 1977) suggests a direct assimilation of the dissolved organic matter.

Shirayama (1984) found close correlations between density of nematodes and factors of food conditions, *e.g.* surface productivity and organic carbon content of the sediment, but the calcium carbonate content of the sediment had the strongest influence on meiobenthic density. Besides indicating a large sedimentation rate, a high $CaCO_3$-content may increase interstitial space. Diversity in deep-sea clayey-silts is significantly higher than diversity in shallow-water clayey-silts. Environmental predictability is probably important in regulating diversity. The increasing diversity with depth is also confirmed by Dinet (1977) down to 4000 m, by Vivier (1978), by Soetaert (1983), and by Tietjen (1984). Tietjen (1984) found in an extensive investigation of the nematodes in the Venezuela Basin that the abundances of the nematodes are directly related to depth and geographic position; the distribution of the trophic types is similar to that found in many coastal regions. Even so, sediment heterogeneity and organic content are important in governing species diversity.

The deep sea is a rather stable environment. Waste disposal and mining for mineral resources have started recently; large-scale industrial operations could disturb this fragile ecosystem rather drastically. A better knowledge and understanding of the deep-sea ecosystems is urgently needed.

SPATIAL DISTRIBUTION

VERTICAL DISTRIBUTION IN THE SEDIMENT

The vertical distribution of meiofauna in general and nematodes in particular in sediments has attracted much attention since the development of the sulphide system concept (Fenchel & Riedl, 1970). As originally defined, this sulphide system is the anaerobic environment typically established under a cover of oxidized aerobic sediments; it occurs world-wide except along surf-stressed beaches and has been postulated to be the environment where early metazoan evolution took place. This sulphide biome or thiobios, as it was subsequently renamed by Boaden & Platt (1971) is bounded at its top by the redox discontinuity layer where oxidized processes become replaced by reducing processes. The RPD layer is sharply defined in protected beaches and occurs there to within a few mm of the sediment surface. Values of the redox potential drop from +400 mV to −50 and −250 mV. The lower limits of the sulphide system can reach metres down and correspond to the deeper burial stage, defined by the termination of bacterial activity and accumulation of poisonous compounds, close to the H_2O/H_2 fence where partial pressure of H_2 reaches 1 atm (Fenchel & Riedl, 1970).

The Metazoa in this system are represented nearly exclusively by the Platyhelminthes (Turbellaria and Gnathostomulida) and the Aschelminthes (mainly Nematoda, although some Gastrotricha and Rotatoria occur). Nematodes extend farthest down into the sulphide system and often have population densities of 1000 ind.·10 cm^{-2} in the system.

This concept of thiobios has recently come under discussion since Reise & Ax (1979) demonstrated that most of the so-called thiobiotic meiofauna occurred consistently in greater numbers near oxygen islands in the anaerobic environment. These oxygen islands are formed around the burrows of macrofaunal animals, such as large polychaetes, nemertines, and amphipods on tidal flats or large decapods in deeper water.

The paper of Reise & Ax (1979) can serve to demonstrate the typical vertical distribution of nematodes on intertidal sand-flats. On this *Arenicola*-flat the colour of the sediment changes abruptly from brown (due to hydrous ferric oxides) in the upper layers to black (due to the presence of FeS) at a depth of 5 mm in summer and 20 mm in autumn. At greater depths, the sediment colour changes again to grey, due to pyrite sulphur FeS_2. In summer, away from the *Arenicola*-burrows about 66% of the meiofauna (82% without the nematodes) occurred above 6 to 8 mm depth. All meiofauna, except for a few nematodes, in the black zone were restricted to its upper fringe and did not occur below 50 mm depth. In autumn the sand turned black at a depth of 17 mm and nematode abundance was highest at 2–3 cm depth.

In the vicinity of the *Arenicola*-burrows the vertical distribution of the meiofauna changed entirely (Fig. 8). Alongside the burrows meiofaunal abundance was much higher than in the surrounding sediments and nematodes, in particular, were ten times more abundant. At 5–15 cm depth away from burrows, the average abundance of nematodes was only 39·10 cm^{-3}.

In salt marshes in Louisiana, nematodes occur in greatest density in the upper 2 cm of the sediment (Sikora & Sikora, 1982), and disappear nearly completely below 6 cm. Oxygen is completely (?) absent below 0·5 cm. Sikora & Sikora (1982) make a good point when remarking that below a redox potential of +350 mV there is no free oxygen, and that the dark zone in many sediments is the zone (125 mV) where ferric iron is reduced to ferrous iron, not the dividing line between oxic and anoxic sediments.

That nematodes do occur in greatest numbers close to the surface has been demonstrated in many studies, but in shallow waters there are often subsurface maxima as well. Tietjen (1969) demonstrated clear seasonal changes in fine shallow subtidal sands, where in spring most individuals were concentrated in the top centimetre whereas they occurred deeper in autumn, despite the fact that the RPD layer did not change position. Only in coarse sands do nematodes remain most abundant in the top layer. Skoolmun & Gerlach (1971) also found relatively high numbers of nematodes in the deeper layers of intertidal fine sands in the Weser estuary in Germany, especially in winter when most nematodes occurred in the 5–8 cm layer (sample depth was 8 cm). Even in spring, when maximum density was reached, still 20% of the nematodes was found beneath 5 cm depth.

Seasonal changes in vertical distribution were also observed by Platt

(1977b) on a fine sand beach in Northern Ireland. In this situation, the RPD layer changed position and the seasonal variation of penetration depth of the nematodes was correlated with this. In Platt's study, species were distinguished. Fifty-three percent of them were virtually restricted to the top centimetre while another 40% had their maximum abundance there but penetrated deeper. Only 7% of the species had their maximum abundance below the top centimetre layer. There was also a change in feeding types, the surface layer comprising about 30% epigrowth-feeders whereas the next centimetre only had 17% epigrowth-feeders, this percentage increasing again in deeper layers with a corresponding decrease in deposit-feeders. The deeper occurring genera were *Microlaimus*, *Monoposthia*, *Paracanthonchus* and *Pomponema* whereas the surface epigrowth-feeders were mainly Chromadoridae, and it may be that the feeding behaviour of these groups are different.

The nematode fauna of the intertidal area studied by Reise & Ax (1979) was analysed by Blome (unpubl.) who found that of 70 species present, 31 were found in the oxic surface sediments, 29 were mainly there but extended into the suboxic zone and only 8 species were found exclusively in the black suboxic sediment layers. In another study (Blome, 1983) on the island of Sylt, clear seasonal changes in vertical distribution were observed with deeper penetration in winter. *Theristus pertenuis*, where it was most abundant, had maximum densities in the 5 to 10-cm layer and in the 10 to 15-cm layer its abundance was nearly ten times higher than at the sediment surface. *Paracanthonchus caecus* had the same vertical extension at the same station in October and occurred consistently with large numbers even below 20 cm in

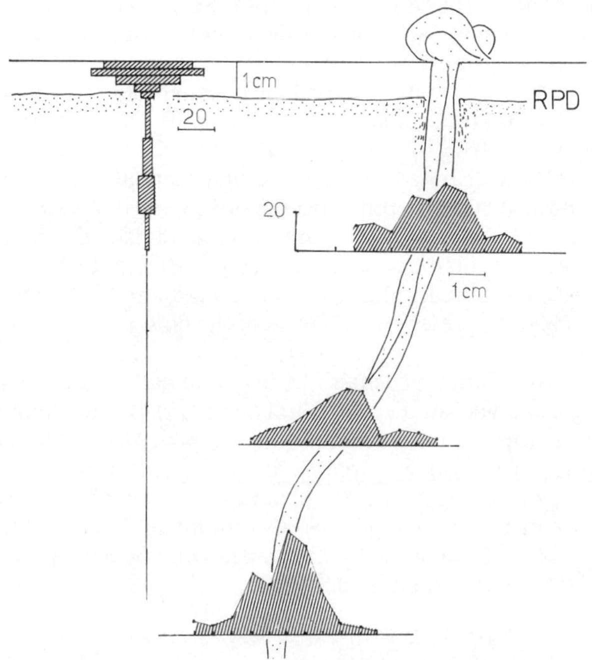

Fig. 8.—Distribution of meiofauna in the vicinity of *Arenicola*-burrows: scale bar is 20 ind.·1 cm^{-3}; data are from Reise & Ax (1979).

January. *Odontophora setosa* was even completely absent from the top 5 cm in that month.

Blome's (1983) paper clearly demonstrates the deep penetration of certain nematode species on intertidal sand-flats, but for nematodes in general this has been known for a long time; *e.g.* McLachlan (1977) reports nematodes 140 cm deep in the sand. Even in shallow subtidal sands the penetration depth may be high. McLachlan, Winter & Botha (1977) used long cores along four transects from 5 to 30 m depth in medium to fine sands off Algoa Bay, South Africa, and found meiofauna abundant to at least 35 cm down in the sediment. Oxygen was present at this depth (about 7% saturation) because in this area wave energy is important. In a stagnant marine lake in the Netherlands, Heip, Willems & Goossens (1975) found nematodes to at least 25 cm depth in a sandy station, but here also oxygen was present. In 1977 and 1978 the RPD layer, however, moved up from −20 cm to −2 cm with a concurrent compression of the nematodes in the surface layer, although significant numbers remained present in the deeper layers (Willems, Sharma, Heip & Sandee, in press).

In deeper waters, penetration of nematodes into the sediment diminishes. In mud stations off the Belgian coast, 93–99% of the nematodes were found in the upper 4 cm, but in a sandy station penetration was much deeper and maximum density was found between 6 and 10 cm (Heip *et al.*, 1979). Boucher (1980a) found most nematodes in the surface layers of a subtidal (−19 m) fine sand in Brittany in spring (59% in the upper 2 cm) and summer (51%), but in autumn and winter nematodes migrated down; only 42% was present in the upper 2 cm in autumn, only 13% in winter. In spring and summer, subsurface secondary maxima were found around 5–10 cm depth. Seven out of the 17 dominant species showed seasonal variations in their depth distribution; six out of 40 species were restricted to the deeper layers.

At depths around 100 m in the North Sea, nematodes remain important to at least 10 cm depth in the sediment (Heip *et al.*, 1979; Faubel, Hartwig & Thiel, 1983). The same is true in deep-sea sediments. Coull *et al.* (1977) found nematodes evenly distributed through the first 6 cm at 400 m depth, with a surface maximum at 800 m depth and a subsurface maximum between 2–3 cm at 4000 m depth. Vivier (1978) found Araeolaimidae, Desmodorida, and Desmoscolecida living in the surface layers, together with Enoplida (especially Oxystominidae), Chromadoridae, Cyatholaimidae, and Choanolaimidae. The Comesomatidae were present in all layers of the fluid red muds studied (168–580 m).

The vertical distribution of nematodes in twelve bathyal to hadal stations in the western Pacific was studied by Shirayama (1984). Meiofauna was most abundant in the top centimetre and declined exponentially with depth in the sediment. At seven stations no organisms were found beneath 12 cm, but at one station nematodes together with rhizopods were found as deep as 30 cm. A surface maximum for meiofauna was also found by Soetaert (1983) in five stations from 350–1600 m in the Mediterranean and by Rutgers van der Loeff & Lavaleye (1984) in the Iberian Deep Sea.

VERTICAL MIGRATION AND HORIZONTAL DISPERSAL

Tidally induced vertical and horizontal displacements have been observed by Rieger & Ott (1971) in the Adriatic Sea near Venice. Four nematode species showed clear changes in distribution over a tidal cycle. *Daptonema leviculum* migrated upwards at high tide and downwards at low tide while *Microlaimus criminalis* had just the opposite behaviour. Two other monohysteroid species (*Theristus pictus* and *Daptonema curvispiculum*) are periodically concentrated and dispersed again by the shifting sands, moving towards the high water line during flood.

Gerlach (1977b) summarizes the few data existing on meiofauna dispersal. Drifting materials, such as driftwood, coconuts, ice and algae may transport meiofauna over long distances. The ballast of sailing vessels mainly consisted of heavy material such as stones and sand, and much material was transported in this way across the oceans.

Nematodes are frequently found in plankton samples in shallow waters (Gerlach, 1977b; Hagerman & Rieger, 1981). Sibert (1981) found nematodes in the water column to be most abundant just above the sediment–water interface; at 5 cm they were two to five times more abundant than at 30 cm above the bottom. Chandler & Fleeger (1983) found that nematodes colonize defaunated sediments either *via* the water column or *via* the sediment, but that even after 29 days their densities were less than in the surrounding sediment.

That nematodes can swim has been observed in cultures (*e.g.*, Jensen, 1981b) but it must be much more common in the field than is generally thought. Gerlach (1977b) cites Schneider who found many females of the nematode *Tripyla* in the gut of the pelagic plankton-feeding fish *Coregonus* and postulates that ripe females of this nematode swim up into the water column to spawn. Quite the opposite happens in the viviparous species *Anoplostoma viviparum*, where ripe females migrate down in the sediment to the anoxic layers before spawning; then the hatched juveniles migrate to the flocculent layer and are dispersed by water currents (Surey-Gent, 1981).

Colonization by nematodes after disturbances may take anything from hours to years, depending on the environment and the kind of disturbance. Sherman & Coull (1980) disturbed a 9 m^2 area of intertidal mud-flat by hand-turning. Total nematode density was restored after only one tidal cycle. Two *Ptycholaimellus* species had significant lower densities the night after the disturbance was created, which may indicate migration. Re-population was thus very quick and may have occurred mainly by advection *via* the flocculent layer. Other recolonization rates reported for meiofauna are much lower, of the order of months (Coull, 1969) and even years (Pequegnat, 1975; Rogers, 1976).

A particular example of a natural disturbance are the pits created by stingrays (*Dasyatis sabina*) in shallow subtidal sands at night. These pits have been studied by Sherman, Reidenauer, Thistle & Meeter (1983) who found reduced nematode densities until three days after the formation of the pit. No single species seemed to exploit the disturbance better than others during recolonization, which was ascribed to the fact that nematodes lack sufficient dispersal capability to arrive quickly at newly opened space, although 48 h after the pits had been made re-population was effective.

By studying the recolonization of tube-caps of the polychaete *Diopatra cuprea*, Bell & Coen (1982) found rapid recolonization by nematodes through the water column and *via* the sediments.

SPATIAL PATTERN

The small-scale spatial distribution or pattern of populations has received considerable attention since space has been recognized as one of the important niche dimensions along which species segregate (*e.g.* Schoener, 1974). In general, three types of pattern are distinguished: regular, random, and aggregated. In benthos studies, as in much terrestrial work, only the two-dimensional horizontal plane is considered. To discriminate these types many authors use the variance/mean ratio $s^2/\bar{x}$ or some measure derived from it such as $(n-1)s^2/\bar{x}$, distributed as χ^2, or the index of clumping $s^2/\bar{x}-1$ (David & Moore, 1954). Discrimination is based on the fact that a Poisson distribution, describing the occurrence of rare, random events, is completely determined by only one parameter since $s^2 = \bar{x}$. When for a given species $s^2/\bar{x} = 1$, its spatial pattern is considered random; when $s^2/\bar{x} > 1$, the pattern is aggregated. Based on such measures nematodes have generally been found to be aggregated (Vitiello, 1968; Gray & Rieger, 1971; Arlt, 1973; Gerlach, 1977a).

There are, however, some problems with this approach. Basically, spatial pattern is a property of populations, not of communities or taxocenes. Other statistical distributions, such as the binomial and normal, also describe the occurrence of random events but without the restriction that the event (presence or absence) must be rare. Empirically it has been found that the variance is often a function of the mean, as in Taylor's power law $s^2 = a\bar{x}^b$ (Taylor, 1961). Since b is often near to two, the variance/mean ratio is often a linear function of the mean $s^2/\bar{x} = a\bar{x}$ and the coefficient of variation $s/\bar{x}$ is often constant (Heip, 1975), at least for $\bar{x}$ large enough. For these reasons the $s^2/\bar{x}$ ratio is of only limited usefulness.

Detailed studies of spatial patterns of marine nematodes are rare. Olsson & Eriksson (1974) studied horizontal distribution of the meiofauna from the Swedish west coast at a depth of 22 m on a clayey bottom. A contiguous grid with cells of 9·3 cm^2 surface was used for subsampling a box corer on deck. For nematodes two adjacent patches with high density were observed on a scale of about 10 cm. The coefficient of variation $s/\bar{x}$ was 0·515 and the number of samples required to detect a 20% difference in means with 95% confidence was estimated as seven. Findlay (1981) studied a mud and a sand station using four different core sizes to sample approximately the same surface area. There was a marked influence of core size but nematodes were only slightly aggregated with the highest values of the index used (Green's index of dispersion $(s^2/\bar{x}-1)/(N-1)$ with N the total number of individuals) at a core size of 5 cm^2 in the mud station and the sand station in February, but no aggregation at all in the sand station in September. These data relate to the total density of the nematode taxocene and the significance of the weak departure from randomness at 5 cm^2 core size is unclear.

Samples taken at the sand site in February with drinking straws were re-analysed in a second paper (Findlay, 1982b). Eighty-one drinking straws in a 9×9 array were taken covering an area of 0·02 m^2 (15×15 cm). From these densities a contour map was drawn which was then sampled by simulation with three different core sizes. Each sampling scheme had the same efficiency for estimating the original abundance, indicating that sampling scale has a negligible effect on abundance estimates. Larger scale sampling schemes tended, however, to obscure the heterogeneity of the distribution. Larger numbers of smaller cores gave a better resolution of the pattern and Findlay (1982b) concludes that roughly fifty 0·5 to 1·0-cm^2 cores is a reasonable optimum sampling scheme for intra-community research on spatial pattern.

Hogue & Miller (1981) also studied total nematode density but in a system with macroscopic heterogeneity, a sand-flat with ripples. Using autocorrelation they demonstrated that a significant periodicity in density existed at the wavelength of the ripple crests. This is unexpected, since organic material concentrates in the troughs, but it was thought that this material is buried, after drying, by the moving sediments so that after several tidal cycles it would be under the crests. Of course this requires that ripples would move over distances equal to half the wave length. Also, samples were taken at low tide and the pattern may change over a tidal cycle, and only to 4 cm depth, and the nematodes may have been deeper.

In a study of sandy sediments on the coast of Oregon, U.S.A., Hogue (1982) took 72 straw samples in two parallel transects 3 cm apart from a box corer at a station with water depth 28 m. Neighbouring samples were correlated over distances of approximately 2–3 cm and there was no periodicity in density flucuations, *i.e.* aggregations were not spaced at constant intervals. The two transects were also correlated, so that the physical dimensions of patches of high density were estimated to be about 3 cm in size.

Using 1·9 cm cores, the pattern of 21 dominant species of the same site was also studied. For all species, the index of dispersion indicated aggregation in May, but in January only five out of eleven superficially occurring species were aggregated, although total density had a very high value for the index of dispersion. Another interesting result of Hogue's (1982) study was that the dominant species were not correlated spatially, they occurred independently. This, and the predominantly random distribution of nematodes in winter, points to biological factors such as food availability and reproductive processes as the most likely cause for small-scale patchiness in marine nematodes (Hogue, 1982).

Similar conclusions were reached by Delmotte (1975) in a study of a 10-cm deep brackish water pond in Belgium. He studied spatial pattern of nematodes from a 7 × 7 grid of 0·8-cm^2 cores, each core 5 cm apart, in summer. All species were aggregated and their frequency distribution could, but for two species, be fitted with the negative binomial distribution. Most species were distributed independently, and juveniles tended to co-occur with adults of the same species.

These few existing studies on small-scale horizontal distribution tend to agree that patches exist on a scale of centimetres and that species tend to be distributed independently. The consequence is that a large number of small samples are required to study spatial pattern but also that a small number of large samples may give good estimates of density. If indeed patch size would be as small as 2–3 cm one can expect that 6 cm diameter cores will already be reliable for this purpose.

LARGER HORIZONTAL SCALES

The studies of Delmotte (1975), Findlay (1981, 1982b), and Hogue (1982) indicate that important variability exists on the scale of a few centimetres. Variability on a larger scale seems to be much less important, except when strong gradients exist, as on beaches (Platt, 1977b; Blome, 1983). Again Hogue's (1982) study is very illustrative of this point. He compared the data from three cores taken within three quadrants (10 cm scale) of a box corer (25 cm scale) at three stations 30 m apart. Only about one-quarter of the total variance was due to between-boxes differences and this difference was even only 9% when species living in the first 5 cm of the sediment were considered. For these species, 67% of the variance occurred within quadrants (10 cm scale) and 24% between quadrants (25 cm scale).

When six stations were compared (km scale), it was found that only 40% of the variance was due to between-station differences, even though two distinct communities could be discerned. The same was found by Heip *et al.* (1979) for eighteen stations covering an area of nearly 1000 km^2 in the Southern Bight of the North Sea.

Variability on a scale larger than the nematode patches may, however, exist when environmental gradients exist or where disturbance is common. Powell, Bright, Woods & Gitting (1983) studied the meiofauna from a very interesting and uncommon system, the East Flower Garden Brine Seep. This consists of a brine seep at 72 m depth in the Gulf of Mexico. The brine, rich in sulphide, flows by gravity into a basin located about 60 m from the bank's edge into a brine lake. The overflow from that lake flows into a 96-m deep canyon. Nematodes were absent from the brine lake and occurred only in low abundance in most stations: between 2–75·10 cm^{-2} in the stations closest to the brine lake where sulphide concentrations were in the range of 60–80 μg-at.·l^{-1}. Then a big peak occurred, together with an even bigger and remarkable peak in the density of gnathostomulids, at a station where sulphide remained high and oxygen was even lower than in the previous station of the brine seep. At stations further from the brine lake, nematode density dropped again despite the fact that oxygen levels increased and sulphide decreased. Why this is so was not explained, although an abundant bacterial growth at this interface between 'thiobiotic' and 'oxybiotic' conditions is to be expected.

LIFE CYCLES

SEASONAL CYCLES

Yearly fluctuations in nematode density and/or species composition have been studied by a number of authors but sampling is in most cases with a low temporal resolution and in some cases not really quantitative. Data exist for the Baltic (Arlt, 1973; Arlt & Holtfreter, 1975; Möller, Brenning & Arlt, 1976), the Swedish west coast (Nyholm & Olsson, 1973), intertidal or brackish water areas along the North Sea (Skoolmun & Gerlach, 1971; Smol, Heip & Govaert, 1980; Bouwman, Romeyn & Admiraal, in press), subtidal areas of the North Sea (Warwick & Buchanan, 1971; Lorenzen, 1974; Juario, 1975), intertidal (Harris, 1972), and subtidal (Boucher, 1980a) areas of the English Channel, the Scottish west coast (McIntyre & Murison, 1973), the western Mediterranean (Soyer, 1971; de Bovée & Soyer, 1974); estuaries (Tietjen, 1969), salt marshes (Coull & Bell, 1979), and sea-grass communities (Hopper & Meyers, 1967b) along the U.S. east coast and some information from the Canadian west coast (Sharma & Webster, 1983).

Skoolmun & Gerlach (1971) studied an intertidal sand-flat in the Weser estuary in Germany. Total density of nematodes, which was very low overall, had a clear peak in May–June and another high value in October. The dominant species had different density patterns: *Hypodontolaimus setosus* had a peak in summer, *Tripyloides marinus*, *Sabatieria vulgaris*, and *Chromadorita tentabunda* had a peak in winter or spring. Two *Theristus* species and *Enoploides spiculohamatus* had several peaks over the year whereas *Viscosia viscosa* and *Oncholaimus brachycercus* had two abundance peaks and may have two generations annually.

Another oncholaimid, *O. oxyuris*, has been studied over several years in a brackish water pond in Belgium (Smol *et al.*, 1980). This species has either two generations annually or two generations in three years, in a strange pattern in which overwintering juveniles belong to two different generations. Large juveniles become adult in spring and produce a second generation in late summer whose offspring overwinter as small juveniles. Small overwintering juveniles become adult in summer and their offspring overwinters as large juveniles.

The Ems-Dollard estuary, on the Dutch–German border, has been studied by Bouwman *et al.* (1983). The Dollard receives a heavy load of organic matter in autumn. Maximum abundance in two stations on a transect near the outfall reached a peak in June, whereas maximum abundance in the third station, further from the outfall, was reached in August. In the first station, closest to the outfall, a second peak was reached in September. Maximum density reached was highest in the intermediate station (close to 19×10^6 ind.$\cdot$m^{-2}) and lowest in the far station (close to 10×10^6 ind.$\cdot$m^{-2}), but the number of species was highest there (between 12 and 20 compared with between 6 and 10). The whole transect had 27 species, with only two dominants near the outfall: *Eudiplogaster pararmatus* and *Dichromadora geophila*, whereas the far station was dominated by *Sabatieria pulchra*, *Ptycholaimellus ponticus*, and *Leptolaimus papilliger*. Similarity between samples from different months was almost as high as between replicates, but significant changes in the nematode community structure took place in spring at the three stations. At this time the density of all the dominant species increased.

Eudiplogaster pararmatus appeared to reproduce all through the year, gravid females and juveniles being always present. *Dichromadora geophila*, however, had only juveniles from April to August and they had a conspicuous peak in May, so that reproduction appears to be restricted to a short period in spring. *Ptycholaimellus ponticus* started reproduction in spring and continued to reproduce during the whole summer. *Leptolaimus papilliger* seemed to reproduce continuously, although more gravid females were present in summer. *Sabatieria pulchra* also had continuous reproduction and juveniles dominated in all seasons. Spring and summer peaks in nematode density in intertidal habitats have also been found by Harris (1972) in a beach in Cornwall, but Schmidt (1969) observed no distinct peaks in Sylt (Germany).

A very thorough analysis of nematode fluctuations on a flatfish nursery ground on the Scottish west coast (McIntyre & Murison, 1973) from monthly samples over seven years taken near M.L.W.N. shows that on average there is a peak in June and another in August. Lowest densities were reached in January, and the average February value was slightly higher than the subsequent March value, but whether a third peak occurs in late winter is uncertain. Statistically, the winter values are lower than the densities from May onwards. At 6 m depth subtidally on the same beach annual fluctuations were much smaller, with a peak in July and another one in January.

Temperature and food are the most obvious factors explaining these density changes. As nematodes belong to different feeding types, one expects changes correlated with the availability of different food items. This has been observed in two shallow estuaries in New England by Tietjen (1969). In these estuaries again one or two annual peaks in density were observed, a spring peak in April–May and a summer peak in July–August. The spring peak was due to an increase in epigrowth-feeders, mostly *Nudora lineata, Tripyloides gracilis, Spirinia parasitifera, Monoposthia costata,* two *Chromadora,* and six *Hypodontolaimus* species. All these species declined rapidly after the summer, maybe due to predation by juvenile polychaetes. A summer increase for *Hypodontolaimus* sp. and *Spirinia parasitifera* was also observed by Hopper & Meyers (1967b) for sea-grass beds in Florida and Wieser & Kanwisher (1961) found increased numbers of *S. parasitifera* in June when compared with November in a salt marsh near Woods Hole.

Deposit-feeders and omnivorous predators did not exhibit such marked seasonal variations in density, but deposit-feeders tended to reach maximum numbers in fall (autumn), winter or early spring, due to the incorporation of dead *Zostera* leaves and other vegetation into the sediment. This incorporation coincided with peaks of *Paralinhomoeus, Theristus, Paramonhystera,* and *Sabatieria.* Oncholaimids were the dominant omnivorous predators and their seasonal distribution closely followed that of the deposit-feeders.

The seasonal cycles of nematodes on algae have been studied by Warwick (1977) for red algae on the Scilly Isles, by Kito (1982) for the fauna on *Sargassum confusum,* and by Trotter & Webster (1983) for the fauna on *Macrocystis integrifolia.* On this last algae, only three species of nematodes accounted for more than 90% of the fauna. They occurred in all months, but *Prochromadorella neapolitana* peaked in summer, when epiphytic diatoms were abundant, and *Monhystera refringens* peaked from July till October; *M. disjuncta* was relatively common throughout the year. All species occurred in greatest abundance on the lower and middle blades of *Macrocystis* in the deep end of the kelp bed.

The nematodes from *Sargassum confusum* show two peaks in population density, one in September and one in May. Three out of four dominant species had both peaks, whereas *Chromadora heterostomata* had only the spring peak. Gravid females and juveniles occurred in all months, except for *C. heterostomata* which was absent from August to October. Of the five most abundant nematodes from algae on the Scilly Isles (Warwick, 1977), *Oncholaimus dujardinii* has peak dominance in spring and early summer, and then makes up for nearly 80% of the total nematode density, *Theristus acer* has a similar though somewhat more irregular pattern, while *Symplocostoma tenuicolle, Euchromadora striata,* and *Cyatholaimus gracilis* have peaks in late summer and winter. These peaks, however, relate to relative dominance; absolute densities in the algae are not known.

The reproductive cycle of *Leptosomatum bacillatum,* a nematode living in sponges, was studied at Texel, The Netherlands, by Bongers (1983), where it lives in *Halichondria panicea.* It has an annual life cycle with small juveniles appearing in July and growing slowly over the winter, the first becoming adult near March.

Whereas spring and summer peaks appear to be common in intertidal and shallow subtidal areas on an annual basis, little is known on long-term temporal variability. Two density peaks were found in two subsequent years by Harris (1972) at the same time of the year but with different heights. McIntyre & Murison (1973) compared mean spring densities over seven years and found differences of about 5:1, the minimum value being 328 ind.·10 cm^{-2} in 1970 and the maximum value 1767 ind.·10 cm^{-2} in 1969. Coull & Bell (1979) describe the density fluctuations from monthly samples from 1972 to 1977 in a shallow subtidal muddy creek, and demonstrated important variations from year to year: from 1972 to 1974 maximum densities were around 700–800 ind.·10 cm^{-2}, from 1975 to 1976 two years with maximum densities around 3500 ind.·10 cm^{-2} followed, while in 1977 maximum density was around 1500 ind.·10 cm^{-2}.

Whereas seasonal and annual variation in nematode density are important in intertidal or shallow subtidal estuarine areas, this is much less the case for subtidal marine areas. Here nematode populations appear to be much more stable. In the German Bight, two studies exist (Lorenzen, 1974; Juario, 1975). In a fine sand station in a TiO_2 dumping ground a summer maximum around August and a spring maximum around May were found and an exceptionally high value in December, but summer densities were statistically not different from winter densities. Dominance of most nematode species did not change appreciably over the year but some species had higher dominance in summer (*Daptonema leviculum*, *Metadesmolaimus heteroclitus*, and *Axonolaimus helgolandicus*). Juveniles made up 50% of the community over the whole year, but for several species an increase in reproduction could be established; *A. helgolandicus*, *Metadesmolaimus heteroclitus*, *Mesacanthion diplechma*, *Paracyatholaimus pentodon* and *Paracanthonchus caecus* all had a maxima in reproductive activity in spring. *Mesacanthion diplechma* only occurred in June to July as mature adults and probably has only one generation annually. This is a large species and related to *Enoplus communis*, which also has only one generation annually (Wieser & Kanwisher, 1960). The other species all had adults and juveniles present throughout the year.

The station studied by Juario (1975) at 35-m depth in the German Bight of the North Sea had maximum nematode densities in April and August, but winter values were not significantly different from summer values and there was no change in the relative abundance of feeding types. The most abundant species, *Molgolaimus turgofrons*, had continuous reproduction throughout the year, and this was also true for all 11 dominant species. Juveniles were found in high densities all over the year. Species composition of the nematode assemblage remained very stable.

Nearly the same features have been observed by Warwick & Buchanan (1971) at an 80-m deep station off the Northumberland coast in the North Sea; there were no significant changes in population structure (juveniles being always dominant), species composition, density and biomass of the nematodes during the year. Boucher (1980a) again came to the same conclusions for the nematodes in a 19-m deep fine sand in the Bay of Morlaix in Brittany. Density variations over one year were not significant when either the four seasons or summer–winter were compared. Only three species out of 67 had significant seasonal fluctuations: *Prochromadorella ditlevseni*, *Actinonema celtica*, and *Theristus bastiani*. Juveniles again were abundant in all seasons and formed a constant proportion of the population. Gravid females were, in general, more abundant in spring and maturing females more in autumn.

In the deeper waters of the North Sea, however, Faubel, Hartwig & Thiel (1983) found peaks in nematode density in March at one station (134 m) and in December when all stations were averaged. In this area bottom water temperature is highest in December due to the influx of Atlantic water.

In the western Mediterranean seasonal changes in density have been studied by Soyer (1971) and de Bovée & Soyer (1974). At a 35-m deep station, there was a clear summer peak in August in 1964 but it was much retarded, in October, in 1965, and earlier, in July, in 1966. This was related to water temperature and the persistence of warm water masses in the area. These seasonal changes disappeared in deeper water; from 75 m to 550 m the average abundance was the same in the four seasons.

LABORATORY STUDIES

Most laboratory studies on life cycles of marine nematodes have focused on the effect of temperature and/or salinity on development. In these studies either embryonic development or generation time have been used with different meanings, which makes comparison difficult. For example, generation time was measured as the time between first egg deposition in successive generations (Gerlach & Schrage, 1971; Tietjen & Lee, 1972, 1973; Warwick, 1981a); as the time between the appearance of gravid females in successive generations (Vranken, Thielemans, Heip & Vandycke, 1981); the experiment sometimes starts with gravid females or eggs and runs until either the first gravid female or egg is detected or until 50% of the daughters become gravid or 50% of the eggs hatch. In any case, all these definitions are only approximations of the mean generation time, and for fast-developing, iteroparous species these time periods are approximations of the development time or minimum generation time T_{min} (Vranken & Heip, 1983). In the rest of this paper we will, unless otherwise stated, only refer to minimum generation time or development time T_{min}.

Development times of marine nematodes may vary between a few days for the rhabditids and several months for the large oncholaimids. In the earlier papers, generation times of approximately one month have been reported for several monhysterids and chromadorids (Chitwood & Murphy, 1964; Hopper & Meyers, 1966a; Tietjen, 1967; von Thun, 1968). We now know that some of these species have been grown suboptimally, resulting in too long generation times; *e.g.* for *Monhystera disjuncta*, Chitwood & Murphy (1964) found T_{min} = 30 days at 22 °C, whereas Gerlach & Schrage (1971) found T_{min} = 12 days at 17–22 °C.

The effect of body weight

An important generalization relating development time with body mass of the species states that small species will have short life cycles, large species mature later and have longer life cycles. This has been shown true for many animal groups (Fenchel, 1974; Steele & Steele, 1975; McLaren, 1966). For the meiofauna, especially for nematodes and harpacticoids, an inverse relationship between P/B and adult body weight was postulated by Banse & Mosher (1980) and demonstrated by Heip, Herman & Coomans (1982). When life-cycle turnover is constant, this relationship may be reduced to a relationship between the annual number of generations and body weight, and it is then evident that a relationship between development time and body weight should exist. One complication arises because reproductive activity is limited by temperature: at the low end of the range the basal temperature or biological zero T_0, below which no reproduction occurs and at the other, the upper temperature limit (see Wieser, 1975).

In general, a direct relationship between development time and biomass (weight) is to be expected. We have checked this using all published data on generation time and added unpublished results from Vranken and co-workers. The relationship between minimum generation time T_{min} at 20 °C and biomass (dry weight) is given in Figure 9. The different corrections made are

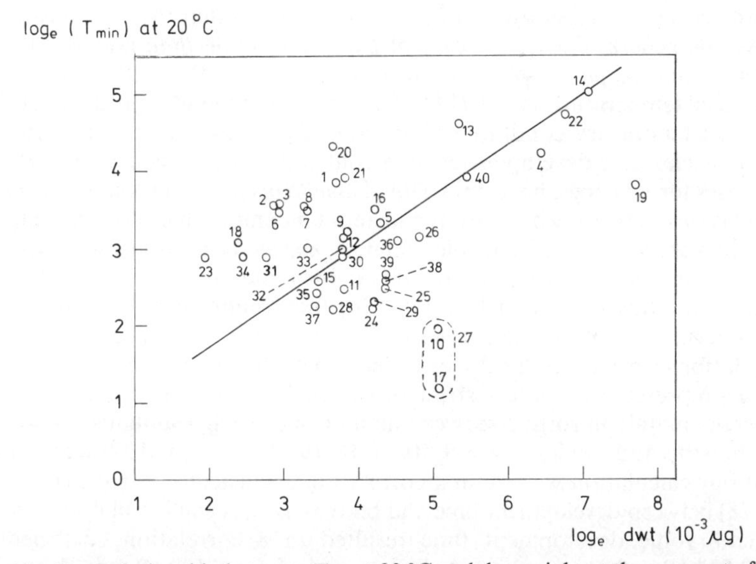

Fig. 9.—Relationship between T_{min} at 20 °C and dry weight at the moment of maturation: corrections to 20 °C were made, using Krogh's normal curve or by equation $T_{min} = aT^b$; code numbers of the species are listed in Tables VII and VIII; 30 is *Chromadorina germanica*; 31, 32, 33, and 34 are consecutively *Monhystera* sp.; *Atrochromadora denticulata*, *Monhystera parva* and *M. multisetosa* (Garcia, 1982); dry weight is 15% of wet weight, which was estimated with Andrássy's formula using a combination of the smallest length and width of a data-set; most morphometric data are from the original experimental papers; the others are from De Coninck & Schuurmans Stekhoven (1933) for 22, 35, 37, and 40; Schuurmans Stekhoven (1935) for 3; Chitwood (1951) for 18; Gerlach (1951) for 21; Timm (1952) for 15; Hopper & Meyers (1967b) for 16 and 30; Lorenzen (1969) for 20; Boucher (1976) for 36; Platt & Warwick (1983) for 19; Herman, Vranken & Heip (1984) for 28; from growth curves (Vranken, pers. obs.) for 23, 27, 29, and 38.

summarized beneath the figure, the coded species in Tables VII and VIII. There exists a weak but significant correlation ($r = 0{\cdot}338$, $0{\cdot}05 > P > 0{\cdot}01$; $n = 40$) between body weight and generation time. There is, however, substantial scatter around the line, calculated by GM-regression (model II, Sokal & Rohlf, 1981) as $\log_e T_{min}$ (days) $= 0{\cdot}430 + 0{\cdot}651 \log_e$ dry wt (10^{-3} μg) (95% CI slope = 0·450–0·853). Part of this scatter can be attributed to the following factors: (a) suboptimal cultivation, resulting in an over-estimate of T_{min} (species 1, 2, 3, 9, and 33); (b) differences in acceleration rate and basal temperature between species inhabiting different climates (species 18, 19, 20 *versus* the others); and (c) the nature of the temperature response of extreme opportunists (species 10, 17, 23, and 27).

To illustrate: *Monhystrella parelegantula* (23) has a very high basal temperature $T_0 = 14$ °C, hence the effective temperature for this species at 20 °C is only 6 °C. Such an increase is not high enough to arrive at $T_{min} = 5{\cdot}5$ days, as predicted by the general equation, even though the species has a very high $-b = 3{\cdot}11$.

The warm-water species cultivated by Hopper, Fell & Cefalu (1973) also have very high basal temperatures (around 12–15 °C) and also a high $-b > 2$, but the effective temperature of 20 °C is too low, so that they are positioned above the general line in the case of *Diplolaimella ocellata* and *Haliplectus dorsalis*. *Enoplus paralittoralis* is, however, under the line, although it has a high basal temperature as well (12·0 °C) and was not found reproducing below 18 °C in laboratory conditions. That above the basal temperature, warm-water species may develop faster than cold-water species is demonstrated in two cases for which we have data: *Diplolaimella ocellata* and *Rhabditis marina* populations of warm water develop faster at a comparable effective temperature than the same species in cold water. *R. marina* is in both cases below the general line, which is expected since it is known that saprobic nematodes such as the rhabditids are extremely fast developers (Tomlinson & Rothstein, 1962; Grootaert, 1976; Schiemer, 1982b). We excluded this species from our calculations since it can hardly be called typical for the marine environment, being a representative of terrestrial forms which happen to tolerate salt water (it occurs mainly in rotting seaweed high on the beach) (Sudhaus, 1974).

When this and species (1, 2, 3, 9, 10, 17, 18, 19, 20, 23, 27, and 33) are excluded from our calculation we obtain a correlation coefficient $r = 0{\cdot}681$ ($P < 0{\cdot}01$, $n = 28$) between development time and body weight. A similar analysis (Fig. 10) on embryonic development time resulted in a correlation coefficient of $r = 0{\cdot}524$ ($P < 0{\cdot}01$, $n = 30$) for all data and of $r = 0{\cdot}599$ ($P < 0{\cdot}01$, $n = 28$) with *R. marina* excluded. As the correlation found between embryonic development time and biomass is not significantly different from the correlation found between generation time and biomass, prolongation in post-embryonic development due to suboptimal cultivation cannot be proved statistically.

Influence of temperature

Temperature has a profound effect on minimum generation time T_{min} in all nematodes studied (Gerlach & Schrage, 1971; Tietjen & Lee, 1972; Hopper, Fell & Cefalu, 1973; Bergholz & Brenning, 1978; Heip, Smol & Absillis, 1978; Warwick, 1981a; Garcia, 1982; Vranken, Herman & Heip, unpubl.). At high

TABLE VII

Mean embryonic duration (E_{min}) and minimum generation time (T_{min}) of free-living brackish-water nematodes in laboratory conditions: ( ), species code number; SD, standard deviation; N_e, number of eggs studied; N, number of females studied; +, sampled from the Dievengat (Knokke, Belgium); + +, sampled from the Sluice Dock (Ostend, Belgium); Medium I, bacto-agar (DIFCO) + 1% Vlasbom-medium + 0·5 − 1% $Na_2SiO_3 \cdot 9H_2O$ (concentration in stock-solution is 15 $g \cdot l^{-1}$); Medium II, bacto-agar (DIFCO) enriched with phosphorus modified Walne-medium (5 parts) and ES medium of Provasoli (1 part) and 0·5–1% $Na_2SiO_3 \cdot 9H_2O$ (15 $g \cdot l^{-1}$) (for constitution see Vranken et al., *1984a); Medium III, modified Killian-nutrient agar after von Thun (1966)*

Species	Medium	Salinity (°/₀₀)	Temp. (°C)	E_{min} (days)	T_{min} (days)	Authority (unpubl.)
Monhystrella parelegantula^{++} (23)	I	30	20	3·8 (SD = 0·75; N_e = 181)	18·1 (SD = 2·87; N = 275)	Vranken
Monhystera parva^{+} (37)	II/III	20	20	2·7 (SD = 0·76; N_e = 163)	11·5 (SD = 1·41; N = 75)	Vranken
Monhystera parva^{++} (35)	III	30	22	3·7 (SD = 1·09; N_e = 53)	8·8 (SD = 1·64; N = 229)	Vranken & Coppieters
Chromadora nudicapitata^{++} (38)	III	30	22	3·3 (SD = 1·17; N_e = 313)	9·7 (SD = 0·96; N = 148)	Vranken & Dua
Chromadora nudicapitata^{+} (39)	II	20	20	—	14·0 (SD = 1·40; N = 108)	Van Brussel & Vranken
Neochromadora poecilosomoides^{++} (36)	II/III	30	20	5·3 (SD = 1·57; N_e = 204)	21·7 (SD = 4·10; N = 115)	Vranken
Paracanthonchus caecus^{+} (40)	II	20	20	—	51·1 (SD = 4·8; N = 66)	Van Brussel & Vranken

TABLE VIII

Life history of free-living marine nematodes: a, number of eggs in the uterus; b, after this time-period approximately 60% of the females died; c, generation time from ♀♂ to ♀♂, data between brackets are the shortest minimum and the longest maximum; d, only 1 individual has been studied; e, only 2 individuals have been studied; f, maximum number of eggs produced; g, generation time measured as the time elapsed between first egg-depositions; h, Sudhaus' population of R. marina *(Kiel, FRG) is viviparous, therefore this figure = postembryonic development time; i, data read from Bergholz & Brenning's fig. 3; j, generation time (T_{min}) calculated as $T_{min} = (0{\cdot}228T + 5{\cdot}573) \times E_{min}$, with T = temperature (°C) and E_{min} = embryonic development time (days); k, daily egg-production·female^{-1}, read from Warwick's Figs 2 and 3; l, number of juveniles·female^{-1}* (E. paramatus *is viviparous) at unspecified temperature; m, generation of* M. disjuncta *cultivated monoxenically on* Alteromonas haloplanktis *(ISC_2); n, mean fecundity (= potential capability of a nematode to produce eggs) of 12 ♀♀ (3 °C) and 9 ♀♀ (12 and 17 °C); *, mean adult longevity, or time at which 50% of the adult females studied, died, mean total longevity can be obtained by adding T_{min} of the corresponding temperature, N = 106 at 17 °C; N = 125 at 12 °C and N = 105 at 3 °C; N, number of females studied during the generation time experiments; SD, standard deviation; SE, standard error; ♂, male; ♀, female*

Species (Code no.)	Salinity (°/₀₀)	Temp. (°C)	Egg number (average and/or min., max.)	Average min. generation time (days) (min.–max.)	Average life span (days)	Reference
Diplolaimella schneideri (1)	Sea water	20–24	—	40	—	Chitwood & Murphy, 1964
Monhystera disjuncta (2)	Sea water	20–24	—	30	—	,,
Acanthonchus cobbi	Sea water	—	—	29	—	Hopper & Meyers, 1966a
Euchromadora gaulica	Sea water	—	—	35 (30–40)	—	,,
Monhystrella parelegantula	Sea water	—	—	30	—	,,
Monhystera filicaudata (3)	Estuarine	20–25	8–20[a]	29·5 (24–35)	—	Tietjen, 1967
Adoncholaimus thalassophygas (4)	15	20–22	17 (14–22)	63 (55–72)	78	von Thun, 1968
Chromadorita tenuis (5)	15	20–22	20 (16–28)	26 (19–34)	43	,,
Diplolaimella ocellata (6)	15	20–22	22 (12–26)	29 (22–39)	56	,,
Diplolaimelloides oschei (7)	20	20–22	36 (24–42)	29 (23–35)	54	,,
Diplolaimelloides islandica (8)	15	20–22	22 (16–25)	31 (24–39)	50	,,
Monhystera disjuncta (9)	5	20–22	16 (14–20)	23 (18–28)	33	,,
Rhabditis marina (10)	25	25	85 (70–100)	4·5–5	♀♀: ±10[b]; ♂♂: up to 19	Tietjen *et al.*, 1970

Monhystera disjuncta (11)	32	26	—	no juvenile growth	—	Gerlach & Schrage, 1971
	32	17–22	37	12 (8–15)[c]	61[d]	,,
	32	13–15	—	15 (9–20)[c]	—	,,
	32	9–12	—	17 (13–24)[c]		,,
	32	7	—	22 (14–32)[c]		,,
	32	0–2	—	78 (77–81)	—	,,
	32	(−1)–(+1)	—	131 (128–134)[c]	—	,,
Theristus pertenuis (12)	32	17–22	—	23 (19–26)[c]	—	,,
	32	13–15	—	42 (35–47)[c]	—	,,
	32	9–12	—	47 (37–54)[c]	—	,,
	32	7	at least 30	72 (61–90)[c]	208[d]	,,
Oncholaimus brachycercus (13)	32	7	6[f]	399	♀♀ (541 & 610)[e] ♂: 580[d]	Gerlach & Schrage, 1972
Desmodora scaldensis (14)	32	7	8[f]	603	♀: >800[d] ♂: >750[d]	,,
Halichoanolaimus robustus (A)	32	7	—	>20 months	at least 855	,,
Monhystera denticulata (15)	13	5	8–19	197 (173–204)[g]	at least 330	Tietjen & Lee, 1972b
	13	15	12–21	36 (30–41)[g]	69	,,
	13	25	16–22	20 (18–24)[g]	37	,,
	26	5	10–17	180 (163–196)[g]	at least 330	,,
	26	15	18–24	18 (15–20)[g]	53	,,
	26	25	18–23	10 (8–12)[g]	34	,,
	39	15	15–23	34 (27–38)[g]	57	,,
	39	25	15–23	17 (15–22)[g]	29	,,
Chromadora macrolaimoides (16)	26	25	10 (9–18)	22 (18–25)[g]	45 (35–54)	Tietjen & Lee, 1973
Rhabditis marina (17)	15	35	—	1·75 (1·5–2)[g]	—	Hopper *et al.*, 1973
(at present called *Pellioditis marina* (Bastian, 1865) Andrássy, 1983 n. comb.)		33	—	1·5 (1–1·5)[g]	—	,,
		30	—	2 (1–2)[g]	—	,,
		24	—	2·25 (2–3)[g]	—	,,
		21	—	2·5 (2–3)[g]	—	,,
		18	—	4 (2–6)	—	,,
		12	—	8 (6–9)	—	,,
Diplolaimelloides sp.	15	37	—	7 (7–8)	—	,,
		35	—	4 (3–5)	—	,,
		33	—	4 (4)	—	,,
		30	—	4·5 (4–5)	—	,,
		24	—	7 (6–8)	—	,,
		21	—	9·5 (8–10)	—	,,
		18	—	14 (13–15)	—	,,
		12	—	30 (24–35)	—	,,

TABLE VIII—*continued*

Species (Code no.)	Salinity (°/oo)	Temp. (°C)	Egg number (average and/or min., max.)	Average min. generation time (days) (min.–max.)	Average life span (days)	Reference
Diplolaimella ocellata (18)	15	35	—	8 (7–11)	—	Hopper *et al.*, 1973
		33	—	6·5 (5–10)	—	,,
		30	—	6 (5–7)	—	,,
		24	—	11·5 (8–16)	—	,,
		21	—	12 (10–14)	—	,,
		18	—	43 (33–60)	—	,,
Enoplus paralittoralis (19)	15	31	—	27 (27)	—	,,
		28	—	21 (21)	—	,,
		24	—	22 (19–24)	—	,,
		21	—	41 (27–59)	—	,,
Oncholaimus sp.	15	33	—	23 (16–34)	—	,,
		30	—	20 (17–28)	—	,,
		24	—	29 (25–40)	—	,,
		21	—	39 (31–44)	—	,,
		18	—	86 (80–94)	—	,,
Haliplectus dorsalis (20)	15	35	—	35 (28–42)	—	,,
		33	—	26 (26)	—	,,
		30	—	27 (21–31)	—	,,
		24	—	34 (30–37)	—	,,
		21	—	70 (60–74)	—	,,
		18	—	112 (109–114)	—	,,
Rhabditis marina		room T	128 (70–260)	3–4[h]	—	Sudhaus, 1974
Rhabditis marina (B)	5	5	17[i]	57 (49–64)	—	Bergholz & Brenning, 1978
		9	27[i]	48 (42–57)	—	,,
		16	31[i]	30 (24–39)	—	,,
		25	43[i]	20 (14–25)	—	,,

Prochromadora orleji (21)	5	5	7[i]	112	—	„
		9	5[i]	87	—	„
		16	13[i]	54	—	„
		25	14[i]	46	—	„
Oncholaimus oxyuris (22)	18–23	5	13·5 (SE = 13·4)	570[j]	—	Heip *et al.*, 1978
		10	18·5 (SE = 6·0)	285[j]	—	„
		15	34·3 (SE = 10·9)	153 (SE = 4·3)	—	„
		20	35·6 (SE = 7·4)	114 (SE = 1·4)	—	„
		25	36·8 (SE = 8·3)	102 (SE = 2·9)	—	„
Monhystrella parelegantula (23)	30	25	—	8·9 (SE = 0·04; *N* = 539)	—	Vranken *et al.*, 1981
Diplolaimelloides bruciei (24)	26	5	—	—	—	Warwick, 1981a
		10	0·7[k]	>64	—	„
		15	3·7[k]	13·5	—	„
		20	4·9[k]	9	—	„
		25	5·2[k]	7·5	—	„
		30	7·1[k]	5·5	—	„
	1·75	20	1·2[k]	—	—	„
	8·95	20	2·3[k]	13	—	„
	17·5	20	2·6[k]	10·2	—	„
	26	20	4·9[k]	9	—	„
	35	20	2·9[k]	12·5	—	„
Chromadora nudicapitata (25)	?	15	8	19	28	Warwick, 1981b
		20	50	13	19	„
Eudiplogaster pararmatus (26)	5	12	8–10[l]	45	—	Romeyn *et al.*, 1983
		21		21		
Rhabditis marina (27)	20	25	600	4·5 (4–5; *N* = 47)	—	Vranken & Heip, 1983
Monhystera disjuncta (28)	30	17	—	♀♀: 10·9 (SD = 2·36; *N* = 226)	—	Vranken *et al.*, 1984a
		17	—	♂♂: 11·0 (SD = 2·68; *N* = 90)	—	„
Monhystera microphthalma (29)	20	20	—	♀♀: 10·2 (SD = 1·20; *N* = 113)	—	„
		20	—	♂♂: 11·2 (SD = 1·61; *N* = 107)	—	„
Monhystera disjuncta	30	17	—	♀♀: 8·8 (SD = 0·96; *N* = 209)[m]	—	„
		17	—	♂♂: 8·6 (SD = 0·87; *N* = 140)[m]	—	„
Monhystera disjuncta	30	3	180[n] (SE = 18·5)	♀♀: 52·3 (SD = 8·41; *N* = 287) ♀♀: 123*	—	Vranken *et al.*, unpubl.
		12	218[n] (SE = 31·9)	♀♀: 17·2 (SD = 4·53; *N* = 662) ♀♀: 49*	—	„
		17	187[n] (SE = 10·4)	♀♀: 10·9 (SD = 2·36; *N* = 226) ♀♀: 38*	—	„
		20	—	♀♀: 9·3 (SD = 2·21; *N* = 291)	—	„

temperatures T_{min} may be as low as 1·5 days (*R. marina* at 33 °C; Hopper, Fell & Cefalu, 1973) whereas at low temperatures development time may be more than 100 days (Gerlach & Schrage, 1971; Tietjen & Lee, 1972). Some species are adapted to low temperatures, *e.g. Monhystera disjuncta* develops at 0 °C and reproduces at 3 °C (Vranken, Herman & Heip, unpubl.). Extremely long generation times have been reported for *Desmodora scaldensis* and *Oncholaimus brachycercus*; at 7 °C the former species needs 566 days to reach sexual maturity, the latter approximately 300 days (Gerlach & Schrage, 1972), whereas *Halichoanolaimus robustus* could not even complete its life cycle within a period of 20 months. Some of these experiments, however, may have been in suboptimal conditions because the authors did not observe the nematodes feeding.

For many species development is more temperature-dependent in the lower temperature range than near the optimum. This is illustrated in Table IX, which shows Q_{10} values for different temperature intervals. As an example, *Monhystera disjuncta* has a $Q_{10} = 7$ in the interval 0–10·5 °C and a $Q_{10} = 1{\cdot}5$ in the interval 10·5–19·5 °C (Gerlach & Schrage, 1971). Most species show this higher response, but some have a more or less constant Q_{10}: *Theristus pertenuis* (Gerlach & Schrage, 1971), *Rhabditis marina* (Tietjen *et al.*, 1970; Bergholz & Brenning, 1978), and *Prochromadora orleji* (Bergholz & Brenning, 1978).

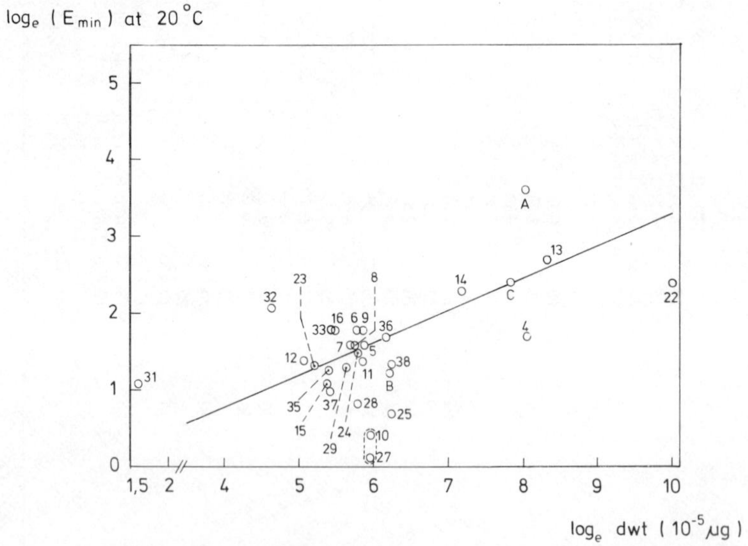

Fig. 10.—Relationship between embryonic development (E_{min}) at 20 °C and dry weight of the eggs (15% of wet weight): the weight of the cylindriform eggs was estimated by using Andrássy's (1956) formula, that of globular eggs was calculated by multiplying the volume (sphere) by the density 1·08 proposed by Andrássy (1956); morphometric data not quoted in the original experimental publications are from Ditlevsen (1911) for 22; Meyl (1954b) for 12; Man (1889), Rachor (1969), and Belogurova (1978) for 13; Luc & De Coninck (1959) for 14; Smol (unpubl.) for A; Vincx (unpubl.) for C = *Anticoma pellucida*; Vranken (unpubl.) for 23, 28, 29, 35, 36, 37, and 38.

Development rate reaches a maximum near some optimum temperature beyond which generation time is prolonged. This is the typical response for most meiofauna (Vernberg & Coull, 1981). The optimum temperature is near to 33 °C for the warm-water species cultivated by Hopper, Fell & Cefalu (1973) and near to 20 °C for *Chromadora nudicapitata* at 20°/₀₀ *S* (Heip *et al.*, 1985). To describe the relationship between temperature and development time the power equation $D = aT^b$ has been used by Heip, Smol & Absillis (1978) for *Oncholaimus oxyuris*, by Warwick (1981a) for *Diplolaimelloides bruciei* and by Heip *et al.* (1982) for reviewing the available information on life cycles. Heip *et al.* (1982) obtained a modal value of $b = -1{\cdot}8$, which is considerably higher than the *b*-values obtained for other groups such as calanoids and harpacticoids.

In Figure 11 all *b*-values for the nematode species cultivated up to now are compiled, with some omissions when the cultivation method was judged to be inadequate, and some of the unpublished data obtained by Vranken and co-

TABLE IX

Temperature coefficient, $Q_{10} = (v_1/v_2)^{10/(T_1-T_2)}$, calculated from development rates $v = 1/T_{min}$ of free-living marine nematodes, for different temperature intervals (figures in parentheses): temperature, T, is in °C and $T_1 < T_2$

Species	(°/₀₀)	Q_{10} (temperature interval)		Reference
Monhystera disjuncta	32	6·99 (0–10·5)	1·47 (10·5–19·5)	Gerlach & Schrage, 1971
Theristus pertenuis	32	2·16 (7–14)	2·99 (14–19·5)	Gerlach & Schrage, 1971
Monhystera denticulata	13	5·47 (5–15)	1·80 (15–25)	Tietjen & Lee, 1972
	26	10·00 (5–15)	1·80 (15–25)	Tietjen & Lee, 1972
	39	—	2·00 (15–25)	Tietjen & Lee, 1972
Rhabditis marina	15	3·64 (12–21)	1·28 (21–30)	Hopper *et al.*, 1973
Diplolaimella sp.	15	3·59 (12–21)	2·29 (21–30)	Hopper *et al.*, 1973
Rhabditis marina	5	1·79 (5–16)	1·57 (16–25)	Bergholz & Brenning, 1978
Prochromadora orleji	5	1·94 (5–16)	1·20 (16–25)	Bergholz & Brenning, 1978
Oncholaimus oxyuris	18–23	3·73 (5–15)	1·50 (15–25)	Heip *et al.*, 1978
Diplolaimelloides bruciei	26	>7·11 (5–15)	1·64 (15–25)	Warwick, 1981a
Monhystera disjuncta	30	3·44 (3–12)	2·16 (12–20)	Vranken *et al.*, unpubl.

workers. The b-values are highly heterogeneous ($F = 7.27$; d.f. $= 18, 48$; $P < 0.001$), so that the use of a single b-value for marine nematodes is invalid on statistical grounds (see Zaika & Makarova, 1979). The bars around the b-values indicate 95% comparison intervals (Gabriel's T-method in Sokal & Rohlf, 1981), when intervals do not overlap the b's are significantly different. Some interesting features emerge from this figure. *Monhystera disjuncta*, a truly marine species, has a much lower b-value than the other species, most of which are estuarine or brackish-water forms. At the other extreme we find the tropical species cultured by Hopper, Fell & Cefalu (1973). These have statistically higher b-values than those inhabiting temperate regions. The monhysterids, with the exception of *Monhystrella parelegantula*, have intermediate b-values, between -2.01 for *Monhystera microphthalma* at 30‰ S and -1.67 for the same species at 11‰ S. *Monhystrella parelegantula* occupies a peculiar position; the influence of temperature on development time is as strong as for the warm-water species. *M. parelegantula* is a brackish-water species with a wide salinity tolerance and one of the dominant species in inland salinas in Germany (Meyl, 1954a; Paetzold, 1955, 1958). Its basal temperature $T_0 = 14$ °C and a high developmental acceleration is necessary to achieve a high productivity and, therefore, survival in the temporary habitats where it lives (Vranken *et al.*, 1981). Temperature appears to have little effect on *Monhystera disjuncta*; as mentioned, it reproduces at 3 °C and was found in Antarctica (Viglierchio, 1974). *Rhabditis marina* has an intermediate b-value

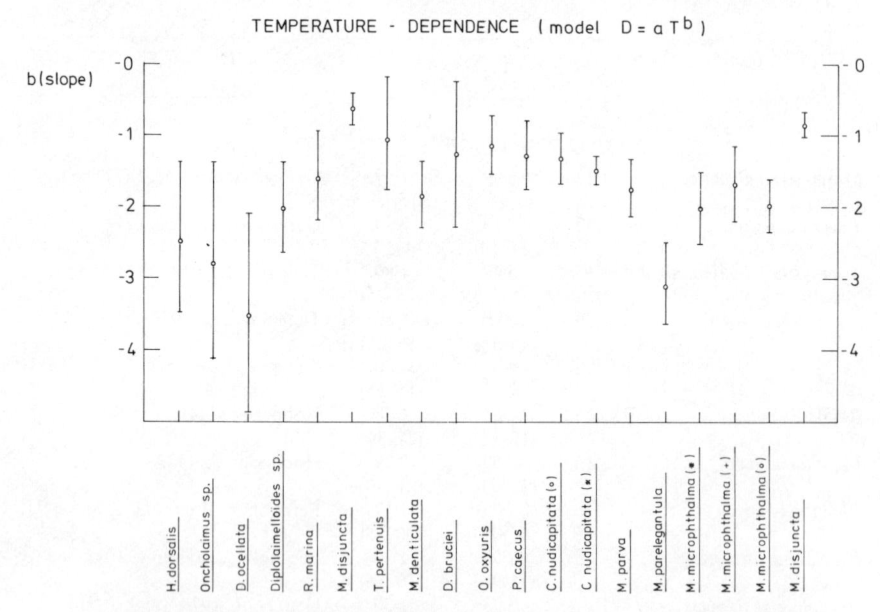

Fig. 11.—Values of b in the equation $T_{min} = aT^b$, with T_{min} = minimum generation time and T = temperature (°C): sources are quoted in Table XI; for *R. marina*, the b-value obtained from the data of Hopper, Fell & Cefalu (1973) is shown; (+) = 11‰ S, (○) = 20‰ S; (*) = 30‰ S; b-values for *P. caecus* and *C. nudicapitata* are from Heip *et al.* (1985); the unquoted data are unpublished results from Vranken and co-workers.

(−1·59). It has a short development time in its lower temperature range and its acceleration rate is limited. The remaining species, mainly chromadorids and *Oncholaimus oxyuris*, have smaller (statistically not significant) *b*-values than the monhysterids and much smaller *b*-values than the warm-water species. Some relationship between temperature-dependency and geographic occurrence appears to exist. The warm-water species have the highest *b*-values.

Some precaution, however, is necessary when comparing *b*-values of different species. This is because the power equation as a temperature model is a simplification of Bělehrádek's function, $T_{min} = a(T - T_0)^b$, with T_0 the basal temperature. In the power equation T_0 is fixed at zero degree Celsius. Therefore, if Bělehrádek's function is the underlying temperature model, it is to be expected that *b* obtained in the power equation highly correlates with T_0. This makes comparison of the *b*-values (power equation), without knowledge of T_0 somewhat difficult. Nevertheless when regressing T_{min} against the effective temperature, extreme opportunists tend to have the higher *b*-values (Vranken, unpubl.).

Influence of salinity

Only a few studies (Tietjen & Lee, 1972, 1973; Warwick, 1981a) report on the combined effect of temperature and salinity on generation time. In *Monhystera denticulata*, generation time was almost doubled by decreasing and by increasing the optimum salinity of 26‰ *S* by 13‰ *S*, both at 15 and 25 °C. At 5 °C, development rates were nearly equal at 13‰ *S* and 26‰ *S*. For *Chromadorina germanica*, population growth, measured as *r*, was maximal at 26‰ *S* and, within the optimum temperature range of 20–30 °C, a drop to 13‰ *S* caused a doubling of the generation time. Outside the optimum temperature range, the difference increased. Tietjen & Lee (1977a) concluded that both species are better adapted to the middle and upper regions of their salinity tolerance range than to the lower part. Mortality, however, was higher at high salinity than at low salinity.

Croll & Viglierchio (1969) claim that *Deontostoma californicum* is able to osmoregulate in hypertonic solutions but not in hypotonic solutions. These results have been questioned by Wright & Newall (1976, 1980) who report on the osmoregulation of *Enoplus communis* and *E. brevis*; especially the latter, brackish-water dwelling species, is able to regulate its body volume in both hyper- and hypotonic solutions. As calcium is required to maintain the normal permeability of membranes, the single salt solutions used by Croll & Viglierchio (1969) and Viglierchio (1974) cause artifacts. Evidence for active osmoregulation in nematodes is absent (Wright & Newall, 1980).

That generation time increases with salinity has been shown for *Diplolaimelloides bruciei* by Warwick (1981a) at both sides of the optimum of 26‰ *S* for this species; Vranken & Van Brussel (unpubl.) found the same effect for *Monhystera microphthalma* on both sides of the optimum 20‰ *S*, although the slope (*b*) did not change.

The number of annual generations

A further step in many papers dealing with the influence of temperature and salinity on development is to estimate the number of generations the population would produce in the field given comparable situations would exist. As many nematodes have continuous reproduction it is often impossible to obtain this information from field studies.

For a number of chromadorid and monhysterid species estimates obtained from laboratory experiments exist. The following annual numbers of generations have been calculated: 1·6 for *Oncholaimus oxyuris* in the Dievengat, a brackish-water pond in Belgium (Heip, Smol & Absillis, 1978); 5 to 5·5 generations for *Monhystrella parelegantula* in the Sluice Dock of Ostend, Belgium (Vranken *et al.*, 1981); 5 generations for *Prochromadora orleji* in the Barther Bodden, G.D.R. (Bergholz & Brenning, 1978); a maximum of 7 and probably nearer to 5 generations for *Theristus pertenuis* (Gerlach & Schrage, 1971); 10 generations for *Rhabditis marina* in the Barther Bodden, G.D.R. (Bergholz & Brenning, 1978); 13 generations for *Chromadorina germanica* in North Sea Harbor, New York (Tietjen & Lee, 1972); an average of 15 and a maximum of 21 for *Monhystera denticulata* in North Sea Harbor, New York (Tietjen & Lee, 1972); 17 generations for *Diplolaimelloides bruciei* in the Lynher estuary (calculated by Heip, Herman & Coomans, 1982, from Warwick, 1981a); and a maximum of 17 and a mean of 12 generations for *Monhystera disjuncta* in the North Sea (Gerlach & Schrage, 1971).

In fact, all these calculations over-estimate the mean annual number of generations. Nematodes cultured in the laboratory immediately start reproduction and reproduce continuously over a relatively long time in comparison with their pre-reproductive life. The time elapsed between two identical stages in two successive generations 'egg to egg' or 'female to female', roughly equals development time and the time necessary to produce the median egg has to be added to obtain a good estimate of mean generation time (Laughlin, 1965). The above calculations are valid estimations of the number of juvenile periods realized each year (see Taylor, 1981, for a definition).

As most species cultivated belong to the group of fast-growing nematodes it remains premature to speculate upon a model minimum generation number of nematodes in the field. Moreover, the data in many papers in this field are sometimes difficult to interpret and calculations erroneous. Important information that is often withheld is replication of the experiment, mortality in cultures, the number of individuals studied; even such basic statistics as the standard error or deviation are not always found in the paper.

Reproductive potential

The intrinsic rate of natural increase r_m (also called Malthusian parameter and the innate capacity for increase) is the actual rate of increase when the population has attained a stable age distribution, in the absence of competition or predation. When this stable age distribution is not reached, every actual rate of increase will depend on age structure of the population; therefore, only r_m is a good measure of the capacity of the species to increase when conditions are favourable and only r_m should be used in interspecies comparisons.

Three features of the life history are of paramount importance in determining reproductive potential: development rate, fecundity and longevity. We have shown how generation time is dependent on temperature and salinity. Food is of obvious importance, but there are no data for marine nematodes.

For the soil-dwelling *Caenorhabditis briggsae* development time increased three times at food densities two orders of magnitude higher (Schiemer, 1982b), and food quality was also important.

Temperature has an important effect on fecundity. The number of eggs produced per female is often higher at higher temperatures (Tietjen & Lee, 1972; Bergholz & Brenning, 1978; Heip, Smol & Absillis, 1978; Warwick, 1981a). For two species, *Oncholaimus oxyuris* and *Diplolaimelloides bruciei*, a linear relationship between daily egg production and temperature has been demonstrated (Table X). Similarly, in *Monhystera microphthalma* daily egg production increased from 2·4 eggs·fem.$^{-1}$·day^{-1} at 15 °C to 12·8 eggs·fem.$^{-1}$·day^{-1} at 25 °C (Vranken, unpubl.). At 30 °C the species still produced 11·7 eggs·day^{-1}. These data also indicate that the effect of temperature is dependent on species; *M. microphthalma* is very temperature-dependent, *Diplolaimelloides bruciei* intermediate and *Oncholaimus oxyuris* is only little affected.

Most literature on longevity is casual. Many authors do not even mention how it was measured and often a very small number of individuals is followed, producing results that have no basis for generalization. Again, temperature has a profound influence on longevity, and Tietjen & Lee (1972) report for *Monhystera denticulata* 330 days at 3 °C compared with 30–40 days at 25 °C. In *M. disjuncta*, mean longevity of adult females decreased from 175 days at 3 °C to 49 days at 17 °C (Vranken, Herman & Heip, unpubl.).

Temperature, salinity, and food thus have a profound influence on the reproductive potential (Table XI) (Tietjen & Lee, 1977a; Heip, Smol & Absillis, 1978; Alongi & Tietjen, 1980; Warwick, 1981a; Vranken *et al.*, 1984a). In four species, a linear increase of r_m with temperature has been observed (Table XII), with a large slope for *Diplolaimelloides bruciei, Chromadorina germanica* and *Monhystera microphthalma* and a small slope for the large *Oncholaimus oxyuris*. Beyond a certain optimum temperature, the reproductive potential is depressed (Tietjen & Lee, 1977a). As for temperature, there exists a salinity optimum with depression of the reproductive potential at both higher and lower salinities (Tietjen & Lee, 1977a; Warwick, 1981a; Romeyn, Bouwman & Admiraal, 1983). The importance of food has been demonstrated by Alongi & Tietjen (1980): r-values ranging between 0 and 0·12 have been reported from *Chromadorina germanica* and *Diplolaimella* sp. grown on different diets; with *Monhystera disjuncta* the range was somewhat smaller. On the other hand, Findlay (1982a) found that the carrying capacity K was a straightforward function of ration and type of detritus given to *Diplolaimella chitwoodi*.

TABLE X

Relation between daily egg-production and temperature

Species	a	b	R^2	Reference
Oncholaimus oxyuris	−0·3050	0·0834	0·9727	Heip *et al.*, 1978
Diplolaimelloides bruciei	−1·429	0·286	0·951	Warwick, 1981a

TABLE XI

Reproductive potential, r (per day) of free-living marine nematodes as a function of temperature, salinity and food: a, r-values read from original figure

Species	Temperature (°C)	Salinity (‰)	Food	r (per day)	Reference
Chromadorina germanica[a]	13	26	*Chlorococcum* sp. and *Cylindrotheca closterium*	0·038	Tietjen & Lee, 1977a
	17·7	26	,,	0·116	,,
	21	26	,,	0·164	,,
	23	26	,,	0·179	,,
	30	26	,,	0·185	,,
	33	26	,,	0·075	,,
	19·5	39	,,	0·115	,,
	24·8	39	,,	0·164	,,
	30	39	,,	0·137	,,
	31·2	39	,,	0·070	,,
	19·5	13	,,	0·051	,,
	26	13	,,	0·120	,,
	30	13	,,	0·078	,,
Oncholaimus oxyuris	5	18–23	*Panagrellus redivivus* and	0·003	Heip, Smol & Absillis, 1978
	10	18–23	bacteria, algae	0·009	,,
	15	18–23	,,	0·015	,,
	20	18–23	,,	0·022	,,
	25	18–23	,,	0·029	,,
Chromadorina germanica	23	26	*Pseudomonas* sp. 1	0·064	Alongi & Tietjen, 1980
	23	26	*Pseudomonas* sp. 2	0·058	,,
	23	26	*Dunaliella* sp.	0	,,
	23	26	*Nitzschia* sp.	0·007	,,
	23	26	*Cylindrotheca closterium*	0·118	,,
Diplolaimella punicea	23	26	*Pseudomonas* sp. 1	0·096	Alongi & Tietjen, 1980
	23	26	*Pseudomonas* sp. 2	0·086	,,
	23	26	*Dunaliella* sp.	0·117	,,
	23	26	*Nitzschia* sp.	0	,,
	23	26	*Cylindrotheca closterium*	0	,,

Monhystera disjuncta	23	26	*Pseudomonas* sp. 1	0·099	Alongi & Tietjen, 1980
	23	26	*Pseudomonas* sp. 2	0·093	„
	23	26	*Dunaliella* sp.	0·098	„
	23	26	*Nitzschia* sp.	0	„
	23	26	*Cylindrotheca closterium*	0	„
Diplolaimelloides bruciei	10	26	Bacteria (unidentified)	0·034	Warwick, 1981a
	15	26	„	0·172	„
	20	26	„	0·231	„
	25	26	„	0·244	„
	30	26	„	0·311	„
	20	1·75	„	0·056	„
	20	8·95	„	0·107	„
	20	17·5	„	0·118	„
	20	26	„	0·231	„
	20	35	„	0·135	„
Diplolaimella chitwoodi	20	25	Detritus from *Ulva fasciata*	0·280	Findlay, 1982a
	20	25	Pablum (mixed cereal)	0·230	„
	20	25	Detritus from *Gracilaria foliifera*	0·280	„
	20	25	Detritus from *Spartina alterniflora* (S-13)	0·093	„
	20	25	Detritus from *Spartina alterniflora* (S-1)	0·302	„
	20	25	Detritus from *Thalassia testudinum*	0	„
Eudiplogaster pararmatus[a]	17	0·5	*Navicula salinarum*	0·052	Romeyn *et al.*, 1983
	17	2·5	„	0·060	„
	17	5	„	0·057	„
	17	10	„	0·028	„
	17	20	„	0·008	„
Rhabditis marina	25	20	Bacteria	0·914	Vranken & Heip, 1983
Monhystera microphthalma	15	20	Bacteria	0·102	Vranken *et al.*, 1984a
	20	20	„	0·277	„
	25	20	„	0·268	„
	30	20	„	0·350	„
Monhystera disjuncta	3	30	Bacteria	0·102	Vranken *et al.*, 1984a
	12	30	„	0·121	„
	18	30	„	0·131	„

Banse (1982) studied the relationship between r_m and body weight for small metazoans, including nematodes. The optimal growth rates of meiobenthic nematodes were markedly lower than those of other nematodes (including *Caenorhabditis briggsae* and *Plectus palustris*) and other small invertebrates. Banse predicted an allometric relationship between r_m and wet weight, with an exponent "well below −0·25". On the other hand, we have the general value of this exponent −0·274 calculated by Fenchel (1974) for heterotherm metazoans. To re-examine the case for nematodes, we have re-calculated this expression using all literature data on r (Table XIII) (not r_m because most values published are only approximations of r_m for reasons given above). Only the figures for *Diplolaimelloides bruciei* (Warwick, 1981a), *Chromadora nudicapitata* (calculated by us from data in Warwick, 1981b), *Caenorhabditis briggsae* (Schiemer, 1982b), and *Rhabditis marina* (Vranken & Heip, in press) are real r_m values. The values quoted from von Thun (1968) relate to capacities for increase r_c (Laughlin, 1965) and are under-estimations, the more so since in recent years it has become clear that marine nematodes may produce a much higher number of eggs than previously assumed. *Monhystera denticulata* produces 60 eggs in a life time according to Tietjen & Lee (1972) whereas *M. disjuncta* lays approximately 200 eggs (Vranken, Herman & Heip, unpubl.); *Rhabditis marina* produces 600 eggs per female in only six days (Vranken & Heip, 1983), and a single observation of 800 eggs per female for this species exists (Vanderhaeghen, pers. comm.). For these reasons we excluded the values from von Thun (1968); the rhabditids were excluded for the same reasons as given previously (see p. 442), although some of the monhysterids also have high r-values: *Monhystera microphthalma*, $r = 0{\cdot}350{\cdot}\text{day}^{-1}$ at 30 °C and 20‰ S and *M. disjuncta*, $r = 0{\cdot}260{\cdot}\text{day}^{-1}$ at 15 °C and 30‰ S (Vranken, unpubl.).

There are also some problems with the other literature data, but we calculated nevertheless a linear least square regression line (model I) through

TABLE XII

Relation between the intrinsic rate of natural increase and temperature: CI, 95% confidence interval

Species	a	b	R^2	Reference
Oncholaimus oxyuris	−0·0042	0·0013 (CI:0·00005)	0·9996	Heip *et al.*, 1978
Diplolaimelloides bruciei	−0·067	0·0134 (CI:0·0042)	0·953	Warwick, 1981a
Chromadorina germanica	−0·1453	0·0144 (CI:0·0051)	0·988	Tietjen & Lee, 1977a
Monhystera microphthalma	−0·0815	0·0147 (CI:0·0217)	0·8196	Vranken *et al.*, 1984a (*logistic function*)
Monhystera microphthalma	−0·1018	0·0146 (CI:0·0157)	0·896	Vranken *et al.*, 1984a (*exponential*)

TABLE XIII

Relation between r-value and biomass: a, r-value computed as r_c (Laughlin, 1965); b, obtained by graphic interpretation; c, corrected to 20 °C assuming a linear relationship between r_m and temperature (Heip, 1977) with $r_m = 0$ at 0 °C; d, real r_m calculated with Lotka's formula

Species	$r \cdot day^{-1}$ at 20 °C	Wet weight (μg)	Reference
Adoncholaimus thallasophygas[a]	0·032	9·8	von Thun, 1968
Chromadorita tenuis[a]	0·072	1·3	von Thun, 1968
Diplolaimella ocellata[a]	0·068	0·2	von Thun, 1968
Diplolaimelloides oschei[a]	0·085	0·2	von Thun, 1968
Diplolaimelloides islandica[a]	0·068	0·2	von Thun, 1968
Monhystera disjuncta[a]	0·076	0·85	von Thun, 1968
Chromadorina germanica[b]	0·15	0·28	Tietjen & Lee, 1977a
Oncholaimus oxyuris	0·022	20	Heip *et al.*, 1978
Diplolaimella punicea[c]	0·118	1·13	Alongi & Tietjen, 1980
Monhystera disjuncta[c]	0·116	1·76	Alongi & Tietjen, 1980
Plectus palustris	0·28	1·5	Schiemer *et al.*, 1980
Diplolaimella chitwoodi	0·27	0·67	Findlay, 1982a
Diplolaimelloides bruciei	0·231	0·45	Warwick, 1981a
Caenorhabditis briggsae[d]	1·136	0·50	Schiemer, 1982b
Eudiplogaster pararmatus[c]	0·071	1·1	Romeyn *et al.*, 1983
Rhabditis marina[c]	0·731	1·1	Vranken & Heip, 1983
Monhystera microphthalma	0·277	0·6	Vranken *et al.*, 1984a
Monhystera disjuncta	0·116	0·32	Vranken, unpubl.
Chromadora nudicapitata[d]	0·209	0·71	Warwick, 1981b
Mesodiplogaster lheritieri[c]	1·158	7·3	Grootaert, 1976
Labronema vulvapapillatum[c]	0·067	10·7	Grootaert & Small, 1982

the existing data, which resulted in the following equation: $\log_e r_m = -1{\cdot}959$ (SE $= 0{\cdot}154$)$-0{\cdot}429$ (SE $= 0{\cdot}122$) $\log_e$ wet wt, WW (μg) ($F = 12{\cdot}4$, d.f. $= 1, 10$; $0{\cdot}005 < P < 0{\cdot}01$). This shows that weight-dependency of r_m in marine nematodes is about 1·6 times higher than calculated by Fenchel (1974) (although the values are statistically not different). A high correlation, $r = -0{\cdot}74$, exists between body weight and reproductive potential, large nematodes have a lower reproductive potential than small nematodes. The equation demonstrates that indeed the reprodutive potential of nematodes is lower than that of similarly sized heterothermic animals. Only the rhabditids realize approximately the r right for their size, the large *Oncholaimus oxyuris* has r about 20 times too low.

Calculation of the intrinsic rate of natural increase

As mentioned, the calculations above suffer from the fact that the literature data are not very precise. The exact calculation of r_m requires the construction of an age-specific fecundity (m_x) and survival (l_x) table from which r_m can be

TABLE XIV

Life-history parameters of Rhabditis marina *and* Chromadora nudicapitata: *all parameters related to generation time (T_c, T, $\bar{T}$ and T_{min}) are in days (d); the capacity of increase r_c and the intrinsic rate of natural increase, r_m, are in reciprocal days (day^{-1}); R_0 is the net reproductivity* (i.e. *the multiplication rate per generation*)

	Species	
	Rhabditis marina	*Chromadora nudicapitata*
Temp. (°C)	25	20
R_0	400	26
r_c	0·837	0·205
r_m	0·914	0·209
T_c	7·2	15·9
T	6·6	15·5
$\bar{T}$	6·1	15·1
T_{min}	4·5	12·5
Reference	Vranken & Heip, 1983	Warwick, 1981b

calculated with Lotka's formula:

$$\sum_{0}^{\infty} e^{-r_m x} l_x m_x = 1$$

with x the pivotal age of the different age groups (Ricklefs, 1973; Krebs, 1978). r_m can also be calculated from growth experiments when the initial population structure is the stable age distribution, *i.e.* when all the different age groups occur in the appropriate proportions (see Krebs, 1978, for the calculation of the stable age distribution). Then r_m can be calculated from $r_m = 1/t \ln N_t/N_0$, provided the stable age distribution is maintained (no heterogeneity around the regression line).

There exist only two life tables for two brackish water nematodes: *Rhabditis marina* (Vranken & Heip, in press) and a graphic one for *Chromadora nudicapitata* (Warwick, 1981b). They are shown in Table XIV, where several measures of generation time are calculated (see Pielou, 1977; Southwood, 1978). These parameters all give substantially longer values than the minimum generation time or development time T_{min}.

ENERGY FLOW

FEEDING

Much of the literature on feeding and indeed general ecology of marine nematodes is based on a paper by Wieser (1953) who classified nematodes in feeding types according to the morphology of their buccal cavity. Indeed, although the organization of free-living nematodes is quite uniform, a great diversity in buccal structures exists which bears on ecological relationships and the niche of the species.

Wieser made a division into four feeding types in two groups; Group 1 contains those species in which the buccal cavity is unarmed, Group 2 species have an armed buccal cavity. Type 1A (Fig. 12) contains those species where there is no real buccal cavity, although some species where it is small are included. These species supposedly pick up small food particles, such as bacteria, selectively. Type 1B (Fig. 12) contains those species in which the unarmed buccal cavity is wide and they supposedly feed non-selectively on deposits. Type 2A (Fig. 12) contains species which are presumably herbivorous.

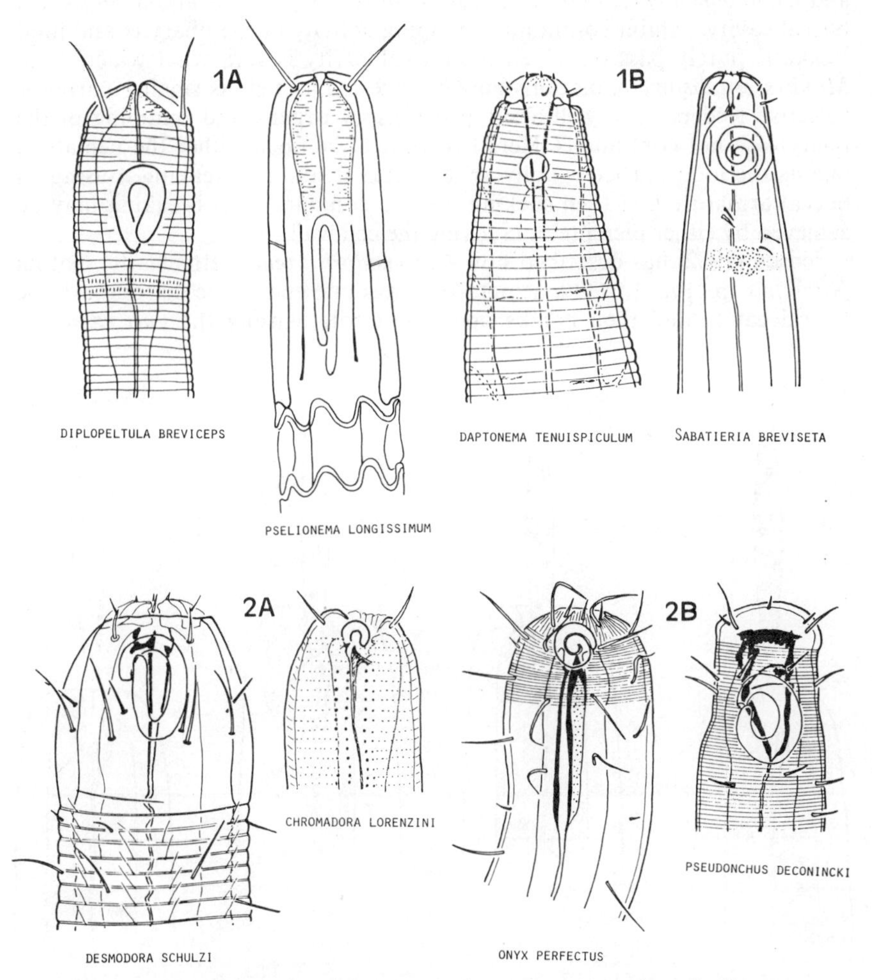

Fig. 12.—Heads of some marine nematode species arranged per feeding type (see text p. 459): note *C. lorenzini* should read *C. lorenzeni.*

Type 2B species (Fig. 12) have wide buccal cavities and glands opening on teeth, and they are supposedly carnivores or omnivores. This rather simple scheme has been used over and over again and has obtained the status of a paradigma in marine nematology. Wieser (1953) himself, however, warned against such over-simplications: ". . . [das Man] ein Einteilung mit ökologischer Zielsetzung niemals ausschlieslich auf morphologischen Fakten aufbauen kann." When writing in 1953, Wieser remarked that there were no observations on solid particles in the guts of 1A animals, but that diatoms had often been observed in 1B and 2A animals. Since then anecdotal observations on the gut contents of nematodes have been numerous; most earlier data were summarized by von Thun (1968). The scheme itself has often been discussed but seldom questioned and only since Romeyn & Bouwman (1983) has some original contribution appeared. They considered oesophageal pumping rates as relevant to the problem. Non-selective deposit-feeders, such as *Monhystera* and *Diplolaimelloides*, that have very small sensory organs and a very small buccal cavity, exhibit continuous pumping activity of the pharynx and food intake is purely passive. When the buccal cavity is somewhat wider, as in *Monhystera disjuncta*, diatoms may be ingested as well as smaller particles. Selective feeders have well-developed sensory organs and pumping of the pharynx is not continuous. Here two routes are open: either the nematode ingests the food particle wholly or it attacks it first by breaking it using its buccal armature and then sucking out the contents. This breaking may be achieved by either piercing or cracking the cell wall.

Jensen (1982) has described how *Chromadorita tenuis* attacks the diatom *Nitzschia* sp. (Fig. 13). The nematode brings one end of the diatom into the buccal cavity and then breaks open the girdle causing the two valves to

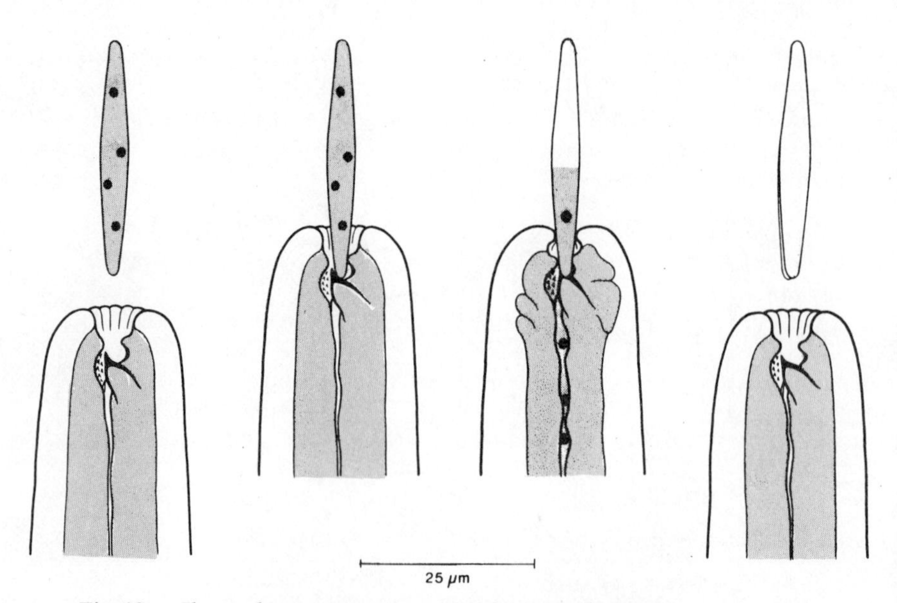

Fig. 13.—*Chromadorita tenuis*: sequence of events when the nematode feeds on the diatom *Nitzschia* sp.; after Jensen (1982).

separate, after which it sucks out the content by one or two pumping movements of the oesophagus. The gut never contained fragments of frustules of *Nitzschia*.

A disadvantage of the Wieser scheme is that it confines nematode species to a single trophic status. Non-selective feeders are in a sense omnivores; although the particle size of the food is restricted, it may include bacteria, non-living aggregates, small flagellates *etc.* Many selective feeders, either with an unarmed or armed buccal cavity, seem to be diatom-feeders, but diatoms are only numerous in shallow sediments and very little is known about the feeding of the abundant deep-sea nematodes.

An example of a 2B species is *Oncholaimus oxyuris*. Older juveniles and adults were fed with another nematode, *Panagrellus silusiae*, in cultures by Heip, Smol & Absillis (1978). It also behaves as a scavenger feeding on dead *Hydrobia ulvae*, moving into the shells of the gastropod. Juveniles and adults also concentrate around the tracks made by the polychaete *Nereis diversicolor*, where organic matter is concentrated and there is a dense growth of microorganisms. Besides a scavenger and bacterivore, it is a predator and adults, especially females, attack in small groups of 2–4 individuals oligochaetes several times longer than themselves. They pierce through the body wall and move into the body of their victim (Vranken, pers. obs.).

Bacteria and algae

Feeding of nematodes in cultures has been very important in obtaining more reliable information on their food requirements. Many species can be cultured on bacteria and on algae. Tietjen & Lee (1973) tested 20 species of algae as food for *Chromadora macrolaimoides* but only 11 were ingested in sufficient quantities and only five were able to sustain growth for many generations. Only two sustained growth indefinitely. That nematodes may be highly selective was also demonstrated in other experiments by Lee, Tietjen, Mastropolo & Rubin (1977) who offered pure cultures of different algal species to nematodes in the field and to selected nematode species in the laboratory in so-called cafeteria experiments (see also Trotter & Webster, 1984). Both these experiments again clearly demonstrated that selective recruitment to patches of particular algae took place.

For bacteria the same selectivity holds. Three species of bacteria out of sixteen tested were not taken up at all in an experiment by Tietjen & Lee (1973). Vranken *et al.* (1984a) report how of 11 species tested only one permitted optimal growth of a *Monhystera* species.

The few data existing on meiofauna grazing on microflora have been summarized by Tietjen (1980a). These studies indicate that the consumption of bacteria and algae is in the order of 0·1 to 10 μg dry wt·animal^{-1}·day^{-1}. For three nematode species ingestion varied from 14×10^{-2} μg C·day^{-1} (*M. disjuncta*), 40×10^{-2} μg C·day^{-1} (*Chromadorina germanica*), and 60×10^{-2} μg C·day^{-1} (*Rhabditis marina*). The ingestion rate of the diatom-feeder *Eudiplogaster pararmatus* was studied by Admiraal, Bouwman, Hoekstra & Romeyn (1983) by radioactive labelling of diatoms. Six or seven diatoms were taken each hour, amounting to a total daily carbon intake of 0·17 μg C. Each nematode, with a carbon content of 0·10 μg C, thus ingests about double its body carbon each day.

The effects of nematodes grazing on bacteria and algae have also been evaluated by Tietjen (1980b) and Admiraal *et al.* (1983). Direct effects on bacterial or algal biomass are difficult to estimate: a nematode community with a standing stock of 0·3 g C·m^{-2} would consume about 0·6 g C·m^{-2}·day^{-1}, or about 220 g C·m^{-2}·yr^{-1}. This is a very large figure compared with the annual C-input into most benthic systems. Admiraal *et al.* (1983) concluded that herbivorous nematodes are unable to decrease algal populations on tidal flats in The Netherlands. These benthic diatoms had a maximum production of nearly 2 g C·m^{-2}·day^{-1} of which at most 100 mg C was grazed by the meiofauna. This estimate may be too low, since daily carbon intake by nematodes was estimated as 0·17 μg C. A standing stock of 1×10^6 herbivorous nematodes would thus consume 170 mg C·day^{-1}.

The effects of nematodes on bacterial dynamics were measured by direct heat flow measurements by Pamatmat & Findlay (1983). When food was absent, sand with 7000 nematodes showed barely detectable heat production. When bacteria were added, heat production started to increase after 6 hours from a basic level of 0·1 mJ·s^{-1} to a peak of 0·7 mJ·s^{-1} 7 hours later. Then heat production levelled off again to a slightly higher level than the production with bacteria alone. The peak level of 0·7 mJ·s^{-1} was also reached without nematodes, but four hours earlier. Nematodes thus seem to affect bacterial population growth by prolonging the lag phase and by reducing the total amount of energy flow, although some inhibition may have been due to metabolites excreted by the nematodes.

Indirect effects of grazing have been summarized by Tietjen (1980b). Burrowing may induce advection of nutrients and oxygen within the sediments (Cantelmo, 1978, cited in Tietjen, 1980b). Tietjen (1980b) provides evidence that meiofauna may contribute to nutrient regeneration; the addition of nematodes to experimental flasks in which algae were grown axenically provoked significant increases in the levels of orthophosphate in the absence of bacteria. The secretion of mucus by producing slime trails may attract and sustain bacterial growth (Riemann & Schrage, 1978; Warwick, 1981b).

When fresh sediments are observed through a stereo-microscope nematodes can be seen gliding along an intricate network of thread-like burrows reinforced by mucus secretions (Cullen, 1973). This mucus may trap organic particles and adsorb macromolecules and thus act as substrate for the growth of microorganisms (Riemann & Schrage, 1978). This mucus-trap hypothesis has been considered as a gardening mechanism by Gerlach (1978), and this has been confirmed by Warwick (1981b) who found that the trails of *Praecanthonchus punctatus* induced the growth of very dense monospecific populations of the non-motile phase of the flagellate *Tetraselmis*. The presence of the nematode induced the formation of these non-motile cells; in cultures without nematodes all the cells remained motile. It is possible that such mechanisms are widespread but nothing is known of their occurrence in nature. More commonly observed is the growth of algae and bacteria on the body wall or within the cuticular ornamentation of the nematode itself.

Of particular interest is the occurrence of mouthless and gutless nematodes in anoxic or microxic environments. Hope (1977), Hope & Murphy (1969), and Vitiello (1970) described the occurrence of bacteria-like structures in deep-sea nematodes. Ott, Rieger, Rieger & Enderes (1982) describe the mouthless and gutless nematode *Astomonema jenneri* from polychaete tubes of an intertidal mud-flat in North Carolina. These animals contained two types of micro-organisms in the rudiment of the gut and they occurred in near anoxic or microxic sediments rich in organic matter and hydrogen sulphide.

Detritus

The utilization of detritus by nematodes has been only rarely studied. The input of higher plants and macroalgae is an important carbon source in coastal systems. The subsequent mineralization of this detritus is predominantly a bacterial process and the impact of nematodes grazing on bacteria has been postulated to exert an effect on mineralization rates in the benthos (Johannes, 1965). This has been studied by Findlay & Tenore (1982) who investigated the rates of carbon mineralization in the presence or absence of the nematode *Diplolaimella chitwoodi.* They looked at detritus derived from a vascular plant, *Spartina,* and a red algae, *Gracilaria.* In controls without nematodes the mineralization rate amounted to 25 μg C·h^{-1}, with less than 6% of the initial dry weight being mineralized in one week. With ten nematodes ·10 cm^{-2} the maximum mineralization rate was doubled and with 50 to 1500 nematodes·10 cm^{-2} it was more than three times higher, and 9–10% of the initial dry weight was mineralized within one week. This increase in mineralization rate occurred independent of particle size. *Spartina*-detritus appeared to be more difficult to attack and, although the increase in mineralization rate in the presence of nematodes was about 50%, the amount mineralized after one week was not much different without and with nematodes present (4·8 and 5·8%, respectively). The effect of nematodes on mineralization rates thus seems to be significant, but nematodes only account for 4% of this mineralization directly. Apparently their main action is to increase bacterial metabolism. The impact of this on the benthic system could vary: for directly available detritus, such as derived from seaweeds, increased mineralization would decrease overall production of the system. For more refractory detritus the effect of nematodes may be to enhance production.

Detritus as a food source has been studied for the same species *Diplolaimella chitwoodi* by Findlay (1982a). The detritus was either dead *Spartina* or *Thalassia* from the field or *Gracilaria, Ulva,* and *Spartina* from cultures, and was inoculated with mixed cereals containing bacteria. Two population parameters were tested, the rate of increase *r* and the carrying capacity *K*. The rate of increase of the nematode proved to be less dependent on detritus quality, but the carrying capacity *K* is a straightforward function of detrital type and ration and nitrogen input was a good predictor of *K*. Findlay (1982a) hypothesized that systems receiving pulses of directly available detritus should exhibit population fluctuations closely tied to the rate of food supply. Conversely, systems receiving pulses of not directly available detritus should exhibit fewer sporadic fluctuations. Many detritus systems are nitrogen-limited and nitrogen input seems to be the best measure of food available to the benthos.

Uptake of dissolved organic matter

Uptake of dissolved organic matter has been reported for two oncholaimids, *Pontonema vulgaris* (Chia & Warwick, 1969) and *Adoncholaimus thalasso-*

phygas (Lopez, Riemann & Schrage, 1979). *A. thalassophygas* consistently failed to incorporate labelled microorganisms and consistently succeeded in incorporating labelled glucose. It was suggested that hatched juveniles feed primarily on dissolved organic matter released by microbial activity and that adults retain this ability but supplement their diet by scavenging and predation. *Rhabditis marina*, however, neither absorbed dissolved organic matter through its body wall nor did it ingest this material through the gut (Tietjen & Lee, 1975). Uptake of the label occurred when particulate material (small latex spheres) was added to the culture, but high levels of organic matter were used in this experiment.

PRODUCTION

There does not exist a single estimate of the production of a marine nematode in the field. Studies on the energy flow through nematode populations (Faubel, Hartwig & Thiel, 1983; Heip *et al.*, 1984; Witte & Zijlstra, 1984) are based on indirect estimations of production either by respiration or by using an annual production/biomass ratio $P/B = 9$ as proposed by Gerlach (1971). This estimate is based on the life cycle of the herbivore *Chromadorita tenuis* as studied in cultures by von Thun (1968). The wet weight of an adult varies between 0·8 and 1·3 μg. Eggs, with a weight of 25 ng, are deposited singly over 5–11 days, with an average total of 20. The embryos develop in 4–5 days and the adult stage is reached after another 12–15 days, at 20–22 °C. Five days later the first eggs are deposited. With a mortality scheme holding the population constant, the average standing stock is 3·3 μg wet weight and production is 9·8 μg. The life-cycle turnover is thus three. The same calculation was done for *Monhystera disjuncta*, with a life-cycle turnover rate of 2·2 and a daily egg production of 2–3 eggs during 8 days.

M. disjuncta has been studied extensively in our laboratory by one of us (G.V.) and its productivity was discussed by Herman, Vranken & Heip (1984). For this species complete life-tables were determined and the intrinsic rate of natural increase r_m, the stable age distribution at exponential growth, the birth rate, and several measures of generation time are known. When a stable age distribution is reached, $P/B = b$, the birth rate (Zaika, 1973). For *M. disjuncta* $P/B = 0{\cdot}18 \cdot \text{day}^{-1}$ or $65 \cdot \text{yr}^{-1}$.

The egg production of *M. disjuncta* in our cultures is very much higher than has been previously observed. Eggs are laid at a nearly constant rate of $5{\cdot}1 \cdot \text{day}^{-1}$ until, after 40 days, senescence starts rather abruptly. So, more than 200 eggs per female are produced. Another species, *Rhabditis marina*, lays between 300–800 eggs (Vanderhaeghen, pers. comm.) in cultures. These figures relate to optimum conditions, but they demonstrate that the productivity of those nematodes that have been successfully cultivated in the laboratory is very high indeed. If this high productivity is realized in the field, it is difficult to see from the existing field data. It seems that, in order to obtain the best possible estimates of actual production of nematodes in the field one may follow one of two ways: either measure the number of generations actually produced and use a life-cycle turnover of three, as this appears to be a good constant figure (Herman, Vranken & Heip, 1984); or use the adult body weight and predict production from Banse & Mosher's (1980) regression equation relating body weight and annual P/B.

The first approach requires knowledge of the annual number of generations in the field. For many species, especially smaller ones, it has been shown that production is nearly continuous and juveniles occur throughout the year (see p. 437). Only for a few, large species are generation times known and they appear to be very long, from less than one to two generations per year (Wieser & Kanwisher, 1961, for *Enoplus communis*; Smol, Heip & Govaert, 1980, for *Oncholaimus oxyuris*; and Gerlach & Schrage, 1972, for *O. brachycercus*). Indirect estimates for the annual number of generations are often based on development time in the laboratory and should be treated with caution. Several authors (Tietjen & Lee, 1973) postulate about 10–15 generations annually for species such as *Chromadorina germanica* and *Monhystera disjuncta*.

The available data have been summarized by Zaika & Makarova (1979). They calculated daily production rates C (or specific production) based on the fact that $P/B = b$, the birth rate. To evaluate the birth rate a life-table has to be constructed and to equate the birth rate with the rate of increase, as the authors do, seems hazardous. From the available literature the equation, $C = 0{\cdot}008\ T^{0{\cdot}96}$ was found. At $T = 10$ °C, specific production would thus be $0{\cdot}07{\cdot}\text{day}^{-1}$ or $27{\cdot}\text{yr}^{-1}$.

Another indirect approach to calculate production has been used by Warwick & Price (1979). They used the empirical relationship between production and respiration of short-lived poikilotherm animals proposed by McNeil & Lawton (1970), $\log P = 0{\cdot}8262 \log R - 0{\cdot}0948$ (in $\text{kcal}{\cdot}\text{m}^{-2}{\cdot}\text{yr}^{-1}$). By measuring the respiration of 16 species they arrived at a value of $P/B = 8{\cdot}4$, surprisingly close to a value of $P/B = 8{\cdot}7$ calculated by accepting a net growth efficiency of $P/P+R = 0{\cdot}38$, as found by Marchant & Nicholas (1974) for the freshwater nematode *Pelodera*.

RESPIRATION

Respiration of individual or batches of nematodes has been measured as oxygen consumption using electrodes (Atkinson, 1973) or cartesian diver respirometry (Lasserre, 1976). Respiration is most commonly expressed in nl $O_2{\cdot}h^{-1}$ per individual or unit body weight or volume (in nl) and described in a power relation $R = aV^b$. As in other poikilotherms $b = 0{\cdot}75$ on average, although Atkinson (1975) suggested a somewhat higher value $b = 0{\cdot}79$ and Zeuthen (1953) even believed that metabolism of nematodes is weight-independent. This is certainly not true but it is interesting to note that Laybourn (1979) found a varying value of b according to temperature in the freshwater nematode *Anonchus*, with b approaching one at temperatures of 20–25 °C.

The value of a is considered to be an indication of metabolic intensity (Schiemer & Duncan, 1974); it represents the repiration of a weight or volume unit nematode and is thus not identical to respiration per unit weight, a confusion often found in the literature. As an example, a nematode with dry weight 0·3 μg and a respiration of 0·6 nl $O_2{\cdot}h^{-1}{\cdot}\text{ind.}^{-1}$ has a weight-specific respiration of 2 nl $O_2{\cdot}h^{-1}{\cdot}\mu g^{-1}$ and a metabolic intensity of 1·48 nl $O_2{\cdot}h^{-1}{\cdot}\mu g^{-1}$.

The literature on nematode respiration prior to 1979 has been reviewed by Warwick & Price (1979), who added many new data (Table XV). The value of b

TABLE XV

Nematode respiration rates at 20 °C, denoted by log a' values: n = number of determinations; after Warwick & Price (1979); S Dep, *selective deposit-feeders;* Non S Dep, *non-selective deposit-feeders;* Epig, *epigrowth-feeders;* Pred/omn, *predators and omnivores;* S, *saltmarsh;* M, *mudflat;* A, *algae;* FS, *fine sand (anaerobic);* Ter, *terrestrial;* FW, *freshwater; 1, Wieser & Kanwisher (1961); 2, Teal & Wieser (1966); 3, Wieser* et al. *(1974); 4, Scheimer & Duncan (1974); 5, Warwick & Price (1979)*

Species	Log *a'*	*n*	Feeding group	Habitat	Reference no.
Oncholaimus campylocercoides	0·386	6	Pred/omn	S	2
Oncholaimus paralangrunensis	0·378	5	Pred/omn	S	2
Axonolaimus spinosus	0·370	3	Non S Dep	S	2
Panagrolaimus rigidus	0·361	40	Non S Dep	Ter	4
Tripyloides gracilis	0·343	10	Non S Dep	S	2
Metoncholaimus pristiurus	0·334	13	Pred/omn	S	2
Bathylaimus 'kanwisheri'	0·328	1	Non S Dep	S	2
Viscosia viscosa	0·188	55	Pred/omn	M	5
Bolbellia tenuidens	0·183	2	Pred/omn	S	2
Enoplus communis	0·128	80	Pred/omn	A	4
Mesotheristus setosus	0·113	16	Non S Dep	S	2
Axonolaimus spinosus	0·105	6	Non S Dep	S	1
Odontophora 'papusi'	0·057	5	Non S Dep	S	2
Aphelenchus avenae	0·053	22	Plant parasite		4
Sphaerolaimus hirsutus	0·027	38	Pred/omn	M	5
Theristus 1	0·019	12	Non S Dep	S	1
Axonolaimus paraspinosus	0·002	26	Non S Dep	M	5
Innocuonema tentabundum	−0·012	5	Epig	M	5
Halichoanolaimus dolichurus	−0·030	4	Pred/omn	S	2
Anticoma litoris	−0·042	6	S Dep	S	2
Mesotheristus setosus	−0·047	40	Non S Dep	M	5
Paracanthonchus caecus	−0·048	7	Epig	S	2
Mesotheristus erectus	−0·062	8	Non S Dep	S	1
Ptycholaimellus ponticus	−0·081	5	Epig	M	5
Praeacanthonchus punctatus	−0·091	17	Epig	M	5
Odontophora setosa	−0·092	2	Non S Dep	M	5
Sphaerolaimus balticus	−0·112	5	Pred/omn	M	5
Sphaerolaimus 2	−0·121	6	Pred/omn	S	1
Dichromadora cephalata	−0·133	5	Epig	M	5
Atrochromadora microlaima	−0·142	5	Epig	M	5
Cylindrotheristus normandicus	−0·150	5	Non S Dep	M	5
Mesotheristus erectus	−0·154	16	Non S Dep	FS	3
Terschellingia longicaudata	−0·170	4	S Dep	M	5
Terschellingia sp.	−0·188	1	S Dep	S	2
Sabatieria pulchra	−0·197	1	Non S Dep	M	5

Odontophora setosoides	−0·199	4	Non S Dep	S	1
Spirinia parasitifera	−0·203	6	S Dep	S	2
Theristus 3	−0·204	3	Non S Dep	S	1
Hypodontolaimus 1	−0·204	4	Epig	S	1
Hypodontolaimus geophilus	−0·218	3	Epig	S	1
Tobrilus gracilis	−0·261	30	Non S Dep	FW	4
Terschellingia communis	−0·277	4	S Dep	M	5
Sphaerolaimus 1	−0·278	5	Pred/omn	S	1
Metachromadora vivipara	−0·424	24	Epig	M	5
Trefusia schiemeri	−0·448	6	S Dep	FS	3
Spirinia hamata	−0·600	5	Epig	S	1
Nannolaimoides decoratus	−0·885	10	Epig	FS	3
Spirinia gnaigeri	−1·081	5	S Dep	FS	3

was assumed constant ($b = 0{\cdot}75$) and all data were re-calculated using this value. The conversion factors used are those of Wieser (1960a), who estimated density equal to 1·13 and a dry weight/wet weight relationship of 0·25. An animal 1 mm in length and 35 μm wide all over has a volume of 1 nl, a wet weight of 1·13 μg, and a dry weight of 0·28 μg. To translate metabolic intensity log a, based on volume, to dry weight, 0·41 has to be subtracted from the values given by Warwick & Price (1979).

The values of metabolic intensity on a volume basis vary between 0·386 for *Oncholaimus campylocercoides* and −1·081 for *Spirinia gnaigeri, i.e.* from 2·43 nl $O_2{\cdot}h^{-1}{\cdot}nl^{-1}$ and 0·08 nl $O_2{\cdot}h^{-1}{\cdot}nl^{-1}$. The distribution of log a values is given in Figure 14. Most species have a log a value between −0·24 and −0·03. When transformed to respiration per unit body weight, two groups can be found (Fig. 15): one with a respiration between 0·14 and 0·43 nl $O_2{\cdot}h^{-1}{\cdot}\mu g$ dry wt^{-1} and one with a repiration between 0·86 and 1·00 nl $O_2{\cdot}h^{-1}{\cdot}\mu g$ dry wt^{-1}. The mean respiration for all nematodes is 0·412 nl $O_2{\cdot}h^{-1}{\cdot}\mu g$ dry wt^{-1}. Fast respiring species include two *Oncholaimus* species, *Axonolaimus spinosus*, and *Tripyloides gracilis*. The slowest respirers are *Spirinia hamata, S. gnaigeri*, and *Nannolaimoides decoratus*.

According to feeding types metabolic intensity differs although no significant difference exists between types 1A and 2A and between 1B and 2B (Table XVI). The average per species is −0·081, the average over the four feeding types is −0·121. This corresponds to a metabolic intensity of 0·76 nl $O_2{\cdot}h^{-1}{\cdot}nl^{-1}$ or 0·69 nl $O_2{\cdot}h^{-1}{\cdot}\mu g$ wet wt^{-1}. To obtain log respiration 0.75 log V (or W) has to be added to log a. For a nematode with a body weight between 0·5 and 1·5 μg wet wt, total respiration would be between 0·41 and 0·94 nl $O_2{\cdot}h^{-1}{\cdot}ind.^{-1}$ or between 10–22 nl $O_2{\cdot}day^{-1}{\cdot}ind.^{-1}$ and 3·6–8·2 μl $O_2{\cdot}yr^{-1}{\cdot}ind.^{-1}$. All these values hold at 20 °C.

From this value an estimate of total community respiration would be between 7·2 and 5·5 litres $O_2{\cdot}yr^{-1}$ per g wet wt. Wieser & Kanwisher (1961), Teal & Wieser (1966), and Warwick & Price (1979) calculated total community respiration of nematodes at 20 °C and arrived at a similar figure of 6·0 l $O_2{\cdot}m^{-2}{\cdot}yr^{-1}$ per g wet weight. A correction for field temperatures is necessary when applying this figure but the available information suggests that it could

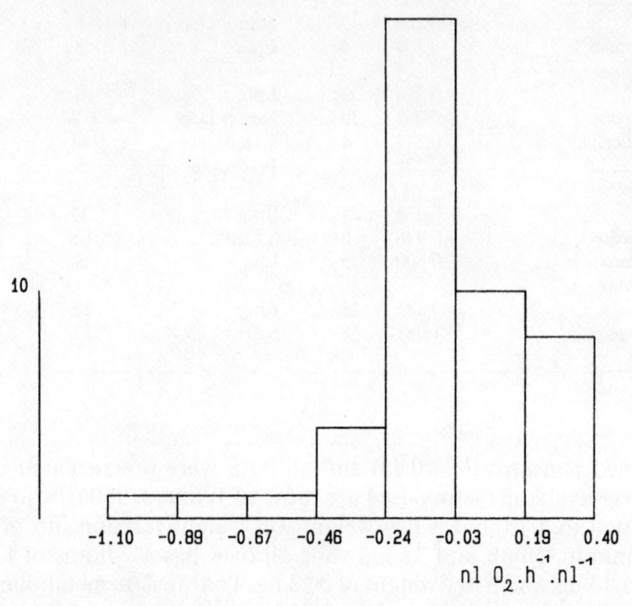

Fig. 14.—Values of metabolic intensity on a volume basis: distribution of log *a*-values.

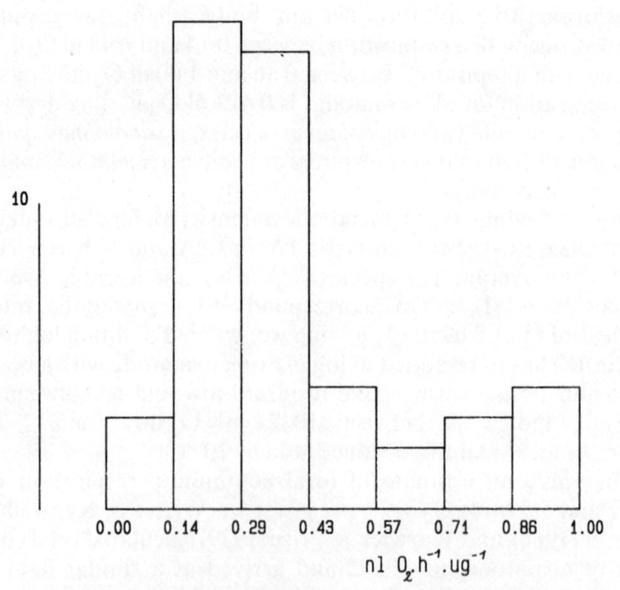

Fig. 15.—Values of metabolic intensity on a volume basis: distribution of weight-specific respiration.

TABLE XVI

Metabolic intensity of feeding types in marine nematodes: values of log a in nl $O_2 \cdot h^{-1} \cdot nl^{-1} \pm SE$ based on n species

	log *a*	*n*	*V* (nl)	Wet wt (μg)	Dry wt (μg)
Selective deposit-feeders	−0·344±0·095	7	0·45	0·50	0·13
Non-selective deposit-feeders	0·018±0·059	18	1·04	1·14	0·28
Epistratum-feeders	−0·258±0·076	11	0·55	0·61	0·15
Predators and omnivores	0·101±0·076	11	1·26	1·38	0·34

be used for making approximate budgets when original respiration data are not available (Warwick & Price, 1979). It must be borne in mind, however, that all data are from intertidal habitats and extrapolation to other biotopes seems risky.

The effect of temperature on respiration has been studied by Price & Warwick (1980) for *Sphaerolaimus hirsutus* and by Wieser & Schiemer (1977) for *Trefusia schiemeri* and *Trichotheristus floridianus*. Q_{10} values of 1·17 were found for *Sphaerolaimus hirsutus* between 5° and 25 °C and about 2 for *Trefusia schiemeri*. For *Sphaerolaimus hirsutus* respiration is related to temperature as $\log R = 0{\cdot}180 + 0{\cdot}0069\ T$, with R in nl $O_2 \cdot h^{-1} \cdot ind.^{-1}$. Price & Warwick (1980) tentatively proposed that animals living in habitats where food supply is stable should have a low Q_{10}, around 1, whereas animals living in habitats with a variable food supply may have a higher Q_{10} of around 2. Too few data exist at present to see if this generalization really holds.

Conversion factors

To convert respiration into other units some difficulties exist. Sikora, Sikora, Erckenbrecker & Coull (1977) determined some conversion factors for marine nematodes which are summarized in Table XVII. To convert oxygen consumption to carbon metabolized most often 1 l O_2 is considered equivalent to 0·4 g C. This conversion factor depends on assumptions concerning the respiratory quotient $RQ = \Delta CO_2/\Delta O_2$. For the respiration of fats $RQ = 0{\cdot}71$, for proteins $RQ = 0{\cdot}71$ and for carbohydrates $RQ = 1$. With an $RQ = 0{\cdot}71$, the conversion factor would be $8{\cdot}4/22{\cdot}4 = 0{\cdot}375$, when carbohydrates dominate the food the conversion would be 0·535. When fermentation or anaerobic respiration is predominant, the respiratory quotient would be larger than 1 and the use of a conversion factor 1 l O_2 = 0·4 g C would no longer be justified.

Anaerobic respiration

Many benthic animal species have evolved very efficient anaerobic pathways yielding as much net energy as aerobic respiration (Pamamat, 1980). Howarth

TABLE XVII

Conversion factors for energy budgets in marine nematodes

Factors				Reference
Wet weight in μg = volume in nl × 1·13				
Wet weight	100%			
Dry weight	25%	100%		Wieser, 1960a
Ash-free dry weight	20%	80%	1 μg	Sikora *et al.*, 1977
Carbon	10·6%	42%	0·53 μg	Sikora *et al.*, 1977
Calorific equivalent			6·12 mcal	Sikora *et al.*, 1977
Calorific equivalent			25·58 mJ	

& Teal (1980) state that in salt marshes sulphate reduction is many times more important in the degradation of organic matter than oxygen respiration and denitrification combined. Microbial production is as high as in aerobic sediments and nematodes are the most important grazers of this production.

Although these anaerobic pathways are certainly very important in many environments, little information on nematodes is available (Wieser, Ott, Schiemer & Gnaiger, 1974).

EXCRETION

Nothing is known about excretion in marine nematodes. For terrestrial nematodes Wright & Newall (1976) found that nitrogen excretion is dependent on body size with $b = 0{\cdot}75$, and a nematode of 1 μg wet wt would excrete 1·92 μmol $N{\cdot}h^{-1}{\cdot}g^{-1}$. In mammals, 1 g of nitrogen excreted corresponds to 5·94 l O_2 respired.

In a soil microcosm significantly more N was mineralized, as ammonium, in the presence of nematodes than with bacteria alone (Coleman *et al.*, 1977). Soil nematodes have high C:N ratios (8:1 to 12:1) in comparison with bacteria (3:1 to 4:1) and thus must release N as a waste product (Anderson *et al.*, 1983). If the same is true in marine systems, the impact of nematodes on the nitrogen cycle may be correspondingly more important than on the carbon cycle.

ENERGY FLOW THROUGH NEMATODES

The energy flow through nematode populations can be written as $C = P + R + U + F$ (Crisp, 1971), in which all terms must be written in the same units, and where C is consumption or ingestion, P is production and can be divided into production due to growth and gonad output, R is respiration, and U and F are soluble and particulate waste products, respectively. A first trial to establish a carbon budget was that of Tietjen (1980b) for females of three species (Table XVIII). As this budget was termed preliminary, we shall not discuss it in detail and shall only use it to clarify some problems. First, assimilation efficiency seems to be low, and much lower for *Chromadorina germanica*, a herbivore,

TABLE XVIII

A carbon budget for females of three nematode species (from Tietjen, 1980b): all values in ng C·day^{-1}

	Chromadorina germanica	*Monhystera disjuncta*	*Rhabditis marina*
Consumption C	400	144	600
Assimilation A	25	26	155
Respiration R	5·3	5·3	5·3
Production P'	19·5	21	150
Growth P	6·7	5	54
Egg production G	12·8	16	96
Efficiencies (%)			
A/C	6·2	18·3	25·8
P'/A	78·7	79·8	96·5
P/A	26·8	19·2	34·8

than for bacterial feeders. As herbivores often do not, however, ingest the cells wholly, this may be an under-estimate. The second point is that production efficiency is very high, and that most of the energy is channelled into egg production.

Warwick (1981a) studied production efficiency of the monhysterid *Diplolaimelloides bruciei*, and the influence of temperature and salinity on this. Daily P/B was measured as the rate of increase r (instead of the birth rate, giving a slight under-estimation) and respiration was also expressed in units of body weight. r was a rather complex function of temperature which was linearized as $r = -0{\cdot}067 + 0{\cdot}0134\ T$. r was highest at 26‰ S and about twice higher than at both higher and lower salinities. Respiration increased exponentially with temperature between 5° and 25 °C as $\log R = -0{\cdot}7238 + 0{\cdot}05957\ T$. Fecundity, expressed as the number of eggs produced daily by a female, was also a linear function of temperature $f = -1{\cdot}429 + 0{\cdot}286\ T$. The proportion of adults in the population was independent of temperature (13·6%) and the sex-ratio was in favour of males (65·6%). From these figures the biomass of eggs produced per unit biomass of the entire population was calculated as $Pr = -0{\cdot}00826 + 0{\cdot}00165\ T$.

The linear relationship between r and temperature, or between production and temperature, was also found for *Chromadorina germanica* (Tietjen & Lee, 1977a) and *Oncholaimus oxyuris* (Heip, Smol & Absillis, 1978). Whereas the slope is similar for *Chromadorina germanica* and *Diplolaimelloides bruciei* (0·0145 and 0·0134, respectively), it is about ten times lower for *Oncholaimus oxyuris* (0·0013).

The energy budget for *Diplolaimelloides bruciei* is given in Table XIX, with the same symbols as for Table XVIII. The same features are apparent as in Tietjen's (1980b) estimates. Production efficiency is very high, between 0·7 and 0·9 between 10 and 30 °C. In females, all of this production is due to eggs and

TABLE XIX

Energy budget for Diplolaimelloides bruciei *(total population): values in J (J^{-1}·day^{-1}) (after Warwick, 1981b)*

	Temperature (°C)		
	5	15	25
P growth (*P*)	0	0·151	0·214
P eggs (*G*)	0	0·021	0·030
P total (*P′*)	0	0·172	0·244
R	0·007	0·026	0·101
$P'/(P'+R)$ (%)	0	87	71
$P/(P+R)$ (%)	0	45	45
$G/(G+R)$ (females)	0	90	77

efficiencies were again very high. Taking into account that females represent only a small fraction of the population, growth efficiency must, however, also be very high in the juveniles. Reproductive effort, $G/(G+R)$, is in the order of 77–90%, and was calculated from the ratio of egg volume on female volume. It may be (Herman, Vranken & Heip, 1984) that the energy content of eggs is much lower than that of adults, since the weight of neonates is often much lower than the egg weight and considerable respiration may occur during the egg stage.

A detailed analysis was done by Schiemer (1982a,b; 1983) on the freshwater species *Plectus palustris* and the soil-dwelling *Caenorhabditis briggsae*, a small and extremely rapid developing species. The weight of an adult female is about 0·4–0·5 μg wet wt. The respiration rate of this species depends on food concentration: at a density of 5×10^7 cells·ml^{-1}, oxygen consumption was 1·23 nl·h^{-1}·μg^{-1}, at a food level of 10^8 cells·ml^{-1} oxygen consumption was 1·64 and at higher food concentrations oxygen consumption remained equal, between 3·2 and 3·7 nl O_2·h^{-1}·μg^{-1}. Production of the species and the P/R ratio vary throughout its life cycle and the P/R ratio is also dependent on food concentration; it varies from 0·62 at 5×10^8 cells·ml^{-1} to 1·77 at 10^{10} cells·ml^{-1}. Over the whole life cycle, from hatching until the end of the reproductive stage, the total amount of energy assimilated is more than double at 10^{10} cells·ml^{-1} than at 5×10^8 cells·ml^{-1} (Table XX).

In geneıal, respiration increases linearly with food supply but production increases hyperbolically. This is probably not due to an increased intake of food only, as ingestion appears to be proportional to food availability (Nicholas, Grassia & Viswanathan, 1973), but to a decrease in assimilation efficiency.

The comparison between these two species yields some interesting results (Schiemer, 1983). Over its whole life cycle, *Plectus palustris* has P = 161 mJ and R = 33 mJ, so that production efficiency is 83%. In the shorter living *Caenorhabditis briggsae* P = 170 mJ and R = 100 mJ, production efficiency

TABLE XX

Energy budgets for Caenorhabditis briggsae *at three food concentrations: values in μJ·10 h⁻¹ (after Schiemer, 1982b)*

	Food concentration (cells/ml)		
	5×10^8	10^9	10^{10}
P growth (%)	11·6	10·2	10·0
P eggs (%)	36·6	43·0	53·0
P total (%)	48·2	53·2	63·0
R (%)	51·8	46·8	37·0
$A = P+R$ (μJ·10 h^{-1})	11·9	19·3	27·2

thus being only 63%, which is still very high. Reproduction accounts for 84% of total production in *C. briggsae*, for 94% in *Plectus palustris*, values that are again astonishingly high.

Whether this enormous productivity is realized in the field depends strongly on food availability. The dependence of assimilation on food concentration can be described by a Michaelis–Menten equation. The food concentration at which assimilation was half its maximum was 0·75 mg dry wt·ml^{-1} for *Caenorhabditis briggsae* and 0·10 mg dry wt·ml^{-1} for *Plectus palustris*. The threshold at which $A = R$ and $P = 0$ was about 0·1 mg dry wt·ml^{-1} for *Caenorhabditis briggsae* and 0·025 mg dry wt·ml^{-1} for *Plectus palustris*.

Oligotrophic lakes have bacterial biomass lower than 0·01 mg dry wt·ml^{-1}, whereas in eutrophic lakes this is between 0·02 and 1 mg dry wt·ml^{-1}.

Joint (1978) showed numbers on an intertidal mud-flat from 1×10^6 to 1×10^9 cells·g^{-1} wet sediment, or about twice these values for the amounts per ml of interstitial water. These densities represent anything from threshold conditions to abundant food supply and much more study is necessary.

Fallon, Newell & Hopkinson (1983) found bacterial densities between 0·97 $\times 10^9$ cells·ml^{-1} wet sediment in subtidal stations 15 km offshore and $8{\cdot}1 \times 10^9$ cells·ml^{-1} wet sediment in a salt marsh. Production, as measured by thymidine uptake, was 36·5 g C·m^{-2}·yr^{-1} 15 km offshore and 296 g C·m^{-2}·yr^{-1} 250 m offshore. There is still much uncertainty in these figures and more study is needed; as marine nematodes are, however, less productive than *Caenorhabditis briggsae*, bacterial production in the shallow areas of the sea may be of the order of magnitude at which nematode production is not seriously limited.

The apparently very high production efficiencies have not been taken into account in existing estimates of carbon flow through nematode populations.

We shall revise here the estimates given by Heip, Herman & Coomans (1982) and Heip, Herman & Vincx (1984). These authors discuss two contrasting situations in the Belgian coastal waters of the Southern Bight of the North Sea. One is a linear sandbank in a highly dynamic environment (the Kwintebank), the other a series of stations in shallow, polluted waters with a high input of organic material. Production estimates were based on respiration, amounting to 2·3 g $C{\cdot}m^{-2}{\cdot}yr^{-1}$ in coastal waters and 1·06 g $C{\cdot}m^{-2}{\cdot}yr^{-1}$ on the Kwintebank. When assuming that production efficiency is 80%, then production in the coastal waters would be 9·2 g $C{\cdot}m^{-2}{\cdot}yr^{-1}$ since food is probably not limiting there. Since assimilation efficiency is again low, around 20% for bacterivores, the total consumption of bacteria by these nematodes would amount to 57 g $C{\cdot}m^{-2}{\cdot}yr^{-1}$, a very significant part of the total input into the benthos. Conversely, when assuming that each nematode eats about double its weight per day (see p. 461), the total consumption would be 1·4 g dry $wt{\cdot}m^{-2}{\cdot}dry^{-1}$ or 511 g dry $wt{\cdot}m^{-2}{\cdot}yr^{-1}$ or about 200 g $C{\cdot}m^{-2}{\cdot}yr^{-1}$. This figure is even higher.

Much more study is necessary before such figures can be more than order of magnitude indications; but as such they show that nematodes are very important components of benthic systems. This has also been shown by Warwick & Price (1979) for the Lynher estuary, where respiration plus production of nematodes is minimum 29·7 g $C{\cdot}m^{-2}{\cdot}yr^{-1}$ and consumption, assuming that a nematode eats about double its body weight each day, would amount to an astonishing 600 g $C{\cdot}m^{-2}{\cdot}yr^{-1}$.

That the extrapolation of general regression equations may yield contrasting results is shown in the study of Witte & Zijlstra (1984) on an intertidal flat in the Wadden Sea. Nematodes number on average 2117 ind.·10 cm^{-2} with an average standing stock of 0·60 g dry $wt{\cdot}m^{-2}$. Respiration and production were estimated using some general equations. For respiration Banse's (1982) regression between biomass and respiration for larger invertebrates was used, $R = 5{\cdot}4\ W^{0{\cdot}75}$ at 20 °C. This yields a value of 2·08 nl $O_2{\cdot}h^{-1}{\cdot}ind.^{-1}$, compared with the value of 0·75 nl $O_2{\cdot}h^{-1}{\cdot}ind.^{-1}$ we calculated from the data in Warwick & Price (1979). For the nematodes total this would amount to 7·5 g $C{\cdot}m^{-2}{\cdot}yr^{-1}$. In Witte & Zijlstra's further calculations a *P*/*B*-ratio of 8·5 for nematodes (also based on a figure of Warwick & Price, 1979) gives a nematode production of 5·1 g dry $wt{\cdot}m^{-2}{\cdot}yr^{-1}$ or 2·14 g $C{\cdot}m^{-2}{\cdot}yr^{-1}$. Total assimilation thus calculated is 9·6 g $C{\cdot}m^{-2}{\cdot}yr^{-1}$. If we assume a production efficiency of 80%, total assimilation would be 37·5 or 13·5 g $C{\cdot}m^{-2}{\cdot}yr^{-1}$ and total consumption would be 180 or 67 g $C{\cdot}m^{-2}{\cdot}yr^{-1}$ (depending on which respiration estimate is used). These discrepancies strongly indicate the need for more research in this area and caution against the use of general equations which may not be at all applicable to nematodes.

POLLUTION STUDIES

FIELD STUDIES

The influence of pollution on nematodes has received little attention until the last few years. Oil pollution in intertidal and shallow subtidal areas has been studied most intensively. A decrease in nematode density after contamination with hydrocarbons has been demonstrated on several beaches (Wormald, 1976; Giere, 1979; Boucher, 1980b), but not on others (Green, Bauden, Gretney & Wono, 1974) and not in sublittoral sands (Elmgren, Hansson & Sundelin, 1980; Elmgren *et al.*, 1983; Boucher, 1980b). An obvious decrease in nematode abundance after an oil spill has often been followed by explosive development of some few opportunistic species within one year (Wormald, 1976; Giere, 1979). After the Amoco Cadiz spill (Boucher, 1980b), however, there was no such explosive development on the beaches.

After an oil spill in La Coruña (northern Spain), *Enoplolaimus litoralis* became extremely dominant; many specimens had ingested oil droplets covered with bacteria. In the intestine of *Bathylaimus* sp. and *Tripyloides* sp., oil particles were found surrounded by clouds of bacteria (Giere, 1979). After the Amoco Cadiz spill, nematode diversity in the sublittoral sands in Morlaix Bay decreased significantly, and most obviously 9 to 12 months after the accident happened (Boucher, 1980b). This was due, on the one hand, to an increase of *Anticoma ecotronis, Sabatieria celtica, Paracyatholaimus occultus,* and *Calomicrolaimus montrosus,* species normally abundant in silty sands; and, on the other hand, to a decrease of *Ixonema sordidum, Monoposthia mirabilis, Rhynchonema ceramotos, Chromadorita mucrocaudata, Xyala striata, Viscosia franzii,* and *Rhynchonema megamphidum,* species normally dominant in clean sands.

Renaud-Mornant, Gourbault, de Panafieu & Helléouet (1981) also examined the same polluted area (mainly Roscoff Beach and Bay of Morlaix) and found that mortality 10 days after the oil input was not important. After one month, density decreased; mortality was especially important in the surface sand layers while in the deeper layers meiofauna was found in the process of spring reproduction. After six months, nematodes became extremely dominant and accounted for 90% of the meiofauna.

In the Adriatic Sea, the density and distribution of sublittoral meiofauna is not influenced by raw domestic sewage (Vidakovic, 1983). The effects of heavy metal pollution have been studied by Lorenzen (1974), Tietjen (1977, 1980a) and by Heip *et al.* (1984) in monitoring field studies and in the laboratory by Howell (1982a,b, 1983, 1984). Lorenzen (1974) found no short term effects on the nematode fauna in a region of the German Bight of the North Sea subjected to industrial waste disposal (containing 10% H_2SO_4 and 14% $FeSO_4$).

Tietjen (1977) found that heavy metals (expressed in mg l^{-1} by weight of the total heavy metal concentration; *i.e.* Cd, Cr, Cu, Hg, Mn, Pb, Zn) did not affect nematode populations in Long Island sublittoral muds, although a slight decrease in diversity was obvious. Tietjen (1980a) examined the nematodes from the New York Bight Apex, a sandy sediment area with high heavy metal content and organic carbon loads. The amounts of heavy metal contamination were between 3·0–302·0 mg $Cr \cdot l^{-1}$; 3·0–361·0 mg $Cu \cdot l^{-1}$; 3·0–47·0 mg $Ni \cdot l^{-1}$; 8·5–141·5 mg $Pb \cdot l^{-1}$; 7·5–580·0 mg $Zn \cdot l^{-1}$. High concentrations of the contaminants in medium sands may result in lowered abundance of the nematode families which normally live in this kind of sediment: Chromadoridae, Desmodoridae, and Monoposthiidae. Other species, such as *Sabatieria pulchra,* which are normally associated with finer sediments, increase in abundance. This species is already adapted for living under low dissolved oxygen concentration and/or high organic content (it lives mostly in muddy sediments).

Heip *et al.* (1984) examined the composition and density of the meiofauna of the Belgian coastal waters (North Sea). The impact of the Western Scheldt river, a highly polluted stream, is reflected in a decrease in diversity on all taxonomic levels. In the coastal zone, nematodes are dominant and represent more than 90% of the meiofauna on all stations; they are the only animal group that survives in normal or even greater abundance, albeit with lower diversity. Nematode richness (number of species) is significantly correlated with heavy metal content (*e.g.* 2–20 mg $Cu \cdot l^{-1}$; 4–17 mg $Mn \cdot l^{-1}$; 15–124 mg $Pb \cdot l^{-1}$; 41–154 mg $Zn \cdot l^{-1}$). In the highly polluted part of the Belgian coast (most close to the mouth of the Western Scheldt) only non-selective deposit-feeders are the meiofaunal component which survive in high density, but with few species per station (2–5): *S. breviseta, S. vulgaris, Daptonema tenuispiculum, Metalinhomoeus* n. sp., and *Ascolaimus elongatus.*

A matter of debate in the last few years has been the use of nematodes (and meiofauna in general) as a possible tool for detecting pollution. Marine nematodes have been suggested as possible pollution indicators as they possess some characteristics such as a short life-span and high diversity which makes them potentially useful in ecological monitoring (Heip, 1980). Three tools are commonly used with meiofauna as pollution indicators: (1) the nematode:copepod ratio; (2) the log-normal distribution of individuals over species; and (3) diversity indices or graphical methods (*k*-dominance curve).

Raffaelli & Mason (1981) indicate that copepods may be more sensitive to environmental stress than nematodes, so that a high nematode:copepod ratio may be indicative of polluted situations. The ratio of nematodes to copepods increases also with decreasing particle size, but ratios from polluted sites were always extremely high and it is proposed that the ratio is a tool for monitoring organic pollution of sandy beaches. Ratios from clean beaches were low and always less than 100, even for muddy sites; all intertidal sites (fine as well as coarse) with ratios exceeding 100 were polluted with organic material (sewage). An increase in the abundance of deposit-feeding nematodes (which profit from the organic material associated with the sewage) and a decrease in copepods which appear generally more sensitive to environmental stress (McIntyre, 1977) are the probable cause for the increase of the ratio. Some sublittoral ratios from unpolluted sites were high, but never approached the very high values characteristic of polluted intertidal sites. The sublittoral ratios also increased with depth. It is obvious that this ratio must be used with caution as this index is also largely affected by sediment granulometry. Coull, Hicks & Wells (1981) dispute that a single ratio is appropriate to describe the very complex meiofaunal community structure. Factors which enormously influence this ratio are horizontal distribution and seasonal variability.

Warwick (1981c) proposes refinement of the ratio based on the trophic dynamic aspects of the meiofauna. He assumed that food is the factor which limits energy flow through the nematode and copepod community; in that case, the total number of copepods should be proportional to the number of type 2A nematodes only, as only 2A nematodes are dependent on the same food source as the copepods. If copepods are indeed more sensitive to the effects of pollution than nematodes, then changes in the proportion of copepods relative to type 2A nematodes might be a useful indicator to separate the effects of pollution from any changes or differences in sediment type. Warwick (1981c) suggests that pollution might be indicated by ratios around 40 for fine sediments and 10 for sands. These values are considerably lower than the values of over 100 proposed by Raffaelli & Mason (1981). Vidakovic (1983) even found a totally opposite trend in the ratio of nematodes to copepods. In Adriatic sublittoral stations, which are constantly influenced by sewage, the number of copepods increases more than the number of nematodes. This controversy indicates that one should be very careful when using the ratio between nematodes to copepods, without further information on the natural variability and the situation before pollution.

Gray & Mirza (1979) suggested an intrinsic probability-plotting method to detect pollution-induced disturbances. This method is based on the frequency distribution of species abundances; deviation from a log-normal distribution of individuals per species may indicate a disturbed or polluted assemblage. They propose the probability paper method of estimating goodness-of-fit to the log-normal distribution. This method has some restrictions; samples should be very large (to prevent effect of patchiness) and this may be problematic for marine nematodes.

Shaw, Lambshead & Platt (1983) and Lambshead, Platt & Shaw (1983) proposed another method for detecting differences among assemblages of marine benthic species (*e.g.* nematodes). Shaw *et al.* (1983) suggested that ranked species abundance curves (RSA-curves) are a sensitive tool in detecting disturbances in the community, as is the index *d* (*i.e.* the proportional abundance of the most abundant species). Where no single species shows overwhelming dominance, it is also interesting to consider the combined dominance of the two, three, . . . *k*, most abundant species (Lambshead *et al.*, 1983). By plotting *k*-dominance (% cumulative abundance) against *k* (species rank) in a so-called *k*-dominance curve, it is possible to 'describe' the diversity pattern of the community. When the *k*-dominance curves of two communities intersect, the communities are not comparable in terms of intrinsic diversity. For an extensive discussion of the last method, we refer to Lambshead *et al.* (1983). As an example of several methods used, we show the data of Platt (1977a), discussed by Shaw *et al.* (1983) and by Lambshead *et al.* (1983).

Figure 16 (A–D) compares the log-transformed species abundances (Gray & Mirza, 1979) from the nematode assemblages in Strangford Lough, Northern Ireland at high (H), mid (M) and low (L) tide level (Fig. 16 A and B) with the RSA curves and *k*-dominance curves of the same data (Fig. 16 C and D). The different kinds of presentation of the data all show the same trends in the nematode assemblages from high to low tide. The dominance curves have the advantage that only a minimum sample size of about 150 individuals is required. For transformation of the abundances, very large samples are necessary.

A combination of trophic diversity (expressed in a trophic index $\Sigma\theta^2$) (θ = percentage of each feeding type) and species richness, provides for the highly polluted Belgian coast a good indication of the influence of pollution (Heip *et al.*, 1984). The relation between the number of species and the trophic index is shown in Figure 17. When non-selective deposit-feeders dominate, the number of species is always low. On moderately polluted stations, type 1B is already the dominant group, but species number still drops when the heavy metal content of the sediment increases. The less polluted sandy stations are always more diverse, with trophic indices approaching 0·25. The high dominance of non-selective deposit-feeders is, however, also correlated with a high silt content. Effects of pollution and effects of sediment granulometry are very often difficult to distinguish.

This review of marine pollution monitoring studies using nematodes shows the difficulties and controversies in the interpretation of observed changes. It is

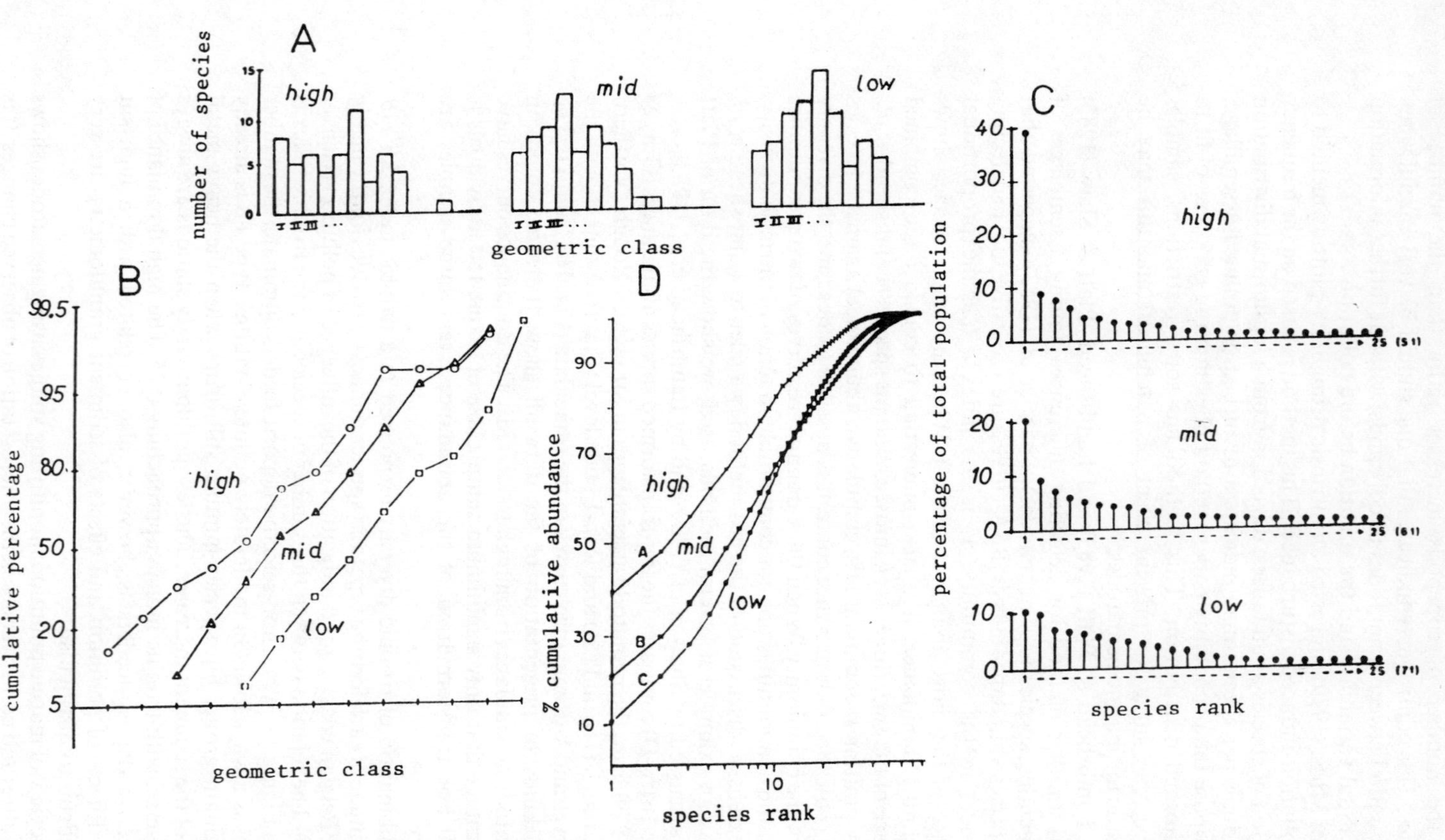

Fig. 16.—A, log-transformed species abundances from nematode assemblages at high, mid and low tide level in Strangford Lough, N. Ireland (after Platt, 1977a in Shaw, Lambshead & Platt, 1983). B, probability plot of the same data (Platt, 1977a): *X*-axis indicates × 2 geometric class; first point of each curve represents class I (*i.e.* the curves are staggered); after Shaw *et al.* (1983). C, RSA (rank species abundance) curve of the 25 most common species from Strangford Lough: data with the total number of species shown in parentheses; after Platt (1977a), Platt & Warwick (1980), and Shaw *et al.* (1983). D, *k*-dominance curves for the three nematode assemblages (same data from Platt, 1977a, in Lambshead *et al.*,

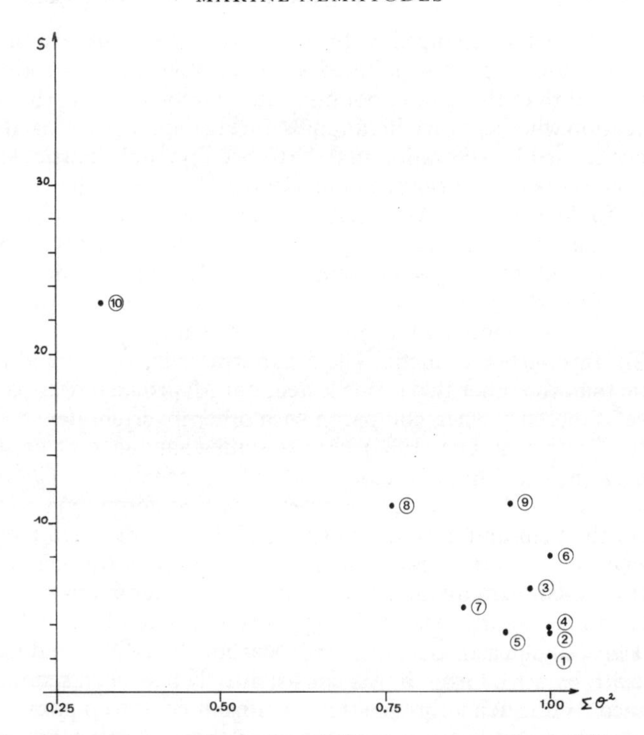

Fig. 17.—The relation between the number of species (S) and the trophic index ($\Sigma\theta^2$): number in circle indicates degree of heavy metal pollution, 1 highest → 10 lowest; after Heip, Herman & Vincx (1984).

very hard to distinguish pollution induced from natural changes, as in most cases the pre-pollution situation is not well known. Density in general is not much affected by pollution, whereas diversity generally seems to decrease. Pollution is often accompanied by general changes in sediment characteristics; the lethal effect of a pollutant or the change in habitat texture may be responsible for the observed changes. We know that some nematode species are resistant to high levels of pollution and anaerobiosis. The effect of *e.g.* heavy metals on nematode population dynamics, however, can only be studied in the laboratory.

LABORATORY STUDIES

Despite their ecological importance, and their significant rôle in all marine and brackish-water sediments, only a few experimental pollution studies, *e.g.* assays with nematodes as model organisms such as the traditional LC_{50} experiments, are available at the present time. As far as we know, only one paper, studying the acute toxicity of pollutants (heavy metals) to marine nematodes, has been published to date (Howell, 1984). From this study it appeared that the susceptibility to copper, lead, and mercury was higher than to zinc and cadmium, with both *Enoplus brevis* and *E. communis*, and that the toxicity of all metals changed with exposure time. Furthermore, it was suggested that animals from polluted stations were less susceptible to the toxicants tested than those of unpolluted sites. Unfortunately, Howell (1984) did not mention whether he fed his animals during the experiments. If not, then interference caused by starvation may have been possible, particularly when exposure time was long. Vranken *et al.* (1984b) dealt with the acute toxicity of mercury to the nematode *Monhystera disjuncta* and the mortality response of the three different life-stages (eggs, juveniles, and adults) was monitored for three different mercury compounds (Hg_2Cl_2, $HgCl_2$, and CH_3HgCl). Mortality during the post-embryonic stages was the most sensitive index for acute stress and developmental inhibition could not be used as a toxicity-criterion for this particular metal. The organic mercury compound was much more toxic than the other two forms tested, but *M. disjuncta* was particularly insensitive to mercury when compared with other invertebrates.

Recently, Tietjen & Lee (1984) used a somewhat different experimental approach to measure the effect of wastes. Two nematodes, *Chromadorina germanica* and *Diplolaimella punicea* were grown in contaminated sediment taken from the field and a number of potential toxicants were measured in these sediments. The intrinsic rate of natural increase (but see p. 452) was used as an indicator of sediment quality. The worms grew better in low contaminated sediments and concentrations of 270 μg polychlorinated biphenyl (PCB)$\cdot l^{-1}$ and 8700 μg polynuclear aromatic hydrocarbon (PAH)$\cdot l^{-1}$ reduced population growth by 50%. Growth was consistently better when sediments were more diluted. With such approaches it remains difficult to pinpoint the most toxic compound and to determine non-toxic levels for individual contaminants.

A somewhat similar idea of experimentation is found in Cantelmo & Rao (1978a,b). These authors used aquaria containing sand and provided with a continuous supply of sea water, to study the effect of the biocide pentachlorophenol (PCP). They found that high concentrations of PCP caused a compositional shift in the nematode assemblages in such a way that epistratum-feeders, which are dominant in conditions of low stress, were outnumbered by selective deposit-feeders in conditions of high stress. An identical experimental set-up was used by Cantelmo, Tagatz & Rao (1979) to study the effect on meiofauna of barite ($BaSO_4$) a major constituent of drilling muds used in marine oil drilling operations. Some groups (Rotifera, Foraminifera, Hydrozoa, Turbellaria, Ostracoda, Polychaeta, and Bivalvia) were unaffected by barite, and a mixture of barite and sand, in some proportions, increased sediment heterogeneity, which resulted in increases in densities of the meiofauna. On the contrary, barite deposited on the surface of the sediment caused a significant reduction in meiofaunal density.

Howell (1982a) studied the occurrence of copper, zinc, lead, cadmium, and mercury in two enoplids, *Enoplus brevis* and *E. communis*, sampled from three estuaries on the northeastern coast of Britain. It was concluded that metal concentrations found in nematodes are in agreement with values reported for benthic organisms with comparable life histories, feeding behaviour and habitus. Howell (1982b) found that the acid mucopolysaccharide (mucus) secreted by the two enoplids (see above) possesses metal-binding properties; this mucus may play an important rôle in heavy metal uptake and loss. The uptake, loss, and tissue distribution of copper and zinc in the same two enoplids was studied by Howell (1983) who suggested that surface adsorption is very important and that most likely the uptake of heavy metals from sea water is *via* the cuticle. Besides this route, uptake *via* the gut, pharynx, and other tissues (hypodermis, muscle layers, and reproductive structures) is significant. In another study (Howell & Smith, 1983) the existence of two heavy-metal binding proteins, binding copper and cadmium, was reported in *E. brevis*. One of the two was probably the collagen-like component of the cuticle. The identity of the other is not yet known.

This review of ecotoxicological work with free-living marine nematodes shows that our knowledge on this topic is very poor and that much more work remains to be done.

ACKNOWLEDGEMENTS

We are grateful to Prof. Dr A. Coomans and Dr P. Herman for a critical reading of parts of the manuscript. This research is supported by the Belgian National Fund for Scientific Research (N.F.W.O.) and the Ministry of Scientific Policy (Concerted Actions Oceanography). Part of this work was conducted under Contract No. ENV.566B of the Environmental Programme of the C.E.C.

REFERENCES

Admiraal, W., Bouwman, L. A., Hoekstra, L. & Romeyn, K., 1983. *Int. Revue ges. Hydrobiol.*, **68,** 175–191.

Alongi, D. M. & Tietjen, J. H., 1980. In, *Marine Benthic Dynamics*, edited by K. R. Tenore & B. C. Coull, University of South Carolina Press, Columbia, pp. 151–166.

Anderson, R. V., Gould, W. D., Woods, L. E., Cambardella, C., Ingham, R. E. & Coleman, D. C., 1983. *Oikos*, **40,** 75–80.

Andrássy, I., 1956. *Acta zool. Hung.*, **2,** 1–15.

Andrássy, I., 1983. *A Taxonomic Review of the Suborder Rhabditina (Nematoda:Secernentia)*. L'office de la recherche scientifique et technique outremer, ORSTOM, Paris, 241 pp.

Arlt, G., 1973. *Wiss. Z. Univ. Rostock*, **22,** *math.-naturw. R.*, 685–692.

Arlt, G., 1977. *Wiss. Z. Univ. Rostock*, **26,** *math.-naturw. R.*, 217–222.

Arlt, G. & Holtfreter, J., 1975. *Wiss. Z. Univ. Rostock*, **24,** *math-naturw. R.*, 743–751.

Ash, C., Crompton, D. & Keymer, A., 1984. *New Sci.*, **101,** 17–20.

Atkinson, H. J., 1973. *J. exp. Biol.*, **59,** 255–266.

Atkinson, H. J., 1975. *J. exp. Biol.*, **62,** 1–9.

Banse, K., 1982. *Mar. Ecol. Progr. Ser.*, **9,** 281–297.

Banse, K. & Mosher, S., 1980. *Ecol. Monogr.*, **50,** 355–379.

Bell, S. S. & Coen, L. D., 1982. *J. exp. mar. Biol. Ecol.*, **61,** 175–188.

Belogurova, L. S., 1978. *Russk. zool. Zh.*, **56,** 1587–1604.

Bergholz, E. & Brenning, U., 1978. *Wiss. Z. Univ. Rostock*, **27,** *math.-naturw. R.*, 393–398.

Bilio, M., 1966. *Int. Revue ges. Hydrobiol.*, **51,** 147–195.

Blome, D., 1983. *Mikrofauna Meeresboden*, **88,** 517–590.

Boaden, P. J. S. & Platt, H. M., 1971. *Thalassia jugosl.*, **7,** 1–12.

Bongers, T., 1983. *Neth. J. Sea Res.*, **17,** 39–46.

Boucher, G., 1972. *Cah. Biol. mar.*, **13,** 457–474.

Boucher, G., 1974. *Vie Milieu*, **23B,** 69–100.

Boucher, G., 1976. *Bull. Mus. natn. Hist. nat.*, Paris, **352,** 25–61.
Boucher, G., 1980a. *Mém. Mus. natn. Hist. nat., Sér. A. Zool.*, **114,** 3–81.
Boucher, G., 1980b. *Mar. Pollut. Bull.*, **11,** 95–111.
Bouwman, L. A., 1983. *Zool. Jb.* (*Syst.*), **110,** 345–376.
Bouwman, L. A., Romeyn, K. & Admiraal, W., in press. *Estuar. cstl Shelf. Sci.*
Brenner, S., 1974. *Nature, Lond.*, **248,** 785–787.
Brenning, U., 1973. *Oikos*, Suppl. 15, 98–104.
Burnett, B. R., 1973. *Deep-Sea Res.*, **20,** 413–417.
Cantelmo, F. R. & Rao, K. R., 1978a. *Mar. Biol.*, **46,** 17–22.
Cantelmo, F. R. & Rao, K. R., 1978b. In, *Pentachlorophenol* (*1978*), edited by K. R. Rao, Plenum Publishing Corporation, New York, pp. 165–174.
Cantelmo, F. R., Tagatz, M. E. & Rao, K. R., 1979. *Mar. environ. Res.*, **2,** 301–309.
Capstick, C. K., 1959. *J. Anim. Ecol.*, **28,** 189–210.
Chandler, G. T. & Fleeger, J. W., 1983. *J. exp. mar. Biol. Ecol.*, **69,** 175–188.
Chia, F. S. & Warwick, R. M., 1969. *Nature, Lond.*, **224,** 720–721.
Chitwood, B. G., 1951. *Tex. J. Sci.*, **3,** 671–672.
Chitwood, B. G. & Murphy, D. G., 1964. *Trans. Am. microsc. Soc.*, **83,** 311–329.
Cobb, N. A., 1917. *Contr. Sci. Nematol.* (Baltimore), **5,** 117–128.
Coleman, D. C., Cole, C. V., Anderson, R. V., Blaha, M., Campion, M. K., Clarholm, M., Elliot, E. T., Hunt, H. W., Schaefer, B. & Sinclair, J., 1977. In, *Soil Organisms as Components of Ecosystems*, edited by U. Lohm & T. Persson, Ecol. Bull. (Stockholm), 25, 299–309.
Coull, B. C., 1969. *Limnol. Oceanogr.*, **14,** 953–957.
Coull, B. C., 1970. *Oecologia* (*Berl.*), **4,** 325–357.
Coull, B. C. & Bell, S. S., 1979. In, *Ecological Processes in Coastal and Marine Systems*, edited by R. J. Livingston, Plenum Press, New York, pp. 189–216.
Coull, B. C., Ellison, R. L., Fleeger, J. W., Higgins, R. P., Hope, W. P., Hummon, W. D., Rieger, R. M., Sterrer, W. E., Thiel, H. & Tietjen, J. H., 1977. *Mar. Biol.*, **39,** 233–240.
Coull, B. C., Hicks, G. R. F. & Wells, J. B. J., 1981. *Mar. Pollut. Bull.*, **12,** 378–381.
Coull, B. C. & Wells, J. B. J., 1981. *N.Z. Jl mar. freshwat. Res.*, **15,** 411–415.
Crisp, D. J., 1971. In, *Methods for the Study of Marine Benthos*, edited by N. A. Holme & A. D. McIntyre, Blackwell Scientific Publications, Oxford, pp. 197–279.
Croll, N. A. & Viglierchio, D. R., 1969. *Proc. helminth. Soc. Wash.*, **35,** 1–9.
Cullen, D. J., 1973. *Nature, Lond.*, **242,** 323–324.
David, F. N. & Moore, P. G., 1954. *Ann. Bot. Lond. N.S.*, **18,** 47–53.
de Bovée, F. & Soyer, J., 1974. *Vie Milieu, sér. B*, **24,** 147–157.
de Bovée, F. & Soyer, J., 1977. *C.N.F.R.A.* (*Com. natl Fr. Rech. Antarct.*), **42,** 237–265.
De Coninck, L. A. P. & Schuurmans Stekhoven, Jr., J. H., 1933. *Mém. Mus. r. Hist. nat. Belg.*, No. 58, 163 pp.
Delmotte, G., 1975. M.Sc. thesis, State University of Ghent, Belgium, 21 pp.
Dinet, A., 1973. *Mar. Biol.*, **20,** 20–26.
Dinet, A., 1977. *Bull. Mus. natn. Hist. nat., Paris, Zool.*, **348,** 1165–1199.
Dinet, A., 1979. *Ambio Special Report*, **6,** 75–77.
Dinet, A. & Vivier, M. H., 1977. *Cah. Biol. mar.*, **18,** 85–97.
Dinet, A. & Vivier, M. H., 1979. *Cah. Biol. mar.*, **20,** 109–123.
Ditlevsen, Hj., 1911. *Vidensk. Meddr. dansk naturh. Foren.*, **63,** 213–256.
Dougherty, E. C., 1960. In, *Nematology. Fundamentals and Recent Advances, with Emphasis on Plant Parasitic Forms*, edited by J. N. Sasser and W. R. Jenkins, University of North Carolina Press, Chapel Hill, North Carolina, pp. 297–318.
Dye, A. H. & Furstenberg, J. P., 1978. *Zoologica Afr.*, **13,** 19–32.
Elliot, J. M., 1971. *Freshwater Biological Association Ferry House, U.K.*, Sci. Publ. 25, 144 pp.
Ellison, R. L., 1984. *Hydrobiologia*, **109,** 131–148.
Elmgren, R., 1978. *Kieler Meeresforsch.*, Sonderh. 4, 1–22.

Elmgren, R., Hansson, S., Larsson, U., Sundelin, B. & Boehm, P. D., 1983. *Mar. Biol.*, **73,** 51–65.
Elmgren, R., Hansson, U. & Sundelin, B., 1980. In, *The "Tsesis" oil spill*, Report to the first year scientific study (October 26, 1977 to December 1978), edited by J. J. Kineman, R. Elmgren & S. Hansson, Boulder, Colorado, U.S. Dept of Commerce. Office of Mar. Pollut. Assessment, pp. 97–126.
Fallon, R. D., Newell, S. Y. & Hopkinson, C. S., 1983. *Mar. Ecol. Progr. Ser.*, **11,** 119–127.
Faubel, A., Hartwig, E. & Thiel, H., 1983. *"Meteor" Forschungsergeb*, Reihe D, No. 36, 35–48.
Fenchel, T., 1974. *Oecologia (Berl.)*, **14,** 317–326.
Fenchel, T. & Riedl, R. J., 1970. *Mar. Biol.*, **7,** 255–268
Findlay, S. E. G., 1981. *Estuar. cstl Shelf Sci.*, **12,** 471–484.
Findlay, S. E. G., 1982a. *Mar. Biol.*, **68,** 223–227.
Findlay, S. E. G., 1982b. *Estuaries*, **5,** 322–324.
Findlay, S. E. G. & Tenore, K. R., 1982. *Mar. Ecol. Progr. Ser.*, **8,** 161–166.
Fricke, A. H. & Fleming, B. W., 1983. In, *Sandy Beaches as Ecosystems*, edited by A. McLachlan & T. Erasmus, Junk, The Hague, pp. 421–431.
Fricke, A. H., Hennig, H. F.-K. O. & Orren, M. J., 1981. *Mar. environ. Res.*, **5,** 59–77.
Ganapati, P. N. & Rao, G. C., 1968. *Mar. Biol. Ass. India*, **4,** 44–57.
Garcia, N., 1982. Thèse présenté à l'université d'Aix-Marseille-II, 82 pp.
Geraert, E., Reuse, C., Van Brussel, D. & Vranken, G., 1981. *Biol. Jb. Dodonaea*, **49,** 148–154.
Gerlach, S. A., 1951. *Kieler Meeresforsch.*, **8,** 106–132.
Gerlach, S. A., 1953. *Z. Morph. Ökol. Tiere*, **41,** 411–512.
Gerlach, S. A., 1954. *Kieler Meeresforsch.*, **10,** 121–129.
Gerlach, S. A., 1971. *Oecologia (Berl.)*, **6,** 176–190.
Gerlach, S. A., 1977a. *Ophelia*, **16,** 151–165.
Gerlach, S. A., 1977b. *Mikrofauna Meeresboden*, **61,** 89–103.
Gerlach, S. A., 1978. *Oecologia (Berl.)*, **33,** 55–69.
Gerlach, S. A. & Riemann, F., 1973/1974. *Veröff. Inst. Meeresforsch. Bremerh.*, Suppl. 4, 1–404 (1973) and 405–734 (1974).
Gerlach, S. A. & Schrage, M., 1971. *Mar. Biol.*, **9,** 274–280.
Gerlach, S. A. & Schrage, M., 1972. *Veröff. Inst. Meeresforsch. Bremerh.*, **14,** 5–11.
Giere, O., 1979. *Cah. Biol. mar.*, **20,** 231–251.
Gourbault, N., 1981. *Cah. Biol. mar.*, **22,** 65–82.
Govaere, J. C. R., Van Damme, D., Heip, C. & De Coninck, L. A. P., 1980. *Helgoländer wiss. Meeresunters.*, **33,** 507–521.
Gray, J. S., 1976. *Estuar. cstl mar. Sci.*, **4,** 653–676.
Gray, J. S. & Mirza, F. B., 1979. *Mar. Pollut. Bull.*, **10,** 142–146.
Gray, J. S. & Rieger, R. M., 1971. *J. mar. biol. Ass. U.K.*, **51,** 1–19.
Green, D. R., Bauden, C., Gretney, W. T. & Wono, C. S., 1974. *Pacif. mar. Sci. Rep.*, **74** (9), 1–39.
Green, R. H., 1979. *Sampling Design and Statistical Methods for Environmental Biologists.* John Wiley & Sons, New York, 257 pp.
Grootaert, P., 1976. *Biol. Jb. Dodonaea*, **44,** 191–202.
Grootaert, P. & Small, R. W., 1982. *Biol. Jb. Dodonaea*, **50,** 135–148.
Guille, A. & Soyer, J., 1968. *Vie Milieu*, **19,** 323–359.
Hagerman, G. M. & Rieger, R. M., 1981. *Mar. Ecol. Progr. Ser.*, **2,** 245–270.
Harris, R. P., 1972. *J. mar. biol. Ass. U.K.*, **52,** 389–403.
Heip, C., 1974. *Nematologica*, **20,** 266–268.
Heip, C., 1975. In, *Proc. 9th European Marine Biology Symposium*, edited by H. Barnes, Aberdeen University Press, Aberdeen, pp. 527–538.
Heip, C., 1977. *Mikrofauna Meeresboden*, **61,** 105–112.
Heip, C., 1980. *Rapp. P.-v. Réun. Cons. int. Explor. Mer*, **179,** 182–187.

Heip, C. & Decraemer, W., 1974. *J. mar. biol. Ass. U.K.*, **54,** 251–255.
Heip, C., Herman, R., Bisschop, G., Govaere, J. C. R., Holvoet, M., Van Damme, D., Vanosmael, C., Willems, K. A. & De Coninck, L. A. P., 1979. *ICES, CM/L9*, 133–163.
Heip, C., Herman, P. M. J. & Coomans, A., 1982. *Academiae Analecta*, **44,** 2, 1–20.
Heip, C., Herman, P. M. J., Smol, N., Van Brussel, D. & Vranken, G., 1985. In, *Biological Processes and Translocations, Vol. 3*, edited by C. Heip & P. Polk, Min. Sci. Policy, Brussels, pp. 11–40.
Heip, C., Herman, R. & Vincx, M., 1983. *Biol. Jb. Dodonaea*, **51,** 116–170.
Heip, C., Herman, R. & Vincx, M., 1984. *Rapp. P.-v. Réun. Cons. perm. int. Explor. Mer*, **183,** 51–56.
Heip, C., Smol, N. & Absillis, V., 1978. *Mar. Biol.*, **45,** 255–260.
Heip, C., Smol, N. & Hautekiet, W., 1974. *Mar. Biol.*, **28,** 79–81.
Heip, C., Vincx, M., Smol, N. & Vranken, G., 1982. *Helminth. Abstr., Ser. B*, **51,** 1–31.
Heip, C., Willems, K. A. & Goossens, A., 1975. *Hydrobiol. Bull.*, **11,** 35–45.
Herman, P. M. J., Vranken, G. & Heip, C., 1984. *Hydrobiologia*, **118,** 21–28.
Herman, R., Vincx, M. & Heip, C., 1985. In, *Biological Processes and Translocations, Vol. 3*, edited by C. Heip & P. Polk, Min. Sci. Policy, Brussels, pp. 41–63.
Hogue, E. W., 1982. *J. mar. Res.*, **40,** 551–573.
Hogue, E. W. & Miller, C. B., 1981. *J. exp. mar. Biol. Ecol.*, **53,** 181–191.
Holme, N. A. & McIntyre, A. D., 1971. *Methods for the Study of Marine Benthos*, Blackwell Sci. Publ., Oxford, 334 pp.
Hope, W. D., 1977. *Mikrofauna Meeresboden*, **61,** 307–308.
Hope, W. D. & Murphy, D., 1969. *Proc. biol. Soc. Wash.*, **82,** 81–92.
Hopper, B. E., Fell, J. W. & Cefalu, R. C., 1973. *Mar. Biol.*, **23,** 293–296.
Hopper, B. E. & Meyers, S. P., 1966a. *Helgoländer wiss. Meeresunters.*, **13,** 444–449.
Hopper, B. E. & Meyers, S. P., 1966b. *Nature, Lond.*, **209,** 899–900.
Hopper, B. E. & Meyers, S. P., 1967a. *Bull. mar. Sci.*, **17,** 471–517.
Hopper, B. E. & Meyers, S. P., 1967b. *Mar. Biol.*, **1,** 85–96.
Howarth, R. W. & Teal, J. M., 1980. *Am. Nat.*, **116,** 862–872.
Howell, R., 1982a. *Mar. Pollut. Bull.*, **13,** 396–398.
Howell, R., 1982b. *Nematologica*, **28,** 110–114.
Howell, R., 1983. *Mar. Pollut. Bull.*, **14,** 263–268.
Howell, R., 1984. *Mar. environ. Res.*, **11,** 153–161.
Howell, R. & Smith, L., 1983. *Nematologica*, **29,** 39–48.
Hulings, N. C., 1971. Editor, *Proc. first int. Conf. on Meiofauna. Smithson. Contr. Zool.*, No. 76, 205 pp.
Hulings, N. C. & Gray, J. S., 1971. Editors, *A Manual for the Study of Meiofauna. Smithson. Contr. Zool.*, No. 78, 83 pp.
Hulings, N. C. & Gray, J. S., 1976. *Mar. Biol.*, **34,** 77–83.
Ito, T., 1978. *J. Fac. Sci. Hokkaido Univ. Ser. 6, Zool.*, **21,** 287–294.
Jansson, B. O., 1967. *Ophelia*, **4,** 173–201.
Jansson, B. O., 1968. *Ophelia*, **5,** 1–72.
Jensen, P., 1974. M.Sc. thesis, State University of Ghent, Belgium, 107 pp.
Jensen, P., 1981a. *Cah. Biol. mar.*, **22,** 231–241.
Jensen, P., 1981b. *Mar. Ecol. Progr. Ser.*, **4,** 203–206.
Jensen, P., 1982. *Nematologica*, **28,** 71–76.
Jensen, P., 1984. *Hydrobiologia*, **108,** 201–217.
Johannes, R. E., 1965. *Limnol. Oceanogr.*, **12,** 189–195.
Joint, I. R., 1978. *Estuar. cstl mar. Sci.*, **7,** 185–195.
Juario, J. V., 1975. *Veröff. Inst. Meeresforsch. Bremerh.*, **15,** 283–337.
Kinne, O., 1977. In, *Marine Ecology, Vol. 3, Cultivation, Part 2*, edited by O. Kinne, Wiley, London, pp. 579–1293.
Kito, K., 1982. *J. Fac. Sci. Hokkaido Univ. Ser. 6 Zool.*, **23,** 143–161.
Krebs, C. J., 1978. *Ecology. The Experimental Analysis of Distribution and Abundance.* Harper and Row, New York, 2nd edition, 678 pp.

Lambshead, P. J. D., Platt, H. M. & Shaw, K. M., 1983. *J. nat. Hist.*, **17**, 859–874.
Lasserre, P., 1976. In, *Benthic Boundary Layer*, edited by N. McCave, Plenum Publishing Co., New York, pp. 95–142.
Lasserre, P., Renaud-Mornant, J. & Castel, J., 1975. In, *10th Europ. Symp. Mar. Biol., Vol. 2*, edited by G. Persoone & E. Jaspers, Universa Press, Wetteren, Belgium, pp. 393–414.
Laughlin, R., 1965. *J. Anim. Ecol.*, **34**, 77–91.
Laybourn, J., 1979. *Oecologia (Berl.)*, **41**, 329–337.
Lee, J. J., Tietjen, J. H., Mastropolo, C. & Rubin, H., 1977. *Helgoländer wiss. Meeresunters.*, **20**, 135–156.
Lee, J. J., Tietjen, J. H., Stone, R. J., Muller, W. A., Rullman, J. & McEnery, J., 1970. *Helgoländer wiss. Meeresunters.*, **20**, 135–156.
Lopez, G., Riemann, F. & Schrage, M., 1979. *Mar. Biol.*, **54**, 311–318.
Lorenzen, S., 1969. *Veröff. Inst. Meeresforsch. Bremerh.*, **11**, 195–238.
Lorenzen, S., 1974. *Veröff. Inst. Meeresforsch. Bremerh.*, **14**, 305–327.
Luc, M. & De Coninck, L. A. P., 1959. *Archs Zool. exp. gén.*, **98**, 103–165.
Man, J. G. de, 1889. *Mém. Soc. zool. Fr.*, **2**, 1–10.
Marchant, R. & Nicholas, W. L., 1974. *Oecologia (Berl.)*, **16**, 237–252.
McIntyre, A. D., 1961. *J. mar. biol. Ass. U.K.*, **41**, 599–616.
McIntyre, A. D., 1964. *J. mar. biol. Ass. U.K.*, **44**, 665–674.
McIntyre, A. D., 1968. *J. Zool.*, **156**, 377–392.
McIntyre, A. D., 1969. *Biol. Rev.*, **44**, 245–290.
McIntyre, A. D., 1977. In, *Ecology of Marine Benthos*, edited by B. C. Coull, University of Columbia Press, Columbia, N. Carolina, pp. 301–308.
McIntyre, A. D. & Murison, D. J., 1973. *J. mar. biol. Ass. U.K.*, **53**, 93–118.
McLachlan, A., 1977. *Zool. Afr.*, **12**, 33–60.
McLachlan, A., 1980. *Trans. R. Soc. S. Afr.*, **44**, 2, 213–223.
McLachlan, A., Winter, P. E. D. & Botha, L., 1977. *Mar. Biol.*, **40**, 355–364.
McLachlan, A., Wooldridge, T. & Dye, A. H., 1981. *S.-Afr. Tydskr. Dierk.*, **16**, 219–231.
McLaren, I. A., 1966. *Biol. Bull. mar. biol. Lab., Woods Hole*, **131**, 457–469.
McNeil, S. & Lawton, J. H., 1970. *Nature, Lond.*, **225**, 472–474.
Meyers, S. P., Feder, W. A. & Tsue, K. M., 1963. *Science*, **141**, 520–522.
Meyers, S. P., Feder, W. A. & Tsue, K. M., 1964. *Devl. ind. Microbiol.*, **5**, 354–364.
Meyers, S. P. & Hopper, B. E., 1966. *Bull. mar. Sci.*, **16**, 142–150.
Meyers, S. P. & Hopper, B. E., 1967. *Helgoländer wiss. Meeresunters.*, **15**, 270–281.
Meyl, A. H., 1954. *Abh. braunschw. wiss. Ges.*, **6**, 84–106.
Meyl, A. H., 1955. *Arch. Hydrobiol.*, **50**, 568–614.
Möller, S., Brenning, U. & Arlt, G., 1976. *Wiss. Z. Univ. Rostock*, **25**, *math.-naturw. R.*, 271–281.
Montagna, P. A., Coull, B. C., Herring, T. C. & Dudley, B. W., 1983. *Estuar. cstl shelf Sci.*, **17**, 381–394.
Moore, P. G., 1971. *J. mar. biol. Ass. U.K.*, **51**, 589–604.
Munro, A. L. S., Wells, J. B. J. & McIntyre, A. D., 1978. *Proc. R. Soc. Edinb.*, **76B**, 297–315.
Muus, B. J., 1967. *Meddr. Danm. Fisk.-og. Havunders.*, **5**, 3–316.
Nicholas, W. L., Grassia, A. & Viswanathan, S., 1973. *Nematologica*, **19**, 411–420.
Nichols, J. A., 1980. *Int. Revue ges. Hydrobiol.*, **65**, 249–257.
Nixon, S. W. & Oviatt, C. A., 1973. *Ecol. Monogr.*, **43**, 463–498.
Nyholm, K. G. & Olsson, I., 1973. *Zoon*, **1**, 69–76.
Olsson, I. & Eriksson, B., 1974. *Zoon*, **2**, 67–84.
Ott, J., 1972a. *Int. Revue ges. Hydrobiol.*, **57**, 645–663.
Ott, J., 1972b. In, *5th Europ. Symp. Mar. Biol.*, edited by B. Battaglia, Piccin Editore, Padova, pp. 275–285.
Ott, J., Rieger, G., Rieger, R. & Enderes, F., 1982. *Mar. Ecol. P.S.Z.N.I.*, **3**, 313–333.
Paetzold, D., 1955. *Wiss. Z. Martin-Luther-Univ. Halle-Wittenb. Math-Nat.*, **4**, 1057–1090.

Paetzold, D., 1958. *Wiss. Z. Martin-Luther-Univ. Halle-Wittenb., Math.-Nat.*, **8,** 17–48.
Pamatmat, M. M., 1980. In, *Marine Benthic Dynamics*, edited by K. R. Tenore & B. C. Coull, University of South Carolina Press, Columbia, S.C., pp. 69–90.
Pamatmat, M. M. & Findlay, S., 1983. *Mar. Ecol. Progr. Ser.*, **11,** 31–38.
Pequegnat, W. E., 1975. *Estuar. Res.*, **2,** 573–583.
Pielou, E. C., 1977. *Mathematical Ecology*. John Wiley & Sons, New York, 385 pp.
Platt, H. M., 1977a. *Estuar. cstl mar. Sci.*, **5,** 685–693.
Platt, H. M., 1977b. *Cah. Biol. mar.*, **18,** 261–273.
Platt, H. M. & Warwick, R. M., 1980. In, *The Shore Environment, Vol. 2, Ecosystems*, edited by J. H. Price, D. E. G. Irvine & W. F. Farnham, Academic Press, London, pp. 729–759.
Platt, H. M. & Warwick, R. M., 1983. *Free-living Marine Nematodes. Part I. British Enoplids.* Cambridge University Press, Cambridge, 307 pp.
Powell, E. N., Bright, T. J., Woods, A. & Gitting, S., 1982. *Mar. Biol.*, **73,** 269–283.
Price, R. & Warwick, R. M., 1980. *Oecologia (Berl.)*, **44,** 145–148.
Rachor, E., 1969. *Z. Morph. Tiere*, **66,** 87–166.
Rachor, E., 1975. "*Meteor*" *Forsch.-Ergebnisse*, **21,** 1–10.
Raffaelli, D. G. & Mason, C. F., 1981. *Mar. Pollut. Bull.*, **12,** 158–163.
Rees, C. B., 1940. *J. mar. biol. Ass. U.K.*, **24,** 195–199.
Reise, K. & Ax, P., 1979. *Mar. Biol.*, **54,** 225–237.
Remane, A., 1933. *Wiss. Meeresunters.* (Abt. Kiel), **21,** 161–221.
Renaud-Debyser, J. & Salvat, B., 1963. *Vie Milieu, Sér. A*, **14,** 463–550.
Renaud-Mornant, J., Gourbault, N., de Panafieu, J.-B. & Helléouet, M.-N., 1981. In, *Actes coll. Intern. COB, Brest, CNEXO, Paris*, pp. 551–561.
Renaud-Mornant, J. & Serène, P., 1967. *Cah. Pacif.*, **11,** 51–73.
Ricklefs, R. E., 1973. *Ecology*. Nelson & Sons, London, 861 pp.
Rieger, R. & Ott, J., 1971. *Vie Milieu*, Suppl. 22, 425–447.
Riemann, F., 1966. *Archs Hydrobiol.*, Suppl. 31, 1–279.
Riemann, F., 1975. *Int. Revue ges. Hydrobiol.*, **60,** 393–407.
Riemann, F., 1980. *Veröff. Inst. Meeresforsch. Bremerh.*, **17,** 213–223.
Riemann, F. & Schrage, M., 1978. *Oecologia (Berl.)*, **34,** 75–88.
Rogers, R. M., 1976. In, *Shell Dredging and Its Influence on Gulf Coast Environments*, edited by A. H. Boume, Gulf Publishing Co., Houston, pp. 327–334.
Romeyn, K. & Bouwman, L. A., 1983. *Hydrobiol. Bull.*, **17,** 103–109.
Romeyn, K., Bouwman, L. A. & Admiraal, W., 1983. *Mar. Ecol. Progr. Ser.*, **12,** 145–153.
Rutgers van der Loeff, M..M. & Lavaleye, M. S. S., 1984. Internal Report 1984-2, Neth. Inst. Sea Res.
Saad, M. A. H. & Arlt, G., 1977. *Cah. Biol. mar.*, **18,** 71–84.
Schiemer, F., 1982a. *Oecologia (Berl.)*, **54,** 108–121.
Schiemer, F., 1982b. *Oecologia (Berl.)*, **54,** 122–128.
Schiemer, F., 1983. *Oikos*, **41,** 32–43.
Schiemer, F. & Duncan, A., 1974. *Oecologia (Berl.)*, **15,** 121–126.
Schiemer, F., Duncan, A. & Klekowski, R. Z., 1980. *Oecologia (Berl.)*, **44,** 205–212.
Schiemer, F., Jensen, P. & Riemann, F., 1983. *Ann. zool. Fenn.*, **20,** 277–291.
Schmidt, P., 1969. *Int. Revue ges. Hydrobiol.*, **54,** 95–174.
Schoener, T. W., 1974. *Science*, **185,** 27–39.
Schuurmans Stekhoven, J. H., 1935. *Nematoda:Systematischer Teil/Nem. errantia—Nem. parasitica*; Tierwelt Nord-u. Ostsee, Lief. 28, 1–173.
Schuurmans Stekhoven, J. H., 1950. *Mém. Inst. r. Sci. nat. Belg.* **37** (2), 1–220.
Seinhorst, J. W., 1959. *Nematologica*, **4,** 67–69.
Sharma, J. & Webster, J. M., 1983. *Estuar. cstl Shelf Sci.*, **16,** 217–227.
Shaw, K. M., Lambshead, P. J. D. & Platt, H. M., 1983. *Mar. Ecol. Progr. Ser.*, **11,** 195–202.
Sherman, K. M. & Coull, B. C. 1980. *J. exp. mar. Biol. Ecol.*, **46,** 59–71.

Sherman, K. M., Reidenauer, J. A., Thistle, D. & Meeter, D., 1983. *Mar. Ecol. Progr. Ser.*, **11**, 23–30.
Shirayama, Y., 1983. *Int. Revue ges. Hydrobiol.*, **68**, 799–810.
Shirayama, Y., 1984. *Oceanol. Acta*, **7**, 113–121.
Sibert, J. R., 1981. *Mar. Biol.*, **46**, 259–265.
Sikora, J. P., Sikora, W. B., Erckenbrecher, C. W. & Coull, B. C., 1977. *Mar. Biol.*, **44**, 7–14.
Sikora, W. B. & Sikora, J. P., 1982. In, *Estuarine Comparisons*, edited by V. S. Kennedy, Academic Press, New York, pp. 269–282.
Skoolmun, P. & Gerlach, S. A., 1971. *Veröff. Inst. Meeresforsch. Bremerh.*, **13**, 119–138.
Smidt, E. L. B., 1951. *Meddr Kommn Danm. Fisk.-og Havunders., Ser. Fiskeri*, **11**, 1–151.
Smol, N., Heip, C. & Govaert, M., 1980. *Annls Soc. r. zool. Belg.*, **110**, 87–103.
Soetaert, K., 1983. M.Sc. thesis, State University of Ghent, Belgium, 133 pp.
Sokal, R. R. & Rohlf, F. J., 1981. *Biometry*, Freeman, San Francisco, 859 pp.
Southwood, T. R. E., 1978. *Ecological Methods*. Chapman & Hall, London, 524 pp.
Soyer, J., 1971. *Vie Milieu, Ser. B*, **22**, 351–424.
Steele, D. H. & Steele, V. J., 1975. *Int. Revue ges. Hydrobiol.*, **60**, 711–715.
Stripp, K., 1969. *Veröff. Inst. Meeresforsch. Bremerh.*, **12**, 65–94.
Sudhaus, W., 1974. *Faun.-Ökol. Mitt.*, **4**, 365–400.
Surey-Gent, S. C., 1981. *Mar. Biol.*, **62**, 157–160.
Tarjan, A. C., 1980. *An Illustrated Guide to the Marine Nematodes*. Florida Institute of Food and Agricultural Sciences, 135 pp.
Taylor, F., 1981. *Am. Nat.*, **117**, 1–23.
Taylor, L. R., 1961. *Nature, Lond.*, **189**, 732–735.
Teal, J. M. & Wieser, W., 1966. *Limnol. Oceanogr.*, **11**, 217–222.
Thiel, H., 1971. *Ber. dt. wiss. Kommn. Meeresforsch.*, **22**, 99–128.
Thiel, H., 1972a. *Verh. dt. zool. Ges., Helgoland*, **65**, 37–42.
Thiel, H., 1972b. "*Meteor*" *Forschungsergeb.*, Reihe D, No. 12, 36–51.
Thiel, H., 1975. *Int. Revue ges. Hydrobiol.*, **60**, 575–606.
Thiel, H., 1979. *Ambio Special Report*, **6**, 25–31.
Thun, W. von, 1966. *Veröff. Inst. Meeresforsch. Bremerh.*, **2**, 277–280.
Thun, W. von, 1968. *Autökologische Untersuchungen an freilebenden Nematoden des Brackwassers*, dissertation, Kiel, 72 pp.
Tietjen, J. H., 1967. *Trans. Am. microsc. Soc.*, **86**, 304–306.
Tietjen, J. H., 1969. *Oecologia* (*Berl.*), **2**, 251–291.
Tietjen, J. H., 1971. *Deep-Sea Res.*, **18**, 941–957.
Tietjen, J. H., 1976. *Deep-Sea Res.*, **23**, 755–768.
Tietjen, J. H., 1977. *Mar. Biol.*, **43**, 123–136.
Tietjen, J. H., 1980a. *Estuar. coast. mar. Sci.*, **10**, 61–73.
Tietjen, J. H., 1980b. In, *Microbiology 1980. VIII. Conference of the American Society of Microbiology on aquatic microbial ecology, 7–10 February 1979, Clearwater Beach, Florida*. Am. Soc. Microbiol. Washington D.C., U.S.A., 335–338.
Tietjen, J. H., 1984. *Deep-Sea Res.*, **31**, 119–132.
Tietjen, J. H. & Lee, J. J., 1972. *Oecologia* (*Berl.*), **10**, 167–176.
Tietjen, J. H. & Lee, J. J., 1973. *Oecologia* (*Berl.*), **12**, 303–314.
Tietjen, J. H. & Lee, J. J., 1975. *Cah. Biol. mar.*, **16**, 685–693.
Tietjen, J. H. & Lee, J. J., 1977a. *Mikrofauna Meeresboden*, **61**, 263–270.
Tietjen, J. H. & Lee, J. J., 1977b. In, *Ecology of the marine benthos*, edited by B. C. Coull, University of South Carolina Press, Columbia, pp. 21–35.
Tietjen, J. H. & Lee, J. J., 1984. *Mar. Environ. Res.*, **11**, 233–251.
Tietjen, J. H., Lee, J. J., Rullman, J., Greengart, A. & Trompeter, J., 1970. *Limnol. Oceanogr.*, **15**, 535–543.
Timm, R. W., 1952. *Contr. Chesapeake biol. Lab.*, **95**, 3–70.
Tomlinson, G. A. & Rothstein, M., 1962. *Biochim. biophys. Acta*, **63**, 465–470.
Trotter, D. B. & Webster, J. M., 1983. *Mar. Biol.*, **78**, 39–43.

Trotter, D. B. & Webster, J. M., 1984. *Mar. Ecol. Progr. Ser.*, **14,** 151–157.
Van Damme, D., Herman, R., Sharma, J., Holvoet, M. & Martens, P., 1980. *ICES, C.M./L*, **23,** 131–170.
Vanfleteren, J. R., 1978. *Ann. Rev. Phytopathol.*, **16,** 131–157.
Vernberg, W. B. & Coull, B. C., 1981. In, *Functional Adaptations of Marine Organisms*, edited by F. J. Vernberg & W. B. Vernberg, Academic Press, New York, pp. 147–177.
Vidakovic, J., 1983. *Mar. Pollut. Bull.*, **14,** 84–88.
Viglierchio, D. R., 1974. *Trans. Am. microsc. Soc.*, **93,** 325–338.
Vitiello, P., 1968. *Recl. Trav. Stn mar. Endoume*, Bull. No. 43, 261–270.
Vitiello, P., 1970. *Téthys*, **2,** 647–690.
Vitiello, P., 1974. *Annls Inst. océanogr., Paris*, **50,** 145–172.
Vitiello, P., 1976. *Annls Inst. océanogr., Paris*, **52,** 283–311.
Vivier, M. H., 1978. *Téthys*, **8,** 307–321.
Vranken, G. & Heip, C., 1983. *Nematologica*, **29,** 468–477.
Vranken, G., Thielemans, L. K., Heip, C. & Vandycke, M., 1981. *Mar. Ecol. Progr. Ser.*, **6,** 67–72.
Vranken, G., Van Brussel, D., Vanderhaeghen, R. & Heip, C. (1984a). In, *Ecological Testing for the Marine Environment, Vol. 2*, edited by G. Persoone *et al.*, State Univ. Ghent & Inst. Mar. Sci. Res. Bredene, Belgium, pp. 159–184.
Vranken, G., Vanderhaeghen, R., Van Brussel, D., Heip, C. & Hermans, D. (1984b). In, *Ecological Testing for the Marine Environment, Vol. 2*, edited by G. Persoone *et al.*, State Univ. Ghent & Inst. Mar. Sci. Res. Bredene, Belgium, pp. 271–291.
Vranken, G., Vincx, M. & Thielemans, L. K., 1982. *Biol. Jb. Dodonaea*, **50,** 93–103.
Ward, A. R., 1973. *Mar. Biol.*, **22,** 53–66.
Ward, A. R., 1975. *Mar. Biol.*, **30,** 217–225.
Warwick, R. M., 1971. *J. mar. biol. Ass. U.K.*, **51,** 439–454.
Warwick, R. M., 1977. In, *Biology of Benthic Organisms, Proc. 11th Europ. Mar. Biol. Symp.*, edited by B. F. Keegan, P. O'Céidigh & P. J. S. Boaden, Oxford, Pergamon Press, pp. 577–585.
Warwick, R. M., 1981a. *Oecologia (Berl.)*, **51,** 318–325.
Warwick, R. M., 1981b. In, *Feeding and Survival of Estuarine Organisms*, edited by N. V. Jones & W. J. Wolff, Plenum Press, New York, pp. 39–52.
Warwick, R. M., 1981c. *Mar. Pollut. Bull.*, **12,** 329–333.
Warwick, R. M. & Buchanan, J. B., 1970. *J. mar. biol. Ass. U.K.*, **50,** 129–146.
Warwick, R. M. & Buchanan, J. B., 1971. *J. mar. biol. Ass. U.K.*, **51,** 355–362.
Warwick, R. M. & Price, R., 1979. *Est. coast. mar. Sci.*, **9,** 257–271.
Wieser, W., 1951. *Öst. zool. Z.*, **3,** 425–480.
Wieser, W., 1953. *Ark. Zool.*, **4,** 439–484.
Wieser, W., 1954. *Acta Univ. lund.*, **50,** 1–148.
Wieser, W., 1959. *Limnol. Oceanogr.*, **4,** 181–194.
Wieser, W., 1960a. *Limnol. Oceanogr.*, **5,** 121–137.
Wieser, W., 1960b. *Int. Revue ges. Hydrobiol.*, **45,** 487–492.
Wieser, W., 1975. *Cah. Biol. mar.*, **16,** 647–670.
Wieser, W. & Kanwisher, J., 1960. *Z. vergl. Physiol.*, **43,** 29–36.
Wieser, W. & Kanwisher, J., 1961. *Limnol. Oceanogr.*, **6,** 252–270.
Wieser, W., Ott, J., Schiemer, F. & Gnaiger, E., 1974. *Mar. Biol.*, **26,** 235–248.
Wieser, W. & Schiemer, F., 1977. *J. exp. mar. Biol. Ecol.*, **26,** 97–106.
Wigley, R. L. & McIntyre, A. D., 1964. *Limnol. Oceanogr.*, **9,** 485–493.
Willems, K. A., Vincx, M., Claeys, D., Vanosmael, C. & Heip, C., 1982. *J. mar. biol. Ass. U.K.*, **62,** 535–548.
Willems, K. A., Sharma, J., Heip, C. & Sandee, A., in press. *Neth. J. Sea Res.*
Witte, J. I. & Zijlstra, J. J., 1984. *Mar. Ecol. Progr. Ser.*, **14,** 129–138.

Wormald, A. P., 1976. *Environ. Pollut.*, **11,** 117–130.
Wright, D. J. & Newall, D. R., 1976. In, *The Organisation of Nematodes*, edited by N. A. Crossl, Academic Press, New York, pp. 163–210.
Wright, D. J. & Newall, D. R., 1980. In, *Nematodes as Biological Models. Vol. II. Aging and other Models*, edited by B. M. Zuckerman, Academic Press, New York, pp. 143–164.
Zaika, V. E., 1973. *Specific Production of Aquatic Invertebrates*, Halsted Press, John Wiley, New York, 154 pp.
Zaika, V. E. & Makarova, N. P., 1979. *Mar. Ecol. Progr. Ser.*, **1,** 153–158.
Zeuthen, E., 1953. *Q. Rev. Biol.*, **28,** 1–12.

Oceanogr. Mar. Biol. Ann. Rev., 1985, **23**, 491–571
Margaret Barnes, Ed.
Aberdeen University Press

CETACEANS IN THE NORTHWESTERN MEDITERRANEAN: THEIR PLACE IN THE ECOSYSTEM

DENISE VIALE

Université Paris VI, Station Zoologique, F-06230 Villefranche-sur-Mer and *Laboratoire d'Ecologie Numérique, Université des Sciences et Techniques de Lille, F-59655 Villeneuve d'Ascq, France*

INTRODUCTION

Humans do not have the same reactions to cetaceans as they do to other marine animals. Considered for a long time as itinerant oil barrels, cetaceans enjoy now public protection due to the prowess of the little Odontoceta in dolphinariums. As to big whales living in Mediterranean, they seem left from a myth: at the time of my first publication on the subject (Viale, 1973), in biologists' circles, speaking about the existence of big whales in Mediterranean provoked doubt or even incredulity. The Mediterranean, an oligotrophic sea, seemed unlikely to contain big cetaceans except by accident. On the contrary, my observations since 1962 lead me to infer that their presence was regular and linked to the oceanography of the area, and to the mechanisms of primary and secondary production. In other respects, since 1972, cetaceans have appeared to me like privileged indicators of chemical pollution in the Mediterranean. Their place at the top of the trophic chains makes them sensitive to the concentrations of pollutants. Because they are mammals and eat the same fishes and cephalopods as man, they are of interest as witnesses of the possible effects of intoxication of the marine environment by industrial effluents.

MATERIAL AND METHODS

Due to the impossibility of collecting material at will, the cetologist is obliged to use varied and different ways of getting information. Such heterogeneous information becomes worthwhile considering the cross-checking it permits.

A result in cetology is always single and historic; it is not foreseeable and is not reproducible in its detail. Cetaceans are scarce animals by comparison with other marine species and, furthermore, largely wandering; so the unit sample has characteristics adapted to the statistics of the rare events, for example, a day of constant observation at sea, or beachings observed on 1000 km of seashore during one month or one year. Every casual event, like an individual or collective beaching, a sighting from the shore, an accidental catch, poaching or an intentional destruction, has to be taken into account in cetological investigations.

In the Mediterranean, exploitation of whales existed, more and less sporadically, between 1921 and 1954. Between 1921 and 1927, 6000 catches of *Balaenoptera* and *Physeter* were reported. There are also reports of beachings or accidental catches spread throughout the literature. These, together with my recent observations and inquiries, amount in a century to 115 observations or beachings and accidents of fin whales, 25 of sperm whales, 57 of *Ziphius*, 62 of *Grampus*, 12 of *Globicephala*, 75 of *Tursiops*, 60 of *Delphinus*, 62 of *Stenella*, 7 of *Orca*, and 1 of *Monodon* (Viale, 1977b). These numbers will appear ridiculous to marine biologists other than cetologists, and show the importance, in this field, of any isolated observation. After statistical treatment when possible, connections may be seen between (1) spatio-temporal localizations of the sightings and catches; (2) analysis of biological conditions, reproductive behaviour, pregnancy, nursing, observation of newborns; and (3) searching for causes of strandings through anatomo-pathological examinations.

Information on strandings and catches used here come from four sources.

(1) Casual scientific publications and departmental archives (detailed in Viale, 1977b).
(2) Collections of the museums around the Mediterranean, of the Museum d'Histoire Naturelle of Paris, of the British Museum, which were partially indexed by Rode (1939), Tortonese (1963c), Paulus (1966), Arbocco (1969), Casinos & Filella (1976), Poggi (1982).
(3) Catching statistics of a whaling industry that existed between 1921 and 1954 in the Gibraltar zone, published by the International Whaling Commission (I.W.C.) (Anonymous, 1942), which were partially rewritten by Aloncle (1964) and by Jonsgård (1966). These data are difficult to use because Mediterranean catches and total catches of Spain and Portugal are often mixed.
(4) Annual stranding reports published by Budker (1968), Duguy & Budker (1972), Duguy (1973, 1974, 1975, 1976, 1977, 1978, 1979, 1980, 1981, 1982, 1983a,b). The commission of Cetology of Barcelona has also published stranding reports since 1972 (Casinos & Filella, 1976, 1981; Casinos & Vericad, 1976).

Such a method of collecting data is of necessity random in nature. The density of information depends on the density of observers. Strandings are aleatory but their detection and reporting are not. The network of observers can become more or less dense due to the sensitivity of the public to the problems of cetaceans. In another way, the working conditions (remoteness of the place of a stranding from a laboratory, *etc.*) limit the possibility of studying the occasional carcases. In fact, the network of observers seems to have been quite stable since 1972; events are regularly reported to the "Centre d'Étude des Mammifères Marins" (C.E.M.M.) of La Rochelle. The number of strandings in summer, during holiday time in coastal areas, does not seem to be systematically larger than the number of winter reports, although the opposite occurs for some species. On the contrary, results before 1971 indicated a larger number of reported beachings and sightings near the big urban centres; this quantity depends on the population of observers. The quantitative analysis of the observation frequencies will, therefore, only take into account the years from 1971 to 1981.

For the open sea observations, there are four sources of information.

(1) The literature data (less numerous).
(2) Reports from the regular boats, centralized by C.E.M.M.
(3) Eighteen personal expeditions in the open sea between 1972 and 1982, from which 11 were aboard the oceanographic vessel KOROTNEFF (C.N.R.S.-PIRO, Marine Station of Villefranche-sur-Mer).
(4) A mensual programme of observations aboard the KOROTNEFF that have been set up since 1980 (see Giordano, 1981 and Palazzoli, 1983).

The present review concerns the following zones: the northern half of the Tyrrhenian Sea, the Ligurian Sea, and the northeastern part of the Algero-Provençal Basin. For the erratic big cetaceans, it is necessary to consider a larger zone between the European coasts and the 40th parallel. The considered time-interval includes a period of ecological and ethological inquiries from 1962 to 1971, then a systematic investigation with statistical methods from 1972 to 1982. The examined material is described in Viale (1977b).

HYDROLOGY AND PLANKTONIC PRODUCTION IN THE AREA STUDIED

CURRENTS

A knowledge of the currents and temperatures is necessary in order to understand the cetological data in the area studied. The western Mediterranean basin is separated from the adjacent basins by, in the west, the Gibraltar threshold and, in the east, the Sicilian-Tunisian and the Tyrrhenian thresholds. The hydrological characters of the western basin are determined by (1) a water balance—there is an entrance of water across the thresholds, compensated for by an evaporation of about 1 m of water by year (Tchernia, 1960), (2) an annual cycle of caloric exchanges between ocean and atmosphere, and (3) a general cyclonic circulation of geostrophic origin.

The principal features of this hydrological regime are given in Lacombe & Tchernia (1960), Tchernia (1960), Gostan & Nival (1963), Le Floch (1963a,b), Gostan (1967a,b,c, 1968a,b) and the Mediprod Expeditions working group (Minas, 1968; Minas & Blanc, 1970; Coste, Gostan & Minas, 1972; Nival, Nival & Thiriot, 1975; Coste & Minas, 1981). There is a surface water layer (0 to 100 or 200 m), an intermediate water layer—warm and of high salinity—(300 to 600 m) originating in the eastern basin, and a deep water layer.

The origin of the surface water is diverse; its principal characteristic is the presence of Atlantic water, which penetrates through the Strait of Gibraltar, forms a current with a speed of 16 to 20 miles per day along the Algerian coast, and then breaks up into two branches: an Algerian-Provençal Current which flows across the western basin and then joins, after flowing to the west of Sardinia and Corsica, the Algerian-Provençal eddy; the other branch turns westwards along the coasts of Sicily and Italy. In addition, a drift leads the surface water from the Esterel to the northwestern coasts of Corsica, in autumn and winter. This conjunction of currents explains some localizations of strandings. At the latitude of Cape Corse, the Atlantic Current divides into three branches: one turns from the Cape southwards and runs along the eastern coast of Corsica; the second washes the Toscan Archipelago, and then reaches the Toscan coasts; and the third turns into the Genoa Gulf, progressively losing its identity, and ends running westwards along the Riviera and the Côte d'Azur. The Tyrrhenian Sea is also occupied by a cyclonic circulation which drifts water northwards, along the Italian coast up to the Strait of Corsica, where this water can reach the Ligurian-Provençal eddy (Gostan, 1967c). The principal features of this circulation are represented in the Figures 1 and 2.

The Atlantic water is characterized by its salinity: from 36.25°/$_{oo}$ at Gibraltar, to 38·28°/$_{oo}$ in the Gulf of Genoa. This water mass gets mixed with a continental water of fluvian origin (essentially water of the Rhône, for the zone concerned here) and forms a current towards the Balearic Islands, and a countercurrent along Roussillon and Languedoc as far as Marseilles. This continental water forms a layer of low salinity, a few metres thick, which gets mixed with Mediterranean water of the Catalanian Current, and dilutes this.

Gostan (1967c) calculated fluxes and speeds perpendicular to a transect from Nice to Calvi, and found a permanent cyclonic current describing a rough ellipse around the point 43° N: 8° E; this point is the centre of one of the most important divergences, where the intermediate water rises to the surface in winter. These average conditions may be affected seasonally or for short periods by a non-negligible variability related to the wind. The Gulf of Genoa is a permanent centre of relative depression and, therefore, is the origin of a cyclonic movement in the atmosphere. These winds are unstable and vary in direction. They have an immediate effect on the surface currents, because they are able to annul or even reverse the geostrophic current. Le Pichon & Troadec (1963) proved that the Mistral (wind from the north) sometimes has an action on the surface hydrology, from the Rhône estuary as far as Nice. On the contrary, along the transect from Nice to Cape Corse, the main winds blow

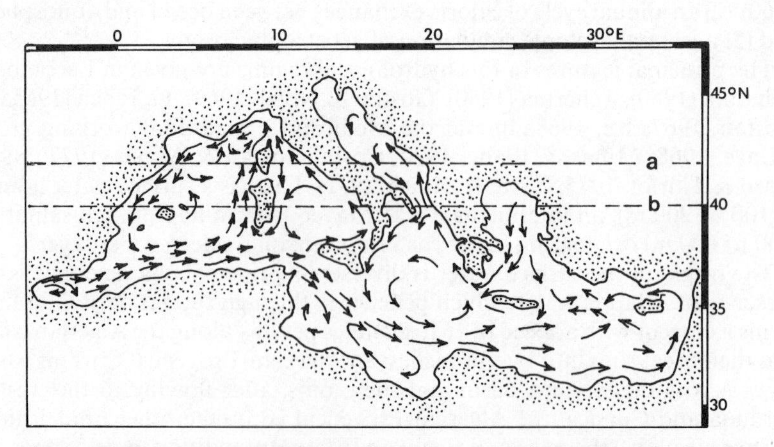

Fig. 1.—Surface currents of the Mediterranean and the southern limits of (a) the "restricted zone" (north from 42° N), and (b) the "enlarged zone" (north from 40° N) (from Anonymous, 1968).

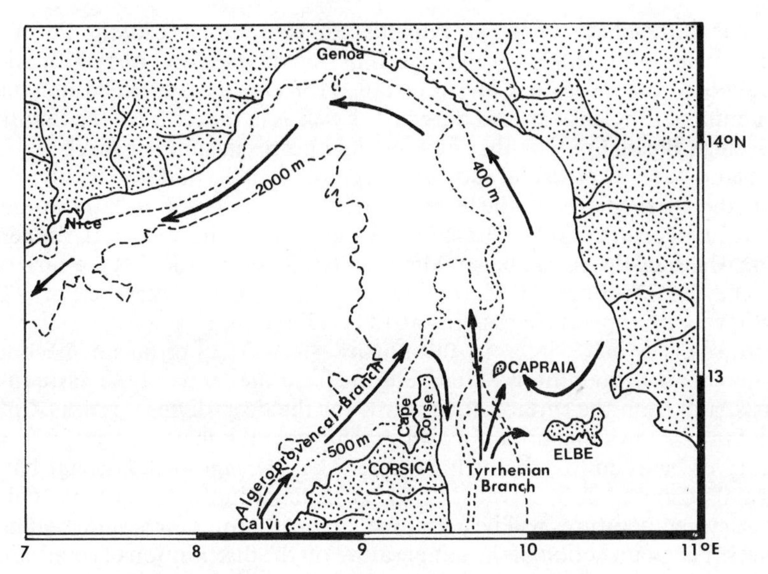

Fig. 2.—Surface currents, detail of the northeastern part of the basin studied (from Gostan, 1967c).

westwards and intensify the geostrophic stream along the Ligurian coast, whereas they annul it along the western coast of Corsica.

The intermediate water is formed in the eastern Mediterranean basin and enters the western basin over the Sicilian-Tunisian threshold. A branch diverges towards Gibraltar, parallel to the North African coast. Another flows northwards along the western coast of Sardinia and Corsica, then turns into the Gulf of Genoa and runs along the Riviera. A third branch, less important, enters the Tyrrhenian Sea, runs along the western Italian coast and enters the Ligurian Sea through the Strait of Corsica, but at a depth of at most 40 m above the threshold, and following an impulsion rhythm. A countercurrent southwards runs along the eastern coast of Corsica, vertically to the isobaths 400 to 500 m (Le Floch, 1963b) and involves the whole water layer; thus, in winter, there is an unvarying temperature of 14 °C from the bottom to the surface.

The deep water layer, homogeneous in the basin (12·7 °C; 38·40°/$_{oo}$) comes from surface waters of the northern part of the basin, which are cooled in winter by continental cold and dry winds, and sink along the continental slope (Bougis & Ruivo, 1954; Lacombe & Tchernia, 1960).

TEMPERATURE

Seasonal temperature variations of the coastal waters have been studied in Villefranche-sur-Mer (Bougis & Carré, 1960), in Marseilles (Leveau, 1965), and in Banyuls-sur-Mer (Thiriot, 1966). The graphs given by Thiriot show a regular and symmetrical oscillation of the surface water temperature. This temperature does not reach 13 °C in the coolest months in the Gulf of Genoa, whereas it reaches 10 °C at Banyuls. In summer, it exceeds 25 °C at Villefranche, but is still lower than 23 °C at Banyuls (see Fig. 10, p. 522). Along a transect from Nice to Calvi (Gostan, 1968a,b), the surface water has a minimum of 13 °C in March at the coast as well as in the centre of the Ligurian-Provençal Basin. On the other hand, temperatures in summer reach 26 °C in Calvi and Nice, whereas they do not exceed 24 °C in the middle of the transect.

On the Tyrrhenian side, the temperature varies from 14 °C in January (equal to the temperature of the intermediate water) to 26 °C in August. In the centre of the Algerian-Provençal basin, Minas (1968) observed 13 °C at the surface in March, 21 °C in June, 23 °C in October, and 14 °C in December; at 300 m depth, water temperature remains at 13 °C all the year.

Cetaceans which winter in the northwestern Mediterranean meet with almost identical conditions of temperature from the centre of the basin to the coasts, and from the surface downwards for the deep-diving species (*Ziphius* and *Physeter*). In summer, surface water temperatures do not exceed 26 °C and will never be a limiting factor for species like *Delphinus delphis*, that have a thermal barrier at 27 °C (Evans, 1976), particularly in the neritic zone. On the contrary, temperatures will be a limiting factor in winter for several delphinid species. The main action of the temperature on the distribution of cetaceans is, however, indirect, by the way of hydrological structures which condition the primary productivity.

THERMOCLINAL STRUCTURES AND UPWELLINGS

The cyclonic circulation leads to divergences. The most important ones are the Ligurian Divergence, centred approximately at 43° N: 8° E, and the Algerian-Provençal Divergence, centred at 42° N: 5° E. Two others are situated in the Catalan Sea and in Alboran Sea. These divergences bring to the surface a mass of intermediate water, hence supplying nitrates and phosphates to the euphotic layer. They have a seasonal character (Gostan, 1967a,b; Minas, 1968).

Winter conditions

In this region, the winters are variable, characterized by the presence of sometimes maritime, soft and wet, sometimes continental, cold and dry air masses. In the first case, the three-layer stratification described above can be seen. In the second case, vertical movements of water masses are observed; the evaporation and cooling induce the formation of a layer with a temperature below 13 °C and a salinity above 38‰, making the stability precarious. In these conditions, winds induce a vertical eddy which homogenize the first and the second water layers in the whole Gulf of Genoa. This phenomenon determines the enrichment of nutrients in the surface layers. In February, a real upwelling situation can be seen, when the thermocline reaches the surface. The intermediate water getting to the surface is caught in the cyclonic circulation and directed towards the coasts.

Summer conditions

Between March and May, the surface 'warming-up' reaches the first 30 m (21 °C, 38·06‰), then 50 m, and later 100 m, recreating the individuality of the surface water layer and the three-layered stratification (Furnestin & Allain, 1962). The thermocline reaches 50 m only in the two central zones and determines a domed separating surface between surface and intermediate waters. Elsewhere, its sinking below the euphotic layer (up to 200 m), particularly in the Tyrrhenian Sea, increases the oligotrophy of the surface water layers, depriving them of new or regenerated production (Frontier, 1978). The thermocline acts as a relative barrier both for diffusion upwards of nutrients, and sedimentation of organic material from the surface layer (Jacques, 1974). Thus, the phyto- and zooplanktonic production is largely determined by the presence of the thermocline inside the eutrophic layer. It follows that seasonal variations of the hydrographic structure have large consequences on primary and secondary productions and, consequently, on the food supply for cetaceans.

PRIMARY PRODUCTION

In the coastal zone, works published before 1960 (see Jacques, Minas, Minas & Nival, 1973) indicate that primary production for French and Spanish coasts varies between 50 and 100 mg $C \cdot m^{-2} \cdot day^{-1}$. The highest values are as high as those in the seas considered as rich. On the contrary, water of the open sea was up to now considered as uniformly poor; more recently, the Medoc and Mediprod Expeditions have defined characters of the primary production in the open sea of the zone being considered. This production is determined by three factors: presence of nutrients, penetration of light, and vertical stability. These factors are not always compatible; vertical stability excludes the vertical movements of mixing, and consequently prevents the nutrient enrichment of the surface layer. Spatial and seasonal variations reflect these antagonist features. The main source of nutrients is related to upwellings depending on the principal divergences. Surface enriched waters are driven towards the coast, where a residual stability subsists at the end of winter and the beginning of spring. It follows that the productivity is, in this period, maximal near the coast, symmetrically on each sides of the divergences, where both nutrients are in enough quantity and stability is still effective (Jacques *et al.*, 1973; Jacques, 1974). Primary production is highest in March and April for the whole region.

At the beginning of summer, the maximum of productivity moves towards the centre of the divergence, as the stratification restores. Simultaneously, the absolute production decreases following the reduction of the nutrient supply. In June, the maximum of productivity is localized at the divergence because at this moment the stratification reaches this zone. After a short time, the nutrient supply does not reach the surface; a productivity dome subsists, reaching the 50-m depth. Surface waters are quickly depleted of nutrients. The oligotrophy, characteristic of the summer conditions, becomes general from the coast to the centre of the divergences, from July onwards. The variations of the primary production follow the sinking of the thermocline; the relative maximum of productivity is at 50 m depth in July, and between 75 and 100 m depth in September. Light then becomes a limiting factor and the production becomes very low.

Other essential factors are the coastal upwellings induced by the winds blowing from land: the Mistral in the Languedoc-Roussillon area, the Tramontane in the Gulf of Genoa, and the Libeccio on the eastern coast of

TABLE I

Comparison between the Provençal region and the Côte d'Azur–Corsica region for chlorophyll and primary production (from Jacques et al., *1973)*

	Provençal region 16 000 km²		Côte d'Azur–Corsica 8500 km²	
	March	April	March	April
Chlorophyll *a*				
mg per m^3	35	98	49	85
total (tonnes)	580	1569	425	724
Daily primary production				
mg C per m^2	311	606	364	819
total (tonnes of C)	4969	9639	3101	6962

Corsica. But the upwelled waters are very poor in nutrients and do not provoke any phytoplankton bloom (Minas, 1974).

A comparative balance of the primary production and of the vegetal biomass of the 0 to 200-m layer, for the Provençal region and for the Côte d'Azur and Corsica region, is given by Jacques *et al.* (1973) (see Table I). The Côte d'Azur and Corsica region is a zone of less intensive upwellings than the Provençal one, because the winter conditions are less drastic. Personal winter observations in 1980, 1981, and 1982 indicate that cetaceans seem to prefer these milder conditions. In spite of this, it can be seen, from Table I, that the total diurnal production, areas being equal, is more important in the first zone. Jacques *et al.* (1973), Minas (1974), and Nival *et al.* (1975) recall the rising of 'blocks' of intermediate water observed during the Mediprod I and II Expeditions to explain the lack of symmetry.

SECONDARY PRODUCTION: ZOOPLANKTONIC AND MICRONECTONIC BIOMASS

Whereas primary production is sometimes high, secondary and tertiary production are always low. Minas (1968) attributes this to the dispersion of the phytoplanktonic cells within the euphotic layer. Zooplankton has, therefore, to expend excessive energy in food collection.

Secondary production of the zone studied is an object of investigation at the Marine Stations of Villefranche and Banyuls. Nival *et al.* (1972, 1975) have estimated the zooplanktonic production and the phyto-zooplankton relations. During the first half of March, the maximum zooplanktonic biomass is about 1600 mg (dry wt)$\cdot m^{-2}$ and is found symmetrically on both sides of the Ligurian Divergence. The poorer coastal zones have a biomass varying between 500 and 1000 mg$\cdot m^{-2}$; the central zone contains 500 mg$\cdot m^{-2}$. On the other hand, Razouls & Thiriot (1973) have studied the horizontal distribution of the biomass and of the production of the mesoplankton in the 0 to 200-m layer in winter, comparing the results of five expeditions. The mesoplankton is essentially composed of copepods, mainly primary consumers. The samples show that the geographical distribution may be considered without taking into account the hour of sampling. Table II summarizes the results. In March and April, during the spring blooms of phytoplankton, the distribution of mesoplankton clearly shows that the enrichment induced by the vertical mixing has repercussions on the secondary production.

The Atlantic Current does not cause any enrichment in the northern part of the basin, whereas in the southern part (*i.e.* off Algeria) the mesoplankton distribution shows important variations related to this current as far as the Strait of Sardinia and along the western coast of Corsica.

In summer conditions, only estimations of biomass are available. Boucher & Thiriot (1973) studied the geographical distribution of these biomasses in the whole western Mediterranean and demonstrated the existence of an important diurnal variation. The mean biomass of micronecton is ten times greater at night than during the day; that of macroplankton is twice as much in the central basin, and three times greater in the Gulf of Genoa. The geographical distribution of macroplankton does not show any pecularities. On the contrary, the most important concentrations of micronecton are found in the Alboran Sea, due to the nutrient enrichment by the Atlantic waters and by a divergence.

The euphausiids are more abundant north from the 40th parallel, particularly in the Gulf of Genoa. A quantitative study of the cnidarians and euphausiids in summer, in the west-north divergence, was made by Goy & Thiriot (1976). At the time of the phyto- and mesoplanktonic blooms, the macroplanktonic, and still more the micronectonic species, are scarce. On the contrary, in May these species reach high concentrations whereas the mesoplanktonic biomass falls. In July, the same authors measured the amount of macroplankton and micronecton along the Nice to Calvi transect (90 miles);

TABLE II

Zooplankton biomass and production in the Provençal region and Côte d'Azur to Corse (from Razouls & Thiriot, 1973)

<table>
<tr><th></th><th colspan="2">Provençal region</th><th colspan="2">Côte d'Azur–Corse</th></tr>
<tr><th></th><th>March</th><th>April</th><th>March</th><th>April</th></tr>
<tr><td>Biomass, mg dry wt per m² at 0 to 200 m</td><td>max. 250
aver. 228</td><td>max. 9560
aver. 3162
min. 1028</td><td>aver. 373·4</td><td>max. 9560
aver. 2754</td></tr>
<tr><td>Daily production, average over 30 days, mg of C</td><td colspan="2">max. 223·8
aver. 96·2
min. 36·8</td><td colspan="2">max. 305·53
aver. 77·99
min. 2·33</td></tr>
</table>

medusans and chaetognaths formed an important fraction of the macroplankton 12 miles off Nice, then disappeared at the level of the divergence, whereas crustaceans constituted 50% of the macroplankton and micronecton (of which 43·5% were euphausiids).

HIGHER TROPHIC LEVELS

It is necessary to examine here the production of the cephalopods, which form the principal food for many cetaceans, and also the production of sardines and other 'blue fish'. Some biomass data will be summarized later (p. 556). Data from Margalef (1967) concerning a coastal zone in the western Mediterranean give an idea of the very low transfer rate from the primary production to the production of pelagic fishes (Fig. 3). The great importance of ciliates, whose production equals that of the other zooplanktonic animals, and which are generally under-considered, is to be noted. The rôle of bacteria has not been estimated.

It may be concluded that there is a lack of information about the biological environment of the Mediterranean. What little there is, however, will permit an interpretation of some features of the behaviour and ecology of cetaceans in this region. It is hoped that international programmes take into account the necessity of undertaking intensive investigations, with the aim of analysing more thoroughly the ecosystems of the Mediterranean.

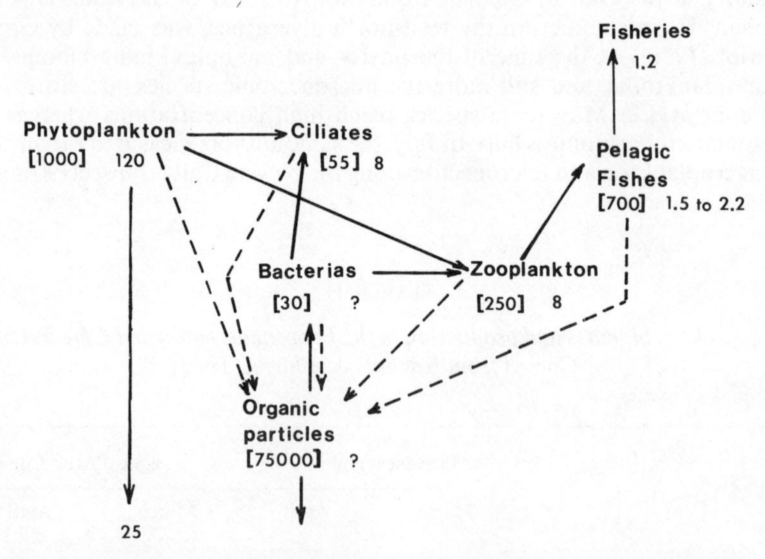

Fig. 3.—Simplified trophic network in a coastal region of the western Mediterranean: numbers in brackets are standing crops in mg of C per m^2, others are daily production rates in mg of C per m^2; from Margalef (1967).

TABLE III

Species of cetaceans sighted or reported in the northwestern Mediterranean

(1) Species with regular distribution, allowing an estimation of abundance.
Mysticeta: *Balaenoptera physalus* Linné, 1758
Odontoceta: *Tursiops truncatus* (Montagu, 1921)
Delphinus delphis Linné, 1758
Stenella coeruleoalba (Meyen, 1833)
Grampus griseus (G. Cuvier, 1812)

(2) Species with a heterogeneous distribution, not allowing any reliable estimation of abundance.
Odontoceta: *Physeter macrocephalus* Linné, 1804
Ziphius cavirostris (G. Cuvier, 1823)
Globicephala melaena (Trail, 1809)

(3) Rare species.
Mysticeta: *Balaenoptera acutorostrata* Lacépède, 1804
Balaenoptera borealis (Lesson, 1821)
Orcinus orca (Linné, 1758)
Pseudorca crassidens (Owen, 1846)

(4) Accidental species.
Odontoceta: *Monodon monoceros* Linné, 1758
Steno bredanensis Lesson, 1821
Mesoplodon bidens (Sowerby, 1804)
Mesoplodon densirostris (Blainville, 1817)

(5) Reported but doubtful species.
Mysticeta: *Balaenoptera musculus* Linné, 1758
Megaptera novaeangliae Borowski, 1781
Odontoceta: *Hyperoodon ampullatus* (Forster, 1770)
Phocaena phocaena (Linné, 1758)

ECOLOGICAL AND ETHOLOGICAL OBSERVATIONS ON CETACEANS OF THE NORTHWESTERN MEDITERRANEAN

We have to distinguish the species considered as frequent, whose distribution is sufficiently homogeneous to allow an estimation of abundance; the species which are present, but with heterogeneous distribution that does not allow an estimation of abundance by way of transect-lines method, although I shall attempt it, but without being able to guarantee the reliability; and finally, some other scarce species, and those whose occurrences are considered as accidental or doubtful. The complete list of the species sighted in Mediterranean is given in Table III.

BALAENOPTERA PHYSALUS

Fin whale, finback whale; in French: baleinoptère, rorqual commun.

B. physalus is the only representative of the Mysticeta regularly found in Mediterranean.

Whaling campaigns

In the past and still today, fin whales constitute the main part of the catches of whaling campaigns off the coast of Spain. In the Strait of Gibraltar, whaling exploitation began in 1921 on the coast of Getares (Algeciras Bay) with an international company directed by a Scandinavian crew. An attempt was made in 1926 to put a whaling factory on the Moroccan coast, but the Rifian people fought it and cut short this project. An offshore whaling factory was then put at Ceuta, after the Spanish colonization of North Morocco. The catches were, however, incompletely reported and are, therefore, difficult to utilize statistically.

The main catches in Getares were fin whales; for example, in May 1924, 15 adults were captured in less than two weeks (Cabrera, 1925; Casinos & Vericad, 1976). Fraser (pers. comm.), who lived a month in Getares, told me that this station was very busy in January and February, 1926, when large whales were captured in the Strait off the Atlantic shore. Aloncle (1964) reported that fin whales were captured in large numbers off Morocco, on the Atlantic side of the Strait, in front of Getares: 153 catches by the shore-station of Benzou between 1949 and 1954 (Anonymous, 1942). Whaling was, therefore, practised on the both sides of the Strait of Gibraltar, *i.e.* on the Mediterranean and Atlantic shores. This means that the catches included those whales entering as well as leaving the Mediterranean. Consequently, these whaling activities should have provided for several years a great deal of information on the movements of the large whales in and out the Mediterranean. But the information is in fact unusable from our point of view because no distinction was made between those entering or leaving. From 1921 to 1927, 6250 large whales were caught on Spanish coasts, mainly in the Strait of Gibraltar (Jonsgård, 1966); other statistics (Anonymous, 1942) show that 6433 (*i.e.* 92%) of 6990 large whales captured are *B. physalus*. According to Ingebrigsten (1929), the main catches in the Strait of Gibraltar were from the end of November up to the beginning of April and in this period, baleen whales were then fattest. In fact, Jonsgård (1966) said that this Strait was a calving area during the end of the autumn and during winter, and later a feeding ground. The exploitation of a large whale population in the very period of mating and breeding had rapid consequences; there was a breakdown of the catches from 1926 onwards, not only in the Mediterranean but also off the western Scottish coast. Jonsgård (1966) infers the hypothesis that fin whales wintering in the Strait of Gibraltar and in Mediterranean go in summer to the northwestern coasts of Scotland and around the Shetland Islands. According to him, the fact strongly in favour of this point is, that the oil extracted from fin whales captured in Shetland in summer was richer in iodine than that extracted from fin whales caught in other parts of the North Atlantic. This feature has to be related with the large amount of iodine in the Mediterranean biomass in comparison with the Atlantic one. This relation suggests that the fin whales come into the Mediterranean for feeding when they come for wintering.

A complete lack of statistical reports between 1927 and 1933 suggests that Mediterranean whaling disappeared or was not very busy; nevertheless, between 1933 and 1939, the exploitation of whales seemed to rise again but was supported exclusively by sperm whales and small species of whales such as *Ziphius*, *Globicephala*, *Balaenoptera acutorostrata* and *B. borealis*. Sperm whales themselves became scarce, as attested by the whaling catches for summer 1934: 66 *Balaenoptera* (all species mixed) and 5 sperm whales (Anonymous, 1942). Further lack of data corresponds to the World War II period of 1939 to 1945; after the war there was whaling, as reported by Aloncle (1964). A whaler who practised in the Strait of Gibraltar from 1950 onwards reported to me that the baleen whales which enter the Mediterranean in May to June, come back again in the autumn; according to him, they come from Mauritania. Jonsgård (1966) too, supposed that the Mediterranean is used by fin whales as a feeding ground in summer.

Whaling ceased after 1960 in the Strait of Gibraltar but not in the north of Spain, such as Biscayan Bay where two factories are maintained with offices in Vigo, and with five whaling boats (1000 horse-power, 16 men in the crew) fishing for them. The catches of these boats from 1966 to 1974 have been analysed (Viale & Ridell, 1975; Viale, 1977a); there was a mean of 18 catches for each boat per year, *i.e.* an average of 90 whales captured each year in the entrance to Biscayan Bay. This catch appeared to Sergeant (1977) very excessive considering the scarcity according to my personal data, of the actual stock in the Biscayan Bay. Another sign of 'overwhaling' was that the females represent 39·3% of the whole catch; this value is significantly different from the normal value 50% at a probability level of 1%. This points to a lack of balance in the sex ratio of the whole population, and hence a decrease of the capacity to restore the stock. This stock is, however, still exploited today with a catching effort asserted to be constant by Aguilar & Lens (1981); but this whaling has spread alarm in the international whale protectionist circles in 1982.

The herd of *B. physalus* summering in the western Mediterranean is probably connected with the stock exploited in the Biscayan Bay; this Bay also constitutes a feeding ground (Jonsgård, 1966). These two herds are sighted at the same time off Gibraltar in May and June, and they are sighted on their respective feeding grounds in July and August. Gruvel (1924), Maigret, Trottignon & Duguy (1976) relate sightings of *B. physalus* off Mauritania between 19 and 21° N, in December, March and April. These sightings are the southernmost for this species in the Northern Hemisphere. Maigret *et al.* (1976) report two strandings of young males (9·5 and 10·2 m) in March 1971 and March 1973 on the Mauritanian coast and suppose, in order to explain them, that there is a migration along the Mauritanian coast. It seems probable that Mauritanian waters constitute a wintering ground for *B. physalus* of the Northern Atlantic, with the possibility of calving and breeding. Each year, when going up along the West African coast towards the summering grounds, a part of this stock enters the northwestern Mediterranean and constitutes the herd observed in the Ligurian Sea in July or in the Algerian-Provençal basin in August. The predation made by whaling in the Biscayan Bay has repercussions on the whole stock and probably explains why the herd of *B. physalus* remains at a very low level, although whaling has been banned in this area at least since 1955.

In order to clarify the summer migrations of *B. physalus*, inquiries were made in summer 1983 by the Oceanographic Research Center of Dakar, Senegal (C.R.O.D.T.). On the one hand the oceanographers and the traditional fishermen were questioned and, on another, the reports from industrial fishing boats were examined for the 1981 tuna-fishing campaign. 1981 was a year of universal effort for studying the effects of fishing on the skipjack

(*Katsuwomus pelamys*). Observer scientists on purse-seine boats supplied information, now centralized in the reports of ICCAT-FAO (Anonymous, 1981). From these reports it was hoped to find out something about cetaceans and above all about sightings of *Balaenoptera physalus*. Reports from eight fishing boats were examined but only one mentioned sightings of *Balaenoptera* that could be identified as *B. physalus* because *B. musculus* was very scarce in this area. These sightings were all located about 0°25′ N, off Cape Lopez, *i.e.* in the intertropical frontal zone. This herd probably belongs to the southern stock. From all these results, a discontinuity in the distribution of this species appears in summer, corresponding to the warm and poor Guinean waters, which extend from south Mauritania to the equator. The Mauritanian stock migrates to the northern feeding grounds: upwelling area off Portugal, Biscayan Bay and the northwestern Mediterranean.

In conclusion, we are led to suppose that there are two different populations of *B. physalus* in northwestern Mediterranean: a winter herd with calving and mating activities and which migrates in summer to the feeding grounds off northwestern Scotland, and a summer herd which migrates along the Mauritanian coast for breeding and mating like the herd that is in summer in the Biscayan Bay. This hypothesis is supported by an analysis of the strandings of *B. physalus* on the Mediterranean coasts.

Strandings

One hundred and forty strandings were reported from 1827 (Parona, 1908) to 1981; the month of stranding is known in 100 cases and the length of the animal in 93 cases.

Strandings occurred all the months, with a maximum in September and in November to December (Fig. 4). The first peak is connected with the mixing of the herd ending its summer stage and the wintering herd beginning to enter the area. The second maximum is not related to an increase of the population abundance in winter; if this were the case, then this increase of abundance would extend throughout the whole winter, *i.e.* the mortality would be maintained in to January and Feburary. The maximum of November and December is related to the incidental mortality at the time of the calving.

The graph (Fig. 4) can be separated, after smoothing, into two curves showing frequencies of stranding corresponding, respectively, to the wintering and the summering herds. This shows that the number of strandings in the

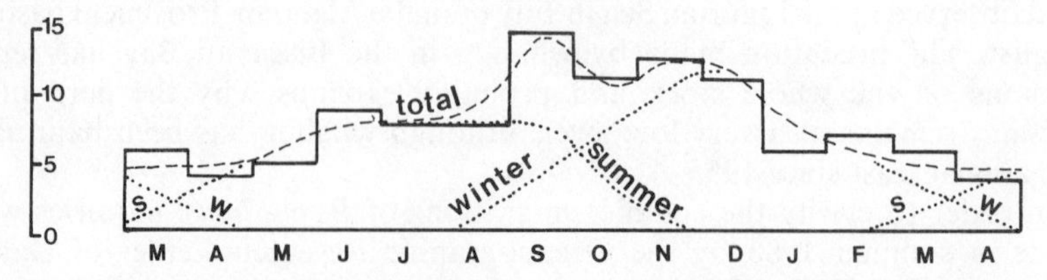

Fig. 4.—Frequencies of strandings of *Balaenoptera physalus* by month: smoothing the histogram (dashed line) and decomposing it into a winter (w) and a summer (s) populations (dotted lines); records from the literature and personal observations.

summering herd remains constant between June and September; during this time there is no biological event in the life of the *Balaenoptera* which would make them more vulnerable. The wintering herd begins to appear at the end of August, provides a maximum of strandings at the beginning of winter and finally leaves this zone in March to April. Births take place from September onwards and may also add to the maximum of strandings shown on the graph.

Reports of strandings of animals more than 22 m long all seem to be before 1910 (probably due to the method of measurement); they were reported between June and November. More recently, according to Figure 5, a quota of individuals between 13 and 20 m seems permanent. On the contrary, the small sizes form two groups that I have tried to interpret using the growth curve (Fig. 6). The size generally reported for new born animals is 5·5 to 6·5 m. For the area studied, the minimum length observed is 5 m (two observations); we have 5 observations of 5·5 m and 6 of 6 m; fin whales of less than 5 m can be considered as premature births (two strandings of 3 and 4 m). A few births take place between April and June, while the great majority are between September and January with a maximum in November. According to Best (1973), calving occurs in the Southern Hemisphere in November to December and Frazer (1973) established from the I.W.C. data that pregnancy lasts from 9·5 to 12 months. All this leads to new born animals being found dead between November and December, in conformity with the above data.

A growth curve of *B. physalus* living in the northern Atlantic or in the Mediterranean is not given in the literature; the size reported at six months is,

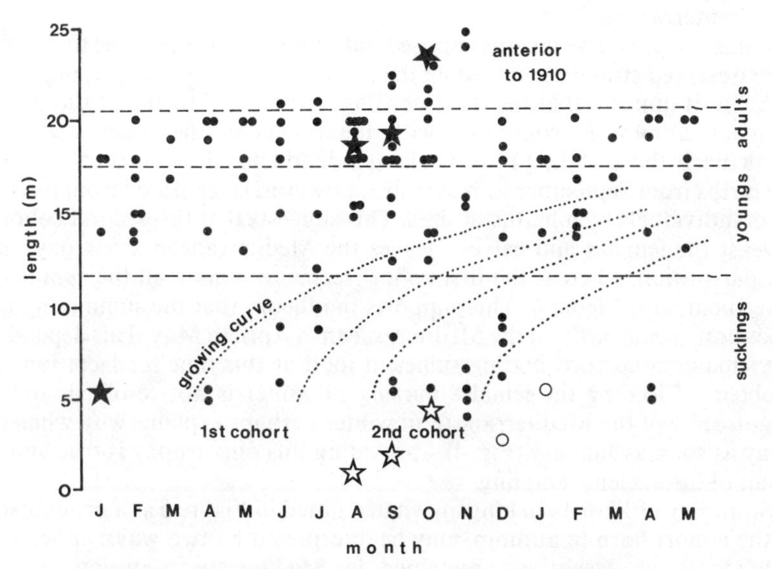

Fig. 5.—Distributions of the accidents at sea and strandings of *Balaenoptera physalus* by sizes and months in the northern Tyrrhenian Sea, Ligurian Sea, and west off Corsica, up to 1982: in order to clarify the interpretation, the first five months have been repeated in the right part of the graph; black stars are pregnant females, and white stars, their foetus; ○, are dead-born or dead immediately after birth; ●, are all other observations.

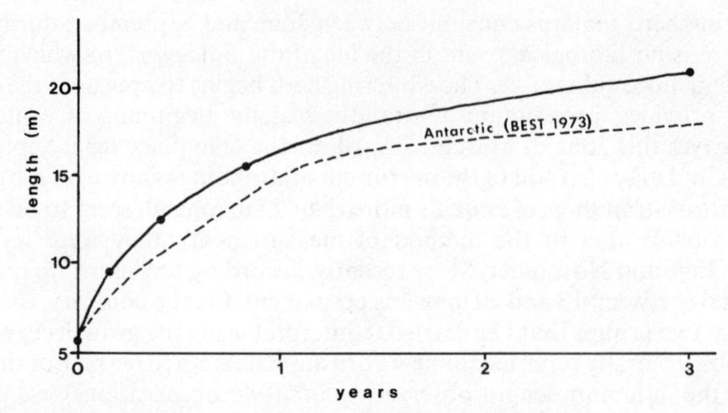

Fig. 6.—Growth curve of *Balaenoptera physalus*: upper line from observations in the Mediterranean; lower line is from observations in the Antarctic from Best (1973).

however, from 12·5 to 13·5 m, that is, the size when exclusive suckling ends. Rayner (1940) reports that a young rorqual tagged at six months and measuring then 12 to 13 m, was recaptured 2·5 years later and was then 20·96 m long; that means a growth of about 8 m between 6 months and 3 years of age. These data are shown in Figure 6 and the shape of the derived graph looks like that given by Best (1973) for the growth of *B. physalus* in the Southern Hemisphere. So Figure 6, upper line, may be considered as the growth curve for Mediterranean.

If this growth curve is superimposed onto Figure 5, it is possible to attribute each observed stranding, following the size of the carcase, to a presumed date of birth. It appears that the dead newborn from April to June form a set of points continuously connected with the points of the other strandings, particularly those of the 13 to 20 m long individuals. The cohort coming from the births from September to November, however, is separated from the 13 to 20 m individuals by a hiatus of sizes. This suggests that this second cohort—at least the females and calves—leaves the Mediterranean a few days after the parturition, whereas the first cohort seems to remain all the summer, in agreement with Figure 4. This supports the theory that the summering herd also shows some births in the Mediterranean in April to May. This depends on the females concerned finding sufficient food at this time for lactation. The problem of feeding the females nursing in winter is not resolved, and the oligotrophy of the Mediterranean in winter perhaps explains why whales go away as soon as January (Fig. 4)—providing this oligotrophy is true and not result of insufficient sampling.

Stranding of 13 or 14 m long fin whales, noted in Figure 5 as a prolongation of the cohort born in autumn, may be interpreted by two ways: either some individuals, as exceptions, remained in Mediterranean during all their suckling time, or some individuals born elsewhere come into this area together with the earliest migrants of the summer herd.

In conclusion, the migration of the species is suggested in order to explain the hiatus in the statistical distribution in time of the strandings on the Mediterranean coasts.

Sightings at sea

Sightings at sea attest to the presence of *B. physalus* throughout the year in the northwestern Mediterranean. When sighting frequencies are reported monthly, a definite maximum appears in August, initiated by an increase in July. This maximum coincides with an increase in the observation effort, *i.e.* an increase of the number of observers (tourists, regular ships, yachts, and so on). An analysis of the observations made between France and Corsica on the regular shipping routes shows an increase in July, but a decrease in August whereas the number of ships remains constant. At the same time, the sightings become more frequent in the area located 40 miles south of Marseilles. Thus, there appears to be a perceptible migration from the Ligurian Divergence area, used by whales at the beginning of summer, to the Algerian-Provençal area used later in the year. The eastern part of the northwestern Mediterranean warms up again before the western part, inducing a shifting of the oceanographical mechanisms between east and west; the fin whales fit them by moving westwards during the summertime. One can note the presence of fin whales off northwestern Sardinia in May to June, then in the Ligurian Sea in July. Many sightings are noted off Cape Corse during the first 15 days of July. At the beginning of September, fin whales gathered in herds of 10 to 15 individuals have been observed in an area 40 to 45 miles south of Marseilles, feeding around the divergence (Viale & Bardin, 1982).

These observations consisted, in 73% of the cases, of isolated animals, 6% of pairs, 14% of herds of 3 to 6 individuals, and at least 7% of schools of 20 to 30 animals. Duguy & Vallon (1976) analysed 478 observations of fin whales in the same area and found completely different percentages. It can be deduced that there is no understandable rule for school structures in this species. Payne & Payne (1972) show that *B. physalus* emits 20 Hz sounds perceptible at 45 miles when transmitted by way of surface water, and at 525 miles by way of the deep-sea waters which are particularly isolated from all the noises due to the boats. According to Payne & Payne, in the ocean even polluted by noise, scattered fin whales should not be considered as isolated but always in contact with their congeners, and thus able to gather themselves at will. It follows, in the northwestern Mediterranean where Corsica and Nice are 90 miles apart, and where two centres of divergence are 180 miles apart, that fin whales scattered for feeding always remain in contact with each other. Therefore, the groups of 2, 5 or 6 fin whales are not significant; such groups move themselves looking for food.

Three typical behaviours have been observed.

(1) Searching behaviour for swarms of euphausiids: the whales move slowly at 3 to 4 knots; 10 or 15-min long dives are separated by three short surfacings for breath. So the animals seem peaceful, may be approached easily and for a relatively long time.

(2) Removing behaviour: when the whales move quickly displacing themselves, probably to join other herds, their normal speed is 12 knots (Budker, 1957) but they are able to swim at 19 knots (Duguy & Vallon, 1976) or even 22 knots (pers. obs.). A young fin whale has been seen swimming at 16 knots. Whatever the speed, this behaviour is characterized by the regularity of the velocity and of the direction. It is impossible to approach them in such conditions.

(3) Feeding behaviour, involving diving: two types of diving for capturing prey have been noted. In early summer, during the day, fin whales dive deeply almost vertically, and reappear practically at the same point after 15 minutes. These observations were made in the Gulf of Genoa where euphausiids live in abundance, daily making 2500 m vertical migrations. At about midday, euphausiids are relatively deep (1000 m). At about 15.00, they will be at 600 or 700 m deep, in the Gulf of Genoa and off western Corsica (pers. obs.). Samples taken then in this area mainly contain *Meganyctiphanes norvegica, Stylocheiron etc.* When feeding at twilight (19.00 to 20.00 in September), fin whales do not dive deeply and emerge very much further away. Hence, they wander over a large area during their shallow diving. For example, I have observed a school of 12 adult fin whales feeding on plankton in an area near the theoretical centre of the divergence; they were very active, having short dives, emerging violently, and rapidly inhaling three or four times between two dives. At the same time our echo sounder detected an intensive Deep Scattering Layer located in the superficial water layers. These whales were moving towards the southwest with the research ship behind them at a speed of 9 knots.

Number summering in the area

The method used to estimate numbers consists of noting the number of whales seen along a transect line, and from that inferring the size of the population (Schweder, 1974; Seber, 1980). This "belt method" was applied for the whole north Atlantic by Mitchell (1975a) and for the Pacific by Nasu & Shimadzu (1970). The maximum distance on both sides of the ship that allows the detection of a large baleen whale is, as admitted, about 1 mile, giving a 2 miles breadth of the sampling 'belt'—although Yablokov (pers. comm.) considers this optimistic—when the weather conditions are good. This is generally the case in the Mediterranean during summer; hence, this method has been used in summertime, but my attempts to estimate the wintering population have up to now failed. This method can be criticized for implying a statistical inference from a sampling pattern which is supposed random sampling; such an optimal sampling pattern is practically impossible to realize at sea, due to the constraints of navigation (Frontier, 1983a). It follows that some parts of the area studied are over-sampled and another one under-sampled, so that the statistical inference is biased.

The results are given in Table IV, which summarizes the surveys of the oceanographic ship KOROTNEFF in 1978 to 1981. They are compared with those obtained in 1973 by Viale (1977b) and with those of Duguy & Vallon (1976). It can be seen that the estimation of the population of *Balaenoptera physalus* in the northwestern Mediterranean, north from 42° N, varies between 500 and 2000 individuals. These upper and lower limits have the same order of magnitude, when considering the approximate character of the sampling procedure. There is, however, a paradox. The figure of 500 individuals came from an intensive sampling of an area where it was assumed that *B. physalus* ought to be more concentrated due to the proximity of a productive zone. The figure of 2000 individuals came from a more aleatory but much lighter sampling of the whole zone studied: the 'belts' sampled were less numerous

TABLE IV

Estimation (hypothetical) of number of Balaenoptera physalus

Cruises	Prospected area (km^2)	Number of sighted animals	Numbers extrapolated to the zones: Between 42° N and isobath 2000 m (120 000 km^2)	Numbers extrapolated to the zones: Between 40° N and isobath 2000 m (300 000 km^2)
Marineland of Antibes, August 1973	1564	9	691	1728
Regular ships, July 1975	42 768	78	219	545
KOROTNEFF, July 1979	1029	7	816	2040
KOROTNEFF, July 1980	1500	15	1200	3000
KOROTNEFF, September 1981	1557	15	1156	2890
Regular ships, summer 1982	18 522	27	175	437
Regular ships, summer 1983	19 894	44	265	664

but better scattered over the whole zone. I have to explain why the first estimation is not greater than the second, in contrast to what was expected. One explanation may be that the frequency of encounters between a whale and an observer not only depends on the abundance of the species, but also on its motility through the area (work on Fractal Theory in preparation in Laboratoire d'Écologie Numérique of the University of Lille). In areas where the food is scarce and scattered, the animals have to travel over long distances to search and collect it, and then the probability of encountering an observer is higher. On the contrary, in the area where a few little extended concentrations of prey exist, whales congregate themselves upon these concentrations to the detriment of the surrounding areas, which become empty, because the *B. physalus* are able to detect euphausiid concentrations at a great distance. On the other hand, the whales probably rapidly exhaust these concentrations of food (see p. 558, where I noted the rapidity of the predation) and, therefore, soon leave the depleted zone. The first sampling proves that fin whales do not stay from day to day at the same place in the productive zone, but are staying during a day. It follows that the species remains for less time, and is less motile, in areas which include little extended food concentrations, than in the others, and that the encounters with an observer ship are less frequent in the first sampling than in the second (Viale, 1984).

Biological data

Only a few results refer exclusively to the Mediterranean, because the sample of 140 carcases available here is too small and covers too long a period to be considered as a representative and homogeneous sample of this population. Some observations can, however, be given.

Pregnancy generally lasts $10{\cdot}8 \pm 1{\cdot}2$ months (Frazer, 1973) but in the Mediterranean seems to be 12 months (see p. 505).

Calving is essentially in October and November in the area, but can begin as soon as September and last to January. Another period of breeding, with much less frequency, takes place in April and May. Thus, the most important period of mating is October-November but can extend to February, in accordance with results from the South Hemisphere (Best, 1973). In the Antarctic, gestation tends to be 9·6 months, whereas in the Mediterranean it tends to be 12 months, which is supported by the superposition of the periods of mating and birth.

The size at birth, generally said to be 6 m, seems to be lower (5·5 m) in the Mediterranean.

There is no information on suckling in the Mediterranean, but according to Jonsgård (1966) and Tomilin (1967) it can be up to 6 months elsewhere.

No statistical data on growth are available from the Mediterranean, but this was discussed on page 506.

In the Atlantic, females are sexually mature when 18·5 m long, whereas in the Southern Hemisphere they are between 20 and 23·7 m long. Two females with their pups came so close inshore at Cape Corse that I could follow them from the coast for an hour. The calves were a third of the length of the mothers; in one case of these two sightings, the new-born calf was found dead, stranded on the coast, the following day; it was 5·5 m long, so that its mother was 17 m long. This observation confirms that of Tomilin (1967).

There are no data on the sex-ratio of the Mediterranean summering fin whales; the sample of 140 stranded animals often does not provide any information on the sex; consequently the sample is too small to give a reliable estimation. It has been established, however, as a hypothesis, that the herd of baleen whales exploited in the Biscayan Bay belongs to the same stock and it contains 40% females (see p. 503). It is not thought that the catching introduced a bias because there is no sexual dimorphism in this species, and the only way to distinguish females during the whaling campaigns is the presence of a calf along the flank of its mother: this peculiarity was mentioned in two cases of 138 reported catches. Best (1973) reports that the behaviour of the whalers in front of a herd of baleen whales including a few calves and their mothers, is to fight the biggest individuals (mother or not). This causes a disorganization of the herd, more severe for the population than the killing of the whole herd.

The estimation of the annual mortality, for the Mediterranean area, arises from the proportion of females, from that of mature females which represent 64% of the females, and from the hypothesis that the calving interval is three years (pregnancy, 1 year; suckling, 1·5 years; and rest, 0·5 year). Consequently, for a set of 100 individuals, there are 40 females of which 25·6 are mature, and among them a third are pregnant, giving 8·5 every year on an average (natality: 8·5% per year). If the population is stable, then the mortality is also 8·5% by year. Eighteen strandings of fin whale were found between 1968 and 1976, which means an average of 2 strandings each year in the "restricted area" (north from 42° N), where the population was estimated to be 400 individuals. Consequently, strandings represent 6% of the dead fin whales in the area.

TURSIOPS TRUNCATUS

Bottlenosed dolphin; in French: grand dauphin.

Only one species is recognized today, with several races defined by variations of body length, of size of the teeth, of the colouring, and of the ecology: *T. aduncus* Ehrenbourg, *T. gilli* Dana, *T. nuuanu* Andrews = *T. catalania* Gray. In the Ligurian-Provençal basin, the *Tursiops* are more massive and more robust than the variety *gilli* of the marinelands. The colour is very dark grey on the main part of the body, while the ventral region is paler. Of the 24 specimens examined from the northwestern Mediterranean, the largest was 3·5 m long. Pilleri & Gihr (1969) consider as doubtful an individual 3·8 m long in the Adriatic Sea, although Duguy & Robineau (1973) give a size of 2 to 4 m for this species on the French coasts.

Sightings and strandings

I have reported eight strandings and 51 sightings at sea around Corsica between 1967 and 1977; from 1978 to 1982, two strandings were found and some systematic observations were made during oceanographic expeditions. Duguy & Budker (1972), then Duguy (1973, 1974, 1975, 1976, 1977, 1978, 1979, 1980, 1981, 1982, 1983a, 1984) report five strandings in the Mediterranean from 1971 to 1976; Casinos & Vericad (1976) report five strandings since 1972 on the Spanish Mediterranean coasts and Balearic Islands; Tortonese (1963c) reports seven in the Gulf of Genoa. In all, 55 strandings and more than 100 observations at sea have been reported.

T. truncatus is a neritic species; I have always found it in the coastal zone, and never offshore during my open-sea expeditions. Furthermore, it has never been reported in the observations from the laboratory buoy of the Commission pour l'Exploration des Océans (COMEXO), 60 miles off Marseilles.

The species is present all the year around Corsica, with much lower densities between September and January than in the spring. The maximum number of sightings around Corsica is in May, which is not the time of the maximum number of observers. The maximum of strandings and accidental deaths on the French coast (Fig. 7) are also in May and June. The abundance of strandings and sightings then decreases regularly up to the end of September, perhaps more rapidly than suggested by my histogram since the number of observers is maximum in July and August (tourists, sailing boats), and this may introduce a bias in the census of the sightings.

A very clear annual cycle appears in the frequency of observations along the coasts, which seems to indicate that the herds make a seasonal migration. The lack of strandings in winter in 15 years of observations in the nothern part of the area studied, shows that the species migrates southwards during this season. A migration towards the open sea is improbable, seeing that the species is neritic, and that the open sea biotope is occupied by two other species of dolphins: *Stenella coeruleoalba* and *Delphinus delphis*. Indeed, Llose (pers. comm.) confirmed the presence of *Tursiops truncatus* during winter off the Algerian coasts. A few individuals come to winter near Corsica: I observed a herd of ten individuals throughout the winter of 1981 and the winter of 1982 near the Cape Corse, and also in December 1980 in the Strait of Bonifacio.

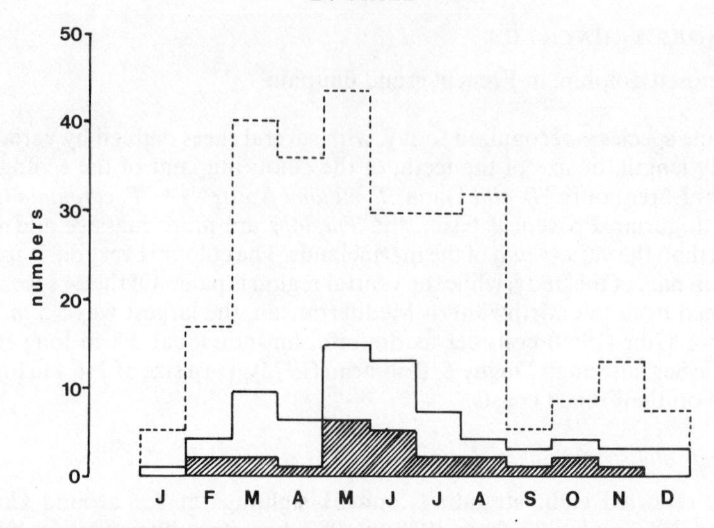

Fig. 7.—Distribution of 68 observations at sea of *Tursiops truncatus* by month (upper continuous histogram) and of 24 strandings (hatched histogram), by months: one observation of any group is here considered as a single observation; upper dotted histogram represents the number of individuals, estimated from the number of observed groups.

In February, *Tursiops* travel northwards, coming from the southern part of the basin, following the Algerian-Provençal branch of the Atlantic Current along the western coasts of Sardinia and Corsica. The exploitation by them on the continental shelf of Corsica is maximum between March and May, then the animals continue their migration northwards, but since 1955 they have never appeared in great numbers in the vicinity of Marseilles nor in the Gulf of Lions. Previously they were frequent in that area so much so that they hindered the fishermen (Peres, pers. comm.). In autumn, the migration southwards brings them back into the Corsican and Toscanian waters.

I have observed groups varying between 2 and 30 individuals. The schooling of individuals seems to vary throughout the year; animals have been observed in small groups or pairs in autumn and winter, and in larger groups from spring onwards. Figure 7 shows the number of individuals and groups observed. The mean number of individuals by group is high between March and August. I infer from this that animals arrive grouped together on the French coast in spring. From March onwards there are very different sized individuals in the groups. Three classes (with different behaviours: see p. 513) can be identified: very young individuals, juveniles near maturity, and adults.

Like Saayman & Taylor (1973), I have seen around Corsica in spring and summer, throughout the day, small groups hunting and searching for food, or playing. By night, I have found them gathered for fishing, in sometimes very well coordinated activities. Thus, it seems that the herds scatter themselves into small groups when seeking for food or playing. In addition, *Tursiops* group themselves, sometimes with another species, around the congeners which have found a fish school. This is the matter of occasional grouping, and not a structured dolphin school.

The structured groups, according to my own observations, belong to three types: (1) groups of two or three adults with very young calves, observed in March to April, and easily identified at sea by the slowness of the large individuals moving themselves following a rectilinear trajectory, whereas the small ones seem very numerous because they jump, going away and coming back very fast around their parents; (2) groups of 10 to 15 dolphins, including three classes of sizes: small ones, recognizable by their playing activities, adults which do not jump, but are regularly together swimming, and juveniles in puberty showing erogenous plays (in spring, from the end of the morning up to the beginning of the afternoon); and (3) groups of 7 or 8 dolphins, all of the same size, showing sexual behaviour: vertical jumps of pairs side by side, and very rapid pursuits, in spring, also around midday.

Fitting a growth curve

Sergeant, Caldwell & Caldwell (1973) drew a growth curve from data from 62 dolphins captured in Florida. Length was 1 m at birth, and asymptotically reached 2·70 m by males and 2·60 m by females. Figure 9 shows that animals from the Mediterranean are distinctly bigger; maximum length is 3·50 m and the smallest observed size is 1·50 m, which implies that the young are not born in the same area, but come into it when one or two months old.

If there is a single birth-period in the year (as asserted by Tavolga & Essapian, 1957, for the Florida coast) and knowing that the small sizes in the Mediterranean are observed at the very beginning of the year, the population observed at a given time in the vicinity of Corsica must be constituted of several cohorts which have year differences between them. It was possible, therefore, to discern various cohorts by modal analysis, corresponding to the age of individuals; it results in as regular a growth curve as possible (Fig. 8). The fitted curve indicates an average length of 1 to 1·20 m at birth, 2·35 m at one year, 2·70 m at two years, and proceeds asymptotically to 3·30 m by males and 2·90 m by females.

The absolute ages of two individuals have been determined through an examination of dentine layers by Kasuya & Miyazaki (pers. comm.): a 3·30 m

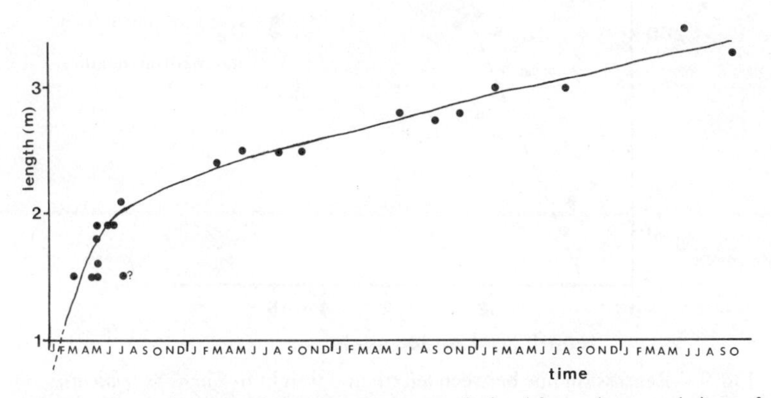

Fig. 8.—Growth curve of *Tursiops truncatus* derived from dates and sizes of strandings.

male was 18 years old and a 2·80 female was 29 years old. McBride & Kritzler (1951) report that *T. truncatus* females can be mature at four years old and calve during their fifth spring. Sergeant *et al.* (1973) noted that *T. gilli* had reached 90% of its definitive size at the time of its sexual maturity. These observations agree with my fitted curve: 3·15 m at about four years old.

Biological data

Five animals have been measured following the norms of Norris (1961) and the results were given in Viale (1977b). The dimensions given by the observers usually follow the standard norms, but weights are generally estimated by eye. A 1·80-m female weighs 180 kg. Five other results concerning the Mediterranean are given by Rode (1939), Pilleri & Gihr (1969), Duguy (1973, 1974, 1976). For the Atlantic, Lilly (1964) gives a regression line on log-log, calculated from seven individuals from Florida, Pilleri & Gihr (1969) add seven results from Florida and two from the Faeroes, and Cadenat (1959b) gives four results from Senegal. All these results are represented in Figure 9 and give a good linear correlation on log-log basis:

$$W = 19{\cdot}824\, L^{2{\cdot}218}$$

where W is the weight in kg and L the length in m. The results from

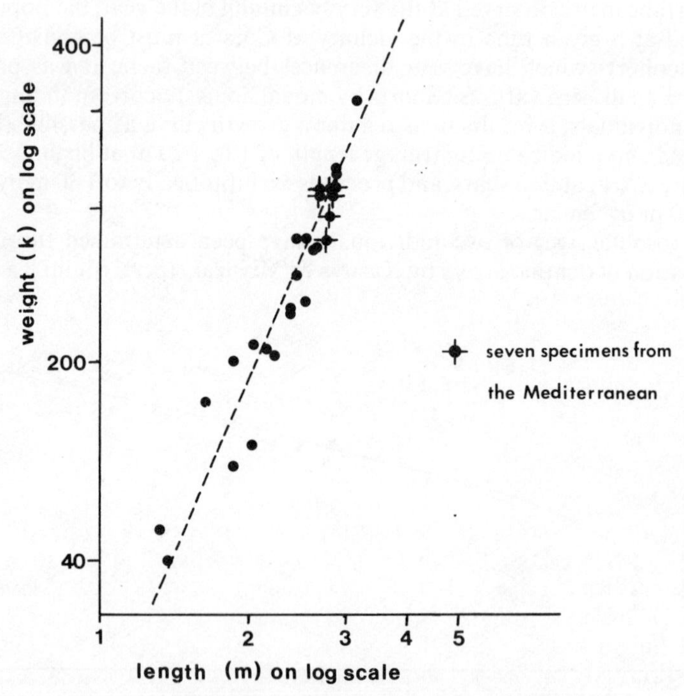

Fig. 9.—Regression line between length and weight in *Tursiops truncatus*, on a log–log scale, according to 26 results from the literature (dots) and personal observations (dots with cross).

Mediterranean are similar to those from Senegal, and well included in the general set of points, which suggests that a single allometric law exists, without any dependence on regions. The regression in Figure 9 distinctly differs from that proposed by Lilly for his seven specimens: following the latter, a 3·50-m adult such as that captured in Bastia ought to weigh 363 kg, instead of 316 kg following my equation; the estimation given by the fishermen was 300 kg. Moreover, individual variations of weight for the same length can be great, as indicated by the stranded individuals, which often reflects physiological (pregnancy, suckling, *etc.*) or pathological variations.

Age at the sexual maturity is still under discussion. For females it is 4 to 5 years according to McBride & Kritzler (1951), or 6 to 7 years according to Harrison & Ridgway (1971) and Harrison, Brownell & Boice (1972), or 12 years according to Sergeant *et al.* (1973) for captive individuals, but Harrison *et al.* suggest that maturity is more precocious in the sea than in a pool. For males it is 10 to 12 years.

The mating and breeding period is from January to the end of March for the majority of animals in the Mediterranean, as well as in Florida (Tavolga & Essapian, 1957). It is in February to June in Britain according to Frazer (1973) and Harrison *et al.* (1972) noted it in the midst of the summer. Pregnancy lasts 12 months according to Frazer (1973). The size at birth is estimated as about 1·10 m in the Mediterranean. In the world, it varies between 0·98 and 1·26 m according to the region (Mitchell, 1975a). No data are available in the literature on the suckling period but 1·50- and 1·60-m individuals stranded on the coast of Corsica in May and June (hence, born in February of the same year) already contained in their stomachs a number of otoliths and vertebres of fishes, which proves mixed feeding.

According to Harrison *et al.* (1972) a female gives birth to eight calves, on an average, during her 30 years of life, giving on an average an interval of three years between two calvings. The average rate of production is, therefore, 0·33 per year with respect to the number of mature females. The population of the Corsican coast being estimated as 225 individuals (see below) in 1976, and assuming 50% are females, and among them 75% are mature, we calculate that there are 28 births every year, *i.e.* 12·5% in respect to the whole population, and an identical mortality if the population is stable. We then can estimate the percentage of dead individuals which beach: noting four strandings by year on an average for the four years preceding 1976, we calculate it as 14%—a greater percentage than in other cetaceans, which agrees with the neritic habits of *Tursiops*.

Ethological observations

Small groups of *Tursiops* can be seen by day, patrolling at 4 to 7 knots, and passing at less than 1 mile offshore. They can detect coastal fishing nets, and when doing this suddenly change their direction, going perpendicularly to their original direction in order to inspect the nets systematically, and to eat fishes captured in them. Another behaviour which proves a high degree of competition with man for fish resources takes place over fishing grounds highly reputed for their richness of fish species of high commercial value. As soon as *Tursiops* enter these grounds fishermen cannot catch any fish; the dolphins eat them all. This behaviour takes place in the morning and at dusk.

By night, *Tursiops* develop another feeding pattern. Besides visiting the nets nightly set off the coast, they congregate around the trawlers and eat at the opening, just in front of the trawl, as proved by the occasional catches of dolphins. They also seem attracted by the small and non-saleable fishes tossed away after every trawl, and eat them with voracity. Their constant presence around the nets and trawlers near Corsica implies a scattered presence on the whole continental shelf, hence a rather homogeneous distribution, at least in spring and at the beginning of summer. This permits an estimation of their abundance.

A trawler which was authorized to fish with light ("lamparo" fishing) in the Strait of Corsica gathered, with the aid of several beater boats, some small schools of pilchards with the aim of making a single big school and to catch it with the trawl net. Dolphins watched the handling; a few days later, they stayed in the neighbourhood during the handling, waiting for the pilchards to be gathered. Then they penetrated this enormous biomass and ate on it to repletion; the fishermen estimated the loss as 30 to 40%. *Tursiops* go so far as to penetrate into the harbours.

Besides observations of *Tursiops* by professional or amateur fishermen (whose hours of activity coincide with those of dolphins), animals have been seen in bays, very near to beaches, during their social activities at the end of the morning or the very beginnings of the afternoon. It is well known, in the Mediterranean, that the greatest number of cetaceans jumping out of the water are seen at the warmest time of the day, around 13.00 hours.

Tursiops have a particular behaviour by day. It is well known that these animals in dolphinariums show a great curiosity and use every opportunity to play. On the east coast of Cape Corse in spring, a group of *Tursiops* have been seen for one and a half hours in a shallow bay on several occasions. During one of these observations, a helicopter flew over the bay at a low altitude: the dolphins all at one time raised their heads out of the water and immobilized themselves to look at the machine. So I was able to see that they were more numerous than had been previously estimated. The combined movement was worthy of note, and had a resemblance with those photographed by Caldwell & Caldwell (1966). *Tursiops* standing up out of the water and leaning with their fins on the edge of the pool in order to look at a gardener mowing the lawn! On the coast of Cape Corse, two fishermen suddenly saw a *Tursiops* standing on its flukes, in an attitude often exploited in the marinelands shows, as if it was trying to see what was happening in the boat (Argivier & Ricci, pers. comm.).

To the north of Cape Corse, men know how to play with *Tursiops*: as soon as a group is detected, the men drive their boat at maximum speed in the direction of the dolphins. The latter immediately come at the stem of the boat and play crossing it at full speed. Does this playing behaviour provide the dolphins with as much pleasure as it provides to men? In all the cases, this game stops as soon as the boat speed falls. The Indians of Columbia also practise this game with the *Tursiops* of the Pacific.

Epimeletic behaviour was first described by Caldwell & Caldwell (1966). It consists of releasing an alarm signal with such a rapidity that the herd can escape the danger. The speed at which the concerted response happens is astonishing, and implies an almost instantaneous communication of dolphins between themselves. Caldwell & Caldwell defined the signal they named as "epimeletic". Busnel & Dziedic (1968) described the movement which gives the alarm in a school of *Delphinus delphis* in Mediterranean. I observed that the Mediterranean *Tursiops* use almost the same posture and movements in the same conditions. In the above-mentioned instance with the helicopter, when it turned rather fast and came close to the sea surface, an adult-sized *Tursiops* rose up entirely out of the water, almost vertically, and fell again fully stretched, heavily, beating the water with a fin. The school of 15 dolphins immediately disappeared and dived, although not very deep since I was able to follow their wakes towards the open sea; they only re-appeared some distance offshore.

In a dangerous condition, the information has to pass quickly to all individuals. Pryor (1973) studied the methods of signalling in pools, both intra- and inter-specific (including behaviour with respect to man), and showed that *Tursiops truncatus*, *Stenella coeruleoalba* and seven other cetacean species use sounds, postures and movements. This does not, however, solve the problem of their organization nor which dolphin gives the information. The group behaviour implies a hierarchy of functions which has not yet been elucidated in cetaceans. Another example shows *Tursiops* in groups able to organize themselves to a common activity, like the intervention they had by night around the fishermen (see above). This method of fishing did not exist in Corsica before October 1975; the behaviour by *Tursiops* is, therefore, an acquired one.

It is impossible to detect any social hierarchy in a pool, but it certainly exists in nature as attested by the collective strandings, and by the examples of concerted action with man in Mauritania as reported by Busnel (pers. comm.) and Maigret, Trottignon & Duguy (1976). Here *Tursiops* and *Feresa* drive fish schools towards the coast; the fishermen provoke the aid of dolphins by beating the water with boards to imitate the noise of fishes. The benefit is reciprocal since the dolphins are then sacred, and so protected.

Saayman & Taylor (1973) report various "social activities" of *Tursiops*; when two groups meet, there appears to be a behaviour described by the authors as a "greeting ceremony", but which seems more like courtship.

Estimation of stock

The damage caused in the nets of fishermen implies a particular spatial distribution of the dolphins, which permits an estimation of the abundance. Such an attempt is justified because of the lack of data at an international level about this species. Gaskin (1976) notes the poorness of our knowledge about these Odontoceta, whereas the stock estimation of the exploited species is given in a large number of publications. Gaskin summarizes the results, and shows how arbitrary and variable they can be.

I infer an estimation of stock from following consideration about the association between *Tursiops* and fishermen. Along the 1000 km of Corsican coast, 160 fishermen are grouped in 15 fishing centres with 15 to 25 nets in each one. There is no industrial fishing. This leads me to the hypothesis of 15 herds of *Tursiops*, one visiting every night the 15 to 25 nets of the fishing centres, along about 70 km of coast. Each herd is of about 15 individuals. I infer that there are about 225 individuals of this species around Corsica. Extrapolating to the northwestern Mediterranean is difficult, because *Tursiops* have almost entirely disappeared from the areas around Marseilles and Genoa; pollution seems to be the cause of their disappearance (Viale, 1977b). Assuming that the whole northwestern basin has an optimal trophic structure in May and June, and that the number of *Tursiops* is proportional to the areas of the continental shelves, we obtain the results shown in Table V.

The southern part of the basin, including Balearian Islands, has a total area of continental shelf little different from that of the north part, but probably is richer because of the Atlantic Current. This makes the hypothesis of a seasonal shift of the *Tursiops* population between north and south plausible.

It is possible to estimate indirectly the total number of *Tursiops* in the northern part of the basin. In Corsica, the average number of strandings and accidents is, at present, 4 by year, *i.e.* 1·78% of the estimated population. In Catalonia, the number of strandings was 20 between 1973 and 1979 that is, on an average, 2·86 by year. Assuming the same percentage of strandings as in Corsica, we calculate a number of 160 *Tursiops* along these coasts, five times lower than the number calculated starting from the area of continental shelf (Table V). Probably the number of animals on the continental shelf of Italy, between Piombino and 40° N, have to undergo a reduction of the same order of magnitude. I finally propose a number of about 2000 *Tursiops* in the northwestern part of the basin, instead of the 3285 calculated above. It is not impossible that the reduction of numbers in the northern part of the basin is compensated by an increase in the southern part. In fact, since 1975, the Algerian Government has complained about the damage caused to fishing nets by dolphins at the end of the winter and beginning of spring, even so far as to contemplate the possibility of eradicating them.

In conclusion, the number of *Tursiops truncatus* in the Mediterranean is low. It is likely that the most observed Odontoceta in my work are the least abundant among the species visiting the Mediterranean. In the northeastern and in the tropical Pacific, *Tursiops* form 7% of the observed and identified cetaceans, whereas they represent 23% of my observations and inquiries between 1967 and 1976. This difference is explained by the rather neritic character of my research, in contrast to the offshore character of the majority of the oceanographic observations.

TABLE V

Estimation of the potential numbers of Tursiops *by zones*

Zones	Areas (km^2)	Numbers of *Tursiops*
Corsican neritic zone	5250	225
Sardinian neritic zone	18 250	780
North Tyrrhenian Sea, Tuscany	9250	400
Gulf of Lions	11 500	490
Riviera, Gulf of Genoa	4750	200
Italy north from Piombino	9250	400
Catalonia up to Valencia	18 000	790
Total	76 250	3285

DELPHINUS DELPHIS

Common dolphin; in French: dauphin commun.

The status of this dolphin in the Mediterranean has completely changed since 1950. Dominant in the past, it has today only a secondary place in the observations at sea because it is replaced in its ecological niche by *Stenella coeruleoalba*, which was considered as a rare species in the Mediterranean up to 1960 (Viale, 1977b, 1980a,b). Taxonomically, *Delphinus delphis* should not be divided in subspecies according to the present state of our knowledge, but it does show some important variations in the length of the rostrum and the zygomatic width (Evans, 1976; Gaskin, 1976). Various authors have described geographical variations (Cadenat, 1959b). Tomilin (1967) reports some significant biometric differences between herds 500 km apart in the Black Sea. Aloncle (1964) notes that the Mediterranean *Delphinus delphis* is characterized by the buff-coloured inferior part of its pectoral fins. The yellow ventral colour is not seen in the individuals from the French Atlantic coast (Duguy & Robineau, 1973), nor in those from the Black Sea, which are clearly distinguished from those of the western Mediterranean by the shape of a white belt along the flank (Tomilin, 1967). A male stranded in Corsica on which I did an autopsy had a black dorsal face with two grey lozenges on the flanks, a white lateroventral surface, and buff-coloured belly. I have photographed at sea some individuals characterized by a yellow ventral colour. Three individuals harpooned in the Gulf of Guinea (Oceanographic ORSTOM Centre of Pointe-Noire, Congo), however, did not show that colour.

Sightings and strandings

Observations before and after 1950 will be considered separately. The data from 1950 to 1970 are still difficult to interpret because, during this period (and even today) observers often confuse the two species of dolphins, which are difficult to distinguish from a ship. The main records refer to “dolphins”, whereas they clearly concern *Stenella*.

Before 1950, the status of *D. delphis* in the Mediterranean according to my questions to fishermen appeared to be as follows: the damage caused in the nets was important in the region of Marseilles, the Gulf of Lions, the Ligurian Sea, and Corsica, where this dolphin always appears to follow the same pattern: a herd of dolphins in pursuit of a fish school would come near the coast and then go offshore again; the herds were always numerous and sometimes described as “streams of dolphins”. In August 1932, a herd of several thousands of dolphins was seen from the top of Cape Corse, in a V-shaped formation surrounding an area where the water was bubbling because of the swarming fishes; the whole was topped by a cloud of sea birds. From 1912 forwards, systematic destruction was requested by the fishermen.

In August 1958, east from Cape Corse, a herd of 200 individuals was noted. In August 1960 and September 1961 at the same place, several herds of 40 to 50 individuals each were seen. Observations from the labotatory buoy of COMEXO, 60 miles south from Marseilles, indicated a dozen *Delphinus* in June and a small group in August. Observations and inquiries, starting in 1962, never recorded such huge gatherings. The herds passing east of the Cape

Corse in June or July were always less than 100. In 1972 and 1973, three cruises organized by the Marineland of Antibes observed some groups of 15 to 50 individuals. From 1972 to 1977, observations at sea were of two kinds: mainly of small groups of two to eight individuals between March and August, contrasting with a few records of herds of approximately 100, which were migrating.

From 1977 onwards, I have made several cruises with the aim of taking a census of *D. delphis*: during seven cruises in spring and at the beginning of summer, *Delphinus* were only observed four times. They were in small groups mixed with *Stenella* in the Strait of Corsica, and in one monospecific herd in the Tyrrhenian Sea off Porto Vecchio (south Corsica). Some sightings were also made off Cape Ferrat, in the Riviera (Palazzoli, 1983). During monthly cruises in the northern part of the Ligurian-Provençal basin in winter 1981–1982, not one *Delphinus* was seen among 154 cetaceans. In 15 monthly cruises in the northern Mediterranean, there were two sightings of *Delphinus* off Villefranche-sur-Mer at the beginning of summer, and in seven cruises (21 days at sea) in spring and at the beginning of summer, only four sightings.

I, in collaboration with J. P. Bardin, Officer on board the regular ship CYRNOS, made repeated observations in the transects from Ajaccio to Marseilles and from Bastia to Marseilles, under the same conditions of effort of the observer, the same state of the sea surface, and for the same time of the day (11.00 to 15.00). Each 90-mile long transect was covered seven times between 3rd July and 8th August 1981 (Viale & Bardin, 1983). *D. delphis* was seen only on the transect from Ajaccio to Marseilles, and only twice among 18 sightings of cetaceans (*i.e.* 11% of the observations); the dolphins were in small groups of four to eight individuals. On the whole, 12 individuals were seen among 218 observed cetaceans, *i.e.* 5·5%. The decrease in number of *Delphinus* is confirmed by these observations at sea, as well as by the decrease of the number of individuals in each group, and particularly the complete disappearance of the very large herds.

This decrease is also evident in the stranding reports. According to the literature and the collections of museums, 30 strandings or catchings were reported between 1890 and 1912, but the high abundance of this species at that time was such that all the strandings were not recorded because fishermen demanded the systematic destruction of dolphins. This implies a hiatus in the information. From 1971 onwards, systematic recording has been organized by P. Budker, and later by R. Duguy confirming the rarity of the strandings of *D. delphis* in comparison with those of *Stenella coeruleoalba*, which are increasing (Table VI).

Seasonal migration and water temperature

The observations at sea in the northern part of the western basin of the Mediterranean are from March to October, with a maximum in July. Evans (1976) showed that, in the eastern Pacific, the geographical range of *Delphinus delphis* is limited by high temperatures (more than 27 °C) and that the maximum abundance is when the water temperature is close to 17 °C. Kasuya (1971), Miyazaki, Kasuya & Nishiwaki (1974) also describe migrations of *Delphinus* from cold to warmer waters, over several hundred miles. Cadenat (1959b) noted the absence of *Delphinus* off Dakar (Senegal) in March to April;

TABLE VI

Number of strandings of dolphins on the French Mediterranean coasts between 1971 and 1983 (data from Duguy & Budker, 1972, and Duguy, 1973, 1974, 1975, 1976, 1977, 1978, 1979, 1980, 1981, 1982, 1983a, 1984)

Year	*Delphinus delphis*	*Stenella coeruleoalba*	*Tursiops truncatus*	Total Cetacea
1971	1	0	0	6
1972	2	2	1	18
1973	2	4	7	27
1974	2	7	1	22
1975	0	10	7	22
1976	1	18	4	29
1977	3	13	3	26
1978	1	10	1	19
1979	1	26	2	42
1980	1	18	1	29
1981	0	12	0	19
1982	1	13	1	22
1983	0	12	0	21

surface isotherms at this time of the year around Cape Vert are below 17 °C (Conand & Fagetti, 1972). Gaskin (1976) shows that the distribution of the species is restricted to lower temperatures by a surface temperature of about 14 °C. Tomilin (1967) describes a seasonal migration between north and south in the Black Sea.

In the Mediterranean too, it seems that the variations of temperature of the surface waters explain the seasonal migration of the species. In Figure 10 the histograms of the numbers of sightings and of strandings, and the curves of the surface temperature at Villefranche-sur-Mer, Marseilles, Banyuls, and at the middle of the Nice to Calvi transect (*i.e.* the centre of the Ligurian Oceanographic Divergence) are superposed. The curves observed in Marseilles and Banyuls are clearly different from the other two; the northwestern part of the basin shows a temperature lower than 14 °C; Corsica and Sardinia are then at 15 °C. On the other hand, the temperature on the coast of north Africa is between 18 and 19 °C up to January. Then, up to March, only the southern part of the Tyrrhenian Sea and the region of the Sicilian-Tunisian Strait have a temperature maintained at 16 °C.

If such a thermal regime of *D. delphis* is also applicable to the Mediterranean population, the species ought to migrate seasonally between the north side of the basin and the Algerian coast. From the latter coast, they would be driven towards Tunisia and the Tyrrhenian Sea before beginning again to move northwards, along Sardinia and Corsica.

During my observations of *Delphinus* in May and June in the northern Tyrrhenian Sea, I verified that the surface temperature was 20·34 to 20·78 °C. In May 1980 in the same area, the surface temperature was 16·29 °C, and at

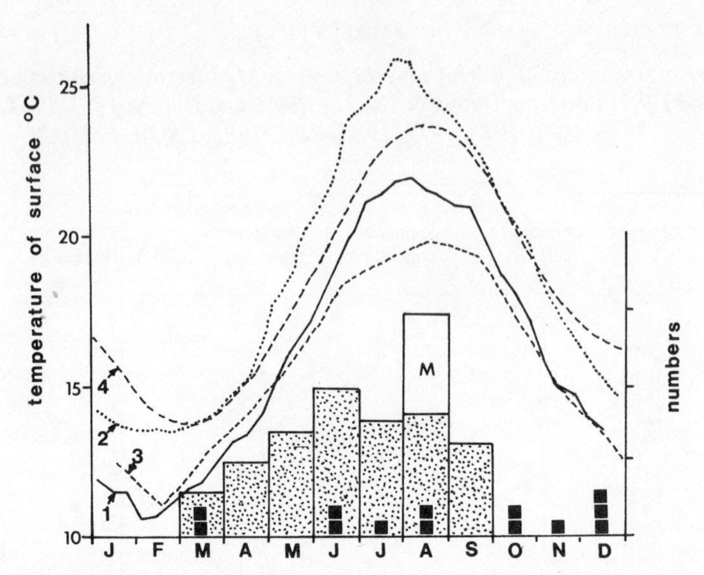

Fig. 10.—Distribution of observations at sea of *Delphinus delphis* by months: any observation of a group is considered as one observation; observations from the cruise of the Marineland of Antibes are considered separately (M); black squares are strandings; curves (from Thiriot, 1966) represent the variations of the surface temperature at (1) Banyuls, (2) Villefranche-sur-Mer, (3) Marseilles, and (4) the centre of the Ligurian Divergence; marks on right hand ordinate are 5, 10, and 15.

10 m depth was 16·30 °C; *Stenella, Grampus*, and *Tursiops* were then observed, but not *Delphinus*. In May 1981, when the surface temperature was 19·09 °C, and at 10 m was 19·00 °C, a monospecific herd of 70 *Delphinus* was seen. Similarly, an expedition of the ship of the Marineland of Antibes towards the coast of Spain and the Balearian Islands in September 1971 encountered 500 *Delphinus* in a week at sea, and noted that the surface temperature at each observation was never lower than 19 °C.

In conclusion, temperature seems to be the determinant factor inflencing the migration of the species. It is, however, possible that the trophic factor, which is strongly associated with temperature in the oceanographic regime of the area studied, also determines this movement of population, because the arrival of *Delphinus* in March off the northwestern coast of Corsica coincides with the maximum primary productivity on the margins of the divergence. Tunas also arrive at this time. Dolphins are more abundant at the end of spring in the north Tyrrhenian Sea and the Ligurian Sea; then the summer oligotrophy reaches a wider and wider coastal belt, and drives the dolphins towards the centre of the Ligurian Divergence, and later towards the Algerian-Provençal Divergence (Viale & Bardin, 1983).

The census of strandings (Fig. 10, ■) shows that they are distributed throughout the whole period when the species is in the area, *i.e.* March to September, followed by a few strandings between October and December, when the species is not sighted. The latter strandings correspond to ill individuals. A stranding in October in Sardinia (the southern boundary of the area studied) confirms the migration, and Richard (1936) noted the beaching of an ill animal in December. Probably two strandings in November and beginning of December were caused by pollution (I noted metallic pollution of organs and skull, and pathologic state of the testicles, see Viale, 1977b, 1978). Strandings in autumn and beginning of winter probably correspond to the mortality of individuals which could not undertake their migration, and could not withstand the winter conditions. The strandings of March were a mother and its calf riddled with bullets.

Number of individuals in groups

In the absence of a programme of aerial observations, as used in the U.S.A., in the U.S.S.R., and in Japan, recourse has to be made to indefinite information about numbers in herds, such as "a dozen", "40 to 50", or "several hundreds" which, of course, does not reflect any accurate counting of animals, but a certain global evaluation of abundance. Two questions then arise: first, what reliability is to be assigned to such qualitative estimations, and secondly, are such estimates (if assumed to be reliable) able to bring to light any ecologically significant features?

Frontier suggested the use of a rapid method of estimating the abundances, hitherto used in plankton work with the aim of obtaining, without any accurate counting of organisms, a semi-quantitative idea of the community (Frontier, 1966, 1969, 1983a) and, more recently used as a statistical approximation for artisanal fisheries (Frontier, 1982, 1983b). The method and its justification are reviewed here.

In an unknown region, planktonic groups may be the objects of qualifications such as "very rare", "rare", "rather rare", "rather abundant", "abundant", "very abundant", *etc.* Annual cycles and geographical distributions then very quickly appeared when fitted on graphs, despite the highly approximate and, above all, apparently subjective character of these estimations by eye. Later, a numerical control of these estimations by means of precise census made evident several points.

(1) The intuitively taken scale corresponds rather amazingly to a geometric series with ratio 1:4·3 by reference to the censused numbers. The author then was led to choose more formally a "quotation of abundance" involving classes such as, Class 1 (1 to 3 individuals in the sample), Class 2 (4 to 18), Class 3 (18 to 80), Class 4 (80 to 350), Class 5 (350 to 1500), Class 6 (1500 to 6500), *etc.* A justification of the ratio 1:4·3 is suggested on the basis of Information Theory (Frontier & Viale, 1977). Experience shows that it is very easy to decide, without accurately counting, if an abundance of individuals belonging to a given category is, for example, included between 80 and 350 (Class 4) or between 350 and 1500 (Class 5); a hesitation only appears when the number approaches 350 and, for convenience, the Class "4·5" (which signifies "about 350") is then adopted. A number of individuals thus is intuitively considered as "clearly" higher than another when it is at least 4·3 times higher.

(2) The coincidence between the results obtained from quotations and from accurate countings, when put on graphs or maps, make it evident that countings do not provide more ecological information than the estimates by classes. The observed differences have the same order of magnitude as the background noise which randomly makes 'fuzzy' the data. Similarly, more

sophisticated treatments such as factorial analysis, indices of diversity *etc.*, show no significant differences between the results obtained from accurate counts and estimates (Frontier & Ibanez, 1974; Devaux & Millerioux, 1976a,b, 1977). Reverting to cetaceans, it is possible to compare the valuation of abundance given by observers, with Frontier's quotations. It appeared that the phrases used to describe herds of cetaceans coincided rather well with the above classes, thus:

Class 1 (1 to 3)	"1 to 4 individuals"
Class 2 (4 to 18)	"about ten, a dozen, a band"
Class 3 (18 to 80)	"several dozens, a herd"
Class 4 (80 to 350)	"100 to 200, a great herd"
Class 5 (350 to 1500)	"several hundreds, a huge herd"
Class 6 (1500 to 6500)	"several thousands, a stream of dolphins".

The examination of observations shows that they generally fit without any ambiguity into one or other of these classes. Any increase in the accuracy (for example, increasing the number of classes in the same range of numbers, every class being smaller) seems difficult because of the difficulty in ascribing an observation to a class.

(3) The results obtained by gathering all observations together seem indicative of ecological phenomena. In Figure 11 all the data are plotted according to the class number and months. In only four observations (among 41), and only between the Classes 2 and 3 ("about 20": Class "2·5") was there any ambiguity. Hence there is evidence of three kinds of grouping for *D. delphis.* (a) The most frequent observations are of "herds of 20 to 80 individuals": Class 3. This kind of gathering is most characteristic of a hunting behaviour behind pelagic fish schools. There are almost the same number of small groups (Class 2), corresponding to the dispersal of the herds during the

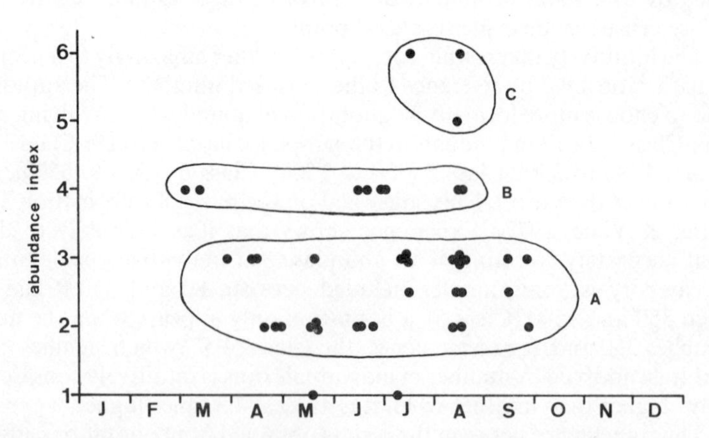

Fig. 11.—Distribution of observations at sea of *Delphinus delphis* by months and by numbers of individuals in groups: numbers in groups are approximately estimated and expressed with the 'quotation' of Frontier (1969); three kinds of grouping can be distinguished and correspond to (A) alimentary behaviour, (B) migratory behaviour, and (C) pairing behaviour.

day, with feeding and diving behaviour if there is no important fish school close to the surface, or to the time around midday when dolphins do not feed but are playing. (b) Gatherings of about 100 individuals (Class 4) corresponding to a social organization in order to undertake migrations, as can be observed on the arrival of the species in the zone studied in March, and the migration in June to August towards other trophic or reproductive areas. (c) Finally, the huge gatherings (Classes 5 and 6) may correspond to sexual pairing in July and August. Nevertheless, in one observation of August 1932, such a gathering was produced by hunting behaviour described on p. 519.

Evans (1976), by statistical analysis starting from aerial observations off California, also found that *D. delphis* lives in "schools" or in "herds", groups varying between 5 and 5000 individuals. Radio-telemetric tagging allowed him to follow a female for 36 h, and he found the same animal again 270 km further south, after she had rejoined a migrating herd of 200 or 300 dolphins.

There are two hypotheses for interpreting huge gatherings of this species (in the western Mediterranean: August 1932, July 1971, August 1973 and that initiated in June 1974 by three herds of 100 individuals). They may represent groupings on the oceanographical domes with high productivity, near the centres of divergences, thus having a trophic motivation. Another hypothesis may be sexually motivated gatherings. In this region, as well as in the neighbouring zones (Black Sea, Senegal), there is a reproductive period in the middle of summer. A 2·05-m male stranded at the end of November and which was autopsied, was mature as indicated by the histology of testicles, but was in a sexual rest period (Harrison, pers. comm.). The only strandings of very young individuals happen in August and September. Since the gestation lasts 10 to 11 months, mating must be in June to July. In Dakar (Senegal), Cadenat (1959a) caught a female ready for calving between October and December. I have no data for winter, but a second reproductive period is not to be excluded at the time when the animals stay near the North African coasts, although it has not been proved. In the area studied, the greatest gatherings of *D. delphis* always coincide with the time of reproduction.

Estimation of stock

A first estimation in August 1972 by the "belt" method (see p. 508) on a 4-day prospecting trip by the ship of the Marineland of Antibes, led to a density of 225 *D. delphis* in 228 square miles, *i.e.* practically one per square mile. Extrapolating to the "restricted area" (see above), between 42° N and the isobath 2000 m (*i.e.*, excluding the coastal belt, where the species do not live) in an area of 120 000 km^2, I estimated a total number of 36 000 dolphins. In fact, this number is probably an over-estimate because the "belt" of the survey was taken in a a region where the species gather at this time of the year in the region of the Algerian-Provençal Divergence. Extrapolating the same data to the "large zone" (*i.e.*, north from 40° N), in a 225 000 km^2 area, I estimated a total number of 65 000 dolphins. This seemed to me a too optimistic estimation, nevertheless Yablokov (pers. comm.) considered it low by comparison with the Black Sea, where estimations by aerial surveys gave a figure of 200 000 individuals in an area one and a half times greater. The trophic conditions and the competition in the Black Sea, however, are very different from those of the western Mediterranean.

TABLE VII

Estimations of abundance of Delphinus delphis *by the 'belt method': *, south from 42° N; others, north*

Cruise	Prospected area (km^2)	Number of observed animals	Mean density (animals per km^2)
May 1978	816	70	0·086
May–June 1979	772	30	0·039
May 1980	463	0	0
May 1981	1166	0	0
May 1981*	960	70	0·073
September 1981	563	0	0

Observations made during cruises of the oceanographic ship KOROTNEFF of the Marine Station at Villefranche-sur-Mer in 1978, 1979, 1980, and 1981 provide the results given in Table VII. Viale & Bardin (1982), on seven 90-mile transects between Marseilles and Ajaccio in August 1981, counted 12 individuals (density, 0·02 by square mile). Observation during a cruise of the ship of the Marineland of Antibes, from Ile Rousse (Corsica) to Gibraltar in September 1972, gave the following densities: 4·17 by square mile in the Alboran Sea, 3·06 between Gibraltar and Ceuta, 0·63 east from Alboran Sea, and 0·25 south from Balearic Islands. The Alboran Sea is specially rich because it receives trophic enrichment from the Atlantic Current, and from a divergence studied by Minas (1968). This author observed here high concentrations of dolphins (Minas, pers. comm.), just as did Pilleri (1967a). So the distribution of *D. delphis* is heterogeneous within the western Mediterranean basin, the southwestern part being clearly richer. In the northern part (north from 40° N), if we take 0·15 individuals by square mile on an average and extrapolate, we get 10 000 individuals today.

The average number of strandings of the French coast between 1971 and 1981 is 1·3 individuals by year, *i.e.* for the European coasts north from 40° N, about four individuals by year as an extrapolation. Assuming a mortality rate of the same order as that of *Stenella* (see below), that is 6% by year, and seeing that *Delphinus delphis* stay in the zone only from March to September (seven months during the year), we estimate that about 1% of the carcases get stranded, proving the oceanic character of this species.

STENELLA COERULEOALBA

Stripped dolphin; in French: dauphin rayé bleu et blanc.

Up to 1970, *S. coeruleoalba* was assumed to be not very common in the Mediterranean (Marcuzzi & Pilleri, 1971), but since then, this species has found a more and more important place in the reports of strandings, until today it constitutes the main part of them (see Table VI, p. 521). In 1976, 90% of the *Stenella* beached in France were Mediterranean, 81% in 1977, 100% in 1978, 81% in 1979, 69% in 1980, 75% in 1981. The species is rarely mentioned in the literature, and sometimes under the name of *Stenella styx* (Gray) or *S. euphrosyne* (Gray). Nevertheless, Van Bree, Mizoule & Petit (1969) report a regular frequentation of the southern part of Gulf of Lions by this species. Fraser (in Busnel & Dziedic, 1968) notes that it is represented in the frescos of the Queen's Palace at Knossos, and I also have recognized it in Carthage. The collections in museums also prove its presence in Mediterranean for a long time: specimens from Genoa and Monaco were harpooned in 1809 and 1912. Van Beneden & Gervais (1868) report one catch at Port Vendres (France) and Robineau (1972) describes some skeletons stranded in Cape d'Ail in 1909. In 1959 and 1967, Busnel (pers. comm.) observed mixed herds of *Stenella* and *Delphinus* in the Alboran Sea at the beginning of summer, and captured a 96-cm *Stenella* which I studied in the collection of the Museum of Paris.

The species is, therefore, undoubtedly present and perhaps even more than previously imagined, but probably was confused with *Delphinus delphis*, as admitted by Casinos & Vericad (1976), and as proved by Duguy & Cyrus (1973) after seeing photographs in the reports of Monaco cruises. I, personally, have stated this frequent confusion in the regional newspapers. Since 1970, numerous observations at sea on board of regular ships include, under the heading "common dolphins", herds of *Stenella*.

Sightings and strandings

One hundred and forty-five strandings were recorded between 1970 and 1983 on the French coasts (Duguy & Budker, 1972; Duguy, 1973, 1974, 1975, 1976, 1977, 1978, 1979, 1980, 1981, 1982, 1983a, 1984; Duguy, Casinos & Filella, 1978); 14 beachings or catches reported in the literature since 1864 should be added. Recent data from the coasts of Spain are given by Casinos & Filella (1976), Grau, Aguilar & Filella (1980), and some Italian data by Marcuzzi & Pilleri (1971). Observations at sea were made by Busnel, Pilleri & Fraser (1968), and five were recorded in the regional newspapers in 1970, 1972, and 1976 (Viale, 1977b). From 1977 onwards, I made 11 oceanographical expeditions and *Stenella* were reported on every one. Among the observations of yachtsmen or officers of regular ships, few are reliable owing to the confusion with *Delphinus*.

Biological data

Stenella coeruleoalba is well known thanks to statistical studies made in Japan by Nishiwaki (1967, 1976), Kasuya (1971, 1972), Miyazaki *et al.* (1974), Kasuya & Miyazaki (1975), Miyazaki (1976), Nishiwaki & Sasao (1976) from the data of an industrial Japanese fishery catching 8000 to 10 000 dolphins each year. I largely refer to their work in what follows and in Frontier & Viale (in prep.).

Pilleri & Busnel (1969), Gihr & Pilleri (1969a,b) give a size–weight relationship for 12 individuals (five males, seven females) all prepubescent; 26 other results are given by Duguy (1975), Tardy (pers. comm.), Besson & Viale (in Duguy, 1982, 1983a, 1984). The fitted line (first principal axis) is represented in Figure 12 and has the equation

$$W = 13{\cdot}25\ L^{2{\cdot}6025}$$

where W is the weight in kg and L the length in m. The correlation coefficient (linear correlation in log–log) is 0·9415. This equation differs from that obtained for *Tursiops*, the shape of which gets thinner with age to a greater degree than does *Stenella*; weight is proportional to the power 2·2 of the length for the first species and 2·6 for the second.

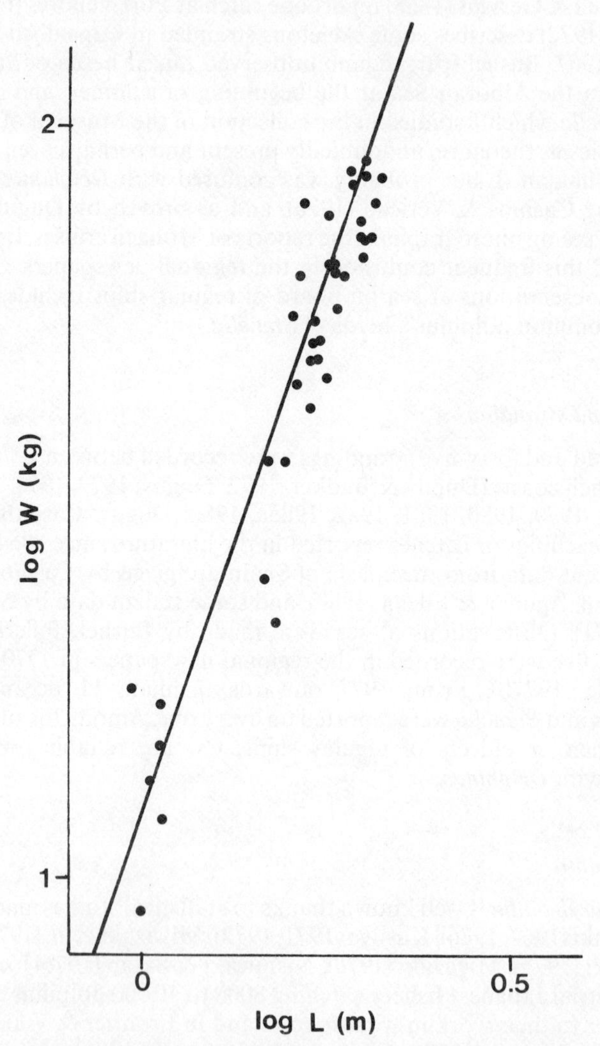

Fig. 12.—Regression line between length and weight, on a log–log scale, in *Stenella coeruleoalba* according to 38 results from the literature and personal observations: fitted line is the first principal axis, the equation of which is: $\log W = 2{\cdot}6025 \log L + 1{\cdot}1292$, and $r = 0{\cdot}9415$.

Growth was studied by the Japanese workers, who gave a length at birth of 1·00 m on the average, length at six months (end of exclusive suckling) is 1·50 m, and maximum length is 2·30 m on the average. From these results, it was possible to calculate a growth equation following the Von Bertalanffy's model:

$$L_t = 2{\cdot}30[1 - e^{-1{\cdot}1817(t+0{\cdot}4828)}]$$

where L is the length in m at age t in years. The curve is given in Figure 13. That equation was used as a size–age key, assuming that growth probably varies little between the Pacific and the Mediterranean populations. It was then possible to infer an approximate age of the stranded individuals, and hence an approximate date of birth of each one, and finally a distribution of births throughout the year (Frontier & Viale, in prep.). It can be seen that there are two distinct peaks of births (Fig. 14): one principal maximum in August to October, and a slighter one in April–May. In Japan, there are three principal periods of reproduction: January–February, May–June, and September–October; but I recognize that my sample is much less important than that used by the Japanese workers.

The percentage of females recorded in the Japanese fishery is 47·7%. In a census material from the Mediterranean, eight strandings out of 46 were without any indication on sex, 19 were males, and 19 were females. In the Pacific, pregnancy lasts 12 months; the suckling time is 18 months, *i.e.* up to a size of 172 cm on the average, but mixed feeding begins at 6 months (158 cm). In the Mediterranean, I dissected a 140-cm female containing in its stomach and gut only milk, and a 143-cm male containing milk, vertebrae of fishes, and beaks of cephalopods; mixed feeding thus begins at a size lower than that indicated by the Japanese workers. A 163-cm male seemed to be fed only with

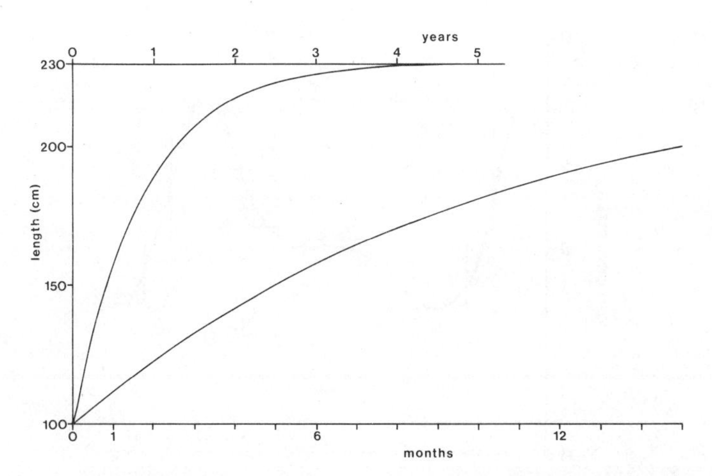

Fig. 13.—Growth curve of *Stenella coeruleoalba*, following the equation:

$$L(t) = 2{\cdot}30\,[1 - e^{-1{\cdot}1817(t+0{\cdot}4828)}]$$

where L is in months and t in years; upper curve is age in years, lower curve in months.

milk, but it bore on its rostrum many very distinct scars of cephalopods suckers, such as those observed in some species of *Grampus*.

The remainder of the growth depends on the sex. The size at puberty is 206 cm in males and 225 cm in females in the Pacific. The size at physical maturity is 222 and 238 cm, respectively. In males, the physical and social maturity are reached at the same age, $13\frac{1}{2}$ years on the average.

The period of sexual rest, from the end of suckling to the beginning of next pregnancy, is 6 months to $2\frac{1}{2}$ years. The interval between two successive calvings varies, depending on the level of exploitation of the population (then on the mortality). In the Japanese fishery in the Pacific, the average rate of natural mortality is about 0·060 per year, and the mortality due to fishery is 0·066. The average calving interval decreased from 4 years at the beginning of the fishery, to 2·26 years in 1974 (Kasuya & Miyazaki, 1975), resulting in an increased rate of natality from 6 to 9% per year (Viale, 1980a,b).

Returning to the stranding data in the Mediterranean, Frontier & Viale (in prep.) studied the statistical distribution of sizes of the 153 individuals whose length was known. Only the salient features are given here. The distribution is not a Gaussian one. Fitting the cumulative curve of frequencies without a probit transformation, there appears a succession of some almost linear segments. Each one results from a combination of the growing curve within a given interval of sizes, and of the mortality curve in the same interval. The breakpoints between intervals are to be interpreted as fast changes of the conditions of growth and/or mortality—perhaps only of mortality of the population, the growth conditions being more characteristic of the species and probably less variable. It may be assumed that the breakpoints within the distribution of sizes of stranded carcases correspond to an equal number of critical periods in the life of the animals, and possibly, seeing the sizes at these

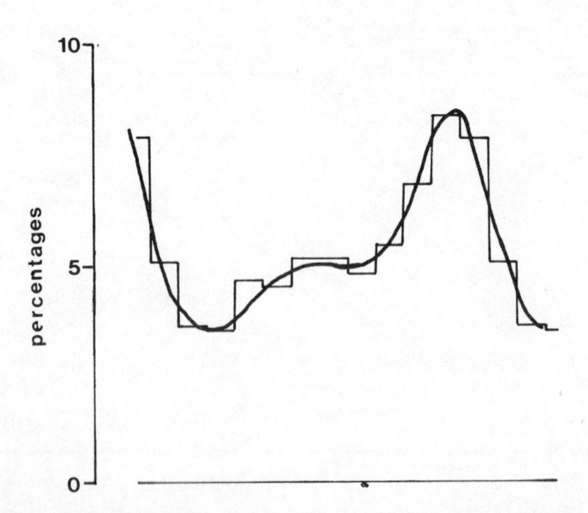

Fig. 14.—Smoothed histogram of distribution of births throughout the year, according to the stranding data in the northwestern Mediterranean: low steps at left of diagram are December and January; for more clarity in the interpretation of the distribution, two months were repeated in the left part of the histogram; from Frontier & Viale (in prep.).

points, to the end of exclusive suckling (145 cm), to puberty (190 cm), and to social maturity (210 cm). Hence, the sizes at these phases of life appear to be smaller than in the Pacific population. Assuming, in addition, that the growth curve changes little with respect to the rates of mortality and reproduction, the size–age key from the Pacific may be used here, resulting in exclusive suckling of $4\frac{1}{2}$ months in the Mediterranean, compared with 6 months in the Pacific, and puberty at 1 year instead of $1\frac{1}{2}$ years by males and 2 years by females.

Kasuya & Miyazaki (1975) determined absolute ages of females according to the number of *corpora albicantia* present in the ovaries. They calculated from the demographic structure of females an average mortality rate of 0·05 per year for immature females, and 0·07 for adults. Combining the mean rate 0·06 with the curve of growth allowed Frontier & Viale (in prep.) to calculate that 5·8% of the dead individuals ought to be between 0 and 1 year old, 5·5% between 1 and 2 years, and 88·7% more than 2 years. The mortality of juveniles is low, and death is restricted to the oldest ages. On the contrary, in the Mediterranean strandings, 66% are between 0 and 1 year, 32·7% between 1 and 2 years, and only 1·3% are more than 2 years.

It is well known that populations respond to a stress (resulting in an increasing mortality) by an increase of natality, often by way of a decrease of the age of first maturity, and a decrease of the interval between two successive calvings (called "calving interval"), as happened in the Japanese fishery. The above reported changes of the reproduction figures in the Mediterranean may be an effect of the huge infantile mortality observed here, of which the cause is not yet understood. The fecundity being restricted, in this species, to one calf at a time, and suckled for a long time, the increase of natality can only occur through a decrease of the calving interval and of the age of sexual maturity. An indication of a shortening of the exclusive lactation period appears to be evident; the stomach contents of the carcases show remains of fishes and squids in individuals having a size corresponding, in the Pacific, to the period of exclusive suckling. Sexual maturity is also reached at a size corresponding to the lower limit in the Pacific.

Estimation of stock

An estimation of stock made in August 1973 by the "belt method" (see p. 508) gave a density of 0·157 individuals per square mile in the neritic zone, and 0·066 in the oceanic zone, from which can be calculated 9250 individuals in the area north from 40° N—an estimation lower than the 65 000 *Delphinus delphis* found at the same time. More recent observations show that the ratio between the two species has been inversed; nevertheless, the ratio between the two species in the stranding statistics cannot be used for an estimation of numbers alive owing to the infantile mortality of *Stenella*. The "belt method" applied during nine oceanographic cruises of KOROTNEFF between 1978 and 1981 (Table VIII) leads to an estimate of 20 000 *Stenella*, *i.e.* a density of 0·34 individual per square mile in the neritic zone, and 0·19 in the oceanic zone. This density represents an average increase of the population of 10% per year.

A natality rate of 10% per year ought to correspond to the following hypothesis: 50% of females in the population, among which 50% are mature, and a 2·5-years calving interval. We thus approach the magnitudes noted in the Pacific for an intensively exploited population of *Stenella*. In fact, the

TABLE VIII

Estimations of density of Stenella coeruleoalba *by the 'belt method', north from 42° N*

Month	Prospected area (km²)	Percentage of that area in coastal zone (%)	Number of observed animals	Mean density (animals/km²)
May 1978	816	100	180	0·22
May–June 1979	772	100	175	0·23
May 1980	463	100	135	0·29
July 1980	1019	22	100	0·10
May 1981	2127	86	146	0·07

natality rate ought to be still higher in order to take into account the natural mortality, and especially the huge infantile mortality, which leads to a quite unrealistic value of the rate of births. Hence, a permanent immigration is necessary to explain the observed increase of population.

The reasons why *Stenella* is now conquering the environment partially abandoned by *Delphinus* remain an open question. An important immigration coming from areas of the Mediterranean where the species was abundant is a quasi-certainty, because no birth rate compatible with the biological figures quoted for the species can explain the speed of its increase in numbers. The immigration must concern relatively young individuals, as proved by the absence of stranded adults. The young population seems to tolerate its new environmental conditions rather well, but the new-born pay a heavy toll due to the environmental change.

A massive increase of the population would be expected to be associated with an increase of adult strandings, but it is not so. It is difficult to say whether the animals observed at sea are adult or young, as the estimation of length by eye from aboard a vessel is difficult. Two hypotheses could explain why there is an increase of the population number and, simultaneously, an increase of juvenile mortality: (a) the population is increasing through a supply regularly coming from other, previously densely populated areas of the Mediterranean, and the influx is principally young animals; (b) there is an increase in the rate of birth, as a response of the population to a stress, that still has to be specified.

Finally the question is, do some new-born survive? Future observations perhaps will continue to report strandings of adults, but nowadays the maintenance of the population through reproduction in the zone, and not only through immigration, is questionable. It is hence hoped that a detailed survey will be continued.

GRAMPUS GRISEUS

Risso's dolphin; in French: dauphin de Risso.

Sightings and strandings

Grampus griseus lives in small groups in areas of more than 200 m depth. Five individuals have been stranded on the Corsican coasts since 1962, and 20 on the French coasts of Mediterranean between 1972 and 1981 (Duguy & Budker, 1972; Duguy, 1973, 1974, 1975, 1976, 1977, 1978, 1979, 1980, 1981, 1982, 1983a, 1984). Paulus (1964) records three strandings and 26 catches before 1934—among these, a herd of 11 individuals at Villefranche-sur-Mer. The Cetological Commission of Barcelona (Casinos & Vericad, 1976; Grau *et al.*, 1980) records three strandings in Mediterranean between 1878 and 1979. In Italy, Tortonese (1963c) and Arbocco (1969) record ten strandings between 1973 and 1977. On the coasts of the zone studied, a total of 62 catches or strandings are recorded between 1811 and 1981, that is, 0·36 individual by year on the average.

Reliable observations at sea are few; I reported five sightings before 1977 and Pilleri (1970) reports two. Among the observations reported by regular ships and centralized in the Centre d'Étude des Mammifères Marins (La Rochelle), few of *Grampus* were noted; this animal is very difficult to identify at the speed of these ships. Since 1978, it has only been observed four times during my cruises in the Corsican Strait; once it was a herd of 15 individuals with calves in May 1980. I have seen them mixed once with *Delphinus delphis*, and once with *Stenella coeruleoalba*. Giordano (1981) reports two sightings. It is, therefore, astonishing to read in faunas that the species is "common in the Mediterranean" (Duguy & Cyrus, 1973; Duguy & Robineau, 1973). Fisher (in Paulus, 1964) affirms having observed several hundred *Grampus* off Gibraltar in July.

Colour

The colour of *G. griseus* is of great interest to the zoologist. Viale (1979a,b) showed that cetaceans excrete the excess salt coming from their diet through the epidermis. A white exudation then hides the real colour of the skin and, in *G. griseus*, the whole body can seem white because of this phenomenon. All the skin surface participates in salt excretion in *Grampus* and at sea this species may appear completely white, as shown by a photograph published by Leatherwood *et al.* (1980) and interpreted by these authors as the "colour of an old adult". Another often observed aspect in *Grampus* is the white marbling or veining on a black background. All the white parts disappear after dying and emersion. Carcases stranded are perfectly black after 24 h (see photographs in Viale, 1984). These colour variations are indications of the intensity of saline excretion according to the nature of the food consumed. Food is more or less rich in salt depending whether it includes more squids or more fishes. Colour, therefore, represents a valuable ecological index (Viale, 1979a) and leads me to the conclusion that *G. griseus*, hitherto considered as exclusively teuthophagous, does in fact consume fish in some circumstances.

Biological data

Little is known of the biology of this species. The size at birth is about 1·50 m, or 1·40 m in the Mediterranean (Paulus, 1964). The maximum size is 4 m. Sexual maturity is acquired at 3 m by males (Pilleri & Gihr, 1969; Harrison, Brownell & Boice, 1972). In the Mediterranean, a 2·95-m male that I dissected was found to be mature, and another of the same size was immature.

Using stranding data

Analysing the stranding frequencies according to months on the French coasts (Fig. 15) indicates a principal period of strandings between April and August, with one maximum in May and another in July, which suggests that the species make migrations: it may arrive in the zone studied in spring or summer, following an annual cycle which is reminiscent of the thermal limitation seen above for *Delphinus delphis.* I made two cruises in December along the Corsican coasts without encountering any *Grampus.* Systematic mensual cruises between Banyuls and Villefranche in 1981 and 1982 allowed observation of individuals from March onwards, and during summer (Palazzoli, 1983).

Some observations of mothers with calves were principally observed in July and August in the northern part of the basin, and at the end of May along the Corsican coasts. In fact, births may occur up to October: a 1·44-m female stranded alive with its umbilical cord was observed on 16th October in Monaco.

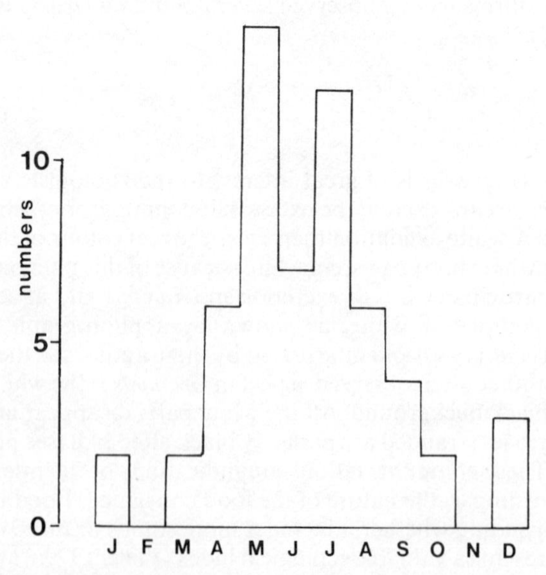

Fig. 15.—Monthly distribution of catchings and strandings of *Grampus griseus.*

Estimation of stock

Using again the "belt" method, 26 *Grampus* were counted in 228 square miles (Marineland Expedition, August 1973). Extrapolating to the "restricted zone" (north from 42° N) and excluding the neritic zone, this gives 2500 individuals. The observed herds varied between 6 and 15 individuals; this means 250 to 300 herds in this zone, and in the "large zone" (north from 40° N), a three times greater population, *i.e.* 7500 *Grampus*, or 0·025 individual by square mile.

This expedition also counted 225 *Delphinus*, that is a ratio of 8:1 with respect to *Grampus*. Comparing the number of strandings recorded in the literature (unhappily not all counted in the same conditions), there were 30 strandings of *Delphinus* between 1890 and 1912 (22 years), *i.e.* 1·36 by year on the average, and 26 strandings of *Grampus* between 1814 and 1964 (150 years), or 0·17 per year. The ratio between species is again 8:1. But the ratio between 1971 and 1977 is not the same: 14 stranded *Delphinus* to 8 *Grampus*. Although these numbers are estimated with a random error which it is impossible to estimate, they seem to indicate an increase of the mortality of the teuthophagous species during the last years—this mortality having no consequences as yet on the number of sighted individuals.

ZIPHIUS CAVIROSTRIS

Cuvier's beaked whale; in French: baleine à bec, baleine de Cuvier.

Taxonomical problems

Van Beneden (1888) listed the synonymy of the different forms described from the Mediterranean: *Delphinus demaresti, D. philippi, Hyperoodon doumetii, H. gervaisi, Ziphius savi.* All are *Z. cavirostris.* This author also questions two other species names: *Z. indicus* and *Petrorhynchus capensis* = *Ziphius capensis.* One species is now recognized (Hershkovitz, 1966; Mitchell, 1975a), in which some observed differences in the skull bones reflect differences of age and not zoogeographical differences (Van Beneden & Gervais, 1868; Fraser, 1942). The species is scattered in all the temperate and subtropical seas.

Two teeth have been described in the lower jaw by Duguy & Robineau (1973) but I have not seen them in any of the seven individuals I have examined: they remained inside the gum. Van Beneden & Gervais (1868) also noted that teeth do not always appear, especially in females. Duguy & Budker (1972), in the first individual they examined, reported this feature as an anomaly. Filella-Cornado (1971) did not see any teeth in two females stranded in Spain.

Colour

The colour described as typical is white on the whole antero-dorsal part, and black on the remainder of the body (Fraser, 1949; Norris, 1966; Duguy & Robineau, 1973), in spite of some statements to the contrary (Van den Brink, 1960, p. 155; Aloncle, 1964; Hershkovitz, 1966; Marcuzzi & Pilleri, 1971).

The first specimen I examined was entirely black, hence my wrong identification, which has been rectified by Budker and by Fraser (pers. comm.)

who admitted that the colour may vary, even may be aleatory, a feature which later was confirmed by the descriptions of Moore (1968). A photograph of the first *Ziphius* I examined was published in the local newspapers, and taken 24 h before my own photograph; it clearly indicated a white anterior part of the body. After dissecting and photographing the entirely black animal, it was difficult to believe that the white colour was more than a mere reflection from the sun! But six years later, I examined a few hours after its death a female that had been stranded alive; it showed a distinct dull white colour on the whole antero-lateral region of body, in conformity with the figure of Fraser (1949). The next day, all the colouration had disappeared. Such observations were confirmed for two stranded individuals in 1974. Mediterranean *Ziphius* thus show, when living, a white colouration on the whole head and obliquely stretched to the dorsal fin. The colour is perceptible for only a few hours after the death of the animal. It is of interest to note that the reconstructed 7-m female (taken at Albisola in 1925) exhibited in the museum of Genoa also shows the white anterior region.

The white part of the skin had not the smooth and shining aspect of the skin of other cetaceans, but was dull and covered with many tiny crystals giving it the appearance of an emery-cloth. It is also iridescent, with bright flashes due to the crystals. Some of the biggest crystals sorted under a stereomicroscope were analysed by the C.E.A. Laboratory at Villefranche-sur-Mer; they were pure sodium chloride (Maetz, pers. comm.). Photographs of these crystals under scanning microscope have been published (Viale, 1977b, 1979a,b). The white region of the animal is apparently entirely covered by a mixture of sodium chloride and crystals of magnesium, calcium and potassium salts. One day after stranding and death, these crystals were washed away by the sea water spray and not replaced, as they would be in the living animal.

Another female, stranded in 1974, also showed crystals in its whole front region. The foetus of a male it was carrying did not show such crystals. Filella-Cornado (1971) described a stranded dying female (5·19 m long) with a white front surface that stretched to the back fin. Omura, Fujino & Kimura (1955) published a photograph of a female whose anterior part was whitish. In contrast, Mitchell (1975a) gives a photograph of a perfectly black *Ziphius* at sea, but does question its identification. Cadenat (1959b) thinks that *Ziphius* must be frequent along the West African coasts, although he never saw them stranded or captured; the species, he says, it easy to recognize thanks to its very anterior back fin, but he makes no mention of the white front part, which must be easily visible at sea. In fact it is possible that, in some physiological conditions, animals do not show this white area (Viale, 1984).

The material

I reported 22 strandings from 1961 to 1981, among which 18 were in Corsica and four on the Riviera; seven were dissected. No sightings were, however, made at sea of this species. From 1850 to 1979, ten strandings on the coasts of Italy and Monaco were reported (Richard, 1936; Rode, 1939; Tortonese, 1957, 1963b; Paulus, 1962; Tortonese & Demr, 1963; Marcuzzi & Pilleri, 1971) and 17 on the Mediterranean coasts of Spain (Cabrera, 1919; Lozano-Rey, 1947; Casinos & Vericad, 1976; Grau, Aguilar & Filella, 1980). On the whole, 60 strandings are recorded in 130 years—less than one in two years on average but, in fact, they beach in groups.

The F.A.O., following the conclusions of the International Commission for Conservation of Nature (U.I.C.N.) classes *Ziphius* among the "rare species", on which essential knowledge is lacking (Anonymous, 1976). Prior to 1972 works are mere osteological descriptions (Cuvier, 1812; Van Beneden & Gervais, 1868; Caldwell & Caldwell, 1971; Caldwell, Rathjen & Caldwell, 1971) or stranding reports. Although the exploitation of *Ziphius* still takes place in the Antilles and in Japan (180 catches per year following Omura *et al.*, 1955), biological data from these regions are scarce (Leatherwood *et al.*, 1980). Observations at sea are even rarer: Busnel (pers. comm.) observed a single herd in ten Mediterranean expeditions. This rarity probably comes from the difficulty of recognizing the animal at sea. Cadenat (1959b) reports observations at sea, but without any photography. Tomilin (1967) published a photograph of the anterior region of a *Ziphius* of the Barents Sea, but it is astonishingly different from those observed in Mediterranean.

Biological data

There is no information in the literature about relationships between weight, length, and age, and my data are still too scarce to be able to give a regression line. Some details of three animals are as follows.

January 1962: male, length 5·10 m, weight 1500 kg. Stomach was empty. Blubber was 5 cm thick indicating that it was in a satisfactory physiological state: it had died accidentally. It was adult, for the diameter of the teeth were greater than 14 mm (Moore, 1968), testicles were 18 cm long and 7 cm wide. Its absolute age, according to dentine layers, was 17 years (Kasuya & Miyazaki, pers. comm.). The skeleton of the pectoral fins (see Fig. 26 in Viale, 1977b) still included cartilaginous parts.

March 1968: female, length 4·50 m, not weighed. Adult, for there was one yellow body in the ovary. Some degree of senescence was indicated by the arteriosclerosis of the vessels of the wide ligaments and by numerous atheromes in the big vessels. Ovary small, 10 cm long and with a smooth aspect.

December 1974: pregnant female; length 5·16 m, not weighed but weight estimated as 1600 to 1650 kg (unhappily the animal was jammed between the rocks, which did not allow it to be weighed in order to derive the weight increase of a pregnant female). The foetus was 20 kg heavy, but perhaps three times more including the amniotic bag and liquid. Blubber was 5 cm thick. The weight of large ligaments which sustains the horns of uterus, and the thickening of all the vessels and tissue of that region, indicate that a female is heavier than a male of the same length, even during the sexual resting period.

Omura (1972) gives maximum size of 7·26 m for males and 7·60 m for females. The maximum length observed in Mediterranean is 7·60 m (female recorded by Casinos & Vericad, 1976). Apart from a 6·50 m female stranded in Frontignan (France), all the strandings in France were of animals less than 5·60 m long.

Ziphius is in the category of the cetaceans for which "biological parameters are unknown and necessitate investigations" following the Symposium at Montreal of the International Whaling Commission (Mitchell, 1975a) and the

Scientific Consultation of F.A.O. in Bergen (Anonymous, 1976). Included here, therefore, is what is available for the western Mediterranean and elsewhere.

Omura *et al.* (1955) consider sexual maturity in males occurs at a length of 5·95 m, "or even smaller". In the Mediterranean, a 5·10-m male with 18 × 7 cm testicles was still immature (histological confirmation by Harrison, pers. comm.). For females, five results from Omura *et al.* (1955) recorded 5·95 m as a minimum length for pregnant females, and 5.27 m immature ones. I observed two mature females of 5·27 and 5·40 m; a 5·30-m female was not pregnant (but the ovaries were not histologically examined). I, therefore, consider that the size of 5·10 m is probably that at which sexual maturity is attained in Mediterranean, both in males and females. I have verified that females are able to grow 2 or 2·50 m more after sexual maturity; the greatest observed size in the Mediterranean strandings was 7·60 m.

In Japan, the maximum number of pregnant females is in August, and at the same time, the smallest (33 cm) and the biggest (321 cm) foetus are observed (the second size is questionable). By contrast, in the Mediterranean, three females in an advanced stage of pregnancy were stranded in November (foetus of 1·30 m), December (foetus of 1·13 m) and March (foetus of 1·42 m). An adult male stranded at the end of January was beginning sexual activity; some spermatogenesis appeared in the seminiferous tubes but the epididyme was empty (Harrison, pers. comm.). Breeding must happen at the end of winter or beginning spring. In June, a 2-m new-born female, stranded alive, was noted (Cetological Commission of Barcelona). The size at birth is less than 2 m, in contradiction with the assessment of a "7 feet" (231 cm) foetus in a 5·95-m female as reported by Omura *et al.* (1955).

Sightings and strandings

The distribution during the months of the strandings records for the northeastern part of the western Mediterranean basin and for the Spanish and Balearian coast are shown in Figure 16. Strandings are scattered throughout the year, but are much more numerous between November and March, excluding an accidental stranding of a herd of 15 individuals in May 1963 (Tortonese, pers. comm.). After searching in the descriptions given in the literature for indications about the state of freshness of the animals, I found the following results:

	October to March	April to September
Living or very fresh carcases	24	4 (+15)
Putrefied carcases	0	4
No information	8	6

Excluding the collective stranding of 15 animals, we note that in the winter 75% of the stranded animals arrive at the coast when dying, or dead but very fresh, and only 30% in the summer. Low temperature, of course, delays the putrefaction, but we can verify that there are six times fewer living or very fresh stranded animals in summer than in winter, which demonstrates that the species comes nearer the coast in winter.

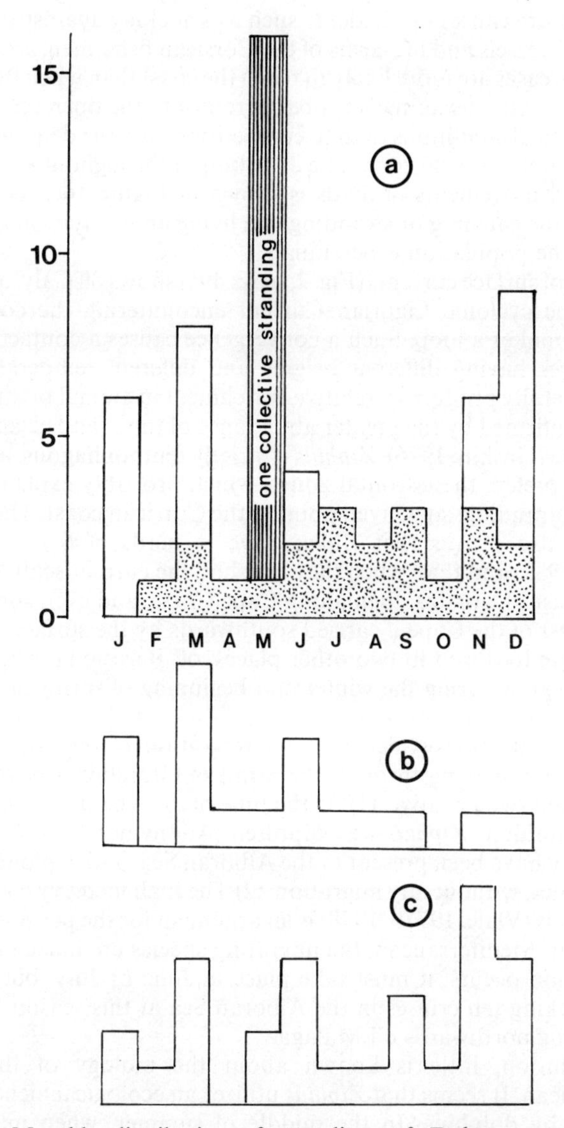

Fig. 16.—Monthly distribution of strandings of *Ziphius cavirostris*: a, northeastern Mediterranean basin, stippling are records of Paulus (1962); b, Mediterranean coasts of Spain and the Balearic Islands; c, supposed distribution of the herds, assuming that a catching or a stranding on an alive animal is a proof of the presence of a herd, since the species is a gregarious one; the collective strandings in (a) concerns a herd of 15 individuals at Genoa in May 1963.

All strandings in Corsica happen in winter, and are along the northwestern coast of the island and on Cape Corse. When approaching the coast, animals encounter more causes of accidents such as knocking against ships, shooting by the naval vessels and fire-arms of the Corsican fishermen, *etc.* On the other hand, the carcases are more likely to reach the coast than when the animals live offshore. In spring the animals probably return to the open sea, less owing to the hydrological conditions than to competition with the delphinids which are then found in the coastal zone. The distribution throughout the year of noted or presumed movements of herds is shown in Figure 16c. As the species is gregarious, the catching or stranding of a living animal indicates the presence of a herd. The population is perennial.

The map of surface currents (Fig. 2, see p. 495) shows off Calvi and Ile Rousse (Corsica) the cyclonic Ligurian Current encountering the counter-current which here makes a loop. Such a convergence causes a contact between two water masses having different origins and different temperatures. Such a contact generally produces a relative enrichment in animal biomass (Frontier, 1978), as confirmed by the greater abundance of tunas and cetaceans near the contact zone (Gaskin, 1976). *Ziphius* is strictly teuthophagous, as is the sperm whale, and prefers these frontal zones, which probably explains the localization of its strandings at a given point of the Corsican coast. The inventory of strandings also proves that *Ziphius* live in herds of 6 to 15 individuals (Sergeant, 1982) and that, when they are shot, the current scatters them along the west coast of Corsica, then along Cape Corse, and even sometimes up to the east coast of the Cape if carried southwards by the surface stream.

Ziphius are localized in two other places: off Barcelona, where there is an upwelling region during the winter and beginning of spring, and south from Balearic islands.

There is conflicting evidence about migration of the species out of the Mediterranean, passing through the Strait of Gibraltar. I have one positive argument and two negative. (1) At the time of the whaling exploitation in the Strait of Gibraltar, *Ziphius* was captured (Anonymous, 1942); however, the animals may have been present in the Alboran Sea, and exploited with sperm and fin whales, without any migration. (2) The high mercury content found in living animals (Viale, 1977b, 1978) is an argument for the permanent residence of animals in Mediterranean; the migrating species are indeed less rich. (3) If this migration occurs, it must take place in June or July, but Busnel (pers. comm.), making ten cruises in the Alboran Sea at this season, only saw one herd, running northwards off Malaga.

In conclusion, little is known about the biology of this species in Mediterranean. It seems that *Ziphius* utilizes an ecological niche left empty in the winter by dolphins. In the middle of summer, when many cetaceans (*Balaenoptera, Delphinus, Grampus*) search for the central zones of the basin, it is possible that *Ziphius* flees towards poorer surface waters and uses its ability to dive deeply, just as do sperm whales (*Ziphius* as well as *Physeter* can stay one hour diving), in order to live at the expense of the mesobathyal biomass.

PHYSETER MACROCEPHALUS

Sperm whale; in French: cachalot.

Sperm whales are now considered as rare in the Mediterranean. The species is, however, very well known because of its presence in many other regions, and all its biological features have been precisely determined. It follows that we are able to interpret the few observations from Mediterranean.

The material

I have recorded few personal observations, almost all in the Tyrrhenian Sea between 1957 and 1982: 12 observations at sea, among which four were of groups, and 10 strandings (seven in Corsica, three in Tuscany). The literature reports three more strandings on the Italian coasts (Marcuzzi & Pilleri, 1971; Ghirardelli, 1944; Pilleri, 1970) and 13 on the Mediterranean coasts of Spain (Casinos & Vericad, 1976; Casinos & Filella, 1976), among which five were before 1900. We shall also use some observations coming from a whaling industry along the Atlantic coast of Spain: 668 observed or captured sperm whales (Viale & Ridell, 1975; Viale, 1977a,b).

Past status and rôle of sperm whale in the whaling industry

The present-day rarity of the species is the opposite of the descriptions of Cabrera (1925) and Aloncle (1964), concerning the Gibraltar region. The catching of sperm whales in the Mediterranean was practised as early as the eighteenth century in the Strait (Barras, 1944). Two onshore factories were functioning on both sides of the Strait: in Getares (Bay of Algeciras) and near Ceuta (Mediterranean coast of Morocco), as described by Jonsgård (1966). Only the catches between 1921 and 1927 have been recorded (Anonymous, 1942). The exploitation was probably revived after World War II. Several whalers operated in the Strait between 1950 and 1960 (Aloncle, 1964; Viale & Ridell, 1975; Viale, 1977b).

From 1921 up to 1927, 489 sperm whales were intercepted in the Strait of Gibraltar (Jonsgård, 1966). In 1927, whaling was reduced as a consequence of the exhausting of the stock of fin whales; it was recommended in 1933, principally with catches of sperm whales and small cetaceans (*Balaenoptera borealis, B. acutorostrata, Ziphius cavirostris, Globicephala melaena*). The catches between 1933 and 1939 for the Spanish and Portuguese coasts are given in Table IX (from Anonymous, 1942). At the same time, the whaling statistics give 66 fin whales and five sperm whales caught in summer 1934.

About the years 1950 to 1960, a whaler operated in the Strait of Gibraltar and reported to me having captured up to 11 sperm whales on the same day during the migration period (Viale, 1977a). The whaling industry of Corcubion (northeastern coast of Spain) maintained itself, and has been transformed into a modern industry in the region of Vigo since 1966 (Viale & Ridell, 1975). I noted 569 catches of sperm whales by a single ship between 1966 and 1974, making an order of magnitude of 300 animals by year for the entire flotilla (five whalers) operating off the Cantabrian coast.

The rarefaction of the sperm whale in the western Mediterranean is

TABLE IX

Catches of cetaceans between 1933 and 1939 on the coasts of Spain and Portugal

	1933	1934	1935	1936	1937	1938	1939	Total
Sperm whales	77	82	136	172	80	0	0	547
Small cetaceans	176	158	140	308	208	388	389	1767

probably not only due to over-exploitation in the Mediterranean, but also perhaps caused by the chemical changes due to industrial pollution (Viale, 1977b). It seems, however, that the main factor for its disappearance is predation by man in the adjacent seas. Doi (1971) estimates that the present rate of catching is greater than the value necessary for the maintenance of the mundial stock, that is, the "Maximum Sustained Yield" (M.S.Y.). Sperm whales are indeed exploited along the whole length of its north–south migration. The exploitation has also been systematically carried out in the Azores (Clarke, 1956); the catches there are too numerous. The disappearance of *Physeter* in the Mediterranean is also to be linked with its intensive predation in the northeastern Atlantic, from whence the Mediterranean stock comes.

Estimation of stock

The estimation of *Physeter* stock is one of today's most studied problems (Doi, 1971; Best, 1975; Holt, 1976; Lockyer, 1976b). These authors have very different opinions about the estimation of maximum sustained yield (M.S.Y.).

Catching statistics of Vigo industry between 1966 and 1974 are available. The area fished includes between 43° and 45° N, 5° and 10° W. Observations of the whalers are recorded in Table X and Figure 17. It can be seen that sperm whales appear in the fishery in May, are most abundant in July and August and disappear in November. Whereas annual catches do not significantly decrease (excepting the two last years), the catches at the time of the maxima (*i.e.* annual peaks of the Fig. 17) are regularly decreasing. This indicates either a spreading of the migration season (which is biologically improbable) or a spreading over the year of the hunting effort; actually the hunting effort was increased during the period of year when the animals were scarce, in the same proportion as the stock was decreasing. The effort during the maximum abundance period was probably maintained. Increasing the fishing effort is a customary reaction of ship-owners facing a decreasing of the stock; hence the preceding statistics, which proved an increase of the fishery effort through spreading in time, also proves a decrease of the stock. In 1974, the establishment captured only 11 sperm whales and one fin whale.

Analysing the frequencies of observations of groups (Fig. 18) it can be verified that, among 280 observations totalling 668 individuals, one third corresponded to isolated animals. Among the total catches realized in 8 years,

Table X

Sightings by months, annual catches and the sex ratio of sperm whales off Vigo (Spain): a, plus two small groups; b, plus herds; c, plus youngs; d, among which some youngs; e, herds of youngs; f, plus two herds; g, plus three herds; h, plus one herd; i, plus some youngs; j, one herd; k, two herds; –, is the absence of whaling activity during the month

Sightings, month	1966	1967	1968	1969	1970	1971	1972	1973	1974	Total
May	0	—	0	6	0	0	5	2	—	13
June	13	7	23	24	22	14[a]	22	7	—	132
July	14	16[b]	22	21[c]	11	25[d]	5	9	—	123
August	9	26	25[e]	8	27[f]	23	19	21	—	158
September	30	24	22	14	22[g]	8[h]	16[i]	7	11	154
October	6	25	3	15	12	19[j]	8	15	0	103
November	2	0	10	0	6[k]	0	0	0	—	18
Total sighted	74	98	105	88	100	89	75	61	11	701
Captured	50	76	72	37	83	87	70	61	11	547
Sex ratio (females/males)	0/50	3/73	3/69	16/21	11/72	20/77	3/67	15/46	0/11	71/476

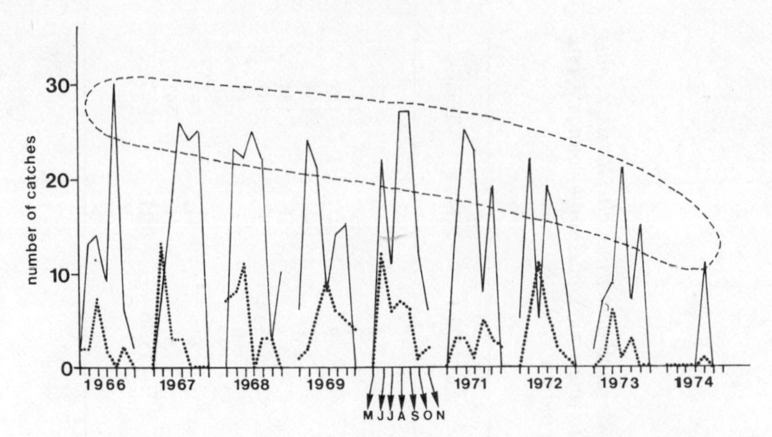

Fig. 17.—Mensual catches of *Balaenoptera physalus* and *Physeter macrocephalus* between 1966 to 1974 off Vigo (Spain): dotted line are *Balaenoptera*; continuous line are *Physeter*; the dashed outline indicates the decrease of sperm whale catches at the time of the year when animals show the highest frequency; only months from May to November are taken into account; from Viale (1977a).

87% were males, what is rather unexpected because the species is considered as polygamous (in the sense of several females for one male, and not the reverse). The sizes of animals unfortunately were not recorded, except for two individuals of 9·40 and 9·60 m caught in 1972. In six cases, the crew noted the "evil character" of the captured animals, which allows us to assume a respectable age and size. This is to be compared with the small size of sperm whales reported in the Mediterranean: no carcase longer than 10 m, except one

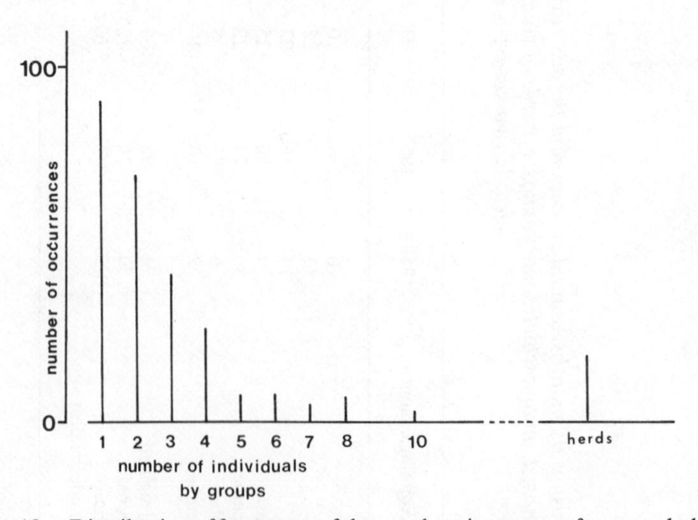

Fig. 18.—Distribution of frequency of the numbers in groups of sperm whales caught off the Atlantic coasts of Spain.

stranded in 1981 on Cape Corse. Sperm whales stranded in the Atlantic are between 13 and 18 m long.

The requirements for the conservation of the species lead the International Whaling Comission (I.W.C.) to fix the minimum size for authorized catches to 9·60 m, and later 10·90 m (Bannister, 1972). Several authors, however, think that these conditions are still insufficient, and an important work on a world scale is in progress in order to organize an efficient management for what remains of the stock (Best, 1975; Holt, 1976; Anonymous, 1981).

Migrations

In opposition to the general opinion, numerous problems still exist about the migrations of sperm whales. Whereas a great movement, from north to south and the reverse, has been proved for balaenopterids, it has not been proved for sperm whales. Big endeavours have been made, using modern techniques such as aerial observations and tagging (Clarke & Paliza, 1972).

In the Mediterranean, Bolognari (in Marcuzzi & Pilleri, 1971) stated that sperm whales arrive during the winter, and the majority leave the next autumn or in the following spring. At Gibraltar, more animals are sighted in May than in October. The migration passes 100 km off Cape Spartel, going northwards. In the Mediterranean, animals perhaps run up to Israel. Thus, it is agreed that sperm whales enter and leave the Mediterranean through the Strait of Gibraltar; a small number (young males) remain in the Mediterranean during the summer. What does this migration mean? Is the Mediterranean a breeding area? Best (1969a,b) and Gambell (1972) think that the majority of matings take place in December off South Africa, and that breeding most frequently occurs in February and March. In the Mediterranean, the only observations of very young individuals were in August (5 m), April (6 m) and 31 December (6·51 m). If we refer to growth data given by Lockyer (1976b) and represented here by Figures 19 and 20, the respective ages must be 6 months, fifteen months, and twenty-one months, then the births take place in February, January and March—without any change in time with respect to the Southern Hemisphere.

These data make reproduction of the species in the Mediterranean

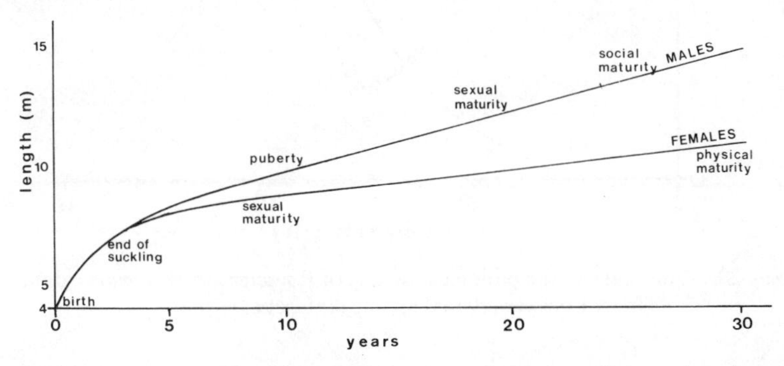

Fig. 19.—Growth curves of *Physeter macrocephalus* according to data from Lockyer (1976b) in the Antarctic.

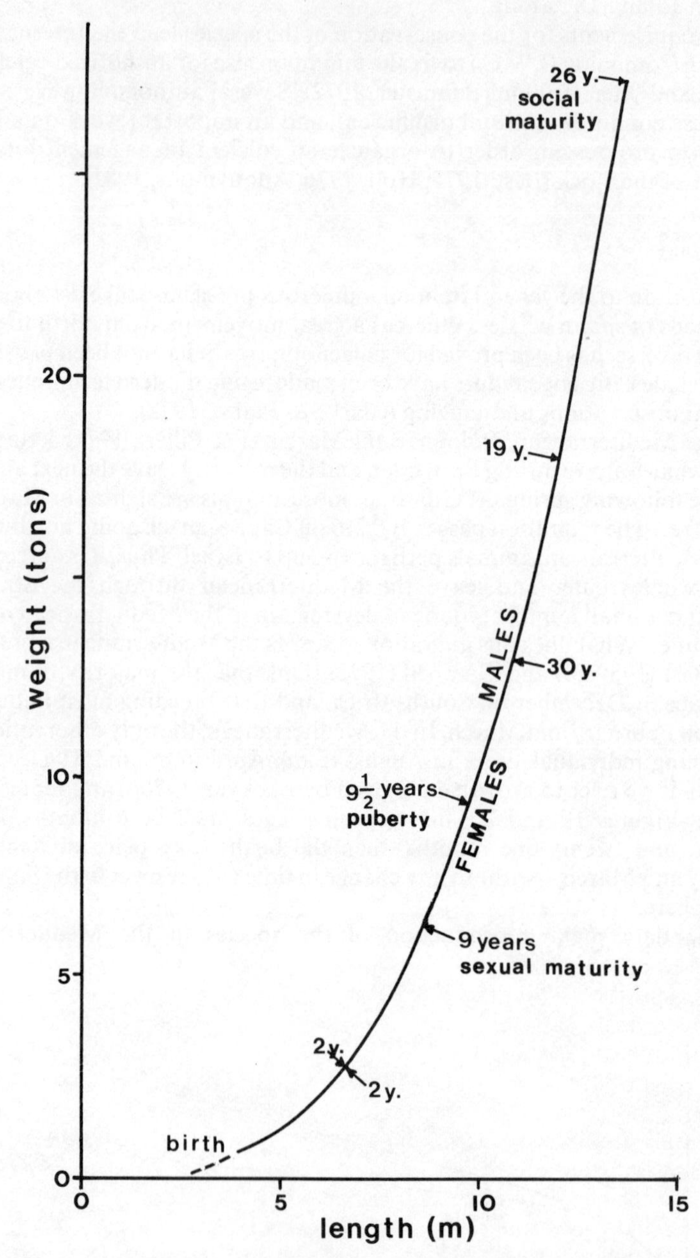

Fig. 20.—Summary of the biological details of *Physeter macrocephalus*; data from Lockyer (1976a) for the Antarctic: y, year.

plausible, since a number of adults stay there during the winter. Nevertheless, only in the first case (specimen born in February and stranded in August), can we be sure that the animal only knew the Mediterranean. As stated by Best (1974), young males wait for sexual maturity (9 years, 9 m long) before undertaking their first migration. I then imagine that the young males born in the Mediterranean stay there and constitute a small permanent population which passes the summer. In fact, summer observations generally are of individuals 6 to 8 m long.

Among the observations made by the whalers off the Cantabrian coast, only seven herds included calves, and a mother and its calf was noted only once (Table X). All these cases were between July and September, that is, during the migration towards the northern feeding grounds. Suckling females do, therefore, migrate with the others.

From the few available data, I conclude that there is a winter migration towards the Mediterranean, which is a breeding area, and a migration in spring and summer towards the northern part of the Atlantic, at least up to 45° N, this zone being (also for fin whales) a feeding area.

Biological data

The growth, mortality, fecundity, *etc.*, have been well known since the works of Clarke (1956), Bannister (1969), Best (1969a,b, 1970, 1973, 1974, 1975), Gambell (1972), and Lockyer (1976b), and are summarized in the Figures 19 and 20.

Observations and hydrological conditions

Important progress has recently been made in the hydrological characterization of the feeding and reproductive areas of sperm whales. In some cases, the distinction between reproductive and trophic areas is hazy, especially for females. Reproduction is related to warm waters; no other condition has so far been established (Gaskin, 1976).

In winter, the whole Tyrrhenian Sea is warmer than the remainder of the Mediterranean, and the only stranding place observed in winter in Corsica is the southeastern coast, at a period when currents run southwards. Among eight strandings, seven were on the Tyrrhenian side. Sightings were also most frequent in the Tyrrhenian Sea.

That part of the Mediterranean situated east from a diagonal line passing through Genoa and the Balearic Islands, to the Algerian–Moroccan frontier, is thermally suitable for the breeding of sperm whales. The western Mediterranean, however, includes only a few zones useful as trophic areas. Studies on the energetic balance (Gambell, 1976; Lockyer, 1976b) show that any pregnant female, in order to supply its needs, has to increase its daily diet by 10% if young, and by 5% if adult, *i.e.* "socially mature" (see Fig. 20). The increase of daily diet for a lactating female reaches 62% for a young one and 33% for an adult. The young one has to fill its stomach four or five times every day (Lockyer, 1976b). According to other authors, females do not feed during the last stage of their pregnancy.

I emphasized, at the beginning of this review, that the Mediterranean is characterized by oligotrophic conditions, especially at the second and third trophic levels. The problem of the food supply of sperm whales, *i.e.* the biomass of cephalopods, is then to be set up. It has to be acknowledged that very little is yet known about the quantities of cephalopods present in the Mediterranean, as well as in other seas. Perhaps suckling has to take place in a richer region; females which breed in the Mediterranean in winter then ought to leave the Mediterranean at the beginning of spring, as confirmed by Bolognari (1950, in Marcuzzi & Pilleri, 1971).

The oceanographic structure of whaling grounds has been studied in Japan by Nasu (1966), in New Zealand by Gaskin (1971, 1973, 1976), and in South Africa by Gambell (1972). Such studies are difficult because of the long- and short-term fluctuations of populations which modified the annual catches and hence the estimation of stock. Gaskin (1976) claimed that only a detailed study, season by season, of the oceanographic conditions in the whaling areas will allow us to understand the true characteristics of the trophic zones. Following thirty oceanographic expeditions in Australia and New Zealand, the most favourable conditions for sperm whales seem to be the simultaneous existence of a horizontal and a vertical gradient of temperature, that is areas where the thermic gradient is about 5 °C for the first 100 m (Nasu, 1966; Gaskin, 1976), and divergence and convergence areas. Do such characterized regions exist in the Mediterranean? Considering the surface monthly isotherms given by Romanovski (1955), and the isotherms at 50 m depth (Furnestin & Allain, 1962, drawn again by Viale, 1977b, see her Fig. 3), we find a few zones with a sharp horizontal gradient 40 to 80 miles southwest from Marseilles, as well as off Catalonia, from September to November. We saw above that the thermocline is located 50 m below the surface at the centre of the Algerian-Provençal Divergence (43° N, 8° E) during the summer, while the surface temperature varies between 20 and 25 °C; this results in a favourable gradient. We have seen that other cetacean species utilize these zones, but observations rarely mentioned sperm whales: only two 7- to 8-m individuals in 1976, and three in 1981 (Viale & Bardin, 1982). The species is thus scarce in the whole northwestern basin.

Sperm whales frequent the abrupt descents at the sides of deep regions. Big isolated males have been observed by Marcuzzi & Pilleri (1971) north from Sicily and around Lipari islands. Another favourable zone is a 2000-m deep hole between Catalonia and Balearic Islands. Among 30 strandings recorded by the Cetological Commission of Barcelona from 1971 to 1976, ten were localized between Barcelona, Valencia, and Majorca. The observers network recently set up by the Commission perhaps causes the results to be biased.

On both sides of the Ligurian cyclonic gyre, off Nice and the Riviera as well as off Calvi and Saint Florent (Corsica), the oceanographic conditions must be favourable at the end of winter and in the beginning of spring. Other species of teutophagous cetaceans also utilize this zone intensively. The Malaga region and the Alboran Sea show an interesting arrangement of isotherms from June to August, and sightings of sperm whales are made during this period (Richard, 1936; Busnel, pers. comm. from his cruises in 1966 and 1967).

In conclusion, although the Mediterranean is reputed to be an oligotrophic sea, it seems that favourable conditions for feeding *Physeter* are only present in certain places. I see a confirmation of this by the fact that Corsican, Provençal and Toscan fishermen agree saying that the species was frequent in the past.

GLOBICEPHALA MELAENA

Long-finned pilot whale; in French: Globicéphale noir.

Two species of *Globicephala* are recognized; *G. macrorhynchus*, or short-finned pilot whale, was not found in the Mediterranean.

The material

G. melaena is a species of the open sea, rarely travelling in the neritic zone. The observations at sea by my observers network, which principally works from the shore, are therefore scarce.

Paulus (1960) records two strandings in the Bouches-du-Rhône (France). Five strandings have been noted on the Corsican coasts since 1955 and Duguy (1974, 1976, 1978, 1979, 1980, 1981, 1982) records eight further strandings on the French coasts. On the coast of Spain, Casinos & Vericad (1976) report four, and Grau *et al.* (1980) one more. Tortonese (1963a) reports four strandings in Ligury, and Richard (1936) six catches in Monaco and one stranding at Beaulieu (Riviera). In total, we have 25 strandings in 150 years on the northwestern coast of the basin studied. I have to add stranded herds, such as the 72 individuals in September 1827 in Calvi (northwest Corsica) and 150 individuals in Majorca in 1860. Since 1977, among ten strandings in France, nine were juveniles: this high infantile mortality has to be compared with that related above for *Stenella coeruleoalba.*

At sea, a herd of 300 individuals was observed in September 1974 in the Gulf of Sagone (northwestern Corsica), another herd in September 1975 northeast of Cape Corse, and the same day a herd surrounding a parturient female off Antibes (Riviera). In my survey in the summer of 1981, during constant observation, four sightings (56 individuals, *i.e.* 14 individuals per group on the average) were noted between Corsica and the French coast (Viale & Bardin, 1982). A similar survey in 1982, in exactly the same observation conditions and season, yielded four sightings (41 individuals). The frequency of *Globicephala* in comparison with other sighted cetaceans was four out of 44 in summer 1981, and four out of 76 in summer 1982.

Vallon, Guigo & Duguy (1976) reported that the main concentrations of this species were located in the western part of the Gulf of Genoa, and they analysed observations at sea by the merchant vessels during the summer of 1974: 12 sightings, totalling 400 individuals between Corsica and the coasts of France and Italy. From May to December 1975, 800 individuals were counted in 500 sightings. The average number of individuals by group was 16 during the winter and 37 during the summer; in the absence of information about the random variability, it is impossible to know whether this difference is significant or not.

Desportes (1983) studied the distance of the species from the coast according to the contents of its stomach.

Biological data

Sergeant (1962) gives some numerical values for the Newfoundland region. The maximum size is 6 m for males, 5 m for females. Sexual maturity is reached in 12 years by males and six years by females. Pregnancy lasts 16 months according to Sergeant, but Frazer (1973) estimates it as 12 months ($\pm 1{\cdot}8$ months). Suckling time is 30 months. Sergeant estimates that the mortality of males is higher than that of females, which results in a polygamy of the population. Breeding takes place throughout the year, but is more frequent in summer. In the Mediterranean, Vallon *et al.* (1976) report a breeding in September and describe a group behaviour around a parturient female: a female of the group assisted the labour and took the new-born in its mouth by the flukes.

Data on the weight-size relationship are scarce. A specimen stranded in 1955 was 5 m long and weighed 600 kg when entirely eviscerated. A 4·85-m female weighed 510 kg (Duguy, 1975); a pregnant female was 3·73 m long and weighed 1100 kg (Richard, 1936). A 4·92 m male stranded in June 1977 could not be weighed. Besson (in Duguy, 1981) records a stranded young male of 2·23 m and 160 kg.

Behaviour, and the effect on the exploitation

The gregariousness of *Globicephala* is well known (Sergeant, 1982); it leads them to get stranded or captured in groups. The exploitation of this species was intensive in the whole Northern Hemisphere, and still is in the Faeroes but Mitchell (1975a) reports that this exploitation is decreasing. In the Mediterranean, exploitation was sporadic: catches were recorded in Gibraltar (Anonymous, 1942), Malaga and Monaco (Richard, 1936), and Majorca (Casinos & Vericad, 1976).

Herds of *Globicephala* can be mixed with other species of Odontoceta: *Tursiops* (Aloncle, 1972); *Stenella* (Vallon *et al.*, 1976), and *Balaenoptera physalus* (photographs in local newspapers, 1975). The last association is the most astonishing, because fin whales are planktophagous whereas pilot whales principally eat squids (with a fish complementation, such as cods in Newfoundland). I conclude that such species assemblages do not always have an alimentary motivation.

Estimation of stock

The absence of data about the variability of the importance of the herds, and about seasonal variations, makes it difficult to estimate the total number of *Globicephala*, starting from the data of Vallon *et al.* (1976) and from the observations of merchant vessels along the Marseille–Corsica and Nice–Corsica regular lines. Estimating approximately the area covered between May and December 1975 by these vessels as 40 000 square miles (137 000 km^2), assuming that Odontoceta are detected at a distance of 0·5 miles from the ship ("belt method"), and extrapolating to the zone north from 40° N, we estimate about 1400 individuals.

According to Mitchell (1975a) and to the data of the International Whaling Commission, the stock of pilot whales from Newfoundland has been destroyed, but should be reconstituted. Other mundial stock do not seem to have been directly attacked but the squids and other cephalopod fisheries, which are developing, may introduce a heavy competition against this species.

BALAENOPTERA ACUTOROSTRATA

Minke whale; in French: rorqual de Rudolphi.

There was a stranding of *B. acutorostrata* in Camogli (Genoa) in 1916 (Toschi, 1965); it was, therefore, included in the Italian fauna. Arbocco (1969) records three strandings, whose carcases are conserved in the museums of Genoa, Rome and Naples. Duguy & Robineau (1973) mention it as an exceptional species. It appears, however, in the list of "small cetaceans" of the Gibraltar fisheries (Anonymous, 1942). A young animal more than 5 m long was caught in a net in June 1977 off Bandol, France (Duguy, 1978), and another juvenile (3·63 m, 160 kg) was captured in a net at Saint Raphael, France (Duguy, 1983a). My network of observers record it more and more frequently. There was one sighting of the species among 44 observed cetaceans in summer 1981, and three among 76 in summer 1982 (Viale & Bardin, 1982). In the observations of summer 1983, the species represents 4% of the observations of large whales, that is practically as often as sperm whales.

Their behaviour allows them to be distinguished from small fin whales. They are observed in pairs, jump easily out of the water and let themselves fall flat; such a behaviour is rare in *B. physalus*.

BALAENOPTERA BOREALIS

Sei whale; in French: petit rorqual.

This species is mentioned among the catches at Gibraltar but is rare. A stranding has been recorded in the same region (Filella-Cornado, 1974).

ORCINUS ORCA

Killer whale; in French: orque.

It is a species from boreal and austral waters, sporadically appearing in the Mediterranean. On the coasts of Corsica, only one dead animal has been observed in 15 years; it was floating off Ajaccio. Several observations of isolated or grouped animals were made by the oceanographic vessel KOROTNEFF of the Marine Station of Villefranche-sur-Mer, and by the ship of the Marineland of Antibes, which made a survey of the basin in 1972 and in 1974.

Richard & Neuville (1897) report three catches in May 1896; two of these were 5·90 and 4·10-m females, and the third was not classified. A specimen in the museum of Monaco came from Gibraltar and was a 4·70-m, 1200-kg female. Duguy & Cyrus (1973) report three strandings in the Herault and the Pyrénées Orientales. Tortonese (1963c) reports none from Italy. Marcuzzi & Pilleri (1971) consider this species as scarce on the Mediterranean Spanish coast, whereas 200 individuals were captured altogether at the end of 1760 on the Atlantic coast of Spain. A killer whale became stranded in 1941 in Majorca when pursuing a herd of dolphins.

PSEUDORCA CRASSIDENS

False killer whale; in French: faux-orque.

Tortonese (1963a) reports a stranding of a 4-m female near Genoa in December 1893. Paulus (1963) records seven in the Mediterranean between 1893 and 1951. Pilleri (1967b) studied the behaviour of a herd of 50 individuals (adults and young) in September 1966 near Malaga. Pilleri & Knuckey (1968) made a morphological study of captured animals. Massutti (1943) indicates a relationship between weight and size: 4·41 m and 750 kg for an animal caught near Majorca. Casinos & Vericad (1976) report a stranding in Malaga. Finally, this species is principally observed in the Alboran Sea, up to Majorca, and rarely appears in Ligury.

BALAENOPTERA MUSCULUS

Blue whale; in French: baleine bleue.

Duguy & Cyrus (1973) relate that among 20 observations recorded in the literature, 14 were in fact *Balaenoptera physalus* wrongly identified. Six doubtful records remain.

Pouchet (1885) mentions the stranding in Corsica of a "110 foot baleen" in 1620 and of a "137 foot" one in 1827 (the information came from a local newspaper, *Journal de la Corse*, 1827). It is difficult to imagine a fin whale of these sizes, hence the animals may have been blue whales. The "foot" mentioned could, however, be a local measurement different from 33 cm. Paulus (1966) includes these two specimens in a list of *Balaenoptera physalus* with a size indication "33 m" for the first, and no indication for the second.

Aloncle (1964) reports five catches of *B. musculus* at Gibraltar. I have never observed this species between Corsican and European coasts.

MEGAPTERA NOVAEANGLIAE

Humpback whale; in French: mégaptère.

This is a boreal and austral species, whose distribution in the Southern Hemisphere has been very well studied. It does not reach the latitude of Gibraltar, nevertheless a (doubtful?) stranding has been recorded in the Var, France (Duguy & Cyrus, 1973).

MESOPLODON BIDENS

An animal stranded in 1790 on Sicily is described by Rafinesque (in Pilleri, 1967a) and assigned to this species. The identification was confirmed by Van Beneden (1888). Its presence is also confirmed in the fauna of Italy (Toschi, 1965) and in the fauna of Europe (Van den Brinck, 1960). Van Bree (1975), however, contests this identification and affirms the absence of the species, as well as of *Lagenorhynchus acutus*, in the Mediterranean.

MESOPLODON DENSIROSTRIS

A stranded individual is recorded by Casinos & Filella (1981) in Catalonia.

HYPEROODON AMPULLATUS

Its presence is affirmed by Van den Brinck (1960). It seems to have been found in the Gard (France), but Duguy & Cyrus (1973) express a doubt. Furnestin (in Aloncle, 1964) records having observed this species on the Algerian coast. Hershkovitz (1966) confirms its presence in the Mediterranean, although it is clearly a boreal species. It is possible to confuse it with *Ziphius cavirostris*, as I myself did when seeing *Ziphius* for the first time, owing to the wrong description of the colour recorded in the literature (see above, p. 535).

MONODON MONOCEROS

Narval.

The stranding of an animal described as "provided with a long twisted sword in the side of the head" has been recorded on the east of Corsica (Viale, 1974). This boreal species only accidentally appears in the Mediterranean.

STENO BREDANENSIS

One specimen was caught in a net, near Toulon (Duguy & Cyrus, 1973). This species is tropical and regularly observed on the west African coasts (Rode, 1939). It must accidentally have penetrated into the Mediterranean.

PHOCAENA PHOCAENA

Harbour porpoise; in French: marsouin.

Its presence is reported in the Balearic Islands (Barcelo, 1975 in Casinos & Vericad, 1976). Companyo (1963, in Casinos & Vericad, 1976) describes some *Phocaena*, adding that it is common in the French Mediterranean waters; Casinos & Vericad express a doubt. An observation off Ajaccio was reported at the Symposium of Les Embiez (Anonymous, 1975) and I in turn expressed a doubt (Viale, 1976). It is possible that porpoises come into Mediterranean in the Atlantic Current, and can lose their way in the Algero-Provençal Current, up to Corsican coast, but that remains to be proved.

An endemic population of this species does exist in the Black Sea (Tomilin, 1967).

STATUS OF CETACEANS IN THE ECOSYSTEM

THE TROPHIC NETWORK OF PLANKTOPHAGOUS CETACEANS

Balaenoptera physalus is the only planktophagous cetacean present in the Mediterranean. In fact, it occasionally consumes pelagic fishes. Other

Mysticeta are strictly planktophagous and are localized in regions where plankton is abundant. I believed this eclectic diet allowed *B. physalus* to inhabit the Mediterranean, but my analysis of stomach contents did not support this hypothesis.

Analysis of stomach contents

The autopsy of an adult killed near the Ligurian Divergence at the beginning of July 1972 had in its stomach an orange-coloured 'mash' looking identical to that found in stomachs of basking sharks, also planktophagous and which were caught near Cape Corse in April and June of the same year. The examination of the 'mash' under microscope showed it to consist of rostra and extremities of appendages of crustaceans, as well as pieces of shells mixed with long muscular bundles. Two bags of the stomach were filled. Beyond the pylorium, the homogenized 'mash' was coloured by the bile. In the small intestine, the 'mash' was finely homogenous and still preserved its 'cooked shrimp' colour; so the whole alimentary canal contained the same planktonic 'mash'. Two other fin whales autopsied by Tardy (pers. comm.) during the summer of 1973 also contained only a planktonic 'mash'. The first stomach bag contained some still recognizable euphausiids. I have no data about stomach content in winter.

Jonsgård (1966) stated that the species may be ichtyophagous during winter and spring, and the remainder of the year prefers planktonic curstaceans: euphausiids and *Calanus finmarchicus*. Ingebrigsten (1929) confirms this exclusive planktonic diet during the summer. Jonsgård (1966) examined the contents of 435 stomachs of summer catches from the northeastern Atlantic: 296 contained plankton, 22 contained fishes, 21 had mixed contents, 2 had milk and 94 were empty. Among crustaceans, *Meganyctiphanes norvegica* constituted only 13% of the plankton, the remainder were *Thysanoessa inermis* and *Calanus finmarchicus*. Mitchell (1975b) found the same diet in the northwestern Atlantic. Gaskin (1976) added fishes living in great schools, myctophids that are micronectonic fishes less than 10 cm long, and micronectonic shrimps. He said that copepods are consumed by the young *Balaenoptera*, owing to the dimensions of their filtration apparatus; Lockyer (1976a) indeed reports that adults only capture 20- to 43-cm euphausiids.

The trophic potentialities of the Mediterranean

The main euphausiid biomass of western Mediterranean belongs to the species *Stylocheiron brevicorne* (14 to 23 mm), *Euphausia kronhi* (11 to 22 mm), *Nematoscelis megalops* (14 to 23 mm), and *Meganyctiphanes norvegica* (7 to 30 mm) according to Casanova-Soulier (1974) and Goy & Thiriot (1976). The last species is most abundant, both juvenile and adult, near the centres of divergences, and shows the highest fecundity but its reproduction stops during the summer. It has an important diel migration between surface (by night) and deep water (by day); it can reach a depth of 2500 m. Franqueville (1970) notes two preferential layers: 500 to 900 m, and the vicinity of bottom.

These data shed light on the different types of behaviour of *Balaenoptera physalus* mentioned on page 507. Goy & Thiriot (1976) and Pustelnik (1976) studied the relative abundance of euphausiids in the Ligurian Sea: *Euphausia krohni* is principally coastal during the winter, and is present in summer up to the divergence. *Meganyctiphanes norvegica* shows its maximum abundance in the Ligurian Divergence zone between February and September and, in the winter, at half way between the coast and the divergence. It is impossible to determine the absolute abundance of euphausiids in the Mediterranean: the sampling techniques indeed are bound to fail owing to the high rate of avoidance of nets by swimming animals (avoidance estimated as 50% on the average for an Isaac-Kidd midwater trawl, but in fact highly variable, and 92% for a classical plankton net, according to Laval, 1974). On the other hand, these organisms form swarms with a heterogeneous distribution in space, from which an adequate sampling method never has been used. The highest estimations recorded in the literature are extremely low, and indicate a biomass which in no way could explain the existence of *Balaenoptera physalus*.

Franqueville (1970) stated that the average weight of an adult of *Meganyctiphanes norvegica* is 0·12 g. If we refer to biomass measurements made with the Isaac-Kidd midwater trawl, we obtain an average biomass of euphausiids and natantiid shrimps of about 0·5 g/m^3. The mesozooplankton biomass is estimated as 0·5 to 1·5 g/m^3, and the macrozooplankton biomass 0·5 to 1·5 g/m^3. Several authors (in Lockyer, 1976a) come to the same conclusion: it is impossible to feed whales with such a low density of plankton. Mackintosh (1972), however, estimates the density of euphausiids inside a swarm as 2 kg/m^3, and Bougis (1967) records quantities reaching 16 kg/m^3. It clearly appears that only the phenomenon of swarming permits the alimentation of whales; the latter are especially adapted, on the one hand to detect the swarms by echo location, and on the other, to engulf the swarms speedily without fear of avoidance or scattering of these. Whales are formidable predators in comparison with our trawls, which are passive filtrating apparatus, the catches of which are unavoidably biased by the high avoidance rate and over-dispersion of animals. Whales are able to exploit the deep scattering layer, that oceanographers have never yet accurately sampled.

A correct study of the deep scattering layer could yield information, because euphausiids constitute an important part of it (D'Arcangues, 1976). Vu Do (1978) emphasizes the importance of Natantia in the samples taken in January–February in the Ligurian Sea. Three species form the main biomass for the whales: *Sergestes arcticus*, *Pasiphae sivado*, and *Gennadas elegans*. These are 3- to 6-cm long shrimps, but perhaps unable to form sufficiently great swarms to be used by fin whales.

Interactions with other elements of the trophic network

Meganyctiphanes norvegica is known, in winter and spring, feeding entirely on phytoplankton which at this time is forming blooms; the remainder of the year, it is omnivorous (Pagano, 1976). It follows that fin whales are a consumer of a third trophic level during half the year, and of a fourth level during the other half. Its competitors are numerous; tuna show in their stomachs a high percentage of *Meganyctiphanes*; euphausiids also constitute the main part of the stomach contents of small selacians, such as *Galeus melanostomus* and *Etmopterus spinax* which are very abundant in the Mediterranean bathyal zone, as well as of basking sharks and of epi- and bathypelagic cephalopods. Lagardere (1971) showed the important rôle of *Meganyctiphanes* in the food of bathyal shrimps such as *Parapenaeus longirostris*, *Aristeus antennatus*, *Aristeomorpha violacea*, *Plesionika martia*, and *P. heterocarpus*, which probably only by night are detritivorous and suspension-feeders. Finally, competition with planktophagous pelagic fishes, *i.e.* "blue fish" (pilchards, *etc.*) should be considered. Thus the trophic network of planktophagous cetaceans is by no means independent of that of ichtyophagous and teuthophagous cetaceans.

THE TROPHIC NETWORK OF TEUTHOPHAGOUS CETACEANS

Strictly teuthophagous cetaceans

Three species belonging to this trophic group are found in the western Mediterranean namely *Ziphius cavirostris*, *Physeter macrocephalus*, and *Grampus griseus*.

Analysis of stomach contents of some *Ziphius* stranded on Corsica indicate exclusive feeding on cephalopods: crystalline lens, ocular capsules, inferior and superior mandibles are found. The mandibles permit an indentification of the prey. Using the key given by Mangold & Fiorini (1966) for 18 Mediterranean species, it was possible to identify three ommatostrephids in the stomach contents: (1) *Todarodes* (=*Ommatostrephes*) *sagittatus* can reach 30 cm dorsal length of mantle (standard length) and it is one of the biggest squid in the Mediterranean. It lives over the 300- to 600-m grounds. Whereas trawlers yield individuals of all sizes, and few mature ones (Wirz, 1958), specimens found in the stomachs of *Ziphius* were all mature with a standard length of more than 24 cm. In the Atlantic and north European zones, *Todarodes sagittatus* is the one most frequently found in the stomachs of teuthophagous cetaceans (Sergeant, 1962, 1969). (2) *Ilex coindetti* (=*I. illecebrosus coindetti*) is smaller than the preceding species and lives over 50- to 300-m muddy grounds (Wirz, 1958). It is characteristic of the eastern Atlantic waters and takes here the place of the American species *I. illecebrosus*, which constitutes the main part of the stomach contents of the American teuthophagous cetaceans such as *Globicephala*, *Delphinapterus* (Sergeant, 1962) and *Delphinus* (Evans, 1976). The beaks found in the autopsied cetaceans of Corsica belong to 18-cm individuals. (3) Beaks belonging to *Todoropsis eblanae* of about 16-cm standard length were also found. (4) A fourth find of large sized mandibles was also in the examined stomachs, and were even the most abundant. They belong to an octopode which is not included in Mangold & Fiorini's list, which proves that *Ziphius* catches cephalopod species different from those caught in trawls.

There are no precise data on *Physeter* from the area studied. Four specimens stranded on Corsica could not be dissected; another one was still being suckled, and a specimen autopsied in Livourne had an empty stomach. The observations of sperm whales in the northern part of the Gulf of Genoa, where there are deep grounds, led me to believe that they were also searching for deep-water cephalopods.

One *Grampus* captured in August 1973 and two stranded in May 1978 contained in their stomachs a number of beaks, among which those of *Loligo vulgaris* were identified in great number, and also some of two unidentified octopodids. Data are lacking on the mundial scale about the diet of this species, although it has been maintained alive in captivity fed principally with cephalopods (Mitchell, 1975a).

Preferentially teuthophagous cetaceans

The two main species in this group are *Stenella coeruleoalba* and *Globicephala melaena*.

Eight autopsies of *Stenella* provided the following stomach contents. (1) Female, 140 cm, November 1946, only milk. (2) Male, 143 cm, December 1974, milk and several kinds of beaks of cephalopods. (3) Male, 163 cm, February 1983, still entirely suckled, but showed on its rostrum numerous traces of suckers of cephalopods of a good size (some were 15 mm in diameter) similar to those observed on a stranded *Grampus*. (4) Male, 169 cm, June 1977, milk and a number of beaks. (5) Male, 169 cm, June 1973, uniform yellow-green stomach content without any trace of cephalopods. (6) Male, 180 cm, March 1975, otoliths of fishes, beaks of cephalopods, small chitinous disks of diameter 3 to 4 mm, and a chitinous tube probably belonging to *Hyalinacia tubicola*—a polychaete abundant on muddy grounds at 300 m on the middle part of the Strait of Corsica; its presence may indicate that *Stenella*, which do not dive deeply, feed on benthophagous cephalopods or fishes (it is an example of 'contamination' of the stomach content of the predator by that of the prey). (7) Male, 192 cm, November 1982, contained beaks of cephalopods and remainders of fishes. (8) Male, 210 cm, March 1983, had its stomach filled with a 'mash' of fishes, with otoliths and crystallines lenses, no beaks of cephalopods; it is to be noted that this individual was ill: coronary vessels were irregularly dilated, and heart abnormally flaccid, thus this individual possibly had difficulty in catching cephalopods.

The precise examination of the remains found in these animals indicated, in addition to unidentified otoliths of fishes, the following species of cephalopods: *Loligo vulgaris*—a 22-cm standard length individual (*i.e.* more than 30-cm total length), and some juveniles of 8- to 12-cm standard length; *Sepiola robusta*—small specimens of 2-cm standard length; and an unidentified species characterized by three well distinct hulls on the crest of the inferior mandible. The identified cephalopods belong to the infra- and circalittoral layers. *Loligo vulgaris* lives, in the Mediterranean, in the vicinity of the 50- to 100-m grounds throughout the year (Wirz, 1958); I, however, have caught it with trawl up to a depth of 200 m.

The stomachs of three *Globicephala melaena* captured in the Mediterranean by Richard (1936) indicate the following contents. (1) Male, 4·92 m, 1200 kg, first bag of the stomach filled with "squids" without identification. (2) Female, 3·86 m, pregnant (foetus of 12 cm), stomach empty. (3) A third only indicated "large squids" (hence, ommatostrephids) having left traces of suckers on the skin of the cetacean. In addition, I autopsied in June 1977 a 4·92-m male, which had in its stomach one otolith and a pair of jaws perhaps belonging to a very small, unidentified cephalopod.

Occasionally teuthophagous cetaceans

Delphinus delphis and *Tursiops truncatus* are in this group.

Three *Delphinus* examined in May and June (Richard, 1936) contained remains of fishes in their stomachs. An animal captured in September contained fishes, cephalopods, and shrimps. One stranded in December 1972 contained beaks of cephalopods and lenses of eyes of fishes. One stranded in November 1973 and an empty stomach.

Evans (1976) studied the diet in relation to size and showed that, during the period of mixed feeding (suckling plus predation), the stomachs only contain milk and squids. Fishes only appear in the stomachs from a size of 1·30 m and bigger. In California, squids constitute 37% of the stomach contents of adults in summer and 25% in winter. In the Mediterranean, fishes form the main part of the contents (Besson & Tardy, pers. comm.). On the West African coasts Cadenat (1959b) also concluded that fishes constitute the main part of the diet, cephalopods only being a complement. Tomilin (1967) records an especially diversified collection of cephalopods from the stomach of a *Delphinus* captured near Corsica: *Pyroteuthis margaritifera* (4 individuals)—an epi- and bathypelagic species, *Chiroteuthis veranyi*—a bathypelagic one, *Loligo vulgaris, Todarodes sagittatus, Heteroteuthis dispar*—nectobenthic ones, *Ancistroteuthis lichtensteini*—macroplanktonic, and finally a new species, described as *Ctenopteryx cyprinoides.*

An adult male *Tursiops truncatus* of 3·30 m captured in October 1975 in Bastia contained squids (*Loligo vulgaris*, 15 to 25 cm standard length) and a 40-cm *Octopus vulgaris*, mixed with a 'mash' of digested fishes, and some entire pilchards.

Trophic potentialities of the Mediterranean

An estimation of potentialities can be made by analysing the yields of commercial trawls, and by referring to the biocoenosis existing between Corsica and Tuscany and in the Gulf of Genoa, well known since, for example, the work of Péres & Picard (1964) and the Polymède expeditions (Anonymous, 1972, in Mangold-Wirz, 1973). Only the species showing important biomasses have any interest as the food of cetaceans. I have studied documents concerning the fisheries of Bastia, Nice and Beaulieu, and the results of Italian scientific cruises (Sara, 1972; Matta, 1973).

Expeditions carried out by the Central Hydrobiological Laboratorium of Rome aboard the commercial trawler MERLO LI tried to estimate "quali-quantitatively" (Matta, 1973) the benthic community north from Corsica, and especially the biomass of cephalopods. In 1973 the cruise, in which I participated, determined the distribution of the species of importance to cetaceans between the depths of 50 and 800 m, from Corsica to the Tuscany Archipelago. The results fell into three groups.

(1) Pelagic cephalopods: *Ommatostrephes* sp., big squid sometimes reaching 1 m long but most frequently 50 cm, living in groups and making important vertical migrations; *Calliteuthis reversa* and many other small species (a few cm long) belonging to the macroplankton—but it is difficult to know whether they contribute to the diet of cetaceans.

(2) Benthic cephalopods from the littoral and circalittoral zones: *Octopus vulgaris*, which is trawled up to 200 m; *Sepia officinalis*, caught with bottom trawl. In Corsica, according to the Maritime Administration of Bastia, the latter species is principally caught in winter, when it comes

close to the coast (November to February); at the other times of the year, a few individuals are caught in the trawls at a depth of 80 m.

(3) Benthic cephalopods from the circalittoral and bathyal zones: *Eledone aldrovendi* (=*cirrosa*), 25 to 30 cm long, is trawled between 50 and 500 m principally on muddy grounds; *Rossia macrosoma*, *Octopus salutii*, and various *Sepiola*, among which is *S. rondeleti*, obtained from 100 to 500 m; *Alloteuthis subulata*, a small 15-cm squid, is caught on the 100- to 400-m grounds, as well as some other small species, probably of little importance to cetaceans.

These results can be compared with those gathered by the Smithsonian Institution (Anonymous, 1972, in Mangold-Wirz, 1973) southwest of Corsica, up to 1000 m. The most sampled cephalopod was *Heteroteuthis dispar*, bathypelagic and living in big schools. In the Tyrrhenian Sea, the most characteristic and the most abundant species in the samples was *Onychoteuthis banksii*, a cosmopolitan pelagic species. On the whole, cephalopods constituted 5% of the total catch of the expedition of June 1972 of the MERLO LI, and 10% in 1973, the average yields being 45·8 and 93·8 kg/km^2 in 1972 and 1973 (prospected areas, respectively, 88 and 83 km^2). The bathymetrical distribution of this cephalopod community is given in the Table XI, in three zones north from Cape Corse: Zone A, from the coast up to 20 miles, Zone B, between 20 to 40 miles from the coast, and Zone C, between 40 to 60 miles from the coast.

In the Tyrrhenian Sea, northeast from Cape Corse, in June 1973, biomasses of cephalopods per hectare for each trawl were: 110–190 m, 0·11 kg/ha; 138–150 m, 0·59 kg/ha; 370–380 m, 0·35 kg/ha; 370–380 m, 1·50 kg/ha; 380–450 m, 0·30 kg/ha; 440 m, 0·83 kg/ha; 500–595 m, 0·14 kg/ha; 700–800 m, 0·10 kg/ha; general average 0·50 kg/ha. These values are increased by 10% by Matta (1973) in order to take into account any loss for avoidance. All these biomasses are very low.

TABLE XI

Percentages of cephalopods with respect to total biomass sampled with a trawl (from Matta, 1973): cruises of the MERLO LI; —, *is the absence of sampling*

	Zone A		Zone B		Zone C	
Depth (m)	1972	1973	1972	1973	1972	1973
50–150	3·2	17·1	—	—	—	—
150–250	—	—	—	—	—	—
250–350	6·6	16·8	—	—	—	—
350–450	5·5	9·5	6·6	7·7	6·6	15·7
450–550	6·9	8·3	7·0	7·8	9·9	12·4
550–650	3·5	2·7	3·1	1·0	3·8	4·4
650–750	—	—	—	—	—	0·4
750–850	0·0	0·8	—	—	—	—

Statistics for cephalopod fisheries available in Corsica and Tuscany are difficult to utilize; catches of cephalopods and other molluscs are not distinguished. The total catch of molluscs varies between 10 000 and 40 000 tons by year for Tuscany. In Corsica, squids appear in the account books of the Maritime Administration only between November and February, with quantities of 5 to 10 kg by fisherman per day: *Octopus* appears throughout the year but *Sepia* and *Sepiola* are recorded very irregularly.

These quantities seem far too low to support a population of teuthophagous cetaceans which, however, does exist. Small Odondoceta daily consume 3% to 13% of their own weight (Andersen, 1965; Sergeant, Caldwell & Caldwell, 1973), but the alimentary needs do not increase proportionately with the size: *Ziphius*, *Physeter*, and *Grampus* daily consume only 3 to 6% of their own weight daily, *i.e.* 45 to 90 kg for a *Ziphius* adult. *Ziphius* can dive for about 15 min. Assuming that the animal needs 2 to 4 min to reach 200–400 m depth, and the same time for rising up again, it has 7 to 11 min to explore a favourable ground at a speed of 7 km/h, hence it describes a trajectory of about 1 km long. Assuming that it prospects (by a movement of the head described by Mitchell, 1975a) a 2 m wide belt, swallowing the cephalopods it encounters, then the raked surface is about 0·2 hectare and would provide, if the cetacean were not more efficient than a trawler, about 100 g of cephalopods! Such a supply, of course, is insufficient to balance the energy expended during a 15-min dive.

In fact, just as fin whales prospect swarms of euphausiids, *Ziphius* have a hunting behaviour. They follow the travelling squids coastwards in winter, or towards the centre of divergence in summer, and over all, they search for and detect the schools. Tomilin (1967) established that the Mediterranean is a favourable region for teutophagous cetaceans. The trawl-net probably gives a very wrong idea of the availability of cephalopods. Three factors are still beyond our sampling techniques: (1) the possibility of the cephalopods to form huge schools, particularly the ommatostrephids; (2) the swimming speed of these, much greater than the speed of a trawler but, however, low with respect to the attacking speed of a cetacean; and (3) the possibility of detecting squids by echo location, and the existence of a specialized hunting behaviour.

Interactions with competitive species

The northern Tyrrhenian Sea is very rich in small selacians, whose stomach contents include (besides macroplankton, see p. 555) cephalopods and principally *Heteroteuthis dispar*, a school-living species sought after by cetaceans. The biomass of these competitors, as estimated by the trawlers, is important. On the other hand, the competition by man, even with inefficient techniques, represents in Tuscany and Ligury 150 000 to 200 000 tons by year of cephalopods.

THE TROPHIC NETWORK OF THE ICHTYOPHAGOUS CETACEANS

No cetacean of the western Mediterranean is strictly ichtyophagous. The species which are principally ichtyophagous are *Delphinus delphis* and *Tursiops truncatus*. *Stenella coeruleoalba* and *Balaenoptera physalus* are occasionally ichtyophagous.

Stomach contents

Delphinus is the most ichtyophagous species among all the Mediterranean dolphins. It is considered to be a consumer of the epi- and bathypelagic fishes, nevertheless it will sometimes be attracted into the neritic zone. The circumstances in which groups of *Delphinus* are observed provide an indication of the alimentary preferences: on the Corsican coast, they appear from April to June, at the same time as schools of *Sardinia pilchardus*, or in the beginning of spring, behind schools of *Engraulis encrassicolus*, and less frequently in pursuit of *Scomber scombrus*. This status has, however, changed since 1950, and such observations are no longer frequent (Viale, 1980a,b, see also p. 519). In the littoral and circalittoral zones, various species of *Mugil*, such as *M. auratus* and *M. cephalotus* also form schools attracting the dolphins near the coast, as well as the species which are caught by night with surface nets in all the area studied: *Trachurus trachurus*, *Maena smaris* and *M. maena*. Although the Corsican and Tuscan fishermen link the presence of dolphins with that of 'blue fish' (*i.e.* epipelagic fishes), examination of stomachs before 1950 revealed that the species also fed on deep and sedentary fishes. Schmidt (1923) and Frost (1924) partially analysed a stomach content and counted 15 000 otoliths, the majority belonging to two species of the genus *Scopelus*, and the others to *Scombresox saurus* and to macrurids. Macrurids are numerous in the western Mediterranean, for example *Coelorhynchus coelorhynchus*, but their hard skin make them less attractive as food. Fitch & Brownell (1968) undertook a study of otoliths sampled in 1903 in a *Delphinus* stomach from California and identified the majority as myctophids. Evans (1976) also analysed the contents of thirty stomachs from California and found that in winter *Engraulis mordax* constitutes 92% of the consumed fishes, *Merluccius productus* and *Loligo opalescens* being the greater part of the remainder. In summer, the diet was more diversified: 70% fishes, 23% cephalopods, and 7% crustaceans; among the 19 fish species, there were numerous micronectonic ones including *Argentina*, *Diaphus*, *Lampedena*, *Scopelogadus*.

It is possible that, in the area studied, *Delphinus* finds an alimentary supply in the benthic and micronectonic fishes. In the Black Sea, the alimentary network of *Delphinus* is rather well known (Tomilin, 1967): the main food is anchovies and sprats, the remainder being *Gadus euxinus*, *Mullus barbatus*, *Trachurus trachurus*, *Temnodon saltator*, and *Pelamys sarda*. Anchovies constitute 99% of the stomach contents in spring, and 22% in winter. Some concentrations of dolphins in the open sea, close to the regions enriched by divergences, consume 99% of sprat. Tomilin explains the substitution of anchovies by sprat, suggesting the possibility of feeding about 500 000 dolphins (estimation from 1967) throughout the year.

Comparing these data from literature with my own observations, *Delphinus delphis* appears as a very opportunistic species, feeding on all species of littoral, epi- or bathypelagic fishes, provided that they live in schools; it travels following the appearance and migrations of such schools. As a result, it seems not impossible that its quasi-disappearance from the area studied, as emphasized on page 520, and especially the complete disappearance of very big schools of dolphins, is linked with the disappearance of big schools of pelagic fishes.

Reports on the stomach contents of *Tursiops truncatus* are very scarce in the literature. Tomilin (1967) noted in the stomach of one female a conger eel and remains of cuttle fishes; he also noted that dolphins steal *Smaris vulgaris* and *S. alcido* from fishing nets. On the other hand, in the Black Sea, a 3-m captured adult contained 16 kg of *Mullus*. In the Mediterranean, Tardy (pers. comm.) found a 3-m *Tursiops* containing a 1-m conger eel and some dorados. My own analyses gave the following results. Female, 2·80 m, June 1967, only a few vertebral disks—this specimen probably had regurgitated its stomach content as a result of the stress of being caught. Male, 2·36 m, March 1973, contained tar pellets and a few otoliths. Male, 1·60 m, May 1973, contained remains of vertebral disks and entire vertebrae, tar, and *Posidonia* leaves certainly swallowed during the stranding on a beach covered with these dead weeds. Male, 1·50 m, June 1974, still suckled young but already containing otoliths, lenses of fish eyes, fish scales and very small vertebrae. Male, 3·30 m, October 1975, accidently captured in a trawl net, and containing *Sardinia pilchardus*, 25-cm *Loligo vulgaris*, 10-cm *Octopus*, a 'mash' of fishes and various vertebrae.

The identification of the otoliths I sampled, kindly made by Aboussouan (comm. pers.), gave the species *Myctophum punctatum* and *Notoscelis elongatus* (myctophids), *Merluccius merluccius*, and *Pagrus vulgaris*. The lenses, fish scales and vertebrae could not be identified. I am indebted to Sardou (pers. comm.) for indications on the ecology and abundance of these species. Myctophids: *Myctophum punctatum*, a species provided with photophores, are in rather large numbers in the samples, and come to the light by night; *Notoscopelus elongatus* are even more frequent; other myctophids bearing photophores are frequent in the deep samples in the area studied, namely *Benthosoma glacialis* (70% of the sampled biomass), *Diaphanus rafinesquei*, and *Lampanyctus crocodilus*. *Argentina sphyraena* (size 10 cm) form schools and are only taken with trawls (I trawled them to the north of Corsica, over 350, 420 and 580 m grounds); they probably have an important rôle in the alimentation of delphinids. The alternative species *Argentina sialis* was found in the stomach contents of *Delphinus* of the Pacific (Evans, 1976). Sparidae constitute a very important part of the diet of *Tursiops*; the consumed species *Sparus pagrus*, *S. aurata*, *Dentex dentex*, *Pagellus centrodontus*, *P. erythraeus*, *P. bogavarec*, and *Oblada melaena* live in schools. These species are also favoured by man and professional fishermen believe that *Tursiops* feed principally on rockfishes and strongly compete against the catching by nets and by long- and short-lines.

In Corsica, where fishing with a short-line or "palangrotte" is practised between 1 to 4 miles from the coast, over 40 to 80-m grounds, very numerous observations of *Tursiops* are made when this kind of fishing begins to be profitable, that is, at the season when *Pagellus* and *Boops* begin to rise up. The *Tursiops* approach, and a few minutes later the fishermen do not even get a nibble; the dolphins then go away. Furthermore, groups of *Tursiops* patrol along the coast, systematically examining the trammel-nets installed perpendicularly to the coast or in circles. I observed groups systematically visiting all the nets over several kilometres along the east coast of Cape Corse. In the nets, they first choose the "good species". They gave proof of their chemical sensitivity in the following experiment: in Bastia, a fisherman hoped to deter the dolphins by hooking in his nets a lot of fishes impregnated with an antiseptic substance; the dolphins did not touch them, but stole all the fresh newly caught fish. The chemical sensitivity of dolphins has not been studied as has the acoustic or optical one; recent work on *Tursiops* in dolphinariums, however, reveals the fineness of this sense: a few cubic centimetres of certain substances put in the pool can modify the speed of executing acts to which animals are trained.

Tursiops truncatus has a very euryphagous behaviour, which allows it to adapt to very diversified conditions. Principally neritic, it is able temporarily to reach offshore regions, sharing fish schools and cephalopods with other cetacean species.

Stenella coeruleoalba is a principally teutophagous species (see p. 557), whose stomach nevertheless contains some remains of fishes. The species is now, as I emphasized, taking over an ecological space partially abandoned by *Delphinus delphis* in the whole western Mediterranean basin and, moreover, encroaching on the ecological niche of *Tursiops*, what implies a partially ichtyophagous diet.

Balaenoptera physalus is, as we saw (p. 554), planktophagous in summer in the western Mediterranean. It may consume fishes at other times of the year, as in the North Atlantic; the species probably eats benthic and nectonic fishes of the circalittoral and bathyal zones in winter. Jonsgård (1966) found, in the stomach contents he analysed, remains of *Micromesistus poutassou* and *Phycis phycis*, species also found in the western Mediterranean, where *P. phycis* and *P. blennoides* are very common in the day catches with trawls over 400- to 700-m grounds, mixed with small selacians. It is admitted that fin whales do not consume selacians, but probably vertical migrations separate the two populations, which facilitates the consuming of *Phycis* by the whales, which never dive very deep. My cruises in 1972 and 1973 allowed me to calculate relative quantities of teleosteans in comparison with other sampled groups (Table XII). *Phycis* and *Micromesistus* constitute 15 to 37% of the sample and represent a biomass of about 1·5 kg/ha, *i.e.* three times that of the cephalopods. On the other hand, Sardou (pers. comm.) stated that considerable quantities of

TABLE XII

Percentages of selacians, teleosteans, and crustaceans with respect to total biomass trawled during the cruises of the MERLO LI (*from Matta, 1973*): –, *is the absence of sampling*

Depth (m)	Selacians 1972	Selacians 1973	Teleosteans 1972	Teleosteans 1973	Crustaceans 1972	Crustaceans 1973
50–150	55·7	50·2	41·1	32·7	—	—
150–250	—	—	—	—	—	—
250–350	32·4	17·3	60·1	63·9	0·9	2·0
350–450	25·9	19·3	60·0	59·0	8·6	12·2
450–550	39·5	30·2	44·4	45·8	9·2	15·7
550–650	44·0	57·3	47·2	24·3	5·3	15·7
650–750	—	76·6	—	9·7	—	13·3
750–850	63·6	85·4	34·2	6·1	2·2	7·7

micronectonic fishes such as *Cyclothone braueri* and *C. pygmaea* (3- to 4-cm long gonostomids) form important schools located between 500 and 1000 m by day, but making diel migrations; they must be important in the formation of the D.S.L. and are present in all the deep samples. I located them with an Issac-Kidd Midwater Trawl, at 11.00 h, in December, 500 m deep, 10 miles southeast of Corsica, in the Tyrrhenian Sea; *Cyclothone* were collected with *Argyropeliscus* and *Meganyctiphanes*; at 600 m depth, in May, 5 miles southwest of Cape Corse, *Cyclothone braueri* constituted the main biomass of the sample.

It seems that the importance of these fish species is not to be neglected in the nutrition of fin whales, which are adapted to catching this kind of prey whose size do not differ from that of macroplanktonic crustaceans; moreover, this biomass is available throughout the year.

Interactions with competitive species

In the absence of big selacians in the zone studied, the competitive predators are principally tuna, swordfish, sea birds, man, and also, which concerns bathypelagic myctophids, big cephalopods. The predation by man is important. The fishery statistics for 1970 indicate, for the Tyrrhenian region, Tuscany, and Corsica, 180 000 tons of fish caught. An approximately equal quantity is caught on the Provençal coast and a still greater quantity in the Ligurian Sea. For pelagic fishes only, we, thus, have a total estimation of over 500 000 tons caught in one year.

TOTAL PREDATION BY CETACEANS IN THE AREA

Taking into account all the above information and estimations, the trophic needs of the cetacean populations have been assessed in conformity with the data of Lockyer (1976a,b) and Sergeant (1969). These data concern cold regions very different from the Mediterranean Sea, but the oligotrophy of this sea imposes on cetaceans the need to migrate, and the energy so expended must counterbalance the energy expended against cold, so I used the same values. The demographic structures and the sharing of the diet among the three groups of prey (fishes, cephalopods, and zooplankton) have also been given. Table XIII is a summary of the results. The numbers, of course, only represent some orders of magnitude. They can be compared with existing data on biomass and production known in the area.

Zooplankton biomass in the neritic zone is about 1 ton by square kilometre, *i.e.* 75 000 tons for my "large zone" (north from 40° N). The oceanic part of this zone has 188 000 tons of zooplankton, with a daily production of 5 to 10% of the biomass, that is, 9400 to 18 000 tons by day. *Balaenoptera physalus* consumes between $\frac{1}{5}$ to $\frac{1}{10}$ of this production!

For teuthophagous cetaceans, the consumed biomass is about 600 tons by day, or 229 000 tons by year, corresponding to about 2·3 times the predation by man between Valencia and Naples (from fishery statistics for 1970). Now, the human population of the north half of the Mediterranean consumes $\frac{1}{10}$ of the world catch of cephalopods: the Mediterranean human population are very teuthophagous and exert an important competition against cetaceans, to which the not negligible competition of benthic selacians should be added. The

TABLE XIII

Summary of biomass of standing crop and consumed biomass between 40° N and the Mediterranean coasts of Europe (300 000 km^2)

Species	Numbers	Mean individual weight (tons)	Biomass (tons)	Consumption (tons per year)		
				Macroplankton and Micronecton	Cephalopods	Fishes
Balaenoptera physalus	1200	40	48 000	330 000	0	?
Physeter macrocephalus	600	50	30 000	0	115 000	0
Ziphius cavirostris	1250	1·5	1875	0	28 000	0
Globicephala melaena	1500	1·5	2025	0	22 000	0
Grampus griseus	7500	0·3	2250	0	49 000	0
Delphinus delphis	10 000	0·1	1000	0	}	14 600
Stenella coeruleoalba	20 000	0·07	1400	0	} 15 000	35 000
Tursiops truncatus	2000	0·2	400	0	}	8500
Total	44 050		86 950	330 000	229 000	58 100

total fish consumed by cetaceans represents $\frac{1}{10}$ of the human consumption in the same area. According to Margalef (1967) the production of pelagic fishes in the area is 2 400 000 tons by year; humans take $\frac{1}{4}$ of this production, and cetaceans $\frac{1}{40}$.

On the continental shelf, the annual fishery is 270 000 tons by year. The quantity of fishes taken by *Tursiops truncatus* is 8500 tons by year, that is, about $\frac{1}{30}$ of the human yield. We can imagine that *Tursiops* are maintained but only in few numbers by such a competition, which explains why they fall back upon myctophids, which are not exploited by man.

In conclusion, the place of cetaceans in the northwestern Mediterranean ecosystem is far from negligible because of the predation they exert, as well as their rôle in the transference of substances (especially heavy metals: see Viale, 1977b, 1978, 1985), in the mechanisms of control of the production, and in the transferance of biomass from zones of high productivity towards oligotrophic ones, which become enriched through the regeneration of nutrients. Cetaceans certainly also play an important rôle in the spread of biomass and nutrients between the depth and the surface. They are the "biggest living particles" Bougis (pers. comm.) of the ocean and participate in the functioning of the marine ecosystem, in competition with man, who ought to make the effort to understand better their biology and precise status.

Some recent works on cetacean populations on a world scale (in Kanwisher & Ridgway, 1983) claimed to estimate these populations at a level much higher than it was usually stated. For example, the world population of sperm whales could be one million or one million and a half individuals, consuming each year hundred millions tons of deep water squids! Populations of small delphinids could be between 50 and 100 million individuals, consuming between 150 and 300 million tons of fish yearly. Knowing that the human world catch is 67 million tons per year for fish and 1 165 000 tons for cephalopods according to FAO reports (Anonymous, 1979), pressure by cetaceans on marine resources are much higher than by man, and the ecological rôle of cetaceans seems much more important than that was described above. But, of course, these numbers are to be confirmed.

REFERENCES

Aguilar, A. & Lens, S., 1981. *Int. Whaling Comm. Rep. Comm.*, **31,** 639–643.
Aloncle, H., 1964. *Bull. Int. Pêches marit. Maroc*, **12,** 21–42.
Aloncle, H., 1972. *Z. Säugetierkd.*, **37,** 180–181.
Andersen, S. H., 1965. *Vie Milieu, Sér. A*, **16,** 799–810.
Anonymous, 1942. *Int. Whaling Stat.*, **16,** 78–86.
Anonymous, 1968. *Instructions nautiques. Côtes sud de France.* Service Hydrographique de la Marine, sér. D, 2, 190 pp.
Anonymous, 1975. Colloque de Cétologie, Les Embiez, Novembre 1976, C.I.E.S.M., 4 pp.
Anonymous, 1976. *FAO Doc.* ACMRR/MM/SC 3, 80 pp.
Anonymous, 1979. *Les pêches mondiales et le droit de la mer. Résumé du programme Z.E.E.*, FAO, 30 pp.
Anonymous, 1981. *Atlas for the living resources of the sea.* FAO.
Arbocco, G., 1969. *Ann. Mus. civ. Stor. nat.*, **77,** 668–670.
Bannister, J. L., 1969. *Int. Whaling Comm. Rep. Comm.*, **19,** 70–76.

Bannister, J. L., 1972. *Aust. Fish.*, **32,** 4–8.
Barras, F. De Las, 1944. *Bull. Soc. esp. Hist. nat.*, **42,** 321–323.
Best, P. B., 1969a. *Investl Rep. Div. Sea Fish. S. Afr.*, No. 72, 20 pp.
Best, P. B., 1969b. *Investl Rep. Div. Sea Fish. S. Afr.*, No. 78, 12 pp.
Best, P. B., 1970. *Investl Rep. Div. Sea Fish. S. Afr.*, No. 79, 27 pp.
Best, P. B., 1973. *Int. Whaling Comm. Rep. Comm.*, **23,** 115–126.
Best, P. B., 1974. In, *The Whale Problem: a Status Report*, edited by W. E. Shewill, Harvard University Press, pp. 257–293.
Best, P. B., 1975. FAO Doc. ACMMR/MM/EC 8, 35 pp.
Boucher, J. & Thiriot, A., 1973. *Mar. Biol.*, **15,** 47–56.
Bougis, P., 1967. *Le Plancton.* P.U.F., Paris, coll. "Que sais-je", 124 pp.
Bougis, P. & Carré, C., 1960. *Cah. océanogr.*, **12,** 627–635.
Bougis, P. & Ruivo, M., 1954. *Bull. Inf. C.O.E.C.*, **6,** 147–154.
Budker, P., 1957. *Baleines et Baleiniers.* Horizons de France, Paris, 193 pp.
Budker, P., 1968. *Norsk Hvalfangst-Tid.*, **57,** 17–19.
Busnel, R. G. & Dziedic, A., 1968. *Annls Inst. océanogr., Paris*, **46,** 109–144.
Busnel, R. G., Pilleri, G. & Fraser, F. C., 1968. *Mammalia*, **32,** 192–203.
Cabrera, A., 1919. *Bol. r. Soc. Esp. Hist. nat.*, **19,** 468–470.
Cabrera, J., 1925. *Trab. Mus. nac. Ciencias nat. Madrid, ser. Zool.*, **52,** 5–48.
Cadenat, J., 1959a. *Bull. Inst. Fr. Afr. noire, Sér. A*, **21,** 1137–1141.
Cadenat, J., 1959b. *Bull. Inst. Fr. Afr. noire, Sér. A*, **21,** 1367–1409.
Caldwell, D. K. & Caldwell, M. C., 1971. *Q. Jl Fla. Acad. Sci.*, **34,** 157–160.
Caldwell, D. K., Rathjen, W. F. & Caldwell, M. C., 1971. *Bull. Sth. Calif. Acad. Sci.*, **70,** 52–53.
Caldwell, M. C. & Caldwell, D. K., 1966. In *Whales, Dolphins and Porpoises*, edited by K. S. Norris, University of California Press, Berkeley, pp. 755–788.
Casanova-Soulier, B., 1974. Thèse, Université Aix-Marseille II, Rapp. Comm. int. Mer Médit., multigr. 62 pp.
Casinos, A. & Filella, S., 1976. Commun. 25th Congr. C.I.E.S.M., Split, October 1976, 5 pp.
Casinos, A. & Vericad, J. R., 1976. *Mammalia*, **40,** 267–289.
Casinos, A. & Filella, S., 1981. *Säugetiere Mitteilungen*, **29,** 61–67.
Clarke, R., 1956. *'Discovery' Rep.*, **28,** 237–298.
Clarke, R. & Paliza, O., 1972. *Hvalradets Skr.*, Nr 53, 106 pp.
Conand, F. & Fagetti, E., 1972. *Cah. O.R.S.T.O.M., Sér. Océanogr.*, **9,** 293–318.
Coste, B., Gostan, J. & Minas, H. J., 1972. *Mar. Biol.*, **16,** 320–348.
Coste, B. & Minas, H. J., 1981. *Thalassia jugosl.*, **17,** 103–108.
Cuvier, G., 1812. *Annls Mus. natn. Hist. nat. Paris*, **19,** 1–16.
D'Arcangues, C., 1976. Thèse, 3e Cycle, Université Paris VI, 160 pp.
Desportes, G., 1983. Deuxième Réunion Statutaire C.I.E.S.M., Goteborg 1983, 4 pp.
Devaux, J. & Millerioux, G., 1976a. *C.-r. Acad. Sci. Paris, Sér. D*, **283,** 41–44.
Devaux, J. & Millerioux, G., 1976b. *C.-r. Acad. Sci. Paris, Sér. D*, **283,** 927–930.
Devaux, J. & Millerioux, G., 1977. *J. français Hydrol.*, **8,** 37–44.
Doi, T., 1971. *Bull. Tokai Reg. Fish. Res. Lab.*, **66,** 89–183.
Duguy, R., 1973. *Mammalia*, **37,** 669–677.
Duguy, R., 1974. *Mammalia*, **38,** 545–555.
Duguy, R., 1975. *Mammalia*, **39,** 689–701.
Duguy, R., 1976. *Mammalia*, **40,** 671–681.
Duguy, R., 1977. *Annls Soc. Sci. nat. Charente-marit.*, **6,** 308–317.
Duguy, R., 1978. *Annls Soc. Sci. nat. Charente-marit.*, **6,** 333–344.
Duguy, R., 1979. *Annls Soc. Sci. nat. Charente-marit.*, **6,** 463–474.
Duguy, R., 1980. *Annls Soc. Sci. nat. Charente-marit.*, **6,** 615–632.
Duguy, R., 1981. *Annls Soc. Sci. nat. Charente-marit.*, **6,** 803–818.
Duguy, R., 1982. *Annls Soc. Sci. nat. Charente-marit.*, **6,** 969–984.
Duguy, R., 1983a. *Annls Soc. Sci. nat. Charente-marit.*, **7,** 121–135.

Duguy, R., 1983b. Editor. *Les Cétacés des Côtes de France. Annls Soc. Sci. nat. Charente-marit.*, special issue, 112 pp.
Duguy, R., 1984. *Annls Soc. Sci. nat. Charente-marit.*, **7,** 189–205.
Duguy, R. & Budker, P., 1972. *Mammalia*, **36,** 517–520.
Duguy, R., Casinos, A. & Filella, S., 1978. Commun. 26th Congr. C.I.E.S.M., Antalea, December 1978, 3 pp.
Duguy, R. & Cyrus, J. L., 1973. *Rev. Trav. Inst. sci. tech. Pêches marit.*, **37,** 151–158.
Duguy, R. & Robineau, D., 1973. *Annls Soc. Sci. nat. Charente-marit.*, Suppl. June 1973, 93 pp.
Duguy, R. & Vallon, D., 1976. Commun. 25th Congr. C.I.E.S.M., Split, October 1976, 4 pp.
Evans, W. E., 1976. *FAO Doc.* ACMRR/MM/SC, 18, 71 pp.
Filella-Cornado, S., 1971. *Misc. Zool.*, **3**(1), 1–7.
Filella-Cornado, S., 1974. *Misc. Zool.*, **3**(4), 1–6.
Fitch, J. E. & Brownell, R. L., 1968. *J. Fish. Res. Bd Can.*, **25,** 2561–2574.
Franqueville, C., 1970. Thèse 3e Cycle Université Aix-Marseille II, 110 pp.
Fraser, F. C., 1942. *Proc. zool. Soc. Lond., Ser. B*, **112,** 21–30.
Fraser, F. C., 1949. In, *Field Book of Giant Fishes*, edited by J. R. Norma & F. C. Fraser, Putnam, New York, pp. 203–360.
Frazer, J. D. F., 1973. *J. Zool.*, **169,** 111–126.
Frontier, S., 1966. *Cah. O.R.S.T.O.M., Sér. Océanogr.*, **4,** 3–37.
Frontier, S., 1969. *J. exp. mar. Biol. Ecol.*, **3,** 18–26.
Frontier, S., 1978. *Annls Inst. océanogr., Paris*, **54,** 96–106.
Frontier, S., 1982. La méthode des cotations d'abondance: essai d'application aux pêcheries artisanales. Rapp. O.R.S.T.O.M., 19 pp.
Frontier, S., 1983a. Editor. *Stratégies d'Échantillonnage en Écologie.* Masson, Paris, 516 pp.
Frontier, S., 1983b. Deuxième mission sur les pêcheries artisanales sénégalaises. Rapp. O.R.S.T.O.M., 18 pp.
Frontier, S. & Ibanez, F., 1974. *J. exp. mar. Biol. Ecol.*, **14,** 217–224.
Frontier, S. & Viale, D., 1977. *J. Rech. océanogr.*, **2,** 15–22.
Frost, G. A., 1924. *Nature, Lond.*, **113,** 310 only.
Furnestin, J. & Allain, C., 1962. *Revue Trav. Inst. Pêches marit.*, **26,** 133–161.
Gambell, R., 1972. *'Discovery' Rep.*, **35,** 199–358.
Gambell, R., 1976. *FAO Doc.* ACMRR/MM/SC 37, 20 pp.
Gaskin, D. F., 1971. *N.Z. Jl Zool.*, **19,** 241–259.
Gaskin, D. F., 1973. *N.Z. Jl mar. freshw. Res.*, **7,** 1–20.
Gaskin, D. F., 1976. *Oceanogr. Mar. Biol. Ann. Rev.*, **14,** 247–346.
Ghirardelli, E., 1944. *Sapere*, **188,** 229 only.
Gihr, M. & Pilleri, G., 1969a. *Invest. Cetacea*, **1,** 15–65.
Gihr, M. & Pilleri, G., 1969b. *Invest. Cetacea*, **1,** 109–126.
Giordano, A., 1981. DEA Université Aix-Marseille II, 37 pp.
Gostan, J., 1967a. *Cah. océanogr.*, **19,** 391–416.
Gostan, J., 1967b. *Cah. océanogr.*, **19,** 469–476.
Gostan, J., 1967c. *Cah. océanogr.*, **19** (suppl. 1), 1–70.
Gostan, J., 1968a. *Cah. océanogr.*, **20,** 37–66.
Gostan, J., 1968b. Thèse, Université Paris VI, 274 pp.
Gostan, J. & Nival, P., 1963. *C.-r. Acad. Sci., Paris, Sér. D*, **257,** 2872–2875.
Goy, J. & Thiriot, A., 1976. *Annls Inst. océanogr. Paris*, **52,** 33–44.
Grau, E., Aguilar, A. & Filella, S., 1980. *Bull. Inst. cat. Hist. nat.* (*Sect. Zool.*), **45,** 167–179.
Gruvel, M. A., 1924. *Bull. Soc. nat. Acclim. France*, **71,** 13–14.
Harrison, R. J., Brownell, R. L. & Boice, R., 1972. In, *Functional Anatomy of Marine Mammals, Vol. 1*, edited by R. J. Harrison, Academic Press, London, pp. 361–429.
Harrison, R. J. & Ridgway, S. M., 1971. *J. Zool.*, **165,** 355–366.

Hershkovitz, P., 1966. *Bull. U.S. natn. Mus.*, No. 246, 259 pp.
Holt, S. J., 1976. *FAO Doc.* ACMRR/MM/SC 61, 5 pp. and ACMRR/MM/EC 29, 11 pp.
Ingebrigsten, A., 1929. *Rapp. P.-v. Réun. Cons. perm. int. Explor. Mer*, **56,** 1–123.
Jacques, G., 1974. *Séminaires Inst. océanogr., Paris*, **1,** 52–76.
Jacques, G., Minas, H. J., Minas, M. & Nival, P., 1973. *Mar. Biol.*, **23,** 251–269.
Jonsgård, Å., 1966. *Hvalradets Skr.*, Nr 49, 62 pp.
Kanwisher, J. & Ridgway, S., 1983. *Science* (French issue), **70 22 F,** 12–22.
Kasuya, T., 1971. *Sci. Rep. Whales Res. Inst. Tokyo*, **23,** 37–66.
Kasuya, T., 1972. *Sci. Rep. Whales Inst. Inst. Tokyo*, **24,** 57–79.
Kasuya, T. & Miyazaki, N., 1975. *FAO Doc.* ACMRR/MM/SC 25, 37 pp.
Lacombe, H. & Tchernia, P., 1960. *Cah. océanogr.*, **12,** 527–547.
Lagardere, J. P., 1971. *Téthys*, **3,** 665–675.
Laval, P., 1974. *J. exp. mar. Biol. Ecol.*, **14,** 57–87.
Leatherwood, S., Perrin, W. S., Le Kirly, V., Hubbs, G. & Dalheim, M., 1980. *Fishery Bull. N.O.A.A.*, **77,** 951–963.
Le Floch, J., 1963a. *Trav. Cent. Rech. Explor. océan.*, **5,** 5–10.
Le Floch, J., 1963b. *Cah. océanogr.*, **15,** 396–403.
Le Pichon, X. & Troadec, J. P., 1963. *Cah. océanogr.*, **15,** 527–539.
Leveau, M., 1965. *Recl Trav. Stn mar. Endoume*, Bull. 37, Fasc. 53, 161–246.
Lilly, J. C., 1964. In, *Handbook of Physiology, Sect. 4: Adaptation to Environment.* Am. Physiol. Soc. Washington, 741–747.
Lockyer, C., 1976a. *FAO Doc.* ACMRR/MM/SC 41, 179 pp.
Lockyer, C., 1976b. *FAO Doc.* ACMRR/MM/SC 38, 32 pp.
Lozano-Rey, L., 1947. *Las Ciencias*, **12,** 263–267.
Mackintosh, N. A., 1972. *Sci. Progr.*, **60,** 449–464.
Maigret, J., Trottignon, J. & Duguy, R., 1976. Commun. 64th Meeting Cons. int. Explor. Mer, Copenhagen 1976, 7 pp.
Mangold, K. & Fiorini, P., 1966. *Vie Milieu, Sér. A*, **17,** 1139–1196.
Mangold-Wirz, K., 1973. *Revue Trav. Inst. Pêches marit.*, **37,** 391–395.
Marcuzzi, P. & Pilleri, P., 1971. *Invest. Cetacea*, **3,** 101–170.
Margalef, R., 1967. *El ecosistema*, in, *Ecologia marina.* Fund. La Salle de ciencias naturales, Carracas, 377–453.
Massutti, M., 1943. *Bol. r. Soc. esp. Hist. nat.*, **41,** 360–361.
Matta, F., 1973. Rilevamenti sulla fauna bentonica del mar Ligure. Rapp. Lab. Centrale Idrobiologia di Roma, 11 pp.
McBride, A. F. & Kritzler, H., 1951. *J. Mammal.*, **32,** 251–266.
Minas, H. J., 1968. Thèse Doctorat, Université Aix-Marseille II, 122 pp.
Minas, H. J., 1974. *Séminaires Inst. océanogr., Paris*, **1,** 1–50.
Minas, H. J. & Blanc, F., 1970. *Téthys*, **2,** 299–316.
Mitchell, E., 1975a. *J. Fish. Res. Bd Can.*, **32,** 875–1242.
Mitchell, E., 1975b. *Int. Whaling Comm. Doc.* SC/26/35, 10 pp.
Miyazaki, N., 1976. *FAO Doc.* ACMRR/MM/SC 73, 3 pp.
Miyazaki, N., Kasuya, T. & Nishiwaki, M., 1974. *Sci. Rep. Whales Res. Inst. Tokyo*, **26,** 227–243.
Moore, J. C., 1968. *Fieldiana Zool.*, **53,** 209–298.
Nasu, K., 1966. *Sci. Rep. Whales Res. Inst. Tokyo*, **20,** 157–210.
Nasu, K. & Shimadzu, Y., 1970. *Int. Whaling Comm. Rep. Comm.*, **20,** 114–129.
Nishiwaki, M., 1967. *Bull. Ocean Res. Inst. Univ. Tokyo*, No. 1, 64 pp.
Nishiwaki, M., 1976. *FAO Doc.*, ACMRR/MM/SC 30, 5 pp.
Nishiwaki, M. & Sasao, A., 1976. *FAO Doc.* ACMRR/MM/SC 124, 8 pp.
Nival, P., Malara, G., Charra, R., Nival, S. & Palazzoli, I., 1972. *C.-r. Acad. Sci., Paris, Sér. D*, **275,** 1295–1298.
Nival, P., Nival, S. & Thiriot, A., 1975. *Mar. Biol.*, **31,** 244–270.
Norris, K. S., 1961. *J. Mammal.*, **42,** 471–476.

Norris, K. S., 1966. Editor. *Whales, Dolphins and Porpoises.* University of California Press, Berkeley, 789 pp.
Omura, H., 1972. *Sci. Rep. Whales Res. Inst. Tokyo*, **24,** 1–34.
Omura, H., Fujino, K. & Kimura, S., 1955. *Sci. Rep. Whales Res. Inst. Tokyo*, **10,** 89–132.
Pagano, M., 1976. *Doc. Centre Rech. océanogr. Villefranche-sur-Mer*, No. 19, 70 pp.
Palazzoli, I., 1983. *Rapp. Comm. int. Mer Medit.*, **28,** 217–218.
Parona, C., 1908. *Atti Soc. Lig. Sci. nat., Genova*, **19,** 173–205.
Paulus, M., 1960. *Bull. Mus. Hist. nat. Marseille*, **20,** 45–52.
Paulus, M., 1962. *Bull. Mus. Hist. nat. Marseille*, **22,** 18–48.
Paulus, M., 1963. *Bull. Mus. Hist. nat. Marseille*, **23,** 29–67.
Paulus, M., 1964. *Bull. Mus. Hist. nat. Marseille*, **24,** 81–124.
Paulus, M., 1966. *Bull. Mus. Hist. nat. Marseille*, **26,** 117–134.
Payne, R. S. & Payne, K., 1972. *Zoologica, N.Y.*, **56,** 158–167.
Péres, J. M. & Picard, J., 1964. *Recl Trav. Stn mar. Endoume*, Bull. 31, Fasc. 47, 1–137.
Pilleri, G., 1967a. *Vie Milieu, Sér. A*, **18,** 355–373.
Pilleri, G., 1967b. *Revue Suisse Géogr.*, **74,** 679–683.
Pilleri, G., 1970. *Invest. Cetacea*, **2,** 21–24.
Pilleri, G. & Busnel, R. G., 1969. *Acta Anat.*, **73,** 92–97.
Pilleri, G. & Gihr, M., 1969. *Invest. Cetacea*, **1,** 66–73.
Pilleri, G. & Knuckey, J., 1968. *Atti Mus. civ. Stor. nat. Trieste*, **26,** 31–76.
Poggi, R., 1982. *Bull. Mus. civ. Hist. nat. Genova*, No. 84, 8 pp.
Pouchet, G., 1885. *C.-r. Acad. Sci., Paris*, **100,** 286–289.
Pryor, K. W., 1973. *Naturwissenschaften*, **60,** 412–420.
Pustelnik, G., 1976. *Doc. Centre Rech. océanogr. Villefranche-sur-Mer*, No. 18, 66 pp.
Rayner, G. W., 1940. *'Discovery' Rep.*, **10,** 245–294.
Razouls, C. & Thiriot, A., 1973. *Vie Milieu, Sér. B*, **23,** 209–241.
Richard, J., 1936. *Résult. scient. Camp. Prince Albert I*, No. 94, 71 pp.
Richard, J. & Neuville, H., 1897. *Mém. Soc. zool. Fr.*, **10,** 100–109.
Robineau, D., 1972. *Mammalia*, **36,** 321–331.
Rode, P., 1939. *Bull. Inst. océanogr. Monaco*, No. 780, 20 pp.
Romanovski, V., 1955. *Doc. Cent. Rech. Explor. océan.*, 2, 21 pp.
Saayman, G. S. & Taylor, C. K., 1973. *J. Mammal.*, **54,** 993–996.
Sara, R., 1972. La pesca a strascico sui fondali della scarpata continentale (Settori de Levanzone Pontelerris). Doc. Minis. Mar. Mercantile, Dir. Pesca maritima, Mem. 21, 65 pp.
Schmidt, J., 1923. *Nature, Lond.*, **112,** 902 only.
Schweder, T., 1974. Ph.D. thesis (Statistics), University of California, Berkeley, 183 pp.
Seber, G. A., 1980. *Some Recent Advances in the Estimation of Animal Abundance.* Washington Sea Grant Program, Tech. Rep. WSG 80/1, 101 pp.
Sergeant, D. E., 1962. *Bull. Fish. Res. Bd Can.*, **132,** 1–84.
Sergeant, D. E., 1969. *FiskDir. Skr. Ser. Havunders.*, **15,** 246–258.
Sergeant, D. E., 1977. *Int. Whaling Comm. Rep. Comm.*, **27,** 460–473.
Sergeant, D. E., 1982. *Sci. Rep. Whales Res. Inst. Tokyo*, **34,** 1–47.
Sergeant, D. E., Caldwell, D. K. & Caldwell, M. C., 1973. *J. Fish. Res. Bd Can.*, **30,** 1009–1016.
Tavolga, M. & Essapian, F. S., 1957. *Zoologica, N.Y.*, **42,** 11–31.
Tchernia, P., 1960. *Cah. océanogr.*, **13,** 184–189.
Thiriot, A., 1966. *Vie Milieu, Sér. B*, **17,** 243–252.
Tomilin, A. G., 1967. *Mammals in the U.S.S.R. and Adjacent Countries, Vol. 9. Cetacea.* Jerusalem, Israel program for scientific translation, No. 1124, 742 pp.
Tortonese, E., 1957. *Res. Ligusticae*, **96,** 1–7.
Tortonese, E., 1963a. *Natura, Milan*, **54,** 31–34.
Tortonese, E., 1963b. *Natura, Milan*, **54,** 120–122.
Tortonese, E., 1963c. *Rapp. P.-v. Cons. int. Explor. Mer*, **17,** 383–386.

Tortonese, E. & Demr, H., 1963. *Rapp. P.-v. Réun. Cons. perm. int. Explor. Mer*, **18,** 575–590.
Toschi, A., 1965. *Fauna di Italia, Vol. 7. Mammalia.* Bologne, 227 pp.
Vallon, D., Guigo, C. & Duguy, R., 1976. Commun. 25th Congr. C.I.E.S.M., Cagliari, October 1976, 2 pp.
Van Beneden, P. J., 1888. *Mém. Comm. Acad. r. Belgique*, **41,** 1–119.
Van Beneden, P. J. & Gervais, P., 1868. *Ostéographie des Cétacés Vivants et Fossiles.* A. Bertrand, Paris, 634 pp.
Van Bree, P. H. J., 1975. *Bull. Mus. civ. Stor. nat. Genova*, **80,** 226–228.
Van Bree, P. H. J., Mizoule, R. & Petit, C., 1969. *Vie Milieu, Sér. A*, **20,** 447–459.
Van den Brinck, F. H., 1960. *Die Säugetiere Europas.* Hamburg & Berlin, French adaptation, Delachaux & Niestlé, 1967, 263 pp.
Viale, D., 1973. *Doc. int. Whal. Comm.* SC/25/31, 8 pp.
Viale, D., 1974. Commun. 26th Congr. Int. Whal. Comm., 10 pp.
Viale, D., 1976. FAO, doc. ACMRR/MM/SC 120, 15 pp.
Viale, D., 1977a. *Mammalia*, **41,** 197–206.
Viale, D., 1977b. Thèse Doctorat, Université Paris VI, 312 pp.
Viale, D., 1978. *Annls Inst. océanogr., Paris*, **54,** 5–16.
Viale, D., 1979a. *J. exp. mar. Biol. Ecol.*, **40,** 201–221.
Viale, D., 1979b. In, *Microscopie Électronique à Balayage, Methode d'Exploration en Biologie*, edited by D. Guillaumin, Arnette, Paris, pp. 49–52.
Viale, D., 1980a. Commun. 27th Congr. C.I.E.S.M., Cagliari, December 1979, 6 pp.
Viale, D., 1980b. In, *Recherches d'Écologie Théorique*, edited by R. Barbault, B. Blandin & J. A. Meyer, Maloine, Paris, pp. 206–216.
Viale, D., 1984. Commun. 8th Sympos. Eur. Assoc. Aquat. Mammals, Manchester 1980 and *Annls Inst. océanogr., Paris*, **60,** 1–9.
Viale, D., 1985. *Commun. Symp. on Endangered Marine Animals and Marine Parks*, Mar. Biol. Ass. India, Cochin, Jan. 1985, 30 pp.
Viale, D. & Bardin, J. P., 1983. *Rapp. Comm. int. Mer Médit.*, **28,** 215–216.
Viale, D. & Ridell, M., 1975. Commun. 63rd Congr. I.C.E.S., Ottawa 1975, 5 pp.
Vu Do, Q., 1978. Thèse Doct. Etat Université Paris VI, 125 pp.
Wirz, K., 1958. *Faune Marine des Pyrénées-Orientales. Fasc. 1: Céphalopodes.* Laboratoire Arago, Université Paris VI, 59 pp.

Oceanogr. Mar. Biol. Ann. Rev., 1985, **23**, 573–597
Margaret Barnes, Ed.
Aberdeen University Press

UTILIZATION BY SHOREBIRDS OF BENTHIC INVERTEBRATE PRODUCTION IN INTERTIDAL AREAS

D. BAIRD
Department of Zoology, University of Port Elizabeth, P.O. Box 1600, Port Elizabeth 6000, South Africa

P. R. EVANS
Department of Zoology, University of Durham, Science Laboratories, South Road, Durham DH1 3LE, England

H. MILNE
Culterty Field Station, University of Aberdeen, Newburgh, Ellon, Aberdeenshire AB4 0AA, Scotland

and

M. W. PIENKOWSKI[1]
Department of Zoology, University of Durham, Science Laboratories, South Road, Durham DH1 3LE, England

INTRODUCTION

It has long been recognized that most shorebirds obtain a substantial proportion of their daily energy requirements in the non-breeding season from intertidal habitats, notably estuaries. In this paper, we shall use the term 'shorebird' to cover not only waders (Charadrii), to which the term 'shorebird' is restricted in North American usage, but also those wildfowl, gulls, and other aquatic birds that regularly use intertidal habitats. The earliest studies of the ecology of shorebirds were mostly of an observational or qualitative nature. Some of the first quantitative studies were published by Goss-Custard (1969) on the redshank, *Tringa totanus*, Heppleston (1971) on the oystercatcher, *Haematopus ostralegus*, and Prater (1972) on the knot, *Calidris canutus*.

Since then, shorebirds have become popular subjects for the study of foraging behaviour, as many species feed in open habitats where their diets may be assessed quantitatively by direct observations. Considerable emphasis has been given to such behavioural questions as the relationships between the density of benthic invertebrate prey species and the density, or food intake rates, of their shorebird predators. Little attention, however, has been paid to the impact of shorebirds on the populations of their prey, except for a short summary by Goss-Custard (1980) of the extent of depletion of certain prey by

[1]Present address: Chief Scientist's Directorate, Nature Conservancy Council, Northminster House, Peterborough PE1 1UA, U.K.

waders wintering in Britain. He concluded that, in their main feeding areas, waders took between 25 and 45% of the autumn standing crops of their prey during a winter, and that the extent of this depletion may significantly reduce the rate at which the birds can acquire food in late winter. Whilst it is clear that the amount and local distribution of the standing crops of prey are the major determinants of the foraging behaviour of predators in the short-term, it could be of more importance to an understanding of the functioning of intertidal (and particularly estuarine) ecosystems in the long-term to examine the impact of waders on the annual production of their prey. This we attempt in the sections that follow. It will become obvious that the data are fragmentary, but nevertheless sufficient to indicate that one of the major pathways of energy flow from intertidal benthic invertebrates lies through shorebirds. Although the distributions of most of these invertebrates extend into the subtidal, and indeed many migrate seasonally between the two zones, we suggest that any attempts to model the dynamics of estuarine or intertidal systems that ignore the extent of predation by birds are likely to be seriously incomplete.

A number of studies illustrating the flow of energy through estuarine ecosystems have been published recently, for example by Milne & Dunnet (1972), Rosenberg, Olson & Ölundh (1977), Dame, Vernberg, Bonell & Kitchens (1977), Warwick, Joint & Radford (1979), and Baird & Milne (1981), but only a very few include quantitative data on energy transfer between invertebrate prey and vertebrate predators.

The aim of this paper is thus to review literature on the utilization of estuarine production by avian predators, with particular emphasis on the rôle of waders. To do this, we consider:

(i) the measurement of food intake by shorebirds;
(ii) methods of estimation of the production of invertebrate prey, and year-to-year variation in productivity;
(iii) factors that determine the number of birds utilizing a particular estuary;
(iv) selected case studies of utilization of benthic invertebrate production by bird predators.

THE MEASUREMENT OF FOOD INTAKE BY SHOREBIRDS

Four methods have been used to estimate the quantities of different prey species removed by shorebirds: two focus on temporal changes in the density of invertebrates and two on daily food requirements of birds.

Changes in the density of prey in a site can be measured by sampling before and after a known period of predation, sampling either from within the same general area or with the aid of exclosures. Because the macrobenthos that form the main foods of shorebirds tend to occur at high densities but often with clumped dispersions, it is necessary to take large numbers of samples, each of relatively small surface area, to a standardized depth, before confidence limits around the average densities of animals become sufficiently narrow to establish statistically significant reductions in density, possibly attributable to predation. Interpretation of such proven changes in density, however, is difficult. If measurements have been taken across a season in which growth of invertebrates has taken place, animals too small to be retained during extraction of samples (*e.g.* by sieving) at the start of the season may have grown sufficiently to be included in the density estimates at the end of the season. Hence the loss to predators would be under-estimated. Losses need not be only to predators, of course. Many intertidal benthic invertebrates are mobile, particularly at certain seasons, and emigration may occur, *e.g.* to subtidal zones before the winter, or from high to low density patches. In addition to such active movements, surface-dwelling benthos may shift as a result of movements of sediments, particularly of finer-grained sediments in areas of high wave energy. Finally, even if all changes in prey density are wholly and solely attributable to predators, the direct methods of measurement do not permit apportioning of losses amongst the different possible predators.

Use of exclosures, however, of different mesh sizes to permit entry and exit of different sizes of predators, can in theory allow the food consumption of particular predators to be estimated. Problems remain in allowing for growth of prey when measuring density changes, and further problems arise as a result of alterations in water flow patterns over the parts of the intertidal zone covered by the exclosures. Even if the mesh of an exclosure remains free from clogging by floating macroalgae, changes in sediment properties within the exclosure often occur, such that finer particles settle there. Certain mobile invertebrate prey species may prefer such finer sediments and redistribute themselves, settling at higher densities within than outside the exclosures, even in the absence of predation. A particularly striking example of this, taken from a study by Millard (1975), is summarized in Table I. The study covered a five-month period during which there was no growth or reproduction of the amphipod, *Corophium volutator*, yet densities increased. Other problems in the use of exclosures, particularly in subtidal areas, have been reviewed by Virnstein (1978) and Peterson (1979).

TABLE I

Results of placement of an exclosure on open mud for five months in winter 1973–1974 (from Millard, 1975): "increase" in Corophium *density during winter inside exclosure is highly significant* ($t_{16} = 5.39$, $P < 0{\cdot}001$); *"decrease" in density outside exclosure is just significant at 5% level* ($t_{16} = 2{\cdot}31$)

	On removal		
	Within exclosure	Outside exclosure	Before positioning
Density of *C. volutator*·m^{-2} (±S.E.)	12 280 ±594	1770 ±299	4725 ±1386
Dry wt filamentous algae·100 cm^{-2} mud	0·116 ±0·063	0·030 ±0·019	
Mean carbon content (%)	3·27	1·43	
Mean silt content (%)	43·6	28·4	
Median particle size (μm)	220	300	
No. of samples (100 cm^2)	10	10	8

Recently, Bloom (1980) claims to have used, with success, exclosures with floating sides to minimize changes to sediment characteristics. At low water, the sides dropped and closed to keep out shorebirds, but at high water they floated open to allow fish and invertebrate predators to enter. We suspect that he was fortunate in working in an area with little wave action and a small tidal range, and would expect that even the refined design he has developed would cause water flow and sediment changes in intertidal regions with large tidal ranges (as in much of western Europe) or on high-energy wave-spending beaches (see Grant 1981). Indeed, a design with even fewer impediments to water flow caused considerable sediment changes on tidal flats in northeastern England (Pienkowski, 1980). Even if no changes in sediment occur, a possibility which has not been checked is that, as densities of benthic organisms outside the exclosures decrease relative to those inside as a result of predation, organisms inside the exclosures may choose to move out to reduce any disadvantages resulting from possible intraspecific competition. This possibility does not seem to have been considered by Quammen (1981) in her experiments using different types of exclosures, set deliberately in areas where water current movement was low to minimize sediment changes. Clearly, the execution and interpretation of exclosure experiments to determine the impact of shorebird predators on benthic invertebrates need to be considered more carefully in future.

The alternative to direct measurement of reduction in density of benthic invertebrates as a result of predation is to determine how much particular predators eat. Direct measurements of the sizes, species and quantities of invertebrates taken by the larger shorebirds can be obtained with reasonable accuracy during the hours of daylight. The feeding rhythms of the birds are, however, governed by the times of exposure of intertidal flats during the tidal cycle, and most species feed by night as well as by day, particularly in mid-winter when the daylight period is brief. Nocturnal feeding may be particularly important for the smaller species of shorebirds, whose diets and rates of food intake are difficult to assess directly (because they contain many small items) even by day. Methods employed for the quantitative estimation of food intake by waders have been reviewed by Pienkowski, Ferns, Davidson & Worrall (1984).

Particularly for the smaller waders, it is difficult to determine daily food intake by direct observation. An alternative approach is to estimate a bird's daily energy requirements, by methods outlined below, and to convert this to an estimate of food requirements by knowing what proportions of the energy content of its prey the bird can assimilate. Daily energy requirements have usually been estimated by establishing time-activity budgets, by direct observations, and then allowing for the differing rates of energy expenditure whilst feeding, flying, roosting, *etc.* These rates of expenditure have usually been expressed as multiples of the species' basal metabolic rate (BMR). Estimates of the daily energy expenditure of a variety of non-shorebird species have lain chiefly between 2 and 4 times BMR (Drent & Daan, 1980), and waders are thought to assimilate 80–90% of the energy content of their prey, so that daily food requirements of between 3·5 and 5 times BMR have been used in calculations of impact of birds on their prey (see *e.g.* Evans, Herdson, Knights & Pienkowski, 1979). BMR has been derived from allometric relationships with body mass (Lasiewski & Dawson, 1967; Aschoff & Pohl,

1970; Kendeigh, Dolnik & Gavrilov, 1977), but it is not always obvious which value should be used for the mass of a particular species, since this varies seasonally and may double before migration (Davidson, 1981; Tuite, 1984).

So far, direct measurement of daily energy expenditure of waders by the D_2O^{18} technique, involving injection of doubly-labelled "heavy" water into free-living birds and their recapture 24 or 48 h later to measure isotope loss, has proved impossible except during the breeding season. Other methods of direct measurement have also been attempted, but are not very reliable (Pienkowski *et al.* 1984).

ESTIMATION OF INVERTEBRATE PRODUCTION AND ITS YEAR-TO-YEAR VARIATION

Methods available for calculation of secondary production of benthic invertebrates have been described by Crisp (1971) in some detail.

One approach involves summation of the growth increments in all members of the population over the period of estimation, or until they die or are eaten. In practice, quantitative samples are taken at regular intervals for at least one year. Individual animals are weighed and measured and age cohorts separated. The products for each cohort of change in weight and numbers during a defined period are summed to give the population's production over that period. Production measured in this way thus reflects growth (*i.e.* increase in weight) and mortality (*i.e.* decrease in numbers) over a specified period for each cohort within the population. There are, however, a number of possible sources of error in this approach. First, the method does not allow for the production of mucus, dissolved organic matter and spawn, and so results in an under-estimation of total production. Secondly, size-selective mortality, through preferential predation of larger animals within cohorts, leads to under-estimation of production of those cohorts.

It has been demonstrated that specific bird predators tend to select particular sizes of prey. For example, oystercatchers preying on the common mussel, *Mytilus edulis* (Goss-Custard, McGrorty, Reading & Durell, 1980; Zwarts & Drent, 1981), knot preying on *Macoma balthica* (Prater, 1972), redshank preying on *Corophium volutator* (Goss-Custard, 1969), and ringed plovers, *Charadrius hiaticula*, feeding on mysids (Schneider, 1981) all took a limited size range of prey from those available. Pienkowski (1983b) found that ringed plovers and grey plovers, *Pluvialis squatarola*, feeding mainly on polychaetes, took large prey when these were available. This depended upon the weather conditions, which affected prey behaviour, and hence the selectivity shown by the predators. The upper limit of size taken is set by the ability of a bird to handle large prey (a limit predictable by foraging theory) or to gain access to large prey. (Larger polychaete worms and bivalves of several species tend to lie deeper in the substratum, out of reach of the bird's beak.) The lower limit of size taken is predicted by foraging theory as that size below which the handling efficiency (weight of food ingested per unit handling time) is less than the average rate of food intake. The lower size limit of acceptability is smaller for smaller predators, *e.g.* only those shore crabs, *Carcinus maenas*, of less than about 5 mm carapace width are ignored by redshank, but of less than about 10 mm are ignored by curlew, *Numenius arquata* (Zwarts, 1980). Within the acceptable size range of prey, actively searching predators should take different sizes in porportion to their available densities; in practice, this does not always happen (see *e.g.* Sutherland, 1982). Furthermore, this prediction does not necessarily apply to 'sit-and-wait' predators, including plovers (see Pienkowski, 1983a).

Size-selection of prey by predators may also lead to erroneous estimates of production if only total densities of prey are measured, rather than densities of each cohort separately. A study of predation by fish and birds on the nereid worm, *Ceratonereis pseudoerythraeensis*, by Kent & Day (1983) demonstrated that, although there was no change in overall worm densities between predated areas and controls, there were significant changes in the worm population structure. In areas of high predation of adult worms, juvenile recruitment was higher than in control areas. In manipulated populations, increased adult density decreased juvenile recruitment, thus supporting the hypothesis that adult–juvenile interactions may counter-balance the effects of adult losses due to predation.

Several predators tend to remove only parts of prey animals, *e.g.* bivalve siphons or polychaete tails. Bar-tailed godwits, *Limosa lapponica*, for example, often fail to extract more than the tails of lugworms, *Arenicola marina*, from their burrows (Smith, 1975). Production of such prey may be under-estimated if calculated by comparing the mean weights of individuals, some of which are regenerating parts after predation, with mean weights earlier in the year, before predation (see also de Vlas, 1981).

An alternative approach to estimation of production of benthic invertebrates involves determination of the change in biomass density from one sampling period to the next, together with an estimate of biomass produced but lost through mortality and gamete production. This approach is subject to errors not only from the sampling of the benthos but also in estimation of losses through predation, as discussed in the previous section.

Most estimates of benthic invertebrate production have been based on a single year's sampling and the assumption made that the ecosystems studied were in a steady state. Thus, annual fluctuations have been ignored in many energy flow studies designed to produce static ecosystem models, of the type popularized by Odum (*e.g.* Odum, 1970). Large fluctuations in annual productivity can, however, occur in the long-term, as shown by Beukema, de Bruin & Jansen (1978) and Beukema (1979, 1981a) in the Wadden Sea and by Baird & Milne (1981) in the Ythan estuary, Scotland, for a large number of benthic invertebrate species. Production estimates for some species may be either exceptionally high or exceptionally low during those years when growth, reproduction and spatfall are favoured or hindered, respectively, by environmental conditions. It must be recognized, however, that practical considerations often prevent measurement of standing crop biomass (and production) regularly over many years. Therefore, the variability in annual production is seldom indicated.

In many studies, invertebrate production (P) is not measured directly but estimated from measurements of mean standing crop biomass ($\bar{B}$) by the use of $P/\bar{B}$ ratios. These are thought to be characteristic of a species or group of similar species. However, because climatic and environmental conditions vary considerably from place to place, the turnover rate may also vary geographically in a widely distributed species. Thus, large errors may arise when applying $P/\bar{B}$ ratios determined in one site to measurements of standing crop biomass in another in order to estimate production in the latter site. Robertson (1979) has drawn attention to the range of variation (0·81–2·07) in $P/\bar{B}$ ratios reported for the bivalve, *Macoma balthica*. These arose chiefly from differences in age structure, since younger age groups have higher $P/\bar{B}$ ratios, and possibly from differences in methods of collecting animals. Some sampling techniques may miss the smallest age groups and hence under-estimate P. Further problems arise in determination of $\bar{B}$, depending upon the time period over which standing crops are measured and the extent of seasonal variation in biomass (Banse & Mosher, 1980).

GEOGRAPHICAL VARIATION IN THE USE OF INTERTIDAL AREAS BY SHOREBIRDS IN THE NON-BREEDING SEASON

Before we examine the total impact of shorebirds on the invertebrate production in different sites, it is necessary to emphasize that many waders that use coastal habitats are migrants. Most of the sandpiper family, Scolopacidae, breed in northern temperate or Arctic areas; so also do several species of plovers (family Charadriidae), although more species of plovers are tropical in distribution and never visit intertidal areas at any time of year. Breeding species from Greenland, Arctic Europe and Siberia migrate not only south in autumn, some travelling as far as South Africa and Australia, but also westwards (from Siberia) and eastwards (from Arctic Canada) to western Europe. Here they settle for the winter on coasts and estuaries around the North and Irish Seas, the Mediterranean and western France. The northern and eastern limits of their wintering distributions correspond to areas where the intertidal zones remain free from ice-cover for all but a few successive days each winter. Thus, it is rare for waders feeding on the benthos of soft sediments to be found within the Baltic in winter, and most travel at least as far as the Dutch Wadden Sea, although they use *e.g.* the Danish Wadden areas extensively during spring and autumn migration (Pienkowski & Evans, 1984).

We have argued elsewhere (Pienkowski & Evans, 1985) that the migration patterns of individuals of most wader species appear to be established in their first winter of life and followed faithfully thereafter, as has been shown for grey plover, *Pluvialis squatarola*, by Townshend (1985). It also appears that, during their first migration, once juveniles have reached a possible wintering area, they travel no further than necessary. As a result, only after breeding seasons with high adult survival and many young raised do juveniles have to 'overflow' from the British and Dutch estuaries to sites farther south, *e.g.* in western France, which are not used so extensively in years of low wader populations (Pienkowski & Evans, 1985). Large year-to-year variation in numbers of birds migrating to distant estuaries, at the end of migration routes, has been recorded also in Tasmania (Thomas, 1970) and South Africa (Robertson, 1981). For this reason alone, one cannot expect the impact of birds on the production of invertebrates in the intertidal zone to be constant from one year to the next.

In the northern parts of the winter range, movements of birds, of a few species of waders and several wildfowl, may also take place in response to the onset of very cold weather. This is known chiefly for plovers (Townshend, 1982) amongst the waders.

On a smaller geographical scale, the distribution of a certain species of shorebird amongst a group of adjacent possible wintering sites may be closely related to variations in the absolute density of their preferred prey amongst these sites. This has been shown, *e.g.* for curlew feeding upon the polychaetes, *Nereis diversicolor* and *Nephtys hombergi*, in the estuaries along the coast of Essex, southeastern England (Goss-Custard, Kay & Blindell, 1977) and in bays along the shore of the Firth of Forth, Scotland (Bryant, 1979). The abundance of curlew in any one site in Essex was, however, more than tenfold greater than in a site in Scotland holding the same densities of prey (Evans & Dugan, 1984). (This may arise from differences in the percentage of prey actually available to the birds, for reasons outlined below.) If such geographical differences also apply to other species of waders, then the impact of birds on the standing crop of a particular prey will also vary from place to place, as will the impact on its annual production, unless the $P/\bar{B}$ ratio varies in the same way and by the same amount as the bird/invertebrate density ratio.

The rate at which shorebirds can capture prey depends chiefly on the density of available prey. This is reduced not only by the action of the predators themselves but also by the responses of the prey to various environmental factors. In general, in European estuaries, as temperatures fall, benthic polychaetes burrow deeper into the substratum and so become less accessible (and, therefore, less available) to birds hunting by probing into soft sediments. Many benthic invertebrates also become less active and so less detectable to birds hunting by sight (Pienkowski, 1983b). Prey activity may also be reduced by the presence of predators, *e.g. Corophium volutator* emerge less frequently from their burrows after redshank have walked close by, than in the absence of birds (Goss-Custard, 1970). Similarly, deeper burrowing by *Macoma balthica* in winter (Reading & McGrorty, 1978) may be in response to the presence of one of their chief predators, the knot, which arrives in late summer or autumn (Evans, 1979). Cold weather does not always reduce the availability of prey to predators; surface-dwelling bivalves such as *Mytilus edulis* may 'gape' in severe weather and thus be more vulnerable to attack by oystercatchers. In many European estuaries, however, the density of available prey for shorebirds is least in mid- to late winter, so that it is at this season that limitations on the use of an estuary by birds could be set. Irrespective therefore of the absolute density of prey, which will be one determinant of the productivity of a site, the proportion taken by birds is likely to be least in sites where mid-winter prey availability is lowest and, thus, the capacity to sustain birds is lowest. Within Europe, such sites should be the most eastern and northern.

It is not known to what extent high temperatures have similar effects on the availability of prey. It could well be that polychaetes burrow deeper to reduce the chances of desiccation, but effects on activity are less easy to predict. Thus, at present, no predictions can be made about the possibility of a seasonal 'bottleneck' in density of prey available to shorebirds in tropical regions.

A further restriction to the use of estuaries by shorebirds is the nature of the supralittoral zone. Waders are generally birds of open habitats. Steep sided, narrow estuaries, or those whose upper tidal levels are covered by mangrove or *Spartina* spp. may not be used as much as their invertebrate resources would predict they might be (Evans, 1976; Millard & Evans, 1984).

CASE STUDIES

One of the objectives of this paper is to examine the extent of energy transfer from benthic invertebrate prey to bird predators. To illustrate this we have selected a few studies from which the necessary information could be extracted. Numerous studies have been reported in the literature on the prey taken in particular estuaries by particular shorebird species, and several on rates of predation and energetic requirements. There are, however, few studies which have examined the impact of a whole bird community on the benthic invertebrate prey community of a single area. Such studies invariably require data not only on bird numbers, diet, and energy intake, but also on invertebrate standing crops, so that reliable estimates of production of the prey organisms can be calculated. We have selected, for the purpose of this review, information on the following coastal and estuarine ecosystems: the Dutch Wadden Sea; the Grevelingen estuary, Netherlands; the Ythan estuary, Scotland; Lindisfarne, Northumberland, and the Tees estuary, both in northeastern England; the Wash, eastern England; and the Langebaan Lagoon, South Africa. Each system is examined in terms of species and numbers of waders (and other birds where information is available), their annual consumption of invertebrate prey and the fraction of the preys' production consumed by the birds. We recognize that birds are not the only predators of the invertebrates and have, therefore, included reference to total predation by fish, birds, and man where sufficient data are available.

THE WADDEN SEA, NETHERLANDS

The Wadden Sea is a shallow coastal sea bordering the Netherlands, the Federal Republic of Germany, and Denmark, and situated behind a series of barrier islands. It covers about 10 000 km^2 and is thus the largest area of coastal flats in Europe (Smit & Wolff, 1980). The Dutch part of the Wadden Sea includes 1300 km^2 of tidal flats, and an equally large subtidal area (Wolff & Smit, 1984). Large numbers of waders, gulls and terns, and wildfowl occur in the Dutch Wadden Sea throughout the year, with maximum numbers (2·5–3·4 million) during August to October. Of these, waders account for between 1·7 and 2·3 million. Bird numbers drop to 1·5 million (0·6–0·8 million waders) in January and February but increase again to 1·3–1·9 million in April and May, including 0·8–1·3 million waders (Smit, 1980a). Smit (1980b) estimated the consumption by all carnivorous birds feeding in the Wadden Sea. The species, mean numbers and mean weight of birds feeding on the intertidal flats are listed in Table II. A substantial amount of research has been conducted on their feeding ecology, for example by Hulscher (1974, 1980, 1982), Zwarts & Drent (1981), Swennen (1975, 1976), and Wolff & Smit (1984). An equal amount of research has been done on invertebrate abundance, distribution, and production (for a comprehensive review and references, see Dankers, Kühl & Wolff, 1981).

Average consumption by birds feeding both inter- and subtidally in the Dutch Wadden Sea has been estimated as 90·0 kJ·m^{-2} by Hulscher (1975), 77·4 kJ·m^{-2} by Swennen (1976) and 85·8 kJ·m^{-2} by Smit (1980b). Wolff & Smit (1984) have suggested that these values are too low, because bird counts

TABLE II

Bird species occurring in the Dutch Wadden Sea and Grevelingen estuary, their mean numbers and mean weights: data for the Wadden Sea from Smit (1980b) and for the Grevelingen estuary from Wolff et al. *(1975); *, species occurring regularly on the Ythan estuary, Scotland (Baird & Milne, 1981)*

Species	Mean number day^{-1}		Mean individual biomass (kg)
	Wadden Sea	Grevelingen	
Species depending on subtidal area			
Great crested grebe (*Podiceps cristatus*)	1000		1·102
Scaup (*Aythya marila*)	4500		0·900
Goldeneye (*Bucephala clangula*)	1800		0·872
Long-tailed duck (*Clangula hyemalis*)	250		0·700
Velvet scoter (*Melanitta fusca*)	250		1·400
Common scoter (*Melanitta nigra*)	8000		1·062
*Eider (*Somateria mollissima*)	62 000		2·118
Red-breasted merganser (*Mergus serrator*)	3900		1·084
Goosander (*Mergus merganser*)	2300		1·624
Herring gull (*Larus argentatus*)	15 000		0·991
*Black-backed gulls (*Larus marinus* & *L. fuscus*)	1500		1·702
Common tern (*Sterna hirundo*)	1300		0·135
*Arctic tern (*Sterna paradisea*)	55		0·110
Little tern (*Sterna albifrons*)	150		0·075
*Sandwich tern (*Sterna sandvicensis*)	900		0·250
Species depending on tidal flats			
*Oystercatcher (*Haematopus ostralegus*)	120 000	16 430	0·497
Grey plover (*Pluvialis squatarola*)	5600	567	0·223
Ringed plover (*Charadrius hiaticula*)	1100	—	0·070
Kentish plover (*Charadrius alexandrinus*)	150	—	0·047
Turnstone (*Arenaria interpres*)	2000	337	0·106
Curlew (*Numenius arquata*)	44 000	2085	0·921
Bar-tailed godwit (*Limosa lapponica*)	30 000	805	0·248
*Redshank (*Tringa totanus*)	10 000	559	0·128
Spotted redshank (*Tringa erythropus*)	700	—	0·149
Greenshank (*Tringa nebularia*)	1100	—	0·180
Knot (*Calidris canutus*)	50 000	2610	0·128
Dunlin (*Calidris alpina*)	120 000	6602	0·057
Curlew sandpiper (*Calidris ferruginea*)	150	—	0·070
Avocet (*Recurvirostra avosetta*)	5800	60	0·340
*Shelduck (*Tadorna tadorna*)	21 000	578	1·237
*Common gull (*Larus canus*)	34 000	2939 (Common gull and Black-headed gull combined)	0·369
*Black-headed gull (*Larus ridibundus*)	50 000		0·241
*Herring gull (*Larus argentatus*)	15 000	449	0·991
*Grey heron (*Ardea cinerea*)	—	5	1·495
Brent Goose (*Branta bernicla*)	—	227	1·382
Teal (*Anas crecca*)	—	16	0·430
*Wigeon (*Anas penelope*)	—	378	0·692
Pintail (*Anas acuta*)	—	55	0·880

have been under-estimates, but they do not, however, indicate the magnitude of the under-estimation. For the purpose of this review, we have used the value of Hulscher (1975).

Beukema (1976, 1981a) has estimated a value for average annual production for all macrobenthic animals living in the intertidal flats of the Dutch Wadden Sea of 619 $kJ \cdot m^{-2} \cdot yr^{-1}$ and a mean annual standing crop of biomass of approximately 556 $kJ \cdot m^{-2}$. If subtidal production is also taken into account, then the average production for the whole of the Dutch Wadden Sea falls to 523 $kJ \cdot m^{-2} \cdot yr^{-1}$ (Beukema, 1981b). Birds, feeding both inter- and subtidally, consume approximately 17% (90 $kJ \cdot m^{-2} \cdot yr^{-1}$) of the total production. Waders, the shelduck, *Tadorna tadorna*, and those gulls feeding only on the intertidal flats (see Table II), again consume nearly 17% (104 $kJ \cdot m^{-2} \cdot yr^{-1}$) of the total intertidal macrobenthic production of 619 $kJ \cdot m^{-2} \cdot yr^{-1}$ (Smit, 1980b).

Of other macrobenthic predators, demersal fish remove about 105 $kJ \cdot m^{-2} \cdot yr^{-1}$ from the whole Dutch Wadden Sea, shore crabs and other invertebrates 62 $kJ \cdot m^{-2} \cdot yr^{-1}$, and fisheries for mussels and cockles another 42 $kJ \cdot m^{-2} \cdot yr^{-1}$, that is an additional 209 $kJ \cdot m^{-2} \cdot yr^{-1}$ (Beukema, 1981b). Total predation then becomes 299 $kJ \cdot m^{-2} \cdot yr^{-1}$ or 57% of the total annual production of the inter- and subtidal areas.

About 50 macrobenthic invertebrate species are found on the tidal flats of the Dutch Wadden Sea. Of these, seven species, namely *Mytilus edulis*, *Arenicola marina*, *Mya arenaria*, *Cerastoderma* (=*Cardium*) *edule*, *Macoma balthica*, *Lanice conchilega*, and *Nereis diversicolor* are responsible for 89% of the mean annual biomass and about 70% of the annual production. Of the bird species, oystercatchers constituted about 44% of the numbers present during 1974–1978 (Prater, pers. comm.), with a mean daily count of 120 000 birds (Smit, 1980b). They feed almost entirely on bivalve molluscs, chiefly *Mytilus edulis*, *Cerastoderma edule* and *Macoma balthica* (Hulscher, 1980, 1982). Oystercatchers alone consume about 36 $kJ \cdot m^{-2} \cdot yr^{-1}$ which is 12·6% of the production (287 $kJ \cdot m^{-2} \cdot yr^{-1}$) of these three bivalve mollusc species on the tidal flats.

GREVELINGEN ESTUARY, NETHERLANDS

The Grevelingen estuary, in the southwestern part of the Netherlands, has a total area of 140 km^2, including 4 km^2 of salt marshes and 55 km^2 of tidal flats. In May 1971 a dam was completed, isolating it from the sea, but in the year immediately before closure of the dam, intensive studies were made (Wolff & de Wolf, 1977). The most important bird species on the intertidal parts of the estuary are listed in Table II. Mean bird days per year, average weights, and annual consumption estimates (per bird species) are given by Wolff *et al.* (1975).

Wolff & de Wolf (1977) computed benthic production for *Hydrobia ulvae*, *Littorina littorea*, *Cardium edule*, *Macoma balthica*, *Arenicola marina*, and *Mytilus edulis* to be in the order of 1200 $kJ \cdot m^{-2} \cdot yr^{-1}$, twice the level in the Wadden Sea. Estuarine birds (zoobenthos feeders) consumed about 72 $kJ \cdot m^{-2} \cdot yr^{-1}$ (Wolff *et al.* 1975) or 6·0% of the total production. Flatfish consumed approximately the same amount (Wolff & de Wolf, 1977).

Approximately 12% of macrobenthic production thus passed to the next trophic level.

THE YTHAN ESTUARY, ABERDEENSHIRE, SCOTLAND

The Ythan estuary is located about 20 miles north of Aberdeen at 57°20′ N: 2°00′ W. It is tidal for about 8 km from the mouth, but has an intertidal area of only 1·85 km^2 and a subtidal area at low water of 0·71 km^2 (Leach, 1971). A substantial amount of ecological research has been conducted during the past 20 years in the estuary with the aim of providing some insight into dynamic relationships within the food web. Earlier results were summarized by Milne & Dunnet (1972). More recently, Baird & Milne (1981) described the utilization and partitioning of net production by the various populations and species in the food web, by means of a steady state energy flow model.

Milne & Dunnet (1972) and Milne (1974) recorded up to 60 bird species in the Ythan estuary and adjacent areas, but only 20 species are present throughout the year or visit it in large numbers. An even smaller number of species (listed in Table II) obtain a substantial proportion of their food requirements from the estuary. Two, the mute swan and the wigeon, are herbivorous, whilst the rest are carnivorous. Details of bird numbers, diet and energy requirements are given by Baird & Milne (1981). These authors have calculated the consumption of benthic macro-invertebrates by all bird predators in the Ythan estuary. The main invertebrate species are *Mytilus edulis*, *Nereis diversicolor*, *Corophium volutator*, *Gammarus* spp., *Carcinus maenas*, *Littorina* spp., *Hydrobia ulvae* and *Macoma balthica*. Their combined annual production amounts to 2450 $kJ{\cdot}m^{-2}{\cdot}yr^{-1}$ of which birds consume an average of 36% (874 $kJ{\cdot}m^{-2}{\cdot}yr^{-1}$). Fish consume an additional 16% (416 $kJ{\cdot}m^{-2}{\cdot}yr^{-1}$) so that the total amount consumed by fish and birds averages 52% (1290 $kJ{\cdot}m^{-2}{\cdot}yr^{-1}$) of the annual production of benthic macrofauna. Baird & Milne (1981) also give values for consumption, by individual species of predators, of individual prey species, as well as information on long-term and short-term fluctuations in standing crops of prey and predators.

LINDISFARNE, NORTHEASTERN ENGLAND

Lindisfarne National Nature Reserve in north Northumberland comprises about 32·4 km^2, mainly of tidal mud and sandflats, saltmarsh, and dunes. Pienkowski (1980, 1982) studied the feeding ecology of the grey plover, *Pluvialis squatarola*, and the ringed plover, *Charadrius hiaticula*, on Holy Island Sands. It is clear from these studies that the capitellid polychaete, *Notomastus latericeus*, and the ariciid polychaete, *Scoloplos armiger*, are the most important prey in numerical terms. Smaller numbers of the much larger lugworm, *Arenicola marina*, form a large component of the diet in energetic terms; various small prey species form only a small proportion of the birds' diet. Smith (1975) investigated the feeding ecology of a larger wader, the bar-tailed godwit, *Limosa lapponica*, at Lindisfarne and concluded that, for this species, the lugworm provided about 94% but the smaller worms only 6% of the energy content of the diet.

Standing crop biomass of *Notomastus* (73 $kJ{\cdot}m^{-2}$) and *Scoloplos* (11·5 $kJ{\cdot}m^{-2}$) at the beginning of summer were measured by Pienkowski (1980) and production was estimated using minimal and maximal $P/\bar{B}$ ratios of 2·5 and 5·0, respectively (McLusky, 1981), yielding annual production values of 182·4–364·7 $kJ{\cdot}m^{-2}{\cdot}yr^{-1}$ for *Notomastus* and 28·8–57·5 $kJ{\cdot}m^{-2}{\cdot}yr^{-1}$ for *Scoloplos*, respectively. The standing crop of *Arenicola* in May was 103 $kJ{\cdot}m^{-2}$ (Smith, 1975) and by applying a $P/\bar{B}$ ratio of 1·21 (Wolff & de Wolf, 1977), production was estimated as about 124 $kJ{\cdot}m^{-2}{\cdot}yr^{-1}$.

During winter ringed plovers removed from a small, but representative, study area of 0·01 km^2 approximately $9{\cdot}3 \times 10^4$ kJ of *Notomastus* (*i.e.* 9·3 $kJ{\cdot}m^{-2}$, between 2·6 and 5·3% of its annual production), $2{\cdot}4 \times 10^4$ kJ of *Scoloplos* (*i.e.* 2·4 $kJ{\cdot}m^{-2}$, between 4·2 and 8·6% of its annual production) and $2{\cdot}1 \times 10^4$ kJ of *Arenicola* (1·7% of its annual production). Grey plovers consumed about $5{\cdot}0 \times 10^4$ kJ of *Notomastus* (1·4 to 2·9% of annual production), $9{\cdot}4 \times 10^3$ kJ of *Scoloplos* (between 1·6 and 3·3% of production), and 13×10^4 kJ of *Arenicola* (10·6% of production). Bar-tailed godwits removed between 0·3 and 0·8% of the production of *Notomastus* and *Scoloplos* but about 38·6 $kJ{\cdot}m^{-2}$ or 31% of the annual production of *Arenicola*. Total consumption, by all three bird species, of *Notomastus* and *Scoloplos* amounts to approximately 19 $kJ{\cdot}m^{-2}{\cdot}yr^{-1}$ (that is between 4·5 and 9·0% of the total production of the two prey species), and of *Arenicola* 54 $kJ{\cdot}m^{-2}{\cdot}yr^{-1}$ (43% of production). This represents between 13 and 22% of the total annual production of *Notomastus*, *Scoloplos*, and *Arenicola*. It should be noted that these values refer to only a single year's data. In addition, a fourth small but abundant wader species, the dunlin, *Calidris alpina*, probably took substantial quantities of the small worms. Because of the different feeding method it used, these quantities could not be measured. (Dunlin at this site did not, however, take *Arenicola*.)

TEES ESTUARY, NORTHEASTERN ENGLAND

The Tees estuary (54°37′ N: 1°12′ W) contains at present an intertidal mudflat of 1·4 km^2, the remainder of an initial area of 24 km^2 which was depleted by reclamation schemes during the nineteenth and twentieth centuries. The mudflat is the only important feeding ground for several species of shorebirds, although a few species also feed on rocks and beaches outside the estuary (Evans *et al.*, 1979). Teesmouth is a regular migration staging post and overwintering area for large numbers of shorebirds, in particular grey plover, bar-tailed godwit, redshank, knot, dunlin, shelduck and curlew.

The main invertebrate prey taken by shorebirds at Teesmouth and estimates of their production in 1973 are given in Table III. It was not possible to measure directly the quantities of each prey species consumed by each species of predator at Teesmouth during the following winter. Evans *et al.* (1979), however, estimated the daily energy requirements for each predator (Table IV), on the basis of 3·5 × BMR (basal metabolic rate) for grey plover and shelduck and 4·5 × BMR for all the other species, and the number of bird-days spent by each species on the feeding area. Total prey production over the whole of this feeding area amounted to 1192×10^6 kJ (Table III), and consumption by the overwintering shorebird populations to 529×10^6 kJ, *i.e.* 44% of the estimated production. Little growth of invertebrates occurred during the winter months when most of the birds were present and feeding. Thus the standing crops were severely reduced by May 1974, as was proved by

TABLE III

*Estimated production of the main prey species of shorebirds on the mudflats at Teesmouth in 1973: *, values taken from Baird & Milne (1981)*

Species	Production over whole area ($kJ \cdot yr^{-1}$)	Average production ($kJ \cdot m^{-2} \cdot yr^{-1}$)	$P/\bar{B}$ ratio used
Hydrobia ulvae	$17{\cdot}2 \times 10^6$	12·3	1·4*
Nereis diversicolor	$59{\cdot}0 \times 10^6$	42·2	3·0*
Macoma balthica	$2{\cdot}5 \times 10^6$	1·7	2·0*
Small polychaetes and *Manayunkia aestuarina*	$552{\cdot}3 \times 10^6$	395	5·0
Small oligochaetes	$560{\cdot}7 \times 10^6$	400	4·0

direct measurement (Evans *et al.*, 1979). The 1973 to 1974 annual production, estimated on an area basis, was 851 $kJ \cdot m^{-2} \cdot yr^{-1}$. Gray (1976) commented on the high productivity, but pointed out that it resulted chiefly from the very high densities of meiofauna and that typical estuarine macrobenthos was restricted in both variety and density.

THE WASH, EASTERN ENGLAND

The Wash is a large rectangle of intertidal land (600 km^2) into which flow four main rivers draining the low-lying agricultural land of the eastern half of southern central England. Proposals in the 1970s to build a series of water storage reservoirs in the upper parts of the intertidal zone led to several thorough studies of the distributions, densities, and diets of the more numerous shorebird species (Goss-Custard, 1977; Goss-Custard, Jones & Newbery, 1977; Goss-Custard *et al.*, 1977), and of their invertebrate prey (*e.g.* Reading & McGrorty 1978). The main shorebird predator–invertebrate prey interactions are listed in Table V.

From the published studies it is unfortunately not possible to estimate average invertebrate production over the whole of the Wash. Estimates of the proportion of the autumn standing crop of several bivalves and polychaetes that were taken by specific predators, however, have been made by Goss-Custard (1977). By use of simple exclosures, he attributed 43% of the fall in biomass of the sedentary polychaete, *Lanice conchilega*, during one winter to predation by shorebirds (see Table V for species). If we assume a steady-state standing crop from year to year in late spring (which may not be a valid assumption—see Beukema, 1979), and a $P/\bar{B}$ ratio of 2·0, then shorebird consumption represented 21% of annual production of this polychaete.

Goss-Custard (1977) found that exclosures were not satisfactory for the estimation of overwinter losses of bivalves. However, by observation of numbers of birds feeding on selected intertidal areas on the southeastern side of The Wash, and calculation of their energy requirements, he estimated the propor-

TABLE IV

Bird species, numbers and estimated energy requirements at Teesmouth, August 1973 to May 1974

Species	Total food requirements during overwintering period (kJ)	Total bird-days at Teesmouth (thousands)	Estimated daily energy requirements (kJ)
Shelduck (*Tadorna tadorna*)	222×10^6	161	1191
Grey plover (*Pluvialis squatarola*)	$6{\cdot}3 \times 10^6$	15	364
Curlew (*Numenius arquata*)	$46{\cdot}0 \times 10^6$	35	1170
Bar-tailed godwit (*Limosa lapponica*)	$16{\cdot}3 \times 10^6$	26	552
Redshank (*Tringa totanus*)	$18{\cdot}4 \times 10^6$	49	334
Knot (*Calidris canutus*)	$16{\cdot}7 \times 10^6$	45	334
Dunlin (*C. alpina*)	203×10^6	1200	150
Grand total	529×10^6		

tion of the November standing crop of *Macoma* and *Cardium* taken by knot and oystercatchers in several main and subsidiary feeding areas of these birds. This varied between 14 and 34% on the main areas. Unfortunately this cannot be related to production.

Calculations from Table 2 in Goss-Custard (1977) allow rough estimates to be made of the proportion of production of all invertebrates taken by all waders. Three estimates are given of standing crop in each of three areas during the period from July 1972 to June 1973. If an average biomass is calculated and an average $P/\bar{B}$ ratio of 2 employed, then it follows that waders removed between one-fifth and one-third of the production in the three areas. If a higher $P/\bar{B}$ ratio were used then the proportion removed would be correspondingly less.

LANGEBAAN LAGOON, SOUTH AFRICA

Langebaan Lagoon (33° S: 18° E) is a large inlet from Saldanha Bay on the coast of the southwestern Cape. It consists of 17·5 km^2 of intertidal sandflats and 6·06 km^2 of saltmarshes (Summers, 1977). Its general ecology has been described by Day (1959), while spatial and temporal variations in densities of the intertidal fauna available to waders have been discussed by Puttick (1977). The numbers and distribution of waders (Charadrii) have been described by Pringle & Cooper (1977) for the whole Cape Peninsula and by Summers, Cooper & Pringle (1977) for the southwestern Cape, while Summers & Waltner (1979) commented on seasonal variations in the body mass of waders in southern Africa. The size of the non-breeding Palaearctic wader populations of Langebaan Lagoon was given by Pringle & Cooper (1975), and

TABLE V

Waders and their principal invertebrate prey at the Wash (from Goss-Custard, Jones & Newbery, 1977)

Bird species	Molluscs *Cardium*	*Mytilus*	*Macoma*	*Scrobicularia*	*Hydrobia*	Polychaetes *Scoloplos*	*Nereis*	*Nephtys* spp.	*Lanice*	*Arenicola*	Crustaceans *Carcinus*	*Crangon* spp.
Oystercatcher	*	*	*									
Knot	*		*		*							
Dunlin					*		*	*				
Redshank			*		*		*				*	*
Bar-tailed godwit			*			*	*		*			
Turnstone	*	*										
Grey plover	*		*				*		*		*	
Curlew	*			*					*	*	*	

Summers (1977) discussed the distribution, abundance and energy relationships of waders there.

Waterbirds occur in large numbers at Langebaan Lagoon; most are waders (Charadrii). Other major groups are terns (Sternidae), gulls (Laridae), and cormorants (Phalacrocoracidae). Table VI gives the mean weight of individuals of each wader species and their summer and winter population sizes. The curlew sandpiper, *Calidris ferruginea*, is the most abundant in terms of numbers and biomass and the grey plover next most important. Puttick (1978, 1979, 1980) investigated the diet, foraging behaviour and energy budgets of the curlew sandpiper at Langebaan.

A check-list of benthic organisms, including their mean monthly biomass and numbers, and production estimates of those benthic invertebrates available to birds, is given by Puttick (1977, 1980). The more important invertebrate species in terms of biomass and production, and importance as prey items to waders, are nereid worms (mainly *Ceratonereis erythraeensis*), *Assiminea globulea*, *Urothoe grimaldi*, *Cleistosoma edwardsii*, *Hymenosona orbiculare*, and stratiomyid larvae (Order Diptera). The mean annual production of these invertebrate species is about 705 $kJ \cdot m^{-2} \cdot yr^{-1}$ (Puttick, 1980), a value comparable with that of benthic production in northern temperate areas.

TABLE VI

Mean biomass and numbers of waders at Langebaan Lagoon during 1975 to 1976 (from Summers, 1977)

Species	Mean weight (g)	Southern winter population (no. birds)	Southern summer population (no. birds)
Haematopus moquini (black oystercatcher)	685	80	32
Arenaria interpres (turnstone)	106	215	968
Charadrius hiaticula (ringed plover)	47		242
C. marginatus (white fronted plover)	45	196	287
C. pallidus (chestnut-banded plover)	36		39
C. pecuarius (Kittlitz's plover)	40	145	123
C. tricollaris (three-banded plover)	32	6	
Pluvialis squatarola (grey plover)	217	433	3615
Hoplopterus armatus (blacksmith plover)	182	38	15
Calidris ferruginea (curlew sandpiper)	56	4	25 357
C. minuta (little stint)	23		346
C. canutus (knot)	143	70	2685
C. alba (sanderling)	55	46	1784
Xenus cinereus (terek sandpiper)	75	40	78
Tringa stagnatilis (marsh sandpiper)	72		34
T. nebularia (greenshank)	200	153	309
Limosa lapponica (bar-tailed godwit)	291		32
Numenius arquata (curlew)	905	32	309
N. phaeopus (whimbrel)	409	143	465
Recurvirostra avosetta (avocet)	324	23	33
Himantopus himantopus (stilt)	160	47	6

Total energy requirements and consumption by all birds feeding in the Langebaan Lagoon, calculated from data in Summers (1977) using a ratio of 5 times BMR (Puttick, 1980), appear to be approximately 142 $kJ \cdot m^{-2} \cdot yr^{-1}$, *i.e.* 20% of the total invertebrate prey production.

In the early 1970s there were between 37 000 and 55 000 curlew sandpipers during the southern summer and 12 000 in the southern winter (Pringle & Cooper, 1975). This species accounts, on average, for about 69% of the total number of waders and 42% of their total biomass (Summers, 1977). It is not only an important numerical constituent of the Langebaan Lagoon avifauna, but also plays an important rôle in the energetics of the system. Curlew sandpipers removed an estimated 87 $kJ \cdot m^{-2} \cdot yr^{-1}$, or 12% from the net annual production potentially available to them (Puttick, 1980). She also observed that the energetic equivalent of the standing crop of benthic invertebrates decreased during a year by approximately 44%, although only 20% of the yearly production was removed by birds.

OTHER AREAS

In addition to the cases discussed above, a number of studies have been reported dealing with certain aspects of energy transfer from invertebrate prey to bird predators. Some refer to a single predator's consumption of a number of prey species; others to the influence of a number of predators on a single prey species; yet others have been limited to a specific area of an estuarine ecosystem. For example, Christy, Bildstein & DeCoursey (1981) investigated the impact of wading birds (herons and egrets as well as shorebirds) on the invertebrates of a saltmarsh in South Carolina. Birds removed approximately 24 $kJ \cdot m^{-2} \cdot yr^{-1}$ but no quantitative information was given on the consumption of individual prey species, except that crustaceans made up about 50%, fishes about 20% and molluscs about 20% of the diet of the entire avian community. Grant (1981) estimated that about 10·4% of the daily energy intake by shorebirds on a beach in South Carolina consisted of the haustoriid amphipod, *Acanthohaustorius millsi*, although this value is heavily dependent on the validity of his sampling programme to determine changes in amphipod density. Schramm (1978) reported a consumption rate of about 8·8 $kJ \cdot m^{-2} \cdot yr^{-1}$ for the grey plover, *Pluvialis squatarola*, in the Swartkops estuary. This represents approximately 0·3% of the production (2950 $kJ \cdot m^{-2} \cdot yr^{-1}$) of the most important benthic invertebrates in the estuary Hanekom & Baird, 1985). Grey plover density in the southern summer was found to be 280 $birds \cdot km^{-2}$ and 30 $birds \cdot km^{-2}$ in winter. Such densities are comparable with those found in north temperate estuaries in Europe.

DISCUSSION

Information from the five most complete case studies is summarized in Table VII. No error margins have been attached to the figures, but several points should be borne in mind. First, for only the Dutch Wadden Sea and the Ythan estuary do the data represent average values, measured over several years, of both consumption by birds and production by estuarine macrobenthos. Secondly, the estimate of consumption is probably too low for the Ythan ecosystem, since it was calculated by assuming a daily energy requirement for birds of only 2–3 times their basal metabolic rates, as opposed to 4–5 for the other studies. Thirdly, invertebrate production at Langebaan is probably higher than the value included in the Table, since Puttick (1980) hints that at least some of the consumption by predators has not been allowed for when calculating benthic production from changes in biomass. Fourthly, estimates of production at Teesmouth by use of $P/\bar{B}$ ratios have relied wholly on values obtained elsewhere; this criticism applies also, but to a much lesser extent, to the Ythan and Langebaan figures.

In spite of these uncertainties, the values for consumption efficiency, representing energy transfer from the total benthic production to shorebirds (as the sole predators for which adequate data are generally available), are high (Table VII) and may even be under-estimates. In general, this ratio represents the degree of utilization of prey by predators at the next higher trophic level and is an index of the pressure of a group of predators on their prey (Ricker, 1968; Krebs, 1978).

The classification of animals into "trophic levels" is, in most cases, an over-simplification based on generalizations concerning the animals' feeding behaviour (May, 1979). This is especially true for invertebrates, in which a single species may behave as a detritivore, a suspension-feeder or a predator, depending on circumstances; but less difficult for, for example, most shorebirds on their wintering grounds. It is valid, therefore, to comment on the efficiency of energy transfer between benthic production and shorebird predators. Table VII clearly illustrates that the efficiency is higher (and if fish and invertebrate predators are added, much higher) than that found in terrestrial ecosystems, where only between 5 and 10% of the primary production is eaten by herbivores (Whittaker, 1975) and about 20% of herbivore production is eaten by carnivores (Krebs, 1978). In such ecosystems, most of the production goes to decomposers rather than to grazers and thence to predators. Other important conclusions from Table VII are that the efficiency of energy transfer

TABLE VII

Summaries of benthic invertebrate production estimates, consumption by shorebirds and consumption efficiencies, for five estuaries

Locality	Energy equivalent ($kJ \cdot m^{-2} \cdot yr^{-1}$) of consumption by birds	Energy equivalent ($kJ \cdot m^{-2} \cdot yr^{-1}$) of invertebrate production	Consumption efficiency, invertebrates to birds
Dutch Wadden Sea (intertidal zone)	103·6	619·2	17%
Grevelingen estuary	71·5	1201·4	6%
Ythan estuary	873·6	2448·1	36%
Tees estuary	367·0	851·0	44%
Langebaan Lagoon	141·6	705·0	20%

from invertebrate prey to bird predators varies widely from place to place and that there is no clear relationship between productivity of prey and the effective utilization of that production by birds. Possible reasons for this were discussed earlier.

Because birds may consume such a high proportion of the annual production of their prey, the question arises as to whether benthic production limits the number of birds using any chosen intertidal area. This is unlikely to be the case, for several reasons. First, production provides only an upper limit to the number of bird biomass-days that could be supported. This has to be translated into bird-days, according to the biomass of each species. Secondly, the food potentially available to birds, *viz.* the standing crop present when the birds arrive from their breeding areas, arises not merely from production in the year immediately preceding this date, but from previous growth of cohorts of invertebrates (especially bivalves) that settled some years earlier. Indeed, the relationship between annual production and utilization cannot be simple, because of the selectivity for size (and therefore age) of prey shown by most shorebirds. High annual production of *e.g. Nereis diversicolor* as first-year worms would be advantageous for small waders such as dunlin which feed on small worms. An equivalent amount of production would not, however, be utilizable by large waders such as curlews unless some of it represented growth from 0-class to 1-class worms, the only ones taken by curlews (Evans *et al.*, 1979). Sequential predation, first by oystercatchers and later by curlews, of a cohort of the clam, *Mya arenaria*, as individual clams grew older and larger, has been described in detail by Zwarts & Wanink (1984).

So far in this discussion, other predators of the benthos have not been considered. It is not only the scale of their impact on benthic production that could affect how much birds take in a particular area, but also the timing of that impact. In shallow European intertidal areas, many fish species graze bivalve siphons and polychaete tails only during the summer months, when water temperatures are relatively warm and the rate of benthic production high. During the autumn, most fish and invertebrate predators, such as the shore crab, *Carcinus maenas*, retreat into the subtidal zone. Migrant shorebirds thus arrive at a time when most of the annual production by the benthos has been achieved and other predators have taken their toll. It may be no coincidence that predation by birds resulted in removal of a higher percentage of benthic production in the two most northern estuaries (Ythan and Tees), where water temperatures are lower than elsewhere, so that the period of use by non-bird predators is likely to be shortest and daily energy requirements of poikilotherms likely to be least. Consumption by birds was highest of all in the Tees where predation by fish was unimportant because of the polluted condition of the river.

In contrast, migrant waders from the Palaearctic, travelling as far as South Africa, reach the Langebaan Lagoon in the Southern Hemisphere spring, and so feed alongside other predators over the period of the year when benthic production is highest. Information about seasonal patterns of prey production in those regions of the tropics where shorebirds are most numerous, *e.g.* the Banc d'Arguin, Mauritania, is unavailable at present, although measurements of the biomass density of the standing crop in January indicate that it is low relative to the biomass density of birds (NOME, 1982). It is known, however, that $P/\bar{B}$ ratios tend to be an order of magnitude higher in Indian tropical areas than in northern temperate areas (Ansell, McLusky, Stirling & Trevallion, 1978), so that the seemingly low food stocks for shorebirds in Mauritania may not be limiting if there is rapid and continuous turnover of the standing crop.

The question of whether bird numbers might be limited by benthic productivity in an estuary has already been mentioned. It should also be remembered that numbers of birds reaching an estuary in autumn may fluctuate from year to year in response to annual variations in adult survival and breeding success. Subsequent survival of birds on an estuary depends, in general, on their ability to find prey fast enough to match their rate of energy expenditure. Although they carry reserves of subcutaneous fat, these can provide energy sufficient for only a few days without food (Davidson & Evans, 1982).

In relation to shorebird–benthic invertebrate interactions, notable progress has been made in understanding and predicting where particular bird species should feed and what sizes and types of prey they should take, by applying and modifying so-called "optimal foraging theory". This considers prey primarily as a source of energy, to be obtained at greater or lesser cost, and predicts that birds should feed where they can maximize their net rate of gain of energy over short periods, taking those items which have high energy content in relation to the time needed to handle them. Many wading birds, however, need not only to obtain an adequate daily energy intake but also to avoid being taken themselves by predators. The extent of predation of waders by raptors and mammals has not been sufficiently appreciated (Page & Whitacre, 1975; Townshend, 1984) and the possibility of predation may affect not only where birds feed (*e.g.* they may choose areas near safe roost sites, even though these do not hold the highest densities of available prey) but also how they divide the time they spend on the feeding grounds between foraging and vigilance. Indeed, if bouts of vigilance are short and frequent, the result is a lower average rate of food intake. Thus, birds may not feed at the maximum possible rate under some circumstances, and energetic considerations clearly are not the only ones determining the impact of birds on benthic invertebrates at a particular site.

Finally, it is worth emphasizing that the range of prey species and sizes taken by a particular species of wader is the energetically profitable sub-set of those available to that wader in a particular estuary. Contrary to the oft-repeated diagrams in text books that link bill-length of the birds to the depth at which particular invertebrates lie within the substratum, and thus predict the diet of each bird, bill-length determines the range of prey species accessible, and therefore available, to only those species of waders that hunt by probing. One group of waders, the plovers, which have short bills, forage only by sight, and rely on benthic invertebrates coming close to the surface to respire or defaecate. For plovers, prey availability depends not only on accessibility but also on prey activity which allows birds to detect them.

CONCLUSION

This review has shown that, albeit from data with substantial but unquantified error margins, a high proportion of the productivity of benthic invertebrates is eaten by birds feeding in the intertidal zone. We make a plea, therefore, for inclusion of birds in models of energy and nutrient transfer in estuarine systems. We suggest, however, that static models of such systems are unlikely to progress much further our understanding of the factors controlling transfer processes in estuaries, not only because estuaries are dynamic and have changed their size and conformation through historical time (even without the intervention of man), but also because variation from year to year in benthic production is sufficiently great that the use of average values distorts the true picture. We applaud the attempts at dynamic modelling of an estuarine system, brought together in book form by Cuff & Tomczak (1983), although it is clear that many gaps remain in prediction and quantification of interactions between the components of such systems.

ACKNOWLEDGEMENTS

One of us (D.B.) thanks the University of Port Elizabeth, the C.S.I.R., and Department of Environmental Affairs of South Africa for financial assistance during the preparation of this review, most of which was completed during a visit to the University of Durham, where facilities were kindly provided by the Department of Zoology through Professor K. Bowler. P.R.E. and M.W.P. thank a variety of organizations for supporting their studies at Teesmouth and Lindisfarne, particularly the Nature Conservancy Council, the Nuffield Foundation, the Natural Environment Research Council and the British Ornithologists' Union.

REFERENCES

Ansell, A. D., McLusky, D. S., Stirling, A. & Trevallion, A., 1978. *Proc. R. Soc. Edinb., Sect. B*, **76**, 269–296.

Aschoff, J. & Pohl, H., 1970. *J. Ornithol.*, **111**, 38–47.

Baird, D. & Milne, H., 1981. *Estuar. cstl shelf Sci.*, **13**, 455–472.

Banse, K. & Mosher, S., 1980. *Ecol. Mongr.*, **50**, 355–379.

Beukema, J. J., 1976. *Neth. J. Sea Res.*, **10**, 236–261.

Beukema, J. J., 1979. *Neth. J. Sea Res.*, **13**, 203–223.

Beukema, J. J., 1981a. In, *Invertebrates of the Wadden Sea*, edited by N. Dankers, H. Kühl & W. J. Wolff, Balkema, Rotterdam, pp. 134–142.

Beukema, J. J., 1981b. In, *Invertebrates of the Wadden Sea*, edited by N. Dankers, H. Kühl & W. J. Wolff, Balkema, Rotterdam, pp. 211–221.

Beukema, J. J., de Bruin, W. & Jansen, J. J. M., 1978. *Neth. J. Sea Res.*, **12**, 58–77.

Bloom, S. A., 1980. *Mar. Ecol. Progr. Ser.*, **3**, 79–81.

Bryant, D. M., 1979. *Estuar. cstl mar. Sci.* **9**, 369–384.

Christy, R. L., Bildstein, K. L. & DeCoursey, P., 1981. *Colonial Waterbirds*, **4**, 96–103.

Crisp, D. J., 1971. In, *Methods for the Study of Marine Benthos* (IBP Handbook No. 16), edited by N. A. Holme & A. D. McIntyre, Blackwells, Oxford, pp. 197–279.

Cuff, W. R. & Tomczak Jr., M., 1983. Editors. *Synthesis and Modelling of Intermittent Estuaries*. Springer Verlag, Berlin, 302 pp.

Dame, R., Vernberg, F., Bonell, R. & Kitchens, W., 1977. *Helgoländer wiss. Meeresunters.*, **30**, 343–356.

Dankers, N., Kühl, H. & Wolff, W. J., 1981. Editors. *Invertebrates of the Wadden Sea*, Balkema, Rotterdam, 221 pp.

Davidson, N. C., 1981. Ph.D. thesis, University of Durham, U.K., 228 pp.

Davidson, N. C. & Evans, P. R., 1982. *Bird Study*, **29,** 183–188.
Day, J. H., 1959. *Trans. R. Soc. S. Afr.*, **35,** 475–547.
de Vlas, J., 1981. In, *Feeding and Survival Strategies of Estuarine Organisms*, edited by N. V. Jones & W. J. Wolff, Plenum, New York, pp. 173–178.
Drent, R. H. & Daan, S., 1980. *Ardea*, **68,** 225–252.
Evans, P. R., 1976. *Ardea*, **64,** 117–139.
Evans, P. R., 1979. In, *Cyclic Phenomena in Marine Plants and Animals, Proc. 13th Europ. Mar. Biol. Symp.*, edited by E. Naylor & R. G. Hartnoll, Pergamon, Oxford, pp. 357–366.
Evans, P. R. & Dugan, P. J., 1984. In, *Coastal Waders and Wildfowl in Winter*, edited by P. R. Evans, J. D. Goss-Custard & W. G. Hale, Cambridge University Press, Cambridge, pp. 8–28.
Evans, P. R., Herdson, D. M., Knights, P. J. & Pienkowski, M. W., 1979. *Oecologia (Berl.)*, **41,** 183–206.
Goss-Custard, J. D., 1969. *Ibis*, **111,** 338–356.
Goss-Custard, J. D., 1970. In, *Social Behaviour in Birds and Mammals*, edited by J. H. Crook, Academic Press, London, pp. 3–35.
Goss-Custard, J. D., 1977. *J. appl. Ecol.*, **14,** 721–739.
Goss-Custard, J. D., 1980. *Ardea*, **68,** 31–52.
Goss-Custard, J. D., Jenyon, J. A., Jones, R. E., Newbery, P. E. & Williams, R. le B., 1977. *J. appl. Ecol.*, **14,** 701–719.
Goss-Custard, J. D., Jones, R. E. & Newbery, P. E., 1977. *J. appl. Ecol.*, **14,** 681–700.
Goss-Custard, J. D., Kay, D. G. & Blindell, R. E., 1977. *Estuar. cstl mar. Sci.*, **5,** 497–510.
Goss-Custard, J. D., McGrorty, S., Reading, C. J. & Durell, S. E. A. le V. dit., 1980. *Devonshire Association*, Special Volume 2, pp. 161–185.
Grant, J., 1981. *Oikos*, **37,** 53–62.
Gray, J. S., 1976. *Estuar. cstl mar. Sci.*, **4,** 653–676.
Hanekom, N. & Baird, D., 1985. *S. Afr. J. Zool.*, in press.
Heppleston, P. B., 1971. *J. Anim. Ecol.*, **40,** 651–672.
Hulscher, J. B., 1974. *Ardea*, **62,** 155–171.
Hulscher, J. B., 1975. *Med. Werkgr. Waddengeb.*, **1,** 57–82.
Hulscher, J. B., 1980. In, *Birds of the Wadden Sea*, edited by C. J. Smit & W. J. Wolff, Balkema, Rotterdam, pp. 92–104.
Hulscher, J. B., 1982. *Ardea*, **70,** 89–152.
Kendeigh, S. C., Dolnik, V. R. & Gavrilov, V. M., 1977. In, *Granivorous Birds in Ecosystems*, edited by J. Pinowski & S. C. Kendeigh, Cambridge University Press, Cambridge, pp. 129–204.
Kent, A. C. & Day, R. W., 1983. *J. exp. mar. Biol. Ecol.*, **73,** 185–203.
Krebs, C. J., 1978. *Ecology*. Harper & Row, London, 2nd edition, 678 pp.
Lasiewski, R. C. & Dawson, W. R., 1967. *Condor*, **69,** 13–23.
Leach, J. H., 1971. *J. mar. biol. Ass. U.K.*, **51,** 137–157.
May, R. M., 1979. *Nature, Lond.*, **282,** 443–444.
McLusky, D. S., 1981. *The Estuarine Ecosystem*, Blackie, Glasgow, 150 pp.
Millard, A. V., 1975. Ph.D. thesis, University of Durham, U.K., 178 pp.
Millard, A. V. & Evans, P. R., 1984. In, Spartina anglica *in Great Britain*, edited by J. P. Doody, Nature Conservancy Council, Shrewsbury, pp. 41–48.
Milne, H., 1974. *Proc. Challenger Soc.*, **4,** p. 40 only.
Milne, H. & Dunnet, G. M., 1972. In, *The Estuarine Environment*, edited by R. S. K. Barnes & J. Green, Applied Science, London, pp. 86–106.
NOME, 1982. *Wintering Waders on the Banc d'Arguin*, Communication No. 6, Wadden Sea Working Group, Leiden, Netherlands, 284 pp.
Odum, E. P., 1970. *Fundamentals of Ecology*. Saunders, London, 3rd edition, 574 pp.
Page, G. & Whitacre, D. F., 1975. *Condor*, **77,** 73–83.
Peterson, C. H., 1979. In, *Ecological Processes in Coastal and Marine Systems*, edited by R. J. Livingston, Plenum Press, New York, pp. 233–264.

Pienkowski, M. W., 1980. Ph.D. thesis, University of Durham, U.K., 386 pp.
Pienkowski, M. W., 1982. *J. Zool.*, **197,** 511–549.
Pienkowski, M. W., 1983a. *Anim. Behav.*, **31,** 244–264.
Pienkowski, M. W., 1983b. *Ornis. Scandinavica,* **14,** 227–238.
Pienkowski, M. W. & Evans, P. R., 1984. In, *Behaviour of Marine Animals, Vol. 6, Shorebirds,* edited by J. Burger & B. L. Olla, Plenum Publishing, New York, pp. 74–124.
Pienkowski, M. W. & Evans, P. R., 1985. In, *Behavioural Ecology* (Symposium of the British Ecological Society No. 24), edited by R. M. Sibly & R. H. Smith, Blackwell, Oxford, pp. 331–352.
Pienkowski, M. W., Ferns, P. N., Davidson, N. C. & Worrall, D. H., 1984. In, *Coastal Waders and Wildfowl in Winter,* edited by P. R. Evans, J. D. Goss-Custard & W. G. Hale, Cambridge University Press, Cambridge, pp. 29–56.
Prater, A. J., 1972. *J. appl. Ecol.*, **9,** 179–194.
Pringle, J. S. & Cooper, J., 1975. *Ostrich,* **46,** 213–217.
Pringle, J. S. & Cooper, J., 1977. *Ostrich,* **48,** 98–105.
Puttick, G. M., 1977. *Trans. R. Soc. S. Afr.*, **42,** 403–439.
Puttick, G. M., 1978. *Ostrich,* **49,** 158–167.
Puttick, G. M., 1979. *Ardea,* **67,** 111–122.
Puttick, G. M., 1980. *Estuar. cstl mar. Sci.*, **11,** 207–215.
Quammen, M., 1981. *Auk,* **98,** 812–817.
Reading, C. J. & McGrorty, S., 1978. *Estuar. cstl mar. Sci.*, **8,** 135–144.
Ricker, W. E., 1968. Editor. *Methods for Assessment of Fish Production in Fresh Waters* (I.B.P. Handbook No. 3), Blackwell, Oxford, 326 pp.
Robertson, A. I., 1979. *Oecologia* (*Berl.*), **38,** 193–197.
Robertson, H. G., 1981. In, *Proc. Symp. Birds of Sea and Shore,* edited by J. S. Cooper, African Seabird Group, Cape Town, pp. 335–345.
Rosenberg, R., Olson, I. & Ölundh, E., 1977. *Mar. Biol.*, **42,** 99–107.
Schneider, D. C., 1981. *Mar. Ecol. Progr. Ser.*, **5,** 223–224.
Schramm, M., 1978. B.Sc. thesis, University of Port Elizabeth, S. Africa.
Smit, C. J., 1980a. In, *Birds of the Wadden Sea,* edited by C. J. Smit & W. J. Wolff, Balkema, Rotterdam, pp. 280–289.
Smit, C. J., 1980b. In, *Birds of the Wadden Sea,* edited by C. J. Smit & W. J. Wolff, Balkema, Rotterdam, pp. 290–301.
Smit, C. J. & Wolff, W. J., 1980. Editors. *Birds of the Wadden Sea,* Balkema, Rotterdam, 308 pp.
Smith, P. C., 1975. Ph.D. thesis, University of Durham, U.K., 138 pp.
Summers, R. W., 1977. *Trans. R. Soc. S. Afr.*, **42,** 483–495.
Summers, R. W., Cooper, J. & Pringle, J. S., 1977. *Ostrich,* **48,** 85–97.
Summers, R. W. & Waltner, M., 1979. *Ostrich,* **50,** 21–37.
Sutherland, W. J., 1982. *Anim. Behav.*, **30,** 857–861.
Swennen, C., 1975. *Vogeljaar,* **21,** 141–156.
Swennen, C., 1976. In, *Proc. Int. Conference on the Conservation of Wetlands and Waterfowl 1974,* edited by M. Smart, IWRB, Slimbridge (U.K.), pp. 184–198.
Thomas, D. G., 1970. *Emu,* **70,** 79–85.
Townshend, D. J., 1982. *Wader Study Group Bull.*, No. 34, 11–12.
Townshend, D. J., 1984. *Wader Study Group Bull.*, No. 40, 51–54.
Townshend, D. J., 1985. *J. Anim. Ecol.*, **54,** 267–274.
Tuite, C. H., 1984. *Ibis,* **126,** 250–252.
Virnstein, R. W., 1978. In, *Estuarine Interactions,* edited by M. L. Wiley, Academic Press, New York, pp. 261–273.
Warwick, R. M., Joint, I. R. & Radford, P. J., 1979. In, *Ecological Processes in Coastal Environments,* edited by R. L. Jefferies & A. J. Davy, Blackwell, Oxford, pp. 429–450.

Whittaker, R. H., 1975. *Communities and Ecosystems.* MacMillan, New York, 2nd edition, 600 pp.

Wolff, W. J., van Haperen, A. M. M., Sandu, A. J. J., Baptist, H. J. M. & Saeijs, H. L. F., 1975. In, *Proc. 10th Europ. Symp. Mar. Biol., Vol. 2, Population Dynamics*, edited by G. Persoone & E. Jaspers, Universa Press, Wetteren, Belgium, pp. 673–679.

Wolff, W. J. & Smit, C. J., 1984. In, *Coastal Waders and Wildfowl in Winter*, edited by P. R. Evans, J. D. Goss-Custard & W. G. Hale, Cambridge University Press, Cambridge, pp. 238–252.

Wolff, W. J. & de Wolf, L., 1977. *Estuar. cstl mar. Sci.*, **5,** 1–24.

Zwarts, L., 1980. In, *Birds of the Wadden Sea*, edited by C. J. Smit & W. J. Wolff, Balkema, Rotterdam, pp. 271–279.

Zwarts, L. & Drent, R. H., 1981. In, *Feeding and Survival Strategies of Estuarine Organisms*, edited by N. V. Jones & W. J. Wolff, Plenum, New York, pp. 193–216.

Zwarts, L. & Wanink, J., 1984. In, *Coastal Waders and Wildfowl in Winter*, edited by P. R. Evans, J. D. Goss-Custard & W. G. Hale, Cambridge University Press, Cambridge, pp. 69–83.

AUTHOR INDEX

References to complete articles are given in heavy type; references to pages are given in normal type; references to bibliographical lists are given in italics.

SYSTEMATIC INDEX

SUBJECT INDEX

References to complete articles are given in heavy type; references to sections of articles are given in italics; references to pages are given in normal type.